Sajid Bashir, Jingbo Louise Liu (Eds.)

Nanochemistry

Also of interest

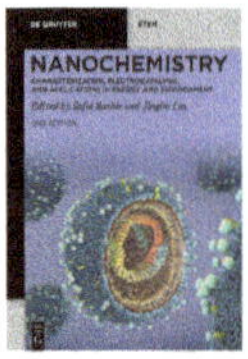

Nanochemistry.
Vol 2: Characterization, Electrocatalysis, and Applications in Energy and Environment
Bashir, Liu (Eds.), 2026
ISBN 978-3-11-225503-2, e-ISBN (PDF) 978-3-11-225504-9,
e-ISBN (EPUB)
978-3-11-225505-6

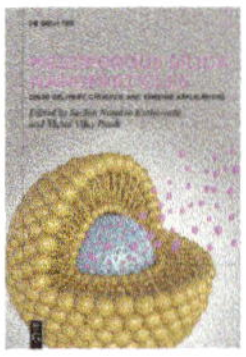

Mesoporous Silica Nanoparticles.
Drug Delivery, Catalysis and Sensing Applications
Sachin Namdeo Kothawade, Vishal Vijay Pande (Eds.), 2024
ISBN 978-3-11-133844-6, e-ISBN (PDF) 978-3-11-133847-7,
e-ISBN (EPUB) 978-3-11-133849-1

Nano-Safety.
What We Need to Know to Protect Workers
2nd Edition
Dominick E. Fazarro, Christie M. Sayes, Walt Trybula, Jitendra S. Tate, Craig Hanks (Eds.), 2023
ISBN 978-3-11-078182-3, e-ISBN (PDF) 978-3-11-078183-0,
e-ISBN (EPUB) 978-3-11-078195-3

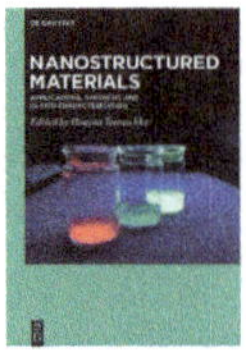

Nanostructured Materials.
Applications, Synthesis and In-Situ Characterization
Huayna Terraschke (Ed.), 2023
ISBN 978-3-11-045829-9, e-ISBN (PDF) 978-3-11-045909-8,
e-ISBN (EPUB) 978-3-11-045833-6

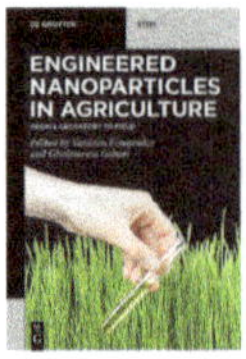

Engineered Nanoparticles in Agriculture.
From Laboratory to Field
Vasileios Fotopoulos, Gholamreza Gohari (Eds.), 2023
ISBN 978-1-5015-2315-1, e-ISBN (PDF) 978-1-5015-2322-9,
e-ISBN (EPUB) 978-1-5015-1552-1

Nanochemistry

Molecular Design and Scalable Synthesis

2nd, Revised and Extended Edition
Volume 1

Edited by
Sajid Bashir and Jingbo Louise Liu

DE GRUYTER

Editors
Professor Sajid Bashir
7534, Beau Terre Court
Corpus Christi
TX 78414-6209
USA
sajid.liu@tamuk.edu

Professor Jingbo Louise Liu
7534, Beau Terre Court
Corpus Christi
TX 78414-6209
USA
jingbo.liu@tamuk.edu

ISBN 978-3-11-162098-5
ISBN 978-3-11-162111-1 (PDF)
ISBN 978-3-11-162103-6 (EPUB)
DOI https://doi.org/10.1515/9783111621111

Library of Congress Control Number: 2026950630

Bibliographic information published by the Deutsche Nationalbibliothek
The Deutsche Nationalbibliothek lists this publication in the Deutsche Nationalbibliografie; detailed bibliographic data are available on the internet at http://dnb.dnb.de.

De Gruyter and Walter de Gruyter GmbH are part of De Gruyter Brill.
www.degruyterbrill.com

Questions about General Product Safety Regulation:
productsafety@degruyterbrill.com

Cover illustration: LumerB/iStock/Getty Images Plus

Contents

Contributing authors

Abedin Ashraf
Formerly at the Cain Department of Chemical Engineering
Louisiana State University
3307 Patrick F. Taylor Hall
Baton Rouge, LA 70803, USA
Now at the National Energy Technology Laboratory
Email: mdashraful.abedin@netl.doe.gov

Alexander Matthew
Department of Chemical Engineering
Texas A&M University-Kingsville
MSC 161, 700 University Blvd.
Kingsville, TX 78363, USA
Email: Matthew.Alexander@tamuk.edu

Alia Telli
Laboratoire de protection des écosystèmes en zone aride et semi-aride
Université de KASDI Merbah
BP 511 la route de Ghardaïa
Ouargla 30000, Algérie
Email: telli.alia@univ-ouargla.dz

Anand Vivek
Department of Chemistry
University Institute of Science
Chandigarh University
Gharuan, Mohali 140413
Punjab, India
Email: vivekanandac88@gmail.com

Asgar Hassnain
Smith School of Chemical and Biological Engineering
Cornell University
Ithaca, NY 14853, USA
Email: ha356@cornell.edu

Bashir Sajid
Department of Chemistry
Texas A&M University-Kingsville
MSC 161, 700 University Blvd.
Kingsville, TX 78363, USA
Email: br9@tamuk.edu

Chava Sai Raghuveer
Department of Chemistry
Texas A&M University-Kingsville
MSC 161, 700 University Blvd.
Kingsville, TX 78363-8202, USA
Email: sai_raghuveer.chava@students.tamuk.edu

Choi Jong-Won
Department of Civil and Architectural Engineering
Texas A&M University-Kingsville
MSC 194
925 W. Avenue B
Kingsville, TX 78363, USA
Email: Jong-Won.Choi@tamuk.edu

Durowoju Emmanuel
Department of Mechanical and Industrial Engineering
Texas A&M University-Kingsville
925 W. Avenue B
Kingsville, TX 78363-8202, USA
Email: durowojues@gmail.com

Gadikota Greeshma
Formerly at the School of Civil and Environmental Engineering
Smith School of Chemical and Biological Engineering
Cornell University
Ithaca, NY 14850/14853, USA
Now at Columbia University
New York, NY, USA
Email: ge2131@columbia.edu

Haik Yousef
Formerly at the Department of Mechanical and Industrial Engineering
Frank H. Dotterweich College of Engineering
Texas A&M University-Kingsville
925 W. Avenue B
Kingsville, TX 78363-8202, USA
Now at the University of Jordan
Amman, Jordan
Email: yhaik@sharjah.ac.ae

 | https://doi.org/10.1515/9783111621111-203

Haq Tanveer ul
Formerly at Sustainable Energy Engineering
Frank H. Dotterweich College of Engineering
Texas A&M University-Kingsville
Kingsville, TX 78363-8202, USA
Now at the University of Jordan
Amman, Jordan
Email: tanveerulhaq341@gmail.com;
thaq@sharjah.ac.ae

He Guanjie
Christopher Ingold Laboratory
Department of Chemistry
University College London
20 Gordon Street
London WC1H 0AJ, UK
and
School of Chemistry
Joseph Banks Laboratories
University of Lincoln
Green Lane
Lincoln LN6 7DL, UK
Email: g.he@ucl.ac.uk

Houf William P.
Department of Chemistry
Texas A&M University-Kingsville
MSC 161, 700 University Blvd.
Kingsville, TX 78363, USA
Email: william.houf@students.tamuk.edu

Huang Hsuan-Yi
Department of Chemistry
Texas A&M University-Kingsville
MSC 161, 700 University Blvd.
Kingsville, TX, USA
Email (Student, Set): hsuan-yi.huang@students.
tamuk.edu

Jiang Zheng
Shanghai Synchrotron Radiation Facility
Zhangjiang Lab
Shanghai Advanced Research Institute
Chinese Academy of Sciences
Shanghai 201210, PR China
Email: jiangzheng@sinap.ac.cn

Kong Dan
Department of Materials Science and Physics
China University of Mining and Technology
No. 1, Daxue Road, Xuzhou
Jiangsu, PR China
Email: dan.kong.14@ucl.ac.uk

Lei Wen
State Key Laboratory of Refractories and
Metallurgy
Wuhan University of Science and Technology
Wuhan 430081, PR China
Email: leiwen@wust.edu.cn

Li Hua
Mechanical and Industrial Engineering
Department
Texas A&M University-Kingsville
MSC191, 700 University Blvd.
Kingsville, TX 78363, USA
Email: Hua.Li@tamuk.edu

Liu Jingbo Louise
Department of Chemistry
Texas A&M University-Kingsville
MSC 161, 700 University Blvd.
Kingsville, TX 78363, USA
Email: jingbo.liu@tamuk.edu; jingbo.liu@tamu.edu

Luckett Robert S.
Department of Social Work
Texas A&M University-Kingsville
700 University Blvd., MSC 123
Kingsville, TX 78363, USA
Email: Robert.Luckett@tamuk.edu

Mei Bingbao
Shanghai Synchrotron Radiation Facility
Zhangjiang Lab
Shanghai Advanced Research Institute
Chinese Academy of Sciences
Shanghai 201210, PR China
Email: meibb@sari.ac.cn

Mishra Roli
Department of Engineering and Physical Sciences
Institute of Advanced Research
Gandhinagar, Gujarat, India
Email: rolimishra2@gmail.com; roli.mishra@iar.ac.in

Mohammed Sohaib
School of Civil and Environmental Engineering
Cornell University
Ithaca, NY 14850, USA
Email: sm2793@cornell.edu

Olaseni Yesirat O.
Department of Chemistry
Texas A&M University-Kingsville
MSC 161, 700 University Blvd.
Kingsville, TX, USA
Email: yesirat.olaseni@students.tamuk.edu

Palakurthi Sushesh Srivatsa
Department of Chemistry
Texas A&M University-Kingsville
MSC 161, 700 University Blvd.
Kingsville, TX, USA
Email: sushesh.palakurthi@students.tamuk.edu

Parkin Ivan P.
Christopher Ingold Laboratory
Department of Chemistry
University College London
20 Gordon Street
London WC1H 0AJ, UK
Email: i.p.parkin@ucl.ac.uk

Tan Yeshu
Christopher Ingold Laboratory
Department of Chemistry
University College London
20 Gordon Street
London WC1H 0AJ, UK
Email: yeshu.tan.19@ucl.ac.uk

Villanueva Luis
Department of Chemistry
Texas A&M University-Kingsville
MSC 161, 700 University Blvd.
Kingsville, TX 78363, USA
Email: luis.villanueva@students.tamuk.edu

Villarreal Pete
Department of Chemistry
Texas A&M University-Kingsville
MSC 161, 700 University Blvd.
Kingsville, TX, USA
Email: Pete.Villarreal@tamuk.edu

Vyas Saurabh
Department of Engineering and Physical Sciences
Institute of Advanced Research
Gandhinagar, Gujarat, India
Email: saurabhvyas.phd2020@iar.ac.in

Wang Xuan
Texas A&M Higher Education Center at McAllen
McAllen, TX, USA
Email: xuan.wang@science.tamu.edu

Wang Zhaohui
Formerly at Texas A&M University-Kingsville
MSC 161, 700 University Blvd.
Kingsville, TX, USA
Currently at North Carolina A&T State University
Greensboro, NC, USA
Email: zwang3@ncat.edu

Zhang Haijun
State Key Laboratory of Refractories and Metallurgy
Wuhan University of Science and Technology
Wuhan 430081, PR China
Email: zhanghaijun@wust.edu.cn;
Tel: +86-15697181003

Zhang Hao
Institute of Functional Nano and Soft Materials (FUNSOM)
Jiangsu Key Laboratory for Carbon-Based Functional Materials and Devices
Soochow University
Suzhou 215123, Jiangsu, PR China
Email: haozhang@suda.edu.cn

Zhang Shaowei
College of Engineering, Mathematics and Physical Sciences
University of Exeter
Exeter EX4 4QF, UK
Email: shaowei.zhang@exeter.ac.uk

Sajid Bashir, Jingbo Louise Liu*

Preface to the second edition

Abstract: This second edition of *Nanochemistry: From Theory to Application for In-Depth Understanding of Nanomaterials* was completed in 2025, 10 years after the publication of the first edition in 2015. The original volume, comprising over 500 pages, was conceived as a comprehensive introduction to nanochemistry at a time when the field was rapidly transitioning from fundamental discovery to early technological implementation.

Over the past decade, nanochemistry has undergone profound expansion in scope, depth, and societal relevance. What was once a specialized scientific discipline has become a foundational pillar of modern materials science, energy technology, environmental remediation, and biomedical innovation. These developments, together with the maturation of scalable fabrication methods, advanced characterization tools, and electrochemical and electrocatalytic applications, necessitated a substantially expanded second edition.

The present edition has grown over 1,000 pages and is therefore presented in 2 complementary volumes. This structure allows for both conceptual clarity and pedagogical coherence while preserving the integrative philosophy of the original work.

Acknowledgments: The research efforts led by Dr Bashir and Dr Liu were supported by the American Chemical Society Petroleum Research Fund (Grant 53827-UR10), federal agencies, including the Office of Naval Research, the Air Force Office of Scientific Research, the Department of Energy, the National Science Foundation, the Department of Education, the Department of Homeland Security (ORISE), and the National Institutes of Health. The Robert A. Welch Foundation (Departmental Grant AC-0006) is gratefully acknowledged for its support of student training. Funding agencies supporting the contributions of other authors are acknowledged within their respective chapters to ensure accurate and appropriate recognition of all sources of support. The authors also acknowledge the technical assistance provided by Texas A&M University-Kingsville and the Texas A&M Energy Institute. The authors further thank their colleagues and collaborators for their valuable discussions and the reviewers for their constructive comments, which have strengthened the quality of this work.

Authors' contributions: S. Bashir and J. Liu collectively conceived the projects and oversaw their progress. S. Bashir was in charge of the submission processing of all chapters. J. Liu wrote the first draft, which all authors edited. All authors agreed to the publication of this chapter.

***Corresponding author: Jingbo Louise Liu**, Department of Chemistry, Texas A&M University-Kingsville, MSC161, 700 University Blvd., Kingsville, TX 78363, USA; Louise Nanoinnnoation North American, 7534 Beau Terre, Corpus Christi, TX 78414, USA; Texas A&M Energy Institute, 1617 Research Parkway, Suite 308, College Station, TX 77843-3372 USA, e-mails: jingbo.liu@tamuk.edu, jingbo.liu@tamu.edu
Sajid Bashir, Department of Chemistry, Texas A&M University-Kingsville, MSC161, 700 University Blvd., Kingsville, TX 78363, USA; Louise Nanoinnnoation North American, 7534 Beau Terre, Corpus Christi, TX 78414, USA; Texas A&M Energy Institute, 1617 Research Parkway, Suite 308, College Station, TX 77843-3372 USA

 | https://doi.org/10.1515/9783111621111-001

Volume 1, entitled *Molecular Design and Scalable Fabrication*, focuses on the principles and methods governing the design, synthesis, and production of nanomaterials. **Volume 2**, entitled *Characterization, Electrochemistry, and Applications in Energy and Environment*, addresses advanced analytical techniques, electrochemical and electrocatalytic processes, and nanomaterial applications relevant to sustainable energy and environmental challenges.

Together, these volumes reflect the editors' long-standing engagement with nanochemistry and their commitment to presenting both established knowledge and emerging directions in the field. While the foundational theory remains central, this edition places increased emphasis on scalability, functionality, and real-world impact. The perspectives and selected highlights presented herein naturally reflect the editors' individual experiences and evolving views, offering readers both continuity and forward-looking insight.

This two-volume work is intended for undergraduate, graduate, and postgraduate students, as well as researchers, technologists, laboratory managers, and faculty seeking a comprehensive and authoritative reference on modern nanochemistry.

Nanotechnology has evolved from a primarily exploratory scientific endeavor into a mature, application-driven discipline with direct relevance to global challenges in energy, environmental sustainability, health, and advanced manufacturing. Since the publication of the first edition, rapid advances in synthesis precision, characterization capability, computational modeling, and data-driven design have reshaped how nanomaterials are conceived, evaluated, and deployed.

This second edition reflects that evolution by expanding its scope beyond discovery-oriented research to encompass technology readiness, scalability, sustainability, and societal impact. New and substantially revised chapters emphasize scalable fabrication strategies, operando and in-situ characterization, electrochemical and electrocatalytic systems, and responsible innovation, positioning nanochemistry as a foundational pillar of modern materials science.

To accommodate the breadth and depth of these developments, the second edition is presented in two complementary volumes:

- **Volume 1: *Molecular Design and Scalable Fabrication*** focuses on the principles, methods, and scalable production of nanomaterials, emphasizing molecular-level design, synthesis strategies, and fabrication pathways.
- **Volume 2: *Characterization, Electrochemistry, and Applications in Energy and Environment*** addresses advanced characterization techniques, electrochemical and electrocatalytic processes, and nanomaterial applications relevant to sustainable energy and environmental systems.

Together, these volumes are designed to serve a diverse readership, including students, researchers, engineers, and practitioners, by integrating fundamental principles with application-oriented case studies and forward-looking perspectives that reflect both the current state and future trajectory of the field.

1 Production of nanomaterials

This book, entitled *Nanochemistry: From Theory to Application for In-Depth Understanding of Nanomaterials*, provides a comprehensive exploration of nanotechnology through the lenses of synthesis, characterization, and application. It covers bottom-up and top-down fabrication strategies, advanced electron microscopy and spectroscopy techniques, and a wide range of functional uses. These include nanomaterials for sustainable energy systems, water remediation, biomedical technologies, and forensic science. The goal of this edition is to offer readers a clear overview of the textbook's scope while highlighting new topics, updated examples, and areas of contemporary relevance that have emerged since the first edition. Naturally, the perspectives and emphases presented here reflect the distinct experiences and viewpoints of each editor.

The text balances theoretical foundations with practical guidance, providing numerous examples across diverse subfields. The opening section on synthesis introduces wet chemistry approaches such as sol–gel, solvothermal, and hydrothermal methods, along with a brief comparison to top-down ball milling. These discussions are particularly valuable for undergraduate, graduate, and postgraduate students who may be familiar with laboratory procedures but seek a deeper understanding of the underlying chemical principles. Technicians and materials technologists will also benefit from the rationale behind these synthetic strategies. At the same time, faculty, researchers, laboratory managers, and directors will find the book a useful compendium of universal techniques and detailed case studies.

Nanomaterials with dimensions below 100 nm offer unique advantages because a large fraction of their atoms reside at or near the surface, thereby enabling enhanced chemical interactions, catalytic activity, and bonding behavior [1]. Materials that are inert in bulk form often become highly reactive at the nanoscale. This phenomenon has fueled rapid advances in the design, synthesis, characterization, and application of nanostructured systems [2]. This textbook examines a wide range of wet chemical synthetic routes for producing nanomaterials, including sol–gel, solvothermal, hydrothermal, and electrochemical techniques. These methods enable the fabrication of metals, metal ions, metal oxides, single- and multi-walled carbon nanotubes, graphene-based nanostructures, graphene oxides, and reduced graphene oxides for catalytic and multifunctional applications [3] (Tab. 1).

2 Evaluation of nanomaterials

The evaluation of nanomaterials is fundamentally rooted in the recognition that reducing material dimensions below approximately 100 nm profoundly alters their physicochemical behavior. At this scale, a substantial fraction of atoms reside on sur-

Tab. 1: Production strategies for nanomaterials.

Synthesis strategy	Representative methods	Typical materials	Key advantages	Scalability considerations
Bottom-up chemical synthesis	Sol–gel, hydrothermal, and solvothermal	Metal oxides, perovskites, and nanoparticles	Precise compositional control and tunable morphology	Batch variability and solvent management
Electrochemical synthesis	Electrodeposition and anodization	Metals, oxides, and nanostructured films	Interface control and direct device integration	Uniformity over large areas
Vapor-phase synthesis	CVD, PVD, and ALD	Thin films, CNTs, and graphene	High purity and conformal coatings	Capital cost and throughput
Top-down fabrication	Lithography and milling	Nanopatterned solids	Device compatibility	Material waste and resolution limits

faces or interfaces rather than within the bulk lattice, thereby enhancing chemical accessibility and reactivity. This surface dominance enables stronger and more diverse chemical interactions, increased catalytic turnover, and tunable bonding environments that are unattainable in conventional bulk materials. Consequently, nanomaterials often exhibit enhanced adsorption capacity, accelerated reaction kinetics, and heightened sensitivity to external stimuli, making them particularly attractive for catalytic, sensing, and energy conversion applications [4].

Materials that are chemically inert or weakly reactive in their bulk form frequently become highly active when engineered at the nanoscale. This transformation arises from a combination of factors, including unsaturated surface coordination, altered electronic structure, quantum confinement effects, and strain-induced lattice distortions. Such features modify ionization energies, electron density distributions, and surface energetics, allowing precise control over reaction pathways and selectivity. The evaluation of nanomaterials, therefore, extends beyond simple compositional analysis to encompass size-dependent electronic, optical, magnetic, and thermal properties, all of which contribute to functional performance [5].

Advances in characterization techniques have been central to this evaluative process. High-resolution electron microscopy, spectroscopy, and in situ analytical methods enable direct observation of nanoscale structure, morphology, and dynamic behavior under operating conditions. When combined with computational modeling and data-driven analysis, these tools provide a comprehensive framework for correlating structure with function and predicting material behavior. As a result, the field has experienced an exponential expansion in the rational design, synthesis, characterization, and application of nanomaterials across diverse technological domains [6]. This integrated approach to evaluation ensures that nanoscale innovations are not only scientifically understood but also optimized for reliability, scalability, and real-world deployment (Tab. 2).

Tab. 2: Evaluation dimensions for nanomaterials.

Evaluation dimension	Representative techniques	Information gained	Relevance to applications
Structural	TEM, XRD, and AFM	Size, morphology, and crystallinity	Performance consistency
Chemical	XPS, FT-IR, and Raman	Surface chemistry and bonding	Reactivity and stability
Electronic	UV–Vis and conductivity	Band structure and charge transport	Optoelectronics and catalysis

3 Applications of nanomaterials

The applications of nanomaterials represent the culmination of advances in synthesis, characterization, and theoretical understanding, translating nanoscale design principles into functional technologies. This textbook highlights a broad spectrum of application-driven nanomaterials derived primarily from wet chemical synthetic routes, including sol–gel, solvothermal, hydrothermal, and electrochemical methods. These approaches enable the controlled fabrication of metals, metal ions, metal oxides, carbon nanotubes (both single- and multi-walled), and graphene-based nanostructures such as graphene oxides and reduced graphene oxides. Owing to their high surface area, tunable electronic structure, and adaptable surface chemistry, these materials have found widespread use in catalytic processes, energy conversion, environmental remediation, and sensing applications [7].

Beyond conventional catalytic systems, the book emphasizes the growing importance of interdisciplinary research in expanding the application space of nanomaterials. In particular, the emergence of novel perovskite materials illustrates how progress at the nanoscale increasingly depends on the integration of chemistry, physics, materials science, and engineering. Perovskite nanostructures exhibit exceptional optoelectronic, catalytic, and ionic transport properties, making them promising candidates for applications in photovoltaics, electrocatalysis, and energy storage. However, realizing their full potential requires a deeper theoretical understanding of structure–property relationships, coupled with precise synthetic control over composition, phase stability, and defect chemistry [8].

The evaluation of nanomaterial applications, therefore, extends beyond demonstrating functional performance to encompass durability, scalability, and compatibility with device architectures. As nanomaterials transition from laboratory demonstrations to components of industrial and commercial technologies, considerations such as long-term stability, reproducibility, and environmental impact become increasingly critical [9]. By presenting diverse application examples alongside their underlying synthetic and theoretical foundations, this textbook underscores how continued advances in nanomaterials research can drive innovation across the global materials engineering and development landscape, ultimately enabling high-performance, societally relevant technologies (Tab. 3).

Tab. 3: Nanomaterial classes and application domains.

Nanomaterial class	Key properties	Primary applications	Performance drivers
Metal nanoparticles	High surface reactivity	Catalysis and sensing	Active site density
Carbon nanostructures	Electrical conductivity	Energy storage and electronics	Defect control
Metal oxides	Redox activity	Catalysis and gas storage	Surface chemistry

4 Future direction: technology readiness

The textbook covers various wet synthetic methods for generating nanomaterials. Techniques include sol–gel, solvothermal, hydrothermal, and electrochemical methods for generating metals, ions, metal oxides, single- and multi-walled carbon nanotubes, graphene-based nanoparticles, graphene oxides, and reduced graphene oxides for multiple catalytic applications [10]. Allied interdisciplinary research is necessary to understand and utilize the field of novel perovskite materials truly. A greater theoretical and synthetic understanding will promote the application of perovskites and, in turn, enhance the field, benefiting the global materials engineering and development market (Tab. 4) [11].

Tab. 4: Nanomaterials across technology readiness stages.

Development stage	Focus	Key challenges	Representative topics
Discovery	Property exploration	Reproducibility	Fundamental synthesis
Optimization	Performance tuning	Stability	Structure–function
Integration	Device compatibility	Interfaces	Electrochemical systems

Nanotechnology (the focus of Chapter 1) has developed into a powerful interdisciplinary field focused on the design, control, and manipulation of matter at atomic and molecular length scales. When materials are engineered with at least one dimension in the nanometer regime, they often display physical, chemical, electronic, magnetic, and biological behaviors that differ fundamentally from those of their bulk counterparts. These size-dependent effects have enabled significant technological advances in areas such as water treatment, renewable energy, healthcare, food production, and other fields that directly affect quality of life. The rapid growth of nanoscience has been supported by progress in high-resolution imaging, advanced structural and property characterization, computational modeling, and precise methods for controlling matter across length scales ranging from approximately 1 to 100 nm [12]. At this scale, materials demonstrate enhanced interactivity and resilience, motivating continued research into efficient, scalable nanomanufacturing for industrial, environmental, and defense applications. This chapter introduces the foundational concepts of nanotechnology and outlines two principal strategies for creating tunable nanostructures: bottom-up and top-down synthesis. Bottom-up approaches, including sol–gel chemistry, microemulsion techniques, hydrosolvothermal processing, and vapor-phase deposition, assemble materials from atoms, ions, or molecules to achieve controlled architectures. Top-down methods, in contrast, rely on physical or chemical means to reduce bulk materials to nanoscale dimensions. Both strategies benefit from modern optimization tools such as artificial intelligence, automated experimentation, and data-driven modeling. Nanomaterials often behave differently from isolated molecules or bulk solids due to size confinement, interfacial dominance, and quantum-mechanical effects. Their shapes, compositions, and architectures critically influence performance, enabling tailored biological, catalytic, electronic, and optical functionalities. Over five decades of research have yielded advanced materials such as carbon nanotubes, quantum dots, thin films, DNA-based assemblies, and hybrid nanostructures. Through functionalization, doping, intercalation, and grafting, scientists continue to engineer materials with unprecedented properties, marking a new era in materials science and technology [13].

Chapter 2 provides a comprehensive examination of how nanoscale dimensions fundamentally alter material properties and enable transformative applications. It begins by explaining that when particle sizes fall below approximately 100 nm, quantum confinement and dramatically increased surface-to-volume ratios lead to behaviors that differ markedly from those of bulk materials. Chemically, these effects modify ionization energies, electron affinities, and surface reactivity, allowing precise control over catalytic activity. Structural transitions, such as the size-dependent lattice rearrangement of gold nanoparticles, illustrate how nanoscale engineering can accelerate reaction kinetics by orders of magnitude. Magnetic properties are similarly size-dependent, with iron oxide nanoparticles exhibiting superparamagnetism only within a narrow diameter range, governed by exchange interactions and orbital effects. Electro-optical behavior arises from quantum-mechanical confinement, where changes in dimensionality reshape electronic density of states, enabling tunable absorption, resonant charge transport, and single-electron phenomena relevant to advanced optoelectronic devices. Thermal transport is also strongly influenced by nanoscale structure, as phonon scattering at grain boundaries and defects reduces thermal conductivity relative to that of bulk solids. Building on these fundamental properties, the chapter highlights five representative applications. Nanocatalysts enable the conversion of plastic waste into hydrogen and graphene, supporting environmental remediation and energy recovery. In biomedicine, nanoparticle platforms enhance cancer immunotherapy, ophthalmic treatments, and trauma care through targeted delivery, reactive oxygen species scavenging, and controlled hemostasis. Finally, nanostructured oxides and metal–organic frameworks demonstrate exceptional gas storage capabilities, offering promising solutions for clean energy storage. Overall, Chapter 2 demonstrates how deliberate control of size, shape, composition, and surface chemistry at the nanometer scale unlocks functionalities unattainable in bulk materials. By linking fundamental properties to real-world applications, the chapter underscores the central role of nanomaterials in addressing critical challenges in energy, health, and environmental sustainability [14].

Chapter 3 advances the discussion from foundational principles to practical synthesis strategies that underpin technology readiness. It focuses on wet chemical routes, including sol–gel, solvothermal, hydrothermal, and electrochemical methods, as versatile platforms for producing nanomaterials with controlled composition, morphology, and surface chemistry. These approaches are highlighted not merely as laboratory techniques, but as scalable and adaptable processes capable of supporting industrial translation. By demonstrating how reaction parameters such as precursor chemistry, temperature, solvent environment, and redox conditions influence nucleation and growth, the chapter establishes a direct link between synthesis control and functional reliability [15].

The chapter further emphasizes that wet chemical synthesis enables the production of a wide range of technologically relevant materials, including metal oxides, carbon nanotubes, and graphene-based nanostructures. These materials are evaluated in the context of catalytic performance, stability, and reproducibility, key metrics for technology readiness. Notably, the discussion acknowledges challenges associated with scale-up, such as batch-to-batch variability and waste management, framing them as engineering problems that can be addressed through process optimization and green chemistry principles. In doing so, Chapter 3 positions wet chemical synthesis as a foundational bridge between nanoscale discovery and deployable material systems [16].

Chapter 4 extends the synthesis discussion toward manufacturing-compatible techniques by examining vapor-phase deposition methods, including physical and chemical vapor deposition. These approaches are presented as critical enablers for integrating nanomaterials into devices and coatings with precise thickness control and uniformity over large areas. Unlike solution-based methods, vapor-phase techniques are directly compatible with semiconductor processing and advanced manufacturing workflows, making them particularly relevant for electronics, sensors, and energy devices [17].

The chapter highlights how vapor-phase synthesis supports technology readiness by enabling reproducible fabrication, clean interfaces, and scalable throughput. Case studies illustrate that thin films and nanostructured coatings produced via vapor-phase routes exhibit enhanced durability and performance stability under operating conditions. By situating vapor-phase deposition within the broader context of nanomanufacturing, the chapter underscores its role in transitioning nanomaterials from experimental prototypes to components of functional devices and systems [18].

Chapter 5 focuses on atomic layer deposition (ALD) as a synthesis strategy uniquely suited to advanced technology readiness. ALD is presented as a method that combines atomic-scale precision with conformal coating capability, enabling the functionalization of complex three-dimensional architectures. This level of control is essential for applications requiring uniformity, reliability, and long-term stability, such as energy storage electrodes, catalytic supports, and protective coatings [19].

The chapter emphasizes that ALD facilitates systematic tuning of interfaces and defect structures, which directly influences device performance and lifetime. By enabling reproducible fabrication across length scales, ALD addresses one of the central barriers to nanotechnology deployment: variability. The discussion positions ALD not as a niche technique but as a cornerstone of next-generation nanomanufacturing, particularly when integrated with automation and data-driven optimization [20].

Chapter 6 reframes technology readiness through the lens of sustainability, arguing that deployable nanotechnologies must be evaluated across their entire lifecycle. The chapter examines green synthesis strategies that reduce energy consumption, minimize the use of hazardous reagents, and improve environmental compatibility. These approaches are presented as essential complements to performance-driven design, ensuring that nanomaterials can be responsibly scaled and adopted [21].

5 Advanced nanomaterials and emerging functional architectures

As nanotechnology continues to mature, increasing attention has shifted toward advanced material classes and architectures that address the long-term performance, stability, and reliability required for real-world deployment. Chapter 17 (renumbered as Chapter 7 in this volume) focuses on this critical stage of nanomaterials development, examining how emerging functional architectures, particularly metallic nanoparticles and complex nanostructures, can be engineered to maintain performance under demanding operational conditions. This chapter highlights the transition from demonstrating nanoscale functionality to ensuring durability and consistency across extended lifetimes, a defining requirement for technology readiness [22].

A central challenge addressed in this chapter is the inherent instability of many nanomaterials. While nanoscale dimensions confer enhanced reactivity and functional performance, they also introduce vulnerabilities, including agglomeration, surface oxidation, and structural degradation. Metallic nanoparticles, in particular, are prone to coalescence and chemical transformation under thermal, chemical, or electrochemical stress. Chapter 17 (renumbered as Chapter 7 in this volume) explores strategies for mitigating these effects through surface passivation, ligand engineering, and the incorporation of stabilizing matrices. By controlling interparticle interactions and surface energetics, these approaches preserve nanoscale features while extending operational lifetimes [23].

The chapter further emphasizes the importance of architectural design in achieving functional robustness. Rather than relying on isolated nanoparticles, advanced nanomaterials increasingly adopt hierarchical and composite structures that distribute functionality across multiple length scales. Core–shell architectures, supported nanoparticles, and hybrid nanocomposites are presented as examples of how structural complexity can enhance stability without sacrificing performance. These architectures enable precise control over active sites while protecting sensitive components from deactivation, illustrating how nanoscale design principles can be leveraged to address macroscopic reliability concerns [24].

Another key theme is the role of interfaces in governing long-term behavior. At the nanoscale, interfaces between particles, supports, and surrounding environments dominate material response. Chapter 17 (renumbered as Chapter 7 in this volume) demonstrates how interfacial engineering through chemical bonding, lattice matching, or controlled defect introduction can stabilize nanomaterials against mechanical and chemical stress. Such strategies are particularly relevant for applications in catalysis, energy storage, and sensing, where materials are subjected to repeated cycling and harsh conditions. By tailoring interfacial properties, researchers can decouple high activity from rapid degradation, enabling sustained performance [25].

From a functional perspective, the chapter highlights how advanced nanomaterials enable emerging applications that demand both precision and resilience. Metallic nanoparticles with controlled size and composition are shown to support plasmonic, catalytic, and electronic functionalities that are sensitive to subtle structural changes. Maintaining these properties over time requires a balance between accessibility and protection, a theme that recurs throughout the discussion. The chapter underscores that achieving this balance is not a matter of incremental optimization, but of deliberate architectural design informed by mechanistic understanding [26].

The chapter also situates advanced nanomaterials within the broader context of manufacturing and deployment. Scalability and reproducibility are identified as critical barriers to adoption, particularly for complex architectures that require multistep synthesis or precise control over interfaces. Chapter 17 (renumbered as Chapter 7 in this volume) discusses strategies for addressing these challenges, including modular synthesis approaches and standardized processing protocols [27]. By aligning material design with manufacturing constraints, the chapter reinforces the notion that advanced nanomaterials must be conceived with deployment in mind from the outset (Tab. 5).

An essential contribution of this chapter is its emphasis on evaluation metrics that extend beyond initial performance. Long-term stability, resistance to degradation, and consistency across production batches are presented as essential criteria for assessing advanced nanomaterials. This perspective aligns with the broader theme of

Tab. 5: Stability strategies for advanced nanomaterials.

Architecture type	Stabilization strategy	Functional benefit	Application context
Core–shell	Surface passivation	Longevity	Catalysis
Supported nanoparticles	Matrix anchoring	Agglomeration resistance	Energy systems
Hybrid composites	Interface engineering	Mechanical robustness	Sensors

technology readiness introduced earlier in the Preface, emphasizing that functional success at the nanoscale must be matched by reliability at the system level. The chapter illustrates how accelerated aging tests, in situ monitoring, and predictive modeling can be used to assess and improve material longevity [28].

In synthesizing these themes, Chapter 17 (renumbered as Chapter 7 in this volume) positions advanced nanomaterials and emerging functional architectures as a bridge between nanoscale innovation and practical technology. By addressing stability, interfacial control, and architectural complexity, the chapter demonstrates how nanochemistry can evolve to meet the demands of real-world applications. This focus on durability and design maturity complements earlier discussions of synthesis, characterization, and application, reinforcing the central message of the second edition: that impactful nanotechnology arises when nanoscale precision is integrated with system-level thinking and long-term performance considerations [29].

Together with the preceding sections, this chapter prepares the reader for the final thematic discussion of societal impact and responsible innovation. By establishing that advanced nanomaterials can be engineered for reliability and scalability, it sets the stage for evaluating how such technologies can be deployed responsibly and equitably, ensuring that the benefits of nanochemistry extend beyond the laboratory to society at large [30].

References

[1] Bell, A. T. (2003). The impact of nanoscience on heterogeneous catalysis. *Science*, 299(5613), 1688–1691.
[2] Savage, N., & Diallo, M. S. (2005). Nanomaterials and water purification: Opportunities and challenges. *Journal of Nanoparticle Research*, 7(4–5), 331–342.
[3] Trasatti, S. (1995). Electrocatalysis in the anodic evolution of oxygen. *Electrochimica Acta*, 40(13–14), 1811–1823.
[4] Roduner, E. (2006). Size matters: Why nanomaterials are different. *Chemical Society Reviews*, 35(7), 583–592.
[5] Smith, A. M., & Nie, S. (2010). Semiconductor nanocrystals: Structure, properties, and band gap engineering. *Accounts of Chemical Research*, 43(2), 190–200.

[6] Weckhuysen, B. M. (2002). Snapshots of a working catalyst: Possibilities and limitations of in situ spectroscopy in the field of heterogeneous catalysis. *Chemical Communications*, 97–110.
[7] Cushing, B. L., Kolesnichenko, V. L., & O'Connor, C. J. (2004). Recent advances in the liquid-phase syntheses of inorganic nanoparticles. *Chemical Reviews*, 104(9), 3893–3946.
[8] Snaith, H. J. (2013). Perovskites: The emergence of a new era for low-cost, high-efficiency solar cells. *The Journal of Physical Chemistry Letters*, 4(21), 3623–3630. https://doi.org/10.1021/jz4020162.
[9] Schmid, G. ed. (2010). *Nanoparticles: From Theory To Application* (2nd ed.). Wiley-VCH. https://onlinelibrary.wiley.com/doi/book/10.1002/9783527631544?msockid=0b479d1851c06266020e8bc150026383.
[10] Jamkhande, P. G., Ghule, N. W., Bamer, A. H., & Kalaskar, M. G. (2019). Metal nanoparticles synthesis: An overview on methods of preparation, advantages and disadvantages, and applications. *Journal of Drug Delivery Science and Technology*, 53, 101174.
[11] Dong, H., Ran, C., Gao, W., Li, M., Xia, Y., & Huang, W. (2023). Metal Halide Perovskite for next-generation optoelectronics: Progresses and prospects. *ELight*, 3(1), 3.
[12] Whitesides, G. M. (2005). Nanoscience, nanotechnology, and chemistry. *Small*, 1(2), 172–179.
[13] Tian, L., & Li, L. (2018). A review on the strengthening of nanostructured materials. *International Journal of Current Engineering and Technology*, 8(02).
[14] Alivisatos, A. P. (1996). Perspectives on the physical chemistry of semiconductor nanocrystals. *The Journal of Physical Chemistry*, 100(31), 13226–13239.
[15] Brinker, C. J., & Scherer, G. W. (1990). *Sol-Gel Science: The Physics and Chemistry of Sol-Gel Processing*. Gulf Professional Publishing. https://collaborate.princeton.edu/en/publications/sol-gel-science-the-physics-and-chemistry-of-sol-gel-processing/.
[16] Ijaz, I., Gilani, E., Nazir, A., & Bukhari, A. (2020). Detailed review on chemical, physical, and green synthesis, classification, characterizations, and applications of nanoparticles. *Green Chemistry Letters and Reviews*, 13(3), 223–245.
[17] Wördenweber, R. (2022). Physical Vapor Thin-Film Deposition Techniques. In *Handbook of Superconductivity* (pp. 348–364). CRC Press. https://www.taylorfrancis.com/chapters/edit/10.1201/9780429183027-26/physical-vapor-thin-film-deposition-techniques-roger-wördenweber.
[18] Bakly, A. A. (2025). Nanocoating techniques of thin films: a comprehensive review. *Journal of University of Babylon for Engineering Sciences*, 33(5), 70–90.
[19] George, S. M. (2010). Atomic layer deposition: An overview. *Chemical Reviews*, 110(1), 111–131.
[20] Puurunen, R. L. (2005). Surface chemistry of atomic layer deposition: A case study for the trimethylaluminum/water process. *Journal of Applied Physics*, 97(12).
[21] Anastas, P. T., & Warner, J. C. (2000). *Green Chemistry: Theory and Practice*. Oxford University Press. https://global.oup.com/academic/product/green-chemistry-9780198506980.
[22] Kamat, P. V. (2011). Graphene-based nanoassemblies for energy conversion. *The Journal of Physical Chemistry Letters*, 2(3), 242–251.
[23] Zhang, H., Jin, M., & Xia, Y. (2012). Noble-metal nanocrystals with concave surfaces: Synthesis and applications. *Angewandte Chemie International Edition*, 51(31), 7656–7673.
[24] Sanchez, C., Belleville, P., Popall, M., & Nicole, L. (2011). Applications of advanced hybrid organic-inorganic nanomaterials: From laboratory to market. *Chemical Society Reviews*, 40(2), 696–753.
[25] Campbell, C. T., & Sellers, J. R. (2012). The entropies of adsorbed molecules. *Journal of the American Chemical Society*, 134(43), 18109–18115.
[26] Maier, S. A. (2007). *Plasmonics: Fundamentals and Applications* (Vol. 1, pp. 245). New York: Springer. https://link.springer.com/book/10.1007/0-387-37825-1.
[27] Whitesides, G. M., & Grzybowski, B. (2002). Self-assembly at all scales. *Science*, 295(5564), 2418–2421.

[28] Jeng, S. L., Lu, J. C., & Wang, K. (2007). A review of reliability research on nanotechnology. *IEEE Transactions On Reliability*, 56(3), 401–410.
[29] Roco, M. C., Mirkin, C. A., & Hersam, M. C. (2011). Nanotechnology research directions for societal needs in 2020: Summary of international study. *Journal of Nanoparticle Research*, 13(3), 897–919.
[30] Kulinowski, K. (2004). Nanotechnology: From "wow" to "yuck"? *Bulletin of Science, Technology & Society*, 24(1), 13–20.

Sajid Bashir, William Houf, Luis Villanueva, Pete Villareal, Jingbo Louise Liu*

Chapter 1 Nanochemistry: design and production of materials

1.1 Introduction

Nanotechnology, a multidisciplinary field, has been systematically investigated to understand, manipulate, and tune materials at either the atomic or molecular levels [1]. The design, construction, and utilization of functional structures are the critical perspectives to tackle at least one characteristic dimension measured in nanometers (nm) [2]. Nanomaterials (NMs) and systems exhibit physical, chemical, electronic, magnetic, and biological properties [3]. These properties and phenomena of NMs en-

Acknowledgments: The research efforts led by Dr Bashir and Dr Liu were supported by the American Chemical Society Petroleum Research Fund (Grant 53827-UR10) federal agencies, including the Office of Naval Research, the Air Force Office of Scientific Research, the Department of Energy, the National Science Foundation, the Department of Education, the Department of Homeland Security (ORISE), and the National Institutes of Health. The Robert A. Welch Foundation (Departmental Grant AC-0006) is gratefully acknowledged for its support of student training. The technical support from Texas A&M University-Kingsville and Texas A&M Energy Institute are also duly acknowledged. The authors also gratefully acknowledge the helpful discussions with colleagues and reviewers' suggestions, which have improved the presentation.

Authors' contributions: S. Bashir and J. Liu collectively conceived the projects and oversaw their progress. S. Bashir was in charge of the submission processing of all chapters. W. Houf, a former undergraduate student, and L. Villanueva, a former graduate student, participated as protégés to obtain material synthesis and characterization knowledge to improve their core knowledge and hands-on experience. Other co-authors have been involved in the related synthesis methods. J. Liu wrote the first draft, which all authors edited. All authors agreed to the publication of this chapter.

***Corresponding author: Jingbo Louise Liu**, Department of Chemistry, Texas A&M University-Kingsville, MSC161, 700 University Blvd., Kingsville, TX 78363, USA; Louise Nanoinnnovation North American, 7534 Beau Terre, Corpus Christi, TX 78414, USA; Texas A&M Energy Institute, 1617 Research Parkway, Suite 308, College Station, TX 77843-3372, USA, e-mails: jingbo.liu@tamuk.edu, jingbo.liu@tamu.edu
Sajid Bashir, Department of Chemistry, Texas A&M University-Kingsville, MSC161, 700 University Blvd., Kingsville, TX 78363, USA; Louise Nanoinnnovation North American, 7534 Beau Terre, Corpus Christi, TX 78414, USA; Texas A&M Energy Institute, 1617 Research Parkway, Suite 308, College Station, TX 77843-3372, USA
William Houf, Luis Villanueva, Pete Villareal, Department of Chemistry, Texas A&M University-Kingsville, MSC161, 700 University Blvd., Kingsville, TX 78363, USA

 | https://doi.org/10.1515/9783111621111-002

able diverse applications across water, energy, medicine, and food, and hence have the most significant impact on human quality of life [4, 5]. Scientists and professionals from different disciplines have revolutionized today's research breakthroughs in lowering energy consumption, developing molecular medicine, and deploying artificial intelligence (AI) systems [6]. Nanoscience, engineering, and nanotechnology involve materials imaging, property measurements, structural characterization, modeling, and the manipulation of matter at the nanoscale. From a size perspective, nanotechnology focuses on understanding and controlling materials at dimensions of roughly 1–100 nm, which are more interactive and resilient [7]. Nanotechnological research and development have been directed toward in-depth understanding and innovation in efficient and large-scale nanomanufacturing. According to end demand, advances in nanotechnology will enable industrial and military applications across diverse fields, such as medicine, catalysis, sustainable energy, consumer products, and electronics [8].

The chapter will discuss several key features of nanotechnology, including bottom-up or top-down approaches to creating tunable structures (Scheme 1.1) [9]. The fabrication variables can be optimized using AI (e.g., ANNs) and laboratory automation techniques, informed by experimental data. These structures in the 1–100 nm range exhibit behavior different from that of isolated molecules or bulk materials [10]. The shapes, architectures, and components of NMs and systems are critical factors in enhancing biological, chemical, electronic, and physical properties and further displaying unique phenomena and processes due to their nanoscale size [11]. Under certain circumstances, NMs' new properties and behaviors may be unpredictable compared to those observed at bulk scales. The most critical changes in the behavior and activities of NMs mainly result from the nanocomponents' observed intrinsic nature, instead of a sole reduction in the order of magnitude size [12]. These observations include size confinement, the predominance of interfacial phenomena, and quantum-mechanical effects correlated with electron configurations and orbital diagrams [13]. After more than five decades of intensive and systematic research in nanoscience, scientists and engineers have revolutionized materials design and production, enabling the development of more resilient materials and system architectures through functionalization, intercalation, doping, and grafting [14]. These approaches allow new features to be added to an engineered material by altering its surface chemistry and other properties [15]. Further decreases in the magnitudes of nanostructures enabled the development of unique properties in a series of newly developed materials, such as carbon nanotubes, quantum wires and dots, thin films, DNA-based structures (to control stimulus responsiveness), and laser emitters (polyhedral oligomeric silsesquioxanes) [16]. These compositions and forms of NMs and devices herald a revolutionary age for science, engineering, and technology [17].

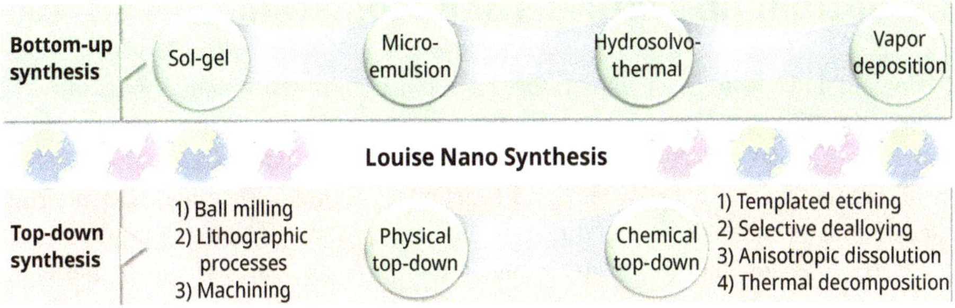

Scheme 1.1: This flowchart outlines six distinct synthesis methods for producing interactive nanomaterials. It categorizes four bottom-up and two top-down approaches, highlighting key technical variables derived from experimental data and literature analysis. Each technique is presented with its operational principles and relevance to advanced material design.

With an emphasis on the dimension and importance of quantum-mechanical effects in NMs, the National Nanotechnology Initiative defines nanotech as "the manipulation of matter with at least one dimension sized from 1 to 100 nanometers" [18]. Nanotechnology has become increasingly essential in modern scientific fields and continues to evolve to meet both commercial and academic interests [19]. The societal need for nanotechnology fosters new research and provides new insights to the scientific community. Nanotechnology can extend beyond conventional device physics to completely new approaches. Research discoveries in molecular self-assembly indicate that nanotechnological advances are opening a new paradigm for SARS-CoV-2 diagnosis, treatment, and vaccine formulation [20]. However, as a double-edged sword, nanotechnology raises problematic issues, including safety, toxicity, and environmental concerns [21]. Buerki-Thurnherr and her team reported that the engineered NMs pose a risk for "pregnancy, fetal development, and offspring health later in life" [22]. The origin and mechanisms induced by NMs will be a cornerstone, and further understanding of their function will guide the design and production of safe and sustainable NMs. The risk posed by NMs to surface water and soil near point sources has drawn significant attention due to NM-induced toxicity, which can be enhanced by sunlight irradiation, natural organic matter, and mineral particles [23]. The other potential impacts of NMs on global economics and speculation about "various doomsday scenarios" have drawn attention from scientists, policymakers, and the public [24]. These concerns have sparked a debate among advocacy groups and governments over whether special regulation of nanotechnology is warranted [25].

1.2 Bottom-up synthesis of nanomaterials

To date, Dr Liu's team and collaborators have synthesized more than 108 distinct NM formulations (Fig. 1.1). These materials span noble metals, metal oxides, metal borides, perovskite ceramics, carbon-based architectures (carbon nanotubes and graphene), metal-organic frameworks (MOFs), and functional polymers. Together, they address challenges in sustainable energy (fuel-cell and battery catalysts, electrolytes, hydrogen generation, carbon capture), water remediation, forensic analysis, cancer theranostics, disinfection, and microwave absorption. Many formulations exhibit exceptional environmental stability, rapid responsiveness to external stimuli, and self-regulating structural properties. In this chapter, we describe four bottom-up synthesis strategies: sol–gel processing, microemulsion templating, hydro(solvo)thermal crystallization, and chemical vapor deposition that deliver highly reactive nanostructures with excellent yields and clear scale-up paths. We also highlight how our group has applied machine learning algorithms to accelerate optimization of key reaction variables, leveraging accumulated experimental data to guide the design of next-generation NMs.

1.2.1 Sol–gel method

The sol–gel synthesis is a chemical process for dispersing solid substances into a liquid medium, forming a continuous 3D network that extends throughout the liquid. The sol–gel method is a bottom-up wet-chemical technique for producing nanostructured materials [26]. During this process, metal alkoxide or ionic compound solutions have commonly been used as starting materials to form a heterogeneous mixture. The above-dispersed substance (known as a sol) typically has a size range of 10^{-7} to 10^{-9} m (100 nm to 1 μm). Due to its high molecular polarity, water or an alcohol–water mixture is commonly used as a dispersing agent. The liquid-phase sol can form a gel (xerogel) under ambient conditions via Brownian motion, driven by hydrolysis and condensation during the sol–gel transition [27]. The high surface tension of these dispersed nanoparticles allows them to "glue" together during collisions. The nanoparticles can either form bonds or agglomerate by electrostatic forces [28]. During these two steps, a continuous inorganic lattice with M–O–M or M–H–M bonding will form as the skeleton for the final products [29]. Upon further polycondensation and solvent removal, three-dimensional, well-aligned networks can form, encapsulating the liquid phase through colloidal particle aggregation and/or flocculation [30]. Upon completion of the sol–gel transition, the viscosity of the 3D gelation approaches infinity. Eventually, immobility will be achieved due to the network's extension following gelation at the gel point [31]. Synthesis variables that affect sol–gel chemistry include activity (pH), solvent, reaction temperature, gelation time, and catalysts [32]. Their influences are briefly discussed below.

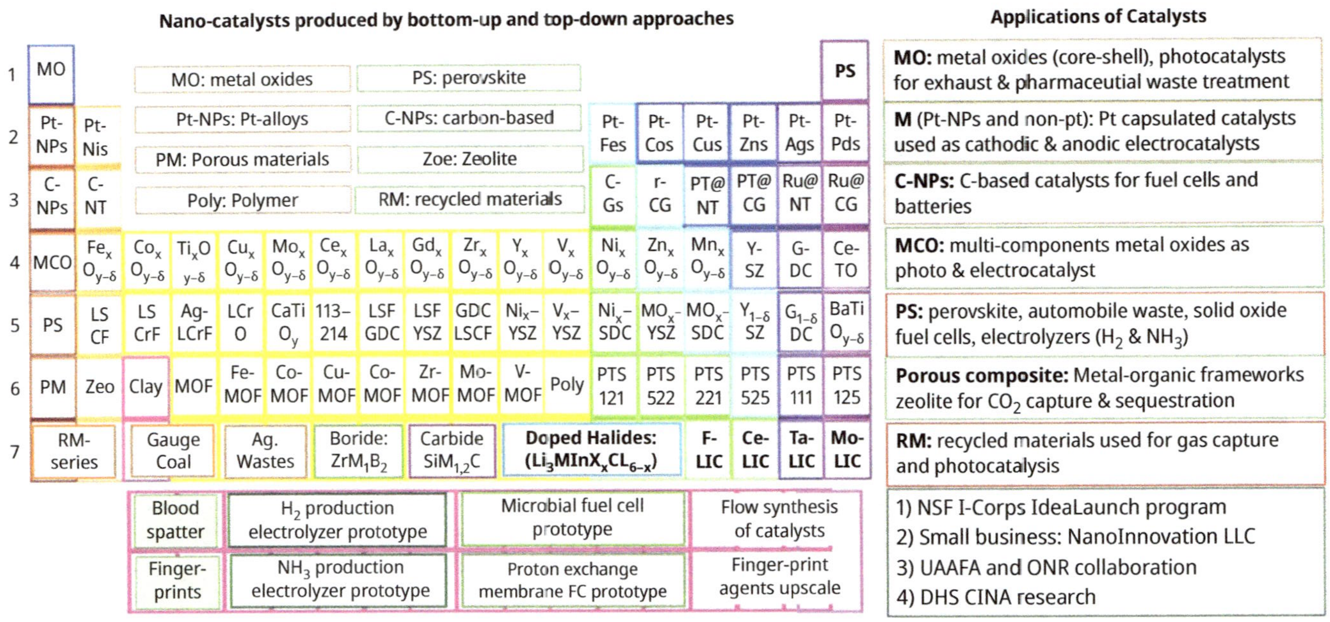

Fig. 1.1: The periodic table of different formulations of nanomaterials, prepared by bottom-up and top-down methods.

Activity effect. Sol–gel colloidal chemistry uses water as a dispersing agent, which is sensitive to pH. The hydrolysis step of most precursors is highly sensitive to acidity. As mentioned above, metal-organic compounds (typically alkoxides) can be used as precursors. In these alkoxides ($M(OR)_n$) structures, M stands for a metal or a metalloid, and R is a hydrophobic alkyl group. The alkoxide compounds are dispersed in a solvent, enabling different reactions under controlled pH conditions. At low pH (2.5–4.5), a general trend indicated that gelation time decreased with increasing pH, temperature, and water content. On the other hand, when pH rises, such as in ammonia conditions, mesoporous xerogels can be produced through cross-linking. Under these conditions, the hydrolysis of alkoxides to form ultrafine nuclei, serving as seed nuclei for gelation, can be the rate-limiting step [33]. A study on polysilsesquioxanes gelation affected by the pH (1.3–12, Fig. 1.2A) was chosen [34]. Slow gelation was observed in the pH range of 4.5–5.2 (Fig. 1.2B). The reaction rate of this system showed an "inverse volcano effect," with a decrease in rate with increasing pH, followed by a rise after a trough at about neutral conditions (Fig. 1.2C). Although the difference between alkyl and aryl groups showed a particular effect, the presence of silanols is the primary factor that causes changes in pH and, consequently, the reaction rate. The NMR data on the degree of condensation of the hexylene-bridged polysilsesquioxane (Fig. 1.2D) indicate a slow condensation rate over a wide pH range [35]. Another observation is that porous xerogels form when the degree of condensation exceeds 75%.

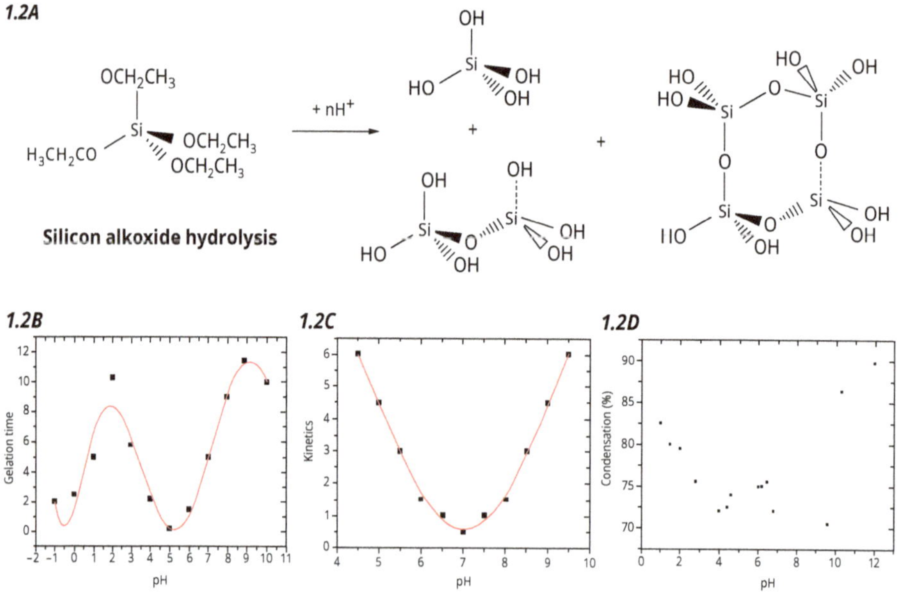

Fig. 1.2: The pH effect on the sol–gel transition of polysilsesquioxanes. A. The silicon alkoxide hydrolysis, B. gelation–pH relationship, C. minimum reaction rate–pH profile, and D. condensation degree dependent on SOH acidity.

Solvent effect. There are two classifications of solvents used in sol–gel chemistry: protic and aprotic. The solvents affect the reaction rate due to their intrinsic properties, such as polarity and molecular complexity. In particular, the rate-determining step of gelation is influenced by the solvent's properties. Protic solvents (water and ethanol) are molecules that typically contain –OH or –NH groups and are capable of forming hydrogen bonds. On the other hand, aprotic solvents (such as chloroform) cannot form hydrogen bonds or act as hydrogen-bond donors. The solvent polarity is a critical intrinsic factor related to its dielectric constant. Polar solvents typically exhibit high dielectric constants. The nonpolar solvent with low dielectric constants is due to the low electronegativity differences between two bonded atoms or the cancellation of bond dipole moments.

During polymerization, the sol–precursors agglomerate into nanoparticles as the solvent is removed. The sol–gel procedures provided advantages for obtaining metastable materials and achieving superior purity and compositional homogeneity in the products. Moderate temperatures and simple laboratory equipment can achieve the above goal by influencing particle morphology during the chemical transformation of the molecular precursor into the final products. A highly polar molecule (H_2O) has been widely used as a solvent in the sol–gel method, where three reaction types (hydrolysis, condensation, and aggregation) occur almost simultaneously. A nonaqueous approach using organic solvents was found to overcome some of the limitations of aqueous methods, provide controllable morphologies, and ensure reproducible final products, as reported by Niederberger [36]. The organic solvent played multiple roles, acting as an oxygen-supplying agent for the metal oxide, providing stabilizing effects, and exhibiting slow reaction kinetics [37]. As the oxygen-supplying agent, the organic solvent strongly influences particle size, shape, surface, and assembly properties (Fig. 1.3) [38]. The slow reaction rates observed with organic solvents were due to the moderate reactivity of the C–O bond. In combination with the stabilizing effect of the organic species, the organic solvent provides oxygen sources for oxide NMs, leading to the formation of highly crystalline products. The characterization results showed that these products exhibited uniform particle morphologies and crystallite sizes. The nonaqueous solvent can donate electrons to the metal d-orbitals, thereby strengthening the metal–oxygen bond. In short, the fundamental role of organic reaction and the intrinsic nature of nonaqueous solvents provide pathways for tailoring particle size, shape, composition, and surface properties.

Temperature effect. The chemical kinetics of the reactions involved in nanoparticle formation and gel network assembly are accelerated by temperature. In general, the sol-to-gel transition (gelation) represented a specific extent of response, which can be described using the Arrhenius equation (1.1A) and/or its logarithm form (1.1B). The gelation time (t_g) is affected by temperature. The temperature increase will raise the kinetic energy of the reactive precursors. The increased kinetic energy leads to more frequent collisions and, in turn, an increase in reaction rate:

$$k = Ae^{-\left(\frac{E_a}{RT}\right)} \tag{1.1A}$$

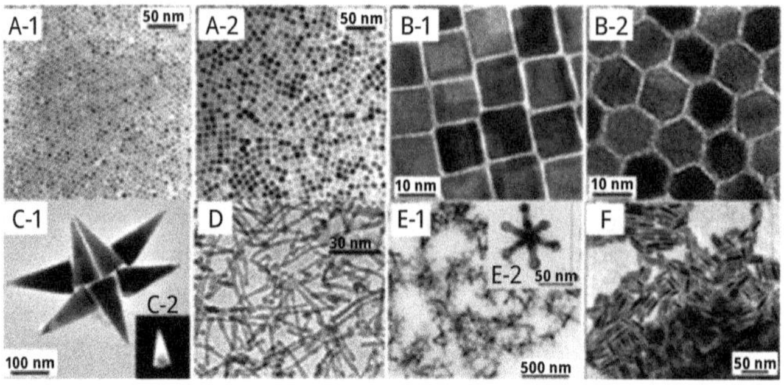

Fig. 1.3: TEM images of different nanomaterials. (A-1) spinel cobalt ferrite ($CoFe_2O_4$) spherical nanoparticles used as seeds (8 nm) and (A-2) cube-like nanoparticles. Reproduced from [39]. Copyright 2004 American Chemical Society. (B-1) cube-like $MnFe_2O_4$ nanoparticles and (B-2) polyhedron-shaped nanoparticles. Reproduced from [40]. Copyright 2004 American Chemical Society. (C-1) ZnO nanocrystals with cone-like and (C-2) zoom-in image of one crystal. Reproduced from [41]. Copyright 2005 Wiley-VCH. (D) Sol–gel derived TiO_2 nanorods with near-monodispersion, single-crystal manganese oxide. Reproduced from [42]. Copyright 2006 Wiley-VCH. (E-1) Single-crystal manganese oxide (MnO) multipods and (E-2) zoom-in image of a hexapod. Reproduced from [43]. Copyright 2005 American Chemical Society. (F) In situ one-pot synthesis of one-dimensional tungsten oxide nanorods. Reproduced from [44]. Copyright 2005 American Chemical Society.

$$\ln k = \ln A - \frac{E_a}{R}\left(\frac{1}{T}\right) \tag{1.1B}$$

where k represents the rate constant, A the frequency factor for the reaction (A, p, z, where p is the orientation probability factor, and Z is the collision frequency); E_a, the activation energy (kJ/mol), exceeds a certain energy threshold for a reaction to happen; T is the absolute temperature (K).

As shown in Tab. 1.1, the low activation energy and high reaction temperature correspond to the high fraction (f) of collisions at a higher temperature. If a reaction shows sufficient energy, equal to or greater than E_a, at a specific temperature, the frequency increases, resulting in an effective collision. As a result, the reaction rate increases. Particles have more energy at higher temperatures and collide more often and more effectively [45].

Tab. 1.1: The relationship between activation energy, temperature, and fraction of collision for the reaction.

Activation energy (kJ/mol) at 298 K	Fraction of collisions	Temperature (K) at 50 kJ/mol	Fraction of collisions
50	1.70×10^{-9}	298	1.70×10^{-9}
75	7.03×10^{-14}	308	3.29×10^{-9}
100	2.90×10^{-18}	318	6.12×10^{-9}

Research data showed that gelation can take weeks or months at low temperatures (Fig. 1.4). In contrast, at high temperatures, the reaction rate is faster, leading to a more rapid formation of the gel network. The rapid gelation typically resulted in a solid precipitate from the liquid due to the formation of large clusters. During gelation, the exothermic reaction favors the growth of nanoparticles and gel networks, leading to a faster reaction and more precipitate. Twej [46] reported that the gelling time of the silica alkoxide obtained by the sol–gel process was shorter at higher temperatures across a range of pH values and hydrolysis ratios (Fig. 1.4A). The large fraction of collisions has enough energy to exceed the activation energy, leading to reaction (Fig. 1.4B). At high temperatures, the high kinetic energy of reactant molecules is correlated to the high frequency of effective collision, meaning a larger fraction of collisions will exceed the activation energy, as shown in the green-colored region in Fig. 1.4B.

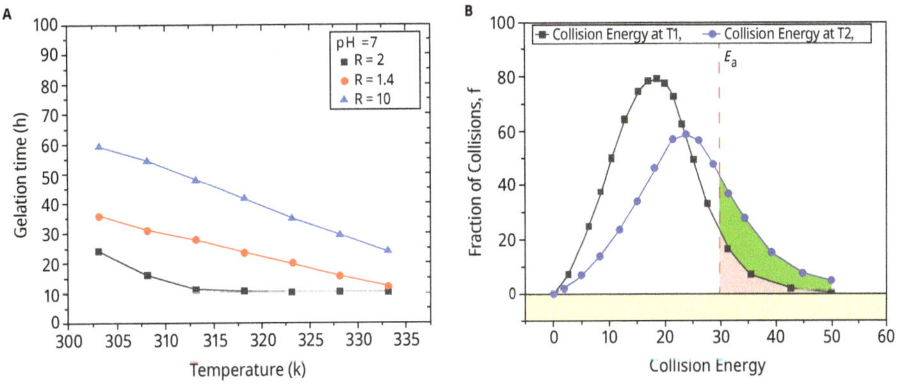

Fig. 1.4: The temperature effect on sol–gel chemistry. A. The gelation time depends on temperature and B. the relationship between the collision fraction and collision energy is strongly temperature-dependent.

Time effect. Generally, a protracted gelation interval promotes more orderly network assembly, yielding a homogeneous gel morphology (Fig. 1.5A). That said, different gel chemistries often progress through discrete kinetic stages that operate on separate time scales. As noted earlier, very fast chemistry usually leads to precipitation because network-forming events occur almost instantaneously. Gelation time is therefore a fundamental control parameter in sol–gel processing: it governs collision frequency, bond formation, and the eventual immobilization of particles into a continuous three-dimensional framework [46]. When gelation unfolds over many hours or even days, individual nanoparticles have sufficient time to diffuse, rotate, and explore configurational space, allowing them to associate with energetically favorable neighbors before irreversible metal–oxygen (M–O–M) or metal–hydrogen (M–H–M) linkages lock the structure. This slow assembly phase encourages the growth of a well-ordered, polymeric network: pore sizes remain consistent, particle domains orient coherently, and internal stress accumulation is minimized. The practical outcome is a monolithic gel whose microstructure translates into reproducible mechanical properties, narrow pore-size distributions, and high optical homogeneity after drying or thermal consolidation.

Conversely, when gelation is driven to completion rapidly by concentrated precursors, potent catalysts, or elevated temperatures, nucleation and condensation events cascade in quick succession. Newly formed nanoparticles bond almost immediately, producing compact aggregates that precipitate before a percolating network can form. Such precipitates characteristically exhibit irregular clustering, broader particle size distributions, and poorly interconnected pore channels. Rapid gelation often coincides with syneresis (solvent expulsion) and crack formation, both of which degrade the material's structural performance [47].

Moreover, many sol–gel recipes display multiple kinetic regimes. For instance, an alkoxide precursor may hydrolyze slowly at first to yield small sol particles, then undergo a critical induction period, followed by a rapid condensation burst that accelerates cluster growth. Organically modified silanes, by contrast, can exhibit separate pre-gel and post-gel phases controlled by different rate-limiting steps (hydrolysis versus polycondensation). By characterizing these time-dependent transitions with rheometry, light scattering, or calorimetry, practitioners can schedule gelation to strike the desired compromise between throughput and microstructural uniformity. In short, precise control of gelation time is the keystone for converting molecular precursors into nanostructured solids whose functionality depends on defect-free, well-aligned networks.

Use of catalysts. Catalysts speed up chemical transformations, and gelation in sol–gel systems is no exception. Both homogeneous acids (H^+) and bases (OH^-) are commonly employed to tune reaction kinetics, and different catalytic regimes can produce distinct formation mechanisms (Fig. 1.5B) [48]. Sol–gel reactions, whether in water or in organic media, are susceptible to pH. Hence, the choice and loading of the catalyst strongly influence the pathway by which precursors convert into a network. By lowering the activation barrier, a catalyst increases the fraction of collisions that lead to reaction and can therefore divert the system onto an alternative kinetic route [49].

Under acid-catalyzed conditions, protonation of metal–alkoxide centers makes the metal site more electrophilic and more prone to nucleophilic attack by water. Hydrolysis (M–OR → M–OH) becomes fast, while subsequent condensation of the newly formed hydroxyls into M–O–M linkages is relatively sluggish. The net effect is the formation of many small, highly branched primary particles that connect into a loose, open network. Gels produced this way tend to exhibit fine, narrowly distributed pores and high optical transparency, although the slower condensation step typically lengthens processing times.

By contrast, base catalysis (OH^-) promotes nucleophilic attack on M–OR groups and accelerates condensation so that bridge formation (M–O–M) can proceed nearly as rapidly as hydrolysis. This favors rapid cluster growth and the formation of denser aggregates within minutes, yielding coarser networks with larger pores. Such behavior can be advantageous when producing thick films or coatings where rapid network buildup is desired, but it makes achieving nanoscale uniformity more challenging [50].

Because both aqueous and nonaqueous sol–gel routes are strongly pH-dependent, practitioners routinely vary catalyst identity and concentration to tune gelation times from under an hour to several days, depending on the target microstructure. In practice, acid catalysts are often chosen when ultrafine porosity and optical quality are priorities (e.g., photonic or sensing materials). In contrast, base catalysts are selected when faster network formation or thicker deposits are required. In all cases, the catalyst's role is the same: reduce activation energies for hydrolysis and condensation, increase the proportion of effective collisions, and open alternative reaction channels that enable scalable, precise control of nanostructure, porosity, and mechanical performance.

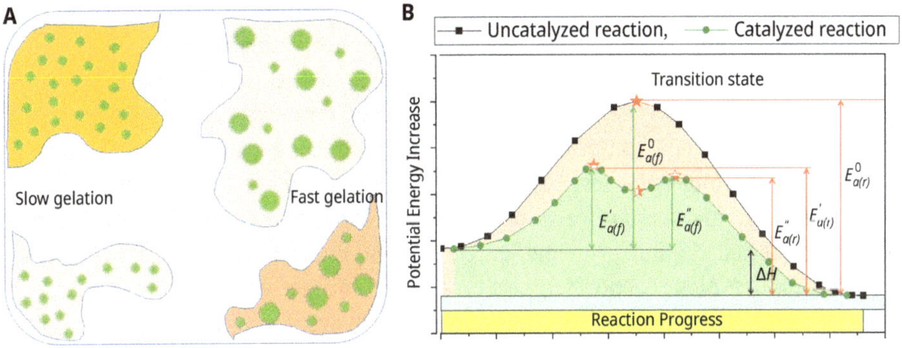

Fig. 1.5: The time and catalyst effects on sol–gel chemistry. A. The gelation time affects particle morphology and size and B shows the schematic of the different reaction pathways when catalysts are used.

Mixing/agitation. Mixing and agitation are essential for achieving a homogeneous solution and guiding the transition from dispersed precursors to a coherent gel. Mechanical or magnetic stirring enhances both mass and heat transfer, improving contact among reactants and evening out concentration and temperature gradients across the vessel. Continuous motion reduces local supersaturation and limits uncontrolled agglomeration by preventing nascent nuclei from growing unchecked, while also influencing the practical chemistry of the components through improved transport [51].

Adequate stirring keeps precursors uniformly distributed, ensuring that hydrolysis and condensation occur evenly throughout the volume. Circulation accelerates thermal equilibration, damping hot spots that could otherwise drive localized, rapid reactions. Enhanced mass transfer reduces concentration gradients and curbs the formation of oversized clusters, which in turn suppresses unwanted agglomeration and helps maintain narrow pore-size distributions.

Mixing rate controls morphology. At moderate agitation, particles have sufficient time to collide, bond, and reorganize into an interconnected network, producing gels with consistent pore architecture, predictable mechanical properties, and reproducible optical clarity. Gentle stirring can be used deliberately to bias particle growth toward larger, well-ordered domains, whereas modestly higher shear tends to favor the formation of finer, more uniform nanostructures.

Excessive shear, however, is detrimental. Vigorous mixing or high-energy ultrasonication can mechanically disrupt growing gel strands, fragmenting incipient domains before they fuse into a continuous matrix. The outcome may be irregular pore geometry, weakened mechanical integrity, or incomplete percolation of the network. In extreme cases, intense agitation prevents gelation entirely, leaving a stable suspension of discrete nanoparticles rather than a monolithic gel.

By tuning agitation intensity and mode (steady stirring, pulsed shear, or controlled sonication), researchers can steer the sol–gel pathway toward uniformity or toward specific morphologies. This parameter, together with stoichiometry and synthesis conditions, enables reproducible scale-up of sol–gel processes for advanced NMs. Sol–gel systems containing multiple components are susceptible to mixing: adjusting the number and composition of sols and optimizing synthesis parameters allows atomic- and molecular-level control over homogeneity and functionality.

Throughout this chapter, we emphasized how small changes in the recipe reshape the final material. The five principal levers, pH (acid vs. base catalysis), solvent choice (aqueous *vs.* organic), reaction temperature, gelation time, and mixing intensity, collectively determine how metal-alkoxide precursors hydrolyze, condense, and interlink into a three-dimensional network. In the chapters that follow, we will examine the hydrolysis of M–OR bonds, derive rate laws for polycondensation (M–OH → M–O–M), and analyze the gelation threshold where viscous sols become solid gels. Mastery of these parameters has enabled Liu's laboratory to translate bench-scale sol–gel chemistry into robust, application-ready NMs, titanium and mixed-oxide photocatalysts for water remediation, multi-metal electrocatalysts for energy conversion, and metal-oxide powders for high-resolution latent-fingerprint development, demonstrating that controlled mixing is as crucial as chemistry in producing functional materials [52].

1.2.2 Microemulsion synthesis

According to the International Union of Pure and Applied Chemistry (IUPAC), a microemulsion is a dispersion composed of water, oil, and surfactant(s) [53]. This system is isotropic and thermodynamically stable, with dispersed domains ranging from 1 to 100 nm in diameter. Usually, these dispersed particles are 10–50 nm in size, and the surfactants are combined with cosurfactants [54]. In the ternary microemulsion, water and "oil" (any non-water-soluble liquids) are present with a surfactant. There are three basic microemulsion families: oil-in-water (o/w), water-in-oil (w/o), and bicontinuous. Unlike the formation of traditional colloids, these microemulsions can be produced by simple mixing of the components without the need for high shear [55]. Microemulsion synthesis has become popular due to its versatility in precisely controlling particle size and properties. A series of monodispersed spherical polymer particles (e.g., polystyrene, poly(methyl methacrylate), and poly(hydroxyethyl methacrylate)) can be prepared using this versatile, scalable microemulsion approach. This feasible wet chemistry enables the production of a wide variety of materials, combined with other synthesis methods developed in Dr Liu's and her collaborators' laboratories. One group of materials is electrocatalysts for green hydrogen production via water electrolysis, nanocomposites for the oxygen reduction reaction, and perovskite-based catalysts for green NH_3 production via solid oxide electrolysis cells. Other materials include photocatalysts for water remediation, flexible ternary metal oxide for emerging energy storage using capacitors, porous materials (e.g., MOF) for carbon capture and conversion, metal oxide catalysts to convert plastic wastes into hydrogen and nanotubes, biocatalysts used in microbial fuel cells, CNT-based sensing elements for in situ COVID-19 detection, nanogold- based materials for cancer imaging and therapy and materials used in lithium battery. This section will discuss the reaction during three key steps: microemulsion, polymerization, cross-linking, and carbonization. The different types of surfactants and their functions will also be addressed to guide their selection.

A simplified flowchart for producing carbon-based materials using the microemulsion is shown in Fig. 1.6A. Three key subsequent steps, polymerization, cross-linking, and carbonization, will occur. A monolayer at the interface between two immiscible phases of water and oil can be formed due to the presence of surfactant molecules. The hydrophilic "tails" of the surfactant allow for interaction within the aqueous phase, while the hydrophobic ends can be dissolved in the oil phase. The mechanism of microemulsion formation can ensure tunable size and geometry, controllable morphology, and homogeneity, as well as a high surface area [56]. The materials produced by microemulsion chemistry are summarized in Fig. 1.6B, and the phase diagram is shown in Fig. 1.6C. The critical factors that affect microemulsion chemistry, pH, surfactants, the catalyst used, and temperature effects follow a similar theory to the sol–gel method and will be discussed in Chapter 2.

As shown in the ternary phase diagram (Fig. 1.6C), the microemulsion domains consist of three components: two immiscible liquids (e.g., water and oil) and a surfactant or cosurfactant. Totally, five regions will be observed: (a) molecular solution (red-colored), (b) water-in-oil (w/o) emulsion (blue), (c) bicontinuous (b/c) structure (yellow), (d) oil-in-water (o/w) emulsion (purple), and (e) conventional two-phase emulsion (green). In the oil–water interfacial layer, the b/c microstructure has zero mean curvature, meaning it is flat on average. The disordered sponge-like structure of microemulsions enables the incorporation of the vinyl monomer into different phases (dispersed, continuous, and bicontinuous) during the polymerization. As a result, this incorporation makes the microemulsion practical to design and prepare versatile microemulsion polymers.

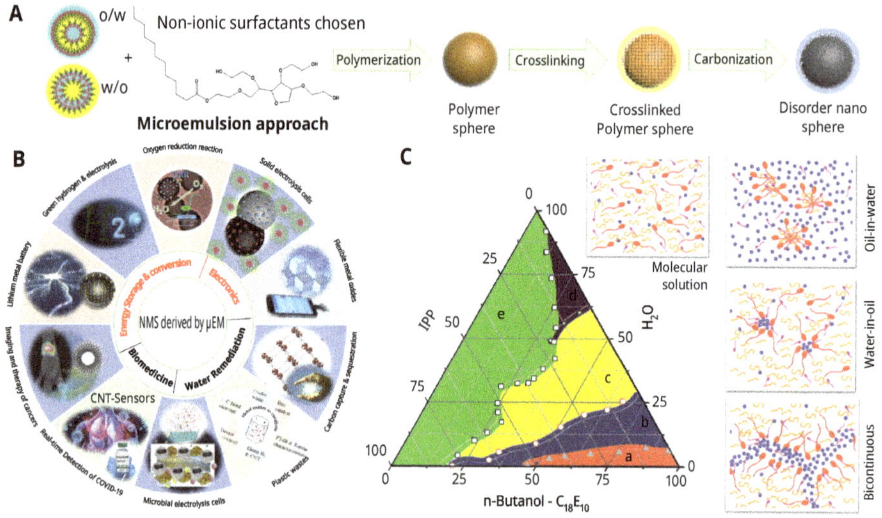

Fig. 1.6: The microemulsion chemistry. A. The simplified flowchart shows three key steps in the microemulsion synthesis of carbon-based materials and B. the summary of NMs produced using microemulsion (bio-imaging graphs were reproduced from [57], Copyright 2010 Elsevier Press. The real-time detection of COVID-19 image was reproduced from [58]. Other data are collected from Liu's group), and C. a typical ternary phase diagram shows the three families of emulsion systems: oil-in-water (o/w), water-in-oil (w/o), and bicontinuous (b/c) phases (data were reproduced from [59]; J. Liu prepared the plot).

The surfactants or cosurfactants are ionic and/or covalent compounds, enabling a synergistic effect for producing stable emulsion systems. When a cosurfactant is used, optimizing the molar ratio between surfactant and cosurfactant is required to form a single component, which is treated as a single "pseudo-component." These three components, water, oil, and surfactant, are found at the triangle's apex, corresponding to the volume fraction of 100%. The component volume fraction decreases as it moves away from the corner. The different points within the triangle correspond to the compositions of the three components in the mixture. The region defined by these points represents the "phase behavior" of the microemulsion system at constant temperature and pressure [60].

Polymerization: Emulsion polymerization is a radical polymerization process in which an emulsion contains water, monomer, and a surfactant [61]. The process usually involves monomers, initiators, a dispersion medium, and a colloid stabilizer, which together form a homogeneous system. Three commonly used emulsions are oil-in-water (o/w), water-in-oil (w/o), and bicontinuous. In the o/w emulsion, the droplets of hydrophobic oil monomer will be emulsified in a continuous water phase in the presence of surfactants. An emulsifier (also called a stabilizer, such as polyvinyl alcohol) prevents particles from agglomerating or coagulating. This polymerization is heterogeneous, and its reaction mechanisms and kinetics differ dramatically depending on the system. This free-radical polymerization mechanism involves three steps: initiation, propagation, and termination, as shown in eqs. (1.2A–C):

Initiation (bond homolysis):

$$I-I \xrightarrow{k_\mathrm{d}} 2I^* \tag{1.2A.1}$$

$$I^* + M \xrightarrow{k_\mathrm{i}} M_1^* + I \tag{1.2A.2}$$

Propagation:

$$M_1^* + M \xrightarrow{k_\mathrm{p}} M_2^* \tag{1.2B.1}$$

$$M_2^* + M \xrightarrow{k_\mathrm{p}} M_3^* \tag{1.2B.2}$$

$$M_2^* + M \xrightarrow{k_\mathrm{p}} M_3^* \tag{1.2B.3}$$

$$M_n^* + M \xrightarrow{k_\mathrm{p}} M_{n+1}^* \tag{1.2B.3}$$

Termination:

$$M_m^* + M_m^* \xrightarrow{k_\mathrm{t,c}} M_{n+m}^* \tag{1.2C.1}$$

$$M_m^* + M_n^* \xrightarrow{k_\mathrm{t,d}} M_{m+Mn} \tag{1.2C.2}$$

where I–I represents the initiator to start the homolytic bond cleavage under light or thermal decomposition; I^*, the initiator radical; M, the reactant monomer; M_1^*, free radical with different monomer units; and M_m, the terminated polymeric chain. The constant values are k_d for the homolysis rate constant of the initiator; k_i for the rate constant of initiation of monomer; k_p for the rate constant of propagation; $k_\mathrm{t,c}$ for the rate constant of combined termination; $k_\mathrm{t,d}$ For the rate constant of disproportionation termination of two different radicals (Fig. 1.7). The green region shows the differences between ideal kinetics and limiting conditions.

Cross-link: In the formation of the microemulsion, the cross-link refers to the intermolecular or intramolecular attraction of two or more molecules through the formation of a covalent bond. Based on the IUPAC definition, the cross-link involves "a small region in a macromolecule from which at least four chains emanate, and formed by reactions involving sites or groups on existing macromolecules or by interactions between existing macromolecules." These small regions may be "an atom, a group of atoms, or several branch points connected by bonds, groups of atoms, or olig-

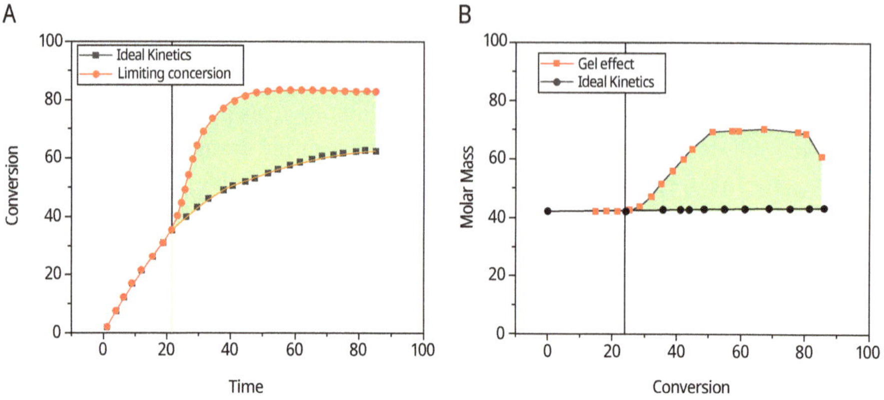

Fig. 1.7: The microemulsion chemistry. A. The time dependence of the conversion during polymerization and B. the molar mass depends on the conversion process (imaging graphs were reproduced from [62]).

omeric chains" (Fig. 1.7A). The cross-link primarily involves covalent bonds, but weaker chemical interactions also occur. The crystallite portions, as well as physical interactions and entanglements, may also be observed [63]. There are two significant types of cross-linking: physical cross-links formed by weak interactions (intermolecular forces) and chemical crosslinks, which are formed through covalent bonding. The oxidative cross-link typically occurs when the reactants are exposed to atmospheric oxygen. Factors such as reaction temperature, pressure, acidity, and irradiation can initiate cross-linking during chemical reactions (Fig. 1.7B). A well-known example is that mixing an unpolymerized resin with cross-linking reagents initiates a chemical reaction. Thermoplastic materials (e.g., C-type cross-linked polyethylene) can be cross-linked using an electron beam [64]. Type A cross-linked polyethylene can be produced when oxidizing agents (such as hydrogen peroxide, H_2O_2) are used during extrusion. The cross-linked membrane (PTS), used as an electrolyte in microbial fuel cells, was chosen to demonstrate the formation of chemical bonds and hydrogen bonds (Fig. 1.8).

Carbonization: An exothermic process that converts organic matter (such as plants and/or animal remains) into carbon. This process is completed through a destructive distillation, an application of pyrolysis. Carbonization is a pyrolytic reaction that decomposes unprocessed material by heating organic matter to high temperatures, typically in the absence of air or other reagents, catalysts, or solvents. Different reactions occur concurrently during this complex process. The reactions "dehydrogenation, condensation, hydrogen transfer, and isomerization" typically occur during this carbonization due to their rapid rates. The degree of carbonization and the amount of the foreign element residual highly depend on the final pyrolysis temperature. The amount of heat applied controls the degree of carbonization and the residual content of foreign elements. Carbonization can be a self-sustaining energy source that does not produce carbon dioxide. For example, glucose carbonization

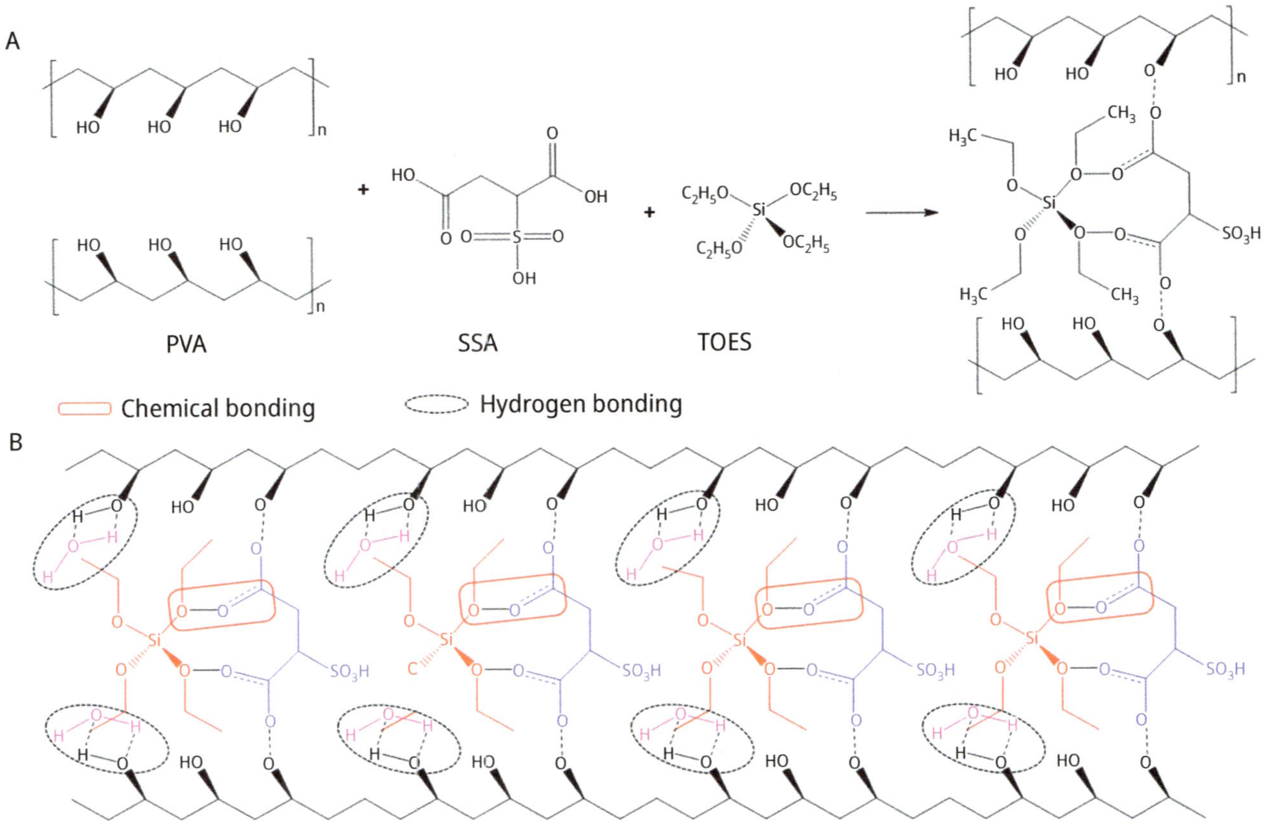

Fig. 1.8: The cross-linking of PTS formation using microemulsion chemistry. A. The monomer formation through covalent bonding and B. the tetramer formation shows chemical bonding (the red-squared region) and intermolecular forces (the black oval area; the imaging graphs were reproduced from [65]).

releases about 237 calories per gram. The intense heat can cause many organic materials to carbonize, turning them into solid carbon quickly when exposed to sudden, searing heat [66]. A series of carbonization processes and a simplified apparatus are listed in Fig. 1.9 [67].

Carbonization occurs under pressure and can be classified into three groups: (a) under pressure resulting from gases decomposed from the reactants, (b) under hydrothermal conditions, and (3) reduction of CO_2 under pressure. The yield and morphology of resultant carbon-based materials are the focus under carbonization conditions. The reaction variables that determine the type of carbon material include temperature, pressure, reactor volume, and the chemical composition of the precursors. Carbonization is a slow exothermic reaction and one of the pyrolysis processes. Typically, carbonization occurs during the heating of the biomass in an oxygen-free or oxygen-limited environment. The final product yield can be maximized and the morphology tuned by varying reaction conditions (Fig. 1.9A). For biomass conversion into highly carbonaceous materials, hydrosolvothermal processing may be a viable approach. Although carbonization has been used to convert dead vegetation into coal, this slow but practical process produces nanocarbon-based materials. It is critical to note that carbonization may not occur during microemulsion synthesis unless carbon coating is required.

Several factors affect emulsion morphology, porosity, and properties. The effects of initiators, the monomer addition methods, and the roles of surfactants will be discussed in the following sections (Fig. 1.9B). Other factors (reactant volume ratio, molar mass, and reaction temperature) will be addressed in Chapter 2.

Effect of initiators: The water-soluble initiator can be used in emulsion polymerization to generate oligomeric radicals in the continuous aqueous phase. These initiators are hydrophilic and often anionic (such as SO_4^{2-} derived from $Na_2S_2O_8$). With the increase in hydrophobicity of these oligomeric radicals, the initiator can enter the monomer, swelling the polymer particles. During the latex particle formation, the negatively charged sulfate end remained on the particle surface. It prevented the entry of free radicals into the interior, thereby encapsulating seed particles. The non-uniform distribution of free radicals in the reaction loci significantly affects the morphology and size of the final products [71]. Unfortunately, the formation of a perfect core/shell structure or uniformly distributed particles cannot be ensured without optimizing other factors, such as reactant polarity and reaction conditions. If an oil-soluble initiator is used, the free radicals distribute more evenly throughout the particles (e.g., latex). Emulsion polymerization will result in a morphological structure in which the post-formed polymer can be present at the latex particle surface and within the composite particle [72]. The different initiators have been investigated to evaluate their effects on the emulsion polymerization of styrene. Examples are potassium persulfate ($K_2S_2O_8$), 2,2′-azobisisobutyronitrile ($[(CH_3)_2C]_2N_2$, alcohol-soluble), and 4,4′-azobis(4-cyanovaleric acid) (water-soluble organic compound, less hydrophilic than $K_2S_2O_8$, $HOCOCH_2CH_2C(CH_3)(CN)N{=}NC(CH_3)(CN)CH_2CH_2COOH$).

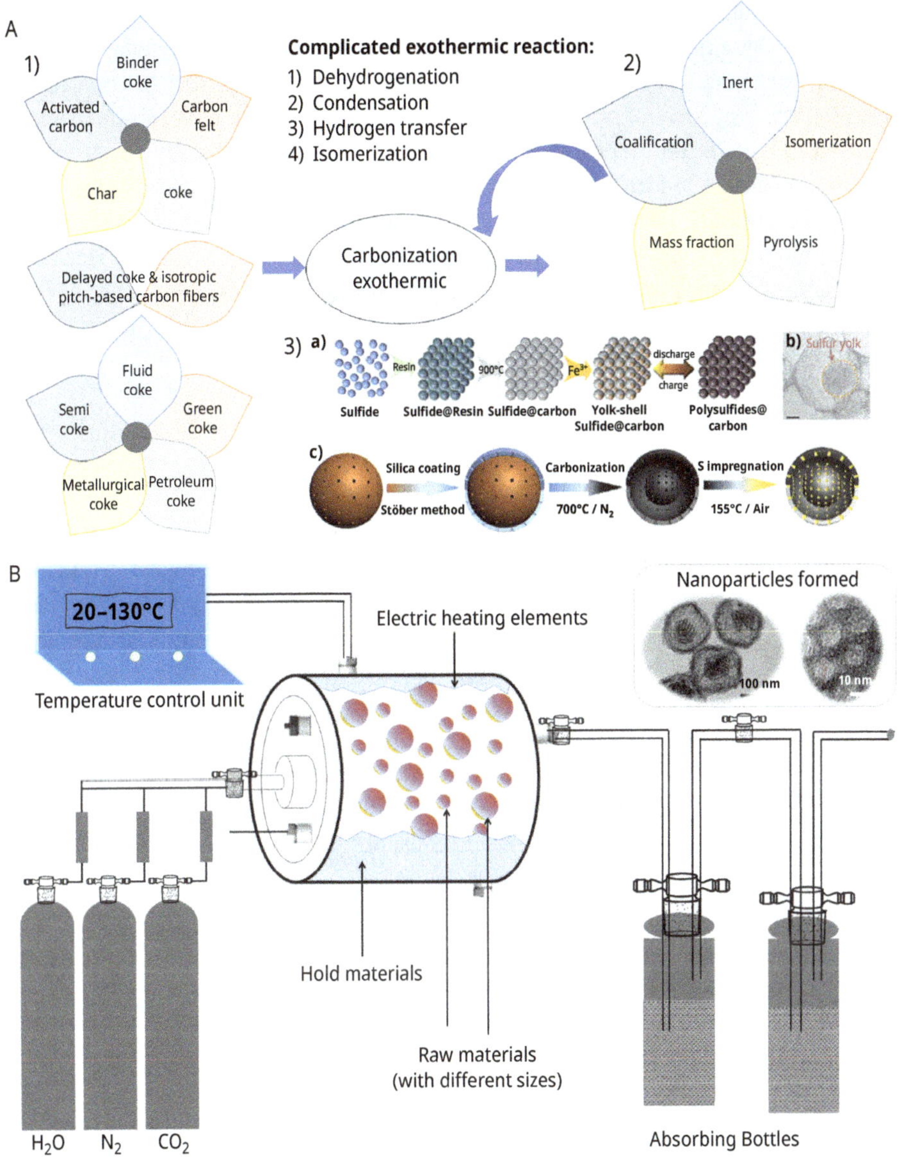

Fig. 1.9: The carbonization process in producing carbon-based nanomaterials. A. Different procedures of carbonization (image in A-1 and A-2) are adapted from [68] (the image in A-3); sulfur nanoparticles encapsulated inside a hollow carbon nanosphere were reproduced from [69]. An apparatus used during carbonization (adapted from [70], and nanoparticle images were copied from [68]).

Potassium persulfate decomposes in the water to form negatively charged sulfate radicals (eq. (1.3A-1)). These free radicals will further react with the monomers (M), which are dissolved in water and form soap-type free radicals, as shown in eq. (1.3A-2). The various morphological structures of latex particles were formed when the potassium persulfate was used to initiate free radicals [73]. These radicals can be associated with dissolved emulsifier molecules to form micelles. They can migrate into existing micelle droplets to form large particles by adding to the growing polymer chains. In this case, polymerization occurs in the water phase due to the solubility of the monomers and initiators. However, the monomers do not provide polymerization sites because the surfactant molecules encapsulate them. The encapsulated monomer droplets, which are negatively charged, are virtually impermeable to negatively charged initiator molecules:

$$S_2O_8^{2-}\,(aq) + heat \rightarrow 2SO_4^{-\,\cdot}\,(aq) \tag{1.3A.1}$$

$$SO_4^{-\,\cdot}\,(aq) + (n+1)M \rightarrow SO_4^{-} + (CH_2 - CX_2)_n\text{–}CH_2\text{–}CX_2. \tag{1.3A.2}$$

On the other hand, inverted core/shell latex particles were obtained when 2,2′-azobisisobutyronitrile (AIBN) or 4,4′-azobis(4-cyanovaleric acid) (ACVA) was used as the initiator, due to their insolubility in water. The thermal- or light-initiation reactions for both AIBN and ACVA are shown in eqs. (1.3B) and (1.3C). The radical formation from AIBN normally occurs at 85 °C within 30 min, while the ACVA occurs at 69 °C within 10 h. The previous study indicated that the initiator concentration and polymerization temperature largely determined the morphology of the subsequent products [74]:

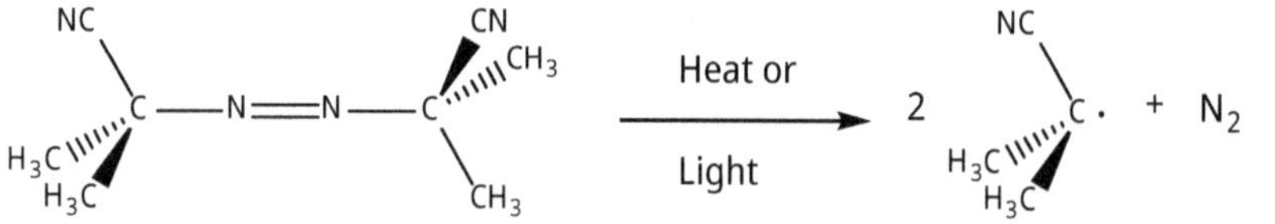

(1.3B)

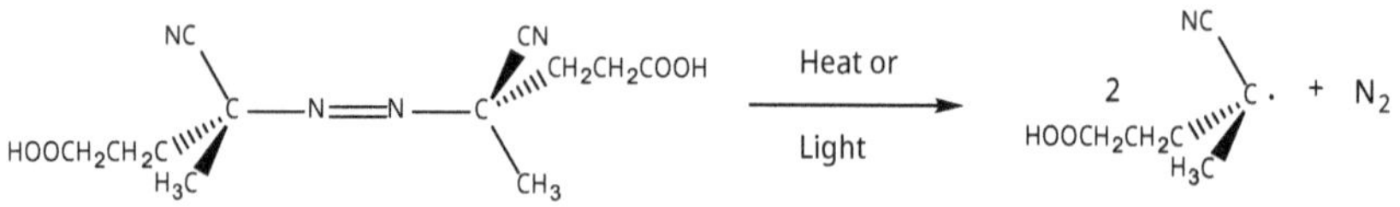

(1.3C)

Effect of monomer addition methods: In emulsion polymerization, the sequence of monomer addition is critical for determining molecular weight, degree of grafting, and the morphology of the final products. It is difficult to control the uniformity of polymerization due to unavoidable irreversible radical termination. As reported by Luo, “over 500 kg mol^{-1} polystyrene with high livingness and low dispersity could be synthesized by a facile two-stage reversible addition-fragmentation transfer emulsion polymerization” [75]. This method enabled the high monomer conversion (90%) within 10 h. Both initiator concentration and the monomer chain transfer constant were significantly lower than the well-accepted values in the conventional method. All the oligomeric reversible addition-fragmentation transfer (RAFT) agents and the initiator were charged in the first stage (initiation to target at 30 kg/mol). A fraction of the styrene monomer was added to the above system. After 100 min of initiation, the remaining styrene monomer, water, and strong base (such as NaOH) were fed for polymerization. The polymerization kinetics and the evolution of molecular weights and dispersities with monomer conversion are shown in Fig. 1.10.

It can be seen that the two-stage synthetic strategy results in a significant reduction of dispersities while the polymerization rates remained comparable to those in the batch cases. This report introduced the styrene monomer in two stages, maintaining the same monomer concentration at each stage. The two-stage method avoids dealing with the low RAFT transfer rate or slow feeding rate. As a result, monomer concentration and propagation rate have been improved compared with the traditional RAFT reaction. Data (Fig. 1.10A) showed that the molecular weight of the high-living polymer exceeded 500 kg/mol, and a low dispersity (ca. 1.5) could be obtained within an adequate polymerization time (Fig. 1.10B). Continuous feed stream, reactor configuration, mean residence time, and precursor concentration are also critical for determining the particle size distribution and for obtaining various morphological structures in the effluent product (Fig. 1.10C).

Surfactant/cosurfactant: In the emulsion system, the incompatibility between oil-soluble and water-soluble phases results in thermodynamic instability. The use of surfactants during emulsion formation can minimize the above issues and prevent nanoparticles from agglomerating. The introduction of surfactants can lower the Gibbs free energy ($\Delta G = \Delta H - T\Delta S$, $\Delta G = g\Delta A - T\Delta S$) since molecules adsorbed on the emulsion droplets result in a decrease in the interfacial tension ($g\Delta A$). Surfactants are amphiphilic organic molecules that simultaneously comprise hydrophilic (head) and hydrophobic (tail) groups to lower surface or interfacial tension [76]. Since surfactants contain water-soluble and oil-soluble components, they can be dissolved in water and oil. Surfactants will diffuse into water and adsorb at the air–water or oil–water interface. These hydrophobic tails may extend into the air or the oil phase. At the same time, the hydrophilic head remains in the water phase (Fig. 1.11A). This schematic diagram shows the formation of a micelle, in which the oil-soluble tails reside in the inner layer. The lipophilicity of the surfactant allows stronger attraction of its tails to the oil phase than to water. The polar heads of the surfactant strongly attract polar water molecules, forming a hydrophilic outer layer. Therefore, a barrier between micelles inhibits oil droplets from agglomerating into larger clusters. In other cases, reverse micelles (Fig. 1.11B) can form when the polar ends face the inner side of the water droplets, with the nonpolar tails on the outer side. These reverse micelles are also water droplets (1–10 nm in size), but the surfactant molecules are organized differently than in micelles.

The polar ends of surfactants facing the inner side can solubilize water, while the nonpolar ends face outside toward the organic phase (Fig. 1.11). Although surfactant molecules have single or double tails, Fig. 1.11A shows only single-tailed schematics. The formation processes of both micelles (Fig. 1.10B) and reverse micelles can be reversible due to the weak intermolecular attractive forces, such as London dispersion

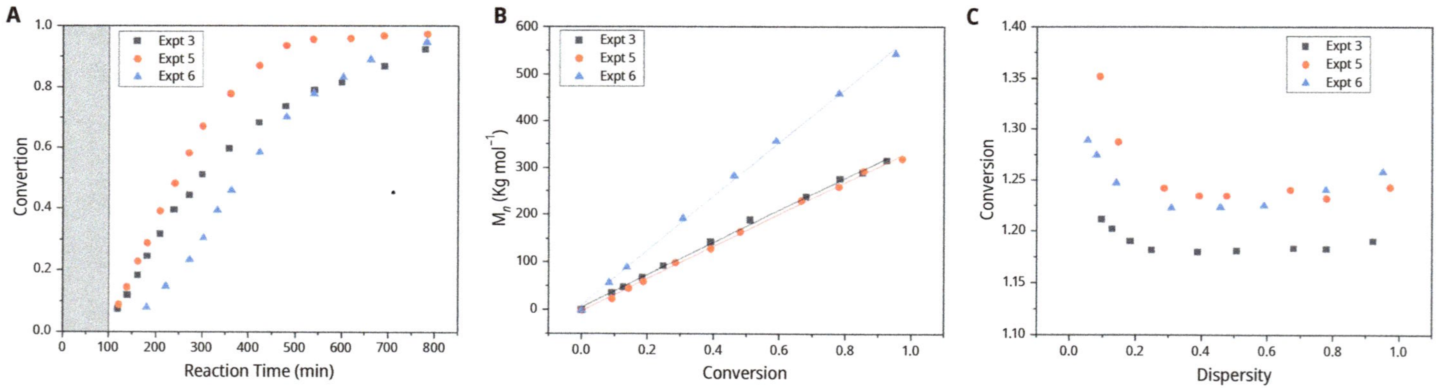

Fig. 1.10: Experimental results of two-stage polymerization of styrene. Experiments 3 and 5 targeted about 350 kg/mol, and 6 at 640 kg/mol. A. Styrene monomer conversion as a function of polymerization time, B. the molecular weight of the polymer as a function of monomer conversion, and C. the dispersity versus monomer conversion. Data were adapted with permission from [75].

(Fig. 1.11C) and dipole–dipole moments (Fig. 1.11D), and hydrogen bond (Fig. 1.11E). The London dispersion can be generally found between nonpolar molecules due to the instantaneous polarization of electron clouds. It is an instantaneous dipole-induced dipole attraction between adjacent atoms. The energy magnitude is relatively low, ranging from 0.05 to 40 kJ/mol [77]. The dipole–dipole moments are attractive forces between polar molecules, from the positive end of one molecule to the negative end of another polar molecule. This attraction is also relatively low, ranging from 5 to 25 kJ/mol [78]. The hydrogen bond (in short, H-bond) is a primarily electrostatic attraction between an H atom and another atom with more electronegativity. The atom can be an oxygen (O), nitrogen (N), or fluorine (F) atom, which bears a lone pair of electrons and acts as an H-bond donor [79].

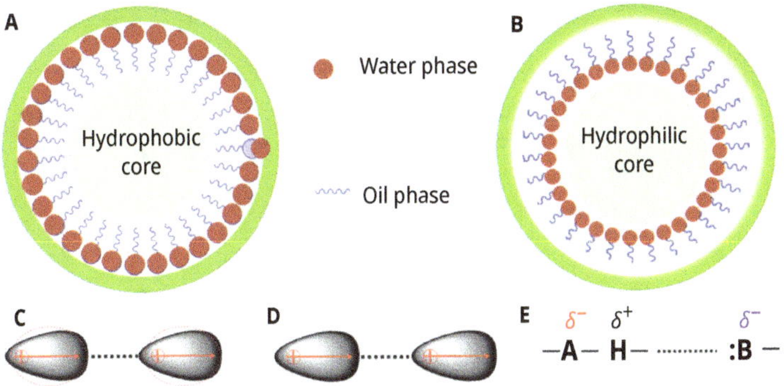

Fig. 1.11: The formation of micelles and reverse micelles through intermolecular forces. A. The schematic diagram of a micelle with a hydrophobic core (with a single tail); B. the diagram of reverse micelles with a hydrophilic core; C. the London dispersion attraction due to polarizable electron clouds between nonpolar molecules; D. the dipole–dipole attraction due to dipole charges between polar molecules; and E. Hydrogen bond formation via a polar bond to H-dipole charges (note: A and B must be one of the N, O, or F atoms). These relatively weak van der Waals attractive forces allow micelles to form and dissociate reversibly.

Most surfactants comprise a polyether chain with a high polar anionic group at the terminal. The polyether groups usually consist of ethoxylated sequences inserted to increase the surfactant's hydrophilicity, due to the polarity of the carbon–oxygen bond. Conversely, the nonpolar sequences in polypropylene oxides can increase the lipophilic character. The surfactants are classified into two groups (ionic and nonionic) based on their polar head groups. There are three types of ionic surfactants: anionic with a negative charge at the polar head, cationic with a positive charge, and zwitterionic (also known as amphoteric) with two oppositely charged heads. The nonionic surfactants contain ether [–$(CH_2CH_2O)_nOH$] and/or hydroxyl [–OH] hydrophilic groups, which are not ionized at any acidity. They are also nonelectrolytes, which are less sensitive to variations in the medium's pH. They can be used to stabilize o/w and w/o emulsions due to their lower sensitivity. They are composed of positively charged heads and can be used in beauty products. Cationic detergents are incredibly harsh and can disrupt the cell membranes of bacteria and viruses. Anionic surfactants consist of negatively charged heads, such as sulfonate (R–SO_3^-), phosphate (PO_4^{3-}), sulfate (SO_4^{2-}), and carboxylate (RCO_2^-). The zwitterionic (also known as amphoteric, shown in Fig. 1.12) surfactants consist of negative and positive charged heads attached to the same molecule. The anionic groups can vary and include sulfonates and betaines, while the cationic functional groups are based on primary, secondary, or tertiary amines or quaternary ammonium cations. These zwitterionic surfactants are sensitive to pH changes and behave as anions or cations at different pH values. The above amphoteric molecules tend to diffuse toward the interface between water and oil phases and reside therein.

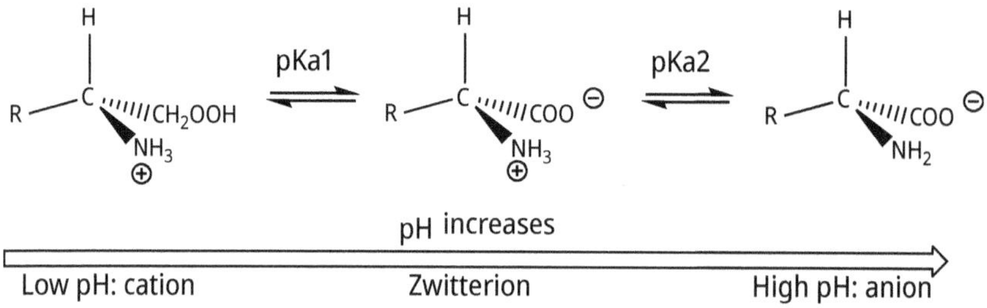

Fig. 1.12: The zwitterionic surfactants and their ionization under different pH values. Amphoteric molecules reach an isoionic point at a specific pH when the positive and negative charges in the hydrophilic portion are internally neutralized.

In general, the surfactant molecules increase the wettability and stability of the microemulsion, although different surfactants have unique functions. The primary function of surfactants is to stabilize the particle nucleation during the early stage of emulsion formation. The further growth and polymerization of particles acquire surfactant molecules from those species adsorbed on the monomer droplets or dissolved in the continuous aqueous phase. As a result, the microemulsion structure can be stabilized using the surfactant/cosurfactant molecules. At the water/oil interface, the stoichiometry of water to surfactant in the microemulsion system is the critical factor in determining micelle size and shape, as discussed above. The examples and definitions of four families of surfactants are tabulated in Tab. 1.2.

The appropriate selection of ionic surfactant enables conformational changes and tunability of the final products by packing these nonionic amphiphiles at the interface of microemulsion droplets. The introduction of an anionic surfactant into a nonionic system results in an attraction between the surfactant and the oil phase, driven by van der Waals forces. This process will lead to the formation of mixed micelles. Commonly used hydrophilic groups can be either ionic or nonionic, for example, SO_4^{2-}, SO_3^{2-}, –COOH, and +$N(CH_3)_3$, –O–$(CH_2–CH_2–O)_nH$. The widely used hydrophobic groups

Tab. 1.2: The summary of compounds and their functions in microemulsion.

Surfactants	**Examples**	**Definition and functions**
Nonionic 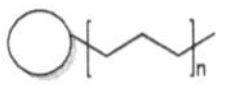	Polysorbates (Tweens®) Sorbitans (Spans®) PEGs Laureth Poloxamer (Pluronics®)	They are charge-neutral to the molecules and are more commonly found in emulsifiers. Nonionic detergents are super harsh and rarely used in skincare. Their functions are: To act as nonionic amphiphiles, induce packing at the interface layer of the microemulsion, and increase the adsorption capacity at the interface.
Cationic 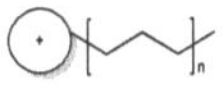	Cetrimonium bromide Benzalkonium chloride (BAC), Cetylpyridinium chloride (CPC) Benzethonium chloride (BZT) Quaternary ammonium cations Alkyltrimethylammonium salts Cetyl trimethylammonium bromide (CTAB) Cetyl trimethylammonium chloride (CTAC)	They are composed of positively charged heads and can be used in beauty products. Cationic detergents are incredibly harsh. Their functions are: To disrupt the cell membranes of bacteria and viruses.
Anionic 	Sodium lauryl sulfate (SLS, $CH_3(CH_2)_{10}CH_2(OCH_2CH_2)_nOSO_3Na$) Alkyl carboxylates (soaps) Sodium lauryl ether sulfate (SLES) Sodium myreth sulfate Dioctyl sodium sulfosuccinate (DOSS) Perfluorooctanesulfonate (PFOS) Linear alkylbenzene sulfonates (LABs) Perfluorobutanesulfonate Alkyl-aryl ether phosphates	Negatively charged, incredibly effective and harsh, higher incidence of irritation, and lathers well and makes foams. Their function is mainly to induce intermolecular attraction.
Zwitterionic 	Cocamidopropyl betaine (CAPB, a mixture of organic compounds derived from coconut oil and dimethylaminopropylamine, $C_{19}H_{38}N_2O_3$) Sodium cocoamphoacetate ($RC(O)NH(CH_2)_2N(CH_2CH_2OH)CH_2COONa$)	Both negatively and positively charged, the final charge depends on the pH values, mild and less irritating, but it foams less: To attract oil and water. To help the different ingredients of a formula blend.
Trade name	Nomenclature/chemical formula	Application [80]
Nonionic		
TritonTM-X-100	Polyoxyethylene glycol octylphenol ethers/$C_8H_{17}-(C_6H_4)-(O-C_2H_4)_{1-25}-OH$	Wetting agents (for coating)

Tab. 1.2 (continued)

Surfactants	Examples	Definition and functions
Nonoxynol-9	Polyoxyethylene glycol alkylphenol ethers/C_9H_{19}–(C_6H_4)–(O–C_2H_4)$_{1-25}$–OH	Spermicide
Polysorbate	Polyoxyethylene glycol sorbitan alkyl esters	Food ingredient
Span®	Sorbitan alkyl esters	Polishes and fragrance carriers
Poloxamers, Tergitol™, Antarox®	Block copolymers of polyethylene glycol and polypropylene glycol	Diversified applications
Anionic		
Pentex 99	Dioctyl sodium sulfosuccinate (DOSS)/ $C_{20}H_{37}NaO_7S$	Wetting agent (coatings and toothpaste)
PFOS	Perfluorooctanesulfonate (PFOS)/ $C_8HF_{17}O_3S$	Scotchguard™ and Skydrol™
Calsoft®	Linear alkylbenzene sulfonates	Laundry detergents and dishwasher detergents
Texapon®	Sodium lauryl ether sulfate	Shampoos and bath products
Darvan®	Lignosulfonate	Concrete plasticizer, plasterboard, and DMSO
	Sodium stearate	Handsoap and HI&I products

include the alkyl chains (–C_nH_{2n+1}) and aryl chains (–C_nH_{2n+1}–C_6H_4–). The commercially available surfactants and their applications are listed in Tab. 1.2. These surface-active agents used in typical emulsion polymerization are usually less than 5%. However, these surfactants play a crucial role in particle nucleation and growth. These surfactants and cosurfactants are critical factors that determine the final product's morphology, properties, and applications. The authors summarized the commonly used surfactants and their functionalities in Tab. 1.2. There are four significant types of surfactants: nonionic, anionic, cationic, and amphoteric, with diverse compositions and polarities.

Giant surfactant: One special case discussed amphiphilic bottlebrush block copolymers (BBCPs), giant surfactants, for producing photonic pigments and their ability to control micellization and self-assembly mechanisms [81]. The amphiphilic BBCPs have been reported to form much larger micelles (a few tens of nanometers) in water than those obtained with traditional surfactants or BCPs. The team led by Parker invented a method using toluene-in-water (t/w) to "induce a controlled swelling of reverse BBCP micelles" to overcome the above limitation. In this method, the (polynorbornene-g-polystyrene)-b-(polynorbornene-g-polyethylene oxide [P(PS-NB)-b-P(PEO-NB)] was dissolved in anhydrous $C_6H_5CH_3$ and emulsified in water. The addition of water during emulsification was evenly absorbed by the toluene droplets, resulting in significant swelling of the BBCP micelles. So-called "soft confinement" within these t/w emulsions resulted in internal aqueous droplets whose size ranges in visible light. Subsequently, highly porous microparticles with tightly packed nanodroplets were formed due to toluene evaporation for about 30 min via self-assembly templation.

The structural coloration of the photonic pigments (Fig. 1.13A) was then created, corresponding to the "short-range order of the pores within the BBCP scaffold." The pore size and colors of the resultant photonic pigments can be tuned by adjusting the amount of water available to the micelles. The different toluene-to-water ratios lead to temporary disruption of the microdroplet interface due to high shear. The reflected color in the visible electromagnetic spectrum can be tuned by increasing the amount of water (Fig. 1.13B and C). It was reported that the wide range of colors, from blue to red, was achieved using this BBCP (MW = 290 kDa, fPEO = 50 vol%, DP = 72). In stark contrast to "lamellar BBCP photonic systems" and the "strongly directional character of the reflection from photonic pigments," the short-order coherent scattering led to the coloration of the resulting pigments, as shown by the optical microscopic analysis upon epi-illumination in reflection mode (Fig. 1.13D). These photonic pigments displayed several key advantages: "suppressed iridescence and improved color purity, removing the need for refractive index matching or the inclusion of a broadband absorber."

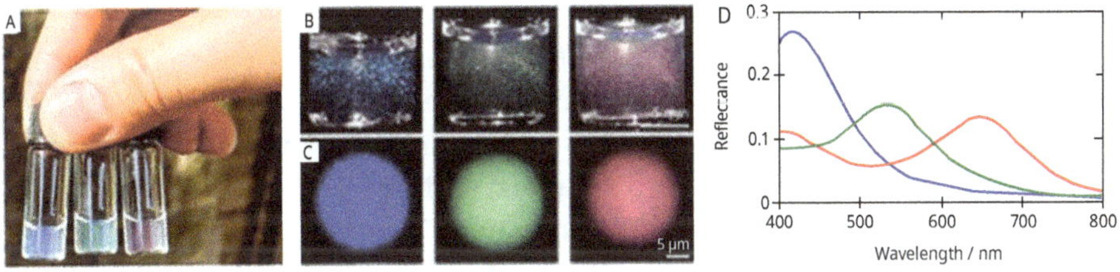

Fig. 1.13: The analyses of the BBCP photonic pigments. A. The aqueous dispersions of three formulations of pigments (blue, green, and red) illuminated under natural sunlight, B. photographs showing the above three dispersions (0.03 wt%) under direct illumination, C. microscopic images of microspheres of these three pigments, and D. reflectance spectra for the microspheres using a Lambertian diffuser and measured from a light intensity from a white light source (images were reproduced from [82]).

1.2.3 Hydrosolvothermal synthesis

The hydrothermal and solvothermal wet-chemistry methods are widely used to provide NMs (such as metal oxides used for supercapacitors [83] and MOFs) [84]. Hydrothermal synthesis is a wet-chemical approach to preparing water-soluble crystalline materials. The reactions usually occur at higher temperatures and pressures in an autoclave composed of steel vessels. The polar molecule water serves as the solvent, providing a medium for nucleation and crystal growth. A temperature gradient can be maintained in the growth chamber, with the hotter regions allowing solute dissolution and the cooler areas promoting crystal deposition [85]. Like hydrothermal synthesis, the solvothermal method uses organic solvents to produce chemical compounds from a nonaqueous precursor solution. Typical solvents include dimethylformamide (DMF), acetone (CH_3OCH_3), acetonitrile (CH_3CN), and dimethylformamide ($(CH_3)_2NCH$), diethyl formamide ($(CH_3CH_2)_2NCH$), ethanol (CH_3CH_2OH), methanol (CH_3OH), and various alcohols [86]. It was reported that almost 70% of MOFs have been prepared by a solvothermal method [87]. In this solvothermal synthesis, a closed system will be used, with escalating pressure and temperature employed above or below the boiling points of the preferred solvents. This solvothermal synthesis is a subcritical synthesis conducted at or below the boiling point [88]. Otherwise, the chemical approach is called the supercritical solvothermal method [89].

When a mixture of water and organic is used as a solvent, the method is known as hydro-solvothermal. The laboratory-scale process can be carried out in the sealed autoclave to sustain various temperatures (up to 300 °C) and pressures (up to 15 MPa). The autoclave consists typically of stainless steel with a Teflon-lined reaction chamber (Fig. 1.14A). The mixed solvent can dissolve the different reactant precursors because of their solubilities in the other solvents. Hydro-solvothermal chemistry has attracted attention due to its potential for flexibility and a green, scalable approach to producing various formulations of products used in supercapacitors and carbon capture [90].

However, reactants under hydrosolvothermal conditions can undergo rapid, unexpected changes, either physical or chemical. Therefore, unexpected reactions may hinder the formation of designed nanoparticles, suggesting that inert gases may be needed to prevent them. Another drawback of this method is that it usually requires high temperature and prolonged duration (300 °C, >72 h). Regardless, well-reticulated single crystals and other tunable structures can be formed using the hydro-/solvothermal reactions [91]. Three key steps in producing final products include dissolving reagents in solvents, nucleating crystalline matter, and crystal growth to form the resulting product (Fig. 1.14B–D). The products' structures, properties, functionalities, and compositions can be achieved by varying and controlling the synthesis parameters, such as temperature, reaction duration, and pressure. The physical parameters of reactants (for example, intermolecular forces and dipole moment), concentration, and acidity are other factors that play critical roles in defining the structure and properties of the end products. The hydrothermal approaches showed high solvation power, high compressibility, and mass transport of these solvents. Therefore, different reactions can result in a series of final products, as listed below [92]:

Synthesis of new phases or stabilization of complexes (such as MOFs)
Crystal growth of several inorganic compounds (oxides and borides)
Preparation of finely distributed materials (nanoparticles)
Leaching of ores in metal extraction
Decomposition, alteration, corrosion, and etching

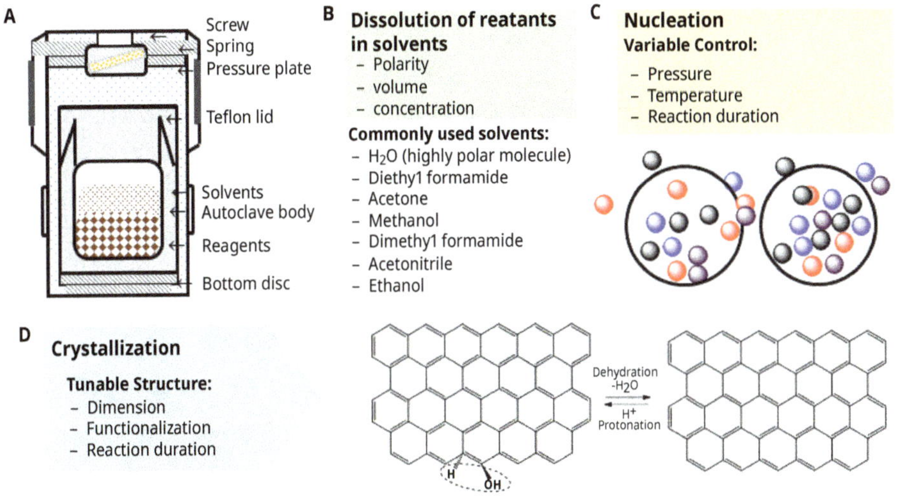

Fig. 1.14: The hydrothermal synthesis: A. the schematic diagram of an autoclave, B. dissolution of solute into the water and other preferred solvents, C. nucleation procedure, and D. crystallization to form a final product (adopted from [93] with permission).

Nucleation: Nucleation is defined as the process by which the crystal reaches a critical size at a threshold to create a new structure or phase in a vapor, solution, or liquid. Beyond this crucial point, the energy benefits of growth exceed the costs. Nucleation is the first spontaneous step in forming a new thermodynamic phase. This new structure starts from metastability, with a higher intermediate-energy state. The time to nucleate, which varies significantly from instantaneous to prolonged, determines the kinetics of nucleation for the new phase to form. Traditionally, two theories have been used to describe this nucleation: the classical and the two-step models (Fig. 1.15A). The classical nucleation theory has been used to explain and quantify nucleation kinetics. It can be broken down into two steps: (1) formation of a crystalline seed due to the collision of ions or molecules and (2) expanding the crystal lattice to form particles when the other solvated ions fall into place [94] (Fig. 1.15B).

The traditional nucleation mechanism posits that nucleation begins with the formation of perfectly spherical nuclei at the threshold size. Based on classical theory, the rate of nucleation can be expressed by eq. (1.4A). This rate expression can be considered the product of the number of nucleation sites and the probability that the threshold mass or size ratio is reached, around which propagation can proceed

(Fig. 1.15C). This threshold is the number of nuclei species at the nucleation barrier, governed by the system's free energy [95]. The probability of a nucleus formation at the loci is inversely exponentially proportional to the Gibbs energy. If the Gibbs energy is significant and positive, the likelihood of forming a nucleus is very low, and nucleation will be slow. At a given time, the nucleation rate will be sluggish or insignificant:

$$R_N = N_s Zjexp\left(\frac{-\Delta G^*}{k_B T}\right) \tag{1.4A}$$

where R_N is the nucleation rate; ΔG^* stands for Gibbs free energy to overcome the nucleation barrier; N_s is the number of nucleation sites; Z is the Zeldovich factor, which gives the probability of forming a new phase rather than redissolving into the solution. j is the rate at which molecules attach to the nucleus; k_B is the Boltzmann constant, and T is the absolute temperature [96].

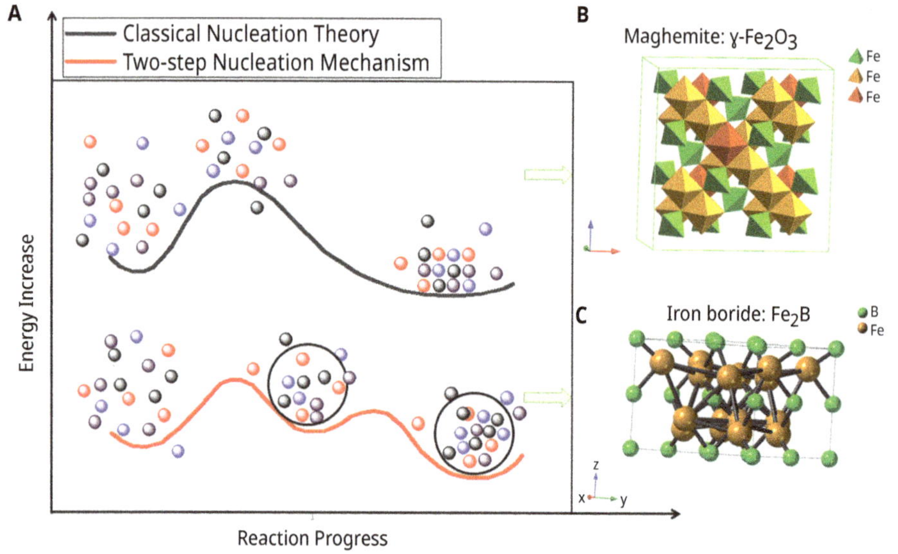

Fig. 1.15: The nucleation mechanisms and select formations of two inorganic compounds: A. the classical and two-step nucleation theories depict crystal formation with energy changes, B. maghemite compound, and C. iron boride.

Recently, a team led by Miao developed four-dimensional atomic electron tomography (AET) to capture the nucleation of FePt nanoparticles (NPs) with a chemically disordered face-centered cubic structure. This 4D AET technology was used to study nucleation at its early stage, spanning three spatial dimensions and one temporal dimension [97]. The research data showed that, according to the traditional approximation, nuclei are irregularly shaped at the early stage. Each nucleus has a core of one to a few atoms with the maximum order parameter. The parameter gradient was defined as the distance from the core to the nuclear boundary. Their study indicated that "growth, fluctuation, dissolution, merging, and/or division" occurred during nucleation. The order parameter distribution and its gradient jointly regulate the structure and dynamics of identical nuclei. The FePt NPs were used as a model system that underwent irregular shape at their early-stage nuclei (Fig. 1.16). The total number of atoms in the FePt NPs was slightly changed (Fig. 1.16A, a–c) under different annealing durations due to the atomic diffusion between NPs. The data also showed that a portion of atoms at the surface and subsurface were rearranged to form L10 phases with no change in the Pt-rich core. The authors proposed that vacancy-mediated atomic diffusion plays a critical role during annealing, creating vacancies on or near the surface because it requires less energy.

Miao's results yield three broad conclusions: 1) That early nucleation is governed by Oh symmetry, that the particles are anisotropic and spherical. When the particles are not spherical, this is due to local inhomogeneity in nucleation forces, anisotropy of physical constraints between the interfacial region and water tension; 2) the nuclei sites are composed of a few atoms, which enable maximum order parameter and order parameter gradient to be optimized from the center to the boundary layer. These boundary conditions result in the generation of a diffuse interface between the nucleation event and the mother liquor or initiatory phase; and 3) the nucleation processes, which propagate, are highly dynamic and consist of fluctuations in atom migrations, dissolution of nucleation bodies, division, or aggregation that are governed by the order parameter distribution and resulting gradient parameter. Each order parameter has a distribution, and the associated gradient represents metastable states between which the nucleus can be distributed and eventually settle, continuing the nucleation process. Peter Ercius from the Molecular Foundry of the Lawrence Berkeley National Laboratory reported that Miao and his team "have gained a never-before-seen view of nucleation, capturing how the atoms rearrange at 4D atomic resolution" [98]. The atomic positions in the NPs were captured using the new AET approach to demonstrate nucleation over different annealing times.

Crystal growth: The conversion of a gas or liquid into a solid, resulting in a phase transition. The growth mechanism depends on the initial phase, the surface state, and the strength of the driving force. During this significant stage of the crystallization process, the crystalline lattice forms as new atoms, ions, or polymer strands are added [99]. This action will yield a single or multiple crystalline phase(s) with highly ordered, close-packed atom and molecular arrangements. Subsequently, an increase in nuclear size will drive further crystal growth toward the critical cluster size. Surface-catalyzed nucleation as an initial stage will occur, followed by the seeded crystallization as shown in Fig. 1.14C and D. During this dynamic process, an equilibrium of solute will be achieved through precipitation out of and redissolution back into solution. Supersaturation is the driving force of crystallization, arising from the solubility of a reactant at an equilibrium point between precipitation and recrystallization. Two different crystallization types are consistent: normal growth (Fig. 1.17A) and nonuniform lateral growth (Fig. 1.17B).

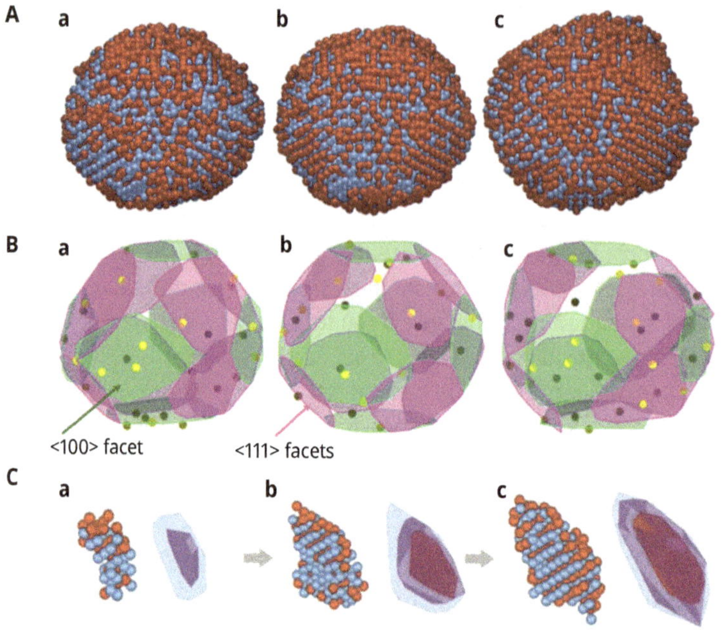

Fig. 1.16: 4D atomic motion captured by AET: A. 3D atomic models (Fe in red and Pt in blue) of FePt NPs with accumulated annealing time; B. heterogeneous nucleation sites, distribution of the nucleation sites with an accumulated annealing time; and C. experimental observation of the identical nuclei undergoing growth, fluctuation, dissolution, merging, or division at 4D atomic resolution (image was reproduced from [97]) (*note*: heat-treatment time: a. 9 min, b. 16 min, and c. 26 min, the lighter colored dots are closer to the front side and the darker dots are closer to the backside of the nanoparticle, and red: Fe and blue: Pt).

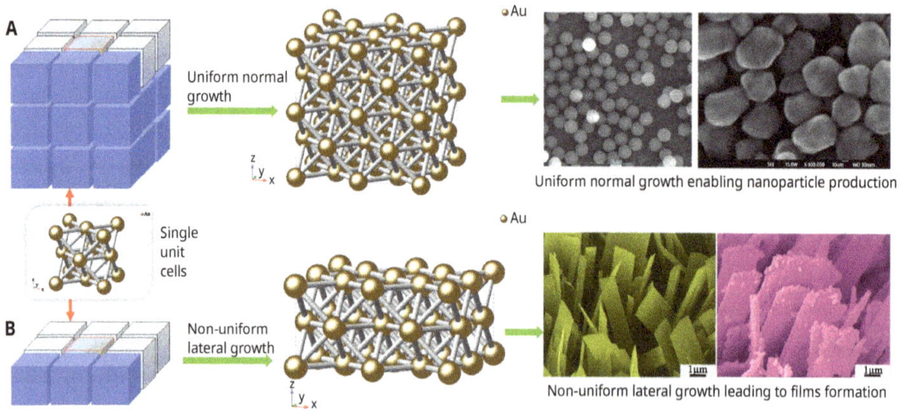

Fig. 1.17: The crystal growth mechanism: A. normal uniform growth to produce nanoparticles (Au NPs were chosen as examples); B. nonuniform lateral growth to produce layered films (MnO_2-grafted V_2O_5 supercapacitor electrodes as examples; copyright is obtained from [100] with permission).

Uniform normal growth: The surface advances isotropically without a stepwise mechanism, a process called "entire surface motion." This uniform "normal growth" is based on the time sequence of an element or atom on the surface. As long as the thermodynamic driving force is sufficient, all different elements or atoms on the surface (red-colored square) can continuously grow and change to advance the interface. During this continuous growth mechanism, surface diffusion may simultaneously require a change via a unit-cell transformation to generate a large-scale superlattice across several successive layers. When crystals grow uniformly over large areas to form a new layer, nanoparticles form to minimize surface tension (Fig. 1.17A).

Nonuniform lateral growth: During nonuniform growth, surface advancement occurs by the lateral motion of each step. These steps can be a single interplanar spacing or an integral multiple of that spacing in height. Without undergoing any changes, some surface atoms will not advance normal growth to themselves, except during the passage of a step. During this lateral growth, two adjacent regions are parallel and show identical configurations, corresponding to a displacement between the areas by an integral number of lattice planes. This nonuniform lateral growth is considered a "geometrical motion of steps" (Fig. 1.17B) instead of an "entire surface motion" [101] (Fig. 1.17A). There is no change in this lateral growth mode except for a step that passes via a continual change. Surface diffusion and the driving force are two criteria used to predict the mechanisms of crystal growth.

A diffuse surface undergoes continuous change from one phase to another, occurring over several atomic planes and being "confined to a depth of one interplanar distance (Fig. 1.17B)" [101]. This step contrasts the "sharp surface" where discontinuous changes in properties occur. The driving force for this diffuse surface is considered a requirement for the appearance of lateral growth. This lateral growth mechanism will be found at the metastable equilibrium of the surface composition in the presence of driving forces. Until the next step is taken, this metastable condition tends to persist. This configuration in the crystalline medium will be kept identical in the following steps until the height step is further advanced. On the other hand, if the surface cannot reach equilibrium conditions, it will continue to expand its regions, leading to nonuniform lateral growth. Therefore, it can be said that the significant driving forces facilitate uniform movement. On the contrary, the heterogeneous nucleation mechanism (or screw dislocation) will favor the presence of driving forces. The diffusiveness of the interface is the critical driving force for crystallization. The crystal properties, including mechanical, chemical, and electrical characteristics, are pertinent to the end products. Thermodynamics and kinetics are the fundamentals for further investigations. In the late nineteenth century, Josiah Willard Gibbs invented the "necessary thermodynamic apparatus" to study heterogeneous equilibrium. Gibbs indicated that surface energy (or surface tension) applies to solids and liquids as a driving force to form large clusters. He also proposed that an anisotropic surface free energy implied a nonspherical equilibrium shape." Thermodynamically, this nonspherical equilibrium can determine particle shape, thereby minimizing the total surface free energy [102].

Water as the solvent: When water is used as the solvent, the method is called a hydrothermal process. This hydrothermal synthesis is the most widely used wet-chemistry method due to its similarity to geological phenomena, its ability to synthesize hydroxylated materials, its low-temperature conditions, and its environmental friendliness. The properties of water, such as thermal conductivity, viscosity, surface tension, and dielectric constant, depend strongly on the composition and type of the materials. Other organic solvents can be selected based on the solubility of the starting materials (reactants) and their compatibility with water. The products' structures and reactivities can be tuned by controlling the synthesis variables, such as solvent volume, solution concentration, reaction temperature, and acidity. The pressure–temperature plots (Fig. 1.18A) show the relationship between the vapor pressure inside an autoclave and temperature as a function of water percentage fill. The graphs indicated that the hydro-solvothermal reaction pressure could be controlled by programming the temperature and altering the solvent volume. Low pressure can be achieved by introducing a small amount of solvent into the autoclave reactor (Fig. 1.14A).

On the other hand, a high solvent volume can create high pressure even at low temperatures in the same reactor. Figure 1.18B shows pressure variation with temperature for different select organic solvents. In general, higher temperatures correspond to higher vapor pressures of solvents and reactants, leading to more effective collisions and further increasing the reaction rate. The organic solvents (e.g., methanol, CH_3OH) showed the highest vapor pressure, while dimethyl formamide ($(CH_3)_2NCH$) produced the lowest pressure under the same reaction temperature. We did not find a clear trend between the vapor pressure and the molecular dielectric constant or the thermal conductivity, due to the molecular structure's complexity and thermal motion. These data provided a guideline for selecting a solvent, which depends on the individual reactions. It is safe to conclude that the less polar molecule with a higher molar mass has a lower vapor pressure under the same conditions.

1.2.4 Chemical vapor deposition

Chemical vapor deposition, abbreviated CVD and also known as vacuum deposition, produces high-quality, high-performance solid materials in an activated environment. Industrial products that use this CVD include cutting tools, filters, optoelectronics, refractory fibers, and semiconductors. As stated by Dr Pierson, "CVD is no longer a laboratory curiosity but a major technology on par with other major technological disciplines" [104]. The CVD facility comprises three key parts (as shown in Fig. 1.19A): (1) a chemical vapor reactant supply system, (2) a CVD reactor, and (3) an effluent gas treatment system. The precursor supply system provides gaseous reactants delivered to the reactor. In the CVD reactor, the substrate will be heated to suitable conditions, enabling the deposition of the final products [105]. The effluent gas treatment system consists of an exhaust gas system and a vacuum system, with pressure, temperature,

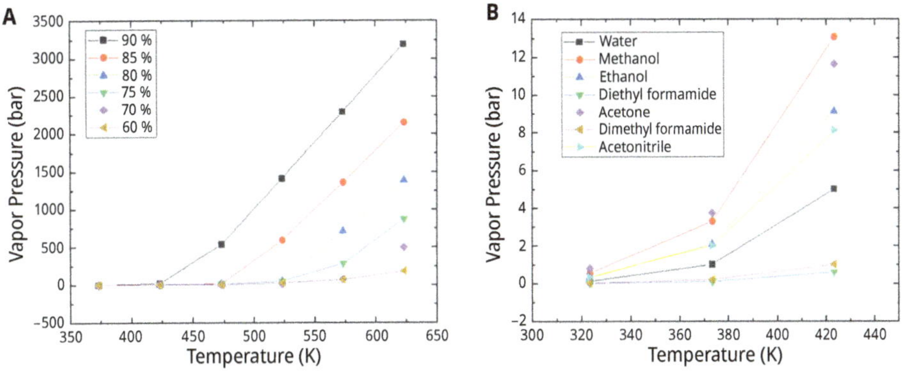

Fig. 1.18: The relationship between vapor pressure and temperature of solvents. A. The phase change of water as a solvent under different humidity (data were derived from the NIST Reference Database Version 9.0) and B. the phase change data for water and six organic solvents as a comparison (pressures were calculated using the Antoine equation ($\log_{10}P = A - B/(C + T)$) The constant values of A, B, and C are cited from the NIST webbook and Figures from [103] with permission.

and vacuum variables that can be controlled during deposition. In this CVD process, the vapor precursors undergo dissociation or chemical reactions under heat, light, and plasma exposure. The CVD method can produce different formulations with high purity, well-controlled structures, and molecular-level compositions. These materials include ultrafine powders (0D), nanowires/nanotubes (1D), nanosheets and wells (2D), and thin films (3D). It was found that ultrafine nanoscale particles can form during CVD in the gaseous state due to their homogeneity [106]. Otherwise, a dense film will form if the reaction occurs near or on the heated substrate, due to the heterogeneous nature of the response [107]. The CVD experimental setup requires complex variables and procedures, including selection of activation methods, precursor choice, and pressure/volume control. The activated methods include heat, plasma, light, and ions. Based on the external energy source, the CVD method can be classified into thermal CVD, plasma-assisted CVD, microwave (MW)-assisted CVD, photo-/laser-assisted CVD, and ion-assisted CVD. Atomic layer deposition (ALD) can operate in either thermal or plasma mode. Those methods typically use high-vacuum systems (Fig. 1.19A).

CVD deposition is a complex process that produces materials with diverse shapes and compositions. The seven key steps [106] are stated below (as shown in Fig. 1.19B):

1. **Vaporization and generation of gaseous reactants:** Evaporation or sublimation occurs when the precursor is exposed to heat, light, or plasma.
2. **Transport of gaseous reactants:** The reactant was then delivered into the reactor, where the chemical reaction occurs. The gaseous phases allow for an even distribution of reactants, leading to a higher effective collision rate.
3. **Gas-phase reactions:** The precursors in the reaction zone produce reactive intermediates at a sufficiently high temperature in the gas phase; intermediates undergo chemical reactions. Stable, solid, fine particles can be made in homoge-

neous gas phases, while by-products can form. At temperatures below the dissociation temperature, the intermediate species will form. They will consequently undergo deposition, diffusion, and removal.

4. **Adsorption of reactants:** The precursors will be absorbed on the heated substrate surface. If heterogeneous reactions occur at the gas–solid interface, deposition and by-products will form.
5. **Surface diffusion:** The deposited substances diffuse or migrate along the heated substrate, allowing for nucleation and crystallization. Film or particle growth will occur due to surface chemical reactions.
6. **By-product removal:** Those gaseous by-products will be removed from the boundary layer via diffusion and convection to improve the quality of products.
7. **Desorption and mass transport:** The by-product and remaining fragments of the unreacted species will be removed from the reaction zone.

Choice of precursors: metal halides and hydrides are commonly used reactants. Metal-organic precursors are widely used in the MOCVD. Deposition pressure: ultra-low pressure is typically required for molecular-beam epitaxy (MBE) and chemical-beam epitaxy (CBE) to deposit products. CVD is a chemical process used to produce high-purity, high-performance materials. CVD involves the thermal decomposition or other chemical reactions of gas-phase species at elevated temperatures ranging from 500 to 1,000 °C. This procedure consists of exposing a quartz substrate to a volatile precursor to deposit the target product.

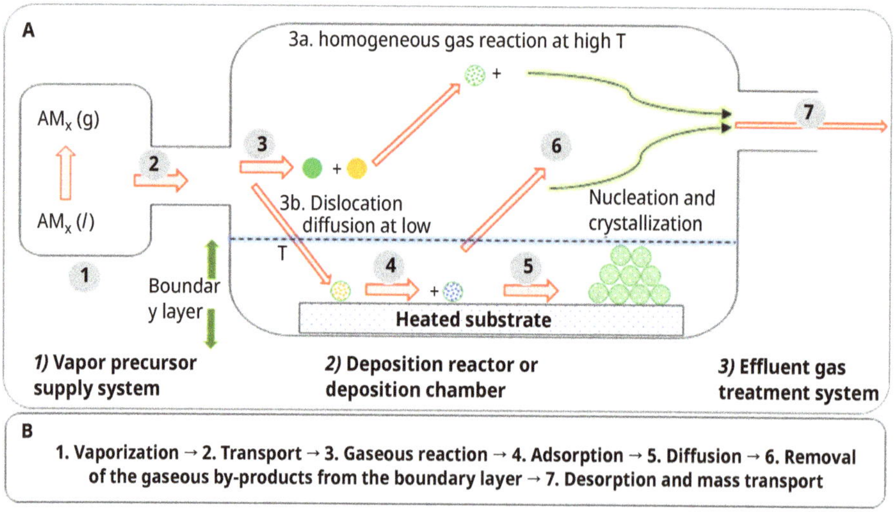

Fig. 1.19: The schematic representation of three key parts in the CVD apparatus: vapor precursor supply system, deposition reactor (chamber), and effluent gas treatment system. The image was adapted from [107] with permission.

The CVD-deposited materials vary from monocrystalline to polycrystalline, amorphous, and epitaxial. Materials with different elemental compositions include silicon, carbon fiber, filament, silica (SiO_2), silicon–germanium (Si–Ge), tungsten (W), silicon carbide (SiC), titanium nitride (Ti_3N_4), various high-K dielectrics, and synthetic diamond [108]. In the conventional CVD technique, the entire substrate is heated, and the reactant gas or vapor is flowed over the substrate surface, where chemical reactions occur and the film is deposited. This process is very slow (film growth rate is 100–1,000 Å/min) and is not energy-efficient, since the entire substrate is heated while only the hot surface is used for film deposition. In this technique, the substrate temperature can reach 1,500 K, depending on the precursors' reaction temperature and the film's properties. Since the diffusion coefficients of the dopants become significant when the temperature exceeds 1,100 K, such high temperatures affect the quality of semiconductor films because of the change in the impurity concentration profiles due to the diffusion of dopants into the film [105]. Also, contaminants inside the CVD chamber or the wafer material can diffuse into the film grown at high temperatures. The volatilities of the various species used for depositing compound semiconductor films differ. Therefore, at high temperatures, the species will evaporate to varying degrees during film deposition, thereby affecting the film's properties. Physical damage to the wafers at high temperatures can also cause defects in the devices produced by the conventional CVD method.

The CVD procedure involves exposing a substrate (also known as a wafer) to a volatile precursor to deposit the target product. The form of CVD deposit materials can be varied from monocrystalline to polycrystalline, amorphous, and epitaxial. Materials produced by CVD are highly diversified, including silicon and carbon nanotubes [109].

Liu and her collaborators employed the two-chambered CVD (Fig. 1.20) method to synthesize the aligned carbon nanotubes (ACNT). In this process, a two-compartment tube furnace was utilized (Fig. 1.20A). The reactants were vaporized in the first compartment and carried into the second, where they deposited ACNT on a glass substrate. This two-furnace process is used to control the length and diameter, which are critical to gas transport and water management in the fuel cell [110] (Fig. 1.20B). The precipitation of carbon leads to the formation of tubular carbon solids with sp^2 bonding, which endows the molecules with their unique strength. Our results show that the impurity of carbon nanotubes can be controlled in a bifurcated furnace (Fig. 1.20C). Moreover, in our study, ferrocene was used as the seeding material for ACNTs (Fig. 1.20D).

Here, the oxygen product partially reacted with the ferrocene ($Fe(C_5H_5)_2$), which decreased the efficiency of the fuel cell due to oxygen loss but maintained ACNT integrity. The xylene–ferrocene solution was used as the precursor, where xylene (or dimethyl benzene $(CH_3)_2C_6H_4$) is the carbon source. Ferrocene provides iron nanoparticles as seed catalysts for nanotube growth. Three 5-cm^2 quartz substrates are placed inside the quartz reaction tube. The tube is placed in a two-stage furnace and tightly sealed from air. The first stage of the stove is at 225 °C, which is sufficient to vaporize the solution. The second stage is held at 725 °C and used to carbonize the vaporized solution, depositing the iron nanoparticles on the quartz substrates and allowing the carbon nanotubes to grow along with the iron seeds [111]. The solution containing the chemicals is injected

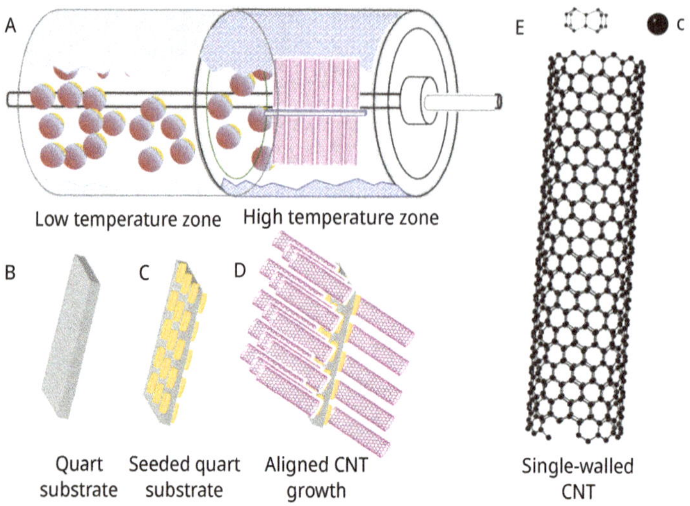

Fig. 1.20: Chemical vapor deposition produces aligned carbon nanotubes used as cathode catalysts for proton exchange membrane fuel cells: A. the schematic diagram of the CVD furnace with two temperature zones, B. substrate used for nanotube growth, C. the metallic Fe seed promotes highly aligned nanotube growth, D. the nanotube growth, and E. carbon unit cells (P1 (triclinic), allowing for chirality; polarity) buckytube.

into the reaction tube during the low-temperature stage, using argon and hydrogen as carrier gases at flow rates of 100 and 50 mL/min, respectively. The chemical injection rate is maintained at 0.225 and 0.250 mL/min, respectively [109] (Fig. 1.20E).

1.3 Top-down synthesis of nanomaterials

The top-down process is usually considered a physical or chemical process that breaks down particles into smaller ones by applying an external force. This synthesis implies that the nanostructures can be obtained by removing crystal planes present on the substrate. In other words, the building blocks are removed or etched out from the substrate to form the nanostructure. Usually, grinding systems (wet and dry) convert regular-sized materials into nanoparticles. Dry systems are obtained by breaking bulk particles using ball mills, jet mills, and roller mills. The reduced particle condensation prevents them from growing and keeps their size below 3 μm. On the other hand, wet grinding can be carried out using the centrifugal fluid mill, vibratory ball mill, flow conduit bead mill, and damp jet mill. When differentiated by both wet and dry methods, the grinding shows advantages in preventing agglomeration and in forming desirable-sized particles to produce nanostructured materials further.

In top-down nanofabrication, the nanoscopic structures can be obtained using physical or chemical-based processes (Fig. 1.21) [112]. The physical top-down approach relies on optical, electron-beam, and ion-beam lithography. This method has been widely used in the microelectronics industry, particularly in dynamic random-access memory (DRAM) devices [111]. The chemical top-down approaches involve chemical etching and heat application reactions. Compared to the most physically intensive top-down methods, top-down chemical techniques do not require highly controlled systems (such as a cleanroom). The inexpensive facility will be sufficient for building and maintaining nanostructures, attracting more attention for in-depth investigation. This approach also offers synthetic versatility, enabling the production of a wide variety of three-dimensional (3D) nanostructures with tunable architectures. These novel nanoscale architectures are promising for diverse applications in biomedicine, catalysis, electronics, energy storage and conversion, and sensors and actuators.

In the last few decades, top-down approaches have been widely used to produce nanostructured materials, optimizing fabrication variables, facility modifications, the affordability of manipulation, and the tunability of product features. The spotlight in this section will be the newly emerging fabrication route with an in-depth understanding of the relationship between nanostructure and preparation parameters. The affordability and ease of use of different top-down methods will also be discussed for future improvement. These final products with novel nanoscale architectures exhibit unique chemical, electronic, and biological properties, enabling diverse applications, as mentioned above. The top-down approaches include physical methods (such as lithographic tools) and chemical-based processes (such as templated etching, selective dealloying, anisotropic dissolution, and thermal decomposition). This section will briefly discuss three bottom up methods: physical top-down (Section 1.31, ball milling, lithography, and machining) and top-down chemical methods (Section 1.3.2). Four different approaches to the chemical process will be included: templated etching, selective dealloying, anisotropic dissolution, and thermal decomposition.

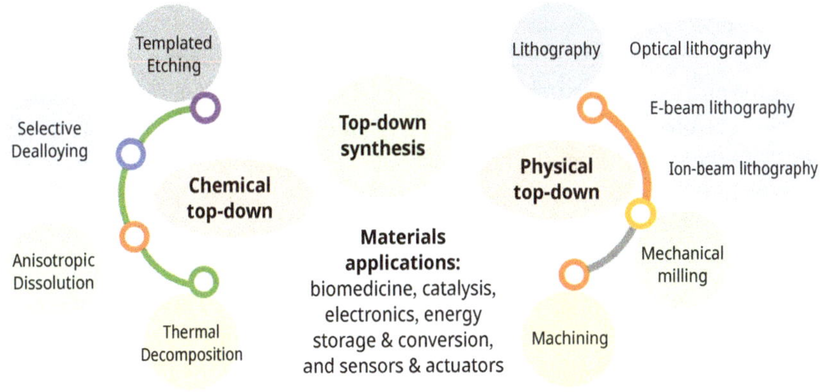

Fig. 1.21: A summary of top-down synthesis, chemical and physical-based processes to produce various materials used in biomedicine, catalysis, electronics, energy storage and conversion, sensors, and actuators.

1.3.1 Physical top-down

The top-down approach has been widely used to fabricate nanostructures used in the microelectronics industry. The significant advantage of top-down synthesis for forming joint circuits lies in creating orderly structures that eliminate the need for additional assembly steps. This section will discuss ball milling, lithography, and machining methods.

1.3.1.1 Ball milling

Ball milling is an operation to grind materials into excellent powders, such as nanoscale. One nanofabrication process, high-energy ball milling, is of industrial importance. This method is also called mechanical attrition or mechanical alloying. In ball milling, coarse-grained powder materials (such as metals, ceramics, and polymers) are mechanically crushed into fine powders in rotating drums lined with hard steel or tungsten carbide (WC) balls. Controlled atmospheric conditions (e.g., the introduction of an inert gas, Ar) can be used to prevent unexpected reactions. Grain size can be reduced via the formation and organization of grain boundaries within particles. Different components can be mechanically alloyed together [112]. Ball milling can also achieve nanosized dispersion of one phase in another. However, the microstructures and phases produced in this way can be thermodynamically metastable and transient. The milling technique can be operated on a large scale, showing industrial interests. The milling of materials has been a significant component of the mineral, ceramic processing, and powder metallurgy industries. The objectives of milling include particle size reduction, mixing or blending, and particle shape changes.

The typical milling method used for these purposes has been the tumbler ball mill. This method uses a cylindrical container that rotates about its axis, with balls impacting the powder charge. The balls may roll down the chamber surface in a series of parallel layers, or they may fall freely and impact the powder and balls beneath them [113]. A vertical rotating shaft with horizontal arms drives the agitator's stirring. A differential movement between the balls and the powder can result from this motion. Therefore, a high degree of surface contact can be achieved in tumbler ball mills. The rotation speed of the tumbler mill cylinder and the attritor shaft are the key factors in determining the kinetic energy imparted to the milling media [114]. Tumbler mills can provide high energy if a large shaft diameter is used and milling operates at critical speed for a short period to "pin" balls to the chamber wall. Another milling method, attrition milling, has been replaced by large-bumbler ball milling to achieve large-scale production of commercial alloys. Another type of mill, the vibratory tube mill, has been used for pilot-scale production [115]. The motion of the balls and particles in a vibratory mill was complicated. The cylindrical container is vibrated, and the impact forces on the powders in the milling chamber depend strongly on the milling rate, the vibration amplitude, and the mass of the milling medium [116]. Our previous study, which used ball milling to produce electrocatalysts for green NH_3 synthesis, indicated improved milling efficiency with two different grinding balls, attributed to increased contact surface area (Fig. 1.22A). The simple, polar molecule methanol (CH_3OH) was found to facilitate ion-dipole intermolecular interactions between reactants; as a result, the electrocatalyst's complexity can be tuned on both intra- and intermolecular levels. During milling, the elastic and plastic deformations were assumed to increase concurrently, thereby generating strong frictional forces. This impact drives grinding the material to smaller sizes, resulting in a crystalline structure with high homogeneity. Three milling methods, the tumble ball mill, the vibratory tube, and the mill attrition mill, will be discussed in Chapter 2.

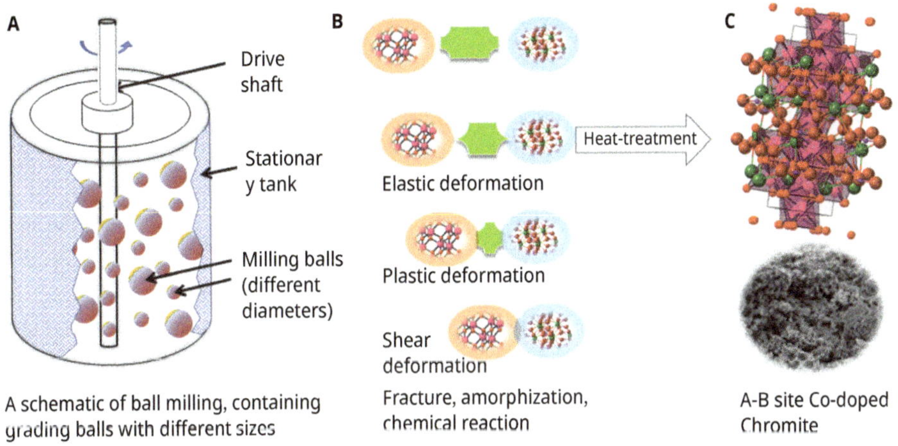

Fig. 1.22: Schematic diagram of ball milling processing: A. a ball milling process, B. shear and impact forces generated in the milling chamber, and C. the catalysts were formed in solid oxide electrolysis cells (adapted from [117] with permission).

High-energy milling forces can be obtained using high vibrational frequencies and small vibration amplitudes. The energy of the milling media depends on the internal mechanics of the specific mill. The power is supplied to drive the milling chamber, and the balls' composition, size, and distribution are determined. Since the balls' kinetic energy is a function of their mass and velocity, dense materials (steel or tungsten carbide) are preferable to ceramic balls. The size and size distribution should be optimized for the given mill [116] (Fig. 1.22B). Too dense packing of balls reduces the mean free path of the ball motion, while a dilute distribution minimizes the collision frequency. Empirically, a mass ratio of 5 to 10 between ball and powder is typically used and effective. One crucial process variable is the mill temperature, as well as the temperature induced in the powders by the kinetic energy of the milling media. This temperature may be critical for the reactions or transformations in the powders during milling. However, the precision of milling temperatures induced by the powder surfaces remains improved. Milling materials has been a significant component of the ceramic processing and powder metallurgy industries. Through milling, particle size can be reduced, mixing or blending can be improved, and particle shape can be changed. High-energy milling is often used to fabricate solid-state alloys mechanically. Benjamin and his colleagues developed this process at the international nickel company and provided a detailed description of the high-energy milling method [118]. This dry, high-energy ball milling process has been used to produce composite metal alloy powders, and the structure and size of the powders can be well controlled. Mechanical alloying is usually carried out in milling equipment capable of high-energy compressive impact forces, such as attrition mills, vibrating ball mills, and shaker mills [119] (Fig. 1.22C). The milling machine will stress the maximum number of individual powders during comminution. The fracture of starting materials will be initiated with minimum energy. The motion of the milling medium and the charge can vary depending on the type of mill. The movement and trajectories of individual balls, the mass of balls, and the degree of energy applied will highly influence shear, attrition, and compression forces acting on powder particles [120].

1.3.1.2 Lithographic processes

Lithography is a "planographic printmaking process" to fabricate nanosized materials or patterns. A design or structure can be drawn onto a flat surface, such as on the surface of stone or a pre-prepared metal plate. Nanoimprint lithography is a novel method. Lithography uses visible or ultraviolet (UV) light, X-rays, electrons, or ions to project an image containing the desired pattern onto a surface [121]. These techniques are generally referred to as photolithography, X-ray lithography, electron beam lithography, or ion beam lithography, depending on the radiation used. The pattern transferring process includes four key steps: 1) solution-based wet chemical etching procedures, 2) dry etching in a reactive plasma, 3) doping using ion implantation techniques, and 4) thin film deposition [122].

Optical lithography is the most widely used top-down synthesis method for creating fine-scale patterning in integrated circuits (Fig. 1.23). UV light transfers the desired pattern from a mask to a photoresist-coated semiconductor substrate (such as a Si wafer) [123]. Subsequently, the predesigned pattern can be transferred to the photoresist substrate, coated with a semiconductor, and often etched through a chemical process (Fig. 1.23A). Projection lenses can enhance the resolution, producing clear images (Fig. 1.23B).

Optical lithography is a top-down method that projects an image onto a silicon wafer using a photon-based technique. This method has been widely used in the semiconductor industry for the production of electronic nanostructures due to its ease of use and feasibility (Fig. 1.23C). The wavelength used in nanoelectronic fabrication determines the minimum feature size that can be achieved. Mercury (Hg)-based lamps emitted near-UV light, which served as the illuminating source during the early years. Moore's law states that "the number of transistors on a microchip doubles about every two years, though the cost of computers is halved" [124]. To keep pace with Moore's prediction, researchers use shorter-wavelength light sources, including excimer lasers, krypton fluoride (KrF) laser (248 nm), argon fluoride (ArF) laser (193 nm), fluorine (F_2, 157 nm), and xenon monochloride (XeCl) laser (308 nm). These lasers produce light with shorter wavelengths in the far-UV region, a technology extensively used for two decades [123] (Fig. 1.23D).

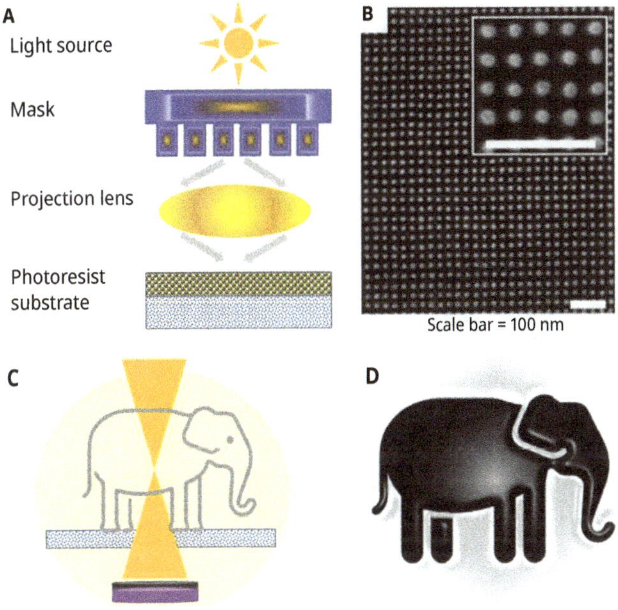

Fig. 1.23: Physical top-down method to fabricate nanostructures: A. simplified schematic diagram showing projection optics for the lithographic process, B. scanning electron microscopic image of silicon (Si) nanodot array produced by electron beam lithography (reprinted with permission from [125]). Copyright 2011, American AIP. C. Simulated scheme of an elephant sculpture fabricated by direct laser, and D. projected 3D image of the elephant (adopted from [126] with permission).

Photolithography can use UV light sources in microelectronic fabrication. Electron-beam (e-beam) and X-ray lithography are two other alternatives that have attracted considerable attention in MEMS and nanofabrication [127]. These methods will be discussed in Chapter 2, Nanochemistry: properties and applications of interactive materials

1.3.1.3 Machining

Machining is one of the top-down procedures to produce patterned and topographically constrained surfaces. Focused ion beams (FIBs) and high-intensity lasers are commonly used to produce and tune materials at micron- to nanoscale levels. In its early stages, starting in 1970, the FIB was used to manufacture electronic components and microdevices by 3D patterned layer-by-layer deposition, ablation, and re-patterning [128]. The patterning is done by ejecting precise liquid droplets using nozzle thrusters of a specific geometry and angle to disperse charged metal droplets. Krohn first documented ion emission from a liquid-metal source in 1961 [129]. His research on liquid metal ion sources (LMISs) for space applications [130] quickly proved helpful across many other areas, especially in semiconductors and materials science. Commercialization of the FIB was not far behind in the 1980s, primarily aimed at the rapidly growing semiconductor industry. The modern FIB system utilizes an LMIS at the top of its column to produce ions (Fig. 1.24). Gallium (Ga) is usually used for its low melting point (30 °C), high mass, low volatility, and low vapor pressure, as well as its easy distinction from other elements. The Ga source is heated to near the boiling point, allowing it to liquefy and flow down the tungsten needle, where it can remain in a liquid state for weeks without further heating, owing to its supercooling properties [131]. The extractor, an annular electrode centered just below the needle tip, is held at −6 kV relative to the source, drawing liquid Ga into a Taylor cone. An electric field on 1,010 V/ m between the tip and aperture induces field evaporation, causing ion emission and acceleration down the column. The source is typically operated at low emission currents of 1–3 μA to produce a stable beam [132]. Another suppressor electrode, working alongside the extractor, maintains a constant beam current at around +2 kV. The suppressor and extractor act as the “fine” and “coarse” controls regulating the ion extraction current (Fig. 1.24A).

As a result of ion generation and ion–ion interactions in the beam, ion–surface interactions are essential considerations to ensure consistency and intricate patterning. The beam flux can be varied by changing the voltage between the electrodes and the rate of liquefaction; the second parameter of ion-surface interaction is more variable, since the final pattern depends on ion energy, ion flux, and the ion–surface interaction angle. Upon collisions, the ions are refracted (slow down within the solid), while some are reflected (backscattered). Since ions are charged atoms, they have a larger radius than electrons (as in an electron beam) and a lower probability of transmission than electrons, but a higher probability of interaction with atoms at the surface. Upon a two-body elastic collision, energy is transferred due to the larger size (~1,850:1) of the Ga^{+} to an *e*. The collision disperses energy but not momentum because the ions’ mass is higher (~400×). The elastic collision process is distinct from surface collisions, which are many-body interactions that result in inelastic collisions and radiative decay [133].

In an elastic collision, kinetic energy and momentum are conserved (during a nuclear collision. When incident ions strike the target atoms, the energy and momentum transferred to the atoms on the material's surface are sufficient to disturb them from their aligned positions. This sputtering effect, which does not occur with an electron beam, allows the FIB to precisely remove atoms from the material's surface in a controlled manner. The incident ions are also scattered, some becoming implanted in the material and others getting backscattered. Furthermore, collisions induced by ion beam bombardment result in numerous secondary processes, including recoil and sputtering of constituent atoms, electron excitation and emission, defect formation, and photon emission (Fig. 1.24) [134]. These collisions lead to thermal- and radiation-induced diffusion, which contributes to various phenomena involving the constituent elements, including phase transformations, crystallization, amorphization, track formation, and the propagation of the process towards material formation. Also, ion implantation and sputtering alter the sample's surface morphology, resulting in craters, facets, pyramids, grooves, ridges, "blisters," or a porous surface. Electronic collisions produce secondary electrons and X-rays, in addition to sputtering effects from elastic collisions. In inelastic ion–electron collisions, the electrons are excited and ionized. Additionally, the incident ions transfer energy to the surface atoms and electrons, producing secondary electrons, ions, and electron ejection. If the incident ion has insufficient energy to penetrate deeply into the atom, the inner-shell electrons screen the nuclear charge; this screening effect must be considered in these atomic collisions [135] (Fig. 1.24B).

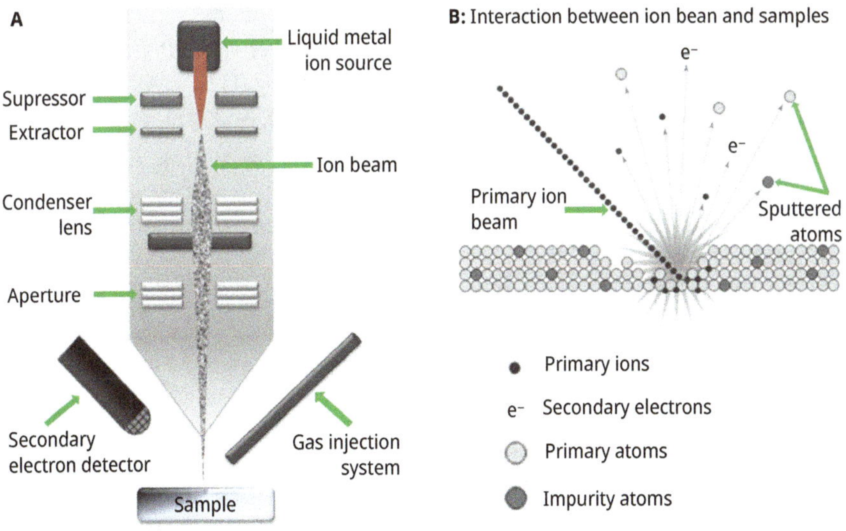

Fig. 1.24: A schematic diagram of focused ion beams (FIBs): A. the generic steps of FIB to produce nanomaterials and B. interaction between ion beams and samples.

FIBs can deliver tens of nanoamps of current to a sample and image the sample with a spot size of a nanometer. FIB is an inherently destructive technique because the specimen is transformed during rasterization. Due to its spattering capability, FIB is used as a micromachining tool [136]. Machining is an operation that produces intricate three-dimensional patterning in a material. FIB and high-intensity lasers have been used to pattern or shape materials directly at the micron and submicron levels. FIB is a semiconductor and materials science technique for site-specific analysis, deposition, and ablation. The focused beam of gallium (Ga is easy to build an LMIS with a concentrated beam of gallium (Ga). Ga metal is placed in contact with a tungsten needle and heated. Gallium wets tungsten, and a high electric field (>108 V/cm) causes ionization and field emission of Ga atoms. FIB can be incorporated into a system with electron and ion beam columns [137].

1.3.2 Chemical top-down synthesis

Top-down photolithography to produce 65-nm commercial transistors has shown significant improvement over the last few decades. Although this method yields precisely controlled shape and size, its manufacturing costs have reached unprecedented levels. Therefore, alternative strategies are needed to produce patterned silicon architectures for academic and industrial uses. One of the new processes compatible with existing Si-based fabrication techniques was the chemical process. The chemical top-down approach has become more attractive in producing nanoscale patterns. This method offers advantages for crafting more sophisticated nanostructures in solids and solutions, as well as for ordered arrays of nanostructures with 3D features. Four different approaches (Figs. 1.21 and 1.25): (1) templated etching (Fig. 1.25A), (2) selective dealloying (Fig. 1.25B), (3) anisotropic dissolution (Fig. 1.25C), and (4) thermal decomposition will be discussed in this section (Fig. 1.25D).

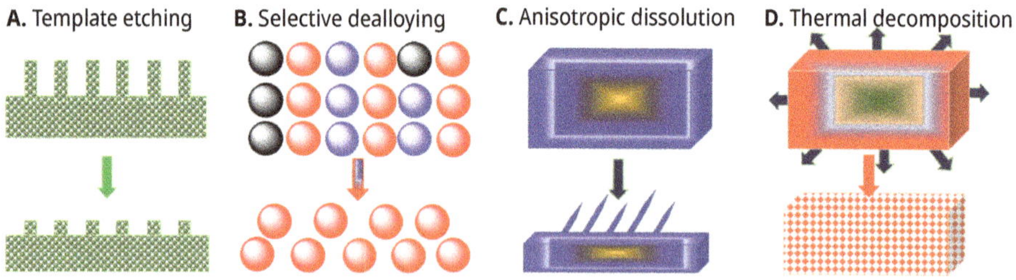

Fig. 1.25: A summary of top-down chemical synthesis: A. the templated etching, B. selective dealloying, C. anisotropic dissolution, and D. thermal decomposition.

1.3.2.1 Templated etching

Templated etching top-down is a technique to produce a nanostructured pattern, which can be crafted on the substrate surface (such as silicon), followed by subsequent template removal. Polymer self-assembly enabled the semiconductor formation of a uniform pattern with large areas and interfaces. The “spontaneous phase separation of block copolymers into nanoscale domains” provides a practical and feasible approach to producing periodic patterns. The template is typically used to direct the substrate etching through a chemical reaction during the templated etching. An amphiphilic block copolymer, poly(styrene-block-4-vinyl pyridine) (PS-b-P4VP, [Mn = 128 400-b-33 500 g/mol]) template (Fig. 1.26A), was used to tune the nanoscaled structure into a 3D etch pit array as reported by Buriak and coworkers [138]. This molecular amphiphilicity is critical in directing surface chemistry in a spatially defined manner. The chemical specificities of different building blocks lead to localized reactions on the substrate surface. Their results indicated that the silicon (Si) surface chemistry could be tuned by controlling the spatial distribution and concentration of the reagents. The “self-assembled block copolymer-based quasihexagonal template arrays” can direct aqueous fluoride anion-based etching of Si surfaces. This processing occurred at ambient conditions within minutes, leading to patterned nanoscale Si-H-terminated etch pits on the Si surface. The surface without etching was “terminated by the native oxide,” suggesting these etch pits and top Si-interface can be further chemically functionalized (Fig. 1.26B). The plane and sectional morphology of the above amphiphilic copolymer-coated Si-surface were evaluated using scanning electron microscopy (Fig. 1.26C). The Si(111) surface was etched in 0.01% HF (aq) for different durations, demonstrating surface changes.

The PS-b-P4VP-coated Si wafer topology varied significantly with exposure duration in HF solution and with single-crystalline orientation. The image indicated that “truncated square and pseudohexagonal-shaped etch pits” were formed, where the center-to-center spacing was averaged at 125 nm. The shape and size were identical to the “parent block copolymer monolayer.” Top- and cross-sectional images of the etched Si surface were collected to evaluate the impact of Si orientation on morphology. The Si(100) orientation was chosen to demonstrate in this chapter. The observation revealed the 3D structures of Si(111) and Si(100) and, possibly, the transitional Si(311) face. The corner rounding was more clearly observed in the images than in samples with a top-surface mask. It was also found that oxygen’s role is minor due to its anisotropic oxidant nature based on the subtle difference in etch pit shapes on Si(100) (Fig. 1.26D).

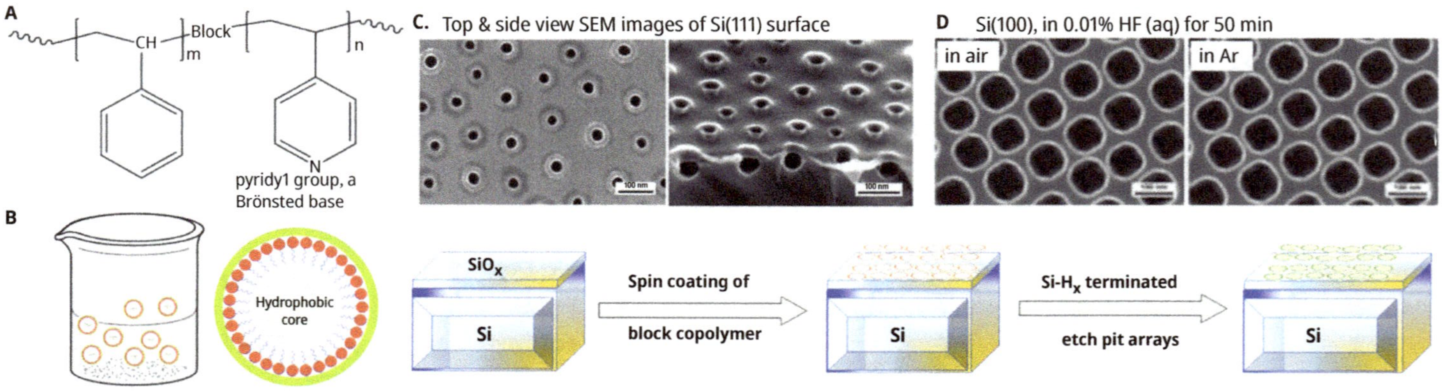

Fig. 1.26: A summary of templated etching top-down synthesis: A. the formula of an amphiphilic block copolymer, poly(styrene-block-4-vinyl pyridine) (PS-b-P4VP), B. schematic diagram to prepare “etch pits on a Si(100) surface” using an amphiphilic templated composed of PS-b-P4VP, C. SEM top and side view of the copolymer template coated Si(111) surface after immersion into 0.01 % HF solution for 40 mins, and D. the impact of Si-crystal orientation (100 was chosen) on geometric shape evolution of the patterned etch pits. Figures B–D were adapted from [138] with permission.

1.3.2.2 Selective dealloying

Selective dealloying (also known as selective leaching) is a procedure for demetalification. This demetalification, as a parting and selective corrosion, can commonly be observed in some "solid solution alloys." A chemical component in the alloys is preferentially leached out from the material under suitable conditions. Due to the redox potential, the more active, or less noble, metal will be removed from the alloy via galvanic redox corrosion. Recently, selective dealloying was used to generate a nanoporous metal from the alloys. The metallic alloy can be selectively "unalloyed" through chemical reactions, removing the most reactive metal constituent. The reactive sequence of different metals is shown in Tab. 1.3.

Tab. 1.3: The reactive sequences of metals and their standard reduction potentials [139].

Reactivity	Metal elements	Ions	Standard reduction potential, $E°$ (V)
1	Lithium (Li)	Li^+	$Li^+(aq) + e^- \rightarrow Li(s)$, $E° = -3.04$
2	Rubidium (Rb)	Rb^+	$Rb^+(aq) + e^- \rightarrow Rb(s)$, $E° = -2.98$
3	Potassium (K)	K^+	$K^+(aq) + e^- \rightarrow K(s)$, $E° = -2.93$
4	Barium (Ba)	Ba^{2+}	$Ba^{2+} + 2e^- \rightarrow Ba(s)$, $E° = -2.92$
5	Cesium (Cs)	Cs^+	$Cs^+(aq) + e^- \rightarrow Cs(s)$, $E° = -2.92$
6	Barium (Ba)	Ba^{2+}	$Ba^{2+}(aq) + 2e^- \rightarrow Ba(s)$, $E° = -2.91$
7	Strontium (Sr)	Sr^{2+}	$Sr^{2+}(aq) + 2e^- \rightarrow Sr(s)$, $E° = -2.89$
7	Calcium (Ca)	Ca^{2+}	$Ca^{2+}(aq) + 2e^- \rightarrow Ca(s)$, $E° = -2.84$
8	Sodium (Na)	Na^+	$Na^+(aq) + e^- \rightarrow Na(s)$, $E° = -2.71$
9	$Mg(OH)_2$	Mg^{2+}	$Mg(OH)_2(s) + 2e^- \rightarrow Na(s) + 2OH^-(aq)$, $E° = -2.69$
10	Lanthanum (La)	La^{3+}	$La^{3+} + 3e^- \rightarrow La(s)$, $E° = -2.38$
11	Magnesium (Mg)	Mg^{2+}	$Mg^{2+}(aq) + 2e^- \rightarrow Mg(s)$, $E° = -2.36$
12	Cerium (Ce)	Ce^{3+}	$Ce^{3+} + 3e^- \rightarrow Ce(s)$, $E° = -2.34$
13	Beryllium (Be)	Be^{2+}	$Be^{2+} + 2e^- \rightarrow Be(s)$, $E° = -1.99$
14	Aluminum (Al)	Al^{3+}	$Al^{3+}(aq) + 3e^- \rightarrow Al(s)$, $E° = -1.68$
15	Uranium (U)	U^{3+}	$U^{3+} + 3e^- \rightarrow U(s)$, $E° = -1.66$
16	Manganese (Mn)	Mn^{2+}	$Mn^{2+} + 2e^- \rightarrow Mn(s)$, $E° = -1.17$
17	Chromium (Cr)	Cr^{2+}	$Cr^{2+}(aq) + 2e^- \rightarrow Cr(s)$, $E° = -0.90$
18	Zinc (Zn)	Zn^{2+}	$Zn^{2+}(aq) + 2e^- \rightarrow Zn(s)$, $E° = -0.76$
19	Chromium (Cr)	Cr^{3+}	$Cr^{3+}(aq) + 3e^- \rightarrow Cr(s)$, $E° = -0.74$
20	Selenium (Se)	Se^{2+}	$Se^{2+}(aq) + 2e^- \rightarrow Se(s)$, $E° = -0.67$ (in 1.0 M NaOH)
21	Iron (Fe)	Fe^{2+}	$Fe^{2+}(aq) + 2e^- \rightarrow Fe(s)$, $E° = -0.44$
22	Cadmium (Cd)	Cd^{2+}	$Cd^{2+}(aq) + 2e^- \rightarrow Cd(s)$, $E° = -0.40$
23	Cobalt (Co)	Co^{2+}	$Co^{2+}(aq) + 2e^- \rightarrow Co(s)$, $E° = -0.28$
24	Nickel (Ni)	Ni^{2+}	$Ni^{2+}(aq) + 2e^- \rightarrow Ni(s)$, $E° = -0.26$
25	Molybdenum (Mo)	Mo^{3+}	$Mo^{3+} + 3e^- \rightarrow Mo(s)$, $E° = -0.20$
26	Titanium (Ti)	Ti^{2+}	$Ti^{2+} + 2e^- \rightarrow Ti(s)$, $E° = -0.16$
27	Tin (Sn)	Sn^{2+}	$Sn^{2+} + 2e^- \rightarrow Sn(s)$, $E° = -0.136$
28	Lead (Pb)	Pb^{2+}	$Pb^{2+} + 2e^- \rightarrow Pb(s)$, $E° = -0.126$
29	Iron (Fe)	Fe^{3+}	$Fe^{3+}(aq) + 2e^- \rightarrow Fe(s)$, $E° = -0.037$
30	Antimony (Sb)	Sb^{3+}	Sb^+ $3H^+ + 3e^-$ SbH3(g), $E° = -0.510$

Tab. 1.3 (continued)

Reactivity	Metal elements	Ions	Standard reduction potential, $E°$ (V)
31	Bismuth (Bi)	Bi^{3+}	$Bi^{3+} + 2e^- \rightarrow Bi(s)$, $E° = +0.317$
32	Copper (Cu)	Cu^{2+}	$Ni^{2+}(aq) + 2e^- \rightarrow Ni(s)$, $E° = +0.34$
33	Thallium (Tl)	Tl^{3+}	$Tl^{3+} + 3e^- \rightarrow Tl(s)$, $E° = +0.742$
34	Platinum (Pt)	Pt^{2+}	$Pt^{2+} + 2e^- \rightarrow Pt(s)$, $E° = +1.2$
35	Mercury (Hg)	Hg^{2+}	$Hg^{+} + 2e^- \rightarrow Hg(l)$, $E° = +0.85$
36	Silver (Ag)	Ag^{+}	$Ag^{+} + e^- \rightarrow Ag(s)$, $E° = +0.78$
37	Gold (Au)	Au^{3+}	$Au^{3+} + 3e^- \rightarrow Au(s)$, $E° = +1.52$
38	Gold (Au)	Au^{+}	$Au^{+} + e^- \rightarrow Au(s)$, $E° = +1.83$

Typically, the elements that undergo selective dealloying or removal are reasonably active metals, such as commonly used aluminum (Al), cobalt (Co), chromium (Cr), iron (Fe), and zinc (Zn). Recently, noble metals have been produced using selective dealloying to create nanoporous structures. Chen and the team reported that nanoporous gold (NPG) was formed by selective removal, named "desilvering" of silver–gold (Ag–Au, $Au_{35}Ag_{65}$) alloys. This desilvering technique was the most feasible nanofabrication approach for tuning pore structures [140]. The SEM images of desilvering NPG from $Au_{35}Ag_{65}$ using nitric acid (HNO_3) showed that the nanopore size increased with dealloying time and temperature (Fig. 1.27A). This team also reported that a series of transition-metal nanopore-containing alloys had been fabricated from their Al-based alloys in alkaline, basic, or acidic solutions under free-corrosion conditions. These elements include three noble metals, palladium (Pd), platinum (Pt), and silver, and one transition metal, copper (Cu). Upon dissolution of Al in either basic or acidic solution, the electron tomography image of the metal showed well-tuned pore structures (Fig. 1.27B). The nanoporous metals showed electrical resistivities one to two orders of magnitude higher than those of their bulk counterparts. Initially, the less-noble metal (LNM) component is dissolved in a "quasi-layer-by-layer mode" to form step edges during the dealloying process. Due to surface diffusion, the more-noble metal (MNM) elements are expected to be etched from their original lattice sites and to aggregate at step edges. As more LNM atoms dissolve, MNM atoms are needed to passivate the steps of the resulting MN-capped hills. The nucleation of MNM clusters will further widen the base perimeters of the above MN-capped hill, allowing for a decrease in the hill spaces [99]. Therefore, the pores extended into the bulk of the materials, forming a porous network within the core–shell structures. The more noble metals were leached out to act as shells with tunable pore structures, as shown in Fig. 1.27B.

Erlebacher et al. [141] have developed a kinetic Monte Carlo model to evaluate the physical mechanisms and porosity evolution during dealloying of a binary metal alloy. The authors proposed a continuum model to demonstrate the intrinsic dynamical pattern formation in nanoporous metal structures. The difference in reduction potential between metals is the driving force behind noble metals aggregating into 2D

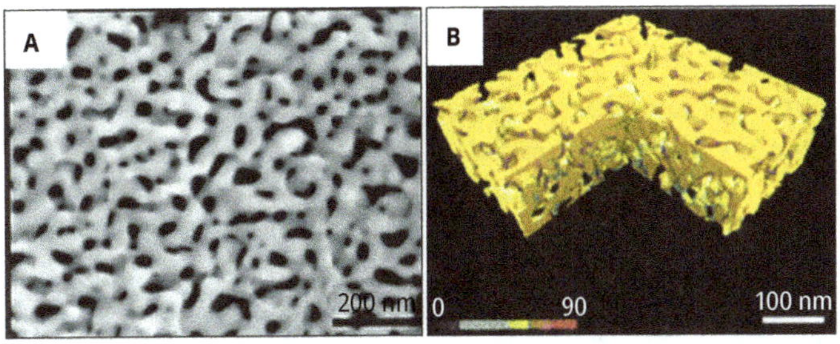

Fig. 1.27: A desilvering method to produce porous nanogold (NPG) from $Au_{35}Ag_{65}$ alloy using HNO_3 under free corrosion conditions: A. SEM image showing the pore size increase with dealloying and temperature and B. electron tomography image, confirming the SEM observation. Reprinted with permission from [140]. Copyright 2008 by APS.

clusters. The phase-separation process (also known as spinodal decomposition) was observed at the solid–electrolyte interface. During chemical etching, the surface area increases, enabling the development of porosity with characteristic length and pore-size dimensions. Through chemical tailoring, the porosity of NPG is anticipated to be suitable for biological sensing applications.

1.3.2.3 Anisotropic dissolution

This is a process in which the materials can be dissolved in a solvent, allowing for different properties in different directions. Unlike nucleation and growth from supersaturated solutions, controlled separation of larger crystals into their constituent components will yield materials with well-defined nanostructure. This procedure occurs typically in "undersaturated solutions" via anisotropic dissolution of preferred crystalline planes and facets (Fig. 1.25C). Yu and coworkers reported that well-aligned nanosheet arrays composed of calcium carbonate ($CaCO_3$, calcite) were formed using the newly developed anisotropic dissolution method [142]. Yu and his team carried out anisotropic dissolution of calcite microcrystals using a 1% aqueous formamide solution, followed by heating at 65 °C for 10 min and 1 day, respectively. Complex calcite, the most stable of $CaCO_3$ polymorphs, was naturally produced through biomineralization, such as protein-directed nucleation and growth, due to the binding between peptide and crystallographic facets of calcite. Yu and his team invented a dissolution method for producing densely packed nanoarrays. Their research (Fig. 1.28A) revealed that stable {104} planes consist of calcium cations (Ca^{2+}) and carbonate anions arranged in an alternating sequence. This plane was found to be more preferentially dissolute than {012} planes. As a result, the higher-energy {012} planes are entirely composed of calcium ions, produced mainly on the nanoarray surfaces. The calcite shows a highly anisotropic crystal structure, including CO_3^{2-} "rotor" groups arranged in layers, while the Ca cation resides between the different layers (Fig. 1.28B). In addition, the high electron density within the CO_3^{2-} layers (purple-colored sub-building units) gives calcite a high refractive index. The low-density component perpendicular to the layers is mainly composed of Ca (indigo, #8de6f5). The intrinsic anisotropic nature of $CaCO_3$, as seen from the unit-cell perspective, lays the foundation for the formation of nanosheet arrays.

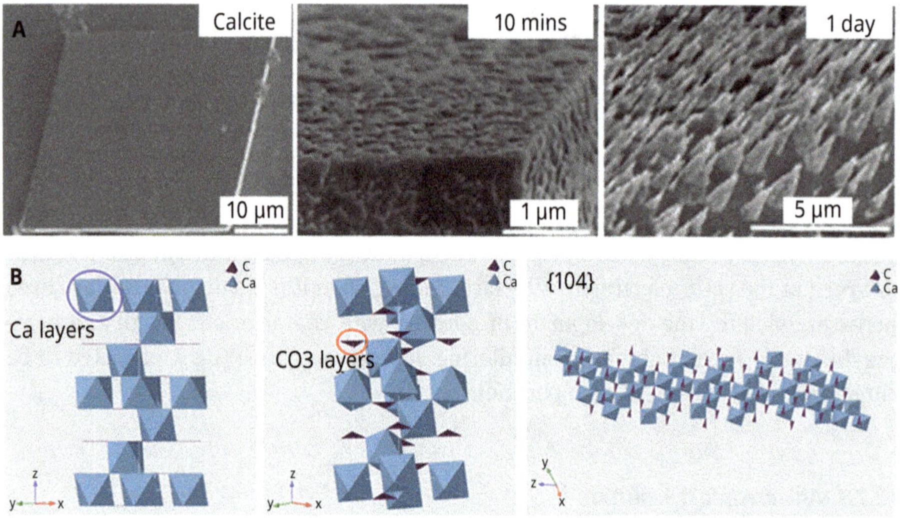

Fig. 1.28: Demonstration of an anisotropic dissolution approach. A. Formation of calcite nanosheet arrays. Reproduced with permission from [142]. Copyright 2010 Wiley-VCH, and B. crystalline structure of calcite (space group: $R\bar{3}c$, trigonal, $a = 4.9900$ Å, $b = 7.0595$ Å, and $c = 367.873$ Å^3).

1.3.2.4 Thermal decomposition

Thermal decomposition is thermolysis, a chemical decomposition when heat is applied to the system. When materials are decomposed at a specific temperature (the decomposition temperature), chemical bonds are cleaved, and new bonds form through rearrangements of electron transfer or sharing. Usually, the thermolysis is endothermic due to the employment of heat treatment. Upon thermal decomposition, the large starting inorganic solids (bulk and microstructure) can be effectively transformed into nanostructured materials with tunable pore sizes and particle sizes. These chemical routes are attractive for their cost-effectiveness in forming NMs with potential for large-scale fabrication. The researchers demonstrated the conversion of calcium carbonate ($CaCO_3$) microcrystals into highly porous materials at 900 °C for 1 h (Fig. 1.29A), as reported by Han and coworkers [143]. It can be seen that $CaCO_3$ crystals ~5 μm in size were thermally decomposed into nanoporous calcium oxide (CaO). These nanoporous CaO particles were subsequently cooled and re-transformed into nanoporous $CaCO_3$ in a carbon dioxide atmosphere. It was found that the large size (~50 µm) of the starting $CaCO_3$ crystals enabled the formation of two distinct pore types: nanopores and micrometer-sized channels. The rapid CO_2 release rate is critical for generating pores of different sizes and shapes. Han and his team from the National University of Singapore reported the "non-crystalline pellets of different calcium compounds." Their research data showed that the calcium content of the starting materials affects the porosity of the final products (Fig. 1.29B and C) [144]. This team employed different resources and compositions of Ca using different compounds, calcium carbonate ($CaCO_3$, 40% Ca), calcium citrate ($Ca_3(C_6H_5O_7)_2$, 21% Ca), and calcium lactobionate ($Ca(C_{12}H_{21}O_{12})_2$, 5% Ca). These compounds were pressed into pellets and heated at 1,000 °C for 1 h to produce porous CaO and release CO_2. The system was cooled down to ambient conditions to reform porous $CaCO_3$. The pore size of the final $CaCO_3$ product varies with the reactants used. It was found that a lower Ca weight percentage in the reactant ($CaCO_3$) correlated with smaller pore sizes, due to a smaller amount of CO_2 released (Fig. 1.29B). The two organic-based compounds, calcium citrate and lactobionate (Fig. 1.29C), each contain many ligands with carboxylate and hydroxyl functional groups, resulting in the release of large amounts of CO_2 and water vapor. The vast, empty spaces will remain open to gas release, leading to large pore sizes. The use of organic compounds enables the control of the porosity or pore size of the final products, which can be used in catalysis, separation, and tissue engineering.

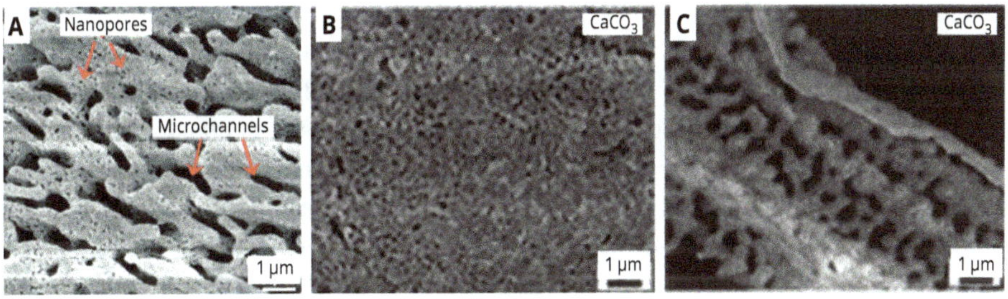

Fig. 1.29: Microscopic image of porous $CaCO_3$ formed using thermolysis. A. SEM image of porous $CaCO_3$ produced from large microcrystals by thermolysis at 900 °C for one hour and subsequently cooled in air. Nanopores and microchannels were formed based on the CO_2 release rate and amount. Adapted from [144]. Moreover, B. SEM image of porous $CaCO_3$ produced using calcium carbonate ($CaCO_3$, Ca%, % = 40) as a reactant with high Ca content, and an SEM image of porous $CaCO_3$ produced using calcium lactobionate ($Ca(C_{12}H_{21}O_{12})_2$, Ca%, % = 5) as a reactant with low Ca content. B and C were adapted from [145].

1.4 Summary

Nanochemistry can be divided into two overarching strategies for building nanoscale materials: bottom-up assembly and top-down deconstruction, through "building up" materials one atom or molecule at a time or "breaking down" bulk solids into nanoscale fragments. In bottom-up synthesis, chemists exploit controlled reactions, such as sol–gel hydrolysis and condensation, microemulsion polymerization and carbonization, hydro(solvo)thermal crystallization, or vapor-phase deposition, to guide molecular precursors into precisely architected nanoparticles, nanorods, and thin films. Top-down techniques, in contrast, use mechanical forces (ball milling, lithography, and FIBs) or chemical etchants (templated etching, selective dealloying, anisotropic dissolution, thermal decomposition) to carve or peel away bulk materials, ultimately yielding nanoscale structures. Together, these complementary approaches give researchers unmatched flexibility to tailor size, shape, composition, and functionality for applications across energy, environment, electronics, and medicine. Together, these complementary routes equip scientists with a versatile toolkit for tailoring the size, shape, composition, and functionality of NMs, from building from molecules to carving from solids, across energy, environmental, electronic, and biomedical frontiers.

In the sol–gel method, metal–alkoxide precursors undergo controlled hydrolysis and condensation to form an interconnected three-dimensional oxide network. By adjusting the solution pH, one can shift the mechanism from acid-catalyzed, slow-condensation pathways that yield uniform, mesoporous gels to base-catalyzed, rapid-condensation regimes that favor dense, microporous particles. Solvent choice (water versus alcohols or nonaqueous media) further tunes nucleation rates and particle crystallinity, while reaction temperature sets the overall kinetics and gelation time. Microemulsion synthesis exploits thermodynamically stable mixtures of oil, water, surfactant, and cosurfactant to create nanoscale reactors. Within these sub-50 nm droplets, free-radical polymerization can be initiated by oil-soluble or water-soluble initiators, producing uniform polymer spheres or core–shell composites. Subsequent cross-linking under oxidative or thermal conditions locks in the network architecture, and a final carbonization step at 700–900 °C transforms the organic matrices into conductive carbon materials with tunable pore sizes. By systematically varying surfactant head-group charge, water-to-oil ratio, and reaction temperature, researchers have synthesized nanocomposites for hydrogen generation, COVID-19 electrochemical sensing, and flexible capacitor electrodes. Hydro(solvo)thermal synthesis drives crystallization under autogenous pressures (up to 15 MPa) and temperatures (150–300 °C) within a Teflon-lined stainless steel vessel. Dissolved metal salts or organometallic complexes nucleate into uniform MOFs or mixed-oxide nanoparticles, with crystal size and phase purity controlled by temperature ramp rates, solvent polarity, and reaction duration (often 12–72 h). A temperature gradient between the hot and cool zones of the autoclave can even direct anisotropic growth into nanorods or nanosheets for supercapacitor and carbon-capture applications. CVD films high-purity materials by transporting volatile metal halides or organometallics into a heated reactor (500–1,000 °C). Precursors thermolyze or fragment under thermal, plasma, or laser activation, depositing atomically smooth films, quantum dots, or ACNTs. In a two-zone furnace, vaporized xylene/ferrocene flows over substrates at 725 °C, where iron seeds catalyze the growth of ACNTs into uniform arrays for fuel-cell cathodes. By selecting

ALD pulses or plasma enhancements, CVD can achieve sub-nanometer thickness control and exceptional conformality on complex three-dimensional supports.

In contrast, top-down methods sculpt or shatter bulk materials into nanoscale features. High-energy ball milling uses densified steel or tungsten carbide balls in rotating drums (5:1 ball-to-powder mass ratio) under an inert gas atmosphere to generate nanoparticles through repeated impact, shear, and cold welding. Lithographic techniques (photolithography, electron-beam lithography, or focused-ion-beam lithography) project patterns with resolutions down to 10 nm onto photoresists, enabling semiconductor and MEMS fabrication. On simpler benches, templated etching with self-assembled block copolymers (e.g., PS-b-P4VP on Si(100)) directs anisotropic HF etching into hexagonally ordered nanoscale pits. At the same time, selective dealloying of $Ag_{65}Au_{35}$ in nitric acid produces nanoporous gold with ligament diameters from 20 to 100 nm. Finally, controlled thermal decomposition of $CaCO_3$ pellets at 900 °C liberates CO_2, yielding CaO frameworks with dual micro- and mesopores, useful for catalysis and tissue-engineering scaffolds.

Dr Liu's laboratory has mastered both strategies to deliver over 108 unique NM formulations. Core–shell metal oxides synthesized by sol–gel and hydrothermal methods serve as highly efficient photocatalysts for degrading automotive exhaust and pharmaceutical contaminants. Platinum-capped nanoparticles and their non-platinum analogues, prepared via microemulsion-templated synthesis and CVD, serve as robust cathodic and anodic electrocatalysts in fuel cells and electrolyzers. Carbon-based nanocatalysts, generated through solvothermal carbonization of polymer templates, boost performance in both hydrogen fuel cells and lithium-ion batteries. Multicomponent metal-oxide spinels combine synergistic photo- and electrochemical activity for water splitting and pollutant degradation. Perovskite ceramics recovered from automotive waste streams have been engineered as solid-oxide fuel cell electrodes to produce clean hydrogen and ammonia. MOFs and their composites grown under hydro(solvo)thermal conditions achieve record CO_2 capture and sequestration capacities. Even recycled materials have been re-fabricated into functional photocatalysts and gas-separation membranes, closing the loop on industrial waste. Beyond the bench, these discoveries are moving toward real-world impact. Through the NSF I-Corps IdeaLaunch program, Dr Liu and her team have transformed laboratory prototypes into market-ready concepts, culminating in the creation of NanoInnovation LLC. Strategic partnerships, such as the collaboration with the Department of Homeland Security's Center for Innovation in Nanotechnology (CINA) on field-deployable sensors, are accelerating the translation of these NMs into resilient, scalable technologies for energy security, environmental remediation, and national defense applications.

References

[1] Khan, Y., Sadia, H., Ali Shah, S.Z., Khan, M.N., Shah, A.A., Ullah, N., & Khan, M.I. (2022). Classification, synthetic, and characterization approaches to nanoparticles, and their applications in various fields of nanotechnology: A review. *Catalysts*, 12(11), 1386.

[2] Al-Harbi, N., & Abd-Elrahman, N.K. (2025). Physical methods for preparation of nanomaterials, their characterization and applications: a review. *Journal of Umm Al-Qura University For Applied Sciences*, 11(2), 356–377.

[3] Nnameka, M., Rahman, M.R., Said, K.A.M.B., Mannan, M.A., & Patwary, A.M. (2023). A review of nanoparticle synthesis methods, classifications, applications, and characterization. *Environmental Nanotechnology, Monitoring & Management*, 20, 100900.

[4] Peng, X., Ai, F., Yan, L., Ha, E., Hu, X., He, S., & Hu, J. (2021). Synthesis strategies and biomedical applications for doped inorganic semiconductor nanocrystals. *Cell Reports Physical Science*, 2(5).

[5] Malik, S., Muhammad, K., & Waheed, Y. (2023). Nanotechnology: a revolution in modern industry. *Molecules*, 28(2), 661.

[6] Sacha, G.M., & Verona, P. (2013). Artificial intelligence in nanotechnology. *Nanotechnology*, 24(45), 452002.

[7] National Research Council, Division on Engineering, Physical Sciences, National Materials Advisory Board, & Committee to Review the National Nanotechnology Initiative. (2006). *A Matter of Size: Triennial Review of The National Nanotechnology Initiative*. National Academies Press.

[8] Serrano, E., Rus, G., & García-Martínez, J. (2009). Nanotechnology for sustainable energy. Renewable and Sustainable *Energy Reviews*, 13(9), 2373–2384.

[9] Abid, N., Khan, A.M., Shujait, S., Chaudhary, K., Ikram, M., Imran, M., & Maqbool, M. (2022). Synthesis of nanomaterials using various top-down and bottom-up approaches, influencing factors, advantages, and disadvantages: A review. *Advances In Colloid And Interface Science*, 300, 102597.

[10] Biswas, A., Bayer, I.S., Biris, A.S., Wang, T., Dervishi, E., & Faupel, F. (2012). Advances in top-down and bottom-up surface nanofabrication: Techniques, applications & prospects. *Advances In Colloid and Interface Science*, 170(1–2), 2–27.

[11] Yadav, T.P., Yadav, R.M., & Singh, D.P. (2012). Mechanical milling: a top-down approach for the synthesis of nanomaterials and nanocomposites. *Nanoscience and Nanotechnology*, 2(3), 22–48.

[12] Pourpoint, V.V., & Skorokhod, V.V. (2008). New dimensionality classifications of nanostructures. *Physica E: Low-Dimensional Systems and Nanostructures*, 40(7), 2521–2525.

[13] Kim, W.Y., Choi, Y.C., Min, S.K., Cho, Y., & Kim, K.S. (2009). Application of quantum chemistry to nanotechnology: electron and spin transport in molecular devices. *Chemical Society Reviews*, 38(8), 2319–2333.

[14] Chen, X., Lou, Y., Dayal, S., Qiu, X., Krolicki, R., Burda, C., & Becker, J. (2005). Doped semiconductor nanomaterials. *Journal of Nanoscience and Nanotechnology*, 5(9), 1408–1420.

[15] Mitin, V.V., Kochelap, V.A., & Stroscio, M.A. (2008). *Introduction To Nanoelectronics: Science, Nanotechnology, Engineering, and Applications*. Cambridge University Press.

[16] Ewels, C.P., & Glerup, M. (2005). Nitrogen doping in carbon nanotubes. *Journal of Nanoscience and Nanotechnology*, 5(9), 1345–1363.

[17] Zhou, H., Ye, Q., & Xu, J. (2017). Polyhedral oligomeric silsesquioxane-based hybrid materials and their applications. *Materials Chemistry Frontiers*, 1(2), 212–230.

[18] Fan, Y., Kou, X., Upadhyaya, P., Shao, Q., Pan, L., Lang, M., & Wang, K.L. (2016). Electric-field control of spin-orbit torque in a magnetically doped topological insulator. *Nature Nanotechnology*, 11(4), 352–359.

[19] Whitesides, G.M. (2005). Nanoscience, nanotechnology, and chemistry. *Small*, 1(2), 172–179.

[20] Rashidzadeh, H., Danafar, H., Rahimi, H., Mozafari, F., Salehiabar, M., Rahmati, M.A., & Mirsaeidi, M. (2021). Nanotechnology against the novel coronavirus (severe acute respiratory syndrome

coronavirus 2): diagnosis, treatment, therapy, and future perspectives. *Nanomedicine*, 16(6), 497–516.
[21] Xuan, L., Ju, Z., Skonieczna, M., Zhou, P.K., & Huang, R. (2023). Nanoparticles-induced potential toxicity on human health: applications, toxicity mechanisms, and evaluation models. *MedComm*, 4(4), e327.
[22] Dugershaw, B.B., Aengenheister, L., Hansen, S.S.K., Hougaard, K.S., & Buerki-Thurnherr, T. (2020). Recent insights on indirect mechanisms in developmental toxicity of nanomaterials. *Particle and Fibre Toxicology*, 17(1), 31.
[23] Hou, W.C., Stuart, B., Howes, R., & Zepp, R.G. (2013). Sunlight-driven reduction of silver ions by natural organic matter: formation and transformation of silver nanoparticles. *Environmental Science & Technology*, 47(14), 7713–7721.
[24] Park, H.G., & Yeo, M.K. (2016). Nanomaterial regulatory policy for human health and environment. *Molecular & Cellular Toxicology*, 12(3), 223–236.
[25] Handy, R.D., & Shaw, B.J. (2007). Toxic effects of nanoparticles and nanomaterials: implications for public health, risk assessment and the public perception of nanotechnology. *Health, Risk & Society*, 9(2), 125–144.
[26] Macwan, D.P., Dave, P.N., & Chaturvedi, S. (2011). A review on nano-TiO2 sol-gel type syntheses and their applications. *Journal of Materials Science*, 46(11), 3669–3686.
[27] Schubert, U. (2015). Chemistry and fundamentals of the sol-gel process. *The Sol-Gel Handbook*, 1–28.
[28] Hu, M.C., Zielke, J.T., Byers, C.H., Lin, J.S., & Harris, M.T. (2000). Probing the early-stage/rapid processes in hydrolysis and condensation of metal alkoxides. *Journal of Materials Science*, 35(8), 1957–1971.
[29] Rankin, S.E., & McCormick, A.V. (2000). Hydrolysis pseudoequilibrium: challenges and opportunities to sol-gel silicate kinetics. *Chemical Engineering Science*, 55(11), 1955–1967.
[30] Rodríguez, R., Rojas, G., Estevez, M., & Vargas, S. (2002). Fractal characterization of silica sol prepared by the sol-gel method: From the sol formation to the flocculation process. *Journal of Sol-Gel Science and Technology*, 23(2), 99–105.
[31] Brinker, C.J., & Scherer, G.W. (1985). Sol→ gel→ glass: I. Gelation and gel structure. *Journal of Non-Crystalline Solids*, 70(3), 301–322.
[32] Hench, L.L., & West, J.K. (1990). The sol-gel process. *Chemical Reviews*, 90(1), 33–72.
[33] Djabourov, M., Leblond, J., & Papon, P. (1988). Gelation of aqueous gelatin solutions. II. Rheology of the sol-gel transition. *Journal De Physique*, 49(2), 333–343.
[34] Bohidar, H.B., & Jena, S.S. (1993). Kinetics of sol-gel transition in thermoreversible gelation of gelatin. *The Journal of Chemical Physics*, 98(11), 8970–8977.
[35] Djabourov, M. (1991). Gelation: A review. *Polymer International*, 25(3), 135–143.
[36] Niederberger, M. (2007). Nonaqueous sol-gel routes to metal oxide nanoparticles. *Accounts of Chemical Research*, 40(9), 793–800.
[37] Rechberger, F., & Niederberger, M. (2017). Synthesis of aerogels: from molecular routes to 3-dimensional nanoparticle assembly. *Nanoscale Horizons*, 2(1), 6–30.
[38] Pinna, N., & Niederberger, M. (2008). Surfactant-free nonaqueous synthesis of metal oxide nanostructures. *Angewandte Chemie International Edition*, 47(29), 5292–5304.
[39] Song, Q., & Zhang, Z.J. (2004). Shape control and associated magnetic properties of spinel cobalt ferrite nanocrystals. *Journal of The American Chemical Society*, 126(19), 6164–6168.
[40] Zeng, H., Rice, P.M., Wang, S.X., & Sun, S. (2004). Shape-controlled synthesis and shape-induced texture of $MnFe_2O_4$ nanoparticles. *Journal of The American Chemical Society*, 126(37), 11458–11459.
[41] Joo, J., Kwon, S.G., Yu, J.H., & Hyeon, T. (2005). Synthesis of ZnO nanocrystals with cone, hexagonal cone, and rod shapes via non-hydrolytic ester elimination sol-gel reactions. *Advanced Materials*, 17(15), 1873–1877.

[42] Li, X.L., Peng, Q., Yi, J.X., Wang, X., & Li, Y. (2006). Near monodisperse TiO2 nanoparticles and nanorods. *Chemistry - A European Journal*, 12(8), 2383–2391.

[43] Zitoun, D., Pinna, N., Frolet, N., & Belin, C. (2005). Single-crystal manganese oxide multipods by oriented attachment. *Journal of The American Chemical Society*, 127(43), 15034–15035.

[44] Seo, J.W., Jun, Y.W., Ko, S.J., & Cheon, J. (2005). In situ one-pot synthesis of 1-dimensional transition metal oxide nanocrystals. *The Journal of Physical Chemistry B*, 109(12), 5389–5391.

[45] Assink, R.A., & Kay, B.D. (1988). Sol-gel kinetics I. Functional group kinetics. *Journal of Non-Crystalline Solids*, 99(2–3), 359–370.

[46] Twej, W. (2009). Temperature influence on the gelation process of tetraethylorthosilicate using the sol-gel technique. *Iraqi Journal of Science*, 50(1), 43–49.

[47] Eshtiagh-Hosseini, H., Housaindokht, M.R., & Chahkandi, M. (2007). Effects of parameters of the sol-gel process on the phase evolution of sol-gel-derived hydroxyapatite. *Materials Chemistry and Physics*, 106(2–3), 310–316.

[48] Lu, Z.L., Lindner, E., & Mayer, H.A. (2002). Applications of sol-gel-processed interphase catalysts. *Chemical Reviews*, 102(10), 3543–3578.

[49] Kritikaki, A., & Tsetsekou, A. (2009). Fabrication of porous alumina ceramics from powder mixtures with sol-gel derived nanometer alumina: Effect of mixing method. *Journal of The European Ceramic Society*, 29(9), 1603–1611.

[50] Alias, S.S., Ismail, A.B., & Mohamad, A.A. (2010). Effect of pH on ZnO nanoparticle properties synthesized by sol-gel centrifugation. *Journal of Alloys and Compounds*, 499(2), 231–237.

[51] Miller, J.B., & Ko, E.I. (1997). Control of mixed oxide textural and acidic properties by the sol-gel method. *Catalysis Today*, 35(3), 269–292.

[52] Wang, X., & Liu, J.L. (2022). Bottom-up synthesis of nanomaterials. *Nanochemistry: From Theory to Application for In-Depth Understanding of Nanomaterials*, 239.

[53] Zhao, M., Wu, D., Wang, J., Zhang, H., Liu, W., & Zhang, R. (2017). Evolution of aqueous micellar solutions in the injection fluids of alkaline-surfactant-polymer and alkaline-surfactant floodings from injection to production. *Asia-Pacific Journal of Chemical Engineering*, 12(4), 651–659.

[54] Svendsen, B., & Hutter, K. (1995). On the thermodynamics of a mixture of isotropic materials with constraints. *International Journal of Engineering Science*, 33(14), 2021–2054.

[55] Gradzielski, M., Duvail, M., De molina, P.M., Simon, M., Talmon, Y., & Zemb, T. (2021). Using microemulsions: formulation based on knowledge of their mesostructure. *Chemical Reviews*, 121(10), 5671–5740.

[56] Paul, B.K., & Moulik, S.P. (1997). Microemulsions: an overview. *Journal of Dispersion Science and Technology*, 18(4), 301–367.

[57] Chang, Y.J., Chang, C.H., Yu, C.Y., Chang, T.J., Chen, L.C., Chen, M.H., & Ting, G. (2010). Therapeutic efficacy and microSPECT/CT imaging of 188Re-DXR-liposome in a C26 murine colon carcinoma solid tumor model. *Nuclear Medicine and Biology*, 37(1), 95–104.

[58] 2020's Top Nanotechnologies for Lifehttps://statnano.com/news/68539/2020-Top-Nanotechnologies-for-Life#ixzz7F80cHrC5;https://statnano.com/news/67755/Real-time-Detection-of-COVID-19-Aided-by-CNT-based-Electrochemical-Sensors; Miripour, Z.S., Sarrami-Forooshani, R., Sanati, H., Makarem, J., Taheri, M.S., Shojaeian, F., & Abdolahad, M. (2020). Real-time diagnosis of reactive oxygen species (ROS) in fresh sputum by electrochemical tracing; a correlation between COVID-19 and virus-induced ROS in lung/respiratory epithelium during this pandemic. *Biosensors And Bioelectronics*, 165, 112435. and the United States Patent Application Publication No.: US 2018 / 0299401 A1 Abdolahad *et al.* Pub. Date: Oct 18 2018.

[59] Chiappisi, L., Noirez, L., & Gradzielski, M. (2016). A journey through the phase diagram of a pharmaceutically relevant microemulsion system. *Journal of Colloid and Interface Science*, 473, 52–59.

[60] Robbins, M.L. (1977). Theory for the phase behavior of microemulsions. In *Micellization, Solubilization, and Microemulsions* (Vol. 2, pp. 713–754). Boston, MA: Springer US.

[61] Fernando, S., & Hanna, M. (2005). Phase behavior of the ethanol-biodiesel-diesel micro-emulsion system. *Transactions of The ASAE*, 48(3), 903–908.
[62] Chern, C.S. (2008). *Principles And Applications Of Emulsion Polymerization*. pp. 1–50 (pp. 154–174). New York: John Wiley & Sons.
[63] Nagao, M., Seto, H., Takeda, T., & Kawabata, Y. (2001). Effects of temperature and pressure on phase transitions in a ternary microemulsion system. *The Journal of Chemical Physics*, 115(21), 10036–10044.
[64] Wiedmer, S., Gellner, D., & Friedrich, K. (2005). Effect of electron-beam radiation on thermoplastic composites. *Plastics, Rubber And Composites*, 34(2), 76–84.
[65] Bashir, S., Houf, W., Liu, J.L., & Mulvaney, S.P. (2021). 3D Conducting Polymeric Membrane and Scaffold Saccharomyces cerevisiae Biofilms to Enhance Energy Conversion in Microbial Fuel Cells. *ACS Applied Materials & Interfaces*.
[66] Zeng, M., & Pan, X. (2022). Insights into solid acid catalysts for efficient cellulose hydrolysis to glucose: progress, challenges, and future opportunities. *Catalysis* Reviews, 64(3), 445–490.
[67] Lewis, I.C. (1982). Chemistry of carbonization. *Carbon*, 20(6), 519–529.
[68] McNaught, A.D., & Wilkinson, A. (2019). Compendium of chemical terminology. IUPAC recommendations (P. 1622); https://goldbook.iupac.org/
[69] Tian, H., Liang, J., & Liu, J. (2019). Nanoengineering carbon spheres as nanoreactors for sustainable energy applications. *Advanced Materials*, 31(50), 1903886.
[70] Li, W., Yang, K., Peng, J., Zhang, L., Guo, S., & Xia, H. (2008). Effects of carbonization temperatures on characteristics of porosity in coconut shell chars and activated carbons derived from carbonized coconut shell chars. *Industrial Crops and Products*, 28(2), 190–198.
[71] Capek, I. (2001). On the role of oil-soluble initiators in the radical polymerization of micellar systems. *Advances In Colloid and Interface Science*, 91(2), 295–334.
[72] Chen, Y.C., Dimonie, V., & El-Aasser, M.S. (1992). Role of surfactant in composite latex particle morphology. *Journal of Applied Polymer Science*, 45(3), 487–499.
[73] Ou, J.L., Yang, J.K., & Chen, H. (2001). Styrene/potassium persulfate/water systems: effects of hydrophilic comonomers and solvent additives on the nucleation mechanism and the particle size. *European Polymer Journal*, 37(4), 789–799.
[74] Abdollahi, M., Yousefi, M.R., Ghahramani, M., Ranjbar, H., & Najafi, F.S. (2017). Synthesis of polybutadiene nanoparticles by emulsion polymerization: The effect of electrolyte and initiator type on particle size and reaction kinetics. *Iranian Polymer Journal*, 26(1), 1–10.
[75] Yan, K., Gao, X., & Luo, Y. (2015). Well-defined, high-molecular-weight polystyrene with high rates and high livingness was synthesized via two-stage RAFT emulsion polymerization. *Macromolecular Rapid Communications*, 36(13), 1277–1282.
[76] Rouse, J.D., Sabatini, D.A., Suflita, J.M., & Harwell, J.H. (1994). Influence of surfactants on microbial degradation of organic compounds. *Critical Reviews In Environmental Science and Technology*, 24(4), 325–370.
[77] Strauss, M.A., & Wegner, H.A. (2019). Molecular systems for the quantification of London dispersion interactions. *European Journal of Organic Chemistry*, 2019(2–3), 295–302.
[78] Talapin, D.V., Shevchenko, E.V., Murray, C.B., Titov, A.V., & Kral, P. (2007). Dipole-dipole interactions in nanoparticle superlattices. *Nano Letters*, 7(5), 1213–1219.
[79] Novoa, J.J., & Sosa, C. (1995). Evaluation of the density functional approximation on the computation of hydrogen bond interactions. *The Journal of Physical Chemistry*, 99(43), 15837–15845.
[80] Hirsch, M. (2021). Classifying Surfactants for Use in Coatings Formulation https://knowledge.ulprospector.com/11661/pc-classifying-surfactants-for-use-in-coatings-formulation
[81] Fenyves, R., Schmutz, M., Horner, I.J., Bright, F.V., & Rzayev, J. (2014). Aqueous self-assembly of giant bottlebrush block copolymer surfactants as shape-tunable building blocks. *Journal of The American Chemical Society*, 136(21), 7762–7770.

[82] Zhao, T.H., Jacucci, G., Chen, X., Song, D.P., Vignolini, S., & Parker, R.M. (2020). Angular-Independent Photonic Pigments via the Controlled Micellization of Amphiphilic Bottlebrush Block Copolymers. *Advanced Materials*, 32(47), 2002681.
[83] An, C., Zhang, Y., Guo, H., & Wang, Y. (2019). Metal oxide-based supercapacitors: progress and prospectives. *Nanoscale Advances*, 1(12), 4644–4658.
[84] Cao, Z., Momen, R., Tao, S., Xiong, D., Song, Z., Xiao, X., & Ji, X. (2022). Metal-organic framework materials for electrochemical supercapacitors. *Nano-Micro Letters*, 14(1), 181.
[85] Ndlwana, L., Raleie, N., Dimpe, K.M., Ogutu, H.F., Oseghe, E.O., Motsa, M.M., & Mamba, B.B. (2021). Sustainable hydrothermal and solvothermal synthesis of advanced carbon materials in multidimensional applications: a review. *Materials*, 14(17), 5094.
[86] Zito, C.A., Orlandi, M.O., & Volanti, D.P. (2018). Accelerated microwave-assisted hydrothermal/ solvothermal processing: Fundamentals, morphologies, and applications. *Journal of Electroceramics*, 40(4), 271–292.
[87] Sun, Y., & Zhou, H.C. (2015). Recent progress in the synthesis of metal-organic frameworks. *Science and Technology of Advanced Materials.*
[88] Sasikala, S.P., Poulin, P., & Aymonier, C. (2017). Advances in subcritical hydro-/solvothermal processing of graphene materials. *Advanced Materials*, 29(22), 1605473.
[89] Devaraju, M.K., Yin, S., & Sato, T. (2009). Morphology control of cerium oxide particles synthesized via a supercritical solvothermal method. *ACS Applied Materials & Interfaces*, 1(11), 2694–2698.
[90] Call, T.P., Carey, T., Bombelli, P., Lea-Smith, D.J., Hooper, P., Howe, C.J., & Torrisi, F. (2017). Platinum-free, graphene-based anodes and air cathodes for single-chamber microbial fuel cells. *Journal of Materials Chemistry A*, 5(45), 23872–23886.
[91] Kemsley, J. (2015). Illuminating crystal nucleation. *Chemical & Engineering News*, 93(2), 28–29.
[92] Byrappa, K., & Yoshimura, M. (2012). *Handbook of Hydrothermal Technology* (Vol. 404). New York: William Andrew.
[93] Bashir, S., Mulvaney, S.P., Houf, W., Villanueva, L., Wang, Z., Buck, G., & Liu, J.L. (2021). Microbial Fuel Cells: Design and Evaluation of Catalysts and Devices. In *Advances In Sustainable Energy* (pp. 681–764). Cham: Springer.
[94] Bøjesen, E.D., & Iversen, B.B. (2016). The chemistry of nucleation. *CrystEngComm*, 18(43), 8332–8353.
[95] Jensen, K.M., Tyrsted, C., Bremholm, M., & Iversen, B.B. (2014). In situ studies of solvothermal synthesis of energy materials. *ChemSusChem*, 7(6), 1594–1611.
[96] Ter Horst, J.H., & Jansens, P.J. (2005). Nucleus size and Zeldovich factor in two-dimensional nucleation at the Kossel crystal (0 0 1) surface. *Surface Science*, 574(1), 77–88.
[97] Zhou, J., Yang, Y., Yang, Y., Kim, D.S., Yuan, A., Tian, X., & Miao, J. (2019). Observing crystal nucleation in four dimensions using atomic electron tomography. *Nature*, 570(7762), 500–503.
[98] Ölfen, H., & Antonietti, M. (2005). Mesocrystals: inorganic superstructures made by highly parallel crystallization and controlled alignment. *Angewandte Chemie International Edition*, 44(35), 5576–5591.
[99] Yu, H.D., Regulacio, M.D., Ye, E., & Han, M.Y. (2013). Chemical routes to top-down nanofabrication. *Chemical Society Reviews*, 42(14), 6006–6018.
[100] Wang, J., Zheng, F., Yu, Y., Hu, P., Li, M., Wang, J., & Liu, J.L. (2021). Symmetric supercapacitors composed of ternary metal oxides ($NiO/V_2O_5/MnO_2$) nanoribbon electrodes with high energy storage performance. *Chemical Engineering Journal*, 426, 131804.
[101] Vance, J.E. (1959). Growth and Perfection of Crystals. *Journal of the American Chemical Society*, 81(13), 3489–3490.
[102] Gibbs, J.W. (1879). On the equilibrium of heterogeneous substances. 300–320.
[103] Bashir, S., Mulvaney, S.P., Houf, W., Villanueva, L., Wang, Z., Buck, G., & Liu, J.L. (2021). Microbial Fuel Cells: Design and Evaluation of Catalysts and Devices. In *Advances In Sustainable Energy* (pp. 681–764). Cham: Springer.

[104] Pierson, H.O. (1999). *Handbook Of Chemical Vapor Deposition: Principles, Technology, And Applications* (pp. 1–505). New York: William Andrew.
[105] Hampden-Smith, M.J., & Kodas, T.T. (1995). Chemical vapor deposition of metals: Part 1. An overview of CVD processes. *Chemical Vapor Deposition*, 1(1), 8–23.
[106] Jones, A.C., & Hitchman, M.L. Eds. (2009). *Chemical Vapour Deposition: Precursors, Processes And Applications* (pp. 1–36). London: Royal Society of Chemistry.
[107] Choy, K.L. (2003). Chemical vapour deposition of coatings. *Progress In Materials Science*, 48(2), 57–170.
[108] Zhou, Y., Bao, Q., Tang, L.A.L., Zhong, Y., & Loh, K.P. (2009). Hydrothermal dehydration for the “green” reduction of exfoliated graphene oxide to graphene and demonstration of tunable optical limiting properties. *Chemistry of Materials*, 21(13), 2950–2956.
[109] Choy, K.L. (Ed). (2019). Chemical Vapour Deposition (CVD): Advances. In Technology, and *Applications* (pp. 1–391). London: CRC Press.
[110] Liu, J., Yuan, Y., & Bashir, S. (2013). Functionalization of aligned carbon nanotubes to enhance the performance of a fuel cell. *Energies*, 6(12), 6476–6486.
[111] Kumar, M., & Ando, Y. (2010). Chemical vapor deposition of carbon nanotubes: a review on growth mechanism and mass production. *Journal of Nanoscience And Nanotechnology*, 10(6), 3739–3758.
[112] Weeber, A.W., & Bakker, H. (1988). Amorphization by ball milling. A review. *Physica B: Condensed Matter*, 153(1–3), 93–135.
[113] Fecht, H.J., Hellstern, E., Fu, Z., & Johnson, W.L. (1990). Nanocrystalline metals prepared by high-energy ball milling. *Metallurgical Transactions A*, 21(9), 2333–2337.
[114] Šepelák, V., Bégin-Colin, S., & Le Caer, G. (2012). Transformations in oxides induced by high-energy ball-milling. *Dalton Transactions*, 41(39), 11927–11948.
[115] Erdem, A.S., & Ergün, Ş.L. (2009). The effect of ball size on breakage rate parameter in a pilot scale ball mill. *Minerals Engineering*, 22(7–8), 660–664.
[116] Mankosa, M.J., Adel, G.T., & Yoon, R.H. (1986). Effect of media size in stirred ball mill grinding of coal. *Powder Technology*, 49(1), 75–82.
[117] Li, R., Liu, X., He, G., Hu, P., Zhen, Q., Liu, J.L., & Bashir, S. (2021). Green catalytic synthesis of ammonia using solid oxide electrolysis cells composed of multicomponent materials. *Catalysis Today*, 374, 102–116.
[118] Zhang, D.L. (2004). Processing of advanced materials using high-energy mechanical milling. *Progress In Materials Science*, 49(3–4), 537–560.
[119] Gavrilov, D., Vinogradov, O., & Shaw, W.J. (1999). Simulation of grinding in a shaker ball mill. *Powder Technology*, 101(1), 63–72.
[120] Shin, H., Lee, S., Jung, H.S., & Kim, J.B. (2013). Effect of ball size and powder loading on the milling efficiency of a laboratory-scale wet ball mill. *Ceramics International*, 39(8), 8963–8968.
[121] Wallraff, G.M., & Hinsberg, W.D. (1999). Lithographic imaging techniques for the formation of nanoscopic features. *Chemical Reviews*, 99, 1801–1822.
[122] Calvert, J.M. (1993). Lithographic patterning of self-assembled films. *Journal Of Vacuum Science & Technology B: Microelectronics And Nanometer Structures Processing, Measurement, and Phenomena*, 11(6), 2155–2163.
[123] Ito, T., & Okazaki, S. (2000). Pushing the limits of lithography. *Nature*, 406(6799), 1027–1031.
[124] Moore, G.E. (2006). Moore’s Law at 40. Understanding Moore’s law: four decades of innovation, 67–84. https://www.alejandrobarros.com/wp-content/uploads/old/4363/Understanding-Moores-Law_Chapter-07.pdf
[125] Lee, M.H., Kim, H.M., Cho, S.Y., Lim, K., Park, S.Y., Jong Lee, J., & Kim, K.B. (2011). Fabrication of ultra-high-density nanodot array patterns (~ 3 Tbits/in. 2) using electron-beam lithography. *Journal Of Vacuum Science & Technology B, Nanotechnology And Microelectronics: Materials, Processing, Measurement, and Phenomena*, 29(6), 061602.

[126] Zhang, Y.L., Chen, Q.D., Xia, H., & Sun, H.B. (2010). Designable 3D nanofabrication by femtosecond laser direct writing. *Nano Today*, 5(5), 435–448.
[127] Karitans, V., Kundzins, K., Laizane, E., Ozolinsh, M., & Ekimane, L. (2012). Applicability of a binary amplitude mask for creating correctors of higher-order ocular aberrations in a photoresistive layer. *Optical Engineering*, 51(7), 078001.
[128] Davim, J.P. Ed. (2008). *Machining: Fundamentals And Recent Advances* (pp. 1–359). Girona: Springer Science & Business Media.
[129] Krohn, V.E. (1961). Liquid metal droplets for heavy particle propulsion. *Electrostatic Propulsion*, 73–80.
[130] Bischoff, L., Mazarov, P., Bruchhaus, L., & Gierak, J. (2016). Liquid metal alloy ion sources: An alternative for focused ion beam technology. *Applied Physics Reviews*, 3(2).
[131] Orloff, J. (1993). High-resolution focused ion beams. *Review of Scientific Instruments*, 64(5), 1105–1130.
[132] Gierak, J. (2009). Focused ion beam technology and ultimate applications. *Semiconductor Science and Technology*, 24(4), 043001.
[133] Grill, V., Shen, J., Evans, C., & Cooks, R.G. (2001). Collisions of ions with surfaces at chemically relevant energies: *Instrumentation and Phenomena. Review of Scientific Instruments*, 72(8), 3149–3179.
[134] El-Hofy, H.A.G. (2005). *Advanced Machining Processes: Nontraditional And Hybrid Machining Processes* (pp. 77–180). New York: McGraw-Hill Professional.
[135] Kubis, A.J., Vandervelde, T.E., Bean, J.C., Dunn, D.N., & Hull, R. (2006). Analysis of the three-dimensional ordering of epitaxial Ge quantum dots using focused ion beam tomography. *Applied Physics Letters*, 88(26), 263103.
[136] Li, J., Malis, T., & Dionne, S. (2006). Recent advances in FIB-TEM specimen preparation techniques. *Materials Characterization*, 57(1), 64–70.
[137] Giannuzzi, L.A., Kempshall, B.W., Schwarz, S.M., Lomness, J.K., Prenitzer, B.I., & Stevie, F.A. (2005). FIB lift-out specimen preparation techniques. In Introduction to focused ion beams (pp. 201–228). Boston, MA: Springer.
[138] Qiao, Y., Wang, D., & Buriak, J.M. (2007). Block copolymer templated etching on silicon. *Nano Letters*, 7(2), 464–469.
[139] Potentials from https://www.nist.gov/system/files/documents/2019/04/02/jpcrd355.pdf; and https://chem.libretexts.org/Ancillary_Materials/Reference/Reference_Tables/Electrochemistry_Tables/P2%3A_Standard_Reduction_Potentials_by_Value
[140] Fujita, T., Okada, H., Koyama, K., Watanabe, K., Maekawa, S., & Chen, M.W. (2008). The minimal electrical resistance of three-dimensional nanoporous gold in external magnetic fields. *Physical Review Letters*, 101(16), 166601.
[141] Erlebacher, J., Aziz, M.J., Karma, A., Dimitrov, N., & Sieradzki, K. (2001). Evolution of nanoporosity in dealloying. *Nature*, 410(6827), 450–453.
[142] Yu, H.D., Yang, D., Wang, D., & Han, M.Y. (2010). Top-Down Fabrication of Calcite Nanoshoot Arrays by Crystal Dissolution. *Advanced Materials*, 22(29), 3181–3184.
[143] Yu, H., Wang, D., & Han, M.Y. (2007). Top-down solid-phase fabrication of nanoporous cadmium oxide architectures. *Journal of the American Chemical Society*, 129(8), 2333–2337.
[144] Yu, H.D., Zhang, Z.Y., Win, K.Y., Chan, J., Teoh, S.H., & Han, M.Y. (2010). Bioinspired fabrication of 3D hierarchical porous nanomicrostructures of calcium carbonate for bone regeneration. *Chemical Communications*, 46(35), 6578–6580.
[145] Yu, H.D., Tee, S.Y., & Han, M.Y. (2013). Preparation of porosity-controlled calcium carbonate by thermal decomposition of volume content-variable calcium carboxylate derivatives. *Chemical Communications*, 49(39), 4229–4231.

Sajid Bashir, William Houf, Luis Villanueva, Pete Villareal,
Jingbo Louise Liu*

Chapter 2
Nanochemistry: properties and applications of interactive materials

2.1 Introduction to properties of nanomaterials

Nanotechnology is the study and application of materials and phenomena at the nanometer scale, where matter exhibits behaviors distinct from its bulk form. At these dimensions, quantum confinement, enhanced surface reactivity, and size-dependent electronic structure become scientifically significant. This field now drives transformative advances across energy, medicine, materials science, and environmental engi-

Acknowledgments: The research efforts led by Dr Bashir and Dr Liu were supported by the American Chemical Society Petroleum Research Fund (Grant 53827-UR10) federal agencies, including the Office of Naval Research, the Air Force Office of Scientific Research, the Department of Energy, the National Science Foundation, the Department of Education, the Department of Homeland Security (ORISE), and the National Institutes of Health. The Robert A. Welch Foundation (Departmental Grant AC-0006) is gratefully acknowledged for its support of student training. The technical support from Texas A&M University-Kingsville and Texas A&M Energy Institute is also duly acknowledged. The authors also gratefully acknowledge the helpful discussions with colleagues, comments, and reviewers' suggestions, which have improved the presentation.

Author contributions: S. Bashir and J. Liu collectively conceived the projects and oversaw their progress. S. Bashir was in charge of the submission processing of all chapters. W. Houf, a former undergraduate student, and L. Villanueva, a former graduate student, participated as protégés to obtain material synthesis and characterization knowledge to improve their core knowledge and hands-on experience. Other co-authors were involved in this project. J. Liu wrote the first draft, which all authors edited, and all agreed to publish this chapter.

***Corresponding author: Jingbo Louise Liu,** The Department of Chemistry, Texas A&M University-Kingsville, MSC161, 700 University Blvd., Kingsville, TX 78363, USA; Louise Nanoinnovation North American, 7534 Beau Terre, Corpus Christi, TX 78363, USA; Texas A&M Energy Institute, 1617 Research Parkway, Suite 308, College Station, TX 77843, 3372, USA, e-mails: jingbo.liu@tamuk.edu, jingbo.liu@tamu.edu
Sajid Bashir, Department of Chemistry, Texas A&M University-Kingsville, MSC161, 700 University Blvd., Kingsville, TX 78363, USA; Louise Nanoinnovation North American, 7534 Beau Terre, Corpus Christi, TX 78363, USA; Texas A&M Energy Institute, 1617 Research Parkway, Suite 308, College Station, TX 77843, 3372 USA
William Houf, Luis Villanueva, Pete Villareal, Department of Chemistry, Texas A&M University-Kingsville, MSC161, 700 University Blvd., Kingsville, TX 78363, USA

 | https://doi.org/10.1515/9783111621111-003

neering. Working at the nanoscale allows researchers to probe and manipulate matter with a level of control that was inconceivable only a generation ago. What continues to motivate scientists is that nanoscale structure–property relationships reveal fundamentally new behaviors, for example, catalytic pathways that accelerate reactions through high-index facets, optical responses that shift with particle size due to quantum confinement, and mechanical strengths that surpass those of bulk analogues because of defect minimization. These characteristics are not academic abstractions; they underpin technologies that increasingly shape global innovation. In environmental systems, engineered nanomaterials serve as high-surface-area sorbents that capture pollutants or convert CO_2 into value-added products and stable mineral phases. In medicine, nanoscale carriers enable targeted delivery, controlled release, and reduced systemic toxicity. In the energy sector, quantum-confined semiconductors and nanostructured electrodes used in batteries, fuel cells, and capacitors are redefining performance limits and accelerating the transition to cleaner technologies. Nanotechnology also serves as a scientific bridge, linking disciplines, cultures, and problem-solving traditions. Its power lies not only in the materials themselves but also in the collaborative, interdisciplinary spirit the field demands. As research continues to evolve rapidly, the potential of nanotechnology remains far from fully realized [1].

These properties include the chemical, electro-optical, magnetic, physical, and thermal characteristics of materials observable at the nanoscale. When particle size decreases to less than 100 nm, materials' properties change significantly compared with those of large-scale bulk materials. At the nanoscale, quantum effects dominate, shaping the materials' properties and performance. Researchers can fine-tune the materials' properties based on the end application. It has been observed that nanomaterial size can affect their fluorescence, porosity, conductivity, and reactivity. This chapter will discuss the chemical, magnetic, electro-optical, and thermal properties of various nanostructured materials. Because of their ultra-high surface area, nanomaterials are more readily exposed to the surrounding environment. Therefore, these materials' chemical reactions (or reactivity) can be significantly enhanced, especially in the field of catalysis. A commonly used catalytic converter in a car contains a nanoscale platinum alloy coated on the surface of porous ceramics. As a result, the gasoline combustion efficiency can be improved. Further, automobile exhaust and air pollution can be reduced or even eliminated. In addition, engineers have fully leveraged the increased reactivity of nanomaterials to design batteries with lighter weight and longer lifespans, fuel cells with enhanced electrochemical performance, and catalysts for emerging energy generation and storage systems [2].

A cubic solid is used to show that particles have phenomenally high surface areas as particle size decreases (Fig. 2.1). We can assume a solid cube with a unit-cell lattice of 1 cm on a side. The total surface area of this "centi-cube" can be calculated to be 6 cm^2 (eqs. (2.1A, B)). If this cube is disassembled into smaller cubes with 1 mm on a side, 1,000 cubes ($10 \times 10 \times 10$) are obtained. Each small "mini-cube" cube has a surface area of 6 mm^2, for a total surface area of 6,000 mm^2 (or 60 cm^2, eq. (2.1C)). With a further decrease in size to 1 nm, the total number of cubes needed will be 1,021 (1 cm = 10^7 nm; $10^7 \times 10^7 \times 10^7 = 10^{21}$). The surface area of each "nano-cube" can be calculated as 6 nm^2. The total surface area of these 10^{21} cubes will be 6 × 1,021 nm^2 or 6 × 107 cm^2. It can be seen that the surface area of cubic nanoparticles can be larger than that of a football field:

$$A_{1\,cm} = 6(1\,cm \times 1\,cm) = 6\,cm^2 \tag{2.1A}$$

$$A_{1\,mm} = 6(1\,mm \times 1\,mm) \times 1{,}000 = 6{,}000\,mm^2 \tag{2.1B}$$

$$A_{1\,nm} = 6(1\,nm \times 1\,nm) \times 10^{21} = 6 \times 10^{21}\,nm^2 \tag{2.1C}$$

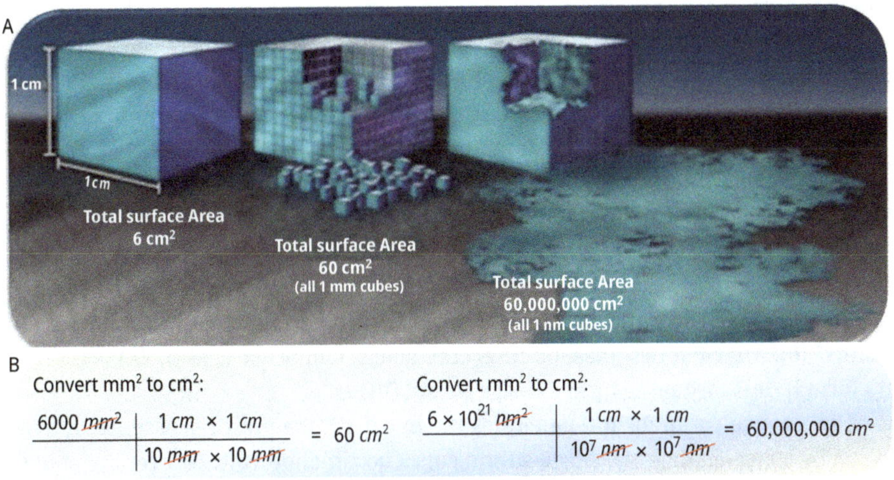

Fig. 2.1: Illustration demonstrating the effect of the increased surface area of nanomaterials. **A.** Nanoscale materials have phenomenally larger surface-area-to-volume ratios than bulk materials and **B.** the conversion of mm^2 and nm^2 to cm^2.

2.1.1 Chemical properties

There is a correlation between chemical structure and electronic properties for nanoscale and microscale materials. Any changes in the structure due to changes in particle size will alter the electronic properties because of the intrinsic connectivity. The minimum energy required to remove an outermost electron is the first ionization energy (IE1), which is typically higher for smaller atomic groups than for bulk materials due to the electron affinity of the atom. Ultra-fine powders used in catalysis can enhance the reaction rate, selectivity, and efficiency of chemical reactions in combustion or synthesis, thereby reducing waste and pollution. For example, gold (Au) nanoparticles generally possess face-centered cubic (fcc) crystalline structure ($Fm\bar{3}m$, cubic,

$a = 4.0786$ Å, cell volume = 67.847 Å^3, 4 sites/unit cell, Fig. 2.2A). However, when the Au size is reduced to 5 nm, Au exists as an icosahedron (Fig. 2.2B) and displays improved catalytic reactivity (Fig. 2.2C) [3]. Scrimin and co-workers reported self-assembly of triazacyclonane ($C_6H_{15}N_3$)-functionalized thiols (R–S–H) on the surface of nanogold to mimic the nuclease systems (Fig. 2.2C). Their data showed that complexation with Zn^{2+} to form a chelating "Au MPC1" tuned into a powerful catalyst [4]. The nanogold enabled transphosphorylation (Fig. 2.2C1) of an RNA model substrate, 2-hydroxypropyl *p*-nitrophenyl phosphate ($C_9H_{11}NO_7P^-$, HPNPP). The "Au MPC1" catalyst showed an improved cleavage activity compared to its monomeric reference complex [5].

Further studies on the catalytic activity of Au MPCs with different formulations indicated that a "low molar fraction of the catalytic unit triazacyclonane-Zn^{2+}" resulted in a decrease in catalytic activity. This observation suggested that this enhanced catalytic reactivity "originates from the cooperative effect between two neighboring catalytic units." The rate constant (*k*) for HPNPP cleavage by Au MPC 1 highly depends on the amount of Zn^{2+} ions (Fig. 2.2C2). A different gold-containing catalyst using an elaborate ligand (a bis-(2-amino-pyridinyl-6-methyl)amine (BAPA) functionalized thiol), named "Au MPC 2," was used for the transphosphorylation of BPNPP. The "Au MPC2" exhibited 100-fold higher efficiency for the cleavage of the bis-*p*-nitrophenyl phosphate (BNPP), shown in Fig. 2.2C3. This improvement is attributed to the "cooperation between metal Lewis acid activation and hydrogen-bonding" by the BAPA–Zn^{2+} complex, allowing for an improved "hydrolytic activity toward phosphate diesters" [6]. After the first plasmid insertion, deoxyribonucleic acid (DNA) is cleaved into form II (less torque) to form a linear strand (III).

A successive cut to form II generates the linear III, the most released (least twist strain), by concatenating 100 single-strand cuts. Coordination with Zn^{2+} results in high catalysis by the "Au MPC2" nanosystem at the bond cleavage site at standard conditions (pH 7.0 and 37 °C for 24 h) with the DNA plasmid chelating with Zn^{2+} and Au MPC2 catalyst to affect cleavage (Fig. 2.2C**4**), while no reaction at all was observed with monomeric-Zn^{2+} complex. At a higher concentration of Zn^{2+} (15 μM), the form III DNA (linear) was 50% greater than nicked DNA (unwinds to form II), taking advantage of the multivalent catalyst MPC 2 and differential chelating and binding of the ligand 6-benzoylacetyl-2-pyridine carboxylic acid (BAPA).

The electronic properties influence topography and dimensionality, with the energy required to remove an electron (ionization potential) higher in small atomic clusters than in the corresponding bulk material. Furthermore, the ionization potential exhibits fluctuations as the cluster size varies. This effect is correlated with chemical reactivity, such as the reaction of Fe_n clusters with hydrogen gas. Nanoscale structures, such as nanoparticles and nanolayers, have very high specific surface area-to-volume ratios and different crystallographic structures. Therefore, the radical alteration in chemical reactivity may result from these changes in either dimension or crystalline structure [7]. Catalysis using ultrafine nanoscale systems can increase reaction kinetics, selectivity, and overall efficiency. This phenomenon is commonly ob-

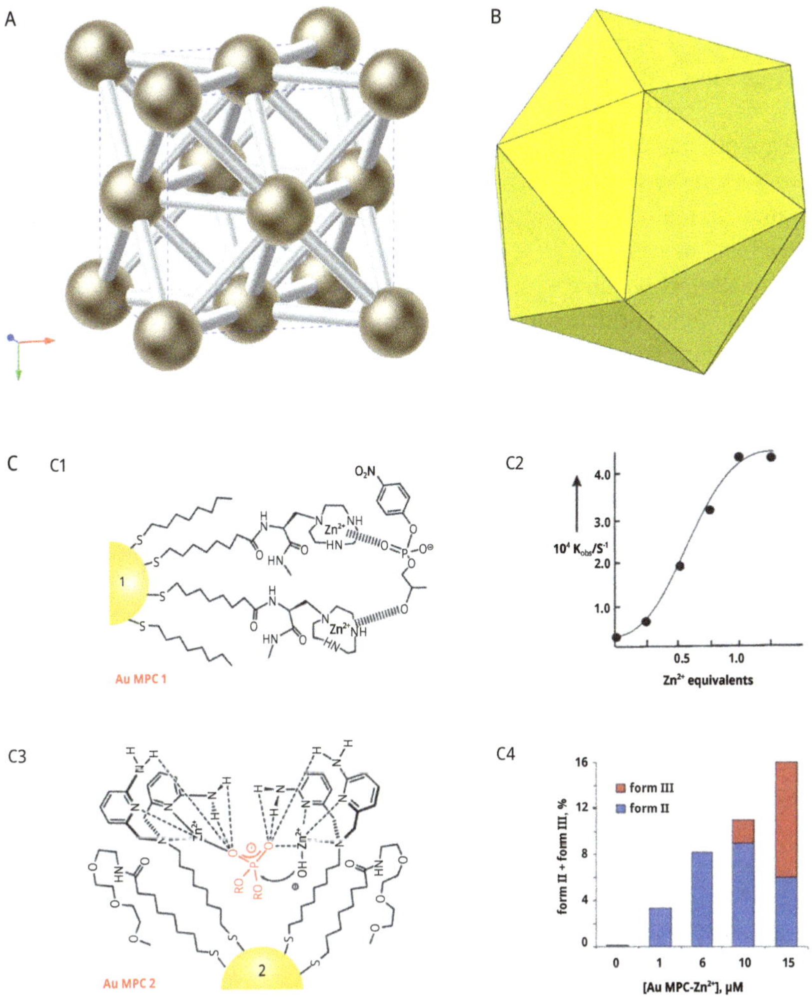

Fig. 2.2: Crystalline structure and reactivity of gold (Au), **A.** face-centered cubic (fcc) crystalline structure (Fm$\bar{3}$ m, cubic, a = 4.0786 Å, cell volume = 67.847 Å^3, 4 sites/unit cell, **B.** icosahedron of nanogold with the size of 5 nm or smaller, and **C.** an improved catalytic reactivity of two families of nanogold catalysts, **C1**: Au catalyst (MPC1) showing an improved cleavage activity, **C2**: rate constant (k) for HPNPP cleavage by Au MPC1 highly depends on the amount of Zn^{2+} ions used for the transphosphorylation of BPNPP, **C3**: Au MPC2 catalyst showing efficiency for the cleavage of the bis-*p*-nitrophenyl phosphate (BNPP), **C4**: catalyst to affect cleavage [5, 6].

served during catalyst-assisted combustion. Consequently, environmental waste and pollution were reduced or mitigated simultaneously. When the size of gold (Au) nanoparticles was reduced to smaller than about 5 nm in diameter, they were known to adopt icosahedral structures rather than the regular fcc arrangement [8]. An extraordinary increase in catalytic activity accompanies this structural change. Furthermore, nanoscale catalytic supports with controlled pore sizes can select products and reactants based on their physical sizes, thereby facilitating their transport to and from internal reaction sites within the nanoporous structure. Additionally, nanoparticles often exhibit new chemistries distinct from those of their larger particulate counterparts. For example, many new medicines are insoluble in water when in micron-sized particles, but dissolve easily when in a nanostructured form [9].

When atoms come into proximity with other atoms in a solid, most electrons remain localized and associated with a particular atom. However, some outer electrons will become involved in bonding with adjacent atoms. The atomic energy-level diagram is modified upon bond formation due to the hybridization of atomic orbitals (AOs). Briefly, the well-defined outer electron states of the atom overlap with those of neighboring atoms and become broadened into energy bands. For a simple diatomic molecule, the two outermost AOs overlap to produce two molecular orbitals (MOs). The formation of MOs can be viewed as a linear combination (addition or subtraction) of the two constituent AOs. A solid can be considered a cluster of many molecules [10]. The electronic energy levels of a one-dimensional (1D) solid (a linear chain of atoms) were affected when the number of atoms within the solid was increased (Fig. 2.3). The electronic band structure of solids can combine the electron waves in a periodic crystalline potential. The Drude–Lorentz free electron model for metals is used for demonstration. In this model, a metallic solid is assumed to consist of a close-packed lattice in which cations are surrounded by an electron sea, or cloud, formed by the ionization of the outer-shell (valence) electrons [11]. Therefore, a classical kinetic gas theory can treat the valence electrons as gas molecules moving inside a container. This model applies to these so-called free-electron metals, the electropositive metals of groups I and II, and aluminum. The model helped explain many fundamental properties of metals, including high electrical and thermal conductivities, optical opacity, reflectivity, ductility, and alloying properties. However, a realistic approach treats the free electrons in metal quantum mechanics and considers their wave-like properties [12].

It is assumed that the free valence electrons are constrained within a potential well to prevent electrons from leaving the metal (the "particle-in-a-box" model) [13]. The box boundary conditions require the wave functions to vanish at the edges of the crystal (or "box"). The Schrödinger equation gives the allowed wave functions, which then correspond to specific wavelengths [14]. For a 1D box of length L (nm), the permitted wavelengths are $\lambda_n = 2\ L/n$, where $n = 1, 2, 3$ is the quantum number of the state; the permitted wave vectors $k_n = 2\pi/\lambda$ are given by $k_n = n\ \pi\ /L$. This simple particle-in-a-box model results in a set of wave functions, and the corresponding energies are given by eqs. (2.2A and B) and Figs. 2.3 and 2.4:

$$Y_n = (2/L)^{1/2}\sin\left(\frac{npx}{L}\right) \tag{2.2A}$$

$$E_n = n^2h^2/8\mathrm{mL}^2 \tag{2.2B}$$

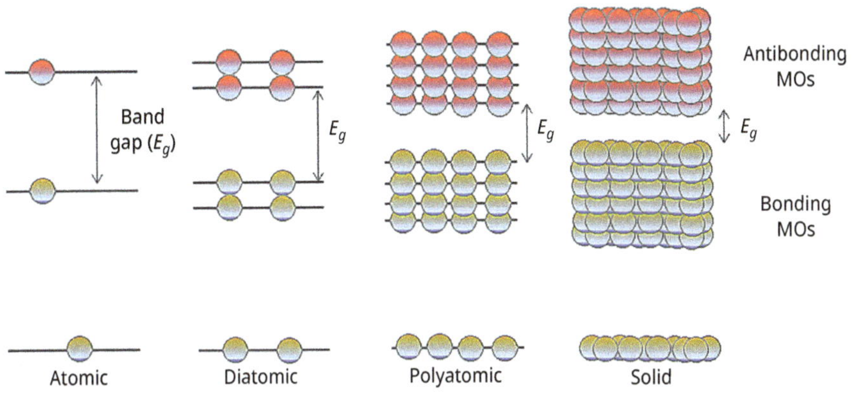

Fig. 2.3: The electron energy-level diagram for a progressively larger linear chain of atoms shows molecular orbitals' broadening into energy bands for a one-dimensional solid [12].

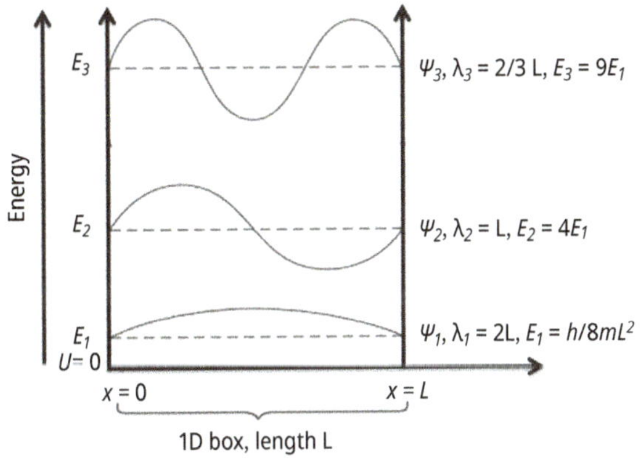

Fig. 2.4: The energy-level diagram also shows some of the allowed wave functions for an electron confined to a one-dimensional potential well [15].

It is well known that nanoscale materials have very high surface area-to-volume ratios. The crystalline structure was also found to differ from the bulk structure. These two findings led to an extraordinary increase in chemical reactivity, selectivity, and reaction efficiency. Recently, nanocatalysis has drawn significant attention due to the use of finely divided nanoscale systems, as discussed above, since this processing can help address environmental pollution [16]. Furthermore, nanoscale catalytic supports with tunable internal and external porosity can selectively accelerate chemical reactions based on their dimensions and functionalities. Importantly, nanoparticles often exhibit unique intrinsic chemistries, distinguishing them from their particulate counterparts at the micro- or macro-scale. Recently, many newly developed medicines have become more water-soluble and more biocompatible than their micron-sized counterparts. Applying bifunctional synthesis methodologies will open a new paradigm for nanomaterial applications [17].

2.1.2 Magnetic properties

Magnetism is a class of physical phenomena resulting from the application of external magnetic forces to other materials, particularly magnetic materials. Magnetism arises from electric currents induced by atomic rotation. When an external magnetic field is applied, the magnetic moments of the substance will be reoriented to align with its direction [18]. Based on the state of electron pairing, we can classify materials as paramagnetic or diamagnetic. A substance with one or more unpaired electrons exhibits paramagnetism. An external magnetic field will attract this material due to its paramagnetism [19]. On the other hand, the species with all electrons paired will repel the applied magnetic field. Therefore, diamagnetism can be observed. Figure 2.5A–C shows the change of apparent mass upon application of the magnetic fields. Nanoparticles that exhibit magnetic properties have extensive applications in catalysis, biotechnology, biomedicine, reverberation imaging, ecological remediation, and information storage [20]. By combining suitable systems of attractive species with different structures, magnetic nanoparticles (MNPs) can be created [21].

This chapter investigated iron oxide nanoparticles (Fe_3O_4 NPs) with ferromagnetic properties [22]. Those materials are intrinsically magnetically ordered and exhibit spontaneous magnetization even in the absence of an external magnetic field. The ordering mechanism is the quantum mechanical exchange interaction [23]. This study focuses on the synthesis of these Fe_3O_4 magnetic NPs. One of the main concerns in producing magnetic particles is size. Ultrafine magnetic particles tend to agglomerate, reducing their vitality due to their high surface tension. However, this vitality is directly linked to the nanoparticles' highly accessible surface area and volume. On the other hand, these metallic nanoparticles or MNPs are synthetically very dynamic; they can be oxidized in air, resulting in a loss of magnetic properties and dispersibility [24]. To broaden the applications of these MNPs, the author treated them with natural product extracts that behave like surfactants, polymers, and other inorganic layers (such as silica or carbon). This treatment also prevents the nanoparticles from degradation and protects them from losing attractive properties. It is also an essential outer layer, or shell, of these nanoparticles, and the shell must be selected based on the functions and applications, as it can be composed of different ligands [25]. The functionalized nanoparticles can be used in a wide range of applications, as mentioned above. This functionalization enhances high scattering, high reactivity, and easy detachment. Now, these magnetic nanomaterials should be blended using techniques that do not involve oxidation or corrosive degradation, as the particles' safety is essential [26]. The Fe_3O_4 MNPs functionalized by natural plants were found to have the potential to imitate wound healing.

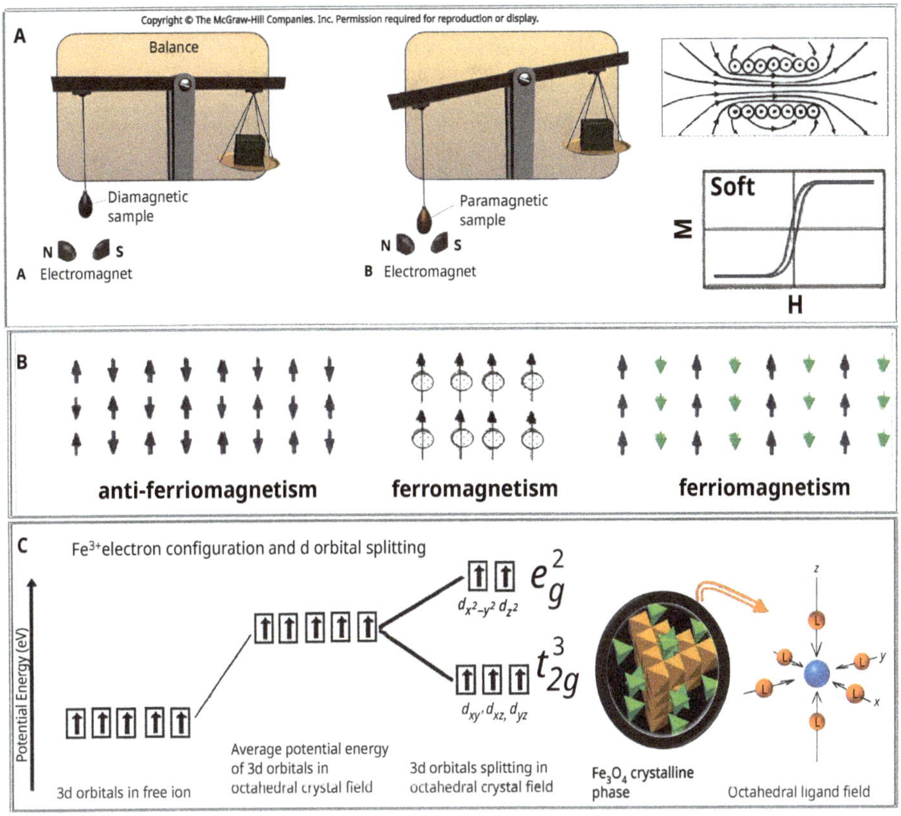

Fig. 2.5: A demonstration of magnetic property. **A.** Apparent mass differences to determine diamagnetic and paramagnetic properties. **B.** The types of ferromagnetism. **C.** The *d*-orbital splitting, resulting in high spin contributing to the magnetism (Fe^{3+} is used as an example) [12].

2.1.3 Electro-optical properties

The Bohr model of the atom depicts the atom as a small, positively charged nucleus surrounded by electrons that travel in circular orbits around it, similar to the solar system, using the hydrogen atom as a model. Nuclei exert electrostatic attraction on the electrons within atoms rather than gravity. The Bohr model suggests that only certain electron orbits, or specific energy levels or quantum shells, are allowed. The shells in a Bohr-type atom occupy three-dimensional (3D) space according to the quantum number, n (not visible) to corresponding spectroscopic labels K, L, M, and N (visible), where the smallest shell closest to the nuclei has a quantum number of 1 ($n = 1$), corresponding to the K transition between the S and next orbital. Similarly, the successive orbitals are labeled by n as 2, 3, 4 [27], and are used to describe chemical stability across the periodic table in terms of half-filled or complete shells. Each Bohr shell can contain $2n^2$ electrons, starting with the K shell ($n = 1$), which can include two electrons, whereas other shells, occupying large volumes, can accommodate more electrons. The next shell, the L shell ($n = 2$), can accommodate eight electrons. Once more than one electron can be occupied, additional information is required to uniquely identify each system, such as angular momentum, spin state, and occupancy, which are quantized at differing energies relative to the vacuum level (zero of the energy scale) that represents the potential energy of a free electron far from the atom [28]. To correspond with atomic emission spectra measured experimentally, the energies of these levels, E_n, are negative (i.e., the electrons are bound to the atom) and are proportional to $1/n^2$, as summarized in Fig. 2.6.

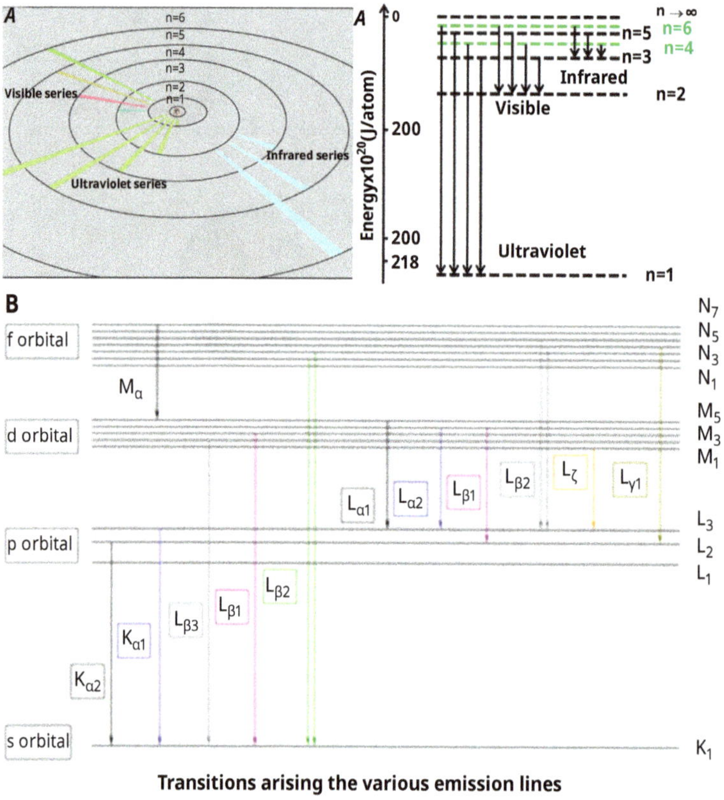

Fig. 2.6: Electron configuration and orbital diagram. **A.** Bohr model depicting the electron distribution around the nuclei and energy distribution of different orbitals. **B.** Demonstration of transitions arising from the various emission lines [12].

Electron occupancy is sometimes described in terms of discrete quanta in the hydrogen atom. However, experimental data also suggest that electrons exhibit many wave-like characteristics due to their subatomic size (Fig. 2.6A). The wave nature of the electrons and the dearth of electrons are correlated with the electronic properties of nanoparticles. These electronic properties change when particle size and shape increase or decrease (Fig. 2.6B). Three-dimensional particles possess completely discrete energies (electrons are confined in all dimensions). Usually, this phenomenon is observed when the particle sizes are comparable to the de Broglie wavelength of electrons [29]. Conductors become insulators below a critical length due to the separation of conduction and valence bands. When the system length scale is reduced, changes in electronic properties are related mainly to the increasing influence of the wave-like property of the electrons and the scarcity of scattering centers based on quantum mechanical effects [30]. When the particle size is comparable with the de Broglie wavelength ($\lambda = h/mv$, *where h represents Planck's constant, m is the mass of the particle,* and *v is* the velocity) of the electrons, the apparent differences between the energy states can be observed, where the electrons are confined in all three dimensions [31]. The electrons can move freely between adjacent nanostructures, a phenomenon known as quantum tunneling, because the barriers are lower than expected under classical mechanics for larger bodies, enabling movement. If an external voltage is applied to the nanosystem, the discrete energy levels can thus be aligned. Therefore, the resonant tunneling effect occurs, leading to a rapid increase in the tunneling current [32] used in the scanning tunneling microscope.

Electrons are highly confined in all dimensions within zero-dimensional (0D) materials (e.g., quantum dots). In this case, the conduction is highly sensitive to the presence of other charge carriers. As a result, the charge state is also sensitive to those carriers. These phenomena are known as "Coulomb blockade effects." In this case, conduction mainly involves single electrons. It requires only a small amount of energy to overcome the energy barrier between the valence and conduction bands, enabling the operation of a switch, transistor, or memory element. Based on the effects, we can develop various devices and systems for electronic optoelectronics. Recent research indicates that these new systems could offer a novel approach to information processing based on resonant tunneling and single-electron transistors [33].

For a large-scale material, the density of states is proportional to $E^{1/2}$. The decrease in the system's dimensionality will highly affect the density of states (Fig. 2.7). This observation results from the associated "reduction of degrees of freedom in wave vector space." In terms of purely dimensional space, 3D, 2D, and 1D systems, the density of states is proportional to k^{n-2}, where E is the state's energy, k is the constant, and n is the number of dimensions in the presence of the nanosystem. For parabolic bands, the density of states is dependent on $E^{(n-2)/2}$ [34]. For a quasi-low-dimensional system, the density of states takes the $E^{(n-2)/2}$ dependence in each subband. These quasi-low-dimensional materials include semiconductor quantum wells or quantum wires, in which a series of confined subbands is formed. There is no continuous distribution of states for a quantum dot due to the confinement in all dimensions. The density of states takes the form of a spectrum of discrete energy values, similar to that found for individual atoms [35].

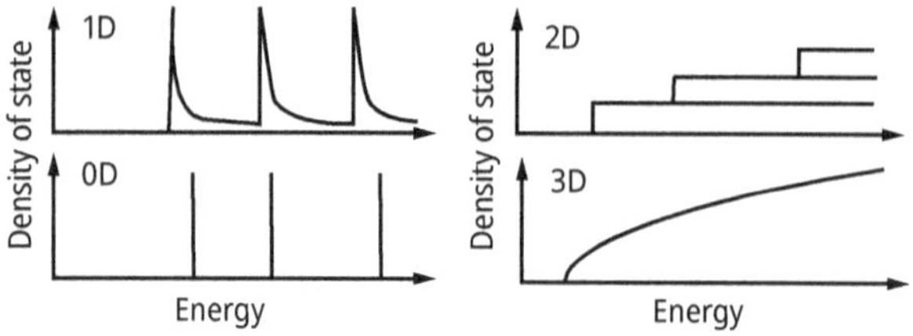

Fig. 2.7: The electronic density of states for a bulk semiconductor (3D), a quantum well (2D), a quantum wire (1D), and a quantum dot (0D).

It was found that the density of states strongly affects the optical and electronic properties of most nanostructured materials. In particular, nanoscale semiconductors exhibit strong dimensionality-dependent behavior. For example, in electrical transport, the number of available states is an essential factor in determining the density of states [36]. It was also found that the strength of the optical transition is proportional to the density of states. This density is correlated with the initial point in the valence band and the final points in the conduction band, creating a joint density of states. The adsorption is found to correspond to the joint density, indicating that the nanostructured materials exhibit different optical and electronic properties [37].

2.1.4 Thermal properties

Thermal conductivity: The theory of heat conduction by lattice waves in solid materials provides insight into understanding how conductivity is influenced by lattice defects (such as point defects and stacking faults), grain boundaries, and extended imperfections. At high temperatures, conduction by electromagnetic radiation within the solid can be controlled. The thermal conductivity (κ, W/K) by mobile carriers, which can be waves or particles, can be expressed as follows (eq. 2.3):

$$\kappa = \frac{1}{3} C_{vl} \tag{2.3}$$

where C is the specific heat per unit volume (J/m^{3}K), v represents the speed of mobile carriers (m/s), and l is the mean free path (nm). Assume the carriers result from waves (lattice or electromagnetic waves) spanning a spectrum of frequencies f. This frequency can be generalized as follows (eq. 2.4):

$$\kappa = \frac{1}{3} \int C(f) v l(f) d(f) v \tag{2.4}$$

where $C(f)df$ is the contribution to the specific heat per unit volume from waves in the frequency interval df, v is the group velocity of the waves, and $l(f)$ is the attenuation length, usually a function of frequency. In eq. (2.4), f_m denotes a practical upper limit to the spectrum because either v or l becomes very small for f/f_m. The f_m is the upper-frequency limit of the acoustic branch. The energy content of waves consists of quanta, phonons, or photons, respectively. In addition, these quanta can be considered as particles to carry and transfer heat. The present treatment emphasizes the wave nature of the carriers of heat [38].

Lattice waves: These are considered either elastic or ultrasonic waves. However, their spectrum can be extended to very high frequencies, f. In this case, their wavelength λ corresponds to atomic dimensions. The relationship between frequencies (f, Hz) And the wavelength (λ, m) must be modified.

At low frequencies, the frequencies are inversely proportional to the wavelength: $\lambda = n/f$. However, at high frequencies, the above relation is no longer applicable due to the discrete atomic structure of the lattice. The displacement field of the wave has the form of a progressive wave; therefore, the displacement $u_{(r,t)}$ varies with position r and time t, as follows (eq. 2.5):

$$m_{(r,t)}\mu \exp(i\,qr - i\,w\,t) \tag{2.5}$$

where the wave vector (q) has magnitude $2\pi/\lambda$. Moreover, points in the propagation direction, the angular frequency (ω), can be expressed by $\omega = 2\pi f$.

The angular frequency (ω) is no longer proportional to the wave vector (q) at high frequencies. The schematic relation between frequency and wave vector for a discrete lattice is shown in Fig. 2.8 [39].

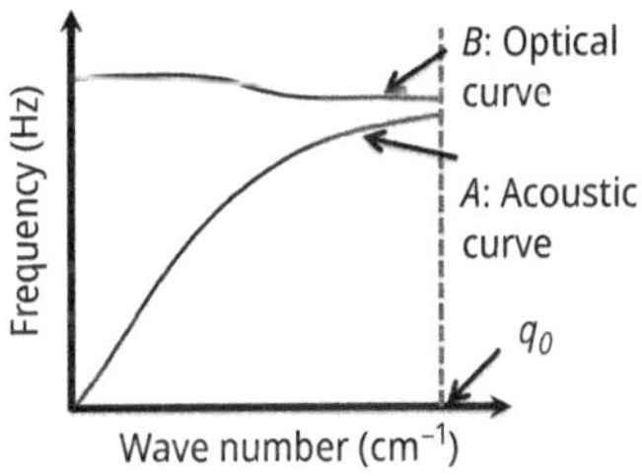

Fig. 2.8: Dependence of angular frequency (ω) on the wave vector (q) for a discrete lattice in the principal direction. **A.** The acoustic branch is the only type in simple lattices. **B.** An optical branch, present in complex crystals and describing the vibration of atoms within molecular units [40].

The slope of the curve (shown in Fig. 2.9) defines the group velocity, which is related to the angular frequencies and wave vector: $v = d\omega/dq$. This relationship indicates that wave speed determines the rate of energy transport. The Debye theory of lattice vibrations states that v is a constant, but the lattice wave spectrum is terminated at the Debye frequency f_D. This frequency is chosen, so that the number of normal modes agrees with the number of atoms. On the other hand, the spectrum can be divided into acoustic modes of the advanced waveform (shown in Fig. 2.9A) if the basic structural unit contains N atoms of varying mass [41]. The spectrum can be divided into optical modes (shown in Fig. 2.9B) if the vibrations of the N atoms in each structural unit are relative to each other. With upper frequency $f_m = f_D/N^{1/3}$, only the former have appreciable group velocity and transport heat. This approximation is justified and can explain the relationship between frequencies and the number of atoms [42].

Another interesting case is the limit $T > hf_m/k$. Under these circumstances, each wave or normal mode reaches thermal equilibrium. The energy equals kT and $C(T) \propto f^2$. Here, k and h are the Boltzmann (m^2/kg s^2 K) and Planck (m^2/kg s) constants. Figure 2.9 schematically demonstrates the difference between acoustic and optical vibrations. Optic vibrations result from the motion of atoms in the lattice. The term "optic" refers to their strong interaction with infrared electromagnetic waves in the case of ionic solids [43].

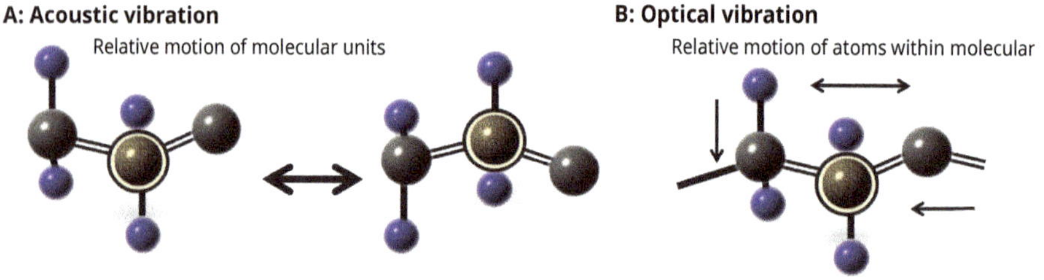

Fig. 2.9: In crystals composed of molecular units, such as ceramic oxides, the lattice waves are composed of **A** acoustic vibration and **B** optical vibration [40].

Interaction processes: In an ideally perfect crystal, it can be assumed that, to be harmonic, structurally perfect, and without external boundaries, the lattice waves would then be normal modes. This approximation indicates that the energy content of each wave would remain constant. Thermal equilibrium between the waves could not be established [44]. However, real solids were found to have deviated from this ideal behavior. The lattice forces exhibit anharmonic components, indicating that linear elasticity has departed. It is also known as strain energy, which is cubic in strain. The lattice defects always exist in real solids, and so do boundaries. These defects cause some energy exchange between the waves. This exchange results in the deviation of the true normal from the ideal modes. The increased number of normal modes results in insufficient randomness in these interactions, leading to thermal equilibrium. To understand their thermal equilibrium properties, we anticipate that interactions between atoms and molecules exist. An in-depth understanding of these interactions will provide insight into transport properties and the estimation of attenuation lengths. Thermal conductivity is limited by various interaction processes that transfer energy between waves and establish thermal equilibrium. Figure 2.10 illustrates a schematic of the interaction process, which corresponds to the different types of departures from ideal behavior (Fig. 2.10A). The intrinsic processes cause energy exchange between triplets of waves (also known as three-phonon interactions). The energy exchange maintains frequency conservation, as in all nonlinear processes (Fig. 2.10B), and wave-vector selection rules. The resulting intrinsic attenuation length is expressed as (eq. 2.6):

$$l_i(f,T) = Bf^{-2}T^{-1} \tag{2.6}$$

where $B\mu ma^3 n\text{¦}m$, μ is the shear modulus (Pa) and a^3 is the volume per atom (nm^3). It is thus inversely proportional to the mean square thermal strain $\kappa T/\mu a^3$ (Fig. 2.10C). The intrinsic conductivity (W/m K) is then expressed as (eq. 2.7):

$$\frac{k_i = \frac{3}{4}N^{-\frac{2}{3}}\ mn^2}{f_D\ T} \tag{2.7}$$

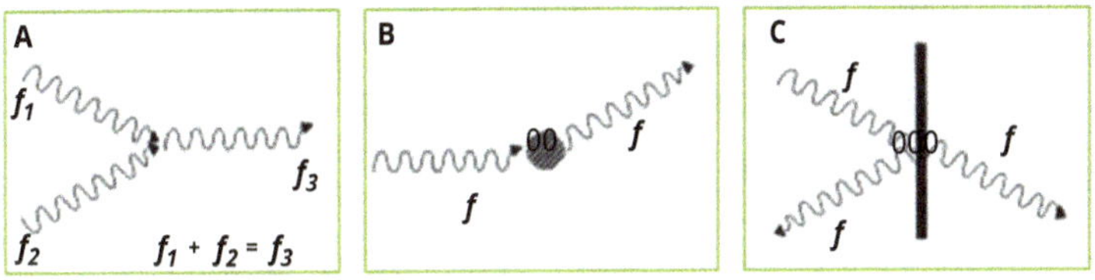

Fig. 2.10: Interaction processes between lattice waves: **A** energy exchange between wave triplets due to anharmonic interactions, **B** scattering of waves by lattice defects, and **C** reflection of waves by boundaries and interfaces [40].

Equation (2.7) well expresses the thermal conductivity of structurally perfect dielectric crystals near and above their Debye temperature $h\,f_D/k$. Factors such as intermolecular forces, large atomic masses, and structural complexity (large N) tend to reduce the intrinsic thermal conductivity.

Note that κ can be expressed as $\kappa - (f)\ df$, indicating the integrand $\kappa_l \propto C(f)\ l_l\ (f)$ is independent of f in the intrinsic case. Therefore, equal-frequency intervals will contribute equally to the inherent conductivity [42]. The thermal conductivity kappa contrasts to the specific heat, which for acoustic modes is mainly due to their highest frequencies, since for $f < f_m$, $C(f) \propto f^2$. This observation is shown schematically in Fig. 2.11.

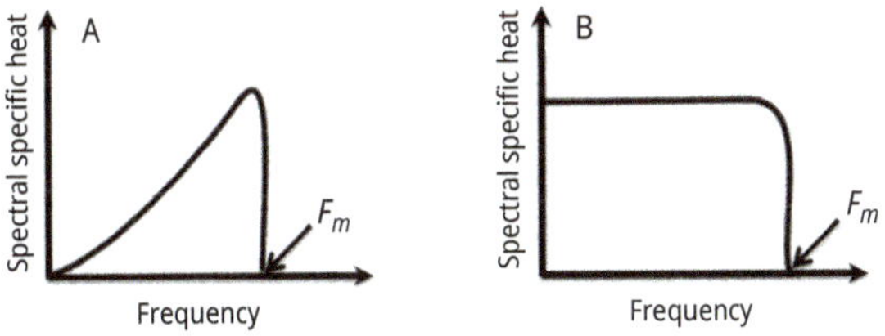

Fig. 2.11: Spectral contributions from lattice waves as a function of frequency: **A** the vibrational specific heat as a function of frequency, and **B** the intrinsic thermal conductivity as a function of frequency [40].

Furthermore, the lattice imperfections will reduce the effective attenuation length. Different imperfections will result in different frequency dependence, accordingly (Fig. 2.11A).It was found that the point defects scatter as the fourth power of frequency, with an inverse attenuation length of $1/l_p$ (f) = Af^4, where A depends on their nature and concentration (Fig. 2.11B). The grain boundaries were found to scatter independently of frequency, with an attenuation length L comparable to the grain size. The inverse attenuation length $1/l(f)$ is composed of the sum of these processes. Since point defects and grain boundaries scatter mainly in different frequency ranges, their conductivity reductions are approximately cumulative, as expressed in eq. (2.8) and shown in Fig. 2.12:

$$k = d_{kl} - d_{kB} - d_{kP} \tag{2.8}$$

where $\delta\kappa_B$ and $\delta\kappa_P$ are the reductions due to boundaries and point defects. Point defects are small volume regions a^3 in which the value of the wave velocity v is locally changed δ_v. Since the intrinsic attenuation length o_f (f, T) has a different dependence on wave frequency f than that due to point defects, it is convenient to define a frequency f_0 at which the two attenuation lengths are equal, i.e., l_i (f_0, T) = l_p (f_0) (eq. 2.9A). Similarly, one can define a frequency f_B such that l_i (f_B, T) = L, where L is the grain diameter (eq. 2.9B):

$$d_{k_B} = k_i \left(\frac{f_B}{f_m}\right) \arctan\left(\frac{f_m}{f_B}\right) \tag{2.9A}$$

$$d_{kP} = k_i \ [1 - \left(\frac{f_0}{f_m}\right) \arctan\left(\frac{f_m}{f_0}\right) \tag{2.9B}$$

Note that the equation for $\delta\kappa_P$ will be applicable if $f_0 > f_m$.

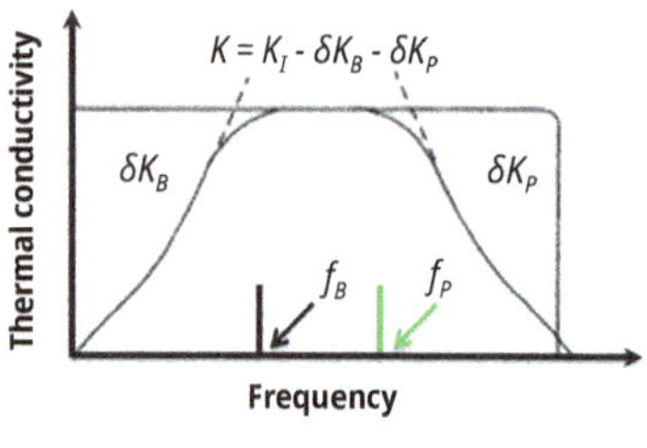

Fig. 2.12: Reduction in the thermal conductivity, which is the area under the curve, at high frequencies due to point defects ($\delta\kappa_P$), and at low frequencies due to grain boundaries and other extended imperfections ($\delta\kappa_B$) [40].

2.2 Applications of nanomaterials

When the grain size of a material decreases to a billionth of a meter (10^{-9}), it is called a nanomaterial or nanocrystalline material. These nanomaterials exhibit desirable and valuable properties, as discussed in Section 2.1. Due to their unique properties, nanomaterials are used in biomedical applications, catalysis, cosmetics, electronics, energy storage, aerospace components, and textiles. In this section, five categories will be discussed to emphasize nanomaterials used for plastic waste remediation (Section 2.2.1), synergistic combination cancer immunotherapy (Section 2.2.2), revolutionization of eye care (Section 2.2.3), treatment of traumatic injuries (Section 2.2.4), and shale gas storage (Section 2.2.5).

2.2.1 Nanomaterials used for plastic waste conversion

Nowadays, the world faces two major problems: the demand for energy and the pollution of the planet due to increased anthropogenic activities. The massive production of plastics used to improve the quality of human life over the past century has resulted in global plastic pollution. Microplastics (MPs) (<5 mm) and nanoplastics (NPLs) (1–1,000 nm) are of special concern due to their harm to ecosystems. Several studies have confirmed MPs' contamination in agroecosystems. A recent study found that MPs up to 2 μm in size are present in edible plants, including vegetables. The Lower Rio Grande Valley in South Texas is heavily dependent upon irrigation water delivered from the Rio Grande River to support the production of staple crops. No studies have been conducted to investigate MPs present in irrigation water or in crops.

The persistent properties of MPs and their release of chemicals/additives have been causing cascading impacts on living organisms worldwide. Natural connectivity among environmental compartments leads to the migration/dispersion of MPs from one sector to another, such as irrigation water to farmland soils. The quantification of MP dispersion across environmental compartments will provide a framework for the remediation of plastic pollutants and their conversion into valuable materials and fuel [45]. This activity will evaluate the properties, distribution, and remediation of MPs collected from soil compartments along the Lower Rio Grande Valley. The morphology can be determined using transmission/scanning electron microscopy (TEM/SEM). High-resolution images will be obtained to determine surface morphology, delicate internal structures, particle shape, and particle size. The chemical constituents, functional groups, and their bonding forces will be analyzed using X-ray photoelectron spectroscopy (XPS), Fourier-transform infrared (FT-IR) spectroscopy, and nuclear magnetic resonance (NMR). The proposed strategy to convert these MPs in irrigation water into "gray" H_2 and graphene will be monitored using a hydrogen fuel cell model car through its electrochemical performance. The evaluation of polyethylene (PE) films and nylon fibers will be discussed below. The chemical congeners/additives (dioxins and polycyclic aromatic hydrocarbons) and endogenous chemicals in MPs will be evaluated using similar approaches. These chemicals consist of hetero- and homo-monomeric units. Representative chemicals include bis(2-ethylhexyl)phthalate, polybrominated diphenyl ethers, bisphenol A, and triclosan.

Area selection and sample collection: Ten farmlands in the Lower Rio Grande Valley of South Texas will be investigated and distributed across the Brownsville, Harlingen, Weslaco, Pharr, McAllen, Edinburg, Mission, San Juan, and Rio Grande City metropolitan areas (Fig. 2.13A). The soil texture type and crop types (corn, fruits, and cotton) will be used for survey purposes. The samples with a point size of 0.3 × 0.3 m^2 were collected from June to August each year to determine the total mean abundance of MPs, mesoplastics, and NPLs. Firstly, crops and significant residues (larger than 5 cm) were removed from the soil surface of the sampling area. The depth of the soil surface was controlled by 0–50 cm. A 500 g sample was collected from different layers at nine other locations. The sample size varied from 200 to 300 and was used for characterization. A saturated aqueous solution of sodium bromide (NaBr) will be applied to extract the MPs. The MP filtrates were digested using a 30% H_2O_2 solution for purification. The abundance of MPs was determined at selected sampling sites in the Lower Rio Grande Valley and decreased from the surface to the deep soil layers. The procedure for MPs purification at different depths is shown in Fig. 2.13B. The team will establish a protocol to collect and measure MPs throughout the Lower Rio Grande Valley during this activity.

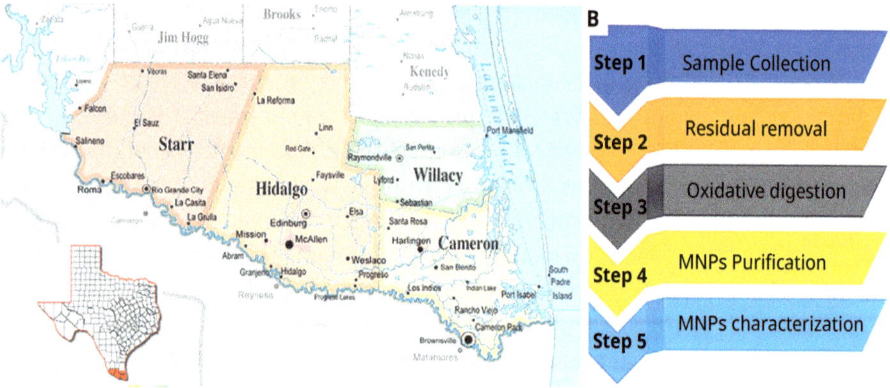

Fig. 2.13: The sample collection and purification. **A.** Geographic sites of MPs along the Lower Rio Grande Valley (image is adapted from [46]). **B.** A procedure for MPs sampling and characterization.

Constituent evaluation: The purified MPs were measured by micro-FT-IR (Fourier Transform Infra Red Spectrscopy) to determine their absorbance or transmittance. The spectrum wavenumber was controlled from 675 to 4,000 cm^{-1}, with a collection time of 3 s for each measurement. The spectra database was used to determine the types of MPs pollutants further. NMR was collected to assess chemical shifts and spin–spin couplings, thereby further identifying the chemical structures of the MPs. The ChemDraw Professional was used to make structure-spectrum assignments. The reported FT-IR and ^{1}H NMR (proton Nuclear Magnetic Resosance Spectrscopy) spectra of PE (Polyethylene) (low Mw) and polyamide (nylon 6/6 pellets) are shown in Fig. 2.14A and B. The abundance of MPs was calculated by mass = ρ (density) × V (volume). The reported MPs abundance by Luo et al. is shown in Fig. 2.14C. This activity will establish the MP's mass density and abundance as a function of sample location and soil layer.

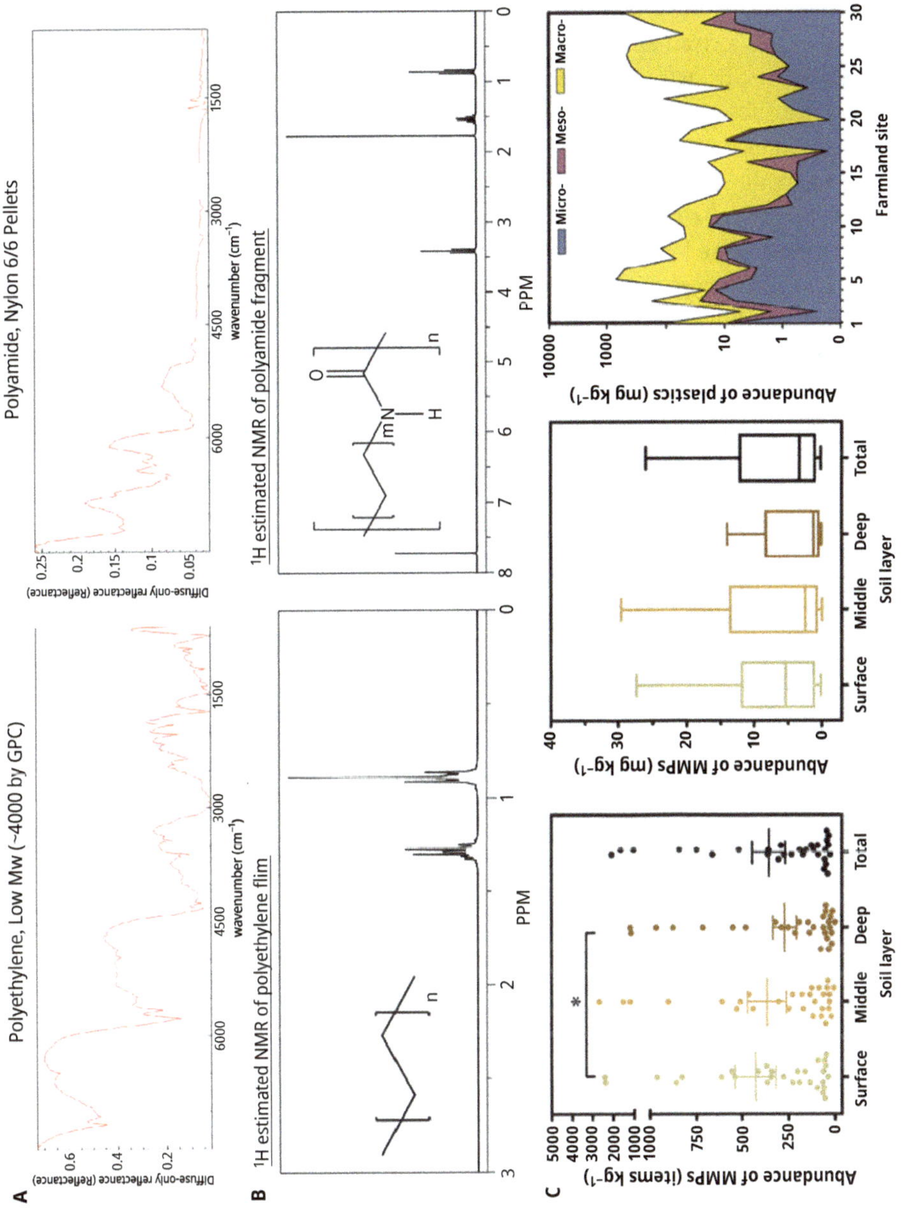

Fig. 2.14: The constituent's measurement of MPs, polyethylene, and polyamide. **A**. The FT-IR spectra (data were adapted from NIST, https://webbook.nist.gov/). **B**. The ChemNMR ^{1}H estimation. **C**. MP's mass density and abundance dependence, which is adapted from [47].

Morphology characterization: The TEM (Transmission Electron Microscopy) and SEM(Scanning Electron Microscopy) will be used to obtain images of MPs, for example, nylon 6 fabrics. Using a carbon-coated copper grid, an electrospinning technique was applied to a substrate to deposit nylon fibers. The surface topography can be observed in surface images and in delicate internal structures. The morphology and diameter of nylon 6 nanofibers examined by Rajendran et al. are shown in Fig. 2.15A and B **[188],** ***demonstrating the nanofibers' homogeneity***. The image and particle distribution will provide a guide for MPs remediation. Using the XPS (X-ray Photon Electron Spectrscopy) and EDX (Energy-Dispersive X-ray Spectroscopy), the elemental composition can be further quantified or semi-quantified, as shown in Fig. 2.15C and D. In this activity, the C_{1s} emission can reflect the elemental content of these collected MPs, while the *O*- and *N*-elemental contents can also be quantified. In addition, the functional groups deposited on the surface of the MPs can be estimated to evaluate the strength of relevant chemical bonds and the formation of new functional groups.

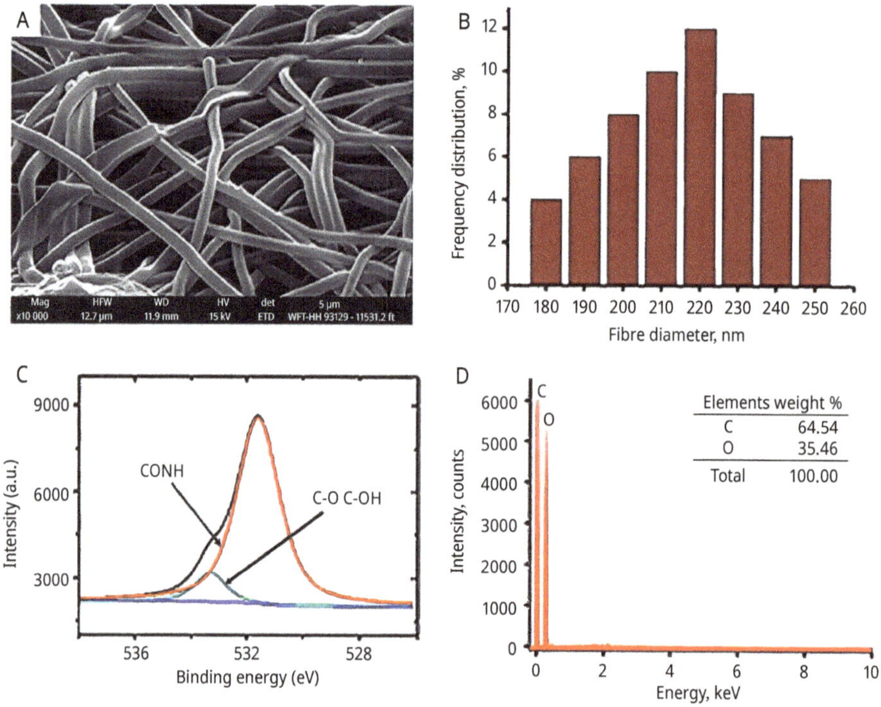

Fig. 2.15: The nanostructural characterization and elemental composition of MPs (nylon 6 fiber as an example). **A.** The TEM image. **B.** The distribution of the nylon fibers. **C.** The XPS elemental analysis and the EDX analysis of the above fiber. (data in **A, B,** and **D** were adapted from [48] and C from [49]).

Remediation strategy: It has been reported that PE and polypropene are converted into hydrogen and porous carbon materials (conductive carbon nanosheets [50]. This research will use metal oxide (V_2O_5–MnO_2) composites as catalysts to convert PE into graphene in three steps: dissolution of PE into toluene, microwave-assisted catalytic pyrolysis, and carbonization (Fig. 2.16A). C–H bond cleavage is proposed to release hydrogen atoms during carbonization. This atomic hydrogen will combine to form hydrogen gas (H_2), which is defined as gray H_2 (GH_2). The so-called GH_2 can serve as a fuel source, and graphene can act as an electrode catalyst to improve the electrochemical performance of microbial fuel cells (Fig. 2.16B).

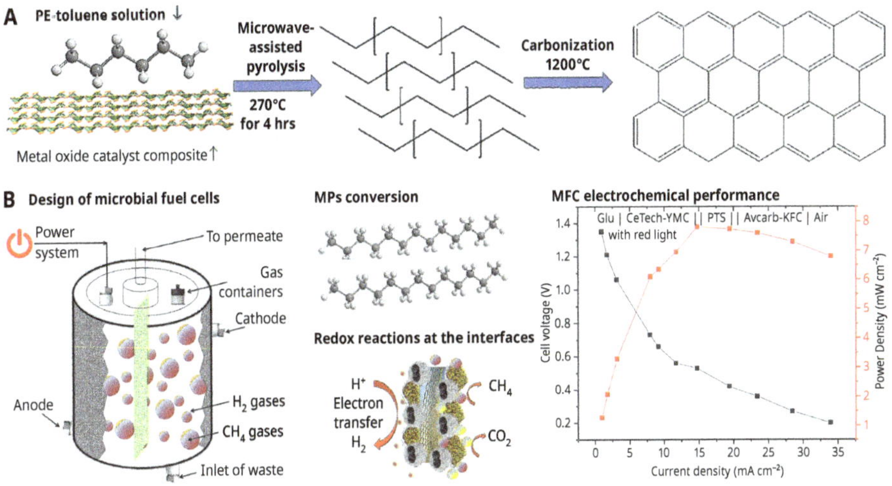

Fig. 2.16: The nonstructural characterization and elemental composition of MPs (nylon 6 fiber as an example). **A.** The TEM image. ***B.*** The distribution of the nylon fibers. **C.** The XPS elemental analysis and the EDX analysis of the above fiber (data in **A**, **B**, and **D** were adapted from [48] and C from [49]).

2.2.2 Nanomaterial-based synergistic combination cancer immunotherapy

Although conventional treatments (e.g., chemotherapy, radiotherapy, and surgery) play critical roles in tumor therapy, cancer immunotherapy becomes attractive due to its advantages in "achieving precision medicine" and "preventing recurrence and metastasis" [50]. The new therapeutic immunotherapy has been used to fight cancer by harnessing the patient's in vivo antibodies [51]. Nanomaterials offer effective, multipronged immunotherapy treatments to enhance patient outcomes [52]. Bioengineered nanomaterials improved the efficacy of combined cancer treatments. This synergistic "cancer combination immunotherapy" demonstrated the potential to inform design strategies for the transformation of versatile clinical applications.

Principle for immune therapy: Nobel laureate in Medicine, Dr. James P. Allison from the University of California, Berkeley, studied the cytotoxic T-lymphocyte-associated protein 4 (CTLA-4). He found that "CTLA-4 functions as a brake on T-cells" and developed an antibody that binds to CTLA-4 to block its function. Allison and co-workers researched mice with cancer and treated them with the "antibodies that inhibit the brake and unlock antitumor T-cell activity" [53]. The team continued their activities to "develop the strategy into a therapy for humans." In 2010, a clinical study found that patients with melanoma (skin cancer) showed significant improvement. Their mechanism (Fig. 2.17) showed that the "activation of T-cells requires that the T-cell receptor binds to structures on other immune cells recognized as *non-self*" (Fig. 2.17A). In addition, a protein that functions as a T-cell accelerator is required to activate T-cells (Fig. 2.17B). Their results suggested that CTLA-4 functions as a brake on T-cells (Fig. 2.17C), inhibiting the accelerator (Fig. 2.17D).

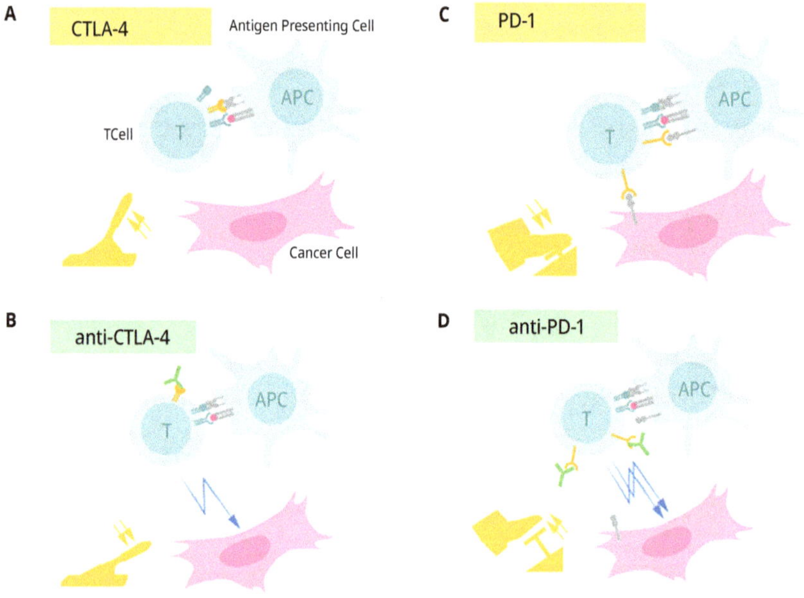

Fig. 2.17: Mechanism of antibody binding to CTLA-4 and blocking its function. **A.** The activation of T-cells. **B.** Antibodies (green) against CTLA-4 block the function of the brake leading to activation of T-cells and attack on cancer cells. **C.** PD-1 is another T-cell brake that inhibits T-cell activation. **D.** Antibodies against PD-1 inhibit the function of the brake leading to activation of T-cells and highly efficient attack on cancer cells [54].

Classification of cancer immunotherapy: In immunotherapy, the patient's immune system is used to trigger the anti-tumor response, further providing personalized vaccines [55]. Cancer immunotherapy includes various treatment modalities (Fig. 2.18A) that lower side effects by targeting cancer cells and regulating the immune system. With its development, immunotherapy strategies were used to harness the anti-tumor immune response for routine cancer management (Fig. 2.18B). Nowadays, this immunotherapy enables rapid mapping of mutations within a genome, the rational selection of vaccine targets, and the on-demand production of a therapy customized to an individual patient's tumor. Since their inception, immunotherapies have become increasingly attractive for activating and boosting the immune response [56]. However, cancers have immune evasion mechanisms to delay, weaken, and even prevent antitumor immunity. The principle of blocking inhibitory surface receptors emphasizes that the immune system is standard and stable. In addition, newly developed immunomodulatory strategies aimed at reversing the tumor microenvironment (TME) are known as vaccine/adjuvant strategies [57]. Figure 2.18A summarizes the common anticancer therapeutic modalities, including chemotherapy, high-intensity focused ultrasound, photodynamic therapy, photochemotherapy, radiotherapy, and gene therapy. Nanomaterials can potentially combine these therapeutic strategies with this new immunotherapy in a TME. A combinatorial approach is used to design, develop, and deploy effective treatments to regulate the immune response and destroy active tumors without compromising the body's innate defense. Five typical immunotherapies are used to treat cancer (Fig. 2.18B). These immunotherapy modalities include cytokines, checkpoint blockade, adoptive T-cell transfer, and antibody–antigen interactions, both mono- and polyclonal. However, they often induce greater side effects. Several small signaling molecules, such as cytokines, are used by cells. These include chemokines, interferons, tumor necrosis factor-alpha, and other related molecules, such as lymphokines [58], which initiate a cellular response; interferons and interleukins stimulate the immune response (Fig. 2.18B-**a**). These innate and programmed responses can be strengthened by therapeutic vaccines, which enhance the programmed response by bolstering the immune response against specific tumors (Fig. 2.18B-**b**) [59].

In an antigen-initiated infection, the immune system generates antibodies, which take time. A method to accelerate the immune response is adoptive cell transfer (ACT), which exhibits specific antigenic epitopes and can enable the biosynthesis of antibodies to be accelerated, with ACS being from the patient (e.g., liver cells) or from a different patient with the same immune disease (Fig. 2.18B-**c**). These ACTs express "T-cells with specific chimeric antigen receptors" and can be stored outside the body in media and transferred as needed, thereby supporting the overall immune response. For traditional checkpoint inhibitors, low selectivity may lead to immune-related toxicities. Synergistic immunomodulation may be more effective at blocking interactions between negative regulators and T-cells. Checkpoint blockade is used to promote self-tolerance and protect cells from immune attack, but can be tailored to recognize and destroy specific tumor cells (Fig. 2.18B-**d**). Monoclonal antibodies bind to specific antigens on the surface of cells, enabling these cells to be better recognized by the immune system for lysis and inactivation. Monoclonals allow a subset of surfaces to be tagged (Fig. 2.18B-**e**).

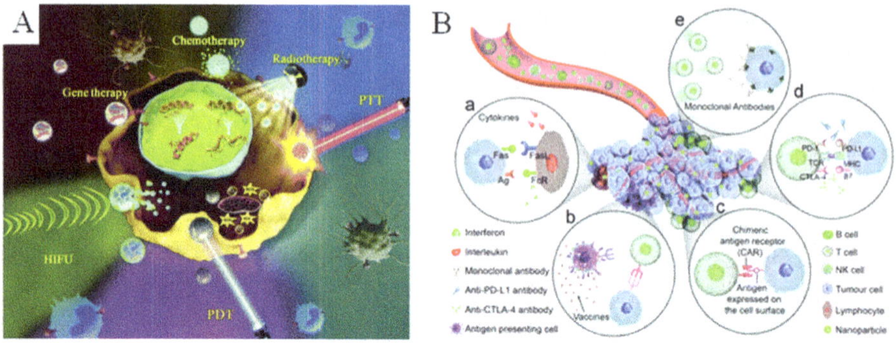

Fig. 2.18: Cancer treatment strategy and typical immunotherapies. **A**. Schematic diagram using common anti-cancer therapeutic strategies. **B**. Five typical types of immunotherapies to treat cancer. They are cytokines (**a**) therapeutic vaccines, (**b**) adoptive cell transfer, (**c**) checkpoint blockade, (**d**), and binding between monoclonal antibodies and tumor antigens (**e**) [52].

Nanomaterials for cancer immunotherapy: Nanomaterials with different formulations have been widely used in cancer diagnosis and therapy [60]. The side effects of nanomaterials play a critical role in avoiding kidney clearance, due to their ability to selectively penetrate tumor tissues. As a result, the drug-loaded inside nanomaterials showed longer blood retention time, decreased toxicity, and improved tumor distribution. Functionalization and modification of nanomaterials enable versatile applications through varying surface chemistry, thereby advancing their uptake by targeted cells [61]. In addition, nanomaterials can be used to deliver cytotoxic drugs or imaging agents, enabling high-precision diagnosis and improved treatment efficacy [62]. Pioneering research indicates that the application of nanomaterials in cancer immunotherapy offers a new platform for both cancer vaccination and modulation of the immunosuppressive TME (Fig. 2.19). When applied in cancer vaccine design, nanomaterials possess several advantages: co-encapsulation of antigen and adjuvant, dendritic cell (DC) uptake and targeting, inherent adjuvant effect, lymph node drainage, and antigen presentation. Antigens are components of foreign cells, composed of proteins and nucleic acids. They can be attached to nanoparticles, which are then injected into a site to initiate and strengthen the immune response. This enhancement can be achieved through innate immune potentiation or by antigen delivery, enabled by the specific physicochemical characteristics of nanomaterials.

Various nanomaterials have been designed to carry active substances, alter the formulations, and modulate the immunological TME. Although nanomaterial-based TME modulation has made significant progress, several limitations must be overcome before it can be applied in clinical trials. These problematic issues relate to limited tumor tissue penetration, immune toxicity, and tumor heterogeneity. The TME is highly tissue and cell-specific. It is often associated with specific tumor cell types that target specific tumors, such as myeloid-derived suppressor cells (MDSCs), tumor-

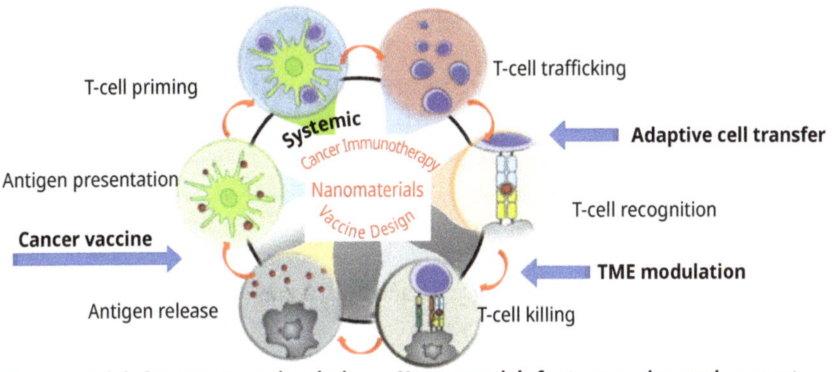

Fig. 2.19: Nanomaterials for cancer vaccine design and tumor microenvironment modulation balance the cancer-immunity cycle. Modified from [63], with permission from Elsevier.

associated macrophages (TAMs), tumor-associated fibroblasts, and soluble small-molecular mediators, which are used as target checkpoints. The nanomaterial can be modified with the appropriate antigen, drug, or ligand to promote cellular disruption of the extracellular matrix, plasma membrane, and tumor vascular systems to enable these tissues to become more porous for easier perfusion of drugs or other nanoparticles, and lower oxygen content to promote hypoxia, slow cell growth, and affect cell death. Modified nanoparticles have been shown to alter the cell populations of cancer-associated fibroblasts by modulating MDSCs, TAMs, and regulatory T-cells, thereby inhibiting cellular proliferation and promoting tumor cell death through targeted approaches. Such approaches have led to the development of more effective chemotherapeutics and fewer adverse immune responses and relapses.

2.2.3 Nanomaterials used to revolutionize eye care

Due to the unique anatomy and physiology of the eye, efficient ocular drug delivery remains a significant challenge for researchers and pharmacologists [64]. Although conventional noninvasive and invasive treatments, such as eye drops, injections, and implants, are commonly used, current therapies suffer from low bioavailability or severe ocular adverse effects. Alternatively, emerging nanoscience and nanotechnology play an essential role in developing novel strategies for ocular disease therapy [65]. Various active molecules have been designed to associate with nanocarriers to overcome ocular barriers and intimately interact with specific ocular tissues. The human eye is a visual spheroidal organ about 24 mm in diameter (Fig. 2.20A). The eye consists of the anterior and posterior segments, including the vitreous humor, which has various biological barriers that protect it from foreign substances. Nanotechnology provides a feasible, rapid, and straightforward eye-care treatment method due to its easy access to the exposed position. Nano drugs must be bio-adherent to prevent being rapidly cleared through the nasolacrimal system [66]. The small amounts of medication delivered to the ocular surface may reduce the risk of blurring, and the possibility exists that the nanoparticles may enter corneal cells, prolonging contact time [67]. The Swiss-designed prototype contact lens sensor (Fig. 2.20B) contains strain gauges that monitor changes in corneal curvature caused by intraocular pressure fluctuations. It can be seen that a microprocessor and an antenna were integrated into the soft contact. This contact lens sensor can collect power and communicate wirelessly. The composition of the nano-drug carriers must be biodegradable and compatible with securing the safety of nanoparticle therapeutics before their potential use. Various biodegradable synthetic polymers and natural polymers (e.g., chitosan, gelatin, albumin, and sodium alginate) are promising for drug delivery. Piedad Calvo and co-workers studied the bioavailability of the indomethacin ($C_{19}H_{16}ClNO_4$)-based nano-drug. This non-steroidal, anti-inflammatory drug was encapsulated in nanoparticles, which showed triple in vitro corneal penetration compared to a conventional solution [68]. Recently, so-called "nanoknives" were fabricated from silicone to make "tiny and precise cuts." This procedure avoided molecular and cellular damage to the surrounding tissue of the treated eye relative to a surgeon's blade [69]. The nano-blades with varying shapes range in size from 100 to several hundred nanometers. The shape of the "nano knife" can be determined by applying individual instruments. The commercial version of the contact lens sensor, called the Triggerfish (Sensimed AG, Lausanne, Switzerland), received CE designation in 2009 and differed slightly from the earlier prototype (Fig. 2.20C) [70].

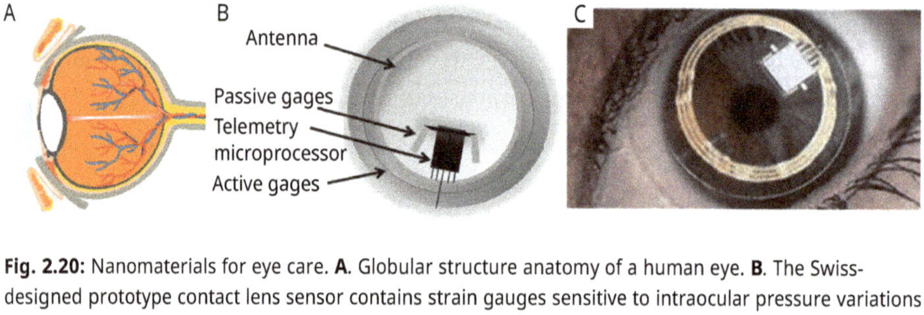

Fig. 2.20: Nanomaterials for eye care. **A**. Globular structure anatomy of a human eye. **B**. The Swiss-designed prototype contact lens sensor contains strain gauges sensitive to intraocular pressure variations and corneal curvature. A microprocessor and an antenna facilitate wireless communication with a computer for readout. **C**. The commercial version of the contact lens sensor, the Triggerfish (Sensimed AG, Lausanne, Switzerland), received the CE designation in 2009. Modified from [70] with permission from Jobson Medical Information LLC.

Nanomaterials have a high surface-area-to-volume ratio. This attribution can help reduce or eliminate reactive species, which are known to cause cataracts and other ocular diseases [71]. Nanoparticles are colloidal systems with several advantages over conventional delivery. They improve delivery efficiency and are particularly useful for chronic diseases such as glaucoma, uveitis, and ocular neovascularization [72]. Molecules such as growth factors can be delivered via nanoparticles. It was found that numerous photoreceptor cells undergo rapid death when exposed to bright light for prolonged periods. To visualize these damaged cells, Chen, McGinnis, and coworkers performed experiments on animals killed 5 days after light exposure. Their eyes were processed for histological analysis using the TUNEL (terminal deoxynucleotidyl transferase dUTP nick-end labeling) assay. This fluorescent assay detects cells damaged by light and undergoing apoptosis (Fig. 2.21). The results demonstrate that exposure to bright light results in TUNEL-positive cells in those retinas from eyes that were either uninfected (Fig. 2.21A-2 and A-3) or injected with saline (Fig. 2.21B-2 and B-3). Nanoparticles (NPs) composed of CeO_2 with different concentrations (0.1, 0.3, and 1.0 μM) were used to protect the retinas and prevent TUNEL-positive cells' appearance (Fig. 2.21C-2, D-2, and E-2) [71]. These data suggest that CeO_2 NPs, even at lower concentrations, effectively inhibit light-induced injury to photoreceptor cells after prolonged exposure. The results of eyes without CeO_2 NPs injection were shown in Fig. 2.21A-1 and B-1, while the injected eyes were evaluated at day 0, exposed to light on day 3, and the experiment ended on day 8 (Fig. 2.21C-1, D-1, and E-1). The CeO_2 NPs with a diameter of 5 nm showed adequate capability to scavenge reactive oxygen intermediates. The high surface-to-volume ratio of nanoparticles allows rapid radical neutralization regeneration without re-addition of the agent. Chen and colleagues have demonstrated that intravitreal injection with CeO_2 NPs effectively prevents light damage in rodents. The nanotechnology invented by Chen shows potential to treat other ocular conditions associated with oxidative damage, such as diabetic retinopathy.

2.2.4 Nanoparticles used to treat traumatic injuries

Our mother nature has created perfect nanoscaled arts of biology over the millennia. Naturally, the inner workings of cells occur at the nanoscale, such as the diameter of hemoglobin, a protein that carries oxygen throughout the human body, which is 5.5 nm (Fig. 2.22A). The strand of the building block of life, deoxyribonucleic acid (DNA), is approximately 2.0 nm. The iron–heme used for oxygen capture is the protein, with a quaternary alpha-beta motif (Fig. 2.22B-**D**). Oxygen binds to the metal axially under high pressure. It is released under low-oxygen conditions in tissues. The natural nanoscale of biology is driving many medical researchers to design tools, develop treatments, and deliver therapies with high precision and accuracy, thereby personalizing treatment compared to conventional methods [72]. Nanopharmacy and nanomedicine are applications of metal within organic or inorganic scaffolds tailored

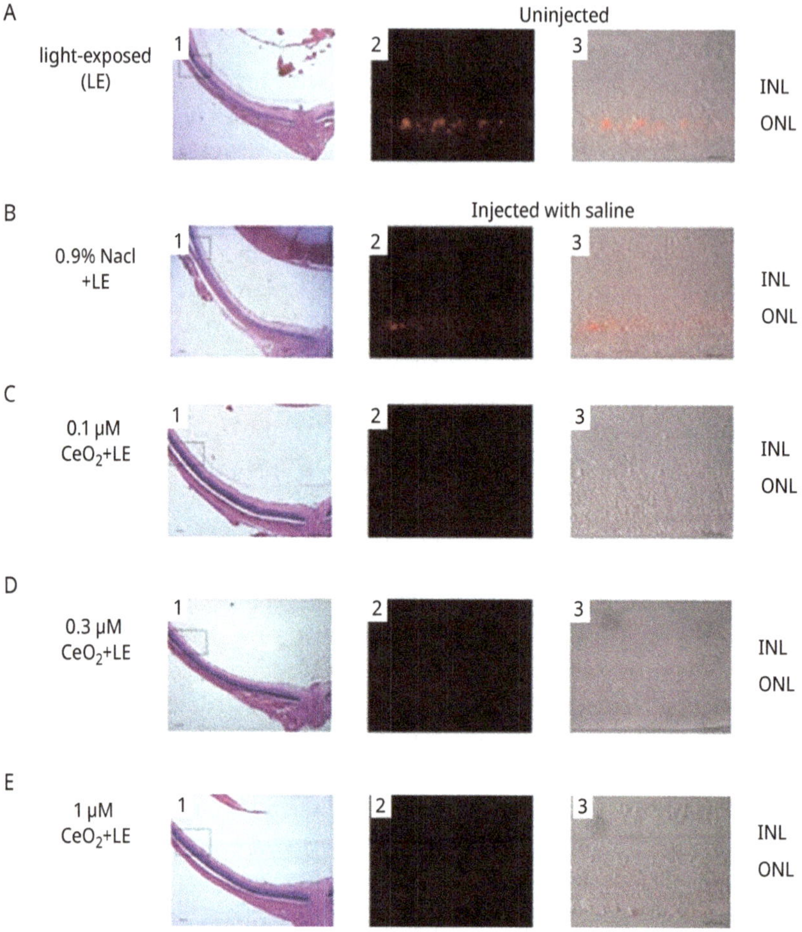

Fig. 2.21: Nanoceria particles prevent the appearance of TUNEL-positive photoreceptor cells, which occur days after exposure to damaging light, resulting in TUNEL-positive cells in retinas that were either uninjected. Images were adapted with permission from [71]. **A–E**. Comparative evaluation of retinal tissues under different experimental conditions. Panels A-1 and B-1 show eyes without CeO_2 nanoparticle (NP) injection, serving as untreated controls. Panels C-1, D-1, and E-1 correspond to eyes injected with 5 nm CeO_2 NPs, assessed at day 0, after light exposure on day 3, and at the end of the experiment on day 8, respectively.

for specific interactions with host tissues, such as the release of agents, scavenging of radicals, generation of reactive oxygen species to inactivate microbes or tumor cells, and antimicrobial properties to reduce the risk of infection. Due to the large number of possible combinations of metal and ligand or coating, nanomaterials can be fine-tuned, unlike their bulk counterparts, for specific biological applications, making their use more ubiquitous.

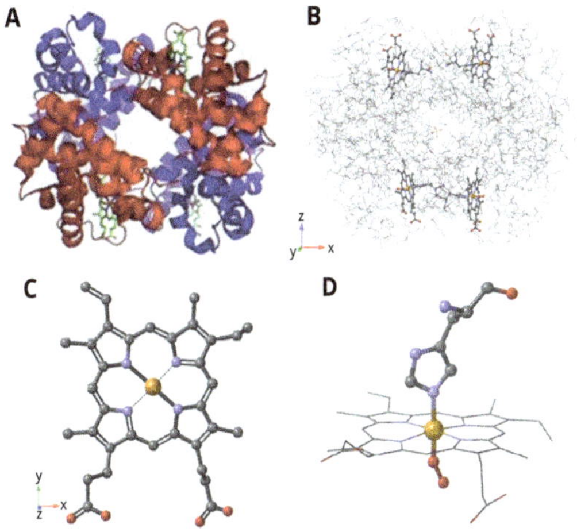

Fig. 2.22: The hemoglobin, a naturally occurring nanoscale protein found in the blood. **A.** Computer simulation of the structure [73]. **B.** The hemoglobin molecule with the four Fe-porphyrin sub-units ($C_{5908}Fe_8N_{1560}O_{2073}P_3S_{24}$). **C.** The Fe-porphyrin subunit ($C_{34}FeN_4O_4$). **D.** Fe-porphyrin with an axial ligand was taken from the complete hemoglobin structure ($C_{40}FeN_7O_7$). Source: Protein Data Bank, ID Code: 1HHO, Revision Level: 1HHOA, Author: B. Shaanan [74].

Blunt trauma is a significant cause of death, especially in individuals under 45 years of age, and is often due to ischemia, loss of blood, and calcium flux. To aid recovery, the potential to inject nanoparticles at the site of the trauma is an attractive option when blood flow is restricted and oxygen is released, enabling the patient to be moved to a more secure location, such as a hospital. However, the role of nanoparticle circulation and toxicity is unknown [75].

A recent study in rats examined the effects of different-sized polymer nanoparticles on their circulation and interaction with platelets, cells that promote blood clotting. The study demonstrated that particles of 150 nm in diameter were the most effective at stopping bleeding. These particles were less likely to travel to other organs or away from the injection site. Particles larger than 150 did travel to other organs, such as the lungs, and would not be effective in promoting blood clotting [76]. "With nanosystems, there is always some accumulation in the liver and the spleen, but we would like more of the active system to accumulate at the wound than at these filtration sites in the body," says Paula Hammond, an MIT Institute Professor, head of the Department of Chemical Engineering, and a member of MIT's Koch Institute for Integrative Cancer Research [77].

Dr Hammond concluded that intravenous nanoparticle hemostats offer a potentially attractive approach to promoting hemostasis, particularly for inaccessible wounds such as noncompressible torso hemorrhage [78]. In their work, particle size was tuned over a range of <100–500 nm, and its effect on nanoparticle-platelet interactions was systematically assessed using in vitro and in vivo experiments. Smaller particles bound a larger percentage of platelets per unit mass delivered, while larger particles resulted in greater particle accumulation on platelet and collagen surfaces. Intermediate particles led to the most excellent platelet content in platelet-nanoparticle aggregates, indicating that they may recruit more platelets to the wound. In biodistribution studies, minor and intermediate nanoparticles exhibited longer circulation lifetimes, whereas larger nanoparticles showed greater pulmonary accumulation. The particles were then challenged in a 2-h lethal inferior vena cava puncture model. Intermediate nanoparticles significantly increased survival and injury-specific targeting relative to saline and unfunctionalized particle controls [79]. The *in vitro* and *in vivo* results suggest that platelet content in aggregates and extended nanoparticle circulation lifetimes are instrumental in enhancing hemostasis [80].

They used a murine lethal injury model to investigate the effects of the nanoparticles. The particles were injected into uninjured mice to assess their circulation lifetimes and organ distributions. The mice were imaged live using the *in vivo* imaging system for 3 h and observed for several days to ensure no deleterious side effects following the injection. Figure 2.23A shows the timeline for blood sampling and imaging postinjection. When injected to achieve an estimated final therapeutic concentration of approximately 1 mg/mL in mouse blood, particles of different sizes were found to have different retention times and organ accumulation (Fig. 2.23B and C): specifically, increased splenic accumulation in the larger particle sizes versus predominantly liver accumulation in the <100 nm size group.

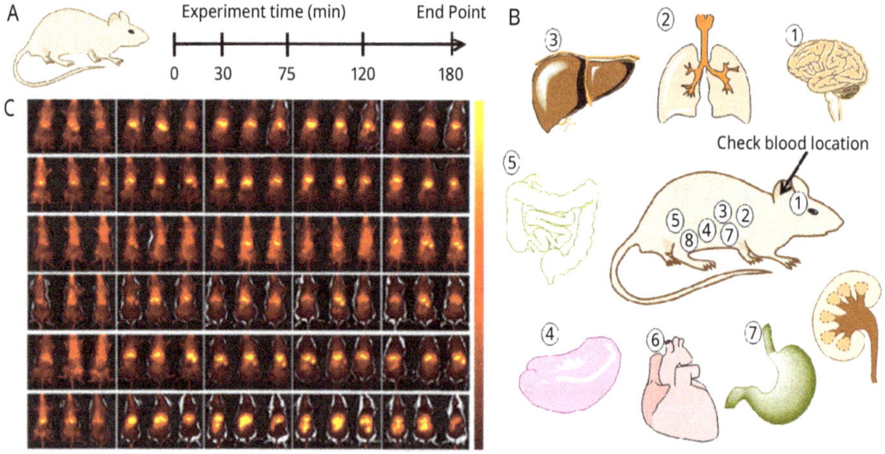

Fig. 2.23: In vivo biodistribution of Cy7-labeled nanoparticles and a free dye control in BALB/c mice. **A.** Dosing timeline. **B.** Organic evaluated. Adapted from refs [74, 80]. **C.** Organ imaging using Cy7-labeled hemostatic nanoparticles and free dye control over 3 h. Dark red/black denotes lower fluorescence levels, while yellow indicates accumulation in a particular area as a function of nanoparticle size.

Size effects: Hemostatic nanoparticles can slow blood flow and promote blood clotting by interacting with platelet co-factors. The nanoparticles are coated with a soft biopolymer, such as polyethylene glycol (PEG)–poly(lactic-co-glycolic acid) (PLGA), to conjugate a surface that reduces nanoparticle toxicity and attracts platelets [81]. The nanoparticles' coating was the initial focus, not their actual size. The group investigated the size and how variation in size affects distribution. "We were trying to look at how the size of the nanoparticle affects its interactions with the wound, which is an area that has not been explored with the polymer nanoparticles used as hemostats before" was reported by Dr Hong [77–78].

While nanoparticles can initiate blood clotting, they also migrate to other organs and potentially cause stress. The group examines low-diameter (<100 nm), intermediate (140–220 nm), and large (500–650 nm) nanoparticles. The stop-flow study examining nanoparticle-to-platelet interactions shows that the smallest nanoparticles bound to platelets at the highest percentage (Fig. 2.24). In a follow-up experiment, they discussed the roles of platelet surface coverage and nanoparticle interaction (Fig. 2.24A) and demonstrated that the largest nanoparticles interacted most effectively with the platelet-bound surface (Fig. 2.24B). The data were converted to the platelet-to-nanoparticle mass ratio, and the highest ratio was observed for nanoparticles of intermediate diameter (Fig. 2.24C1). "If you attract a bunch of nanoparticles and block platelet binding because they clump onto each other, that is not very useful (Fig. 2.24C2). We want platelets to come in," Dr Hong said. Also, "When we did that experiment, we found that the intermediate particle size was the one that ended up with the greatest platelet content" (Fig. 2.24C3) [77, 78].

Stopping the bleeding: In the in vivo model, mice were injected with nanoparticles of three sizes. It was determined that the largest diameter nanoparticles would accumulate in other organs, limiting their on-site circulation time and half-life. Using an internal injury rat model, the team found that intermediate-diameter nanoparticles were optimal for promoting blood clotting and achieving the highest accumulation at the trauma site. Dr Hammond said. "This study suggests that the bigger nanoparticles are not necessarily the system we want to focus on, which was unclear from the previous work." Hammond says, "turning our attention to this medium-size range can open up some new doors." (Fig. 2.24).

Dr Hong commented, "These particles are meant to address preventable deaths. They are not a cure-all for internal bleeding, but they are meant to give a person extra hours until they get to a hospital to receive adequate treatment" [77, 78] (Fig. 2.24).

2.2.5 Nanomaterials used for shale gas storage

Shale gas introduction: As an essential source of natural gas, Shale gas has been studied in many countries [82]. The USA (Fig. 2.25A) and Canada have significant shale gas production, although other countries have many recoverable resources [83]. The Eagle Ford in Southern Texas enhanced the confidence in producing natural gas from shale

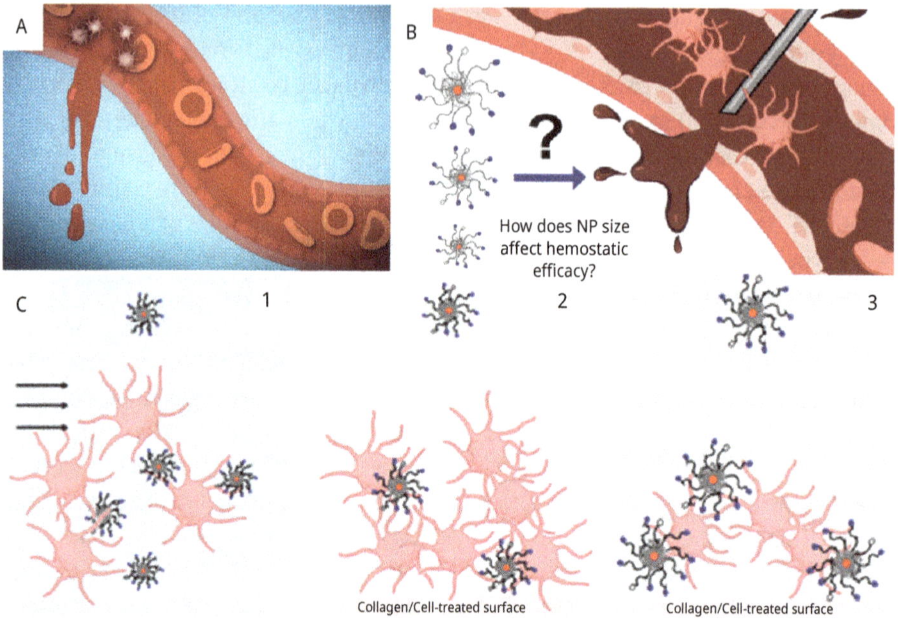

Fig. 2.24: Use of nanoparticles to treat trauma: **A.** a schematic of bleeding; **B.** NP size effect on hemostatic efficacy; and **C.** summary of in vitro data to evaluate NP size on lethal injury model. **C1.** NPs sized <100 nm resulted in the most significant number of specifically bound platelets underflow. **C2.** Intermediate-sized NPs (140–220 nm) resulted in the most considerable number of platelets in a platelet-particle surface aggregate. **C3.** Large NPs (500–650 nm) resulted in the most significant polymer mass accumulation onto platelet-particle surface aggregates. The image was adapted from Biorender [77].

(Fig. 2.25B) [84]. It was found that the Eagle Ford is predominantly composed of organic matter-rich fossiliferous marine shales and marls. It was projected that natural gas from shale resources would grow from 42 BCF/D in 2015 to 168 BCF/D in 2040, a fourfold increase. By the end of the forecast period, shale will account for 30% of world natural gas production. Therefore, it is critical to develop materials to improve shale gas storage [85]·

Porous materials used for shale gas storage: The shale is a fine-grained sedimentary rock that forms from the compaction of silt and clay-size mineral particles. One of the significant components of shale gas is methane (CH_4). Efficient methane storage is the key perspective in this review [86]. To improve methane (CH_4) storage, its physical and electronic properties need to be fully understood. For example, at room temperature, CH_4 is a gas. It can be placed in a liquid-like state under supercritical storage conditions. Another prominent characteristic of methane (T_c = 190.6 K, P_c = 45.8 atm) is that CH_4 cannot be liquefied by compression above T_c. It is challenging to store a large amount of CH_4 in a given volume for portable applications, such as gas-powered automobiles [87]. Therefore, preparing porous materials becomes essential to achieve the above end application of CH_4. This report presented a series of materials (metal-organic frameworks and caged porous composites) prepared by a feasible

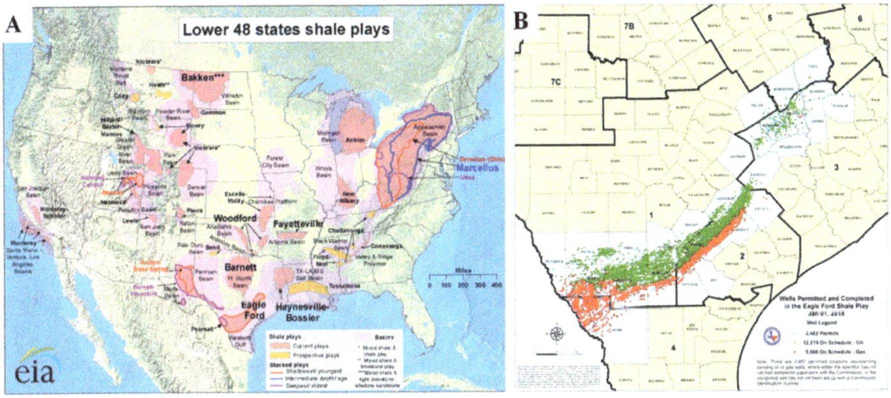

Fig. 2.25: The U.S. Energy Information Administration (EIA) estimates that in 2020: A. the National shale oil and shale gas resources and B. the Texas Eagle Ford shale gas reserve.

wet-chemistry approach. The gas uptake was also measured to evaluate their porosity and surface area. The ideal materials for methane storage need to display the following properties: (1) high surface area, (2) large pore volume, (3) low framework density, and (4) a strong interaction with the adsorbed methane. Recently, the U.S. DOE announced that the volumetric and gravimetric targets for CH_4 onboard storage are 263 cc (STP)/cc and 699 cc(STP)/g, respectively. In reality, a 25% loss in volumetric capacity will occur after the porous materials are packed into the fuel tank [88]. We also implement a strategy by introducing amphiphilic groups to produce caged porous materials to reach this goal. Unfortunately, the pore size was decreased due to this introduction.

Experimental procedure: A feasible wet chemistry was used to produce a series of shielded porous metal oxides and metal-organic frameworks for CH_4 storage. The synthesis method variables were optimized to control the porosity, size, and shape of nanocomposites [89]. Those so-prepared materials were characterized using state-of-the-art techniques (TEM, X-ray diffraction, XPS, and Brunauer–Emmett–Teller (BET)) to evaluate their nanostructure, porosity, and gas uptake. The materials' safety was also assessed through toxicological testing, focusing on cellular assays to determine cellular homeostasis in the presence of ultrafine particles [90].

Materials preparation: The colloidal method produced Ag-MOFs, Fe_3O_4 composites, and ZnO composites. In this method, the monodispersed system is found to be in a high-energy state, corresponding to the free energy required to fabricate the increased surface area due to attractive van der Waals interactions, leading to particle growth [91]. Therefore, amphiphilic dispersing agents were introduced to prevent particle growth by balancing attractive and repulsive forces between the particles. The second method to avoid aggregation resulted from steric stabilization [92]. Herein, the long-chain molecules (e.g., starch and gum arabic) are chemically bonded to or physically attracted to colloidal particles, creating a repulsive force due to steric hindrance. Polymeric surfactant chains attach to more than two particles at low concentrations, resulting in bridging flocculation [93].

Materials characterization: The morphology and elemental composition of the nanocomposite were characterized using TEM (FEI Company, Tecnai F20-G^2, Hillsboro, OR). The TEM (Fig. 2.26A–**C**) images of three series of nanocomposites indicate that the shapes of these fabricated materials (a totally of 89 formulations were prepared) were controlled as rod-shaped crystals with an average diameter of 10 nm and length of 78 nm (gum arabic shielded Fe_3O_4, Fig. 2.26A), cubic (aloe vera extract-modified ZnO, Fig. 2.26B), and fiber (starch-modified ZnO, Fig. 2.26C) depending on the shielding agents and heat-treatment conditions. The silver-based metal-organic frameworks exhibited highly crystalline structures [94]. However, the Cu-, Co-, and Zn-metal -centered MOFs did not show the desired structures or gas uptakes. The amount of MOFs also needs to be scaled up for the gas storage study.

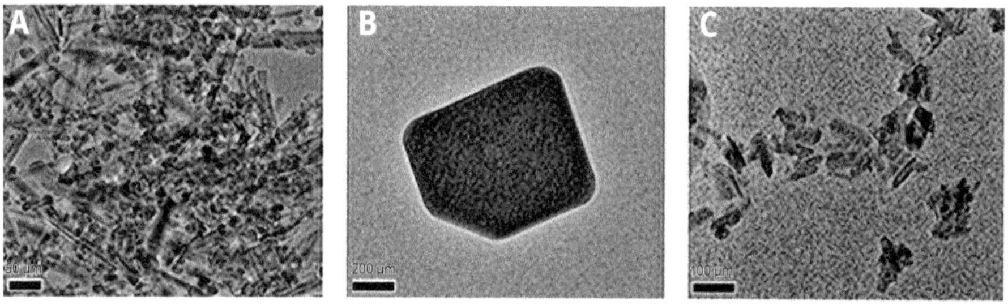

Fig. 2.26: The morphology and elemental composition of the nanocomposite were characterized using transmission electron microscopy: A. gum arabic-shielded Fe_3O_4 nanoparticles; B. aloe vera extract-modified ZnO nanoparticles; and C. starch-modified ZnO nanoparticles.

The study of powder X-ray diffraction (Bruker-AXS D8 Vario X-ray Powder Diffractometer, Madison, WI) indicated that nanocomposites and Ag-MOFs were highly crystalline. The gum arabic-protected Fe_3O_4 (Fig. 2.27A) displayed an inverse spinel structure (PDF 65-3,107, $a = 8.3905$ Å, $\alpha = 90°$). The natural product-modified ZnO nanocomposites (Fig. 2.27B) were tetragonal and well-indexed with the standard (PDF 89-0511 and PDF 05-0664 ($a = 3.249$, $c = 5.205$, $\alpha = 90°$). Briefly, the nanocomposites were generated via colloidal chemistry to produce well-defined, highly crystalline compounds. Nucleation of the crystal units was promoted as solvent molecules evaporated from the solution. This controllable nucleation favored the formation of composites with tunable size and monodispersity. Natural product extracts play a critical role in the shape of nanocomposites [95]·

XPS (Omicron ESCA+, Scienta Omicron, Inc., Denver, CO) determined the elemental composition from photoelectron emission [96]. The chemical state of Ag, Fe, Zn, C, and O atoms can be determined from the binding energy obtained from photoelectron spectros-

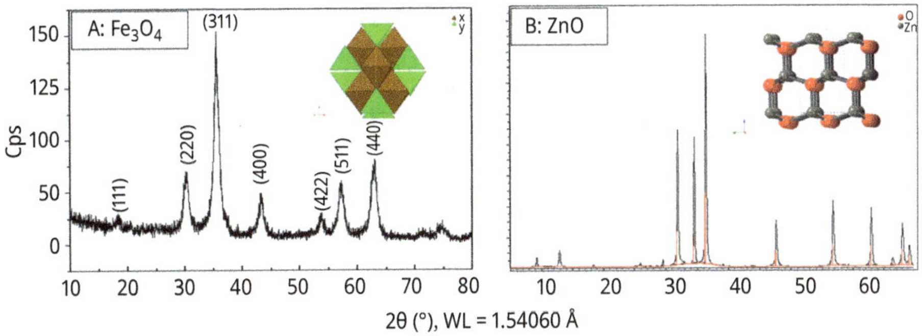

Fig. 2.27: The crystallographic analyses of the nanocomposite obtained using X-ray powder diffraction: A. gum arabic-shielded Fe_3O_4 nanoparticles and B. aloe vera extract-modified ZnO nanoparticles (note: different natural product extracts show no effect on the crystalline phase or crystallinity of ZnO nanoparticles).

copy. XPS can be used as a "fingerprint" to identify the elements in the nanocomposite, based on the bonding information derived from these chemical shifts. The XPS spectra of the nanocomposite (Fig. 2.28A–**C**) indicated the presence of Fe, Zn, and Ag; and O and C elements, which stem from the composite, dispersing, or reducing agents [97].

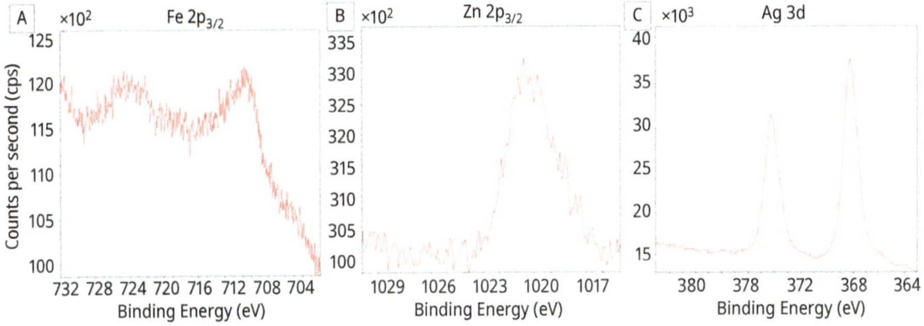

Fig. 2.28: The elemental analyses of the nanocomposite were obtained using X-ray electron–photon spectroscopy (XPS): *A*. Element Fe from Fe_3O_4 nanoparticles. *B*. Element Zn from ZnO nanoparticles. *C*. Element Ag from Ag^0 nanoparticles.

Gas uptake (CH_4, N_2, H_2): The BET absorption model (Micromeritics ASAP 2020, Norcross, GA) and related measurement techniques will be applied. CH_4, CO_2, H_2, and N_2 were used to evaluate the catalyst's selective adsorption capacity for these gases. The measurement temperatures are controlled at 87, 175, and 231 K using liquid argon, a methanol–liquid nitrogen bath, and an acetonitrile-dry ice bath, respectively. Before the adsorption measurement, the samples will be activated using the "outgas" function of the adsorption analyzer for 12 h at 80 °C. These activated samples will then be used for gas adsorption measurements with a purity grade. Table 2.1 shows the CH_4 storage data.

The nanometer-scale pore size distributions allow gases to penetrate effectively. However, the porosity decreased due to the introduction of functional groups (such as alpha acetal from the starch coil structure). The Fe_3O_4 nanoparticles were chosen to showcase the CH_4 uptakes (Fig. 2.29A). It was found that the adsorption reproducibility exceeds our expectations, suggesting a stable structure and stable pores (Fig. 2.29B and C). For the two metal-oxide nanoparticles studied, methane-storage metrics at 87 K are summarized as follows. Fe_3O_4 nanoparticles exhibited an exceptionally high BET surface area of 536.8 m^2/g and a total pore volume of 0.73 cm^3/g. When those pore characteristics are translated into a macroscopic packing density, the material exhibits a porosity of 67.6% and an apparent bulk density of 0.167 g/cm^3. Under these conditions, the excess CH_4 uptake was measured at 4.32 g/L, and the isosteric heat of adsorption (Q_{st}) was 7.54 kJ/mol. ZnO nanoparticles prepared with similar textural properties had the same surface area (ca. 560 m^2/g) and pore volume (0.75 cm^3/g), resulting in identical porosity (70.6 %) and packing density (0.171 g/cm^3). Their CH_4 excess uptake was lower, at 4.50 g/L, while Q_{st} remained 7.54 kJ/mol.

Tab. 2.1: A summary of CH_4 storage using natural product-coated metal oxide nanoparticles.

Samples	CS (m^2/g)	PV (cm^3/g)	P	D (g/cm^3)	EU (g/L)	Q_{st} (kJ/mol)
GA-Fe_3O_4	536.82	0.730	67.62	0.167	4.32	7.54
ST-ZnO	561.73	0.764	70.74	0.175	4.52	7.89
AV-ZnO	560.62	0.752	70.62	0.171	4.50	7.82

CS, calculated surface area; PV, pore volume; P, porosity; D, density; EU, excess CH_4 uptake.

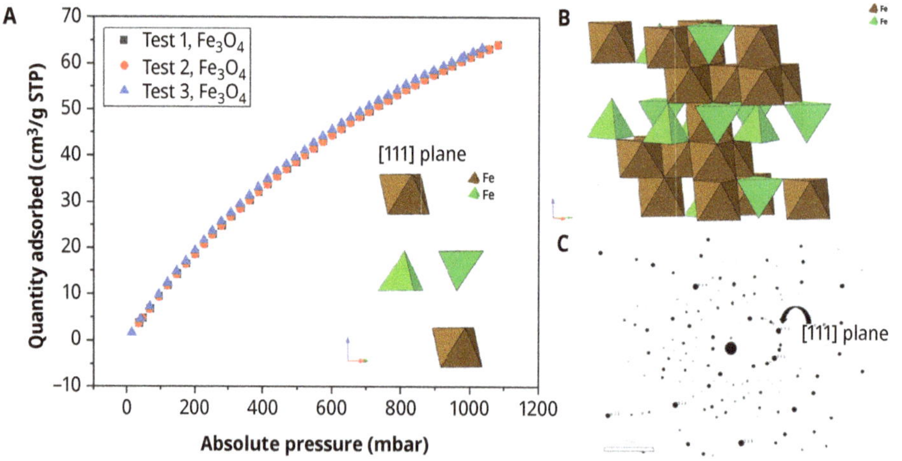

Fig. 2.29: The CH_4 uptake using Fe_3O_4 nanoparticles as a demonstration: A. gas uptake results; B. Fe_3O_4 111 plane showing a stable crystalline structure; C. Laue plane front scattering, showing high crystallinity of the nanoparticles with well-maintained integrity.

2.3 Summary

This chapter opens by explaining that when material dimensions shrink below about 100 nm, quantum confinement, and extreme surface-to-volume ratios radically alter their behavior compared to bulk matter. Chemically, the surface-area-to-volume ratio explodes: a 1 cm cube offers 6 cm^2, whereas dissecting it into 1 nm cubes (about 10^{21} particles) yields roughly 6×10^7 cm^2 of reactive surface area. At this scale, ionization energies and electron affinities shift, finely tuning catalytic activity. For example, gold transforms from its fcc lattice ($a = 4.0786$ Å) into an icosahedral form at particles near 5 nm, which in practice accelerates the rate constant for RNA (Ribonucleic acid)-model phosphate cleavage by more than two orders of magnitude under physiological conditions (pH 7, 37 °C). Magnetically, iron oxide nanoparticles in the 10–20 nm range become superparamagnetic: their saturation magnetization (M_S) and coercivity (H_C) vary sensitively with diameter. At the same time, quantum exchange interactions and *d*-orbital splitting dictate whether individual particles attract or repel applied fields. Surface coatings of silica, carbon, or tailored polymers prevent agglomeration and oxidation, preserving both dispersibility and magnetic performance. Electro-optical properties also emerge from quantum mechanics. Treating a nanoparticle as a "particle-in-a-box" yields allowed wavelengths $\lambda_n = 2\,L/n$ and energy levels $E_n = n^2h^2/(8\ mL^2)$. Changes in dimensionality, from zero-dimensional quantum dots to 3D nanoparticles, reshape the electronic density of states $\rho_{(E)}$ in proportion to $E^{(n-2)/2}$, tailoring absorption spectra and enabling resonant tunneling currents. In quantum dots, Coulomb blockade effects confine single electrons, giving charging energies (E_C) that underpin next-generation single-electron transistors and memory elements. Thermally, phonons carry heat according to $k = (1/3)\,C_{vl}$, where C is volumetric heat capacity, v is the phonon group velocity, and l is the mean free path. Anharmonic three-phonon interactions reduce and scattering at grain boundaries (frequency-independent) plus point defects ($\propto f^4$) further suppress thermal conductivity above the Debye cutoff frequency f_D.

These unique properties translate into five transformative applications. First, environmental remediation employs V_2O_5–MnO_2 catalysts under microwave-assisted pyrolysis to cleave C–H bonds in MPs and NPs, extracted from soils via density separation and peroxide digestion, producing "gray" hydrogen and graphene that serve as high-performance fuel-cell electrodes. Second, in cancer immunotherapy, lipid or polymer nanoparticles co-deliver tumor antigens and adjuvants to DCs, while conjugated checkpoint inhibitors (anti-CTLA-4, anti-PD-1) significantly boost T-cell activation (including IL-2 secretion) with fewer side effects. Third, ophthalmology benefits from 5 nm cerium oxide nanoparticles that scavenge reactive oxygen species, reducing photoreceptor apoptosis by over 90% in preclinical models, and from smart contact lenses that embed strain gauges and micro-antennas in polymer matrices for continuous, noninvasive intraocular pressure monitoring. Fourth, life-saving hemostasis has been achieved with PEG-PLGA nanoparticles sized 140–220 nm, which maximize platelet aggregation and double survival in murine vascular injury models; intermediate particles remain at the wound site with a half-life of about 3 h, avoiding pulmonary or hepatic accumulation. Finally, advances in clean energy storage are exemplified by Fe_3O_4, ZnO, and silver–metal–organic framework nanoparticles. Their ultrahigh surface areas (≈500–560 m^2/g), large pore volumes (≈0.7–0.76 cm^3/g), and low bulk densities (≈0.17 g/cm^3) create porous architectures capable of storing methane at cryogenic temperatures. At 87 K, these materials achieve excess CH_4 uptakes of about 4.3–4.5 g/L, with isosteric heats of adsorption near 7–8 kJ/mol, reflecting stable pore structures and reproducible adsorption behavior. Amphiphilic ligands further tune pore diameters, allowing the frameworks to approach Department of Energy targets of 263 cm^3 (STP)/cm^3 and 699 cm^3 (STP)/g. Together, these textural properties and ligand-controlled pore sizes position oxide and MOF nanoparticles as promising candidates for cryogenic methane storage and broader clean-energy applications.

This chapter reveals how engineering matters at the nanometer scale transforms its fundamental behavior, opening doors that bulk materials cannot unlock. When particle diameters shrink below 100 nm, quantum confinement and vastly increased surface-to-volume ratios profoundly shift chemical reactivity, magnetic response, light–matter interactions, and heat flow. Fine-tuning nanoparticle size and shape alters ionization energies and catalytic rate constants, as seen when 5 nm gold switches from a cubic to an icosahedral lattice and accelerates phosphate-cleavage reactions by over two orders of magnitude. Iron-oxide nanoparticles become superparamagnetic only in the 10–20 nm range, while quantum wells and dots display size-dependent energy levels that underpin next-generation optoelectronics and single-electron devices. Thermal conductivity likewise depends on grain boundaries and phonon scattering that are controlled by crystal structure and defect density. By harnessing these unique properties, researchers have designed catalysts that split biomass into clean hydrogen, engineered nanoparticles that boost cancer vaccines and hemostasis after injury, created nano-enabled contact lenses and retinal protectants, and developed ultraporous oxides and metal–organic frameworks that store methane at cryogenic temperatures. In melding precision nanochemistry with practical applications, this chapter charts a path toward interactive materials capable of addressing urgent energy, health, and environmental challenges.

References

[1] Kolahalam, L. A., Viswanath, I. K., Diwakar, B. S., Govindh, B., Reddy, V., & Murthy, Y. L. N. (2019). Review on nanomaterials: Synthesis and applications. *Materials Today: Proceedings*, 18, 2182–2190.

[2] Serrano, E., Rus, G., & Garcia-Martinez, J. (2009). Nanotechnology for sustainable energy. *Renewable and Sustainable Energy Reviews*, 13(9), 2373–2384.

[3] Lyu, Y., & Scrimin, P. (2021). Mimicking enzymes: The quest for powerful catalysts from simple molecules to nanozymes. *ACS Catalysis*, 11(18), 11501–11509.

[4] Bonomi, R., Cazzolaro, A., Sansone, A., Scrimin, P., & Prins, L. J. (2011). Detection of enzyme activity through catalytic signal amplification with functionalized gold nanoparticles. *Angewandte Chemie International Edition*, 50(10), 2307–2312.
[5] Manea, F., Houillon, F. B., Pasquato, L., & Scrimin, P. (2004). Nanozymes: Gold-nanoparticle-based transphosphorylation catalysts. *Angewandte Chemie*, 116(45), 6291–6295.
[6] Bonomi, R., Selvestrel, F., Lombardo, V., Sissi, C., Polizzi, S., Mancin, F., & Scrimin, P. (2008). Phosphate diester and DNA hydrolysis by a multivalent, nanoparticle-based catalyst. *Journal of the American Chemical Society*, 130(47), 15744–15745.
[7] Jana, N. R., Gearheart, L., Obare, S. O., & Murphy, C. J. (2002). The anisotropic chemical reactivity of gold spheroids and nanorods. *Langmuir*, 18(3), 922–927.
[8] Uzun, O., Hu, Y., Verma, A., Chen, S., Centrone, A., & Stellacci, F. (2008). Water-soluble amphiphilic gold nanoparticles with structured ligand shells. *Chemical Communications*, 2, 196–198.
[9] Guardia, P., Pérez, N., Labarta, A., & Batlle, X. (2010). Controlled synthesis of iron oxide nanoparticles over a wide size range. *Langmuir*, 26(8), 5843–5847.
[10] Sondheimer, E. H. (2001). The mean free path of electrons in metals. *Advances in Physics*, 50(6), 499–537.
[11] Brack, M. (1993). The physics of simple metal clusters: Self-consistent. In Jellium Neth, E. J., Flowers, P., Theopold, K., Langley,R., & Robinson, W. R. 2018. *Chemistry: Atoms First*. OpenStax).
[12] Cummings, F. E. (1977). The particle-in-a-box problem is not simple. *American Journal of Physics*, 45(2), 158–160.
[13] Makri, N., & Miller, W. H. (1989). A semiclassical tunneling model for use in classical trajectory simulations. *The Journal of Chemical Physics*, 91(7), 4026–4036.
[14] LibreTexts (Textmap of Petrucci's Book), H., W., J., Petrucci, R. H., & Mosher, M. D. (1996). *General Chemistry* (Vol. 2). New Jersey: Prentice-Hall.
[15] Mehndiratta, P., Jain, A., Srivastava, S., & Gupta, N. (2013). Environmental pollution and nanotechnology. *Environment and Pollution*, 2(2), 49.
[16] Kim, D., Shin, K., Kwon, S. G., & Hyeon, T. (2018). Synthesis and biomedical applications of multifunctional nanoparticles. *Advanced Materials*, 30(49), 1802309.
[17] Zhang, X., Brynda, M., Britt, R. D., Carroll, E. C., Larsen, D. S., Louie, A. Y., & Kauzlarich, S. M. (2007). Synthesis and characterization of manganese-doped silicon nanoparticles: Bifunctional paramagnetic-optical nanomaterial. *Journal of the American Chemical Society*, 129(35), 10668–10669.
[18] Charpentier, S., Kassiba, A., Emery, J., & Cauchetier, M. (1999). Investigation of the paramagnetic centres and electronic properties of silicon carbide nanomaterials. *Journal of Physics: Condensed Matter*, 11(25), 4887.
[19] Pankhurst, Q. A., Connolly, J., Jones, S. K., & Dobson, J. (2003). Applications of magnetic nanoparticles in biomedicine. *Journal of Physics D: Applied Physics*, 36(13), R167.
[20] Du, X., Yao, Y., & Liu, J. (2013). Structural architecture and magnetism control in metal oxides via surface-grafting techniques. *Journal of Nanoparticle Research*, 15(7), 1–8.
[21] Hong, R., Li, J., Wang, J., & Li, H. (2007). Comparison of schemes for preparing magnetic Fe_3O_4 nanoparticles. *China Particuology*, 5(1–2), 186–191.
[22] Li, Y., Jiang, R., Liu, T., Lv, H., Zhou, L., & Zhang, X. (2014). One-pot synthesis of grass-like Fe_3O_4 nanostructures by a novel microemulsion-assisted solvothermal method. *Ceramics International*, 40(1), 1059–1063.
[23] Kulkarni, S. A., Sawadh, P. S., Palei, P. K., & Kokate, K. K. (2014). Effect of synthesis route on the structural, optical, and magnetic properties of Fe_3O_4 nanoparticles. *Ceramics International*, 40(1), 1945–1949.
[24] Meng, H., Zhang, Z., Zhao, F., Qiu, T., & Yang, J. (2013). Orthogonal optimization design for preparation of Fe_3O_4 nanoparticles via chemical coprecipitation. *Applied Surface Science*, 280, 679–685.

[25] Choi, S. U., & Eastman, J. A. (1995). *Enhancing Thermal Conductivity of Fluids With Nanoparticles* (No. ANL/MSD/CP-84938; CONF-951135-29). Argonne National Lab.(ANL), Argonne, IL (United States).
[26] Haendler, B. L. (1982). Presenting the Bohr atom. *Journal of Chemical Education*, 59(5), 372.
[27] Suto, K. (2009). True nature of potential energy of a hydrogen atom. *Physics Essays*, 22(2), 135–139.
[28] Jacobson, J., Björk, G., Chuang, I., & Yamamoto, Y. (1995). Photonic de Broglie waves. *Physical Review Letters*, 74(24), 4835.
[29] Arndt, M., Nairz, O., Vos-Andreae, J., Keller, C., Van der Zouw, G., & Zeilinger, A. (1999). Wave-particle duality of C60 molecules. *Nature*, 401(6754), 680–682.
[30] Steuernagel, O. (2002). de Broglie wavelength reduction for a multiphoton wave packet. *Physical Review A*, 65(3), 033820.
[31] Halbritter, J., Repphun, G., Vinzelberg, S., Staikov, G., & Lorenz, W. J. (1995). Tunneling mechanisms in electrochemical STM-distance and voltage tunneling spectroscopy. *Electrochimica Acta*, 40(10), 1385–1394.
[32] Lehmann, G., & Taut, M. (1972). On the numerical calculation of the density of states and related properties. *Physica Status Solidi (B)*, 54(2), 469–477.
[33] Lang, J. K., Baer, Y., & Cox, P. A. (1981). Study of the 4f and valence band density of states in rare-earth metals. II. Experiment and results. *Journal of Physics F: Metal Physics*, 11(1), 121.
[34] McMillan, W. L., & Rowell, J. M. (1965). Lead phonon spectrum calculated from the superconducting density of states. *Physical Review Letters*, 14(4), 108.
[35] Wegner, F. (1981). Bounds on the density of states in disordered systems. *Zeitschrift Für Physik B: Condensed Matter*, 44(1), 9–15.
[36] Rohlfing, M., & Louie, S. G. (2000). Electron-hole excitations and optical spectra from first principles. *Physical Review B*, 62(8), 4927.
[37] Klemens, P. G. (1951). The thermal conductivity of dielectric solids at low temperatures (theoretical). *Proceedings of the Royal Society of London. Series A. Mathematical and Physical Sciences*, 208(1092), 108–133.
[38] Cuevas, J. C., & García-Vidal, F. J. (2018). Radiative heat transfer. *ACS Photonics*, 5(10), 3896–3915.
[39] Liu, J. L., & Bashir, S. (2015). *Advanced nanomaterials and their applications in renewable energy* (pp. pp.1.44). New York: Elsevier.
[40] Kieffer, S. W. (1979). Thermodynamics and lattice vibrations of minerals: 1. Mineral heat capacities and their relationships to simple lattice vibrational models. *Reviews of Geophysics*, 17(1), 1–19.
[41] Howell, J. R., Mengüç, M. P., Daun, K., & Siegel, R. (2020). *Thermal Radiation Heat Transfer* (pp. 1–52). Baton Rouge: CRC Press.
[42] Alivisatos, A. P., Harris, T. D., Carroll, P. J., Steigerwald, M. L., & Brus, L. E. (1989). Electron–vibration coupling in semiconductor clusters studied by resonance Raman spectroscopy. *The Journal of Chemical Physics*, 90(7), 3463–3468.
[43] Narisetti, R. K., Ruzzene, M., & Leamy, M. J. (2012). Study of wave propagation in strongly nonlinear periodic lattices using a harmonic balance approach. *Wave Motion*, 49(2), 394–410.
[44] Kumar, M., Chen, H., Sarsaiya, S., Qin, S., Liu, H., Awasthi, M. K., & Taherzadeh, M. J. (2021). A review of current research trends on micro-and nano-plastics as an emerging threat to the global environment. *Journal of Hazardous Materials*, 409, 124967.
[45] https://www.lawyersgunsmoneyblog.com/2021/05/racial-polarization-and-american-politics
[46] Hu, J., He, D., Zhang, X., Li, X., Chen, Y., Wei, G., & Luo, Y. (2022). National-scale distribution of micro (meso) plastics in farmland soils across China: Implications for environmental impacts. *Journal of Hazardous Materials*, 424, 127283.
[47] Dhineshbabu, N. R., Manivasakan, P., & Rajendran, V. (2014). Hydrophobic and thermal behavior of nylon six nanofibre web deposited on cotton fabric through electrospinning. *Micro & Nano Letters*, 9(8), 519–522.

[48] Peng, M., Li, L., Xiong, J., Hua, K., Wang, S., & Shao, T. (2017). Study on surface properties of polyamide 66 using atmospheric glow-like discharge plasma treatment. *Coatings*, 7(8), 123.
[49] Choi, D., Yeo, J. S., Joh, H. I., & Lee, S. (2018). Carbon nanosheet from polyethylene thin film as a transparent conducting film: "Upcycling" of waste to organic photovoltaics application. *ACS Sustainable Chemistry & Engineering*, 6(9), 12463–12470.
[50] Gabrilovich, D. I., & Nagaraj, S. (2009). Myeloid-derived suppressor cells as regulators of the immune system. *Nature Reviews Immunology*, 9(3), 162–174.
[51] Sang, W., Zhang, Z., Dai, Y., & Chen, X. (2019). Recent advances in nanomaterial-based synergistic combination cancer immunotherapy. *Chemical Society Reviews*, 48(14), 3771–3810.
[52] Leach, D. R., Krummel, M. F., & Allison, J. P. (1996). Enhancement of antitumor immunity by CTLA-4 blockade. *Science*, 271(5256), 1734–1736.
[53] https://www.nobelprize.org/prizes/medicine/2018/press-release/
[54] Waldman, A. D., Fritz, J. M., & Lenardo, M. J. (2020). A guide to cancer immunotherapy: From T cell basic science to clinical practice. *Nature Reviews Immunology*, 20(11), 651–668.
[55] Sahin, U., & Türeci, Ö. (2018). Personalized vaccines for cancer immunotherapy. *Science*, 359(6382), 1355–1360.
[56] Khalil, D. N., Smith, E. L., Brentjens, R. J., & Wolchok, J. D. (2016). The future of cancer treatment: Immunomodulation, CARs, and combination immunotherapy. *Nature Reviews Clinical Oncology*, 13(5), 273–290.
[57] Xu, X., Lu, H., & Lee, R. (2020). Near-infrared light-triggered photo/immuno-therapy toward cancers. *Frontiers in Bioengineering and Biotechnology*, 8, 488.
[58] Kulkarni, J. A., Witzigmann, D., Thomson, S. B., Chen, S., Leavitt, B. R., Cullis, P. R., & Van Der Meel, R. (2021). The current landscape of nucleic acid therapeutics. *Nature Nanotechnology*, 16(6), 630–643.
[59] Peer, D., Karp, J. M., Hong, S., Farokhzad, O. C., Margalit, R., & Langer, R. (2020). Nanocarriers as an emerging platform for cancer therapy. *Nano-Enabled Medical Applications*, 61–91.
[60] Chen, Y., Wang, L., & Shi, J. (2016). Two-dimensional non-carbonaceous materials-enabled efficient photothermal cancer therapy. *Nano Today*, 11(3), 292–308.
[61] Irvine, D. J., Swartz, M. A., & Szeto, G. L. (2013). Engineering synthetic vaccines using cues from natural immunity. *Nature Materials*, 12(11), 978–990.
[62] Chen, D. S., & Mellman, I. (2013). Oncology meets immunology: The cancer-immunity cycle. *Immunity*, 39(1), 1–10.
[63] Weng, Y., Liu, J., Jin, S., Guo, W., Liang, X., & Hu, Z. (2017). Nanotechnology-based strategies for the treatment of ocular disease. *Acta Pharmaceutica Sinica B*, 7(3), 281–291.
[64] Sang, W., Zhang, Z., Dai, Y., & Chen, X. (2019). Recent advances in nanomaterial-based synergistic combination cancer immunotherapy. *Chemical Society Reviews*, 48(14), 3771–3810.
[65] Li, Y. J., Luo, L. J., Harroun, S. G., Wei, S. C., Unnikrishnan, B., Chang, H. T., & Huang, C. C. (2019). Synergistically dual-functional nano eye drops for simultaneous anti-inflammatory and anti-oxidative treatment of dry eye disease. *Nanoscale*, 11(12), 5580–5594.
[66] Mobaraki, M., Soltani, M., Zare Harofte, S., L Zoudani, E., Daliri, R., Aghamirsalim, M., & Raahemifar, K. (2020). Biodegradable nanoparticle for cornea drug delivery: Focus review. *Pharmaceutics*, 12(12), 1232.
[67] Calvo, P., Vila-Jato, J. L., & Alonso, M. J. (1996). Comparative in vitro evaluation of several colloidal systems, nanoparticles, nanocapsules, and nanoemulsions, as ocular drug carriers. *Journal of Pharmaceutical Sciences*, 85(5), 530–536.
[68] Chang, W. C., Hawkes, E. A., Kliot, M., & Sretavan, D. W. (2007). In vivo use of a nano knife for axon microsurgery. *Neurosurgery*, 61(4), 683–692.
[69] William D. Townsend, O. D. (2011). How Nanotechnology Will Revolutionize Eye Care, Nov 18, https://www.reviewofcontactlenses.com/issue/november-2011-1691

[70] Chen, J., Patil, S., Seal, S., & McGinnis, J. F. (2006). Rare earth nanoparticles prevent retinal degeneration induced by intracellular peroxides. *Nature Nanotechnology*, 1(2), 142–150.
[71] Pajic, B., Pajic-Eggspuchler, B., & Haefliger, I. (2011). Continuous IOP fluctuation recording in normal-tension glaucoma patients. *Current Eye Research*, 36(12), 1129–1138.
[72] Nanotechnology Initiative, https://www.nano.gov
[73] Perutz, M. F., Fermi, G., Luisi, B., Shaanan, B., & Liddington, R. C. (1987). Stereochemistry of cooperative mechanisms in hemoglobin. *Accounts of Chemical Research*, 20(9), 309–321.
[74] Hong, C., Alser, O., Gebran, A., He, Y., Joo, W., Kokoroskos, N., & Hammond, P. T. (2022). Modulating Nanoparticle Size to Understand Factors Affecting Hemostatic Efficacy and Maximizing Survival in a Lethal Inferior Vena Cava Injury Model. *ACS Nano*.
[75] Bakhaidar, R., O'Neill, S., & Ramtoola, Z. (2020). PLGA-PEG nanoparticles show minimal risks of interference with the platelet function of human platelet-rich plasma. *International Journal of Molecular Sciences*, 21(24), 9716.
[76] Hammond, S. M., Aartsma-Rus, A., Alves, S., Borgos, S. E., Buijsen, R. A., Collin, R. W., & Arechavala-Gomeza, V. (2021). Delivery of oligonucleotide-based therapeutics: Challenges and opportunities. *EMBO Molecular Medicine*, 13(4), e13243.
[77] Hong, C., Olsen, B. D., & Hammond, P. T. (2022). A review of treatments for non-compressible torso hemorrhage (NCTH) and internal bleeding. *Biomaterials*, 121432.
[78] Hammond, P. T. (2004). Form and function in multilayer assembly: New applications at the nanoscale. *Advanced Materials*, 16(15), 1271–1293.
[79] Venkataraman, S., Hedrick, J. L., Ong, Z. Y., Yang, C., Ee, P. L. R., Hammond, P. T., & Yang, Y. Y. (2011). The effects of polymeric nanostructure shape on drug delivery. *Advanced Drug Delivery Reviews*, 63 (14–15), 1228–1246.
[80] Huerta, T. S., Devarajan, A., Tsaava, T., Rishi, A., Cotero, V., Puleo, C., & Chavan, S. S. (2021). Targeted peripheral focused ultrasound stimulation attenuates obesity-induced metabolic and inflammatory dysfunctions. *Scientific Reports*, 11(1), 1–12.
[81] Nicot, J. P., & Scanlon, B. R. (2012). Water use for shale-gas production in Texas. *Environmental Science & Technology*, 46(6), 3580–3586.
[82] Nandlal, K., & Weijermars, R. (2022). Shale well factory model reviewed: Eagle Ford case study. *Journal of Petroleum Science and Engineering*, 212, 110158.
[83] Xu, B., Haghighi, M., & Cooke, D. (2012, March). Production data analysis in Eagle Ford shale gas reservoir. In *SPE/EAGE European Unconventional Resources Conference & Exhibition-From Potential to Production* (pp. cp–285). European Association of Geoscientists & Engineers.
[84] Raterman, K., Liu, Y., Roy, B., Friehauf, K., Thompson, B., & Janssen, A. (2020, December). Analysis of a multi-well eagle ford pilot. In *Unconventional Resources Technology Conference* 20–22 July 2020 (pp. 19–38). URTEC: Unconventional Resources Technology Conference.
[85] Colosimo, F., Thomas, R., Lloyd, J. R., Taylor, K. G., Boothman, C., Smith, A. D., & Kalin, R. M. (2016). Biogenic methane in shale gas and coal bed methane: A review of current knowledge and gaps. *International Journal of Coal Geology*, 165, 106–120.
[86] Howarth, R. W. (2019). Ideas and perspectives: Is shale gas a major driver of the recent increase in global atmospheric methane? *Biogeosciences*, 16(15), 3033–3046.
[87] Makal, T. A., Li, J. R., Lu, W., & Zhou, H. C. (2012). Methane storage in advanced porous materials. *Chemical Society Reviews*, 41(23), 7761–7779.
[88] Eddaoudi, M., Kim, J., Rosi, N., Vodak, D., Wachter, J., O'Keeffe, M., & Yaghi, O. M. (2002). Systematic design of pore size and functionality in isoreticular MOFs and their application in methane storage. *Science*, 295(5554), 469–472.
[89] Bashir, S., Hanumandla, P., Huang, H. Y., & Liu, J. L. (2018). Nanostructured materials for advanced energy conversion and storage devices: Safety implications at end-of-life disposal. In *Nanostructured*

Materials for Next-Generation Energy Storage and Conversion (pp. 517–542). Berlin, Heidelberg: Springer.

[90] Niederberger, M. (2007). Nonaqueous sol-gel routes to metal oxide nanoparticles. *Accounts of Chemical Research*, 40(9), 793–800.

[91] Niederberger, M., & Garnweitner, G. (2006). Organic reaction pathways in the nonaqueous synthesis of metal oxide nanoparticles. *Chemistry – A European Journal*, 12(28), 7282–7302.

[92] Loy, D. A., Mather, B., Straumanis, A. R., Baugher, C., Schneider, D. A., Sanchez, A., & Shea, K. J. (2004). *Effects of pH on the Gelation Time in the Sol-Gel Polymerization of 1, 6-Bis (trimethoxysilyl) Hexane* (No. LA-UR-04-2104). Los Alamos National Lab.(LANL), Los Alamos, NM (United States).

[93] Abdi-Khanghah, M., Adelizadeh, M., Naserzadeh, Z., & Barati, H. (2018). Methane hydrate formation in the presence of ZnO nanoparticle and SDS: Application to transportation and storage. *Journal of Natural Gas Science and Engineering*, 54, 120–130.

[94] Liu, J. L., & Bashir, S. (2015). Advanced nanomaterials and their applications in renewable energy.

[95] Song, Q., & Zhang, Z. J. (2004). Shape control and associated magnetic properties of spinel cobalt ferrite nanocrystals. *Journal of the American Chemical Society*, 126(19), 6164–6168.

[96] Zeng, H., Rice, P. M., Wang, S. X., & Sun, S. (2004). Shape-controlled synthesis and shape-induced texture of $MnFe_2O_4$ nanoparticles. *Journal of the American Chemical Society*, 126(37), 11458–11459.

S. Bashir, Y. Huang, S. Palakurthi, Z. Wang, Y. Olaseni, P. Villarreal, W. Houf, M. Alexander, J. Liu*

Chapter 3 Wet-chemistry-derived nanomaterials and their multidisciplinary applications

3.1 Overview of bottom-up synthesis

Bottom-up synthesis is a foundational approach in nanomaterials engineering that builds complex structures from atomic or molecular precursors. Unlike top-down

Acknowledgments: The research efforts led by Dr Bashir and Dr Liu were supported by the American Chemical Society Petroleum Research Fund (Grant 53827-UR10) federal agencies, including the Office of Naval Research, the Air Force Office of Scientific Research, the Department of Energy, the National Science Foundation, the Department of Education, the Department of Homeland Security (ORISE), and the National Institutes of Health. The Robert A. Welch Foundation (Departmental Grant AC-0006) is gratefully acknowledged for its support of student training. Technical assistance from Texas A&M University-Kingsville and the Texas A&M Energy Institute was invaluable. Finally, we thank our colleagues for their insightful discussions and the reviewers for their constructive comments, which significantly enhanced the rigor and clarity of this chapter.

Author contribution: *Conceptualization and supervision*: S. Bashir and Dr J. Louise Liu jointly conceived the project and oversaw its execution. *Investigation*: RPE characterization and cancer-toxicity assays: S. Bashir; nanostructured photocatalyst design for pharmaceutical-waste treatment: S. Bashir and J. Louise Liu; polymeric membrane synthesis and microbial fuel cell (MFC) assembly: J. Louise Liu; ANN-based fingerprint development and analysis: Z. Wang; water-bioremediation experiments: M. Alexander and J. Ren.

Training and mentorship: Y. Huang, S. Palakurthi, Y. Olaseni, P. Villarreal, and W. Houf participated as protégés in nanotechnology research under project supervision.

Writing and editing: J. Louise Liu drafted the manuscript; all authors contributed to revisions and approved the final version.

All authors have read and approved the submitted manuscript.

***Corresponding author:** **J. Liu,** Department of Chemistry, Texas A&M University-Kingsville, MSC161, 700 University Blvd., Kingsville, TX, USA; Louise Nanoinnovation North American, 7534 Beau Terre, Corpus Christi, TX, USA; Texas A&M Energy Institute, 1617 Research Parkway, Suite 308, College Station, TX 77843-3372 USA, e-mails: jingbo.liu@tamuk.edu, jingbo.liu@tamu.edu

Sajid Bashir, Department of Chemistry, Texas A&M University-Kingsville, MSC161, 700 University Blvd., Kingsville, TX, USA; Louise Nanoinnovation North American, 7534 Beau Terre, Corpus Christi, TX, USA; Texas A&M Energy Institute, 1617 Research Parkway, Suite 308, College Station, TX 77843-3372 USA

Y. Huang, S. Palakurthi, Y. Olaseni, P. Villarreal, W. Houf, Department of Chemistry, Texas A&M University-Kingsville, MSC161, 700 University Blvd., Kingsville, TX, USA

Z. Wang, The Department of Electrical Engineering and Computer Science, North Carolina A&T State University, Greensboro, NC, USA

M. Alexander, Department of Chemical Engineering, Texas A&M University-Kingsville, MSC 161, 700 University Blvd., Kingsville, TX, USA

 | https://doi.org/10.1515/9783111621111-004

methods, which carve nanoscale features from bulk solids, bottom-up techniques rely on chemical reactions that assemble nanoparticles from the ground up, starting with atoms, ions, or small molecules. These precursors undergo controlled decomposition, releasing reactive species that initiate nucleation. Once nucleation begins, atoms or ions aggregate into stable clusters, which grow into monodispersed colloids through processes governed by thermodynamics and kinetic control. This method allows precise tuning of particle size, shape, and crystallinity, often yielding uniform nanostructures with tailored surface chemistry. The resulting materials exhibit enhanced properties, such as quantum confinement, high surface reactivity, and tunable optical or magnetic properties, which are critical for applications in catalysis, electronics, and biomedicine. Sol–gel synthesis, chemical vapor deposition, and hydrothermal growth are standard bottom-up techniques, each offering unique control over morphology and composition. By manipulating reaction conditions, such as temperature, pH, solvent polarity, and ligand environment, researchers can guide the self-assembly of nanoparticles into functional architectures. Bottom-up synthesis thus represents not just a fabrication strategy but also a conceptual shift: from sculpting matter to orchestrating the atom-by-atom emergence of the building blocks of next-generation technologies [1].

Bottom-up synthesis is a powerful strategy in nanotechnology that builds materials from the atomic or molecular level, allowing structures to emerge through self-assembly driven by chemical and physical forces. Rather than carving down bulk materials, this approach starts with fundamental building blocks, atoms, ions, or small molecules that interact and organize into nanostructured arrays. These interactions, often governed by van der Waals forces, electrostatic attraction, or covalent bonding, enable the formation of highly ordered and uniform nanomaterials. One of the key advantages of bottom-up methods is their ability to produce nanostructures with fewer defects compared to top-down techniques. Because the material forms under controlled chemical conditions, researchers can precisely control its composition, crystallinity, and surface functionality. This leads to molecular homogeneity, which is critical for applications in electronics, catalysis, and biomedicine. Moreover, bottom-up synthesis allows fine-tuning of particle size and shape, which directly influences properties such as optical absorption, magnetic behavior, and catalytic activity. Techniques such as sol–gel processing, chemical vapor deposition, and molecular self-assembly exemplify this approach, offering scalable, versatile routes to engineer materials with tailored performance. For both experts and newcomers, bottom-up synthesis is a cornerstone of modern nanoscience, in which complexity arises from simplicity and function begins at the most fundamental scale [2].

The bottom-up approaches evaluate "specific characteristics and micro attributes of an individual stock." Bottom-up synthesis refers to the construction of nanomaterials by assembling atoms, ions, or molecules into larger, organized structures through chemical or physical interactions. Unlike top-down methods that reduce bulk materials to nanoscale dimensions, bottom-up approaches begin with fundamental building blocks and guide their self-assembly into clusters or arrays. This process allows for precise control over the material's composition, morphology, and functional properties. As individual units come together, they form nanoscale products with enhanced performance characteristics, such as increased surface area, tailored reactivity, and quantum effects absent in bulk counterparts. Scientists often evaluate the specific chemical and structural attributes of these precursors to influence the final architecture. The result is a material whose properties can be finely tuned for applications in catalysis, electronics, medicine, and energy [3]. The final products were nanoscale because the assembly of basic units formed clusters that formed larger structures. Bottom-up (also called self-assembly) approaches were used for nanofabrication using chemical or physical forces [4]. This method is widely used to construct desired structures from atom to atom. In the "bottom-up" method, the starting materials were in the gas, liquid, or solid state. The bottom-up processes essentially encompass chemical synthesis, producing a controlled deposition and materials growth [5]. The associated chemical synthesis was carried out via reactions in the solid, liquid, or gaseous state. Solid precursors (or reactants) are mixed by intimate contact with different reactants in solid-state synthesis. However, this bottom-up approach had disadvantages, such as complexity, time-consuming nature, and lower effectiveness. The bottom-up method embraces diverse techniques and a simplified mechanism [6], as summarized in Scheme 3.1. In this section, we briefly discuss the top-down syntheses used in our laboratory.

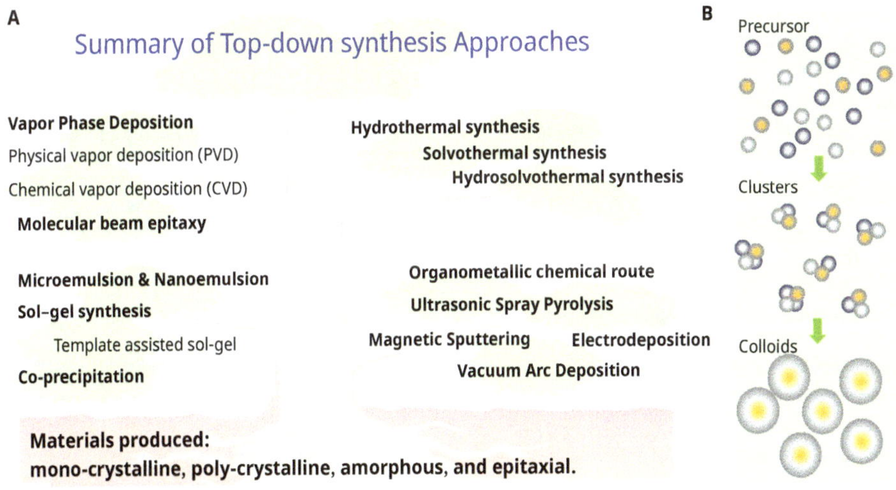

Scheme 3.1: Bottom-up synthesis is used to produce monocrystalline, polycrystalline, amorphous, and epitaxial materials. **A.** Summary of different methods and **B.** a simplified mechanism illustrates the molecular precursors that undergo decomposition, nucleation (re-nucleation), and growth to form clusters and colloids.

3.2 Colloidal synthesis

A colloid is a type of mixture in which tiny, insoluble particles are evenly distributed throughout another substance, known as the dispersing medium. These particles are small enough to remain suspended without settling, yet large enough to scatter light, a phenomenon known as the Tyndall effect, which gives colloids a cloudy or opalescent appearance. Additionally, the particles exhibit Brownian motion, meaning they move randomly due to collisions with molecules in the surrounding medium. Colloids are not uniform like solutions; they are heterogeneous, often showing visible boundaries between phases under certain conditions. Both the dispersed particles and the dispersing medium can be solids, liquids, or gases, allowing for a wide variety of colloidal systems. Based on the physical states of these components, colloids are classified into six major types: aerosols, emulsions, and sols. This versatility makes colloids essential in fields ranging from medicine and food science to nanotechnology and materials engineering (Tab. 3.1). The suspended particles are small, ranging from 1 nm to 1 µm [7]. Colloidal synthesis relies on the precipitation of nanosized particles within a continuous fluid solvent matrix to form a colloidal sol (a misnomer, but commonly used). A finely monodispersed system is in a high-energy state, equivalent to the free energy required to increase its surface area. The colloidal materials tend to aggregate due to the attractive van der Waals forces (Fig. 3.1A). The presence and magnitude of an energy barrier to agglomeration depend on the balance of attractive and repulsive forces between the particles. The energy of a colloidal solution arises from the Brownian motion (Fig. 3.1B) and in the order of K_bT.

Tab. 3.1: A summary of different colloids based on the dispersed phases [8].

Type of colloid	Dispersed phase	Dispersing medium	Examples
Aerosol	Liquid	Gas	Fog
Aerosol	Solid	Gas	Smoke
Foam	Gas	Liquid	Whipped egg
Solid form	Gas	Solid	Marshmallow
Emulsion	Liquid	Liquid	Soy milk
Solid emulsion	Liquid	Solid	Butter and cheese
Sol	Solid	Liquid	Cell fluid
Solid sol	Solid	Solid	Opal and gemstones

3.2.1 Colloidal stabilization

Stabilizing colloidal dispersions against aggregation is essential for applications across diverse fields, such as sustainable energy, the food industry, and personal care products. The aggregation includes coagulation (irreversible particle growth) and flocculation (reversible). Often, the system is an oil-in-water dispersion stabilized by interfacial and ac-

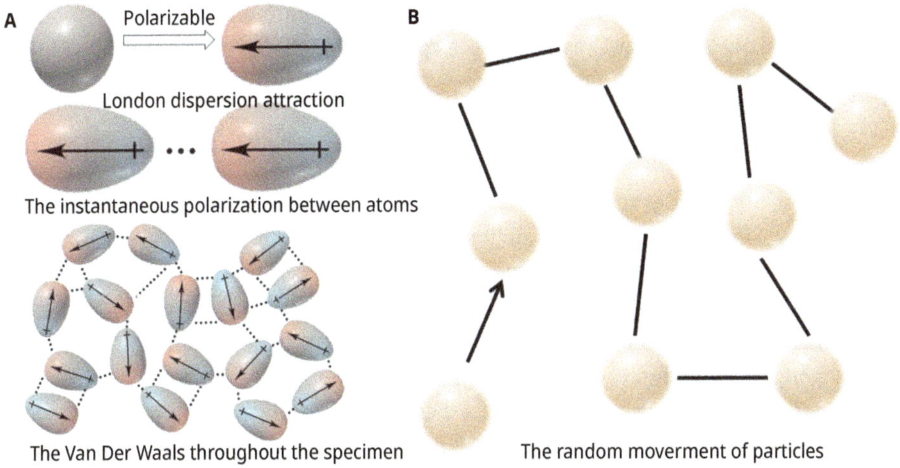

Fig. 3.1: Key characteristics of colloidal systems. **A.** The van der Waals forces from instantaneous polarization of non-polar molecules, leading to weak but cumulative attractions between dispersed particles. **B.** The Brownian motion, driven by thermal agitation, causes erratic particle movement that sustains dispersion and prevents settling.

tive components, such as amphiphiles or proteins [9]. These components segregate to the oil–water interface and stabilize emulsions by reducing interfacial tension. The enhanced rigidity and elasticity of the membrane that forms also help prevent coalescence. Colloidal sols found in paints and pastes also need to be stabilized for extended shelf life; this was achieved in several ways [10]. First, the balance between the repulsive electrostatic and attractive van der Waals contributions to the total potential energy was adjusted for charged colloidal particles in an electrolyte medium. A barrier to aggregation is created. The phenomenon of charge stabilization was analyzed using the Derjaguin–Landau–Verwey–Overbeek (DLVO) theory [11]. The DLVO theory suggests that a particle's stability in solution depends on its total potential energy function, V_T. This theory recognizes that V_T is the balance of several competing contributions (eq. (3.1A)):

$$V_T = V_A + V_R + V_S \tag{3.1A}$$

where V_A and V_R are the attractive and repulsive contributions; V_S is the potential energy due to the solvent, which usually makes only a marginal contribution to the total potential energy over the last few nanometers of separation. Much more important is the balance between attraction and repulsion. In principle, these two forces are much larger and act over much greater distances. The attractive force is associated with particle separation (D) and was determined using eq. (3.1B):

$$V_A = -A/\left(12\ \pi D^2\right) \tag{3.1B}$$

where A is the Hamaker constant and D is the particle separation.

The repulsive potential V_R is a far more complex function, estimated by eq. (3.1C):

$$V_R = 2\ \alpha \varepsilon r \xi^2 \exp(-\kappa D) \tag{3.1C}$$

where r is the particle radius, α is the solvent permeability, κ is a function of the ionic composition, and ξ is the zeta potential [12].

The DLVO theory predicts the stability of a colloidal system, which is determined by the sum of the van der Waals attractive (VA) and electrical double-layer repulsive (V_R) forces, as mentioned above. Both forces exist between particles as they approach each other due to their Brownian motion. This DLVO theory proposes that an energy barrier arising from repulsive forces prevents two particles from approaching one another and adhering (Fig. 3.1A). However, if the particles collide with sufficient energy to overcome that barrier, the attractive force will bring them into contact, where they adhere strongly and irreversibly, resulting in coagulation. Therefore, the colloidal system was stable if the particles had a sufficiently high repulsion to resist flocculation. However, if a repulsion mechanism no longer exists, flocculation or coagulation will eventually take place [13].

There are five postulations in the DLVO theory: 1) infinite flat solid surface, 2) uniform surface charge density, 3) no redistribution of surface charge (surface electric potential remains constant), 4) no change of concentration profiles of both counter-ions and surface charge-determining ions (the electric potential remains constant), and 5) solvent exerts influences via dielectric constant (no chemical reactions between the particles and solvent). These assumptions will lead to deviations when the DLVO theory is used to explain observations. This theory is widely used to provide guidelines for controlling the stability of the colloidal system. By adjusting the repulsive and attractive forces, the colloids were stable when the repulsion was more influential than attraction. Two fundamental mechanisms, steric and electrostatic repulsion, are significant factors in dispersion stability [14].

Steric repulsion: This repulsive force is achieved by adding polymeric dispersing agents to the system, which stabilize the particles. The long-chain polymer was adsorbed onto the particle surface, preventing particles from coming into close contact with one another. If enough polymeric molecules were adsorbed, the coating thickness would be sufficient to overcome van der Waals forces. Therefore, the particles were separated by steric repulsions between the polymer layers (Fig. 3.2A).

Electrostatic repulsion: This mechanism is also called charge stabilization, reflecting the distribution of charge species. These repulsive Coulomb interactions between particles can counterbalance van der Waals forces (Fig. 3.2B). When two charges move along parallel paths, electrostatic repulsion occurs. These moving entities will form two parallel currents and attract each other due to magnetic forces. The magnetic forces reduce the electrostatic repulsion; on the other hand, the space-charge limit was increased [15].

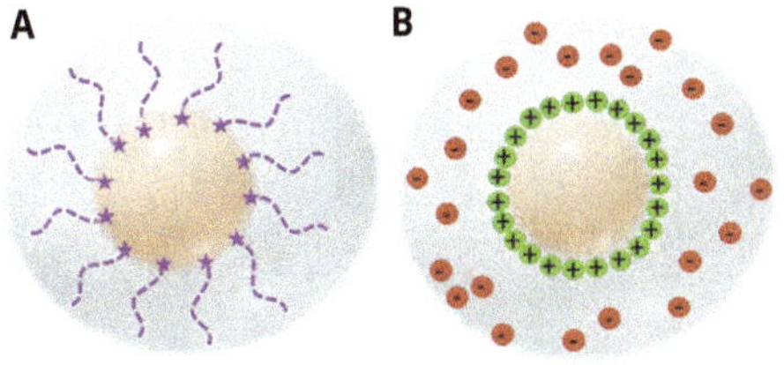

Fig. 3.2: The fundamental mechanisms influencing colloidal stability. **A.** The steric repulsion arises from surface-bound polymers preventing particle overlap, and **B.** electrostatic repulsion results from charged surfaces generating interparticle forces that maintain dispersion and avoid aggregation.

In our research lab, steric stabilization has been widely used to control the size and morphology of the nanoparticles. The fatty acid, a family of long-chain molecules, was chemically attached to or physically adsorbed onto colloidal particles, creating a repulsive force due to steric hindrance (Fig. 3.3A). The chains act as dispersing agents, interpenetrating as particles approach one another. This steric stabilization is highly applicable in nonaqueous media, unlike charge stabilization. A wide range of particles was produced with controllable size distribution and uniformity. The concentration and characteristics of surfactants need to be selected carefully to achieve steric stabilization (see **Chapter 1**, Section **2.2**). At low concentrations, it was found that polymeric surfactant chains can attach themselves to two (or more) particles, resulting in flocculation. At higher polymer concentrations, non-adsorbed polymeric molecules can cause depletion flocculation, as recognized by Asakura and Osawa [16].

In general, the colloids with high zeta potential (ζ) are stable for the long term due to the high degree of electrostatic repulsion between adjacent particles. Consequently, the charged particles in dispersion are separated to avoid close contact or agglomeration (Fig. 3.3B). As an indicator of colloidal stability, the small potential suggests that attractive forces may overcome the repulsion, leading to dispersion collapse and ultimately flocculation. Either a high negative or a high positive zeta potential confers colloidal stability [17]. The schematic diagram (Fig. 3.3C) illustrates the relationship between free energy and particle agglomeration, as predicted by DLVO theory. The net energy was determined by summing the double-layer repulsion and the van der Waals attractions. The particles experience both forces, with the repulsive forces dominating as they approach each other. The net energy increased exponentially as particle agglomeration increased.

Most colloidal dispersions in aqueous media carry an electric charge on the surface; the origin of the charge depends on the surrounding medium or the nature of these particles. There are three types of origins of surface charge: ionization of surface group, ion loss from the crystal lattice, and adsorption of charged species. The ionization of surface groups primarily depends on their acidity or basicity. When the acidic groups on the particle surface dissociate, a negatively charged surface is formed. The particle surface can absorb some surfactants; in this case, if it is a cationic surfactant, then a positively charged surface is obtained. On the other hand, an-

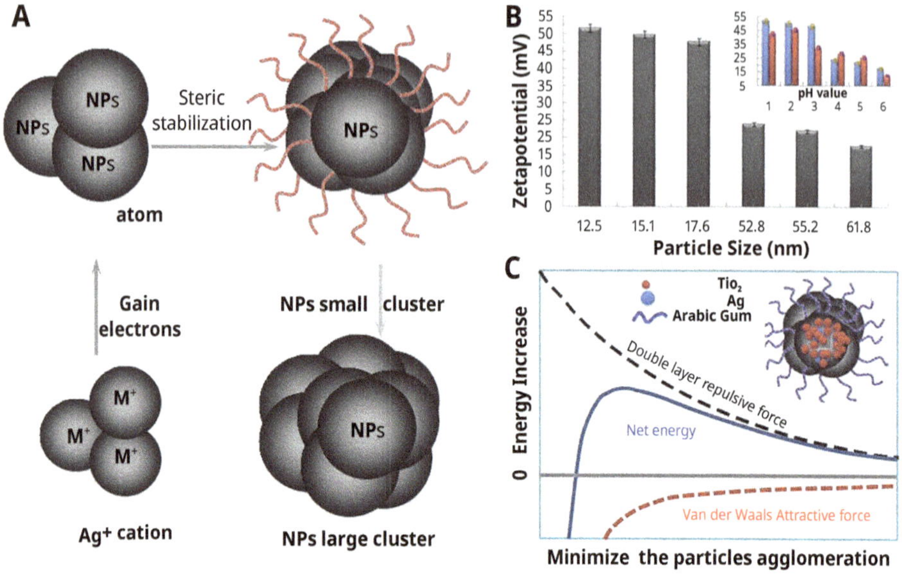

Fig. 3.3: Diagram of stabilizing colloidal dispersions against aggregation by adding interfacially active components. **A**. The steric stabilization, resulting in uniform nanoparticle formation, **B**. zeta potential as a measure of ultrafine particle stability, and **C**. the net energy depends on the balance between attractive and repulsive forces. Image adapted from ref. [18].

ionic surfactants will make particles negatively charged [19]. Theoretically speaking, the zeta potential is the potential difference between the dispersion medium and the stationary layer of fluid attached to the dispersed particle. This electrokinetic potential occurred in the interfacial double layer (DL) [20], particularly at the location of the slipping plane relative to a point in the bulk fluid away from the interface (Fig. 3.4A). The liquid layer around the particles was observed to consist of a Stern layer (inner region) and a Gouy layer (outer diffuse region), as shown in Fig. 3.4B. The ions are bonded to the particles; conversely, they are loosely distributed in the diffusion layer. Only the ions within the boundary move when the particles move, due to gravity. However, ions beyond the boundary stay with the liquid dispersant.

As mentioned above, the zeta-potential (ζ) is the electrokinetic potential at the slipping plane. The magnitude ζ can indicate the stability of particles. Because particles with significant positive or negative ζ charges tend to repel each other, the aggregate behavior will not be present in this situation. The aggregation can easily happen when the zeta potential is small, owing to the absence of a repulsive force. Generally speaking, the absolute value of zeta potential over 30 mV contributes toward a stable colloid [21]. Two factors, the net electrical charge and the location of the slipping plane, determine the magnitude of the zeta potential. The zeta potential is measured at the slipping plane, not at the particle surface, and is a key indicator of

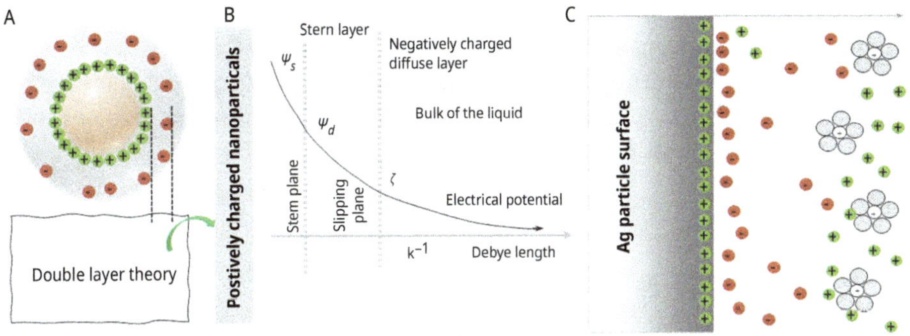

Fig. 3.4: Schematic demonstration showing the ionic concentration and potential difference. **A.** The double-layer theory explains the zeta potential and ion distribution in a colloidal suspension; **B.** the function of the distance from the charged surface of a particle suspended in a dispersion medium; and **C.** a case study of Ag nanoparticle colloidal suspension with a positive surface charge. Images were adapted from ref. [22].

colloidal stability. As shown in Fig. 3.4B, the Stern potential differs from the electric surface potential because of their different locations Image: schematic of surface and Stern potentials. This distinction is important for understanding how electrostatic stabilization controls particle size in colloidal systems.

Electrostatic stabilization of charged colloidal particles is another essential method to control particle size. In this mechanism, four perspectives are: **1)** surface charge density, **2)** electric potential at the proximity of the solid surface, **3)** van der Waals attraction potential, and **4)** Interactions between two particles: DLVO theory (see the section on Derjaguin, Landau, Verwey, and Overbeek theory (DLVO theory).

Surface charge density: When a solid is dispersed in a polar solvent (also known as an electrolyte solution), a charge is developed on the surface through one or more mechanisms. The mechanisms include preferential adsorption of ions, dissociation of surface-adsorbed charged species, isomorphic substitution of ions, accumulation or depletion of surface electrons, and physical adsorption of charged species. The Nernst equation is commonly used to determine the electric potential (E) at the surface of a solid in a given system) (eq. 3.2A):

$$E = E_0 + \left(\frac{R_g T}{n_i} F\right) \ln(a_i) \tag{3.2A}$$

where E_0 is the standard electrode potential (in V); R_g, the ideal gas constant (0.02816 atm L/mol K); T, the absolution temperature in kelvin; F, the Faraday constant (96,485 C/mol, or kJ/mol); and a_i represents the valence state of ions (dimensionless).

In particular, the surface charge in oxides depends on the types of ions adsorbed on the solid surface. This surface charge was derived from ions' preferential dissolution or deposition, which determines the charge of the ions (also called co-ions). Two ions, proton (H^+ or hydronium H_3O^+) and hydroxide (OH^-) are the commonly known determining co-ions in the typical metal oxide system. The surface change and concentrations of the co-ions are intercorrelated. The point of zero charges (PZC) and acidity (pH values) are two key factors determining the surface potential of the typical metal oxide system (eq. (3.2B)). The PZC values of oxide systems in an aqueous solution are tabulated in Tab. 3.2. The system pH was calculated using eq. (3.2C):

$$E = 2.303 R_g T(PZT - \mathrm{pH})/F \tag{3.2B}$$

$$\mathrm{pH} = \log([\mathrm{H_3O}]^+) \tag{3.2C}$$

Tab. 3.2: A summary of zero charge (PZC) points of metal oxides in aqueous solution [23].

Metal oxides	PZC	Metal oxides	PZC
Fe_2O_3	8.6	TiO_2 (calcined)	3.2
ZnO	8.0	TiO_2	6.0
Cr_2O_3	8.4	SnO_2	4.5
ZrO_2	6.7	WO_3	0.5
V_2O_5	1.0–2.0	β-MnO_2	7.3
SiO_2	2.5	δ-MnO_2	1.5
SiO_2 (quartz)	3.7	Al–O–Si	6.0

The electric potential at the interface of the solid surface: When a surface charge density of a solid surface is established, there is an electrostatic force between the solid surface and the charged species in the proximity to segregate positive and negatively charged species. However, Brownian motion (Fig. 3.1B) and entropic forces also act to homogenize the distribution of various species in the solution. Two oppositely charged ions, surface charge-determining ions and counterions, always exist in the solution. Although the system of interest maintains charge neutrality, the distributions of charge-determining ions and counterions are inhomogeneous and vary with proximity to the solid surface. The distributions of both ions are mainly controlled by the following forces: coulombic (also known as electrostatic) force, entropic (or dispersion) force, and Brownian motion [23].

The net ion concentration depends on the magnitudes of the determining ions and counterions. If the surface is positively charged, the ions that determine the surface charge have the lowest concentration near the surface and increase with distance from the surface. On the contrary, the counterions show the highest concentration near the solid surface, decreasing with increasing distance from it. A so-called double-layer structure is formed due to these inhomogeneous ion distributions in the proximity of the solid surface (Fig. 3.4B) [24]. The DL consists of the Stern and Gouy layers (also called the diffuse DLs). The Stern layer is formed between the solid surface and the Helmholtz plane. In this Stern layer, the electric potential decreases linearly through the tightly bound layer of solvent and counterions. Beyond the Helmholtz plane, the Gouy layer persists until the counterions reach the solvent's average concentration. In this Gouy diffuse layer, the counterions diffuse freely, and the electric potential does not decrease linearly with distance. The electric potential (E) change is inversely proportional to the Debye–Hückel screening strength ($1/\kappa$) and was approximated in eq. (3.3A). The Debye–Hückel screening strength, corresponding to the solvent dielectric constant, valence of counterion, and counterion concentrations, was determined using eq. (3.3B):

$$E \propto e^{(-\kappa(h-H))} \tag{3.3A}$$

$$\kappa = \left\{ \frac{(F^2 \sum_i C_i Z_i^2)}{\varepsilon_r \varepsilon_0 R_g T} \right\}^2 \tag{3.3B}$$

where κ^{-1} is known as the Debye–Hückel screening strength; ε_r, the dielectric constant of the solvent; ε_0, the permittivity of vacuum; C_i, the concentration of the counterion of type I; and Z_i, the valence of the counterion of type I.

Based on the above exponential equations (3.3A) and (3.3B), the increase in counterion concentration results in a decrease in electric potential in the proximity of the solid surface. When the valence state (Z_i) increases, the electric potential (E) also exponentially decreases. On the other hand, the increase in the dielectric constant of the solvent ε_r and absolute temperature collectively increases the electric potential (E). The thickness of the DLs also highly depends on the valence and concentration of the counterion and the dielectric constant of the solvents [25]. The Gouy diffuse layer ended at the zero point of electrical potential, where its distance from the solid surface was infinite. However, in reality, the DL is approximately 10 nm thick; otherwise, the system's stability will be compromised. The above discussion assumes that the solid surface is flat and the electrolyte solutions are applicable. If the surface is sufficiently smooth, the above discussion explains the observation of curved surfaces due to their constant surface charge densities. If the particles are spherical, these assumptions hold because they are evenly dispersed in an electrolyte solution. Their considerable separation does not affect each other's particle charge distributions. In reality, the interactions between particles are much more complex than in the above discussion. The electric potential and surface charge are two factors that influence particle interactions, such as electrostatic repulsion (ϕ_R), as approximated by eq. (3.3C). This electrostatic repulsion arises from the electric surface charges of the particles when they come into contact. The double-layer thickness can result in varying attenuation. The DL shows no overlap when the particles are far apart, indicating that electrostatic repulsion is negligible. When particles approach each other, repulsion dominates, keeping them separated and preventing agglomeration (eq. 3.3C):

$$\phi_R = 2\pi\varepsilon_r\varepsilon_0 R E^2 \exp(-S\kappa) \tag{3.3C}$$

Van der Waals attraction potential: If the particles are sized in micrometers or less (known as nanoparticles), they are readily dispersed in a solvent. Under these conditions, van der Waals forces and Brownian motion play important roles [26]. The gravitational influence of these nanoparticles was ignored due to their ultrasmall size. The van der Waals attraction is an intermolecular nonbonding force that is relatively weak compared to the bonding forces. This attractive force exists only over a short distance, typically between adjacent molecules. The combined van der Waals attraction and Brownian motion result in particle agglomeration. Brownian motion is associated with the thermal motion of particles, and van der Waals attraction results from molecular interactions. The total interaction energy was calculated using eq. (3.4A). The negative sign (–) in the equation indicates the intrinsic nature of the attractive forces between two adjacent particles [27]:

$$\phi_A = -\frac{A}{6}\left\{\frac{2r^2}{(S^2+4rS)}+\frac{2r^2}{(S^2+4rS+4r^2)}+\ln\left[\frac{(S^2+4rS)}{(S^2+4rS+4r^2)}\right]\right\} \tag{3.4A}$$

where ϕ_A stands for the total interaction energy or attraction potential; A, a positive constant termed the Hamaker constant (in the order of 10^{-19} to 10^{-20} J, the values of representative materials listed in Tab. 3.3); r stands for the radius of the spherical particles; and S, a distance of two separated particles [28].

A simplified equation (3.4B) was used if the distance of particles is significantly smaller than the radius of the particle with equal sizes ($S \ll r$):

$$\phi_A = -\frac{Ar}{12S} \tag{3.4B}$$

If the two spherical particles have different radii, r_1 and r_2, the equation for the total interaction energy can be simplified to eq. (3.4C):

$$\phi_A = -\frac{Ar_1r_2}{6S(r_1+r_2)} \tag{3.4C}$$

If two parallel particles are flat in shape with a plate thickness of δ, the total interaction energy per unit area was determined using eq. (3.4D):

$$\phi_A = -\frac{A}{12p[(S)^{-2}+(S+2d)^{-2}+(S+d)^{-2}]} \tag{3.4D}$$

If two identical spheres are separated by a distance S, the total interaction energy per unit area is determined using eq. (3.4F) [29]:

$$\phi_A = -\frac{A}{12p(S)^{-2}} \tag{3.4F}$$

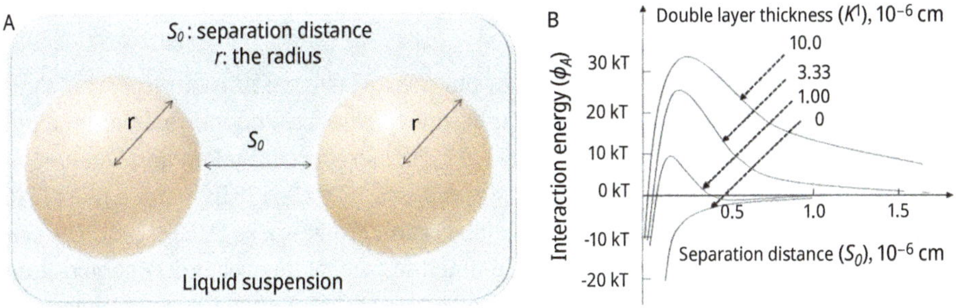

Fig. 3.5: Electrostatic stabilization: **A.** van der Waals interactions between two molecules over two spherical particles with equal size, and **B.** relationship between electric potential and separation distance between two particles.

Tab. 3.3: Hamaker constants (a_i (10^{-20} J)) in a vacuum and across the water (vacuum/water) of inorganic materials interacting against four materials [30].

Materials	Hamaker constants (a_i (10^{-20} J))			
	Silica	**Silicon nitride**	**Alumina**	**Mica**
$BaTiO_3$	10.1/0.62	16.5/4.84	15.2/3.55	12.4/1.98
BeO	9.67/0.95	15.4/3.87	14.8/3.50	11.9/2.06
C (diamond)	13.7/1.71	22.0/7.94	21.1/7.05	17.0/4.03
$CaCO_3$	8.07/0.69	12.9/2.53	12.3/2.17	9.94/1.35
CaF_2	6.70/0.45	10.6/1.17	10.3/1.10	8.26/0.73
CdS	8.03/0.72	13.1/3.12	12.0/2.15	9.86/1.43
KCl	5.94/0.37	9.53/0.73	9.00/0.51	7.31/0.46
$MgAl_2O_4$	9.05/0.85	14.5/3.39	13.8/2.97	11.2/1.79
MgF_2	6.15/0.36	9.74/0.66	9.42/0.69	7.57/0.50
MgO	8.84/0.81	14.2/3.26	13.5/2.79	10.9/1.69
Mica	8.01/0.69	12.8/2.45	12.2/2.15	9.86/1.34
NaC1	6.45/0.44	10.3/1.17	9.77/0.88	7.93/0.66
PbS	5.37/−0.08	8.88/0.64	7.90/−0.20	6.57/−0.03
6H-SiC	12.6/1.52	20.3/7.22	19.2/6.05	15.5/3.54
β-Si_3N_4	10.8/1.17	17.3/5.13	16.5/4.43	13.3/2.61
SiO_2 (quartz)	7.59/0.63	12.1/2.07	11.6/1.83	9.35/1.16
$SrTiO_3$	9.44/0.57	15.4/4.02	14.2/2.98	11.6/1.69
TiO_2	9.46/0.69	15.4/4.26	14.2/3.11	11.6/1.83
Y_2O_3	9.24/0.89	14.9/3.80	14.0/3.11	11.4/1.89
ZnO	7.38/0.58	12.0/2.30	11.1/1.58	9.06/1.08
ZnS (cubic)	9.69/1.02	15.7/4.56	14.6/3.55	11.9/2.19
3Y-ZrO_2	11.4/1.25	18.4/5.89	17.4/4.95	14.1/2.89

Note: Data collected at room temperature, calculated using Lifshitz theory [31].

3.2.2 Kinetic modeling

The colloidal compositional chemistries of nanoparticles have drawn researchers' attention (Fig. 3.5A), both in fundamental and applied research, due to their unique optical, electrical, and catalytic properties (Table 3.3), as reported by Ren and Duan et al. [32] (Fig. 3.5B). These nanomaterials were composed of metals [33], metal oxides [34], ceramics [35], semiconductors [36], and metal–organic frameworks (MOFs) [37]. The colloidal-derived nanoparticles were precisely tuned to control their chemical composition and architectures, including particle size, shape, crystal facets, and surface chemistry [38]. In most cases, ligands with small molecular masses play critical roles in modulating nanomaterial growth by interacting with precursors and nanoparticles. As a result, the structure and morphology of the final multicomponent nanoparticles were well-controlled. The composition and morphology of colloidal inorganic compounds highly depend on kinetics and thermodynamics [39]. For decades, the mechanisms of nanoparticle nucleation and growth have been described by the LaMer burst nucleation (Fig. 3.6A) [40], followed by Ostwald ripening (Fig. 3.6B) [41] to describe changes in particle size. Originally, this model was proposed by Reiss [42], with a well-accepted postulation being developed by Lifshitz–Slyozov–Wagner, LSW theory [43]. This theory was dominantly used to explain nucleation and growth until Watzky and Finke formulated the two-step approach. In 1997, Watzky and Finke proposed a two-step mechanism (Fig. 3.6C) for the formation of iridium (Ir) nanoparticles: slow, continuous nucleation

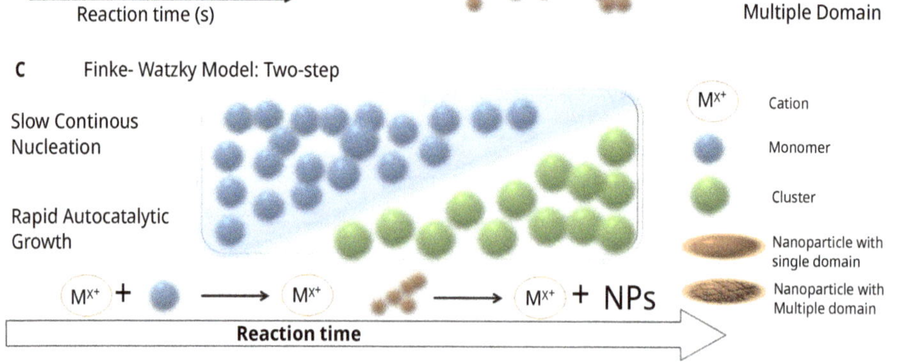

Fig. 3.6: The mechanism of nanoparticle formation: **A.** traditional LaMer burst nucleation model, **B.** Ostwald ripening, and **C.** Finke–Watzky two-step model.

followed by rapid autocatalytic surface growth [44]. The "pseudo-temporal separation of nucleation and growth" was observed because rapid growth occurred relative to slow nucleation. However, this Finke-Watzky two-step model did not consider the role of the ligand in explaining the overlap between nucleation and growth [45].

Mozaffari and coworkers reported a combined in situ small-angle X-ray scattering (SAXS) and kinetic modeling of colloidal metal nanoparticle size control [46]. The palladium (II) acetate ($Pd(CH_3CO_2)_2$, tetragonal structure) and trioctylphosphine ($P(C_8H_{17})_3$, abbreviated TOP) precursor were used for kinetic evaluation.

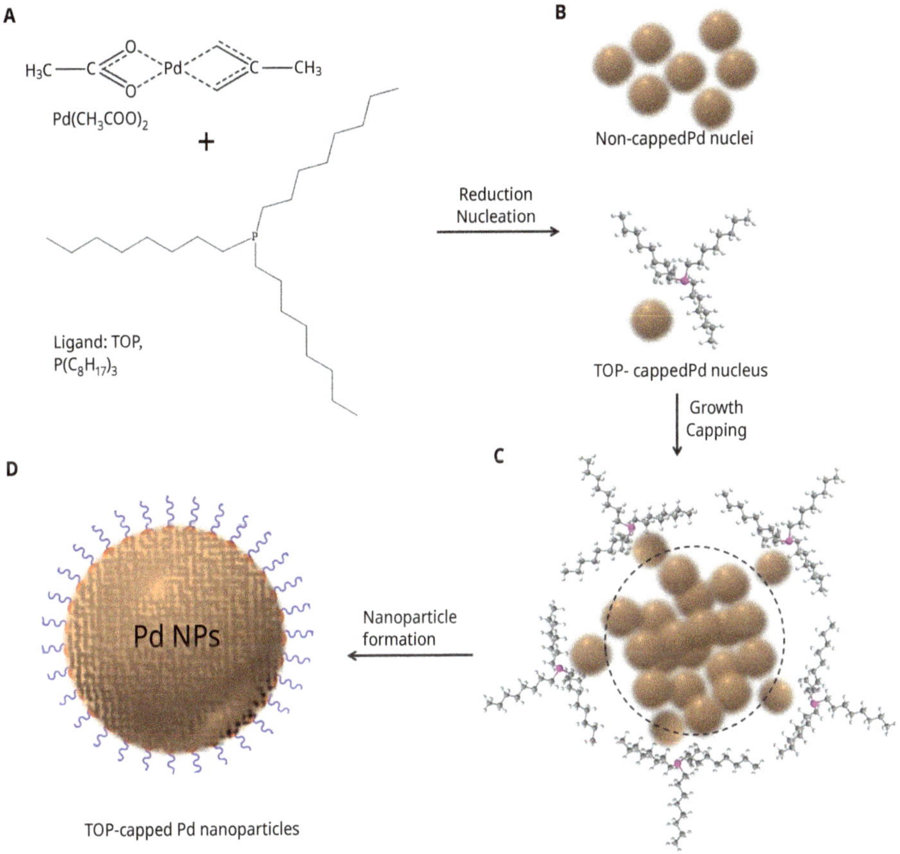

Fig. 3.7: The "ligand-based model" schematic diagram proposed by Mozaffari and coworkers: **A.** the Pd-containing reactant and trioctylphosphine (TOP) as a capping agent; **B.** the non-capped Pd nucleation and TOP-capped Pd nucleation; **C.** the continuous growth, capping, and formation of Pd nanoparticles.

This method quantitatively captured the roles of ligand-metal binding in controlling the nucleation and growth of nanoparticles on the surface. Their research data showed that the kinetic rate constant depends on the particle size and number [47]. The solvents used in the colloidal metal synthesis also affect the metal–ligand binding and the coverage of Pd nanoparticles. It was found that toluene, as a solvent, produced small particles (average 1.4 nm), whereas pyridine produced larger particles (4.3 nm) [48]. Metal concentration, ligand-to-metal ratio, and solvents affect the nucleation and growth kinetics of the metal nanoparticles. The author proposed a "ligand-based model" (Fig. 3.7 and eqs. (3.5A–D)), taking "the metal-ligand binding as reversible reactions" into consideration (Fig. 3.7A). This model is an advancement of the two-step mechanism (Fig. 3.7B) proposed by Finke and Watzky, which has been considered an improved mechanism (Fig. 3.7C) to explain metal nanoparticle synthesis (Fig. 3.7D) compared with the traditional LaMer model [49].

In this "ligand-based model," the authors proposed two additional reversible steps [39] and slow nucleation followed by rapid catalytic growth, as in the Finke-Watsky model. There are three postulations in this ligand-based model: (1) the fraction of atoms that become core atoms is not considered, (2) autocatalytic growth and binding of ligands are independent of nanoparticle size and ligand coverage, and (3) the stoichiometry of the ligand-metal precursor binding is approximated to be unit 1. This four-step model (eqs. (3.5A)–(3.5D)) provides a more accurate and thorough explanation for metal nanoparticle formation [50]. The first two equations ((3.5A) and (3.5D)) are aligned with the slow nucleation and rapid autocatalytic surface growth in the Finke–Watzky model. The reaction rate constant ($k_{1-\text{nul.}}$) in step 1 (eq. (3.5A)) is used to evaluate the combined pseudo-elementary reduction–nucleation. The reaction rate constant ($k_{2-G.}$) in step 2 (eq. (3.5B)) is for the autocatalytic surface growth. Similarly, in step 3 (eq. (3.5C)), the $k_{3-F.}$ is the forward reaction rate constant for the precursor–ligand binding, while the $k_{4-F.}$ It is associated with the ligand-particle binding in the forward reaction (eq. (3.5D)). Two equilibrium constants ($K_{3-\text{eq}}$) correspond to ligand–precursor (eq. (3.5C)) and ligand–nanoparticle (eq. (3.5D)) binding processes, respectively [51]:

$$\text{Nucleation and reduction: } A \overset{k_{1-N.}}{\rightarrow} B \tag{3.5A}$$

$$\text{Autocatalytical surface growth: } A + B \overset{k_{2-G.}}{\rightarrow} C \tag{3.5B}$$

$$\text{Ligand binding to precusor: } A + L \overset{k_{3-F.},\ K_{3-\text{eq}}}{\longrightarrow} A - L \tag{3.5C}$$

$$\text{Ligand binding to NP surface: } B + L \overset{k_{4-F.},\ K_{4-\text{eq}}}{\rightarrow} B - L \tag{3.5D}$$

The authors conducted experiments to examine the relationships between reaction kinetics and the TOP-Pd interaction during Pd nucleation and growth. It was found that the rate of Pd nucleation was fast at its early stage of formation (Fig. 3.8A); within 2,000 s, a rapid decrease was observed, followed by a subsequent slower decrease. During particle growth, the Pd nucleus growth rate increased sharply, reaching a maximum between 1,800 and 2,000 s. Afterward, an exponential decrease and a slight decrease were observed with prolonged reaction time (Fig. 3.8B). The ratio between growth and nucleation showed a trend similar to that of the growth rate (Fig. 3.8C). With the precursor ($Pd(CH_3COO)_2$) consumption, its concentration decreases, decreasing the autocatalytic surface growth rate. The percentage of TOP-capped Pd NPs on the surface is attributed to the ligand's dissociation rate (TOP), which is the reversible reaction of eq. (3.5C). In addition, with Pd formation (denoted as B), its surface percentage increases, leading to a decrease in the growth rate. The TOP capping influenced nanoparticle nucleation and growth (Fig. 3.8D). The "ligand-based model" includes metal–ligand interactions corresponding to the number of free surface sites B. The ligand–nanoparticle binding consequently affects the growth/nucleation ratio. The model enabled the accurate predictions of an increase in both the "fraction of precursor coordinated with TOP (AL) and in the coverage of TOP on the particle surface (BL)," as reported by Mozaffari and coworkers [39].

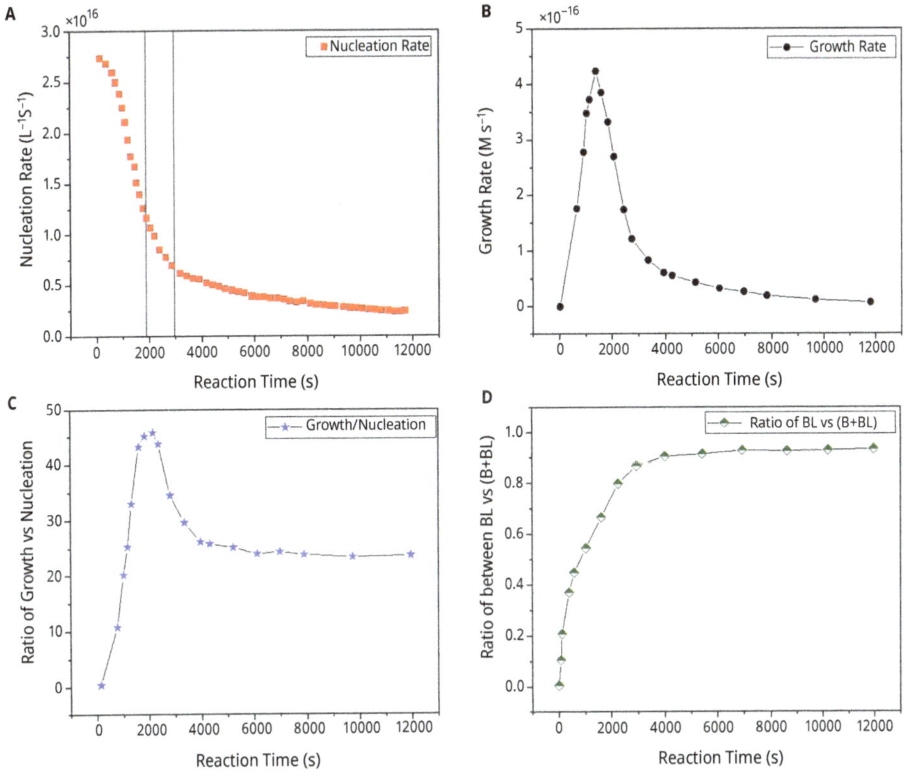

Fig. 3.8: The formation mechanism of Pd nanoparticles: **A**. nucleation rate; **B**. growth rate; **C**. ratio between growth versus nucleation; and **D**. (BL)/(B + BL) ratio (BL: precursor-ligand binding). Reaction conditions: 10 mM $Pd(OAc)_2$ in 50:50 toluene: hexanol, trioctylphosphine: Pd = 2:1, $T = 100$ °C. The images were adapted with permission from the ref. [39].

The shape and architecture of nanoparticles were tuned via colloidal synthesis using different templates and ligands. However, the small-molecule-assisted colloidal synthesis strategy showed intrinsic drawbacks and technical hindrances, such as 1) lattice mismatch in the multicomponent structures, 2) inability to produce unique structures, and 3) lack of stability. Due to lattice mismatch in multicomponent materials, lattice strain forms between neighboring lattice units, hindering the integration of diverse components into large structures [52]. The core-gap-shell nanostructures with highly ordered arrays are not obtainable due to the fixed interparticle distance and the gap between the core and shell [53]. Research indicated that the interparticle distance between two metallic entities affects the optical properties of the designated core–shell metallic structures. Researchers employed functional macromolecules to interact with precursors, both chemically and physically, to address these issues. These attractions between inorganic precursors and nanoparticles enabled the production of nanocrystals and the construction of multicomponent nanostructures [54].

3.2.3 Macropolymer templation

This section summarizes the structures and morphologies produced by colloidal wet chemistry using functional and reactive macromolecules as templating agents (Fig. 3.9). The macromolecules and polymeric colloidal systems exhibit intrinsic reactivity for modulating particle growth. Three families of polymeric agents were chosen to highlight the achievement using "macromolecule-enabled colloidal synthesis of nanomaterials," as reported by Duan and coworkers [32]. Based on the functions of macromolecules involved in the colloidal synthesis, the polymers are categorized into three types: (1) Synthetic block copolymer (Fig. 3.9A) [55], (2) bioinspired multifunctional adhesive polymers (Fig. 3.9B) [56], and (3) biomacromolecules (Fig. 3.9C) [57]. The properties of these functional polymers highly depend on the "pendant functional groups on their monomeric units." The chemical interactions between the monomer and inorganic precursors were similar to those observed in small-molecule binding. The fundamental mechanism of "small molecule-assisted synthesis" was applied to the system, using a functional macropolymer as a template. These macromolecules offered several intrinsic advantages for colloidal synthesis. First, the monomer's covalent coordination between functional groups gives a "multivalency effect," providing multiple reactive sites. These binding sites enable the immobilization of inorganic precursors onto the polymer backbone [58]. The large size of the functional macropolymer and coordination between precursors in the backbone of polymeric groups restrict the precursors' diffusion. The chemical transformation of starting materials (or inorganic precursors) was confined within a nanoscale region. Secondly, the multiple anchoring groups provided a pathway to stabilize the capping ligands, which showed affinity to the surface nanoparticles, as shown in Fig. 3.9C. The diversity of surface chemistry using functional polymers yields diverse morphologies of the final products, including nanospheres, nanorods, nanosheets, core–gap–shell, core–satellite, and nanoframeworks (Fig. 3.9D). In particular, the core–shell structure composed of dissimilar materials was produced using macromolecule-assisted colloidal synthesis. The significant lattice mismatch was overcome by using multifunctional groups to stabilize the nanoparticles' surface reactivity. Thirdly, multicomponent nanomaterials were made by integrating functional groups via different configurations. In this case, the functional macropolymer acts as a directing ligand to adjust the interfacial energy between adjacent monomers [59]. The growth rates of different components were integrated to produce multicomponent nanomaterials. Additionally, the polymeric modulations provide a different pathway to "immobilize inorganic precursors within their backbones." As a result, templated nanoparticle growth yielded single-crystalline nanoparticles. The macropolymer components are integrated into the lattice structure or become parts of the lattice, as observed in monodispersed colloids [60] (Fig. 3.9A).

In the colloidal suspension, the macro-sized functional polymers (Fig. 3.9B) occupy large volumes in collapsed and stretching conditions and act as "a spacing layer." [61] As a result, the multicomponents in the nanomaterials were separated to prevent agglomeration driven by thermal motion. The structural formula of the polymer is the determining factor of the thickness of these "spacing layers," which were adjusted from angstroms and nanoscales, depending on the molar mass and solubility in the medium of these polymers. Smart macro-sized polymers were responsive to external signals, such as pressure, temperature, light intensity, and electrical or magnetic stimuli, to adjust their structure and properties [62]. These functional polymers were used to tune the inorganic nanostructures by manipulating their molecular geometry, surface chemistry, and elemental compositions. Precise, accurate nanosynthesis control was achieved by applying macropolymers as spacing layers. In addition, the functional building blocks provide another pathway for synthesizing nanomaterials by varying their structure via a "living polymerization" strategy (Fig. 3.9C) [63]. These designed polymeric structures and architectures were manipulated into the form of nanospheres, nanosheets, nanofibers, and core shells (Fig. 3.9D). Since 1999, "click" chemistry has been used to engineer nanostructures with different formulations via bioconjugation, joining specific biomolecules via chemical bonding and intermolecular interactions. The intrinsic reactivities of the naturally occurring polymers have been utilized to generate colloidal products by mincing natural evolutions. The "click" chemistry was practical in chemoproteomic, pharmacological, and biomimetic fields, allowing for the detection, localization, and qualification of biomolecules [64].

The macro-sized polymeric templation enabled precise, accurate control of nanomaterials with tailorable catalytic, chemical, electrical, magnetic, and optical properties. The complex polymeric geometries were tailored using the intrinsic properties and functionality of naturally occurring biomacromolecules and natural product extracts. "Click" chemistry integrates macromolecules via chemical bonding or intermolecular interactions to produce structure-controlled nanomaterials with diverse formulations and architectures [65] (Fig. 3.9D).

3.3 Water remediation

The United Nations has designated 2005–2015 as the "Water for Life" decade and estimates that one-fifth of the world's population faces water shortages [66]. This complex problem requires multifaceted approaches to reduce water stress on photosynthesis, improve water-use efficiency, and enhance irrigation-water productivity [67]. The traditional debate on water scarcity focuses on surface water and groundwater (called blue) while assessing the shortage of evapotranspiration (green) water [68]. Both natural and anthropogenic factors collectively contribute to water stress, including population growth, industrialization, and globalization [69]. Green water is the primary source of food, feed, fiber, timber, and bioenergy, although blue water is used for irrigation [70]. Therefore, we must understand the factors causing water stress, with a focus on freshwater scarcity. Three variables, green water footprint (WF_g), maximum sustainable level of green water footprint ($WF_{g,m}$), and green water scarcity (WS_g), are used to express green water availability and appropriation.

The green water footprint (WF_g) reflects the human appropriation of green water for crop production, livestock grazing, wood production, and urban uses (Fig. 3.10A). To support current water consumption and maintain healthy biodiversity, the maximum sustainable level of the green water footprint ($WF_{g,m}$) was approximately 18×10^3 km^3/year [71]. It was found that 56% of the world's green water flow has already been allocated to human activities, leaving only sustainably available water (Fig. 3.10B) [72]. The darker shades of gray (WF_g) were close to ($WF_{g,m}$), meaning that the sustainable avail-

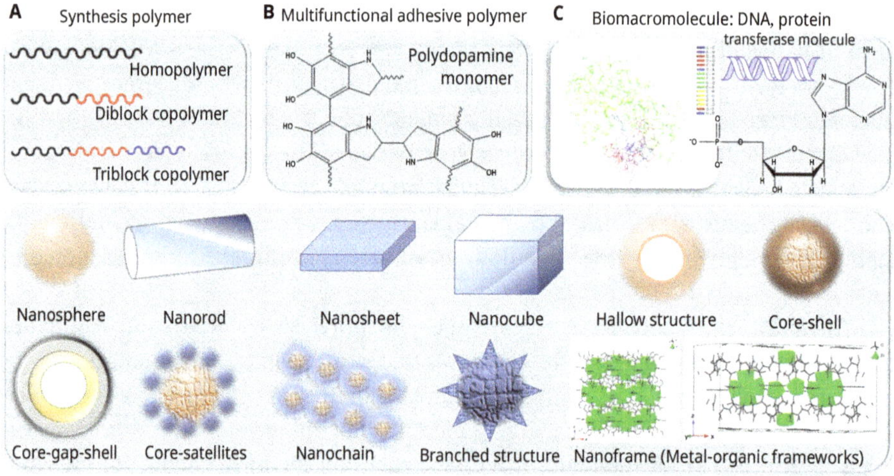

Fig. 3.9: Schematic illustration using functional macromolecule colloidal synthesis to tune the architecture of nanoparticles: **A**. Synthetic polymers (homopolymer, bi-block copolymer, and tri-block copolymer), **B**. multifunctional adhesive polymer providing anchoring effect and multiple binding sites, **C**. naturally existing biomacromolecules, e.g., deoxyribonucleic acid (DNA) and proteins, and **D**. 1D, 2D, and 3D structures including biomimetic structures.

able green water flow has been almost allocated to human use. Furthermore, the lighter shades of gray suggested a local increase in WF_g. Green water scarcity (WS_g) was expressed as the sustainable green water flow available and estimated using the national aggregate $WF_g/WF_{g,m}$ ratio (Fig. 3.10C). In Europe, Central America, the Middle East, and South Asia, green water scarcity is close to 1, indicating that sustainably available green water flow has been fully allocated to human activities. The remaining potential in another part of the country offsets the overshoot in the country of interest. Over half the overshoot occurred in just 10 countries (Fig. 3.10D). The countries, including the United Kingdom, Germany, Indonesia, and New Zealand, are known for ample rainfall [73]. Their $WF_{g,m}$ is limited and mostly or entirely allocated to human activities. In addition, the green water flows are reserved for nature, but are already mainly appropriated for agriculture. In short, humans and nature share the world's limited green water, and its explicit allocation is imperative to maintaining sustainable green water supplies [74].

Contamination of water bodies has become a serious problem for human beings, other living organisms, and other life forms, jeopardizing the sustainable future of humanity and the conservation of ecosystems [76]. Xenobiotics, microplastics, dyes, industrial waste, and discarded pharmaceuticals pose a significant threat to clean water, human consumption, and aquatic life in green and blue water bodies [77]. Exposure to toxic metals (mercury, cadmium, chromium, lead) for a longer time will cause various disorders in the human body, including decreased immunological defenses, disabilities associated with malnutrition, and a high rate of upper gastrointestinal cancer [78]. Therefore, water remediation requires appropriate measures to eliminate heavy metals and metalloids, degrade organic compounds, and disinfect biological contaminants to acceptable levels, thereby achieving a pollution-free environment. Attention has also been focused on developing green, sustainable technologies that must be aesthetically pleasing, eco-friendly, socially acceptable, and economical [79]. Water remediation is any process that improves water quality for specific end uses, including drinking, industrial supply, irrigation, river flow maintenance, and water recreation [80]. Water remediation technology generally involves chemical and physical treatment, biological treatment, and other technologies (such as desalination and portable water purification based on different constituents) [81], as shown in Fig. 3.11.

Recently, green technologies (bioremediation) have drawn significant attention for using microorganisms to transform hazardous contaminants into less toxic or nontoxic forms and further purify water [82]. Bioremediation is a decontamination process using plants, fungi, algae, bacteria, and other microorganisms to improve water quality. Natural or modified products can also alter the treatment process to ensure clean water. Bioremediation uses plants and microorganisms to break down contaminants into less harmful or nontoxic substances [83]. Bioremediation can purify groundwater, soil, and sludge, removing pesticides, hydrocarbons, organic compounds, and pharmaceutical chemicals. Bioremediation involves both autotrophic and heterotrophic methods. The former involves four categories: (1) phytoremediation using living plants to clean up soil, air, and water; (2) rhizofiltration using culti-

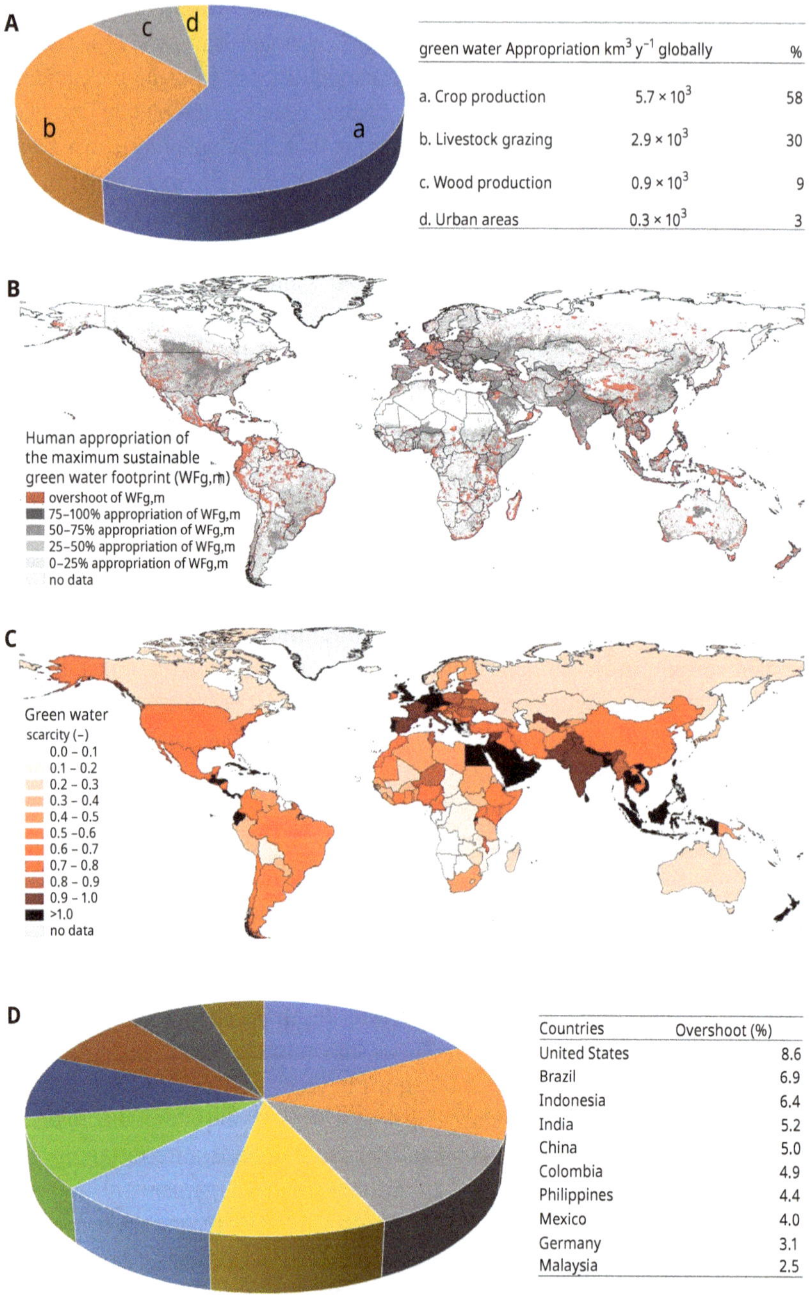

Fig. 3.10: Green water appropriation and maximum sustainable green water footprint. **A.** Human appropriation of the green water flow, **B.** the degree of human appropriation of the $WF_{g,m}$, at 5 × 5 arc-minute grid cell resolution, **C.** green water scarcity (WS_g), and **D.** half the overshoot occurs in just 10 countries (images were adapted from ref. [75] with permission).

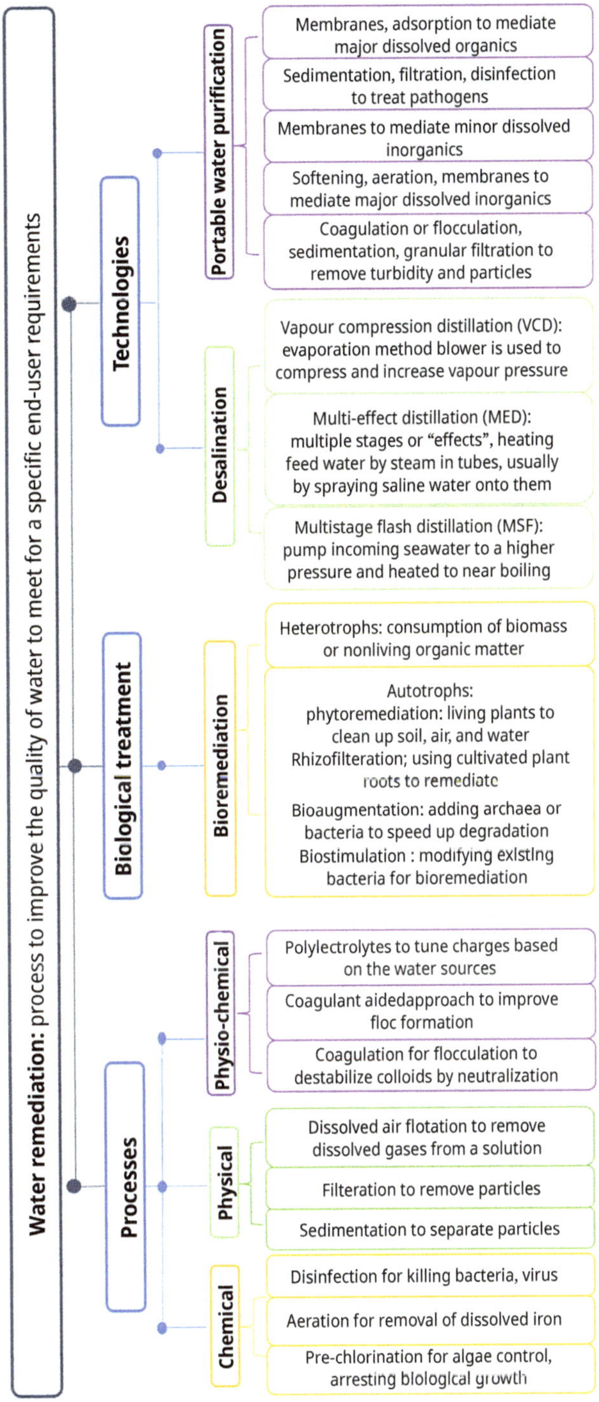

Fig. 3.11: A summary of water remediation technologies using chemical, physical, physicochemical, biological, and desalination methods.

vated plant roots to remediate water; (3) bioaugmentation by adding archaea or bacteria to speed up the degradation of contaminants; and (4) biostimulation through modifying existing bacteria for bioremediation [84].

Our investigations focused on phytoremediation, where the detoxification is carried out through green plants and their biomass as nanoparticle surface coatings. The bioadsorbent-tailored nanoparticles were effective in removing toxic metals by complexation [85]. Periphyton biofilms composed of prokaryotic *cocci* have also been used to treat aquatic ecosystems by removing chemical oxygen demand (COD), limiting aerobic growth, and inactivating cells, similar to 254-nm UV radiation on organic matter. Unlike UV irradiation, periphyton adsorbents can also remove toxic metals [86]. Natural product extracts also emerge as potential agents for the remediation of toxic metals from industrial effluents [87]. We selected natural extracts from polysaccharides, crab shells, and papaya wood. Using these product extracts, the modification of porous materials, such as zeolites, demonstrated selectivity for ammonia, transition metals, bacteria, and radioactive elements [85]. Nanomaterials as potential antimicrobial agents have been widely used to remove metals and organic contaminants. Various heavy metals were extracted using ionic liquids as excellent solvents to achieve metal-free water environments. The surfactant modification for removing oxoanions, such as chromate and oxalate, was effective at removing multiple wastewater contaminants [88]. The inorganic–organic hybrid polymeric materials prepared using the click-chemistry approach were found to have applications for water purification [89]. Such materials will lead to better, more sustainable treatment of polluted water. Most of these disinfection strategies have drawbacks, such as incomplete disinfection [90]. This chapter aims to identify a feasible and effective solution for eliminating impurities from contaminated water using three approaches: photocatalysis (Section 3.3.1), heavy-metal adsorption (Section 3.3.2), and bioinspired remediation (Section 3.3.3).

3.3.1 Photocatalysis in water remediation

Nanomaterials with high surface area, surface-to-volume ratio, and porosity will enhance catalyst reactivity and self-regulate structure and chemical properties at the macroscale of engineering [91]. The potential advantage of responsive nanosystems is that they facilitate treatment at the source of waste generation rather than relying on large, centralized treatment systems for degradation [92]. The leading-edge design of photocatalysts that respond rapidly to external stimuli, beyond current approaches for water remediation and pharmaceutical waste treatment, requires sound knowledge and a data-driven approach to ensure high photocatalytic activity and persistence [93]. A wet-chemistry method can fabricate mesoporous nanomaterials that uniformly “disperse” one component into the supporting substrate under moderate conditions, creating a tunable interface and nanoscale morphology [94]. Photocatalytic functionality is tuned with changes in band structure that govern the reactivity of photocatalysts at a wavelength window, where light energy is used to drive catalysis to yield reactive species (such as •OH radicals) by affecting the reactions between photoelectrons and the primary oxidant (O_2 or O_3) as well as holes at the interfacial region (e.g., H_2O) [95]. Although these complex relationships among synthesis, structures, and functions of engineered nanomaterials have been systematically investigated, the specific parameters for titania-based catalysis that enable self-regulation remain to be improved [96]. Therefore, optimizing fabrication variables and tailoring the band structure of photocatalytic systems while maintaining the structure and properties is critical to establishing the robust relationship between structure and function. It is crucial to develop catalysts that are both persistent and responsive [97]. Given the relationship between the number of charge carriers and their reactivity in photocatalytic reactions, it is essential to select catalytic components and optimize their design and synthesis parameters, using an extended network scaffold, such as graphene, to support them [98].

In collaboration with Dr C. Li, we developed sol–gel chemistry that produced TiO_2 nanoparticles with polyvinylpyrrolidone (PVP) templation and graphene oxides to produce nanocomposites [99]. The crystalline phase of TiO_2 was controlled to be anatase, using a porogen to tune porosity and pore size, thereby achieving high photocatalytic degradation of pharmaceutical waste. These heterogeneous composites were sintered at 400 °C to obtain crystalline TiO_2, while graphene oxide was converted to reduced graphene oxide (rGO) under ultraviolet radiation. Through four steps, hydrolysis, templation, crystallization, and reduction, the uniformly distributed nanocomposites were used to advance the degradation rate of tetracycline. Antibiotics (tetracycline) were used in food and feed to promote animal husbandry and treat bacterial infections. It is essential to note that the natural product extract, citric acid, acts as both a tri-topic ligand and a dispersing agent in the sol–gel process. The use of citric acid facilitated the formation of a coordinative complex between the Ti metal center and carboxylic groups. This coordination successfully prevented the rapid hydrolysis of $Ti(OnBu)_4$, thereby inhibiting particle growth during heat treatment. The water-soluble polymer PVP K30 (PVP-K30) was used to tune the polymerization phase of the Ti-precursor due to its intrinsic hygroscopic nature, adhesive properties, and stable acidity [100]. The five-numbered lactam, K30 strand, with a Guinier length (5.43 ± 0.01 nm) and contour length (34.31 ± 0.51 nm), was chosen due to its intrinsic properties, such as linearity and ease of bonding to a polar bond, O-Ti. At the same time, its molar mass averaged 40,000 g/mol to secure the formation of mesopore sizes. The uniform pore diameters and particle sizes were obtained, attributed to the above intrinsic properties of PVP-K30, particularly the pyrrolidone ring, high polarity, cohesivity, and resultant propensity. The ion-dipole moment between the Ti atom in the precursor and the O atom in PVP resulted in polymer-anchored TiO_2 hybrid materials. A continuous inorganic lattice, such as Ti–O–Ti or Ti–H–Ti [101], was formed to create an oxide matrix through polycondensation of a subatomic precursor [102]. Different amounts of PVP-K30 were introduced through controlled stoichiometry to tune the pore structures. The above "wet-gel" was converted into anatase TiO_2 polycrystals with further drying and heat treatment [103]. Based on the crystal field theory, the five d-orbitals of the Ti atom were subject to an energy increase when Ti was coordinatively bonded with bonding atoms, such as oxygen elements. It was anticipated that a rapid conversion from tetrahedral to octahedral units, driven by a decrease in splitting energy, would yield thermally stable anatase crystalline phases upon heat treatment. The last step was the reduction of graphene oxide (GO) under ultraviolet light, as demonstrated in our previous publication [101]. The Hummer method, derived from graphene, was ideal for incorporating the PVP-templated TiO_2 matrix; however, the functional groups reduced photocatalytic reactivity. The graphene substrate anchors PVP-templated TiO_2 via chemical bonding, primarily through dipole–dipole interactions [104]. This relatively weak attraction was readily overcome during heat treatment, leading to the formation of anatase TiO_2 on the graphene surfaces. Following ultraviolet radiation, the graphene oxide was reduced to remove a controlled amount of functional groups and impurities [105]. During this reduction, the hybridization of the carbon atoms changed from sp^3 to sp^2, creating unoccupied pz orbitals that facilitated electron transfer. The steric hindrance of citric acid and the cohesivity of PVP-K30 prevented the TiO_2 crystal growth at high heating temperatures. The homogeneity and architecture of RGO-supported TiO_2 nanocatalysts were tuned, enabling precise and accurate light photoreceptor activity in the UV-VL regimes. Multiscale texturation and upscaling production were achieved by combining sol–gel chemistry and heat treatment, followed by post-reduction thermal-dynamic treatment.

The crystalline structure obtained using X-ray diffraction showed that the reduced GO-TiO_2 composite and PVP-templated RGO-TiO_2 (Fig. 3.12A) were well-indexed with standard XRD patterns. At the diffraction angle of 13.10°, the diffraction pattern of GO was found, corresponding to the (001) plane. Upon reduction under UV radiation, diffraction at 26.60 was detected due to the formation of reduced GO. Although reduced graphene oxide was obtained in this approach, the XRD signal (~26°) overlapped with the central peak (101) of anatase TiO_2. The XRD results were well-indexed with the standard JCPDS 89-4921 (tetragonal, SG I41/amd (141), lattice parameters: a = 3.777 Å, c = 9.501 Å; α = 90°; cell volume = 136.300 Å^3), in agreement with published data [106]. The decrease in the graphene layer thickness (0.9–0.4) was observed with an increase in the C/O ratio and a reduction in C sp^3 hybridization, in which the sp^3 hybridization of carbon partially recovered to sp^2 [107]. The diffraction peaks of different samples appeared at 25.40° (101), 28.10° (004), 48.20° (200), 54.20° (105), and 62.70° (204), indicating that the tetragonal anatase TiO_2 was obtained with high purity and crystallinity. The in-depth study showed that the (101) diffraction angle and PVP-K30 amount form a volcano relationship. An increase in PVP concentration from 8 mass % to 11 mass % results in a red shift of the central peak on the (101) plane (Fig. 3.12B), which continues up to 12 mass%. This shift of the thermodynamically stable peak of TiO_2 corresponded to the tensile stress (<1.1%), which was found to occur between the boundary of the PVP template and RGO–TiO_2 composite.

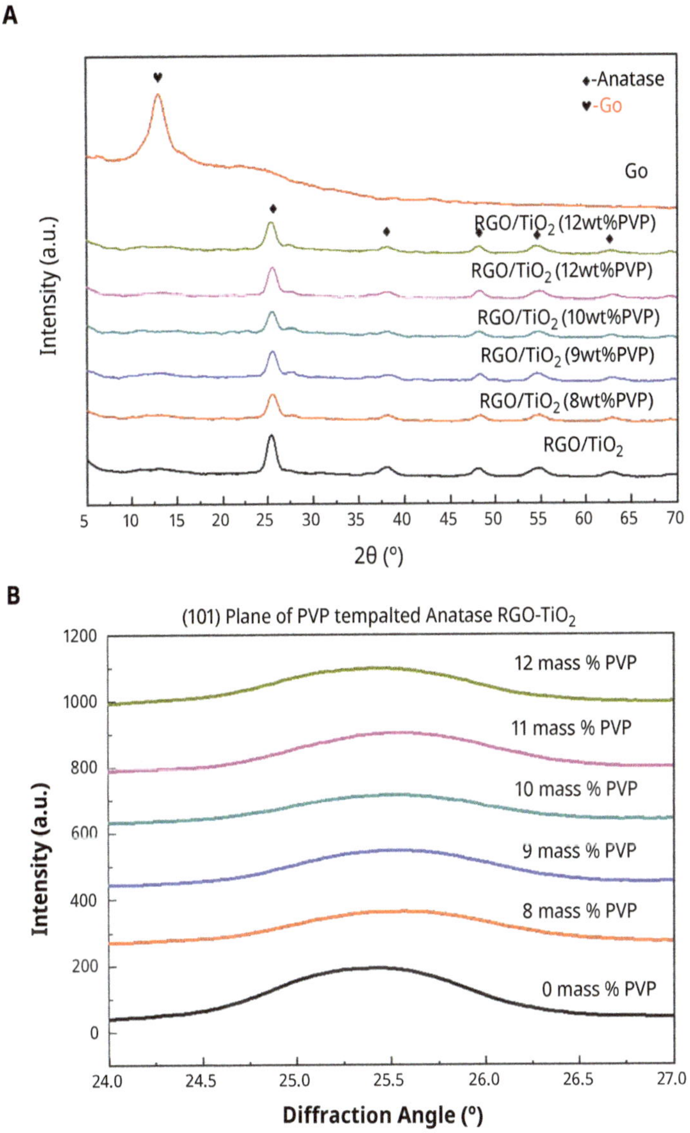

Fig. 3.12: XRD patterns of different TiO_2-based hybrid photocatalysts are mediated with PVP. **A.** Formation of GO, RGO–TiO_2 composite, PVP-K30 templated RGO–TiO_2 composite and **B.** The Williamson–Hall plot is used to estimate the crystallite size and effective strain of the RGO–TiO_2 composite.

The effective strain, ε, was calculated to be 0.964 by the Williamson-Hall plot (Fig. 3.13A), indicating that the lattice expansion occurs along the (101) plane. These defects serve as reactive chemical sites that allow tetracycline decomposition [99]. The positive slope of Fig. 3.13A suggested that the lattice expansion occurred linearly with the increased PVP-K30 amount. Another observation from the XRD data is the broadening of the (101) peak and others, suggesting a decrease in particle size (Eq. 3.6). The Debye–Scherrer (D–S) equation (3.6) was used to calculate the crystallite size of the PVP-templated nanocomposites, assuming negligible strain [108]. The Williamson–Hall (W–H) equation (3.7) calculated the crystallite sizes and lattice strains [109]. It was found that the effective strain is close to 1, indicating that the TiO_2 anatase structure was well-maintained, while the positive slope is associated with cell expansion. The correlation between particle size and the amount of PVP-K30 polymer is shown in Fig. 3.13B, indicating a volcano-shaped relationship. The XRD data on particle sizes also confirmed that a volcano trend of the ratiometry curve, indicative of the optimized amount of PVP-K30 (11 mass%, as shown by the red star in Fig. 3.13B), resulting in the smallest size (12.5 nm for RGO–TiO_2 and 11.0 nm for PVP-templated RGO–TiO_2) and lowest effective strain. With the PVP-K30 amount increase, an exponential increase in particle size (yellow region) was found, followed by a cliff drop. As a result, particle growth of nanosized TiO_2 was prevented by this polarity effect and steric hindrance [110]. With PVP above 11 mass%, the trend was different. Accordingly, this size decrease leads to increased photocatalytic reactivity, as evidenced by tetracycline degradation. Both Debye–Scherrer and Williamson–Hall methods must estimate only the crystallite sizes and effective strain, given the complex dependence of elastic constants and dislocation contrast factors on the polymer-templated TiO_2-based catalysts [111]:

$$D = K\lambda/\beta \cos\theta \tag{3.6}$$

$$\beta \cos\theta/\lambda = 1/D + \varepsilon \sin\theta/\lambda \tag{3.7}$$

where β is the full width at half maximum (FWHM), θ is the diffraction angle, λ is the X-ray wavelength, D is the average crystallite size, and ε is the effective strain. In all cases, the Scherrer formula yields a smaller size than that extracted from the W–H plot, as expected [112].

The high-resolution transmission electron microscopic (TEM) images (Fig. 3.14) showed well-crystallized particles and uniform pore size in the RGO–TiO_2 nanocomposite treated at different PVP concentrations (6–14 mass%). It was found that 11 mass % of PVP was the optimal porogen amount, with 5 mass% of RGO as the support, and all catalysts were tested under these conditions. The RGO–TiO_2 without templating has an average size of 10.5 nm (±0.15 nm), with high integrity, and is distributed on the surface of the RGO. The distinctive motifs of nanocomposites (Fig. 3.14A) were obtained after heat treatment, while some of their edges and corners were observed. These unique regions were found to act as reactive sites and to be polarized by the PVP polymeric template's polarity. The reduced graphene oxides (Fig. 3.14B) were observed as ultrathin layers with corrugated surfaces, which served as hosts for the TiO_2 nanoparticles (Fig. 3.14G). The ring pattern (Fig. 3.14D) was well-aligned with anatase TiO_2 polycrystals, while the selected area electron diffraction patterns (Fig. 3.14E)

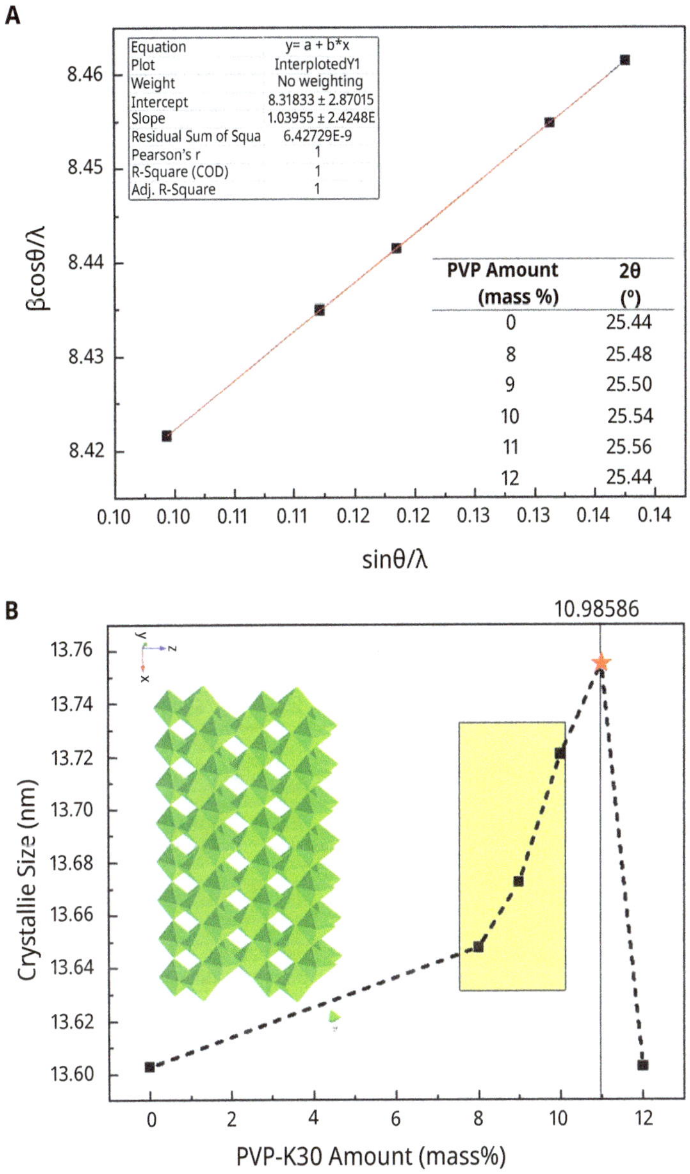

Fig. 3.13: XRD patterns of different TiO_2-based hybrid photocatalysts mediated with PVP. **A**. The Williamson–Hall plot estimates the crystallite size and effective strain of the RGO–TiO_2 composite (the inserted table shows the diffraction angle at which the peak splits). The relationship between crystallite size and PVP-K30 amount (the inserted diagram shows the anatase TiO_2 of 8 × 2 × 2 supercells).

confirmed that highly crystalline catalysts were obtained. Simulated SAED patterns for different sample thicknesses, including Laue front-plate (Fig. 3.14F), back-plate (Fig. 3.14G), and cylindrical scattering, were used to understand the topotactic relationships among the components in the nanocomposites. TEM characterization of delicate structures confirmed that the configurations of anatase RGO-TiO_2 composites with different formulations have undergone minor changes, with detectable local stress and crystallite-size variations induced by polymeric templating. In situ templation with the nonionic polar polymer PVP-K30 successfully tuned the morphology and delicate structures of the final products. Optimizing synthesis variables can regulate the relationship between nanostructures and catalytic reactivity.

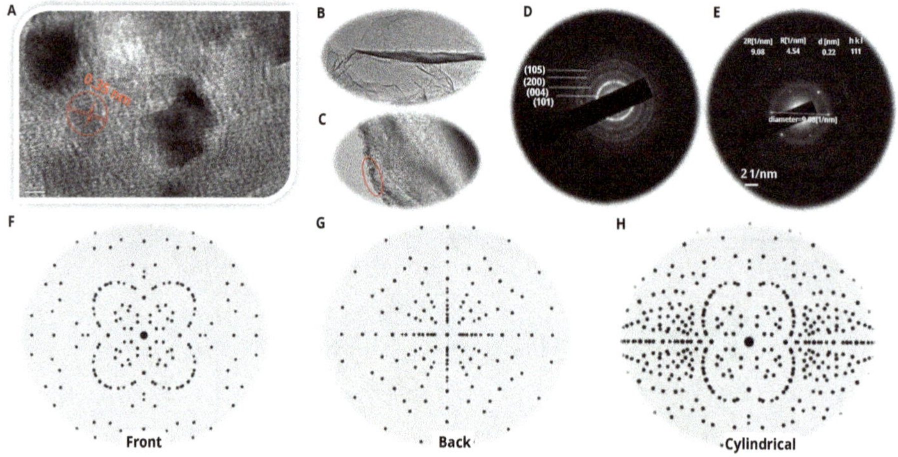

Fig. 3.14: High-resolution TEM images of PVP templated RGO–TiO_2 photocatalysts: **A.** Select RGO-coordinated (RGO_5, 5 mass%) TiO_2, **B.** morphology of RGO thin layer, **C.** TEM image of the templated RGO_5–TiO_2 catalyst, showing TiO_2 distributed along the curvature of RGO, **D.** the ring pattern resulted from TiO_2 in polycrystal cluster, **E.** selected area of electron diffraction patterns, and **F–G.** simulated SAED results, the Laue plate scattering from the front, back, and cylindrical, indicating highly crystallized TiO_2 formation upon RGO coordination and PVP-templating (images were adapted with permission from [99]).

The Barrett, Joyner, and Halenda (BJH) analyses were performed to determine the pore structure of the PVP-templated RGO–TiO_2 photocatalysts. It was evidenced that the in situ templation generated a contiguous network of mesopores [113]. The instrumentation analyses provided an in-depth understanding of the physical and chemical properties of porous nanocomposites as photocatalysts. The CrystalMaker was also used to demonstrate the proposed structure of RGO–TiO_2 composites. The pore sizes (Fig. 3.15A) of the RGO_5–TiO_2 and PVP-K30-templated RGO_5–TiO_2 were about 3.40(65) nm to 4.81(56) nm, respectively, based on the BJH model. The diagram in Fig. 3.15B shows the supercell ($4 \times 2 \times 2$) of TiO_2, with unit cell volume: 2,180.80 Å^3; filled space: 996.93 Å^3 (45.71%); and void space: 1,183.87 Å^3 (54.29%) per unit cell. The graphene layers were used as substrates to control the orientation of the TiO_2 nanoparticles. Based on the BJH analysis, the PVP-K30 porogen successfully enlarged pores, enabling interaction between the photocatalysts and N_2 adsorbates and further enhancing layered adsorption of N_2 molecules. These mesopores provide suitable adsorption sites and channels to promote rapid mass diffusion of the aqueous TCH towards the interfaces, further strengthening the degradation reaction, with a maximum efficiency of 94.22 % when 11-mass% PVP was used under simulated sunlight. The degradation efficiency of the antibiotic tetracycline (TCH) using PVP-K30-templated RGO–TiO_2 photocatalysts was evaluated to identify the optimal operating conditions for pharmaceutical waste treatment. This advancement in the current state of the art of TiO_2-based photocatalysts for the management of pharmaceutical waste under visible light, without additional energy consumption [99].

The PVP-K30-templated RGO_5–TiO_2 composites were evaluated using Raman, photoluminescence (PL), and UV–visible (UV–vis) spectroscopy (Fig. 3.16) to characterize the crystalline phase, crystallinity, and defects. Raman data, in conjunction with XRD analysis, indicated that anatase TiO_2 and RGO were formed (Fig. 3.16A and B). The anatase TiO_2 showed four bands, rising from the (101) O–Ti–O symmetrical stretching and aligned with principal peak E_{1g} (142.20 cm^{-1}) and E_g (636.40 cm^{-1}) of anatase TiO_2 [114]. The B_{1g} (395.40 cm^{-1}) and A_{1g} (512.60 cm^{-1}) bands are well-indexed to the (001) O–Ti–O bending variation. As shown in the green region (Fig. 3.16B), the intensity ratio of the D and G bands (I_D/I_G) has been used as a metric to evaluate graphene disorder, including ripples, edges, charged impurities, and domain boundaries [115]. Our study indicated that (I_D/I_G) decreased from 1.186 (GO) to 1.074 (RGO), indicating a decrease in graphene disorder after partial reduction of graphene oxide. The most important observation regarding this structural deformation of graphene concerns the electrical conductance of RGO–TiO_2 and its chemical interaction [99].

The photoluminescence (PL) spectra were obtained to evaluate the recombination mechanisms and bandgap of RGO–TiO_2 photocatalysts [116]. The date for graphene oxide and reduced graphene oxide-supported TiO_2 (11.00 mass% PVP-K30) and pure TiO_2 (Fig. 3.16C) indicated that broad emission bands were detected in the range of 3.54–2.48 eV (corresponding to 350–500 nm). The anatase TiO_2 displayed three emission lines, where the blue PL intensity centered at 2.658 eV, while the violet PL peak split into two emissions with intensity centered at 3.139 and 3.003 eV. Aggregation-caused quenching (ACQ) of TiO_2 emission was observed upon the introduction of aromatic graphene and the polymeric PVP-K30 [117]. Our data indicated that the recombination of electrons (e^-) and holes (h^+) resulted from the synergistic effect of the graphene support and PVP-K30 templating. Hypothetically, self-trapped holes are located at oxygen atoms in the lattice of the TiO_6 subunit, while trapped electrons occupy mid-gap states positioned below the Fermi level. This approach showed an increase in graphene oxide content, and the PL intensity decreased accordingly, as demonstrated in our previous work [99].

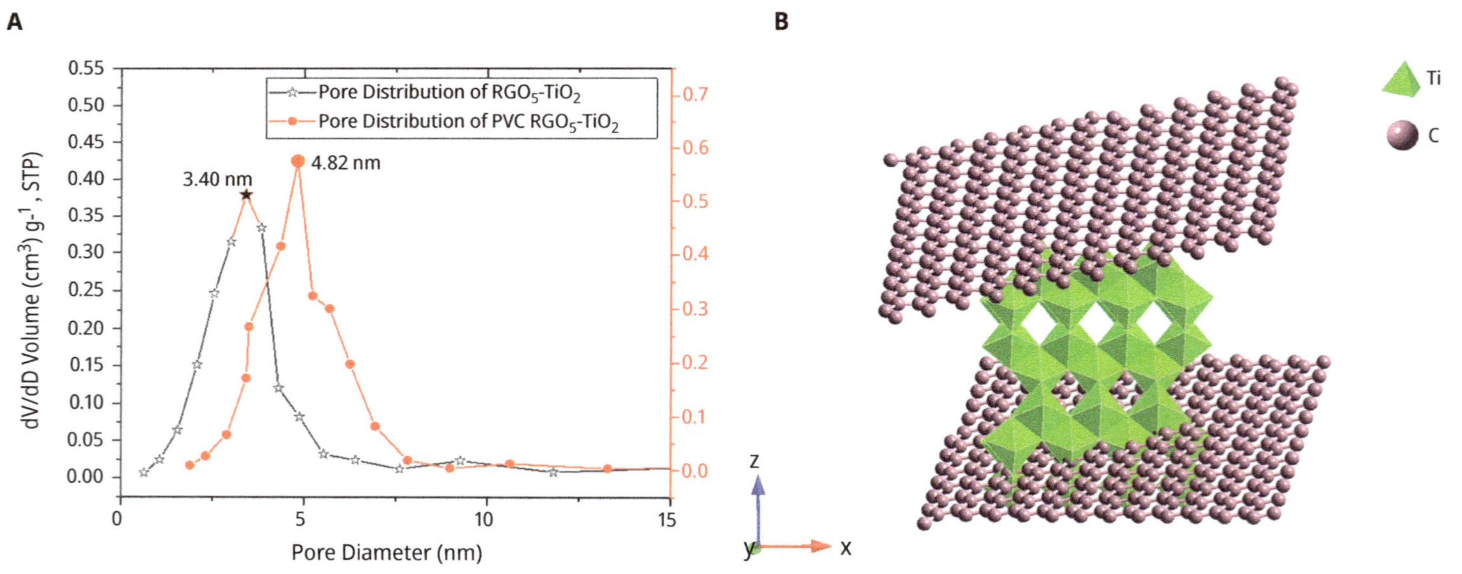

Fig. 3.15: Barrett, Joyner, and Halenda (BJH) analysis of PVP templated RGO–TiO_2 photocatalysts: **A**. Mesopore size distribution of RGO_5–TiO_2 composite with an average size of 3.40 nm and **B**. mesopore size distribution of PVP-K30-templated RGO_5–TiO_2 composite with an average size of 4.82 nm.

As a complementary method to the PL spectroscopy mentioned above, UV-Vis diffuse reflectance spectroscopy (DRS) was used to evaluate the optical bandgap of the PVP-K30-templated RGO–TiO_2. The results (Fig. 3.16D) indicated that both RGO coordination and PVP templating reduced peak intensity. These adsorptions also correlated with the observations from blue PL and violet PL, implying that the lifetimes of charge carriers were extended [118]. The DRS peak broadening was also observed, indicating defects and a decrease in the particle size of RGO–TiO_2. A sharp absorption edge was also observed, which was used to calculate the band gaps of three selected photocatalysts. It was found that the bandgap decreased upon the introduction of both graphene and PVP-K30. The absorption edge (428.2 nm) of RGO_5–TiO_2 nanocomposite showed a red-shift from 417.1 nm (TiO_2 alone), while the PVP-K30 templated RGO_5–TiO_2 showed a further red-shift to 446.8 nm. The corresponding band gap energies were calculated to be 2.97(3), 2.89(6), and 2.77(5) eV for the three photocatalysts, indicating high degradation efficiency under broad-spectrum light. These energy data were attributed to an upward shift of the Ti-O-C band (bending and stretching), resulting in an effective separation of electrons and holes at the interfaces or along the edges of the three components [119].

The antibiotic tetracycline (TCH, $C_{22}H_{24}N_2O_8$) is a highly adaptable molecule that readily undergoes tautomerism in response to different chemical environments [120]. It was chosen as a model molecule to test RGO–TiO_2 photocatalytic activity under ultraviolet (UV) and simulated sunlight (SS) illumination at varying pH values, Ca^{2+} concentrations, and degradation durations. Considering four deprotonations, we can identify 64 diversified tautomers of tetracycline. Among these tautomeric species, two TCH tautomers with highlighted differences (Fig. 3.17A and B) were examined to understand their bonding and geometry upon achieving equilibrium during the degradation process. When these TCH molecules undergo different deprotonation or bond-cleavage pathways, the pK_a can vary, influencing the RGO–TiO_2 photocatalytic reactivity. TCH tautomerism is essential for its chemical behavior under various environmental conditions. This comprehensive molecule's spatial orientation, bond cleavage, and degree of protonation affect the efficacy of deterioration, as discussed in the following sessions.

The degradation rate (η) using the templated RGO–TiO_2 photocatalysts (Fig. 3.18A and B) was investigated under UV and SS conditions. The generalization is that PVP addition increased η by 9.8% under both UV and SS conditions, due to increased pore size and surface area. UV light exposure showed a higher degradation rate (about 3.9%) than SS radiation in degrading tetracycline. The graphene substrate provided "in situ" riveting sites to guide TiO_2 nucleation and facilitated hybridization of the p-orbital from carbon with the d-orbital of titanium [121]. As a result, the electrons were delocalized among neighboring atoms, leading to an extended exciton lifetime. The band gap energy decreased by 0.34 eV from 3.21 eV to 2.87 eV, enabling visible-light adsorption by TiO_2-based catalysts and allowing degradation under visible-light conditions.

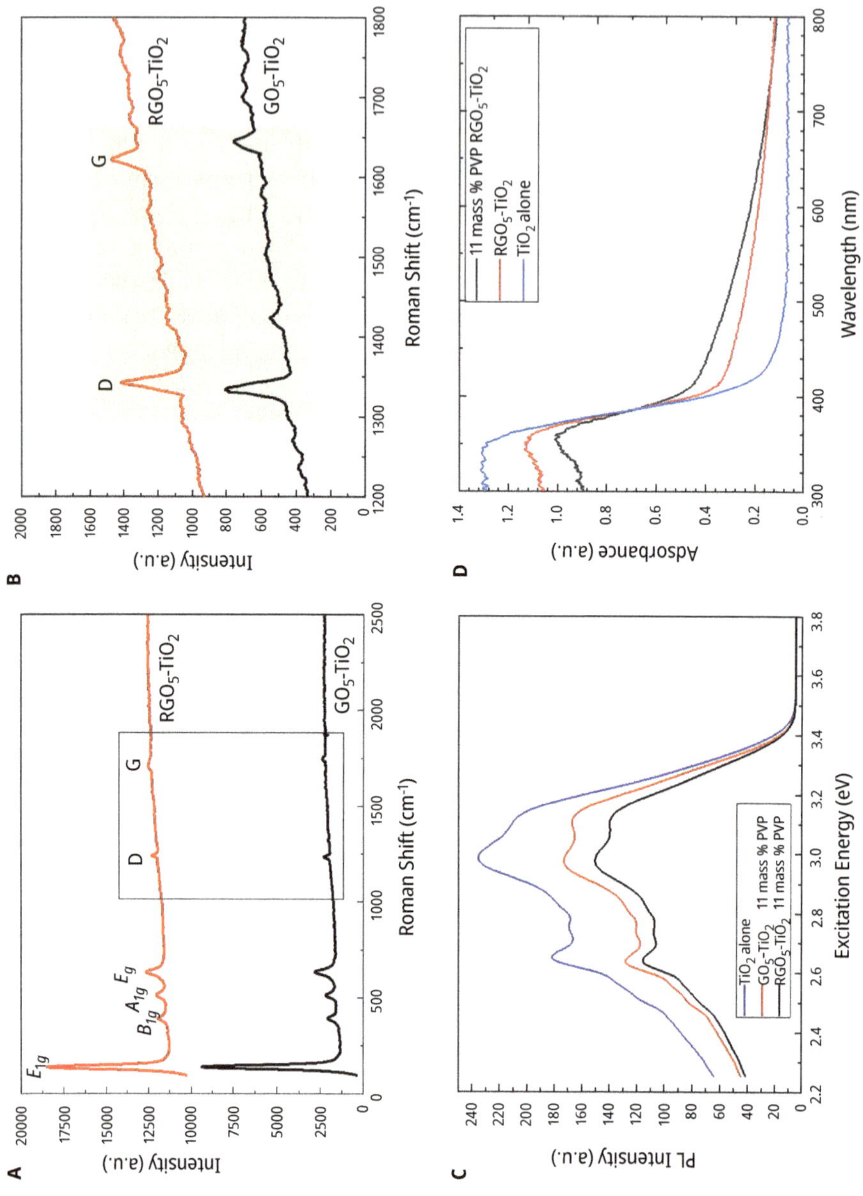

Fig. 3.16: Spectroscopic analysis of TiO_2-based nanocomposites: **A.** Raman spectra of 11 mass% PVP-K30 templated GO_5–TiO_2 catalysts, demonstration of vibration mode of TiO_2; **B.** the D and G bands were attributed to graphene oxide and reduced graphene oxide; and **C.** The photoluminescence of three select TiO_2-based catalysts, and **D.** the ultraviolet visible spectra of three select TiO_2-based catalysts.

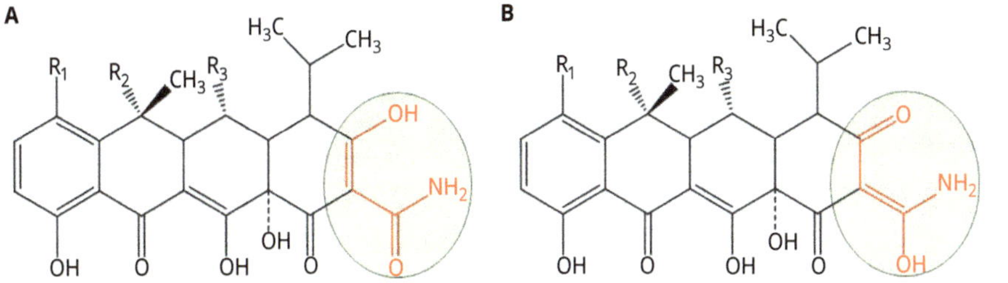

Fig. 3.17: Two tautomers of tetracycline exist, with tautomerization among 64 possible configurations.

The mesoporous PVP template has been used to tune the pore structure and surface area of the RGO–TiO_2 photocatalyst by varying its hydrophilicity, dispersibility, and cohesivity. It was found that the degradation rate depends on the amount of PVP-K30, showing "volcano" relationships under both UV and SS conditions across different radiation durations. Suppose the PVP-K30 is less than 11 mass%, and its degradation rate increases. Otherwise, the higher PVP amount corresponds to a decrease (as shown in the SS data in Fig. 3.19A across different treatment durations). The experimental data showed that RGO–TiO_2 catalysts with 11 mass% PVP-K30 achieved the highest degradation, averaging 93.36% and 91.34% under UV and SS conditions, respectively. The further increase in PVP to 12 mass% resulted in a decrease of approximately 5.29% under both UV and SS radiation. It was proposed that the Ti from the reactants and O from PVP formed ion–dipole intermolecular forces, anchoring the TiO_2 nucleation and particle growth (Fig. 3.19B). This water-soluble PVP templating was critical for tuning pore size and surface area and for improving the catalytic reactivity of these RGO–TiO_2 photocatalysts, thereby further enhancing antibiotic degradation.

The pH values of the TCH solutions ranged from 3, 5, 7, and 9, and a volcano relationship between the degradation rate and pH was also observed (Fig. 3.20A). Under neutral conditions (pH 7), the photocatalytic efficiency reached 93.36% and 91.34% under UV and SS conditions, respectively. This relationship was attributed to the amphoteric nature of TiO_2, with a theoretical isoelectric point of about 6.4, and to the high degree of freedom of the TCH tautomers (64 possibilities) due to protonation. At neutral conditions (pH = 7), there are five tautomers of TCH, among which two are neutral isomers, accounting for about 61 [99]. Under neutral conditions, the reduced repulsion between RGO–TiO_2 and TCH accelerated TCH adsorption, thereby improving degradation. TiO_2 attracted negatively charged hydroxide ions (OH^-) at basic conditions. The repulsion between negatively charged species prevented TCH from being adsorbed, and the degradation efficiency decreased further. The amphoteric propensity of the catalyst and the complexity of tetracycline conformations must be considered when designing photocatalysts.

The concentration of the calcium cation ([Ca^{2+}]) as a water hardness indicator was found to influence the efficiency of TCH degradation (Fig. 3.20B). There is no discernible trend in degradation rate with [Ca^{2+}]. However, with [Ca^{2+}] = 100.00 mg/L, the

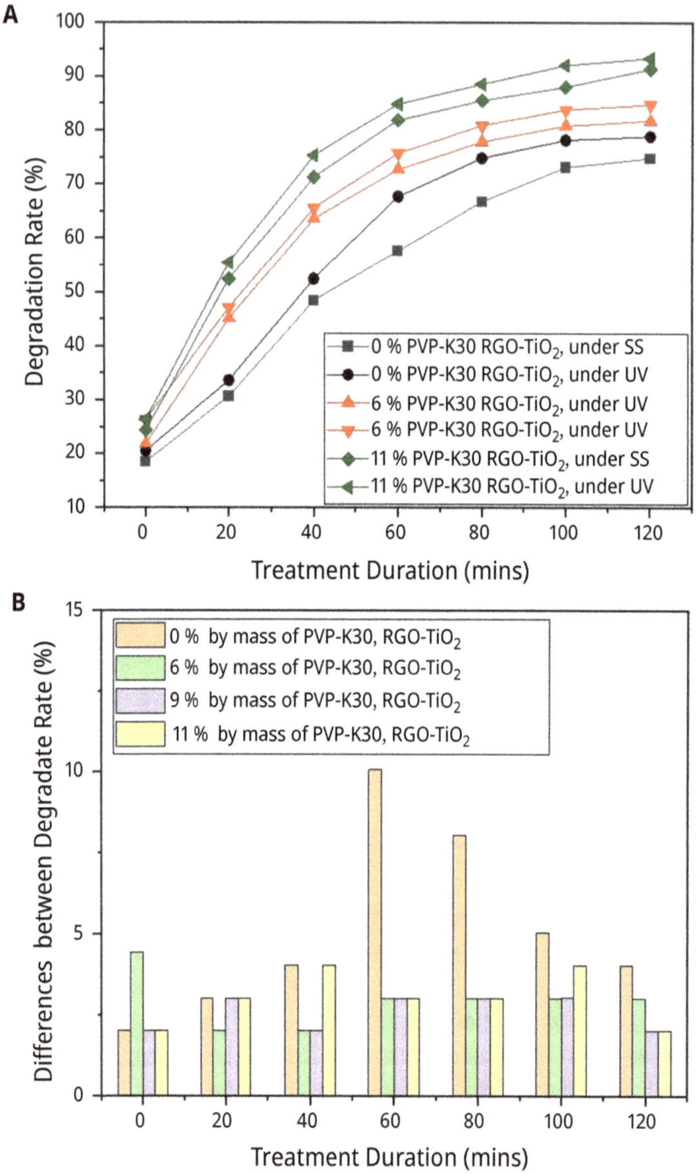

Fig. 3.18: The degradation rate of tetracycline using PVP-K30-templated $RGO_5–TiO_2$ photocatalyst with different adoption amounts of PVP, where three formulations were chosen for comparison. **A**. Degradation rate of catalysts under ultraviolet light (UV) and simulated sunlight (SS) and **B**. differences between degradation rates indicated the UV exposure shaving a higher efficacy; however, the SS provided an applicable photocatalysis for pharmaceutical waste remediation.

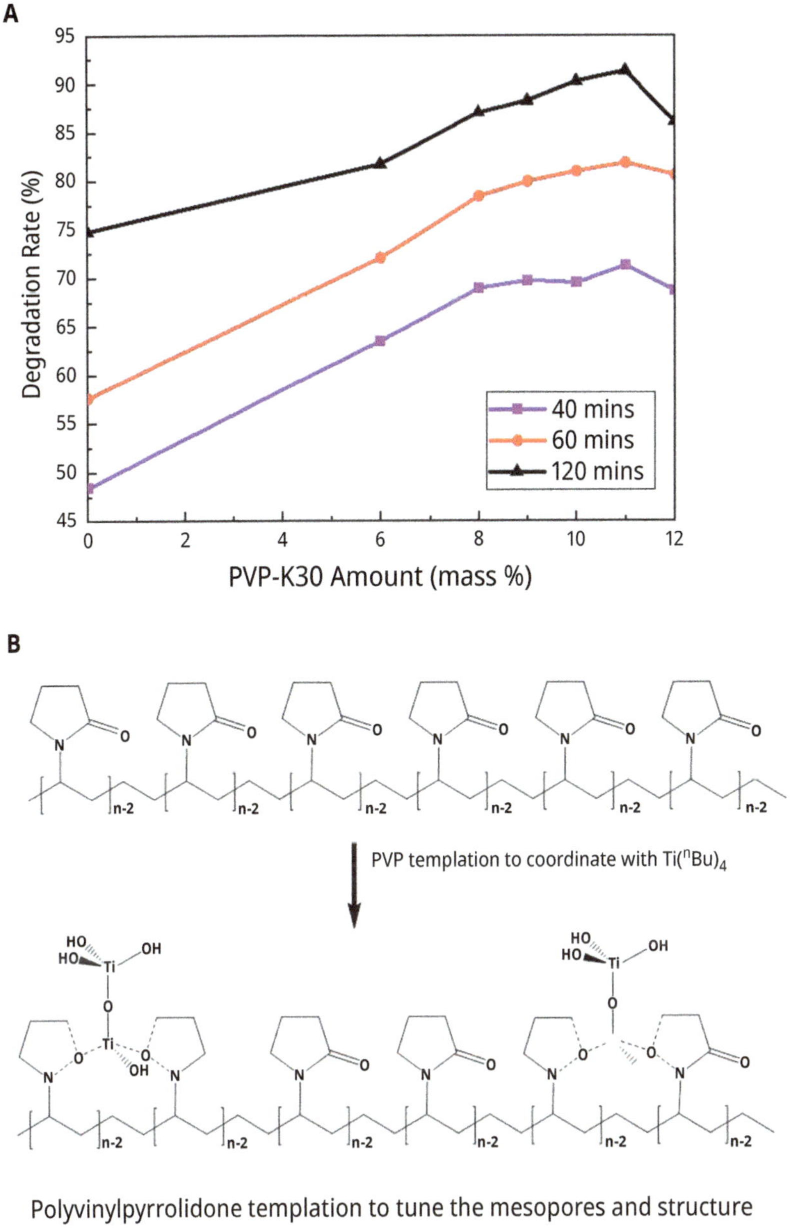

Fig. 3.19: The degradation rate of PVP-K30-templated RGO–TiO_2 photocatalyst with different adoption amounts of PVP: **A.** Degradation rate of catalysts under simulated sunlight and **B.** the PVP-K30 templation through intermolecular forces.

photocatalytic efficiencies were 93.36% and 91.34% under both UV and SS conditions, respectively. These Ca^{2+} cations act as a metallic center to form a complex with the N-end of the 2-dimethylamino and phenolic OH group in TCH, causing internal electron transfer from TCH to electron-deficient cations. At the lower Ca^{2+} concentration range, hydrated $Ca(H_2O)_6^{2+}$ cations were formed, causing spatial hindrance. The hydrated Ca^{2+} cations were prevented from contacting the TCH oxygen-negative pole or from occupying the RGO–TiO_2 reactive sites. As a result, the photocatalytic degradation of TCH was hindered by Ca^{2+} occupying specific sites. On the other hand, concentrations greater than 100.00 mg/L increased TCH degradation. This observation suggested that hard water can facilitate TCH degradation using the RGO–TiO_2 photocatalyst. The TCH molecules (Fig. 3.19) were activated to form a radial structure by losing electrons. These unstable radicals were ready to be further oxidized to small molecules by hydroxyl radicals and other reactive oxygen species (ROS) present in the aqueous solution.

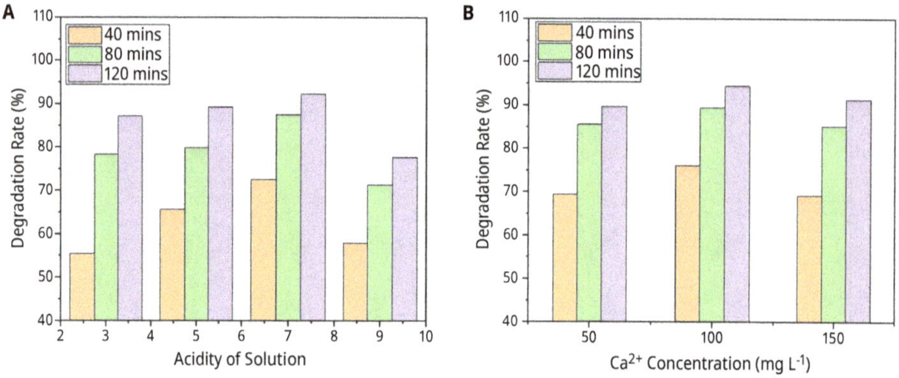

Fig. 3.20: The degradation rate of RGO–TiO_2 photocatalyst templated by PVP-K30 (11 % by mass) under simulated sunlight: **A**. Effect of pH on the degradation rate and **B**. effect of Ca^{2+} concentration on the degradation rate.

3.3.2 Heavy metal absorption

Heavy metal toxic waste has become increasingly abundant due to increased energy demand over the past two decades, causing serious global health and environmental problems. We developed a series of catalysts shielded by natural products to remediate wastewater. The hazardous ions composed of AsO_4^{3-} and CrO_4^{2-} have been absorbed using bio-encapsulated Fe_3O_4 magnetic nanoparticles (MNPs) [85]. These MNPs were encapsulated with natural product extracts (such as pomegranate polyphenols and chitosan from shrimp shells) to improve their biocompatibility and mitigate their toxicity. The results demonstrated reduced oxidative stress and cellular damage in human retinal pigmented epithelium cells when encapsulated MNPs were used. These functionalized NMPs also showed high removal efficiency of sodium hydrogen arsenate (Na_2HAsO_4). We further optimized the molecular structure of a series of MNPs using cost-effective colloidal chemistry, studied their thermodynamics and interactions with heavy-metal ions, and confirmed the mechanism under both electrochemical and physiosorption conditions. Three key perspectives are discussed in this section: (1) material design and characterization to improve the robustness and coverage of MNPs; (2) removal of heavy metal ions from solid or aqueous waste to provide an in-depth understanding of the thermodynamics and interactions between AsO_4^{3-} or CrO_4^{2-} with Fe_3O_4 MNPs; and (3) mechanistic analyses using spectroscopic method(s) to determine the oxidation number change or bond vibration.

These two types of nanocomposites (Fe_3O_4 and $Ca_{10}(PO_4)_6(OH)_2$) were chosen for water remediation. The Fe_3O_4-based nanomaterials were prepared via wet-chemistry methods, with natural product extracts used as dispersing agents to tune particle size, shape, and functionalization. The colloidal chemistry, sol–gel, and inverse emulsion approaches have been employed to control the composition and architecture of nanoproducts at the atomic or molecular level. The structure-diversified core–shell, yolk-shell, Janus, and needle-shaped nanocomposites were obtained [122]. We also aim to produce recyclable nanocomposites with mesoporous, hollow, or multilayered structures to enhance their permeability and adsorption capacity. The structural characterization will provide feedback and guidance to optimize synthesis variables further and improve the stability, configuration, and reactivity of these nanocomposites for remediation applications. Their surface grain, particle size, pore size, wettability, and interactions with contaminants were characterized at the nanoscale [123]. The $Ca_{10}(PO_4)_6(OH)_2$ was studied using CrystalMaker to determine its electron density, crystallographic structure, and functionalization.

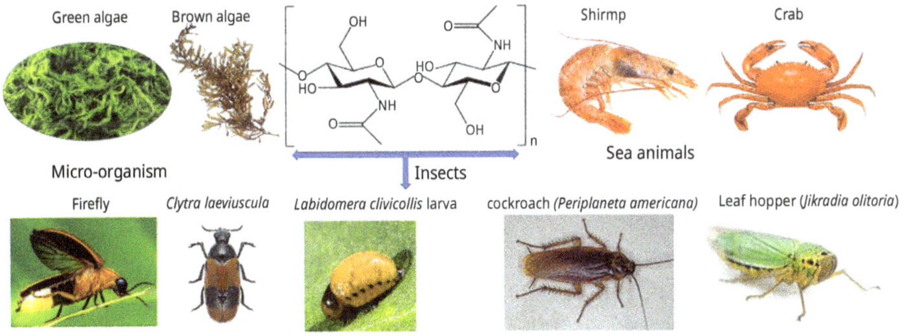

Fig. 3.21: Chitin is an integral structural component present in different organisms. (images were from shutterstock.com with permission).

An inverse emulsion method was used to produce core–shell mixed Fe_3O_4-based magnetic nanocomposites (MIMNs). The starting materials are $Fe(NO_3)_3$ and $Fe(NO_3)_2$ for the synthesis of mixed-oxide magnetic nanoparticles under N_2 protection. Natural product extracts include chitin and chitosan, obtained from the shrimp shell waste. Other compounds, namely citronellal, citrate, and limonene extract from orange peels, are also used to prepare MIMNs. These compounds are chosen for their non-toxicity and ability to recycle natural products, further minimizing energy and resource consumption. Chitin (2-(acetylamino)-2-deoxy-D-glucose, also named β-(1–4)-poly-*N*-acetyl-D-glucosamine), is the second most abundant long-chain polymer, an amide derivative of glucose. This molecule is widely distributed in nature as an amino polysaccharide, second only to cellulose [124]. Chitin occurs as ordered macrofibrils and is the major structural component in the exoskeletons of the crustaceans, crabs, shrimps, and the cell walls of fungi (Fig. 3.21). For biomedical applications, chitin is usually converted to its deacetylated derivative, chitosan [125].

As a nontoxic and linear polysaccharide copolymer, chitosan is the main component of a crustacean's shell. It is composed of β-(1→4)-linked *N*-acetyl-D-glucosamine (acetylated unit, $C_6H_{13}NO_5$) and *N*-acetyl-glucosamine ($C_8H_{15}NO_6$). Many benefits of chitosan were observed, including antibacterial activity and the ability to chelate metal ions [126]. Chitosan was soluble in acidic solutions as a deacetylated form of chitin with varying degrees of deacetylation [127]. Citronellal ($C_{10}H_{18}O$, also called radical) is the monoterpenoid aldehyde and primary reductive compound. It exists in plants, Cymbopogon, as the reason for oranges' special fragrance, giving citronella oil its distinctive lemon scent [128]. In this study, citronellal was extracted from waste orange peel and used as a dispersing agent to improve MIMN dispersion and biocompatibility. Citric acid is another reducing and capping agent used in MIMN's synthesis, inhibiting nanoparticle growth and agglomeration. It exists in a variety of fruits and vegetables. These naturally occurring compounds are ideal candidates for fabricating high-networking polymeric scaffolds for biomimetic materials. Using these natural products includes high porosity, good biodegradability and biocompatibility, predictable degradation rate, and maintainable structure integrity (Tab. 3.4).

More than 20 known iron oxides and oxyhydroxides (Tab. 3.5) play critical roles in geological and biological processes. The structures and cell volumes of iron oxides change significantly with temperature and pressure. For example, iron peroxide showed 24 different phases under different synthesis conditions. Table 3.5 summarizes iron metal oxides' selected crystallographic structure and lattice parameters.

The simulated X-ray powder diffraction patterns of Fe_3O_4 (Hermann Mauguin, Fd3m;[139] Hall F4d23$\bar{1}$d, Point Group, m$\bar{3}$m; Crystal System, cubic). Moreover, Fe_2O_3 (maghemite standard) was shown in Figs. 3.22A and 3.22B. These crystallographic data have been used to guide materials synthesis and to produce highly crystalline structures for heavy-metal removal. It was found that these well-crystallized nanomaterials could be dispersed in aqueous solution and maintain their sub-building units for a prolonged period [140].

Tab. 3.4: A summary of natural product extracts and their properties.

Name	The formula of the natural product extract	Properties
Chitin	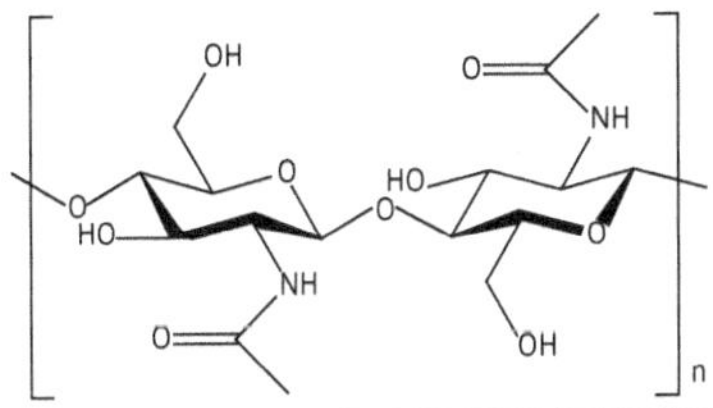	Abundant amino polysaccharide polymer biodegradability lack of toxicity antifungal effects wound healing acceleration immune system stimulation
Chitosan	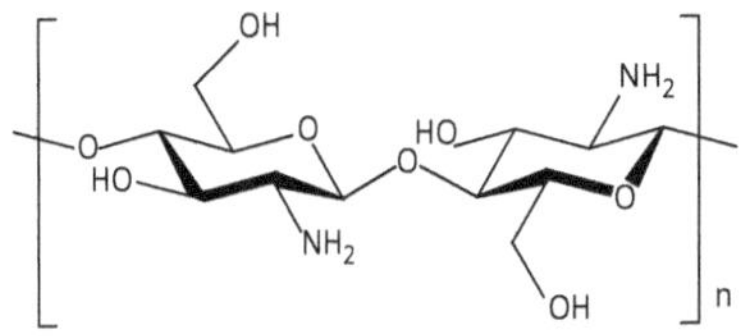	Protonation in neutral solution water-soluble and a bioadhesive binding to negatively charged surfaces making cross-linked polymeric networks enhancing the transport of polar drugs
Citronellal	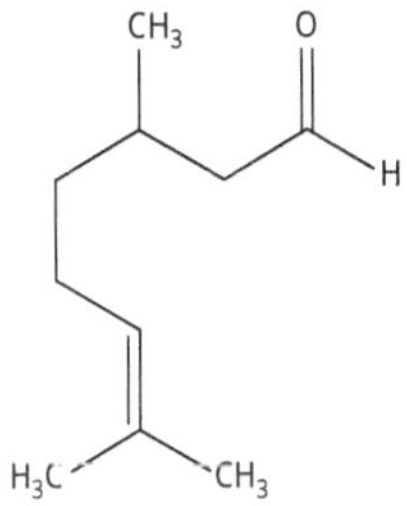	insect repellent properties high repellent effectiveness against mosquitoes strong antifungal qualities low or very low-volatile Combustible substance poorly flammable (flash point >60 °C up to 93 °C)
Citric acid	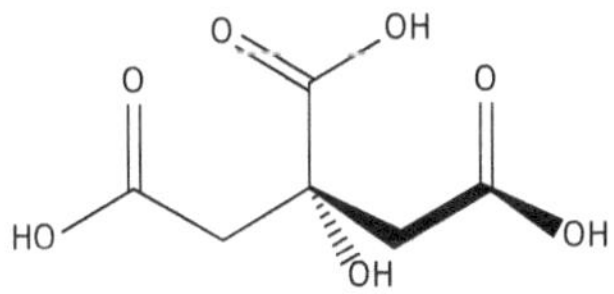	Colorless weak organic acid intermediate in the citric acid cycle acidifier flavoring and chelating agent

Band structure and density of states of Fe_3O_4 and Fe_2O_3 were shown in Fig. 3.23A and Fig. 3.23B, respectively. The electronic band structure of these metal oxides can describe the range of energy levels. The electrons occupied orbitals of lower energy, and those of higher energy were occupied afterward. There is a gap between the highest occupied molecular orbital and the lowest unoccupied molecular orbital, defined as the band gap or forbidden band. Except for the first element, hydrogen (H), all atoms have more than one electron. The electrons occupy the individual atomic orbitals within a single isolated atom with discrete energy levels. When different atoms bond to form a molecule, their nuclear orbitals overlap through hybridization, splitting these discrete energy levels. The adjacent energy levels in nanomaterials are close together and are considered an energy band continuum [142]. Density functional theory (DFT) has been widely used to calculate the ground-state electronic structure. Physi-

Tab. 3.5: The summary of iron oxides and their crystallographic structures, Fe_2O_3 [129].

No.	Nomenclature	Crystalline structure	Lattice parameters
		Oxide of Fe^{2+}	
1	Iron(II) oxide (FeO)	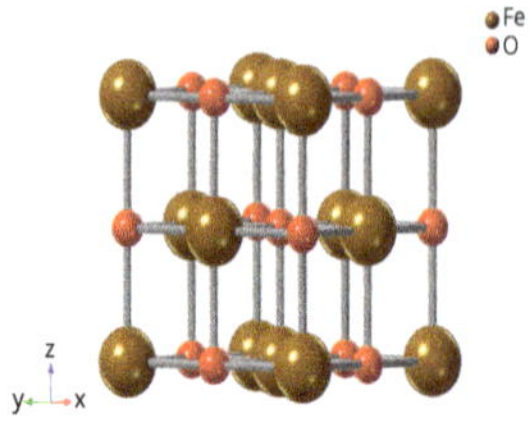	Wüstite[130] Space group: $Fm\bar{3}m$ Crystal system: Cubic *a*: 4.3320 Å *α*: 90° Cell volume: 81.295 Å^3 Rock salt (halite) structure Fe and O in octahedral coordination
2	Iron peroxide (FeO_2)	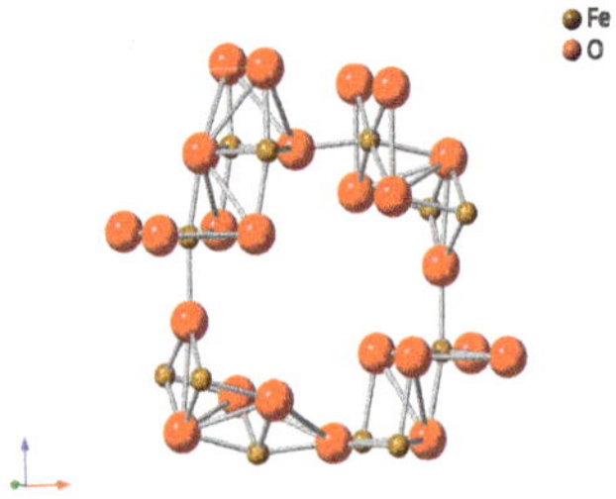	Space group: P 1 Allows chirality; polar Crystal system: triclinic *a*: 10.0830 Å *b*: 2.9695 Å *c*: 10.2054 Å *α* = *γ*: 90.000° *β*: 90.195° Cell volume: 305.563 Å^3
		Mixed oxides of Fe^{2+} and Fe^{3+}	
3	Iron(II, III) oxide and magnetite (Fe_3O_4)	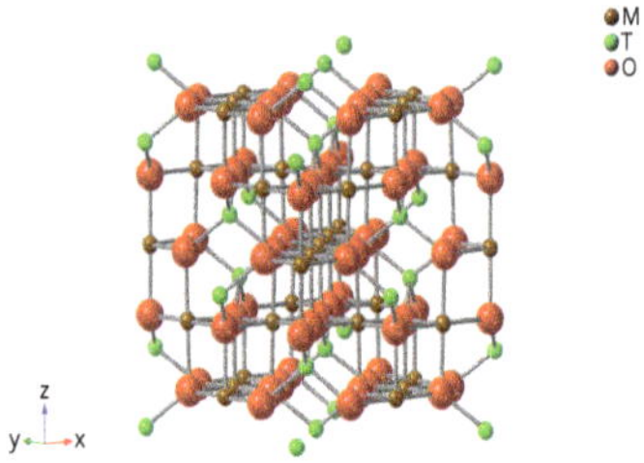	Space group: F d3m[131] Crystal system: cubic *a*: 8.3941 Å Cell volume: 591.456 Å^3 Spinel-type structure O is octahedrally coordinated to Fe_1 Fe_2 is tetrahedrally coordinated by O
3'	Iron(II, III) oxide (Fe_3O_4) Formed at 44 GPa	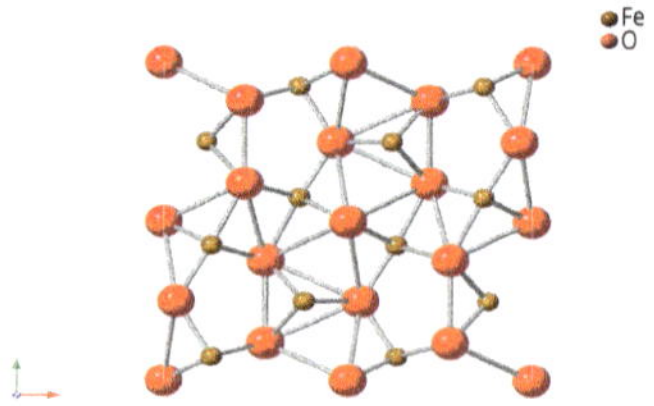	Space group: Bbmm Crystal system: orthorhombic *a*: 9.3090(30) Å *b*: 9.2820(20) Å *c*: 2.6944(9) Å Cell volume: 232.813 Å^3

Tab. 3.5 (continued)

No.	Nomenclature	Crystalline structure	Lattice parameters
4	Fe_4O_5 Formed under high pressure, $P = 10$ GPa	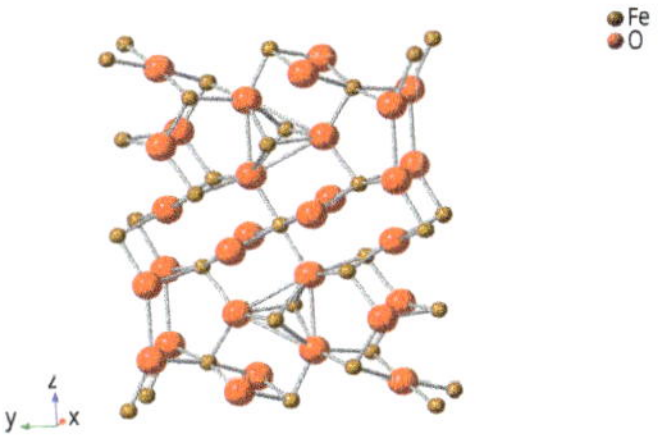	Space group: Cmcm[132] Crystal system: Orthorhombic a: 2.8430 Å b: 9.7000 Å c: 12.2900 Å Cell volume: 338.923 $Å^3$
5	Fe_5O_7 (homologous series nFeO $m$$Fe_2O_3$, formed at 41 GPa)	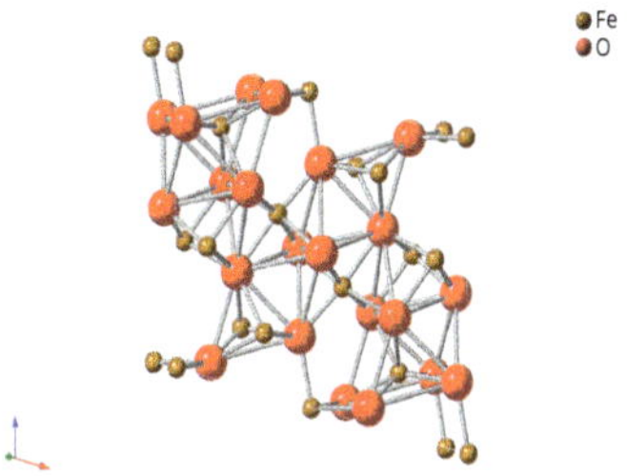	Space group: C2/m Crystal system: monoclinic a: 9.2080(70) Å b: 2.7327(10) Å c: 8.2700(50) Å β: 105.500(80)° Cell volume: 200.527 $Å^3$ Prisms are connected through familiar triangular faces Octahedra connect only via shared edges Triangular face-shared prisms and edge-shared octahedra
6	$Fe_{25}O_{32}$ (formed at 80 GPa)	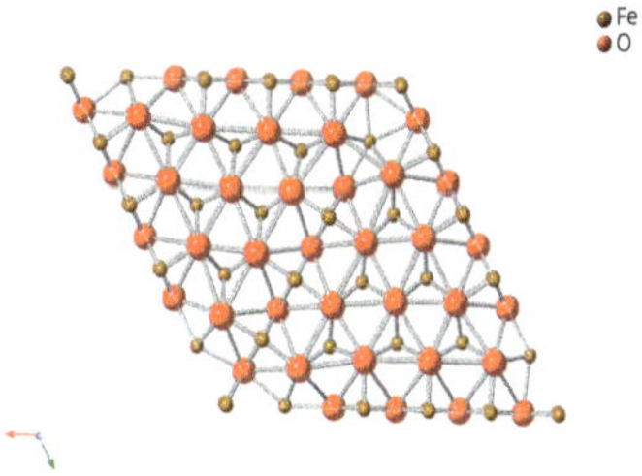	Space group: $P\bar{6}2m$ Crystal system: hexagonal a: 13.4275(16) Å c: 2.6289(4) Å Cell volume: 410.483 $Å^3$ Edge-shared one-capped prisms; belongs neither to the homologous series nor adopts any other known structural motif
7	$Fe_{13}O_{19}$	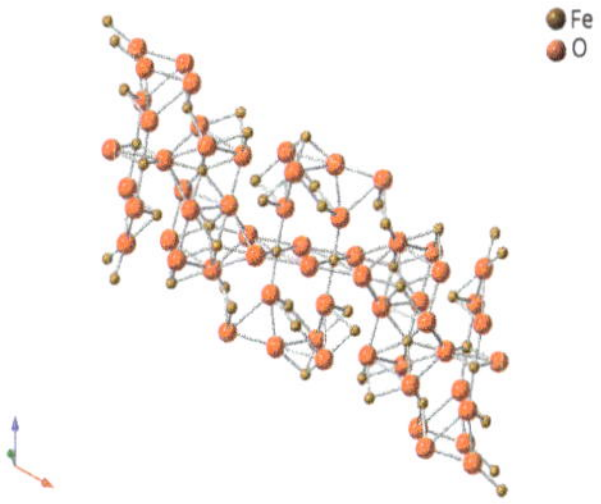	Space group: P 1 Allows chirality; polar Crystal system: triclinic a: 22.1917 Å b: 2.9108 Å c: 11.0205 Å $\alpha = \gamma$: 90.000° β: 118.470°

Tab. 3.5 (continued)

No.	Nomenclature	Crystalline structure	Lattice parameters
		Oxide of Fe^{3+}	
8	α-Fe_2O_3: alpha phase, hematite	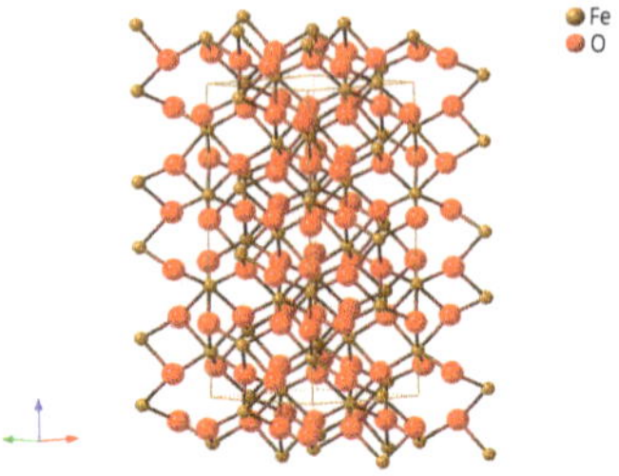	Space group: R $\bar{3}$c Crystal system: trigonal *a*: 5.0380 Å *c*: 13.7720 Å Cell volume: 302.722 $Å^3$
9	$Fe_{1.76}H_{0.06}$ O_3, hematite proto (formed at *T* = 313 °C)	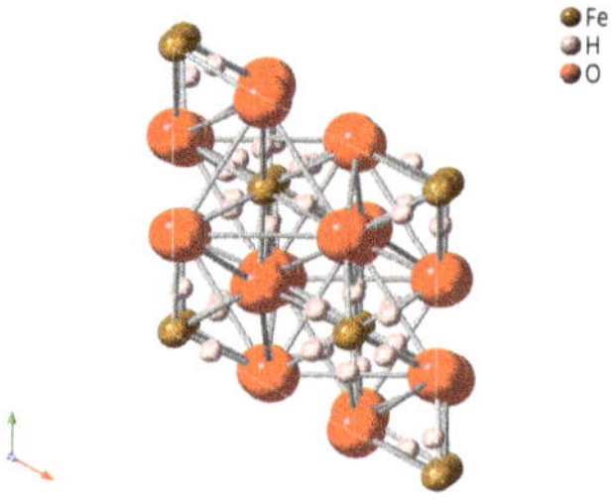	Space group: *R*$\bar{3}$c Crystal system: trigonal *a*: 5.0145 Å *c*: 13.6920 Å Cell volume: 298.162 $Å^3$
10	Fe_2O_3	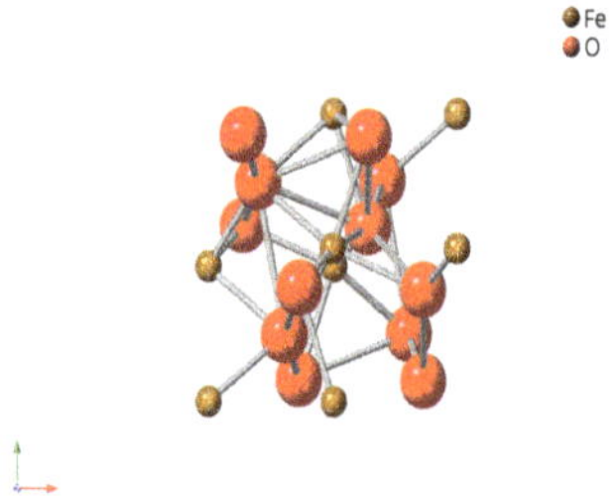	Space group: *P* 1 Crystal system: triclinic *a*: 5.1447 Å *b*: 5.2884 Å *c*: 7.5114 Å $\alpha = \beta = \gamma$: 90.000° Cell volume: 204.363 $Å^3$
11	β-Fe_2O_3: beta phase	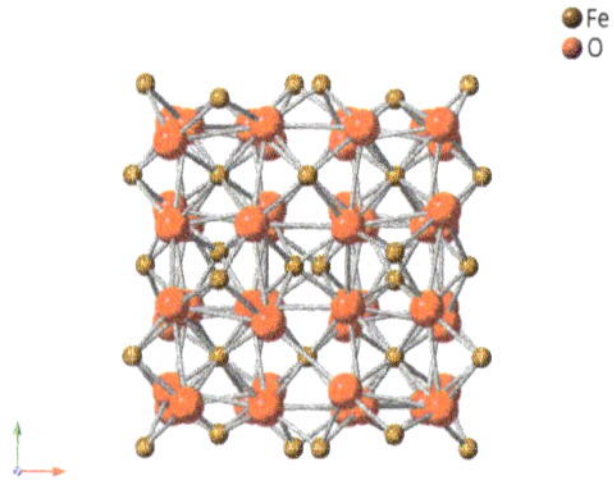	Space group: Ia$\bar{3}$ (P 1) Allows chirality; polar Crystal system: triclinic $a = b = c$: 9.5561 Å $\alpha = \beta = \gamma$: 90.000° Cell volume: 872.658 $Å^3$

Tab. 3.5 (continued)

No.	Nomenclature	Crystalline structure	Lattice parameters
12	γ-Fe_2O_3: gamma phase, maghemite	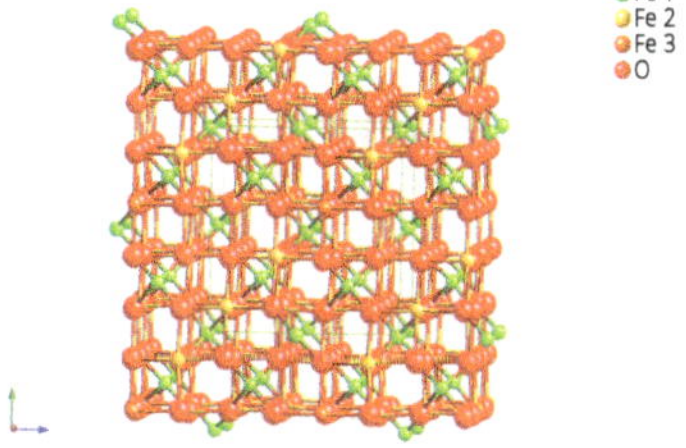	Space group: P 43 3 2[133] Allows chirality Crystal system: cubic *a*: 8.3474 Å Cell volume: 581.639 Å^3
13	η-Fe_2O_3 at 64 GPa	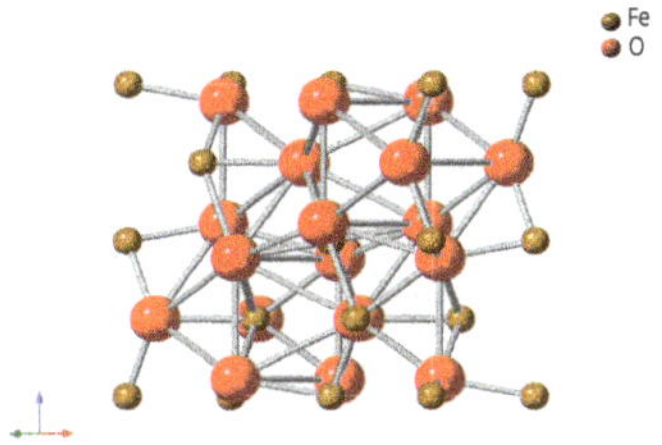	Space group: Cmcm[134] Crystal system: orthorhombic *a*: 2.6400(60) Å *b*: 8.6390(90) Å *c*: 6.4140(140) Å Cell volume: 146.284 Å^3
14	θ-Fe_2O_3 (formed at 74 GPa)	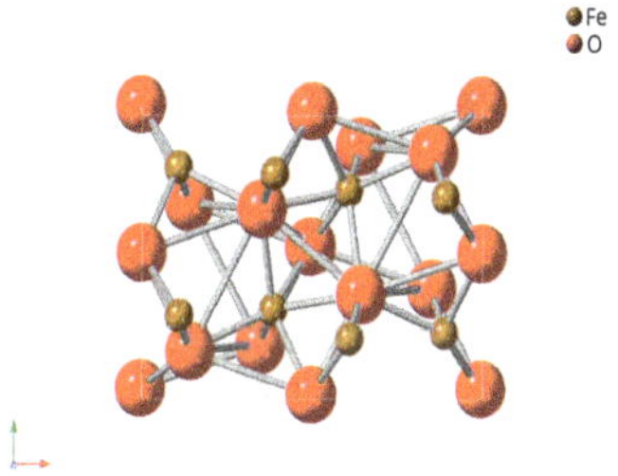	Space group: Aba2[135] Polar Crystal system: orthorhombic *a*: 6.5240(90) Å *b*: 4.7020(30) Å *c*: 4.6030(70) Å Cell volume: 141.201 Å^3
15	τ-Fe_2O_3 (formed at 41 GPa)	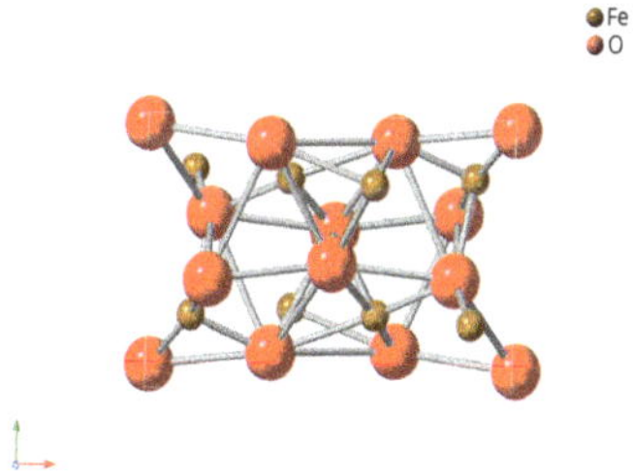	Space group: Pbcn Crystal system: orthorhombic *a*: 7.0620(100) Å *b*: 4.8108(13) Å *c*: 5.0019(8) Å Cell volume: 169.934 Å^3
		Oxide-hydroxides	
16	Goethite (α-FeOOH),	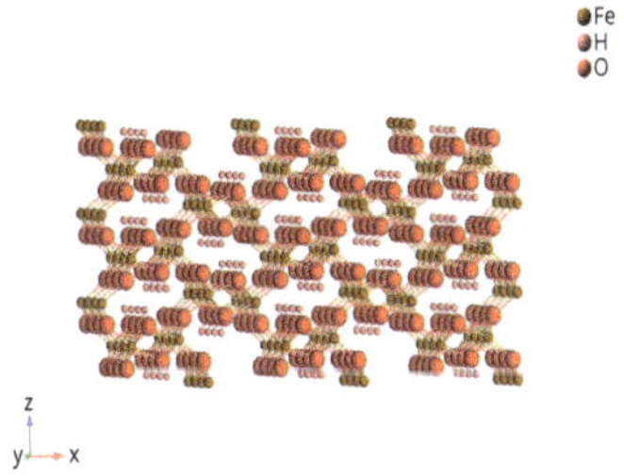	Space group: Pnma Crystal system: orthorhombic *a*: 9.9500 Å *b*: 3.0100 Å *c*: 4.6200 Å Cell volume: 138.367 Å^3 All atoms in 4c positions

Tab. 3.5 (continued)

No.	Nomenclature	Crystalline structure	Lattice parameters
17	Akaganéite (β-FeOOH), an iron oxyhydroxide mineral ($Fe_{7.6}H_{9.28}Ni_{0.4}O_{16}Cl_{1.16}$)		Space group: I 2/m[136] Crystal system: monoclinic *a*: 10.5870 Å *b*: 3.0311 Å *c*: 10.5150 Å *β*: 90.030° Cell volume: 337.429 Å³ Hollandite structure Octahedral chains of FeO_6 square channels along the structure's *y*-axis
18	Lepidocrocite (γ-FeOOH), an iron oxo-hydroxide mineral	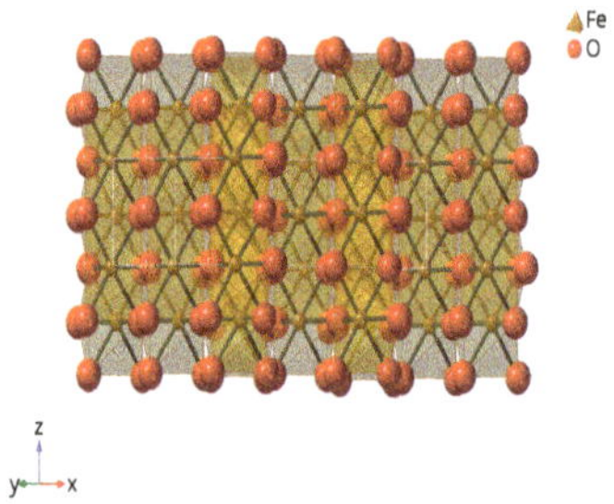	Space group: Bbmm[137] Crystal system: orthorhombic *a*: 12.4000 Å *b*: 3.8700 Å *c*: 3.0600 Å Cell volume: 146.843 Å³ FeO_6 octahedral groups Share edges along the structure's *y*- and *z*-axes Form crenulated octahedral sheets
19	Feroxyhyte (δ-FeOOH), an iron oxyhydroxide mineral	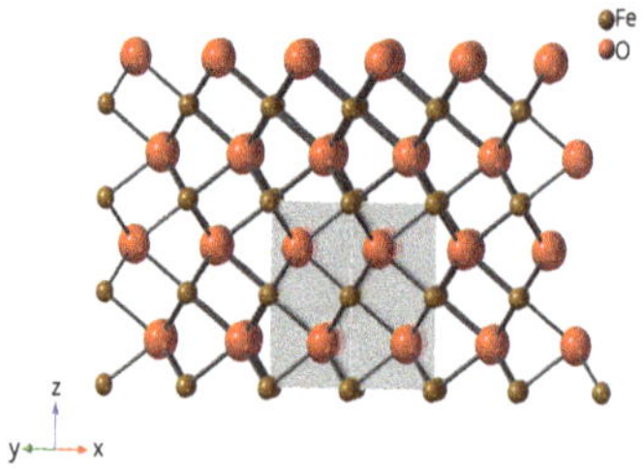	Space group: P $\bar{3}$m1[138] Crystal system: trigonal *a*: 2.9500 Å *c*: 4.5600 Å Cell volume: 34.367 Å³ Edge-sharing FeO_6 groups connected to form hexagonal sheets parallel to (001) Adjacent sheets share octahedral faces along the *z*-axis

cists, chemists, and materials scientists use DFT to investigate many-body systems, especially atoms, molecules, and condensed phases [143]. The DFT has been built using force constants and basis set extensions based on the Bohr atom concept, while accounting for extended forces. These generalized Gradient approximations (GGAs) for exchange-correlation energy expressions as mathematical functionals accurately predict the electron density used to evaluate the properties of a "many-electron system" [144]. Specific basis and extended sets use these functionals for in-depth understanding and design of catalytic processes, drug delivery, electron transport, energy conversion, and other problems in science and technology [145]. One of the functions de-

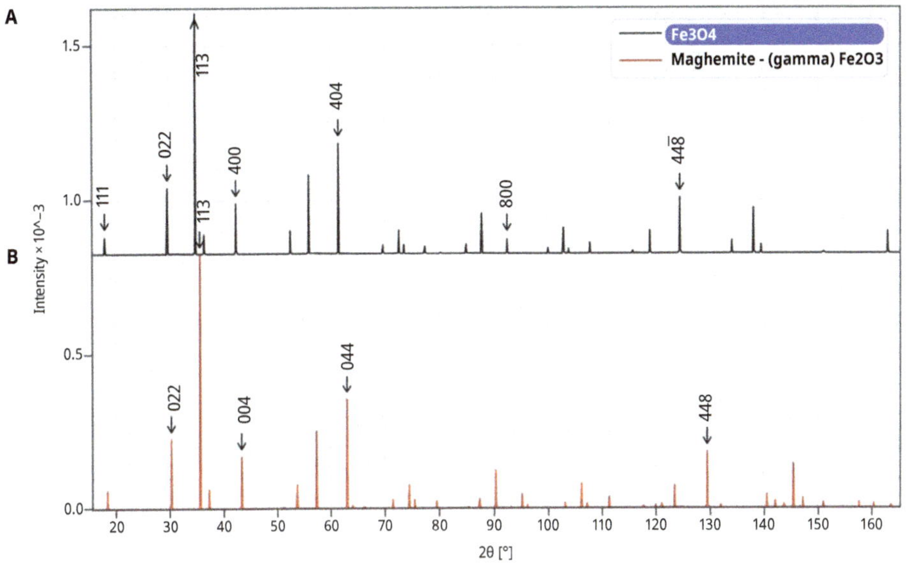

Fig. 3.22: The X-ray powder diffraction patterns were simulated using Crystal Maker to demonstrate the lattice planes of (A) Fe_3O_4 [141] and (B) Fe_2O_3 [133].

pends on the electron density. This computational quantum mechanical modeling method is one of the most popular and versatile techniques available for computational studies [146]. The density of states (DOS) of a system can explain the proportion of states occupied at each energy. The DOS is a mathematical representation of the electronic distribution, expressed as a probability density function [147]. The various states occupied by the system provided this function over the space and time domains. Two figures (Figs. 3.23A and 3.23B) show the electronic band structures and density of states for the Fe_3O_4 and Fe_2O_3 systems [148]. The DOS is directly related to the dispersion relations at a specific energy level. The DOS corresponded to many available states for electrons to be allocated.

X-ray absorption spectroscopy (XAS) is an analytical technique used to evaluate the local geometric or electronic structure of solid-state materials [152]. The experiment is usually performed at synchrotron radiation facilities, which provide intense and tunable X-ray beams (Fig. 3.24A). Samples were in the gas phase, solution, or solid state [153]. XAS data is obtained by tuning the photon energy [153] using a crystalline monochromator to a range where core electrons are excited (0.1–100 keV) [154] (Fig. 3.24B-D).

In our investigation [85], Ms. Huang used gum arabic (*N*-(8-(2-hydroxybenzoyl) amino)caprylate) as a dispersing and capping agent due to its weak acidity. This heteropolymer is composed of a neutral heteropolysaccharide with branched chains. The gum arabic was found in salts consisting of polysaccharides, acids, and cations such as magnesium (Mg^{2+}), sodium (Na^{+}), and potassium (K^{+}). The units of l-arabinose (aldo-

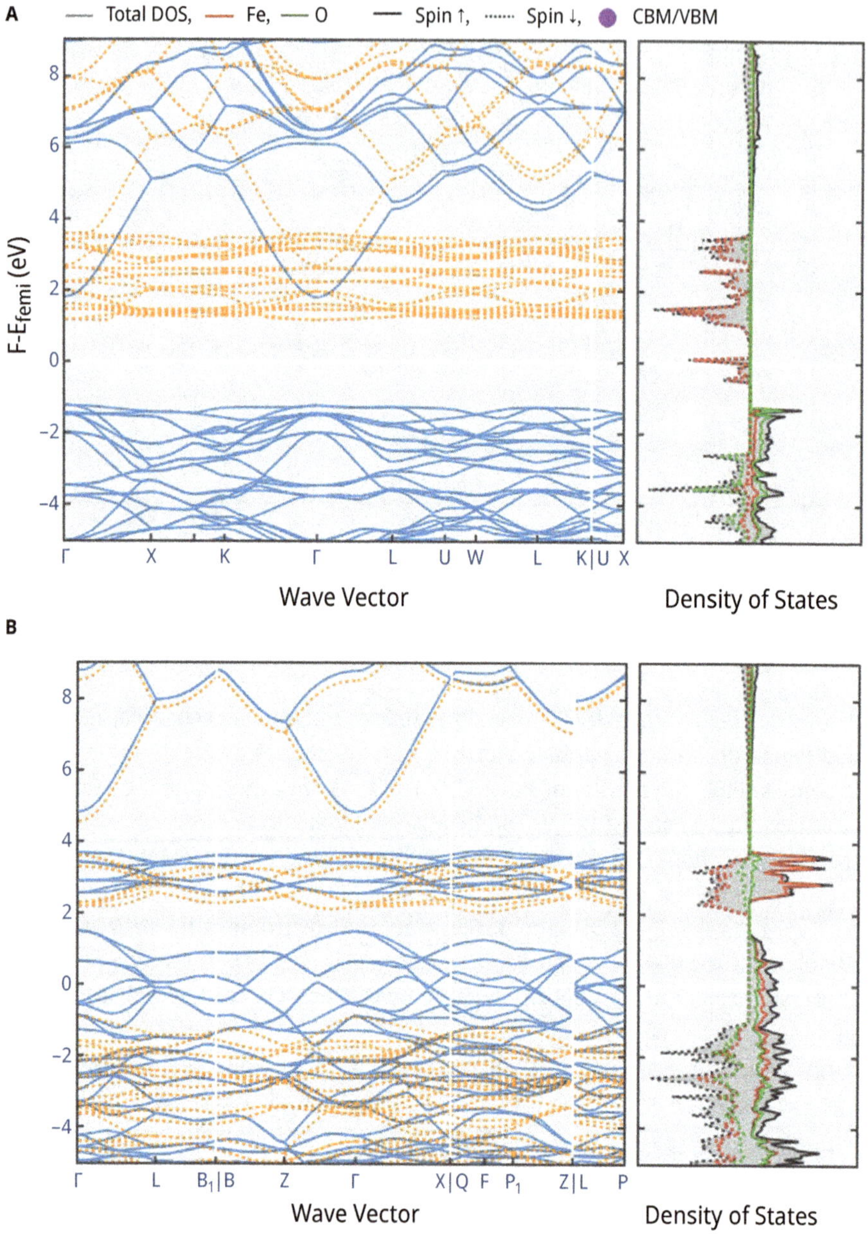

Fig. 3.23: Band structure and density of states using semi-local density functional theory (DFT). **A.** Fe_3O_4 results [149] and **B.** Fe_2O_3 results [150]. (Note: The semi-local DFT method tends to underestimate bandgaps severely.) [151].

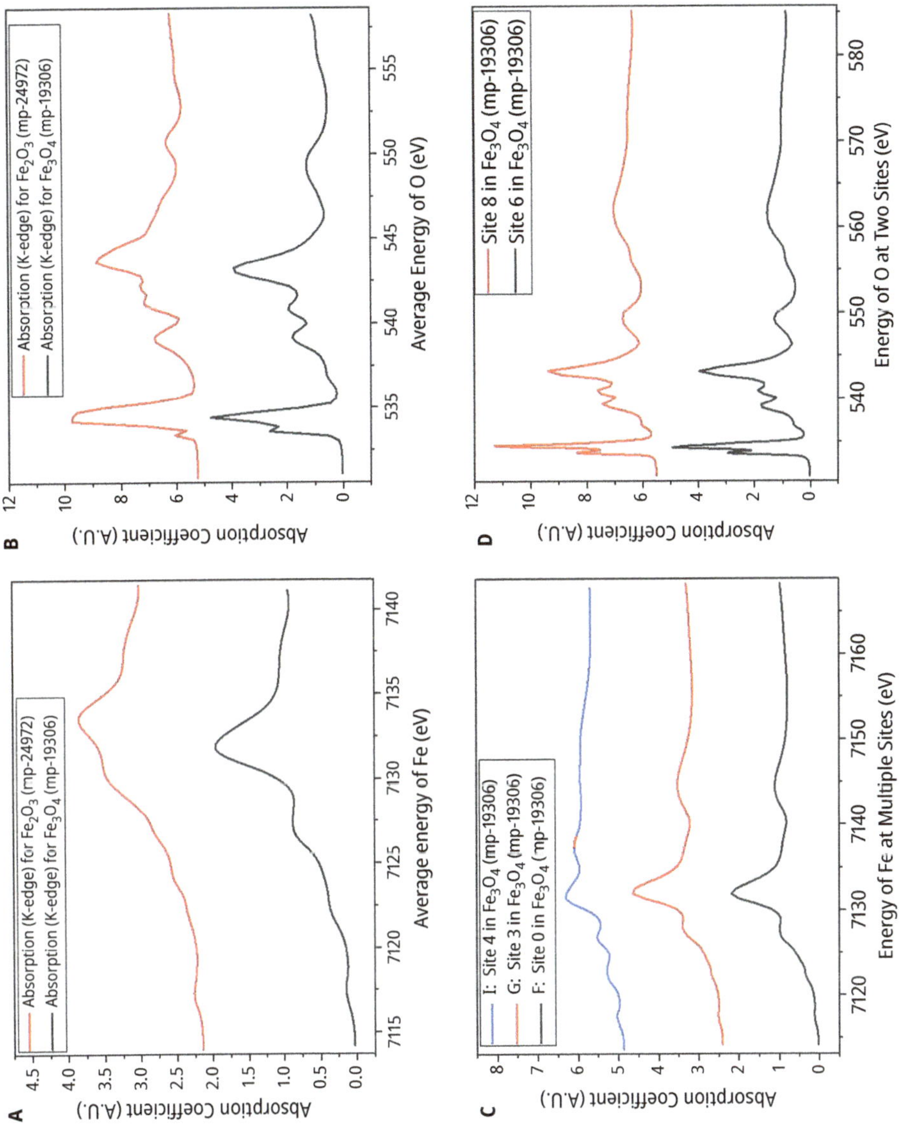

Fig. 3.24: X-ray absorption spectra of Fe_2O_3 and Fe_3O_4 were obtained from the Materials Project, an open-access database. **A.** The average K-edge absorption of the Fe element in Fe_2O_3 and Fe_3O_4, **B**. The average K-edge absorption of the O element in Fe_2O_3 and Fe_3O_4, **C.** K-edge absorption of the Fe element in Fe_3O_4 at three different sites, and **D.** K-edge absorption of the O element in Fe_3O_4 at two other sites (averaged spectra are third-order spline interpolations ranging over the intersection of *E*-axis ranges of site spectra) [156].

pentose, $C_5H_{10}O_5$), l-rhamnose (deoxy sugar, $C_6H_{12}O_5$), d-glucuronic acid (uronic acid, $C_6H_{10}O_7$), and 1,3-linked β-d-galactopyranosyl are the backbones in gum arabic (Figs. 3.24A, B, and 3.25A). It has been used to control the hydrolysis rate and tune the architecture of the final catalysts. A key component, sodium 2-[(7-carboxyheptyl)-C-hydroxycarbonimidoyl]benzen-1-olate ($C_{15}H_{20}NNaO_4$), was used for the demonstration of the templation of the Fe_3O_4 MIMNs (Figs. 3.24B, C, and 3.25B and C). Based on the simulated structural results, it was hypothesized that the O from gum arabic could occupy the lattice sites of the Fe_3O_4 sub-unit building blocks via the common element oxygen, thereby preventing particle growth (Fig. 3.25D). The nanoparticles are evaluated using a Tecnai F20-G2 transmission electron microscope (TEM) equipped with X-ray energy-dispersive spectroscopy to determine morphology. The TEM images showed that uniformly distributed nanoparticles were obtained (Fig. 3.25E). It was also found that the capping agent can result in different shapes of NPs, depending on the other chemical compounds and synthesis conditions (Fig. 3.25F).

X-ray photoelectron spectroscopy (XPS) was used to evaluate the surface and in-depth chemical composition of the core/shell MIMNs (Fig. 3.26). The Fe element's mixed oxidation states (Fe^{2+} and Fe^{3+}) were found based on the principal emission. The characterization data for self-synthesized core/shell Fe_3O_4 MIMNs coated with gum arabic were well-indexed with the standard Fe_3O_4. The Si peak observed in this XPS data was caused by glassware or impurities in the starting materials. Besides the slight binding energy difference between Fe_2O_3 and Fe_3O_4, the existence of a satellite between Fe $2p_{1/2}$ and Fe $2p_{3/2}$ of Fe_2O_3 can also be another distinguishing index. Based on previous research (Fig. 3.26), a satellite peak was observed in the XPS data for Fe_2O_3 (Fig. 3.27A), but none was observed in Fe_3O_4 (Fig. 3.27B) [157]. In a previous study, Ertl et al. reported that the Fe $2p_{3/2}$ peak in the standard Fe_3O_4 specimen lacks a satellite peak [158]. The presence of the satellite peak at 718 eV is used as an indicator of Fe_2O_3 rather than Fe_3O_4 formation. In the Fe_3O_4, the peak positions of Fe $2p_{3/2}$ and Fe 2p1/2 are located at 710.6 (S.D. = 0.05) and 724.1 eV (S.D. = 0.07), respectively. The Fe $^{2+}$:Fe^{3+} ratio should be 1:2 (0.33:0.67) in Fe_3O_4 (also expressed as FeO Fe_2O_3), calculated by the mean relative areas of each constituent peak. Our data align with the published results for Fe_2O_3 (Fig. 3.28A) and Fe_3O_4 (Fig. 3.28B), indicating that the mixed iron oxides were produced using gum arabic as a capping agent under a nitrogen atmosphere.

Atomic absorption spectroscopy was used to evaluate arsenic removal using the MINMs as adsorbents (Fig. 3.28). The "plant-extract and sea-animal extract encapsulated Fe_3O_4" nanocomposites demonstrated that the "biota-on-surface" strategy can improve the sequestration of heavy metals and the bioremediation of pharmaceutical waste. The AsO_4^{3-} removal using Fe_3O_4 NPs coated by natural product extract was increased by 25% compared with the non-shielded ones. The water remediation efficiency reached 100% when the As(V) concentration was below 10%. The result showed that the higher arsenic concentration attained the lower removal capacity in Fig. 3.28A. However, the reason for the lowest water remediation capacity at 10 ppm remains unclear, as it did

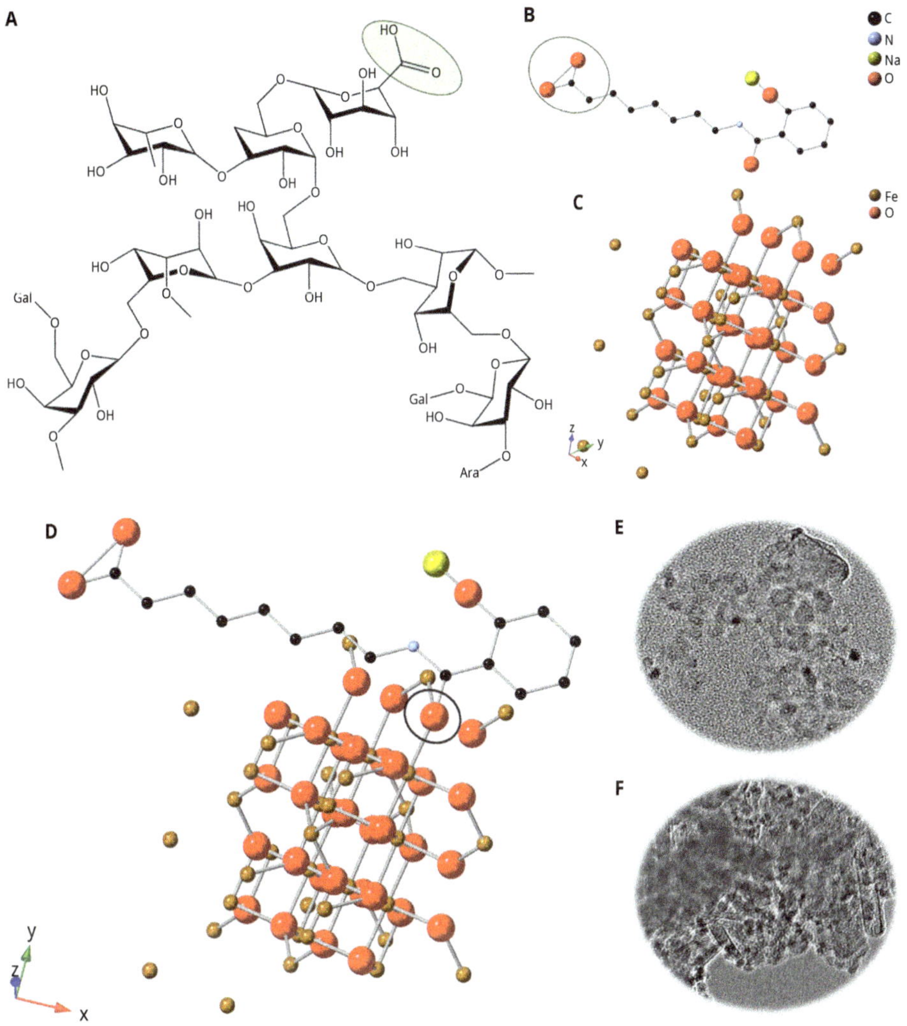

Fig. 3.25: The formation of Fe_3O_4 nanoparticles using gum arabic as a capping agent: **A**. The skeleton of gum arabic, **B**. the key unit in gum arabic chelated to Fe_3O_4 surface oxygen corner, **C**. Fe_3O_4 unit cell, **D**. the conjugation between gum arabic and Fe_3O_4 to form chelated nanoparticles, resulting in spatial hindrance, and **E**. the spherical structure of gum Arabic modified Fe_3O_4 NPs. The rod-shaped morphology was obtained for the Fe_2O_3-FeOOH combination.

not align with the overall efficiency trend. Shrimp shell extract core/shell MNPs obtained slightly more arsenic removal than gum arabic MNPs because chitosan contains an amino group. Based on the introduction in this paragraph, the amino group is responsible for chitosan's ability to absorb various substances. The amount of arsenic oxoanions removed did not show a distinct trend with concentration (Fig. 3.28B).

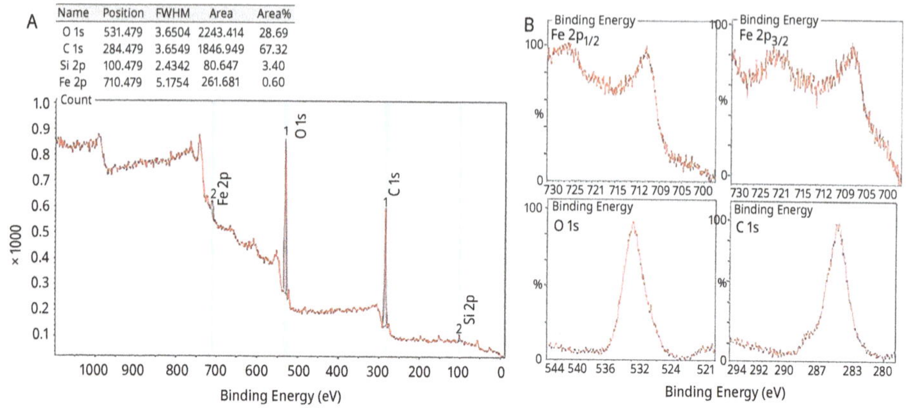

Fig. 3.26: The XPS analysis of gum arabic-modified magnetite nanoparticles: **A**. Survey of the MIMNs and **B**. binding energy of Fe ($2p_{1/2}$, $2p_{3/2}$), O 1s, and C 1s.

3.3.3 Bioinspired remediation: a practical approach

Our objective in bioinspired water remediation was to produce three formulations of interactive nanocomposites (INCs) that respond to variations in environmental stimuli and self-regulate their nanostructures and biochemical properties. These interactive nanomaterials aim to improve the bioremediation efficacy of organic pollutants and a heterogeneous mixture of polycyclic aromatic hydrocarbons (PAHs) and heavy metals. These INCs are biota-anchored porous nanocomposites with hollow core–shell structures and MOFs. The first formulation is composed of a β-cyclodextrin-oriented $CeO_{2-\delta}$ thin film (β-CDC) used for the biodegradation of benzene, toluene, ethylbenzene, and xylenes (BTEX), and trichloroethylene (TCE) under aerobic conditions. The second is iron-centered MOFs with positive surface functionalization for bioremediation of a heterogeneous PAHs and Cr(VI) mixture. The third is Zn bio-MOFs with adverse surface modification to enhance bio-sequestration and degradation of PAHs and Cd(II) mixture. The naturally occurring biota, such as *Pseudomonas aeruginosa*, were aligned along the membrane or framework surface to provide metabolic pathways for the remediation of organic compounds. These "biota-on-surface" structures were tuned to environmental conditions when used as interfacial biocatalysts, with an optimized amphiphilic balance between the shell and core, as well as between the host and guest. As discussed previously, our earlier study on "plant-encapsulated Fe_3O_4" and "citrate-coordinated TiO_2" nanocomposites indicated that this biota-on-surface strategy enhanced the adsorption capacity of heavy metal ions (Hg^{2+} and AsO_4^{3-}) [123] and bioremediation of pharmaceutical waste (tetracycline, TCH) with a dynamic and responsive reconfiguration to adapt to condition changes [99]. It is anticipated that the bacteria-anchored synthetic INCs act as "bioreactors" to provide biota protection, facilitate O_2

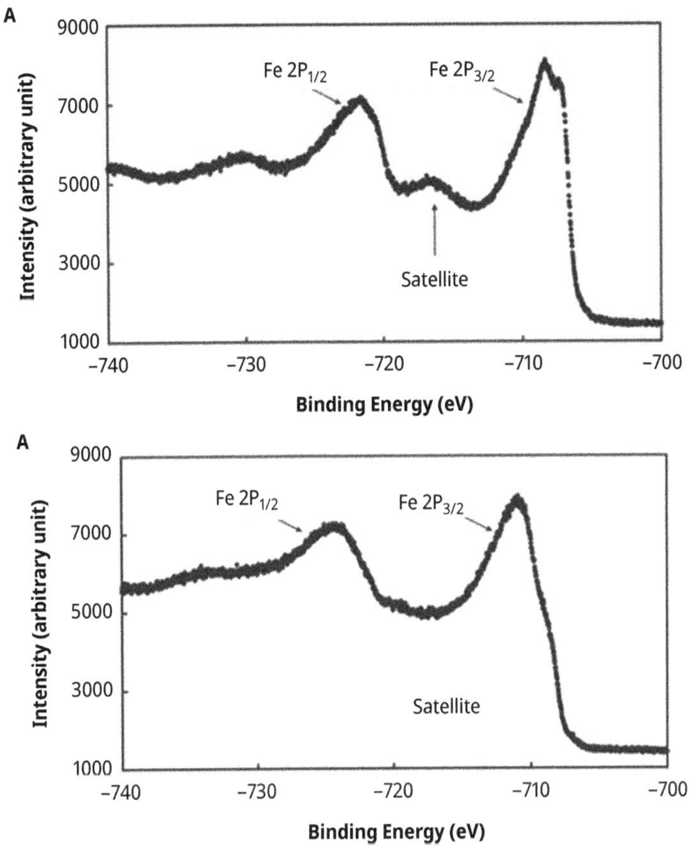

Fig. 3.27: The XPS analysis of iron oxides: **A**. The XPS spectrum of Fe 2p from the fractured surface of the Fe_2O_3 standard sample and **B**. the XPS spectrum of Fe 2p from the broken surface of the Fe_3O_4 standard sample (Images were adapted from [157] with permission).

diffusion, and collectively enhance remediation efficacy. The working hypotheses of these studies lie in that: (1) the reversibility between the two crystalline phases of cerium oxides is desirable for O_2 release via the formation of oxygen vacancies, enabling isometric cubic crystal unit; carried out by an encapsulated cargo (β-CDC) into biological environments; and (2) the functionalization of MOF motifs allows for fine-tuning of the surface charges and internal and external porosity, maximizing bioremediation efficacy of pollutant mixtures via immobilization of charge carriers at the heterojunctions of host–guest systems.

Bioremediation: The bioremediation rates of the CDCs and MOFs were measured in aqueous solutions containing organic compounds and mixtures of heavy metals and PAHs. Aqueous model solutions of contaminants (20 ppm) were prepared, into which INC colloids were introduced. Then the mixture was agitated for 50 min in the

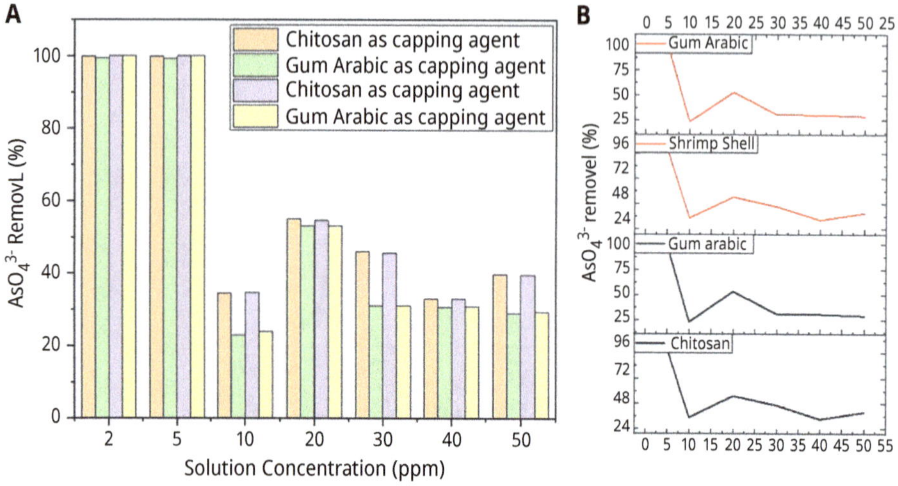

Fig. 3.28: The removal of AsO_4^{3-} using core–shell Fe_3O_4 nanoparticles: **A**. the relationship between removal efficacy and solution concentration and **B**. the comparison of the effectiveness of different formulations of nanoparticles.

darkness to establish contaminant adsorption–desorption equilibrium. The variables include degradation duration, pH, and pollutant concentrations. At 20-min intervals, an aliquot (0.3–1.0 mL) was filtered to remove INCs. The supernatant was collected and analyzed for absorbance or fluorescence using a Tecan M200 Pro to determine the degradation efficiency.

Metal ion sequestration: Both the Langmuir and the Freundlich isotherms were used to describe the adsorption behavior of heavy metals using functionalized MOFs. A self-generated host–guest system was tested to evaluate its bioremediation efficacy. The in situ analyses used pilot data on the reactivity of different formulations of interactive nanocomposites. These various formulations of INCs are responsive to environmental changes and self-regulatory in their structures and performance. Metal oxide-based INCs enhance the bioremediation of organic compounds through spontaneous catabolism. MOF-based INCs address metal sequestration by forming self-assembled host–guest systems that immobilize metallic pollutants. Naturally occurring bacteria are expected to align along the material surface, creating a bioreactor favorable for electron transfer. The nontoxic nature of these materials to neuron (RPE) cells mitigates or eliminates potential secondary pollution from these INCs. Thus, biota-on-surface nanomaterials were developed to improve environmental control of hazardous wastes and the bioremediation of co-occurring contaminants.

Three activities were carried out to improve water quality using different formulations of interactive nanomaterial composites (INCs): (1) fluorescence changes were measured to evaluate the bioremediation efficacy of the BTEX, PAHs, and TCE in a biphysic system composed of water-organic solvents using a Tecan M200 PRO; (2) the adsorption capacity of heavy metal ions by MOFs (host-guest system) and the bioremediation efficacy of the PAHs were measured in the same approach as mentioned above; and (3) the oxidation state, crystalline phase, and morphology were tested using XPS, powder, and single X-ray diffraction and transmission electron microscopy. Based on our literature survey, INCs with self-regulating behavior have not been reported. Their efficacy in bioremediation will open a new paradigm for controlling organic and heavy metal pollutants. A reversible, rapid lattice switch and space-occupancy upon oxidation-state changes between Ce^{4+} and Ce^{3+} are anticipated to enable free-electron transfer, thereby avoiding high lattice distortion. Due to electrostatic or van der Waals attraction, the heavy metal ions existing in wastewater and soil are absorbed by MOFs. Therefore, functionality and porosity must be tailored to improve the adsorption capacity for metal cations and polyatomic anions. The biodegradation efficacy of organic waste and the biological organism is expected to be enhanced by this self-generated host–guest system, driven by synergistic effects.

As we all know, nature has inspired researchers to design and develop materials with functionalization and fine-structural tuning under different reaction conditions [158]. An appropriate selection of solvent molecules, sub-building units, and switchable functional groups enables the preparation of innovative, responsive materials [159]. Water is a widely used solvent to create versatile aqueous solutions due to its high polarity and uneven electron distribution, which enable this simple molecule to dissolve a majority of inorganic and some organic compounds [160]. Its powerful solvation ability will facilitate self-assembly, biomimicry, and the complexity of the materials produced from intra- and intermolecular perspectives [161]. The significant advances in designing and constructing interactive and structure-tunable porous nanomaterials using wet chemistry are cost-effective and user-friendly [162]. The homogeneity and architecture of end products were tuned at the molecular level, enabling precise, accurate fine-tuning [163]. Multiscale texturation and upscaling production were achieved using sol–gel chemistry [164] and self-assembly to obtain thin films, with cyclodextrin as a cargo to disperse CeO_2 uniformly-δ NPs (nanoparticles), as found in our previous study [165]. The combined solvothermal and microwave methods also facilitated rapid nucleation and miniaturization of MOFs [166], followed by a post-functionalization to modify their surface properties [167]. The final products were manifested as particles, tubes, hollow spheres, and thin films [168]. These synthesis methods maintain the intrinsic biological, physical, and chemical properties of starting materials, inorganic compounds, polymeric surfactants, or hybrid matrices [169]. Therefore, the porosity, pore size, surface area, particle size, and distribution were tuned to obtain well-defined porous polymeric networks, interface connectivity, and accessibility to reactive sites [170].

The inner transition metal oxide, nonstoichiometric $CeO_{2,\delta}$ has been chosen as a core for the hollow core-shelled INCs, used for in situ bioremediation of organic compounds due to its advantages, low cost, environmental protection friendliness, and high chemical/structure stability. It was reported that CeO_2 hollow spheres enabled the removal of dyes such as Congo red and acid orange [171]. Adsorption and separation of heavy metal ions and dyes using CeO_2 NPs with high adsorption capacity were also reported, owing to their large pore/cavity volumes and facile mass transport [172]. Moreover, these CeO_2 hollow spheres can be readily regenerated via a centrifugation–calcination procedure, with minimal loss in adsorption performance [173]. Introducing a pendant beta-cyclodextrin (β-CDC) unit into the $CeO_{2\text{-}\delta}$ sol–gel synthesis allowed self-assembly into a Ce-containing hydrogel through complexation between both components through ion-dipole attraction [174]. The redox state of $CeO_{2\delta}$ can alter its binding affinity for β-CDC, thereby seducing polymer chains with variable modulus[175], surface charges, and accessibility to pores [165]. As a result, the O_2 release (1.5×10^{-4} mmol O_2/L h) was modulated to meet the needs of the existing biota. Iron-centered MOFs were selected for their proven ability to sequester toxic heavy metals [176], thereby forming a host–guest system anticipated to synergistically enhance the biodegradation of aromatic compounds (BTEX and PAHs) [177]. The MOFs are crystalline frameworks of metal ions and organic ligands [178]. The pores inside an open MOF were stable and tunable to high surface areas, various topologies, and network types via increasing linker length and size, targeting specific topologies, and further addressing metal-organic vertices to produce bio-MOFs [179]. The carboxylate linkers were widely used to prepare MOFs with microporous and mesoporous cavities [180]. Uniform and tunable pore sizes, along with functionalization readiness, offer great potential for MOF applications [181]. Cavities up to 4 nm have been introduced to hierarchically assembled MOF motifs by adopting an angular, semi-flexible tetra-topic ligand [182]. The resulting MOFs exhibit permanent porosity and show stepwise sorption isotherms for select metal ions by utilizing ion exchange, altering ligand functional groups, and introducing unsaturated metal centers [183]. The functionalized MOFs can prevent dissociation of the host–guest complex by preventing possible ion exchange or covalent cross-linking [184]. A recent study found bacterial attachment to PAHs, which serve as carbon and energy sources, depending on PAH solubility [185]. Since bacteria initiate PAH degradation by intracellular dioxygenases, PAHs were oxidized into cis-dihydrodiols and further into catechols or derivatives by incorporating both atoms of O_2 [186]. This approach demonstrates that harmful metal ions were immobilized via surface adsorption, thereby moderating their toxicity. Consequently, a self-assembled host–guest system was formed to mitigate the damaging effects of biota and provide reactive sites for synergistic PAH degradation.

The "plant-extract encapsulated Fe_3O_4" and "citrate-coordinated TiO_2" nanocomposites indicated that the "biota-on-surface" strategy could improve the sequestration of heavy metals and bioremediation of pharmaceutical waste. The AsO_4^{3-} removal using plant-Fe_3O_4 NPs was increased by 25% compared with the non-shielded iron nanomaterials. Using citrate-TiO_2 NPs (20 ppm), the TCH degradation rate achieved a 92.5% reduction within 2 h under simulated sunlight [99]. Their toxicity was evaluated using retinal pigment epithelium (RPE) cells at the US Air Force Research Laboratory (AFRL), Fort Sam Houston, indicating they are user-friendly [122]. Sol–gel wet chemistry is one of the most feasible and effective methods for developing interactive CeO_2-nanofilms with bacteria aligned along the film surface. The Ce oxidation number was controlled by electrochemically active biota that released O_2 in a controlled manner, with the β-CDC matrix serving as the nutrient source. The synthesis variables for producing different INCs have been optimized in our group, and 12 formulations were found to be biocompatible based on cytotoxicity evaluation in RPE cells. This strategy will modify the current sol–gel synthesis to produce nonstoichiometric $CeO_{2,\delta}$ into which 1–5 mass% of carbon graphene quantum dots is introduced to facilitate the Ce^{4+} to Ce^{3+} redox reaction for controllable release of O_2. There are four key variables to control β-CDC structure using Ce(IV) alkoxides as starting materials: hydrolysis inhibition, hollow sphere templating, crystallization, rapid shift of redox states, and formation of "biota-on-surface." This method provided ultrasmall particle size, a long triple-phase boundary, molecular-scale homogeneity, and O_2 and C sources to activate bacterial spontaneous degradation of BTEX, PAHs, and TCE. This proposed family of innovative and biocompatible INCs is expected to enhance the bioremediation efficiency of in situ hazardous substances in groundwater and contaminated soil by leveraging naturally existing bacteria to construct "biota-on-surface" structures (Fig. 3.29).

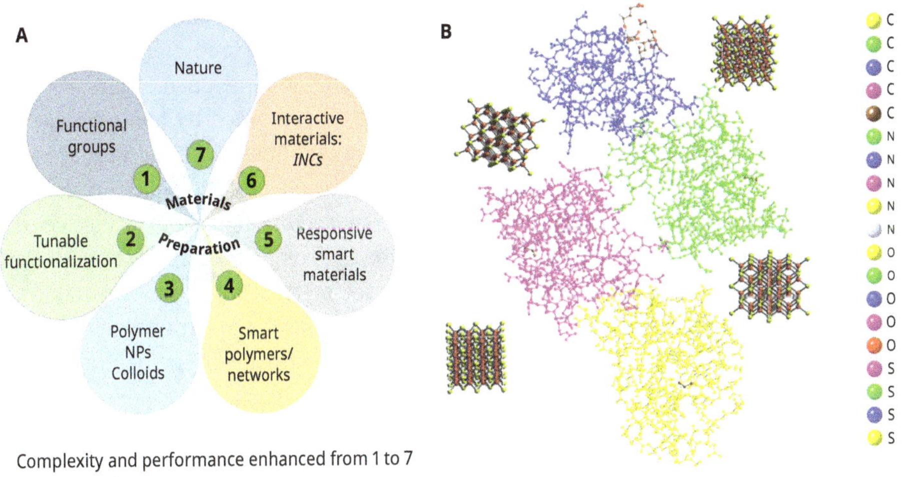

Fig. 3.29: Creation of "biota-on-surface" structure with improved remediation of organic wastes with a sensible response to external stimuli and tunable geometry: **A.** Wet chemistry synthesis to balance complexity and performance of INCs and **B.** "biota-on-surface" lectin–CeO_2 structure offering spontaneous metabolic pathways.

The controllable O_2 release is proposed to occur through vacancy formation within the CeO_2 lattice, while maintaining its crystalline structure. The rapid phase shift will not produce enough O_2 for bacteria to react to unsaturated covalent bonds. Two alternative approaches were also implemented (Fig. 3.29A): (1) carbon quantum dots (1–5 % by mass) were introduced to initiate the redox reaction of $Ce^{4+/3+}$; (2) aloin-encapsulated MO_2 composite was tested to facilitate the spontaneous oxidation of C=C double bond, and C–Cl single bond. Lectin is a family of carbohydrate-binding proteins. Recently, seven lectins have been identified: C-type, L-type, P-type, M-type, fibrinogen-like domain lectins, galectins, and calnexin/calreticulin. These lectins with different structures showed diverse expression patterns and multiple functions in the immune response. The lectin C was chosen to demonstrate biota-on-surface concepts, enabling nanomaterials to be sensitive and responsive to external signal changes and to self-regulate their structures. Although the mechanism by which lectins bind to and communicate with cells remains to be investigated, binding triggers a response to environmental changes (Fig. 3.29B). This example focused on forming lectin-anchored nanoparticles to remove biological pollutants through selective chemical binding.

The functionalized MOFs are proposed for heavy-metal sequestration and PAH biodegradation. These highly porous MOFs are attractive for outer coordination and intermolecular interactions to immobilize harmful contaminants. The proposed MOFs are iron ($Fe^{2+/3+}$) and Zn^{2+} as central elements, dicarboxylic benzoic acid (BDC), BDC derivatives, and adenine as ligands. The Zn-MOFs have been widely studied and produced using a solvothermal method, followed by a post-modification to tune the surface charge and porosity. A mixture of H_2O and DMSO was used as the solvent, and the temperature was maintained at 120 ± 5 °C for 24–72 h. The biocompatible and robust bio-MOF-1 ($C_{104}H_{64}N_{20}O_{25}Zn_8$) produced by An et al. [187] demonstrated large open cavities. In this MOF structure, polyhedral model, benzene rings are represented by black hexagons; green pentagons represent imidazolate (C_3N_2) rings, and pyrimidine (C_4N_2) rings are represented by blue hexagons (Fig. 3.30A). A series of organic pollutants (Fig. 3.30B) was oxidized through adsorption, and heavy metals were removed through ion exchange.

This geometry enables post-modification of the ligands or MOF surfaces with amphiphilic groups. Both protic and aprotic channels for metal adsorption were achieved in the synthesis under mild conditions. Our previous results demonstrate that bioremediation of xenobiotics was achieved using engineered biota-on-surface nanomaterials (Fig. 3.30C). If cytochrome % activity > whole-cell (WC), indicating favorable e-transfer, bioremediation occurs spontaneously (Fig. 3.30D). Thermodynamically, the Gibbs free energy is negative; however, the reaction is slow due to the steric hindrance of the functional groups that limit the kinetics of the bio-remediation of BTEX compounds. We aimed to increase the reaction rate under ambient conditions using a biota-on-surface strategy. The immobility of harmful metal ions using MOF-based materials and their subsequent host–guest formation will be problematic, causing damage to biota and low biodegradation efficacy. Natural products were found to be more resistant to these poisonous cations. We aim to emulate natural processes, enabling spontaneous bioremediation of mixed contaminants in the environment.

Liu and her collaborator conducted a preliminary study on nanobioremediation (NBR) of PAHs in soil under varying levels of heavy metals to improve the effectiveness of in situ remediation. PAHs, widespread environmental pollutants that contain benzene rings arranged in linear, angular, or cluster arrangements, are considered priority pollutants by the U.S. Environmental Protection Agency due to their toxicity,

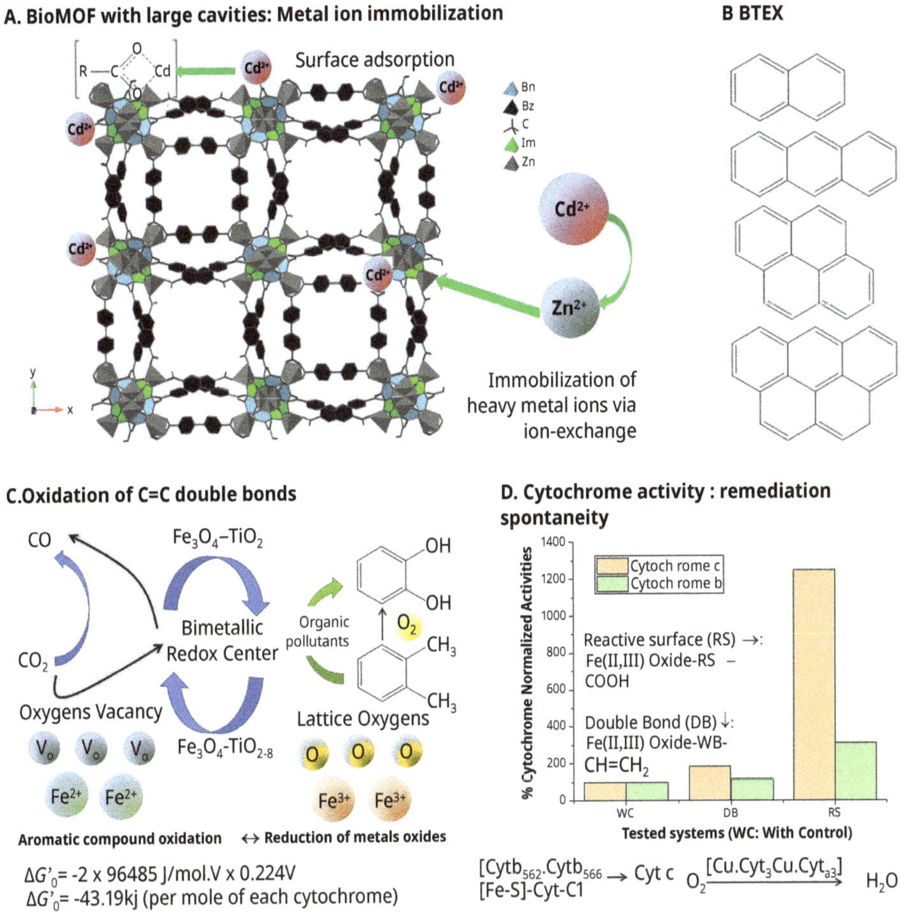

Fig. 3.30: Evaluation of pharmaceutical waste degradation using INCs: **A.** Metallic ion immobilization using bio-MOFs, as bioreactors to break down aromatic compounds and remove heavy metals through ion exchange; **B.** a series of organic compounds used as the model molecules; **C.** proposed redox reaction of organic compounds, and **D.** % cytochrome normalized activity indicating spontaneous bioremediation.

mutagenicity, and carcinogenicity [188]. At many Superfund sites across the United States, heavy metals often co-contaminate PAHs and are reported to impair the biodegradation of PAHs [189]. To reduce PAH abundance in soil at specific locations, more research and investment are needed to develop field-deployable approaches. These methodologies yield real-time tox and species data and can help design nanoparticles to remediate PAH effects in soil [190]. In particular, nanoparticle-enabled in situ remediation techniques represent a key area of contemporary scientific advancement for removing pollutants from various media, such as soils, surfaces, and groundwater. As a result, there has been growing interest in the synthesis and assessment of biomolecule-functionalized nanoparticles (NPs), for example, via surface functionalization, to increase the sensitivity, selectivity, reliability, and practicality of environmental nanoremediation [191]. However, several challenges have been encountered during the in situ deployment of these technologies, including limited mobility, transport, and distribution of these NPs in field applications [192]. In addition, the application of bioremediation in the natural environment is often significantly influenced by field conditions; thus, in situ environmental remediation demands significant effort to achieve a complete understanding of the biochemical mechanisms, ecological constraints, and their by-products to avoid unpredictable adverse remediation outcomes. Given the above technical hurdles, our approach focused on NBR of PAHs in soil in the presence of co-contaminating heavy metals. PAHs often co-occur with heavy metals at Superfund sites. Thus, the successful implementation of PAH biodegradation usually depends on understanding its biodegradation in the presence of heavy metals [193].

Many researchers have highlighted the benefits of bioremediation of combined heavy-metal and PAH pollution. For example, Lu et al. [194] found that adding a moderate dose of pyrene altered the microbial community, promoting microbial growth in soils and thereby relieving metal-induced stress and favoring phytomass yield. However, other researchers, such as Gauthier et al. [195], reported significant deleterious effects of PAH–metal mixtures on microbes. Also, most current studies have focused on the remediation outcomes of combined pollution. The mechanisms underlying interactions among heavy metals, PAHs, and microbes remain poorly understood. Bacteria of the genus *Pseudomonas* are among the microorganisms that can effectively decompose organic pollutants through co-metabolism in natural water and soil environments. Patel et al. [193] concluded that a bacterial consortium consisting of four strains, including *Achromobacter* sp. BAB239, *Pseudomonas* sp. DV-AL2, *Enterobacter* sp. BAB240, and *Pseudomonas* sp. BAB241 degraded naphthalene (1,000 ppm) over a wide pH range, with an excellent degradation rate of 80 mg/h, and other pollutants as well. Brito et al. [196] reported that a bacterial consortium comprising *Pseudomonas* sp. is highly efficient at degrading pyrene. Zhang et al. [197] reported that inoculating a bacterial mixture containing *Pseudomonas stutzeri* (91.7%) and *Candidatus Kuenenia* (2.3%) removed 54% of the added phenanthrene (45 mg/L) under anaerobic conditions. This bacterial mixture has been identified in the functional microbes during the bioremediation of landfill leachate with high-strength nutrients and organic pollutants. The engineering of biotechnological systems that employ these diverse microbial communities plays a key role in the efficient and sustainable treatment of PAH pollution. However, the microbial synergistic mechanisms and interactions underlying the efficient removal of PAHs in heavy-metal co-contaminated environments remain poorly understood [198].

NBR, the integration of traditional bioremediation with NPs, has recently received greater attention as an innovative method for effective, efficient, and eventually sustainable remediation. Recent findings suggest that inoculating multiple microbial communities and biofunctionalized nanomaterials can enhance microbial activity, synergistically improve PAH removal, alleviate heavy-metal stress, and produce high-value materials [199]. Various laboratory and field application studies on NBR in saturated soils or groundwater have been conducted [200]. Most of these studies focused on the use of elemental or zero-valent metals in nanoscale form, such as iron, nickel, and palladium; however, problems involving the reactivity, the useful life, in situ transport processes, clustering tendency, and accelerating deposition of nanoparticles, and their adverse effects on microbes have been reported [200]. In addition, studies on the impact of NP behaviors in the soil pores, adsorption on mineral particles, interaction with soil microorganisms, and biogeochemistry of the contaminated site, such as particle size, moisture, pH, presence of nutrients, organic matter, native microorganism, and type of clay mineral on the NBR have been minimal [189]. Thus, the successful in situ application of this biotechnology is only possible after all these limitations and problems are sufficiently addressed.

Fe_3O_4 NPs coated with chitosan in removing chromium(VI) from wastewater were evaluated by us using batch adsorption experiments. Results have shown that the chitosan-coated Fe_3O_4 NPs have a better adsorption capacity than uncoated Fe_3O_4 NPs reported in the literature, and chromium removal by the surface-coated NPs is highly dependent on solution pH and the NP dosage. The removal of organic refractory chemicals, such as bisphenol-A (BPA) and estrone (E1), in a sequencing batch reactor (SBR) inoculated with anammox bacteria, has also been evaluated by us. Results show that BPA and E1 have negligible effects on microbial activity, while removing 78% and 99% of BPA and E1, respectively.

3.4 Biological and forensic application

Nanomaterials are redefining theranostics by addressing key challenges such as low drug specificity, instability in biological fluids, and rapid clearance from host tissues. MOFs (metal organic frameworks), with their high surface-area-to-volume ratio and crystalline stability, offer precise control over drug release while minimizing side effects. Their nanoscale dimensions reduce the immune response and enable selective capture of signaling molecules, such as nitric oxide (NO), which plays a dual role in cell survival and apoptosis. Nickel-based MOFs further enhance therapeutic potential by generating singlet oxygen and ROS, modulating mitochondrial respiration, and supporting phototherapy via near-infrared/red light (NIR/RL). To improve targeting, MOFs were engineered with silica shells and RDG tags, yielding singlet oxygen yields comparable to those of methylene blue. In parallel, nanotechnology is transforming forensic science. Covalent organometallic polyhedra (COP)-based agents and hyperspectral scanners detect faint fingerprint residues, while artificial neural networks (ANNs) match low-quality prints. These innovations, supported by realistic artificial sweat formulations and specialized training enable high-resolution forensic analysis and advanced suspect identification.

3.4.1 Cancer theranostics

The current therapeutic barriers are low specificity, rapid instability in biological fluids, and removal from the host organs via several biological modalities [201]. It has recently been found that MOFs are ideally suited to address these challenges. MOFs are suitable as drug carriers due to their high surface-area-to-volume ratio, which allows for low drug loading and fewer side effects, as well as their host–guest properties, crystallinity, and stability in biological fluids [202]. The small size of the crystal can minimize an immune response and provide a large surface area for rapid kinetics (controlled drug release or selective capture of messengers such as NO) (nitric oxide) [203]. NO is essential, but it is central to cell survival. High NO levels lead to mitochondrial membrane polarization, calcium overload, energy (adenosine triphosphate) depletion, lactate dehydrogenase, and cytochrome c release [204]. NO cross-talk also leads to caspase-3 activation and apoptosis [205]. Low NO levels promote vasodilation, cell proliferation, enhanced wound healing, decreased phosphorylation of protein kinase B (PKB, also known as Akt; ('thymoma AKT8 virus oncogene')), and control of pro-apoptotic factors [206]. Recent research also suggests that NO levels be attenuated (decreased) or potentiated (increased) by irradiation with NIR/RL (near infra-red/red light), a phenomenon known as phototherapy (PT) [207]. The beneficial effects of NIR/RL are thought to occur via cytochrome c oxidase [208]. The Ni-based MOFs will also enable us to examine this biochemical pathway by monitoring changes in singlet oxygen species (SOS) and ROS (reactive oxygen species) that alter mitochondrial membrane potential and affect respiration through changes in cytochrome c NO reductase activity (as well as its more common oxidase activity) [209].

A potential difficulty is poor drug storage or a lack of interaction at the release site. A silica shell (for improved stability) and amino acid RDG (Arginine -Glycine -Aspartic acid) tag binding were included for site-specificity, and targeted release was attempted. Our preliminary use of MOFs suggests singlet oxygen generation comparable to that of methylene blue (MB), indicating that the MOF architecture can be incorporated to offset poor drug or NO release by promoting SOS generation as supplemental photodynamic therapy (PT) [210]. Other cell lines, such as RPE (ATCC® CRL-4000) (Retinal Pigmented Epithelium), will be used according to standard protocols. Cell morphology was examined using confocal microscopy. Cell proliferation/cytotoxicity assays, such as lactate dehydrogenase, were used for the MTT (3-(4,5-diMethylThiazol-2-yl)-2,5-diphenylTetrazolium bromide) assay, and caspase-3/7 activity was used as an alternative to SOS/ROS [211]. These tests yield information on the mechanism of cell death and the target organelle (cytoplasm, mitochondria, or nucleus) by comparing cell death biomarkers and mapping (EELS (Electron Energy Loss Spectroscopy), confocal microscopy) data, coupled with MOF design [122]. The PT assay was based on the recovery of RPE cells to H_2O_2 wounding by comparing RPE (in the dark, d), RPE + H_2O_2 (d), and RPE + PT, RPE + H_2O_2 + PT, where PT is continuous-wave red light (2.88 J/cm^2).

The multi-metal entities were potentially adaptable as nodes for constructing diverse molecular architectures, and the stability and magnetism in the supramolecular systems resulted in new properties and functions [212]. Potentially, the stability, magnetism, and adaptability of multi-metal entities as nodes for constructing diverse molecular architectures could result in supramolecular systems with new properties and functions. An antineoplastic drug (SN-38, a topoisomerase I inhibitor) can lead to low metabolism, resulting in diarrhea and toxicity in some patients [213]. These adverse effects were reduced through greater drug loading within the MOF superstructure and slower release. Drug metabolism is lower than current loads, leading to reduced glucuronidation and increased toxicity. Nanostructured MOFs effectively encapsulate and release the drug via pH-triggered biodegradation, thermal hysteresis induced by an external magnetic field, or slow release from the target organ [214]. If suitably tuned, the MOFs will be activated by NIR radiation, facilitating photodynamic therapy and CT (Computed Tomography), potentially in the same MOF structures [215].

NMOFs offer improved pharmacokinetic properties similar to those of polymer-encapsulated drugs [216]. A size range between 35 and 70 nm was generated with encapsulation of sugars or silica to improve stability and further post-synthetic modification by functionalization with alpha–beta integrin (RDG) as a delivery vector [217]. The IC_{50} of such functionalized NMOFs is expected to be lower than that of uncoated MOFs. In addition, by functionalizing with a specific peptide vector such as RDG, the NMOF (Nanoscale Metal-Organic Frameworks) can facilitate receptor-mediated endocytosis and prodrug activation (by altering the microenvironment of cancer cells), potentially lowering required doses and side effects [218]. An alternative approach is to engineer the MOF to release NO and promote cell death upon activation with a visible-light laser [219]. The development of functionalized MOF-based probes can provide a new paradigm for understanding cellular mechanisms underlying oxidative stress and for treating cancerous cells. Our approach focused on these three activities: (1) preparation of a series of new MOFs using a feasible wet-chemistry method; (2) characterization of the structures of these MOFs using state-of-the-art instrumentation; and (3) embedding of drugs into the MOFs and using them for cancer theranostics (NO release/adsorption, ROS, and SOS test). The team directed new MOF applications in cancer theranostics and disinfection science [220]. Some of her MOFs were evaluated as NO, SOS, and ROS probes using capabilities at the US AFRL, based at Fort Sam Houston.

Traditional approaches include selectively attaching nanoparticles to cancerous and healthy cells. Drug dose and release gradient, side effects, drug clearance, potential immunogenicity, nonspecific drug release, or low bioavailability are potential issues in any disease treatment [221]. However, MOFs with tunable structures and functionality could minimize many of these issues [222]. The metal within the MOFs will be used for diagnosis or therapy using intrinsic magnetic, optical, or pH-related properties [223]. We aimed to leverage the integrated therapeutic and diagnostic capabilities of prepared MOFs in a theranostic approach targeting cancer cells. Metastatic and drug-resistant cancers, as well as cancer stem cells, pose the most significant challenge to targeted therapy [187]. This research investigates the theranostic application of MOFs to overcome biological barriers to reaching cancerous cells, leveraging their crystalline nature, ultrahigh surface area, tunable porosity, drug-loading capacity, controlled release, and high stability in biological fluids [224]. In addition, the prepared MOFs allow for the integration of therapeutic and diagnostic platforms into this theranostic approach, target cancer cells specifically, and examine the role of NO in any phototherapy (PT) [225]. In particular, metastatic cancers, drug-resistant cancers, and cancer stem cells pose the most significant therapeutic challenge for targeted therapy [226]. In immunotherapy, targeted therapy was achieved using appropriately designed drug-delivery vehicles, such as nanoparticles, adult stem cells, or [thymus-matured] T cells [227]. We implemented a strategy to treat drug-resistant tumors using combination therapeutic agents with distinct mechanisms to achieve synergistic effects and overcome the drawbacks mentioned [228]. In this example, functionalized MOFs were used in cancer theranostics.

To understand the mechanism of interaction between the cancerous cells and nanomaterials: Different cancers affect distinct biochemical pathways, and many cancers "uncouple" cell proliferation from metabolism, allowing cells to divide uncontrollably and consume valuable metabolic resources [229]. MOFs can track biochemical changes and deliver drugs of known efficacy to aid in understanding biogenesis, cell-to-cell communication, and the regulation of apoptosis [230]. The application of NO (a neuromessenger) affects many biochemical pathways identified as essential in low-intensity visible light therapy [231]. Low-light illumination therapy exploits NO deregulation, which mirrors mitochondrial deregulation, characterizing cells in a cancer-like state rather than the normal state, where mitochondria appear tolerant to apoptotic signals [232]. Using MOFs as probes to capture and release NO and then examine other "reporter" ions allows us to analyze the mechanisms [233]. In most studies, because inhibitors and site-specific dyes are used in tandem, MOFs enable the integration of these two distinct molecules into a single MOF with dual or multiple purposes. We designed a quintuple-faceted MOF for *imaging* (virtue of metal center fluorescence), *quantification* of NO (via dichlorodihydro-fluorescein diacetate based loading), *diagnosis* (electron energy loss spectroscopy mapping of MOF within the cell), and *therapy* (SOS or ROS) (or PT (using NIR to RL). Thus, versatile MOFs can effectively treat disease and serve as NO probes to understand better the mechanism underlying observed PT [234].

A feasible solvothermal synthesis has been used to produce a series of MOFs featuring four metallic cations (Fe^{3+}, Ni^{2+}, Cu^{2+}, and Zr^{4+}) as central elements, owing to their magnetic and fluorescent properties. Five groups of ligands (chosen from Zhou's ligand library) with different sizes and bridging angles were selected to tune the MOF's architectures and properties. Amphiprotic groups (such as carboxylic and amine) were added to functionalize the structure and increase its flexibility for binding to cancerous cell surfaces. To construct MOFs for cancer theranostics, we selected pyrene, a stable organic linker that exhibits strong fluorescence and photostability under pH and moisture conditions, as the organic linker. We used pyrene to construct MOFs for PT and charge-transfer (CT) interactions. Based on the "hard and soft (Lewis) acids and bases" theory, Fe^{3+} and Zr^{4+} cations, with high valence and closed-shell electron configurations, can form strong coordinative covalent bonds with carboxylate groups and produce fluorescence, along with their intrinsic magnetism [235].

Similarly, the Ni and Cu MOFs were prepared by solvothermal synthesis to control pore size for NO capture and release. Morphological and crystalline structure analyses were conducted using a transmission electron microscope and atomic force microscopy (Fig. 3.31). Images showing well-defined domains were received (Fig. 3.31A), with the edges (Fig. 3.31B) and corners serving as reactive sites (Fig. 3.31C) for cancer diagnosis and treatment.

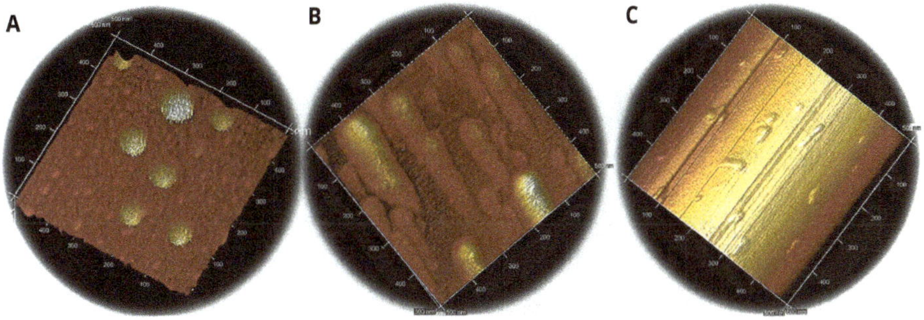

Fig. 3.31: Atomic force microscopy analysis of MOF motifs.

Commercially available SN-38 complexes were embedded into the magnetic MOFs through double injection. Under vigorous agitation, an aliquot of SN38 powder has been introduced into the above MOF's DMSO (dimethoxy sulfoxide) solution. The resulting suspension was then filtered through a 0.45-μm syringe filter to remove the unbound drug. The drug loading was estimated by high-performance liquid chromatography from phosphate-buffered saline (PBS) extracts over 2 weeks, the maximum time SN-38 (Sankyo-Nihon series 38) is expected to remain in the encapsulated form. DAF (Diaminofluorescein) (50 μg in DMSO) was impregnated into the MOF using pressure- and temperature-assisted injection, with an anticipated release rate of 50 μM/h, as measured by fluorescence integration (signal increases over time). In vitro cytotoxicity of Zr-MOFs was tested on ovarian cancer cell line (series A, A2780 line) (Fig. 3.32) and retinal pigmented epithelium (RPE) cells. The 3-(4,5-dimethyl-2-thiazolyl)-2,5-diphenyl-2 H-tetrazolium bromide (MTT) assay measured cell viability. A singlet oxygen sensor green (SOS) [236] and ROS [237] were used to measure oxidative stress [238]. Data were collected using a microplate reader with appropriately set excitation and emission wavelengths. Cell viability after exposure to the MOFs was expressed as a percentage of that in standard growth medium (Fig. 3.32A). The MOF-based materials have applications in healthcare, including as catalysts for water purification and as NO probes for cancer treatment. The MOF motifs exhibit crystalline structures with high surface area and large pore volumes. These factors enable them to adsorb functional molecules or anti-cancer drugs. These substances were either adsorbed on the external surface or within the open channels of the MOFs, or trapped inside the pores of the MOF framework. High efficacy in medical diagnosis, treatment, and imaging was achieved using MOFs alone, even without loading with SN-38 drugs. Loading of 5–15 mass% of drug within the MOFs was achieved by adjusting the open porous structures. The MOFs were stable in PBS and in biological matrices of similar pH, and they will biodegrade at lower pH, typical of rapidly dividing cells (IC_{50} (half maximal inhibitory concentration) < 10 μM, Fig. 3.32A) [239]. The MOFs (Fig. 3.32B) were excreted in the urine, with the Fe recycled by the body, similar to iron supplements. The aim was not to upload ultrahigh doses of the drug, but with moderate loads and slow release. The slow release facilitates a rapid drawdown of the cancerous cell population without common side effects such as diarrhea. The TEM images (Fig. 3.32C) and the schematic diagram (Fig. 3.32D) show a well-defined MOF structure with high porosity, which contributes to improved cancer theranostic efficiency Image: TEM and schematic of MOF structure. Zr-PCN (Porous Coordination Network) MOFs display an isoreticular structure of (P6/mmm group with $a = 39.38$ Å and $c = 16.48$ Å, $\alpha = 90°$) [240]. The secondary building unit comprises a Zr 6-octahedron with eight μ_3-OH, eight normal OH, and eight carboxylate groups. The 3D (three-dimensional) framework with two types of 1D (one-dimensional) channels oriented along the *c*-axis was controlled based on the end application. TEM analysis showed that Zr-MOFs displayed well-defined, highly crystalline coordination compounds. These nanoscale MOFs (NMOFs), equipped with versatile delivery vectors, enable theranostic applications and demonstrate improved efficacy against ovarian cancer cell lines. Percentage cell viability in the regular A2780 cell line (Fig. 3.32E) showed that cells were ~100% viable at concentrations below 1×10^{-8} g/mL. The MOF concentration was controlled from 100 μM to 0.1 nM. After treatment with MOF motifs, the A2780 cells swelled significantly, indicating cell death. Developed nanotechnology platforms exhibit high target specificity with minimal collateral damage and immune response. It is hypothesized that the magnetic properties of Fe-MOFs are crucial in cancer theranostic studies, owing to the hybridization of the metal's d orbitals with the 2p orbitals of oxygen [241], as well as magnetically induced hyperthermia resulting from the spinning of the metal center.

Theranostics combines a diagnostic tool with targeted therapy, thereby expanding our understanding of cell function [243]. Engineered nanomaterials are well-suited for the theranostics approach [244]. In particular, MOFs allow for controlled drug loading. The flexibility of ligands and metal availability in fabricating MOFs, along with the potential for differential drug loading, makes them suitable for imaging, quantification, diagnosis, and therapy [245]. We anticipate that MOFs are neither the cause of oxidative stress (via SOS/ROS) nor cytotoxic (via MTT assays); however, once the drug is released or the MOF bound, cell proliferation is expected to decrease in a dose-dependent manner, resulting in cell death via ROS or inhibition of cell proliferation pathways [246]. Our data (Fig. 3.33) indicated that MOFs cytotoxicity could be tuned (low Fe-ABBT alone to high with H_2O_2) [247]. We expect to design MOFs with steady NO release over the entire exposure period (0–4 h). NO release was determined using an NO-specific dye, 4-amino-5-methylamino-2′,7′-difluoro-fluorescein diacetate (DAF-FM dye), measured using differen-

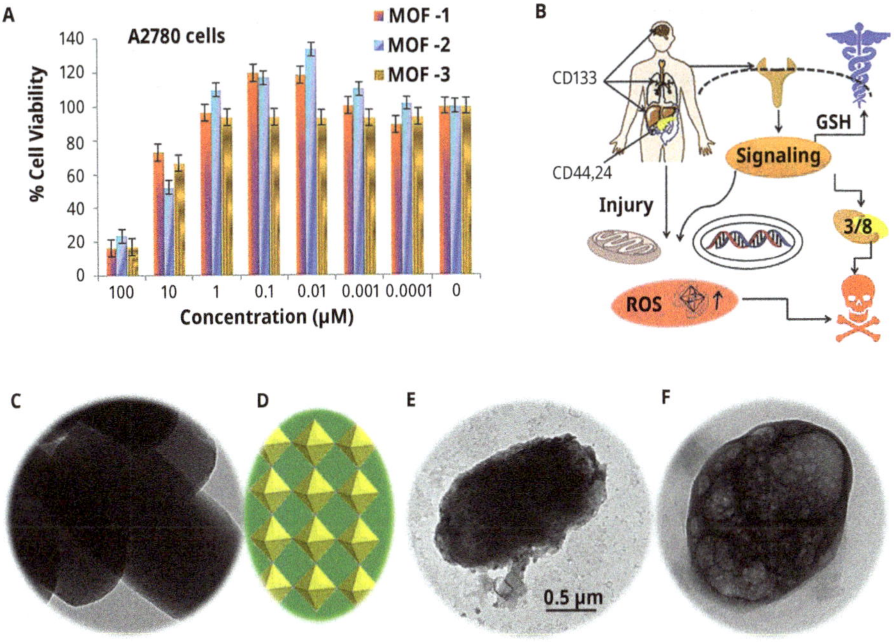

Fig. 3.32: Three metal–organic frameworks used as nanodrugs targeting cancer cells. A. Percentage cell viability at different concentrations of MOFs with regular A2,780 cell line revealed that cells are ~100 % viable below a concentration of 1×10^{-8} g/mL: **A.** viability of Fe-MOFs, **B.** mechanism of inactivation of cancer cells, **C.** TEM image of Zr-MOF, **D.** a diagram showing highly ordered and porous MOF motifs, **E.** intact A2780 cells as control; and **F.** damaged A2780 Cells permission from [242].

ces in nonfluorescent and (unbound) and fluorescent benzotriazole (NO bound) species upon light irradiation using a plate reader (Tecan 200Pro). Both MOFs using ligands of HTTA (2′-hydroxy-[1,1′:3′,1″-ter phenyl]-4,4″,5′-tricarboxylic acid) (Fig. 3.33A) and PDDA (5,5′-(propane-1,3-diylbis(oxy))diiso-phthalic acid) show lower toxicity and low inhibition of cell division (Fig. 3.33B) ($m = 0.01$, $b = 17.55$; $m = 0.01$, $b = 33.57$ respectively, where m is the rate of inhibition and b is the intercept).

The MOF projects summarized in the last segments demonstrate a new paradigm in cancer theranostics using organometallic chemistry as a therapeutic strategy. There have been three broad, innovative research discoveries: (1) The functionalized MOFs display a unique surface structure and internal pores, allowing for drug entrapment and release, and precise epitopic tagging of flexible, tailored MOFs to be selectively conjugated with the cell surface. Through surface-to-surface interactions, the embedded drugs entrapped within the MOF structure were released to the cell membrane of the cancerous cell membranes to promote cell death. (2) NO capture and controlled release provide a better understanding of the mechanism of interaction between MOFs and the cancerous (or oxidized) cell membranes. The progression was

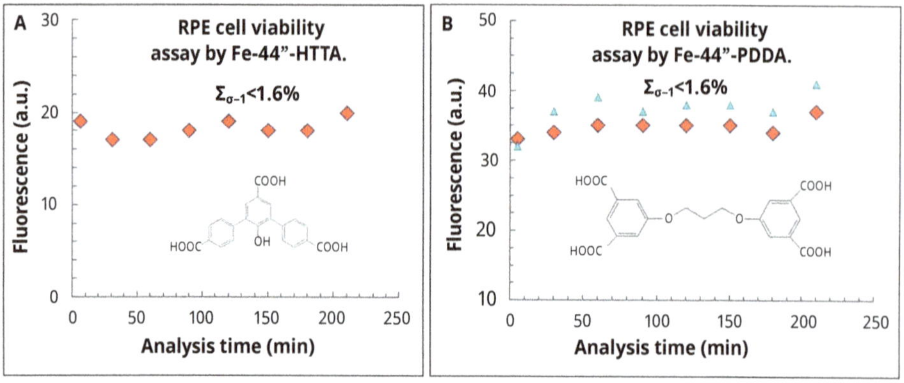

Fig. 3.33: Evaluation of MOF toxicity against RPE cells. MOFs incorporating different drugs exhibit different toxicities. A. RPE cell viability assay with Fe-44-HTTA and B. RPE cell viability assay with Fe-44-PDDA.

inferred due to the release of secondary messengers upon cellular stress and was utilized to measure the development of the treatment with minimal side effects. (3) The potential to construct NO probes with NIR to RL capabilities, which have yet to be documented to assist PT-based therapies by enabling "deep scans" via NIR and therapies via RL. The example shown in Fig. 3.33 demonstrates the efficacy of MOFs against RPE whole cells (WC) at low (1–20 ppm) doses.

The RPE is a single layer of pigmented cells located between the retina and the choroid (Fig. 3.34A). Positioned at this critical junction, the RPE functions both as a metabolic regulator and as structural support for the photoreceptors above. Its cells form a tightly ordered hexagonal mosaic, with apical microvilli extending toward photoreceptor outer segments to enable nutrient exchange and phagocytosis of shed discs. On the basal side, the RPE attaches to Bruch's membrane, creating a selective barrier that controls ion transport, fluid balance, and immune signaling. This dual interface between the neural retina and vascular choroid is essential for maintaining retinal homeostasis, sustaining the visual cycle, and defending against oxidative stress. TEM imaging highlights the RPE's distinctive morphology (Fig. 3.34B), showing polygonal cells that are typically six-sided and tile efficiently with minimal gaps. This geometric arrangement ensures both mechanical stability and functional precision.

TEM imaging revealed notable structural changes in RPE cells after exposure to three MOFs: Fe-HTTA, Fe-ATTA, and Zr-MOF (Fig. 3.34C–E). Untreated cells retained their typical morphology, with distinct nuclear boundaries and well-organized cytoplasm. In contrast, MOF-treated cells displayed clear signs of stress, including swollen membranes, disrupted organelle architecture, and partial loss of atomic definition. These findings indicate that even low concentrations of MOFs can induce cytotoxicity, likely through mechanisms involving oxidative stress or mitochondrial dysfunction. Comparable assays in cancer cell lines produced similar outcomes, with elevated lac-

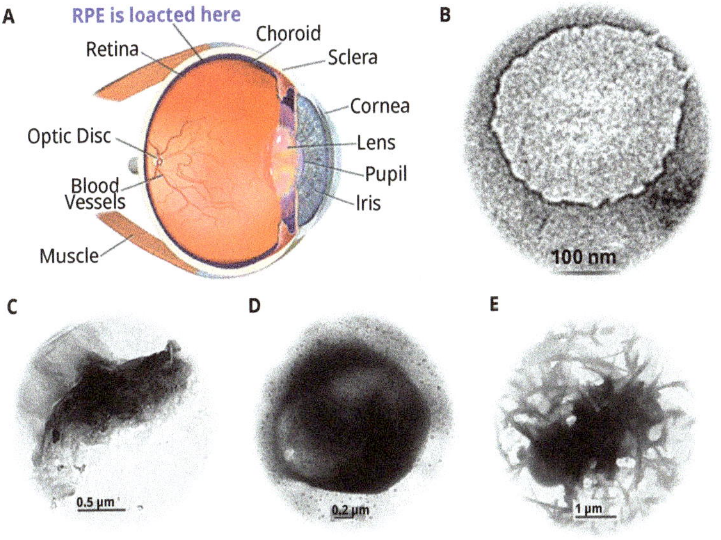

Fig. 3.34: Efficacy of MOFs against retinal pigment epithelium (RPE) whole cells (WC): low (1–20 ppm) doses. **A.** Diagram of the RPE whole cell as a control, **B.** TEM image of the intact RPE whole cell, **C.** TEM image of the damaged RPE WCs by Fe-HTTA, **D.** TEM image of the damaged RPE WCs by Fe-ATTA, and **E.** TEM image of the damaged RPE WCs by Zr-MOFs.

tate dehydrogenase release and increased caspase-3/7 activity consistent with apoptosis. Together, these results suggest that MOFs may act not only as drug delivery vehicles but also as active agents in photodynamic or oxidative therapies against malignant cells. Ongoing studies are focused on linking these dose-dependent structural changes to functional outcomes in cancer models.

A comprehensive review of public health impacts from energy fuel cycles in Europe reported that natural gas use is associated with 1–11 deaths per terawatt-hour, averaging about three deaths per TWh [248]. In the United States, oil and gas extraction has been linked to 10–20 casualties. While some experts, such as Richard A. Muller, have argued that the public health benefits of shale gas outweigh its environmental costs [249], it remains critical to assess the safety of materials employed in gas storage. In our evaluation, wild-type RPE cells were benchmarked against established standards: cells resuspended in phosphate-buffered saline with Ca^{2+}/Mg^{2+} (control, 0% change), hydrogen peroxide (oxidative stress, 3%), and sodium nitroprusside (NO donor, 50 μM). Nanocomposites including ZnO-22 (10 mg/mL), ZnO-28, Fe_3O_4-T2, and the Ag-A19 MOF were tested for toxicity using dimethyl sulfoxide (DMSO) as solvent (Fig. 3.35). Standards were grouped into three categories: control (PBS), oxidizing (HP), and nitric oxide-releasing (NitroP) to provide comparative baselines for evaluating MOF-induced cellular effects.

Compared to the cells without treatment (PBS group), the addition of HP resulted in a relative increase in NO. The addition of NO-releasing agents is complicated by the

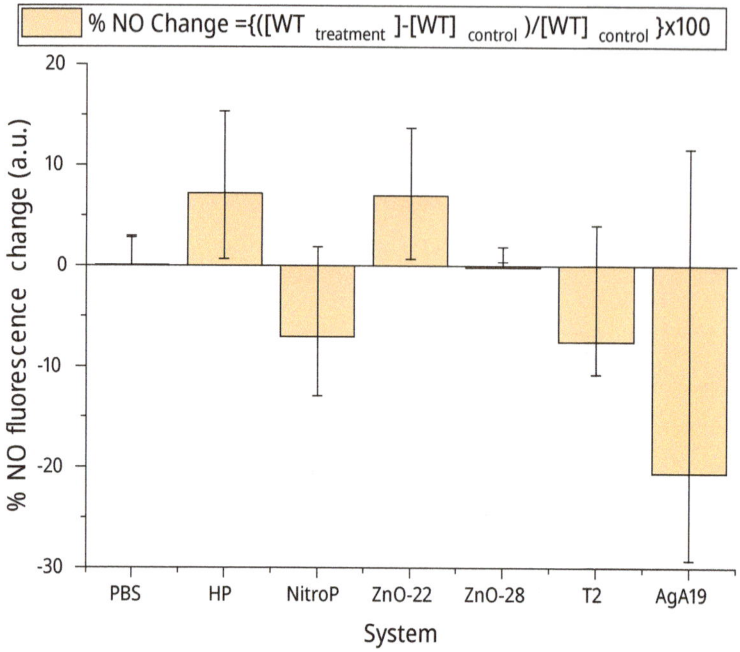

Fig. 3.35: Nitric oxide fluorescence as a difference between WC cells in PBS (control) and treated cells with nanoparticles, with independent error bars ($n = 8$).

NO feedback loop, in which high NO levels trigger a biological switch-off of NO synthase (NOS), the primary enzyme responsible for NO production. The addition of sodium nitroprusside releases NO and, under high NO conditions, activates the negative feedback loop, lowering NO levels. The nanocomposite Zn-28 has almost no effect on NO generation and appears to be nontoxic. ZnO-22 seems to be slightly oxidizing, not sufficient to cause cellular damage, but enough to increase NO levels, which protect against further oxidative stress. The NO levels do not change after the addition of Fe_3O_4-T2, suggesting that the agent can translocate to the mitochondria and directly affect the redox states of the respiratory heme and copper centers, similar to sodium nitroprusside. The Ag-A19 motif destabilized and cross-linked the mitochondrial membrane via the silver center, leading to a drop in NO [250].

3.4.2 Fingerprint development

The initial goal of this research direction is to enhance the capacity, capability, and readiness in the field for the detection and development of latent fingerprints on porous and non-porous surfaces, which are currently problematic. The technical barriers are partial, overlapping, or aged prints, while the administrative hurdles lack sophisticated crime-detection equipment, personnel, and training in using the latest detection techniques. The team has been focused on designing and implementing nanotechnology-based research and artificial intelligence to facilitate latent-print and forensic crime scene analysis and to provide reliable evidence for criminal justice. Three technical activities have been carried out: (1) hyperspectral scanner development using green light to enhance detection of latent fingerprints using chemical fluorescence of chemical components in latent print residue; (2) latent print development agent synthesis using a chemical approach to obtain two novel classes of agents composed of COP; (3) automated fingerprint identification using ANNs and fingerprint aging analysis to enhance decision-making in matching low-quality print images. These applied research methodologies are used to design an expert fingerprint matching and identification system. The matching prints were interfaced with the automated fingerprint identification system (AFIS). The practice aspects were strengthened through specific nanoscience and forensic science courses.

Latent print and forensic crime scene analysis are complex areas of law enforcement that rely on applied and developmental goals to advance the practice. Current barriers to latent print identification include technical factors, such as detecting faint residues using technology or chemical development agents. The evaluation of latent fingerprint age, the resolution of partial and overlapping prints, and the eventual matching of prints to a suspect are the biggest additional obstacles that must be overcome to facilitate criminal investigations and the delivery of justice. These barriers are of particular concern to KCSO, which has identified four areas. These areas are the abilities to: (1) resolve poor quality prints, (2) detect latent prints, (3) determine the age, and (4) identify likely suspects. The first two areas were addressed using technology, whereby the fluorescence arising from the fingerprint residual was measured. Knowledge of latent fingerprint age and synchronicity of a print is based on the chemical decay in the latent print [251]. The inability to resolve partial or overlapping prints affects the ability to confirm or refute a suspect's alibi. Multiple factors [252] influence the outcome of latent fingerprint detection. These are the amount of print residue, the surface texture, and the type of DAs used [253]. The sensitivity factor is a metric of the DA, where the signal intensity limits detection through fluorescence [254]. This low detection sensitivity will be addressed by developing a more sensitive hyperspectral scanner to provide high resolution [255] or by synthesizing nanostructured composites with different functional groups that exhibit higher quantum yields [256].

A hyperspectral scanner would compensate for slight print residue by simultaneously scanning lines and producing a three-dimensional information cube. Each fluorescence wavelength would be linked to a particular chemical component in the latent print residue. An alternative approach is to design agents with amphiphilic topologies that generate high quantum yields using latent print residues from porous and nonporous substrates [257]. Using the DAs with polyhedra will provide more binding sites, thereby improving fluorescence detection and increasing sensitivity to latent print residues [258]. The new latent fingerprint DAs have the potential to replace iodine or ninhydrin and can be used on a variety of surfaces [259].

An ANN-based expert system resolved the last two concerns, even though the latent prints are challenging to develop. This difficulty is due to limited print residue, resulting in low print resolution, incomplete ridges, partial bifurcations, or even a lack of a core, hindering expert print-matching. Automated searching and matching of prints with ridges but without minutiae pose a barrier to correct print identification. An ANN-based expert system will improve the probability of suspect identification by reducing the time required to search the massive database before detailed matching is implemented. The contrast enhancement can reveal ridges and bifurcations, while spatial filtering can remove noise between distinct points. This approach simulates what human experts would do by using additional information not observable in the prints. The last concern was addressed by implementing high-quality training and professional development in nanoscience and forensic science [260]. The tools will enable small police departments to gain expertise in fingerprint-suspect matching, on par with that of experienced police departments. Crime detection requires experience and know-how, which are problematic for latent fingerprint detection and evaluation due to numerous variables in print quality. The ability to train under standardized conditions and develop a toolkit library to meet various scenarios will benefit detectives' in-field operations. This benefit can only be accomplished using standard prints with artificial sweat for print duplication. Current problems with using artificial sweat are two-fold: a) the composition of commonly used sweat does not resemble sweat in healthy adults due to a lack of metal cations, and *b)* the sweat "concentration" is adjusted through dilution of the cartridges with water. This dilution does not reflect latent prints in touch pressure and finger sweat, which determine print homogeneity and uniformity of spread. The latter was addressed by creating more realistic sweat and by tailoring specific courses for law enforcement needs.

The team focused on three activities to improve the development accuracy and precision of fingerprint extractions. ***Hyperspectral scanner development workflow***: The hyperspectral scanner was constructed and evaluated for print detection on porous and nonporous surfaces. Detection without DAs can utilize print and latent print residue(s). The detection threshold can also be compared and contrasted against eccrine secretion DAs. The data were analyzed for probability match using the five-point grade scale (Fig. 3.36A) and examined for statistical significance using ANOVA. Construction: The scanner was constructed based on an existing design with an imaging spectrograph and charge-coupled camera orthogonal to a second monitored image, capturing portions that undergo hyperspectral (HS) imaging (Fig. 3.36B).

A green-emitting diode can illuminate the surfaces. An HS image is generated in which every exposed surface is digitized over the wavelength range of the interesting visible light. The pictures are stored as a pixelated HS dataset. An instrument designed by Mitaka Kohki Co., Japan, has been used as our model [261]. The image is obtained by moving a spectrograph and a charge-coupled device-equipped camera (Pika XC2, Resonon, USA). The spectrograph has a spatial resolution of 1.3 nm × 13 nm slit over

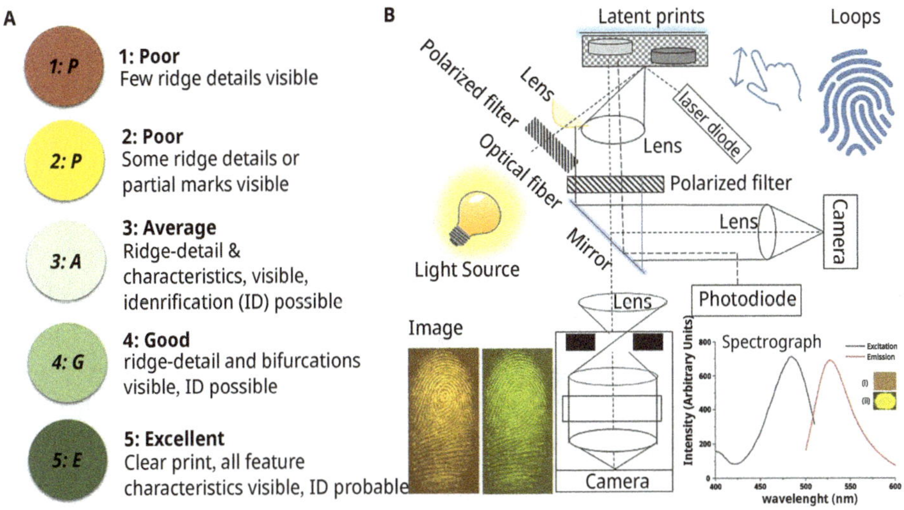

Fig. 3.36: The hyperspectral scanner (HS) development workflow. **A**. Five-point grade scale for latent fingerprints and **B**. plan of HS scanner to detect partial or overlapping prints.

440–800 nm. A surface with latent prints was illuminated and focused onto the spectrograph's entrance slit. The output spectrum is focused onto the camera, enabling all points within the frame (1,600 × 1,600 × 447 line, scan, and spectral pixels) to be recorded at 30 frames per second. Depending on the distance between the surface and the lens, the image area was as small as 30 × 30 mm, with a spatial resolution of about 46 µm, a spectral resolution of 2.0 nm, and an acquisition time of 50 s. The acquisition time can increase with a larger area, while spatial resolution can decrease.

A second orthogonal camera (Pika L, Resonon, USA) was used to assist in capturing part of the image undergoing hyperspectral imaging. The underlying (latent) fingerprint surface was illuminated and filtered using a polarizer (VWR, USA) at 532 nm with 2.0–4.0 W of power. A continuous-wave laser was focused onto the surfaces using a cylindrical lens. The incident light was linearly polarized, and the light was diffused and reflected with the polarization perpendicular to the incident light (Fig. 3.36B). The light is selected by a linear polarizer aligned with the detector, and focusing is automatically controlled by a laser diode operating at 850 nm. Accordingly, a photodiode and long-pass cutoff filter (Sigma Koki, Japan) were selected for the wavelength of interest.

Spectral processing: The fluorescence data for latent prints were processed as a ratio (β) of the intensity difference (ΔI) or I between the acquired and reference spectra, where the reference is the fluorescence reflected from the standard surface. Under high background, fluorescence illuminated between 560 and 660 nm with a background at $I_{600}\beta$ is evaluated using the following calculation (eq. 3.8):

$$b = (I_{560} - I_{600})/(I_{620} - I_{600}). \tag{3.8}$$

The surfaces with low background fluorescence were treated using eq. (3.9) as a first approximation:

$$b = I_{560}/I_{620} \tag{3.9}$$

These proposed expressions can differentiate surfaces with small print residues and overlapping prints. Scanning prints over a wavelength window (Δλ) and processing the resulting intensity ratios as a subtracted intensity ratio can yield a generated hyperspectral dataset.

Quality control procedure: Fingerprint images were ranked using the quality assessment scale (Fig. 3.36A) described by Verdue et al. [262]. The grade "3: A" or higher was examined, where ridgelines and junctions are recognized without laceration or incomplete regions. An expected print and related spectra from ridges and furrows are shown in Fig. 3.37A. While hyperspectral (HS) scanning is helpful for individual prints, the technique will extract valuable information about partial or different prints. This fluorescence is due to the hydrophobic domains of the print residue, which undergo dipole transitions. Each residue would emit energy at a unique wavelength, determined by its atomic arrangement and internal electronic states. If the residue is excited at a particular wavelength, the excitation spectrum reflects the fingerprint components as a fluorescence spectrum. The pressure exerted on the surface results in different residue deposition. This uniqueness, known as the aging of latent fingerprints, is reflected by its fluorescence spectra [263].

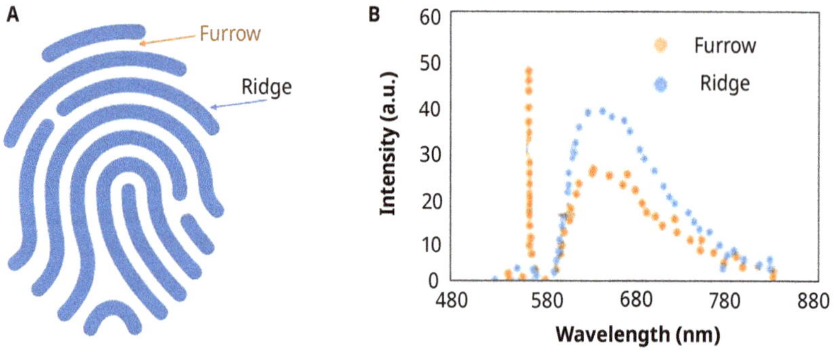

Fig. 3.37: Quality control to improve fingerprint image resolution: **A.** a schematic picture of a fingerprint and **B.** simulated excitation wavelength dependence of fluorescence on latent residue shown in a photo of the ridge (print) and furrow (background) with a green light (bandwidth: 10 nm) with a red filter.

These temporal changes in the fluorescence spectra between "initial" and "final" prints were identified, even when ridge or bifurcation structures are incomplete; the dataset stores the spectral–temporal coordinates (Fig. 3.37B). An individual with the same fingerprint and secretions will result in an incomplete or overlapping print, which cannot be identified by standard means. Each print emits fluorescence due to an electron transition from the ground to an excited state. Through the deconvolution of superposition coordinates of both initial and final prints, the images will be "reconstructed." Therefore, these prints can extract useful information from different dipole transitions, thereby representing an "HS fingerprint" (Fig. 3.38). The HS scanning was successful for unique prints and was practical for reconstructing partial or overlapping prints. Background subtraction via spectral difference (eqs. (3.8) and (3.9)) is expected to resolve some latent prints, depending on the ratio (β) of the intensity difference (ΔI) between the acquired and reference spectra, as stated above. When no print is extracted, the fluorescence spectra can serve as a basis for identity verification.

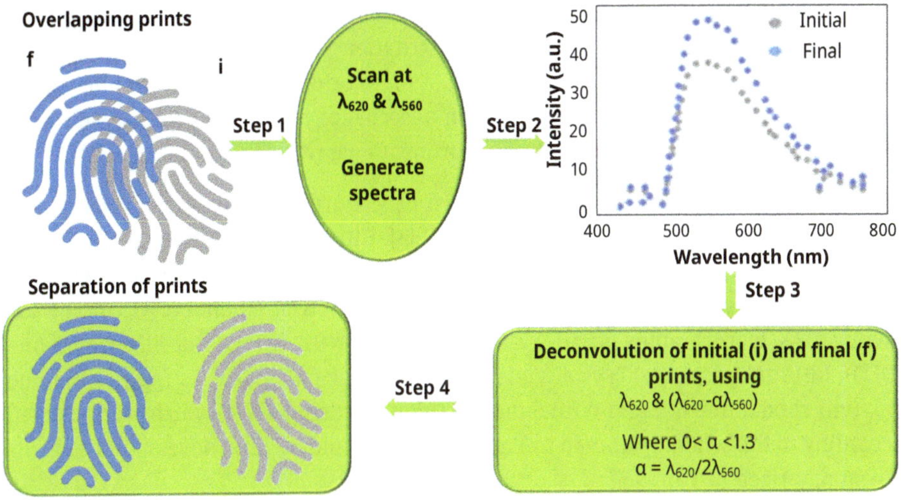

Fig. 3.38: The HS fingerprint schematic depicts the different dipole transitions of residues across four key steps to improve image resolution and separation.

Fingerprint acquisition procedure: The sample collection protocol followed the guidelines of the International Fingerprint Research Group [264]. The purpose of obtaining three types of deposits is to evaluate the HS scanner's sensitivity by examining the fluorescence of each. Eccrine and sebum-rich latent fingermark deposits from three donors were obtained on microscope glass slides, pre-cleaned with ethanol and ultrapure water, then dried under nitrogen. Immediately before sample collection, both hands were thoroughly washed with soap and warm water for 2–3 min. Moreover, eccrine deposits were obtained after 30 min without touching any objects between the handwashing stage and deposition [265]. For sebum-rich deposits, donors can press their fingers together and run the fingertips across their nose and forehead. Afterward, the fingerprints were placed lightly on the glass slides in a downward motion for 5–10 s. The same procedure was repeated to collect fingermarks on the black electrical tape, a polyethylene bag, and a soda can as examples of nonporous surfaces, and copy, gloss papers, and card stock as porous surfaces [266].

Spectroscopic characterization and imaging procedure: Fluorescence spectra of the developed surface were obtained using a spectrophotometer with an average of 10 scans and with excitation and emission slit widths of 10 nm. The surface was photographed in absorbance and luminescence modes. The white balance calibration was set up to obtain the best contrast using white light (WL), ultraviolet (UV) light, or filtered light, similar to actual observations (with safety goggles). The images were captured as fluorescent marks using a suitable light (WL or UV) and colored (blue (B), green (G), yellow (Y)) filters, and saved as an 8-bit grayscale TIFF file. For both modes, the focal length was kept at 60 mm, exposure and white balance set to automatic, and aperture corresponding to f11, mimicking a film sensitivity of ISO 200. In absorbance mode, the shutter speed was set to 1/20 s, and in luminescence mode to 1 s to obtain the highest-quality image on either a dry or a wet exhibit. An excitation wavelength of 490–510 nm was used in luminescence mode with a green (G, 529 nm) or orange (O, 620 nm) filter to improve image contrast, while WL was used in absorbance mode. The prints were scored using grade scoring [267] and processed according to the workflow (Fig. 3.43).

Latent fingerprint visualization: Standard procedures for visualizing fingerprints have been described by Lee and Gaensslen [268]. Ultrapure (18 MΩ) water (H_2O) and another solvent (analytical grade) were utilized as solvents for dissolution. Fluorescence quantum yield was measured using rhodamine 6G in ethanol (C_2H_6O). The following DAs were employed: (i) 1,2-indanedione (IND); (ii) 1,8-diazafluoren-9-one (DFO); (iii) ninhydrin, (iv) crystal violet (CV); (v) silver nanoparticles (Ag); (vi) iodine (I_2); (vii) rhodamine 6G; and (viii) 5-methyl-thioninhydrin (MNT), which were treated according to the scheme outlined in Fig. 3.39. The commonly used development agents are summarized in **Tab. 3.6**.

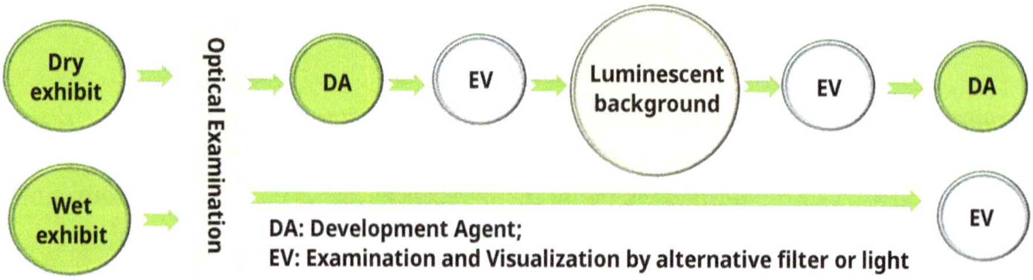

Fig. 3.39: The sequences for enhancing latent fingermarks on porous and nonporous surfaces for the wet or dry exhibit.

Latent print development agent (DA) synthesis workflow: The new DAs were synthesized via solvothermal chemistry, using carboxylic acids or succinic acid as organic linkers and multi-aromatic ring structures as fluorescent dyes. In another approach, mi-

Tab. 3.6: A summary of the development agents used to extract fingerprints.

No.	Nomenclature/properties	The molecular or crystalline structure
1	Rhodamine 6G chloride IUPAC name: 9-[2-(Ethoxycarbonyl)phenyl]-*N*-ethyl -6-(ethylamino)-2,7-dimethyl-3H-xanthen-3-iminium chloride highly fluorescent	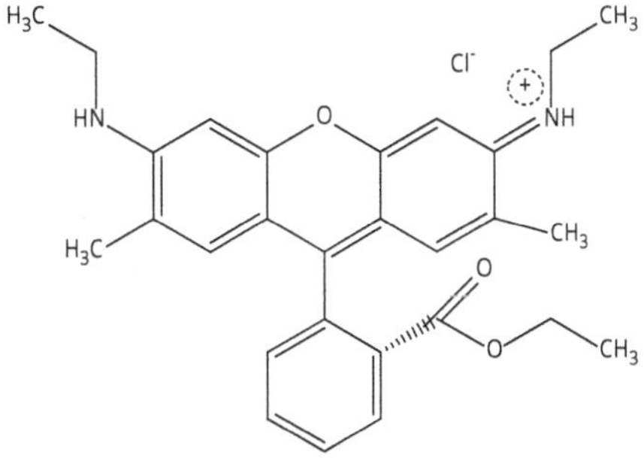
2	1,2-Indanedione (IND) Other names: Indan-1,2-dione 1H-Indene-1,2(3H)-dione 3H-Indene-1,2-dione vicinal diketone	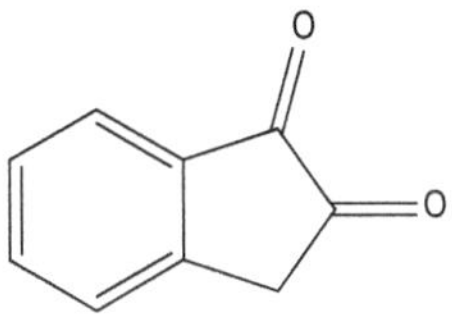
3	1,8-Diazafluoren-9-one (DFO) IUPAC name: 9H-Cyclopenta[1,2-b:4,3-b′]dipyridin-9-one aromatic ketone	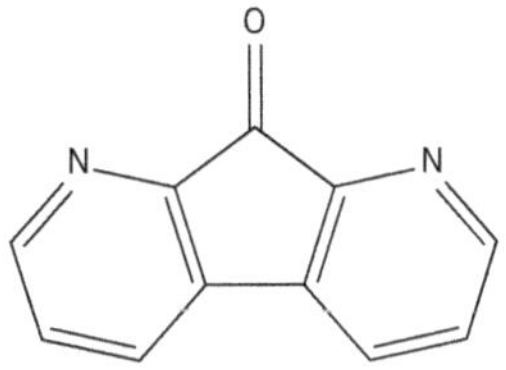
4	Ninhydrin IUPAC name: 2,2-Dihydroxy-1H-indene-1,3(2H)-dione	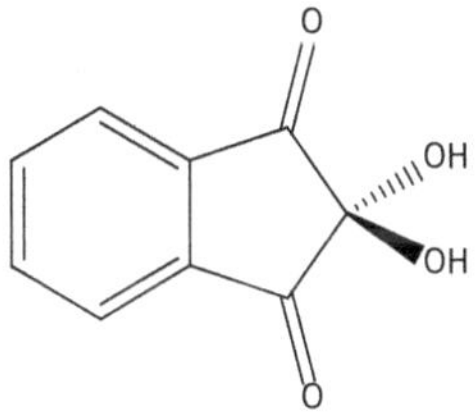
5	Crystal violet IUPAC name: 4-{Bis[4-(dimethylamino)phenyl] methylidene}-*N*,*N*-dimethylcyclohexa-2,5-dien -1-iminium chloride	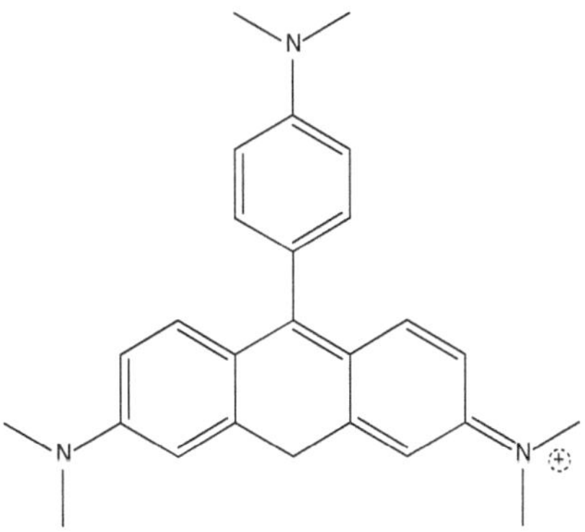

Tab. 3.6 (continued)

6	5-Methyl-thioninhydrin (MNT)	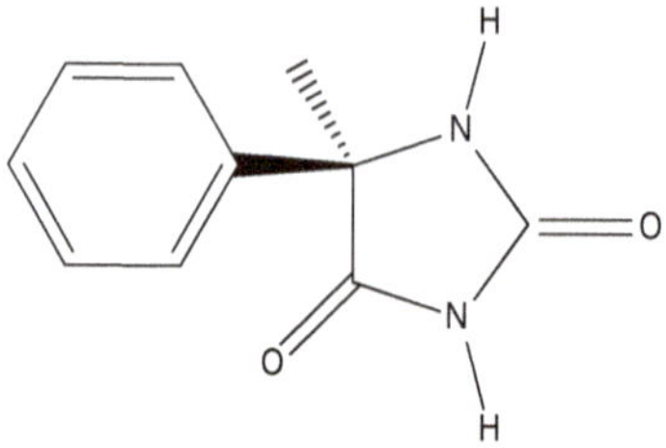
7	Silver (Ag) Space group: Fm3m Crystal system: Cubic *a*: 4.0855 Å Cell volume: 68.192 Å^3 Asymmetric unit: 1 site Unit cell: 4 sites/unit cell Density: 10.5099 g/cm^3 [269].	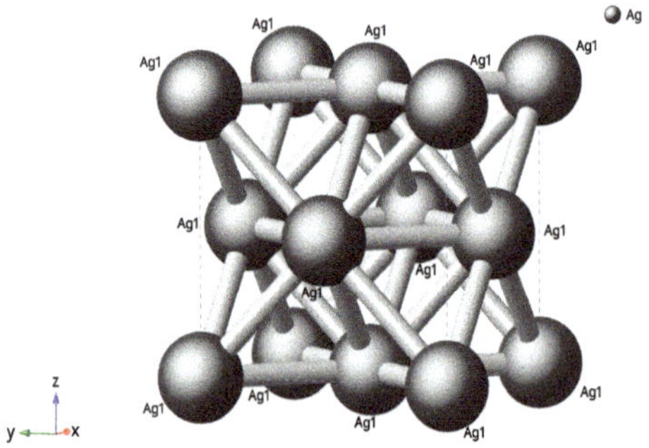
8	Iodine Space group: *Bmab* Crystal system: orthorhombic *a*: 7.2701 Å *b*: 9.7934 Å *c*: 4.7900 Å Cell volume: 341.046 Å^3 Asymmetric unit: 1 site Unit cell: 8 sites/unit cell Density: 4.9430 g/cm^3 [137].	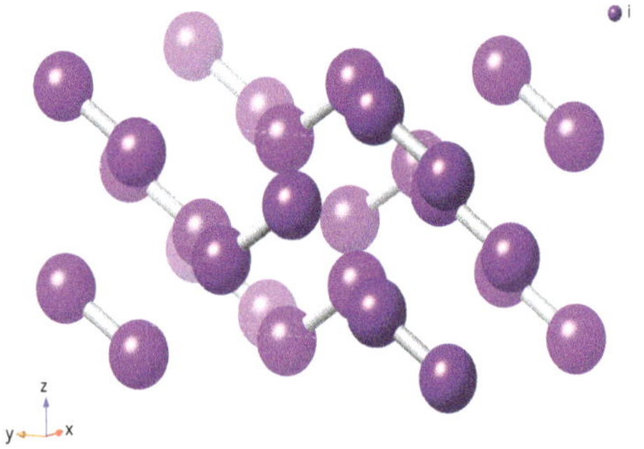

crowave-assisted synthesis was used to generate MOFs as DAs composed of pyrene derivatives. The ligands were coordinated to preferred metals for charge stabilization and enhanced detection, based on the metal d-orbital vacancy in both methods [270]. This activity consists of three specific deliverables used to tailor DAs to detect prints for nonporous and porous surfaces. Three perspectives are discussed in detail. ***Nanostructured covalent organometallic polyhedron (COP) DAs to detect fingerprints from the nonporous surfaces***: A solvothermal chemistry has been widely employed to prepare COPs (Fig. 3.40) [271]. A complex was formed between a metal cation (Mn^+) and multi-coordinate ligands (M-COPs). This coupling can introduce hydrophilic active sizes into the substructure. These substructures are highly stable, enabling the COPs to remain stable in air and water and enhancing print-residue detection and selectivity through various carboxylates that can interact with amino acids in fingerprint residues [272]. The interaction between prints and M-COPs is expected to occur within 5–30 s. The sensitivity and visibility of latent prints are intended to increase by impregnating fluorescent dye into COP's pores after post-modification [273]. This technique allows for the encapsulation of M-COPs by dyes, which emit different colors when "hemiketalized" with the amino acid in the fingerprint residue. The ligands, dyes, and metal ions that can produce COP DAs to improve sensitivity, selectivity, and detection speed are tabulated in **Tab. 3.7.** Dr H.C. Zhou's group produced these ligands with different bridging angles and molar masses.

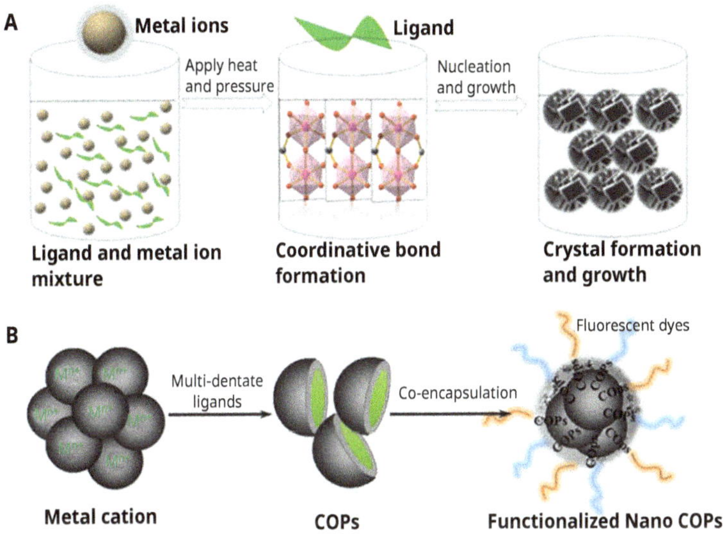

Fig. 3.40: The COP synthesis and post-modification. **A.** A schematic diagram of solvothermal wet-chemistry to produce COPs and **B.** a double-jet approach to preparing dye-encapsulated COPs (increase adhesion between inorganic and organic interfaces).

The solvothermal wet chemistry (Fig. 3.40A) was used to produce metal-centered COPs. The ligand bridging angles (shown in **Tab. 3.7**) varied to tune the pore structure and tomography of different MOFs with transition elements as the center. These binding ligands (also called linkers) were dissolved in organic solvents (such as alcohols or dimethyl sulfoxide), and the metal ions were dissolved in aqueous solution in glass vials. The mixtures were ultrasonicated to ensure that the various raw materials with different solubilities formed similar solutions. The reaction was controlled between 25 and 120 °C for 72 h to form well-reticulated crystals or a colloidal suspension [274].

COP encapsulation: Commonly used fluorescent dyes with different functional groups (**Tab. 3.7**) were dissolved in ethanol or other chosen solvents. Then, these dyes were embedded in the COPs via double injection under intense agitation (Fig. 3.40B) [242]. An ali-

Tab. 3.7: The series of multi-dentate ligands was covalently bonded with three fluorescent dyes using different metal ions.

Bridging angle: ~0°	Bridging angle: ~60°
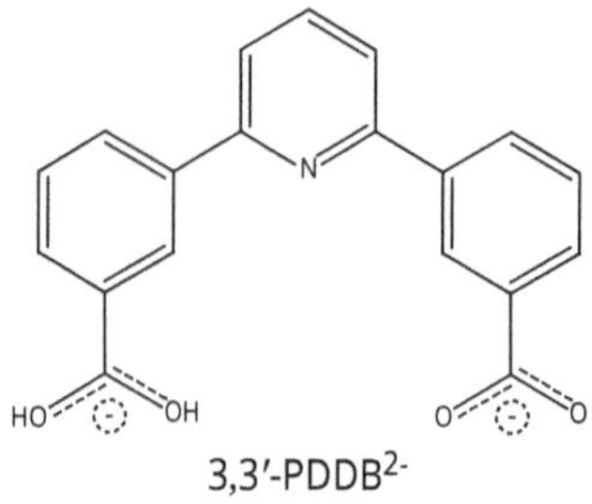 3,3′-PDDB^{2-}	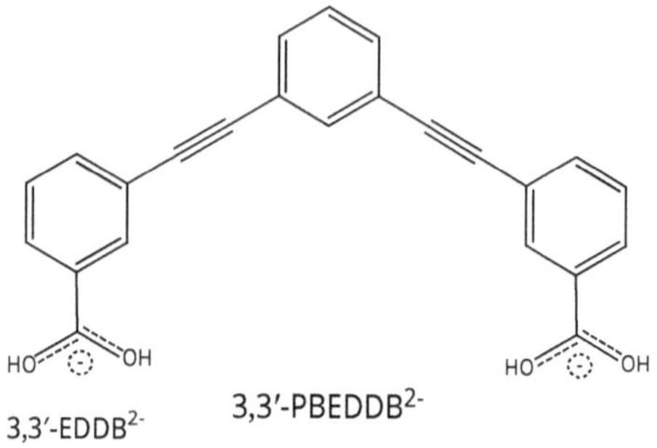 3,3′-EDDB^{2-}
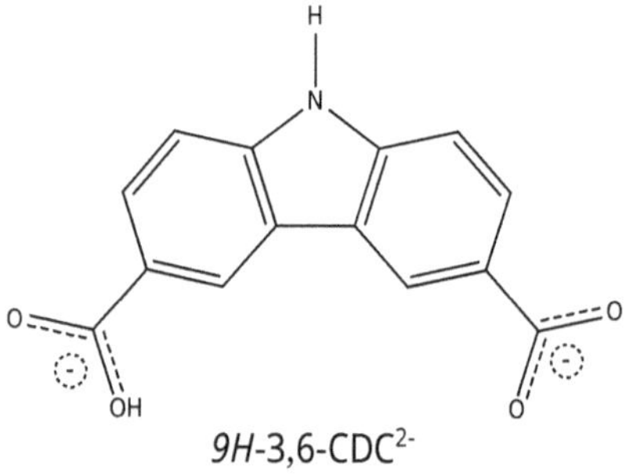 3,3′-PBEDDB^{2-}	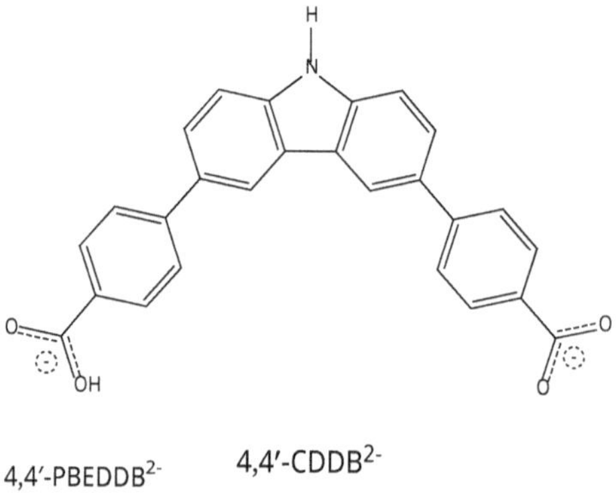 4,4′-PBEDDB^{2-}
Bridging angle: ~90°	Bridging angle: ~120°
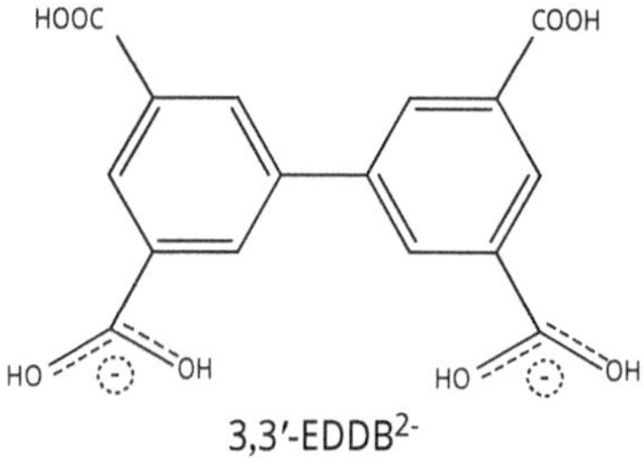 *9H*-3,6-CDC^{2-}	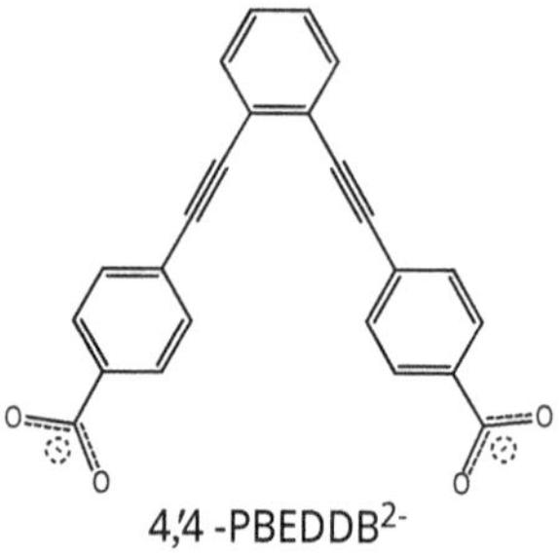 1,3-BDC^{2-}

Tab. 3.7 (continued)

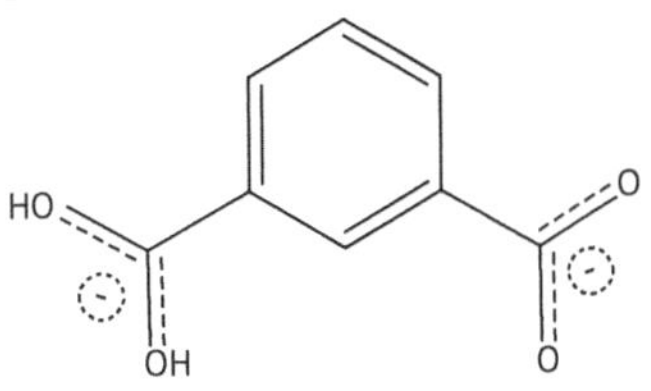 1,3-BDC^{2-}	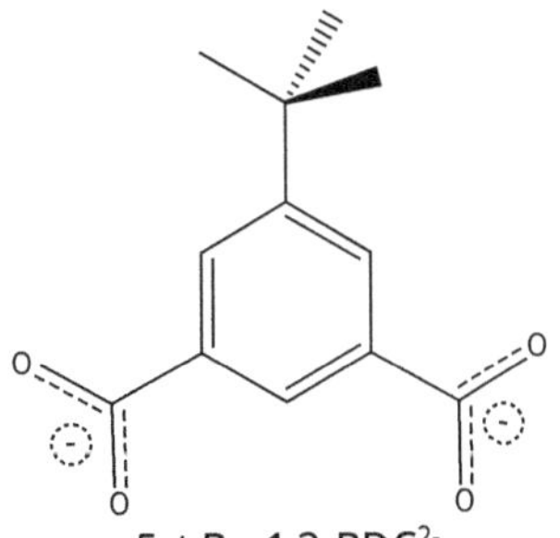 5-*t*-Bu-1,3-BDC^{2-}
4,4′-CDDB^{2-}	5*t*-Bu-1,3-BDC^{2-}

quot of a dye solution with different concentrations (0.05–0.5 M) was used. The resulting suspension was filtered through a 0.45-μm syringe filter to remove the unembedded dye. The filtrate was further diluted with H_2O/CH_3CH_2OH and analyzed for complexation efficiency using ultraviolet-visible (UV-vis) spectroscopy at a wavelength of 200–1,100 nm. The loading efficiency of the dye was quantitatively estimated by high-performance liquid chromatography (HPLC) on a C8 column in reverse-phase mode. Estimating dye-loading efficiency enables us to optimize preparation parameters to produce a suitable dye-encapsulated COP formulation [275].

Fluorescent MOF DAs to detect fingerprints from porous surfaces: DAs composed of PAHs were produced [276]. Pyrene derivatives (Sigma-Aldrich) are chosen as molecular probes to design novel DAs for fingerprint detection. PAH fluorescence is sensitive to solvent polarity [277]. Emission bands are unaffected because the pyrene-excited state has a different transition dipole than its ground state. Our previous study demonstrated that highly stable zirconium (Zr)-MOFs were fluorescent and had negligible toxicity (0.25 ppm) to humans, as assessed by in vitro bioassays using RPE cells. M-MOFs were constructed via chemical synthesis using pyrene linkers with high fluorescence, which were selected as the functional group. For example, a 4,4′,4″4‴-(pyrene-1,3,6,8-tetryl) tetrabenzoic acid (H_4TBAPy) containing pyrene as the backbone was chosen as the binding ligand [220]. The "hard and soft (Lewis) acids and bases" theory suggested that the Zr^{4+} cation with high valence can form a strong coordinative covalent bond with carboxylate.

As a result, fluorescent frameworks were produced. Microwave synthesis can generate MOFs between H_4TBAPy and Zr^{4+} (Fig. 3.41A). Microwaves spanning a broad frequency range (300 MHz to 300 GHz) were used to create thermal excitations. Microwave radiation shortens reaction times, provides a rapid rate of crystal nucleation and growth, and yields high crystal yields with minimal by-products. Our previous study indicated that microwave irradiation for less than 1 min was optimal for producing well-crystallized products. The charge attraction developed during microwave radiation is the driving force for nucleation, leading to crystallization and growth. The fluorescent

MOFs (with different formulations) with highly porous structures (Fig. 3.41B) were generated. Functional groups of $-NH_2$ and –C=O were introduced into the MOF pores. These groups can react with 1° and 2° amines in amino acids present in the papillary exudate, forming hemiketals. The fluorescent emission yield was used to evaluate fingerprint detection thresholds quantitatively [278]. Two MOF motifs were chosen as examples, which are anticipated to improve the accuracy and precision of fingerprint development. The MOF-10 (crystal system: triclinic, a: 17.1470 Å, b: 23.3220 Å, c: 25.2550 Å, $\alpha = \beta = \gamma$: 90.000°, cell volume: 10,099.533 $Å^3$) is an MOF-5 framework, composed of a “strut” that connects the zinc oxide “joints” in this framework structure. The MOF-5 (space group: $Fm^{\bar{3}}m$, crystal system: cubic, a: 25.8247(4) Å, cell volume: 17,222.883 $Å^3$, asymmetric unit: 8 sites, unit cell: 428 sites/unit cell (0.0246 atoms/$Å^3$), density: 0.5938 g/cm^3). Lock et al. [279] reported groups of ZrO_4 tetrahedra interlinked via benzo-ethanoic acid units. An MOF-114 (space group: C 2/c, crystal system: monoclinic, a: 22.2410(20) Å, b: 12.7034(12) Å, c: 17.6543 (16) Å, β: 126.361(2)°, cell volume: 4,016.809 $Å^3$) is composed of the disordered C4S site, which has been shown as an average site, reported by Omar M. Yaghi et al. [280].

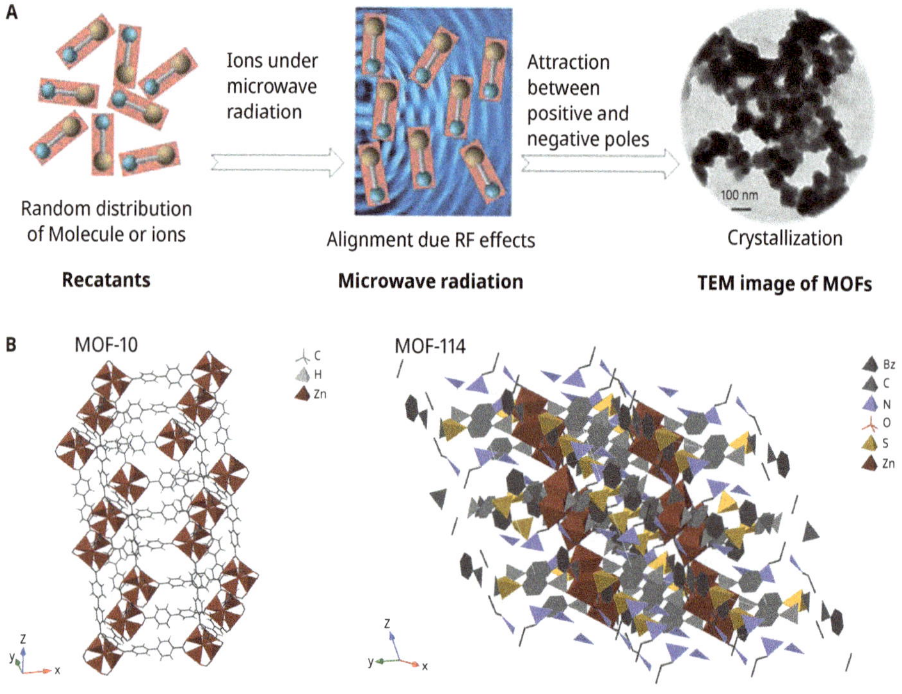

Fig. 3.41: The synthesis and characterization of COPs (MOFs as an example): **A**. The synthesis of MOFs using microwave-assisted solvothermal chemistry and **B**. two existing MOF motifs with different pores and geometries were chosen as examples.

Fabrication of an aerosol system to effortlessly smooth the deposition of DAs on surfaces: To convert COPs and MOFs into an aerosol, we can use the procedure described by Fox [281]. An aerosol composed of MOF with high porosity and surface areas can hold 2–50 mg of the newly prepared DAs suspended in a 100–250 mL water/ethanol mixture (1:1 v/v). Organic solvents, such as ethanol, can facilitate solvent vaporization and accelerate fingerprint development. The argon inert gas with pressure controlled at 2–6 bar was introduced for aerosolization under ambient conditions (Fig. 3.42) [282]. The supercell dimensions with 2 × 2 × 2 (Fig. 3.42A) were generated as multiples of the existing unit cells along *x*, *y*, and *z*. By controlling the solvent–COPs ratio, aerosol was produced to collect figure prints (Fig. 3.42B). The X-ray powder diffraction, X-ray energy dispersive spectroscopy, and porosity map of MOF-114 were simulated and shown in Fig. 3.43 as a guideline for future evaluation of COPs.

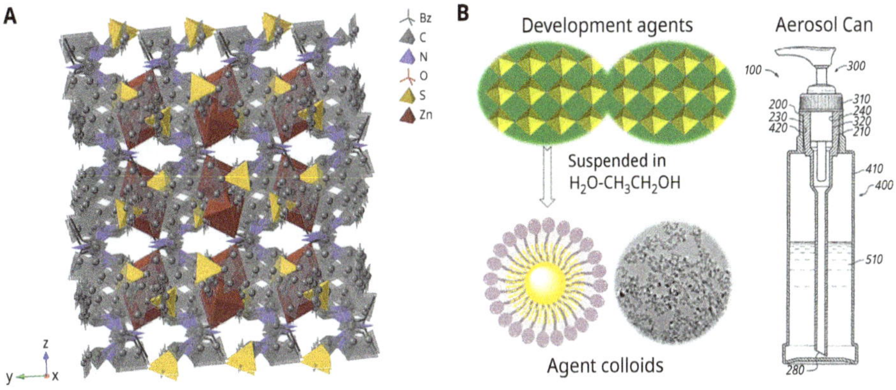

Fig. 3.42: A production flow of an aerosol system. **A.** MOF-114 with enlarged unit cells as development agents (DAs) of fingerprints and **B.** aerosol used to enable smooth depositions of DAs on surfaces.

Automated fingerprint identification workflow: Enhanced fingerprint matching using ANNs was implemented using the NIST Spectral Database 4 as the training module. The Microsoft Visual Studio Professional and MATLAB packages were used to design, implement, and deliver a Windows-compatible graphical user interface (GUI) [283]. This activity has two specific deliverables: latent fingerprint aging analysis via hyperspectral overlay and enhanced matching using ANNs and ML (Fig. 3.43). These two features are discussed below. ***Latent fingerprint aging analysis by hyperspectral overlay***: The fingerprint aging is determined by a series of spectral images. These prints, acquired at variable time lags, were used to improve recognition accuracy. Sweat composition, environmental factors, surface type, fingerprint contamination, resolution, and measured area mainly influence aging [284]. The overlapped print spectral separation depends on the temporal degeneracy difference between spectra [285]. The intensity differences between the "initial" and "final" fingerprints become significant between 560 and 640 nm.

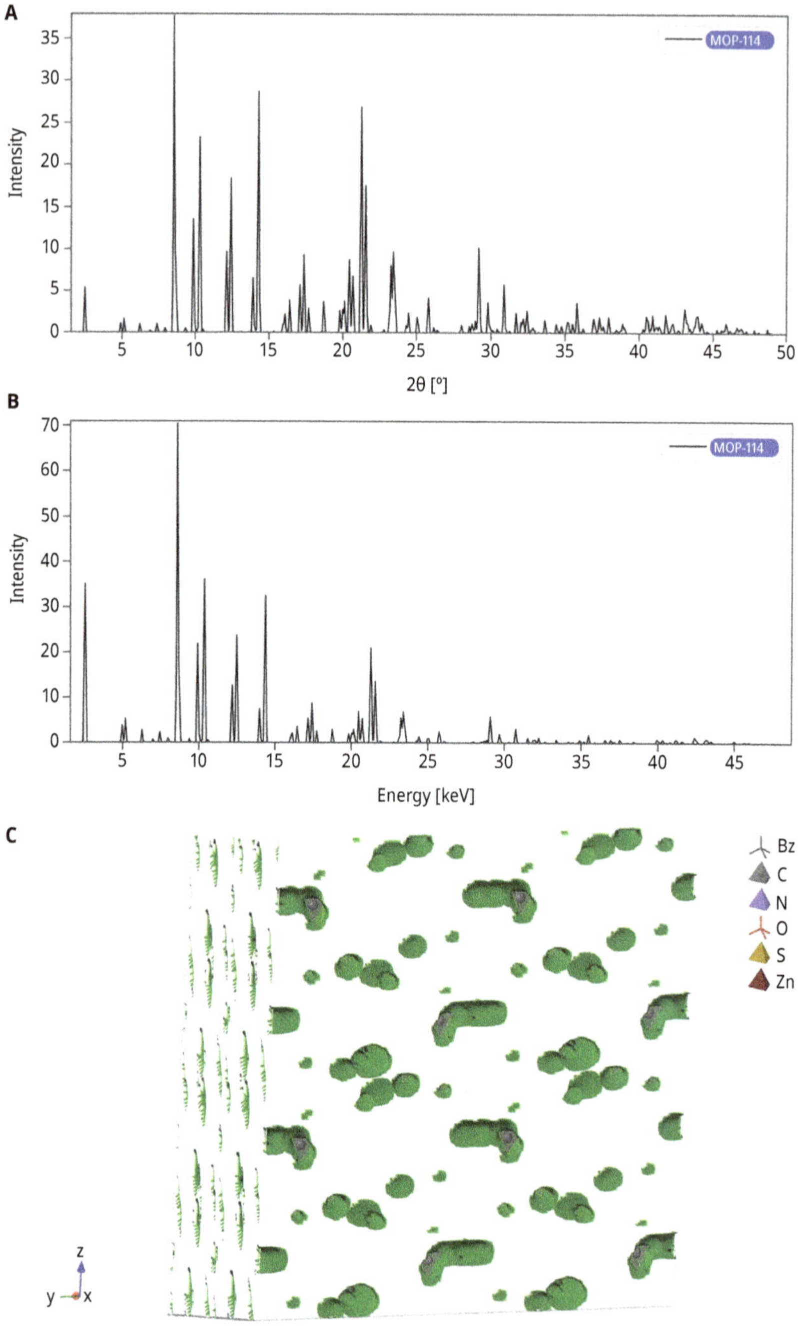

Fig. 3.43: The simulated diffraction pattern and porosity mapping of MOF-114 with supercell dimension of 2 × 2 × 2. **A**. X-ray powder diffraction pattern, **B**. energy dispersive spectroscopic diffraction pattern, and ***C***. porosity mapping.

Therefore, single fingerprint images (Fig. 3.38) were separated using eq. (3.8) or eq. (3.9) from the overlay [286]. The quality of images derived from HS data should be high enough for an ANN to compare with reference images. The first stage in overlay processing is the normalization of the picture and the removal of outliers at the beginning ($t = 0$ ms; eqs. (3.10A)–(3.10C)). The second stage was the remeasurement of fluorescence at different time points until t_{max} was reached. From the t_{max} value, the fingerprinting age was derived. Dr. Wang can apply two additional transforms to the acquired images to achieve an optimal spatial and temporal resolution. The tendency of the ridgeline is improved by dilation of the original pixel (**eq. (3.10A)**):

$$F_1 = \frac{1}{x_{max} y_{max}} \sum_{x=1, y=1}^{x=x_{max}, y=y_{max}} \text{Binarize}(I_{C\_DN}(x, y), \text{thresh} = 0.8) \tag{3.10A}$$

$$F_2 = \frac{1}{x_{max} y_{max}} \sum_{x=1, y=1}^{x=x_{max}, y=y_{max}} \text{Binarize}(I_{C\_SN}(x, y), \text{thresh} = 0.8) \tag{3.10B}$$

$$F_3 = \frac{1}{x_{max} y_{max}} \sum_{x=1, y=1}^{x=x_{max}, y=y_{max}} \text{Binarize}\left(I_{M(x,y)} \cdot I_{C\_DN}(x, y), \text{thresh} = 0.8\right) \tag{3.10C}$$

where I_C is cropping; I_{C_DN} is the dynamic normalization; I_{C_SN} is the static normalization; and I_M = Dilate(i,j)·Binarize(I_{C_DN}, thresh = 0.8) (Eq. 3.10B).

Age determination is currently considered a significant barrier to effective law enforcement. The aging behavior of latent fingerprints was determined with greater certainty through the above processing. Using the bin and sort algorithms, the decay in the spectral pattern was correlated with the approximate age of the fingerprint (Eq. 3.10C).

Enhanced matching using ANNs: One bottleneck in latent fingerprint crime scene detection is the poor quality of generated print images. We can characterize images with grade "2: P" (Fig. 3.36A) using the ANN expert system with varying degrees of intelligence to guarantee picture quality (Fig. 3.37–3.43). These criteria are applied to each extracted fingerprint image in the spectral separation trial [287]. The automated processing of the latent images (Fig. 3.44A and **B**) and the reference images (Fig. 3.44C) determines the regions of the pictures with well-defined ridge flow [288]. Processing steps include the binarization and skeletonization of the images and marking quality areas in the pictures. Quality areas of print are defined, and high-contrast print images are created (Fig. 3.44D). The print image is rendered as a high-contrast image that captures ridge flow (with black ridges and white furrows) based on phase measurements of bands of light and dark patterns, and masks out areas where constant-quality ridge flow is not present. An examiner can manually refine the clarity of the high-contrast image portion of the automated process when the computerized process produces an inadequate result. ANNs are used to determine fingerprint patterns, thereby expediting candidate searches in a massive database. The reference candidates with a high match-ranking of the pattern can continue the warp analysis with the latent image to achieve convergence. An accurate warp of a latent image to the mated or nonmated reference image is automatically created [289].

All ridges and furrows within the high-contrast latent and the reference images are thinned to one-pixel skeletons. Cubic and higher-order Bezier curves can approximate small segments of ridge-and-furrow skeletons. A cubic Bezier curve, defined by two endpoints and two internal control points, was altered by manipulating these four control points. Higher-order Bezier curves have more control points and are used as required to fit more complex curvature on the skeleton [290]. A new Bezier curve segment was added to provide finer coverage in the reference collection, starting at a shorter interval and a skeleton for a reference image than the range for a latent image [291].

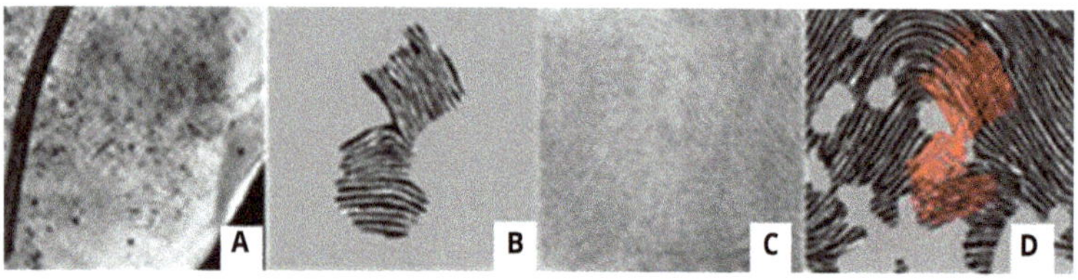

Fig. 3.44: Spectroscopic and image of prints. **A.** Select latent images, **B.** The high-contrast masked image, **C.** The actual mate reference print with poor clarity, and **D.** The best warp of the latent onto the quality-masked actual mate reference print.

The accuracy of warp and selection of the overlay (i.e., "best" warp) between the latent print image and each reference image is quantified at the level of single pixels within the skeletonized latent image. The basis for defining pixel-level quantification is the unique pairing of a skeletonized reference image (Fig. 3.44C) with a slight Bezier curve overlaid on the latent image (Fig. 3.44A). The matching criterion for Bezier curves is a distance measure of their similarity. The reference images identified by the latent print overlays are prioritized. Once an optimal warp provides the overlay between the latent and each reference image, the overlays are compared for accuracy, and the reference images are scored for final ranking (Fig. 3.44D). A scoring algorithm is applied to index and rank the overlays relative to the set of candidate reference images [292]. A competitive score between warps w_i and w_j at the latent skeleton pixel ω is defined as $S_{i,j} = -\log[d_i(\omega)/d_j(\omega)]$. Here, $d_w(\omega)$ is the minimum distance between all slight Bezier curves, containing the skeletonized latent image pixel ω and the warp w, inverted paired short Bezier curves from the skeleton of the reference image. Microsoft Visual Studio Professional was used to develop a GUI package that incorporates ANN algorithms trained on NIST DB 4 (freeware). Its advantage lies in the fact that smaller police departments can automate expertise before AFIS-interfacing and identification are installed. Recognition results (Fig. 3.45A) of the reference images (Fig. 3.45B) were recorded into the database for accurate verification.

Professional development and training workflow: Multiple fingerprint generation is difficult due to variability in touch pressure, sweat production, and the tendency to lose sweat by successive impressions. Artificial sweat and 2D printers were used to cre-

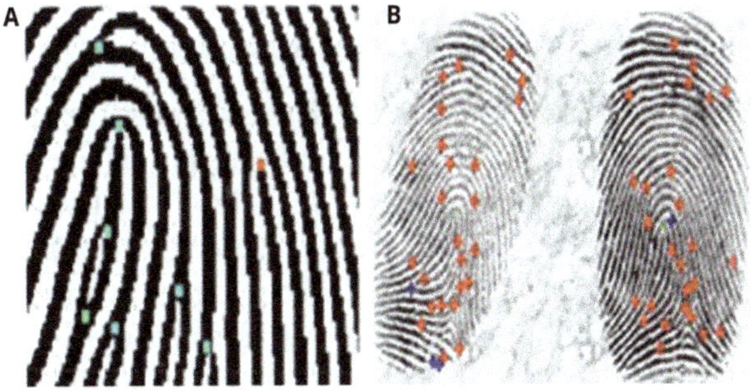

Fig. 3.45: Pattern matching of latent fingerprints: **A.** Pattern recognition and **B.** alignment and matching of the latent and reference fingerprints using ANN-based algorithms.

ate standardized prints to overcome the high barriers to obtaining uniform latent fingerprints. This procedure was utilized in the technician's training to enhance their expertise and experience in crime investigation. This activity has two deliverables, as described below: using a 2D printer with artificial sweat to generate standardized prints and creating a specialty certificate in nanoscience and forensic science.

Use of a 2D printer with artificial sweat to generate standardized prints: We can formulate a sweat composition using the procedure described by Schwarz et al. [293]. This method was found to have two drawbacks: a lack of metal cation and inhomogeneity of sweat concentration. Our approach can generate a realistic sweat composition and control print quality by using copy numbers rather than sweat dilution. Standard fingerprints from Liu's team were obtained for analysis. The visible fingerprints were saved as print image pairs in the 2000 8-bit grayscale TIFF format. The digital TIFF image was made "transparent" to remove artifacts between the ridges. Then, gamma correction (default 2.5) and the image contrast output parameters (default 0) in the picture-processing software were adjusted to achieve a digital image quality grade 5 (excellent). These digital fingerprints are then reproduced on white cotton print paper (A4 size) and evaluated using standard DAs described in the latent fingerprint visualization section. Dr. Bashir cut out the prints previously sprayed with Ag nanoparticles to examine reproduction efficiency. The print cuts were immersed in methanol to release Ag-bound print residues. The Ag excitation was measured by UV-vis spectroscopy from the resulting suspension. Herein, the reproducibility of the latent prints was examined by varying the Ag resonance peak at 410–425 nm.

Artificial sweat composition: All *L*-amino acids were purchased from VWR International. The structure can consist of an equimolar concentration of serine, glycine, alanine, leucine, threonine, histidine, valine, asparagine, lysine, sodium chloride, calcium chloride, and magnesium chloride. The average concentrations of amino acids and salts were controlled at 30 mM and 115 mM, respectively. The prints were evaluated using DFO, ninhydrin, MTN, Ag nanoparticles, and IND. The cartridges were filled with sweat and prints using an Epson Artisan 50 Color Inkjet Printer with T0781 refillable cartridges.

Creation of a specialty certificate in nanoscience and forensic science: The proposed course can cover five topic areas in the nanoscience certificate and six themes for the forensic science certificate at undergraduate and graduate levels for law enforcement personnel. The enrolled students shall select 3 for the graduate certificate or 4 for the undergraduate certificate, which will be transcribed into their academic records. Artificial sweat and 2D printing can generate standardized prints evaluated using different DAs (Fig. 3.46 A–D). Enrollees and volunteers who have contributed their fingerprints have been selected for undergraduate or graduate studies under the supervision of Dr S. Bashir and Dr J. Louise Liu. This professional training can introduce a new paradigm for forensic evidence collection.

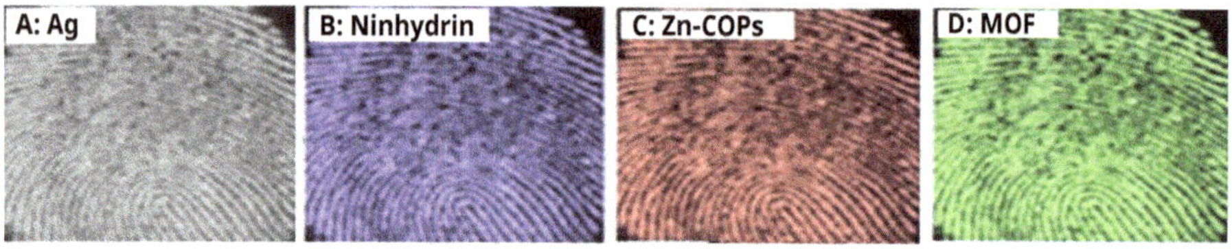

Fig. 3.46: Incorporation of new development agents (DAs) into a class for fingerprint technicians. **A.** Standard print generated using artificial sweat and 2D printing and **B–D.** eccrine development agents were tested on different surfaces and as part of classes for a certificate in materials or forensic science: **B.** ninhydrin as DA; **C.** Zn-COP (MOF-10) as DA; and **D.** MOG-114 as DA.

3.4 Summary

This Chapter begins by framing bottom-up synthesis as the atom-by-atom or molecule-by-molecule construction of nanomaterials, where controlled nucleation and growth produce uniform nanoparticles. Colloidal synthesis emerges as a versatile route: steric and electrostatic stabilization prevent aggregation, kinetic modeling predicts size distributions, and macropolymer templation directs the formation of ordered mesostructures. The chapter then turns to water remediation, emphasizing photocatalysis and heavy-metal sequestration. Photocatalytic semiconductors harness light to generate electron-hole pairs that cleave C–H bonds in organic pollutants, mineralizing dyes and pharmaceuticals into innocuous small molecules. Adsorptive strategies employ biota-anchored nanocomposites, microalgae, or bacterial exopolymers grafted onto iron oxides to capture Pb^{2+}, Cd^{2+}, and other toxic metals with >95 % efficiency. Bioinspired remediation leverages enzyme-mimetic nanomaterials and peptide-coated frameworks to degrade recalcitrant contaminants under mild conditions. MOFs bridge environmental and biomedical realms. Their high surface area and tunable pore chemistry enable both contaminant uptake and drug delivery, while single-atom catalysts embedded in MOF nodes drive selective oxidation reactions. In biological and forensic applications, MOF-based theranostics selectively generate ROS to induce apoptosis in cancer cells, and hyperspectral-active nanoparticles illuminate latent fingerprints on challenging surfaces. Coupling these nanostructures with machine-learning pattern recognition and three-dimensional imaging foreshadows a new era in environmental cleanup, precision medicine, and forensic science.

References

[1] Sabatier, P. A. (1986). Top-down and bottom-up approaches to implementation research: A critical analysis and suggested synthesis. *Journal of Public Policy*, 6(1), 21–48.

[2] Lu, J., Elam, J. W., & Stair, P. C. (2016). Atomic layer deposition Sequential self-limiting surface reactions for advanced catalyst "bottom-up" synthesis. *Surface Science Reports*, 71(2), 410–472.

[3] Langille, M. R., Personick, M. L., Zhang, J., & Mirkin, C. A. (2011). Bottom-up synthesis of gold octahedra with tailorable hollow features. *Journal of the American Chemical Society*, 133(27), 10414–10417.

[4] Perry, J. J. IV, Kravtsov, V. C., McManus, G. J., & Zaworotko, M. J. (2007). A bottom-up synthesis that does not start at the bottom: Quadruple covalent cross-linking nanoscale faceted polyhedra. *Journal of the American Chemical Society*, 129(33), 10076–10077.

[5] Wang, Y., & Xia, Y. (2004). Bottom-up and top-down approaches to synthesizing monodispersed spherical colloids of low-melting-point metals. *Nano Letters*, 4(10), 2047–2050.

[6] De Oliveira, P. F., Torresi, R. M., Emmerling, F., & Camargo, P. H. (2020). Challenges and opportunities in the bottom-up mechanochemical synthesis of noble metal nanoparticles. *Journal of Materials Chemistry A*, 8(32), 16114–16141.

[7] Wang, L., Sun, Y., Li, Z., Wu, A., & Wei, G. Bottom-up synthesis and sensor applications of biomimetic nanostructures. *Materials* 9, no. 1 (2016): 53.

[8] Silberberg, M. (2017). *Chemistry: The Molecular Nature of Matter and Change with Advanced Topics* (pp. 2–514). New York: McGraw-Hill Education.

[9] Bijlard, A. C., Wald, S., Crespy, D., Taden, A., Wurm, F. R., & Landfester, K. (2017). Functional colloidal stabilization. *Advanced Materials Interfaces*, 4(1), 1600443.

[10] Piacenza, E., Presentato, A., & Turner, R. J. (2018). Stability of biogenic metal (loid) nanomaterials related to the colloidal stabilization theory of chemical nanostructures. *Critical Reviews in Biotechnology*, 38(8), 1137–1156.

[11] Lyklema, J., Van Leeuwen, H. P., & Minor, M. (1999). DLVO theory, a dynamic re-interpretation. *Advances in Colloid and Interface Science*, 83(1–3), 33–69.

[12] Ninham, B. W. (1999). On progress in forces since the DLVO theory. *Advances in Colloid and Interface Science*, 83(1–3), 1–17.

[13] Dahirel, V., & Jardat, M. (2010). Effective interactions between charged nanoparticles in water: What is left from the DLVO theory? *Current Opinion in Colloid & Interface Science*, 15(1–2), 2–7.

[14] Missana, T., & Adell, A. (2000). On the applicability of DLVO theory to the prediction of clay colloids stability. *Journal of Colloid and Interface Science*, 230(1), 150–156.

[15] Mishchuk, N. A. (2011). The model of hydrophobic attraction in the framework of classical DLVO forces. *Advances in Colloid and Interface Science*, 168(1–2), 149–166.

[16] Asakura, S., & Oosawa, F. (1954). On the interaction between two bodies immersed in a solution of macromolecules. *The Journal of Chemical Physics*, 22(7), 1255–1256.

[17] Hanaor, D., Michelazzi, M., Leonelli, C., & Sorrell, C. C. (2012). The effects of carboxylic acids on the aqueous dispersion and electrophoretic deposition of ZrO_2. *Journal of the European Ceramic Society*, 32(1), 235–244.
[18] Liu, J. L., & Bashir, S. (2015). *Advanced Nanomaterials and Their Applications in Renewable Energy* (pp.57). New York: Elsevier.
[19] Butler, J. A. V. (1948). Theory of the stability of lyophobic colloids. *Nature*, 162(4113), 315–316.
[20] Hiemenz, P. C., & Rajagopalan, R. (2016). Principles of colloid and surface chemistry. In Revised and Expanded (pp. 461–498). Boca Raton: CRC Press.
[21] Hunter, R. J. (2013). *Zeta Potential in Colloid Science: Principles and Applications* (Vol. 2, pp. 59–141). New York: Academic Press.
[22] Nič, M., Jirát, J., Košata, B., Jenkins, A., & McNaught, A. (2009). *IUPAC Compendium of Chemical Terminology*. IUPAC, Research Triangle Park, NC.
[23] https://depts.washington.edu/solgel/pages/courses/MSE_502/Electrostatic_Stabilization.html
[24] Delgado, Á. V., González-Caballero, F., Hunter, R. J., Koopal, L. K., & Lyklema, J. (2007). Measurement and interpretation of electrokinetic phenomena. *Journal of Colloid and Interface Science*, 309(2), 194–224.
[25] Brinker, C. J., & Scherer, G. W. (2013). *Sol-Gel Science: The Physics and Chemistry of Sol-Gel Processing* (pp. 235–302). New York: Academic Press.
[26] Cao, G. (2004). N*anostructures & Nanomaterials: Synthesis, Properties & Applications* (vol. 7–10). London: Imperial College Press.
[27] Probstein, R. F. (2005). *Physicochemical Hydrodynamics: An Introduction* (pp. 237–273). New York: John Wiley & Sons.
[28] Myers, D. (2020). *Surfactant Science and Technology* (pp. 115–159). New York: John Wiley & Sons.
[29] Pierre, A. C. (2020). *Introduction To Sol-Gel Processing* (pp. 15–64). Switzerland: Springer Nature.
[30] Bergström, L. (1997). Hamaker constants of inorganic materials. *Advances in Colloid and Interface Science*, 70, 125–169.
[31] Ninham, B. W., & Parsegian, V. A. (1970). Van der Waals forces: Special characteristics in lipid-water systems and a general method of calculation based on the Lifshitz theory. *Biophysical Journal*, 10(7), 646–663.
[32] Lu, D., Zhou, J., Hou, S., Xiong, Q., Chen, Y., Pu, K., & Duan, H. (2019). Functional macromolecule-enabled colloidal synthesis: from nanoparticle engineering to multifunctionality. *Advanced Materials*, 31(44), 1902733.
[33] Polte, J. (2015). Fundamental growth principles of colloidal metal nanoparticles-a new perspective. *CrystEngComm*, 17(36), 6809–6830.
[34] Korotcenkov, G., Thomas, S., Sunny, A. T., & Prajitha, V. (2019). *Colloidal Metal Oxide Nanoparticles: Synthesis, Characterization, and Applications* (pp. 15–22). Oxford, UK: Elsevier.
[35] Moreno, R. (2012). Colloidal processing of ceramics and composites. *Advances in Applied Ceramics*, 111(5–6), 246–253.
[36] El-Sayed, M. A. (2004). Small is different: Shape-, size-, and composition-dependent properties of some colloidal semiconductor nanocrystals. *Accounts of Chemical Research*, 37(5), 326–333.
[37] Yang, L., & Wen, J. (2022). Can DLVO theory be applied to MOF in different dielectric solvents? *Microporous and Mesoporous Materials*, 343, 112166.
[38] Xia, Y., Xiong, Y., Lim, B., & Skrabalak, S. E. (2009). Shape-controlled synthesis of metal nanocrystals: Simple chemistry meets complex physics? *Angewandte Chemie International Edition*, 48(1), 60–103.
[39] Mozaffari, S., Li, W., Thompson, C., Ivanov, S., Seifert, S., Lee, B., & Karim, A. M. (2017). Colloidal nanoparticle size control: Experimental and kinetic modeling investigation of the ligand-metal binding role in controlling the nucleation and growth kinetics. *Nanoscale*, 9(36), 13772–13785.
[40] LaMer, V. K. (1952). Kinetics in phase transitions. *Industrial & Engineering Chemistry*, 44, 1270–1277.

[41] Ostwald, W. (1900). Über die vermeintliche Isomerie des roten und gelben Quecksilberoxyds und die Oberflächenspannung fester Körper. *Zeitschrift Für Physikalische Chemie*, 34(1), 495–503.
[42] Reiss, H. (1951). The growth of uniform colloidal dispersions. *The Journal of Chemical Physics*, 19(4), 482–487.
[43] Nakahara, A., Kawakatsu, T., & Kawasaki, K. (1991). Ostwald ripening in open systems. *The Journal of Chemical Physics*, 95(6), 4407–4417.
[44] Watzky, M. A., & Finke, R. G. (1997). Transition-metal nanocluster formation: Kinetic and mechanistic studies. A new mechanism when hydrogen is the reductant: Slow, continuous nucleation and fast autocatalytic surface growth. *Journal of the American Chemical Society*, 119(43), 10382–10400.
[45] Xia, Y., Xiong, Y., Lim, B., & Skrabalak, S. E. (2009). Shape-controlled synthesis of metal nanocrystals: Simple chemistry meets complex physics? *Angewandte Chemie International Edition*, 48(1), 60–103.
[46] Mozaffari, S., Li, W., Thompson, C., Ivanov, S., Seifert, S., Lee, B., & Karim, A. M. (2018). Ligand-mediated nucleation and growth of palladium metal nanoparticles. *JoVE (Journal of Visualized Experiments)*, (136), e57667.
[47] Li, W., Taylor, M. G., Bayerl, D., Mozaffari, S., Dixit, M., Ivanov, S., & Karim, A. M. (2021). Solvent manipulation of the pre-reduction metal-ligand complex and particle-ligand binding for controlled synthesis of Pd nanoparticles. *Nanoscale*, 13(1), 206–217.
[48] Mozaffari, S., Li, W., Dixit, M., Seifert, S., Lee, B., Kovarik, L., & Karim, A. M. (2019). The role of nanoparticle size and ligand coverage in size focuses on colloidal metal nanoparticles. *Nanoscale Advances*, 1(10), 4052–4066.
[49] Watzky, M. A., & Finke, R. G. (1997). Nanocluster size-control and "magic number" investigations. Experimental tests of the "living-metal polymer" concept and mechanism-based size-control predictions led to the syntheses of Iridium (0) nanoclusters centering about four sequential magic numbers. *Chemistry of Materials*, 9(12), 3083–3095.
[50] Morris, A. M., Watzky, M. A., Agar, J. N., & Finke, R. G. (2008). Fitting neurological protein aggregation kinetic data via a 2-step, Minimal/ "Ockham's Razor" Model: The Finke-Watzky mechanism of nucleation followed by autocatalytic surface growth. *Biochemistry*, 47(8), 2413–2427.
[51] Watzky, M. A., Finney, E. E., & Finke, R. G. (2008). Transition-metal nanocluster size vs. formation time and the catalytically effective nucleus number: A mechanism-based treatment. *Journal of the American Chemical Society*, 130(36), 11959–11969.
[52] Feng, Y., He, J., Wang, H., Tay, Y. Y., Sun, H., Zhu, L., & Chen, H. (2012). An unconventional role of ligand in continuously tuning of metal-metal interfacial strain. *Journal of the American Chemical Society*, 134(4), 2004–2007.
[53] Yoo, S., Kim, J., Kim, J. M., Son, J., Lee, S., Hilal, H., & Park, S. (2020). Three-dimensional gold nanosphere hexamers linked with metal bridges: Near-field focusing for single-particle surface-enhanced Raman scattering. *Journal of the American Chemical Society*, 142(36), 15412–15419.
[54] Bi, D., Yi, C., Luo, J., Décoppet, J. D., Zhang, F., Zakeeruddin, S. M., & Grätzel, M. (2016). Polymer-templated nucleation and crystal growth of perovskite films for solar cells with efficiency greater than 21%. *Nature Energy*, 1(10), 1–5.
[55] Jones, M. R., Osberg, K. D., Macfarlane, R. J., Langille, M. R., & Mirkin, C. A. (2011). Templated techniques for the synthesis and assembly of plasmonic nanostructures. *Chemical Reviews*, 111(6), 3736–3827.
[56] Lee, H., Dellatore, S. M., Miller, W. M., & Messersmith, P. B. (2007). Mussel-inspired surface chemistry for multifunctional coatings. *Science*, 318(5849), 426–430.
[57] Liu, N., & Liedl, T. (2018). DNA-assembled advanced plasmonic architectures. *Chemical Reviews*, 118(6), 3032–3053.
[58] Zhulina, E. B., Borisov, O. V., & Birshtein, T. M. (1999). Polyelectrolyte brush interaction with multivalent ions. *Macromolecules*, 32(24), 8189–8196.

[59] Lin, M., Kim, G. H., Kim, J. H., Oh, J. W., & Nam, J. M. (2017). Transformative heterointerface evolution and plasmonic tuning of anisotropic trimetallic nanoparticles. *Journal of the American Chemical Society*, 139(30), 10180–10183.
[60] Pang, X., Zhao, L., Han, W., Xin, X., & Lin, Z. (2013). A general and robust strategy for the synthesis of nearly monodisperse colloidal nanocrystals. *Nature Nanotechnology*, 8(6), 426–431.
[61] Chen, T., Yang, M., Wang, X., Tan, L. H., & Chen, H. (2008). Controlled assembly of eccentrically encapsulated gold nanoparticles. *Journal of the American Chemical Society*, 130(36), 11858–11859.
[62] Gil, E. S., & Hudson, S. M. (2004). Stimuli-responsive polymers and their bioconjugates. *Progress In Polymer Science*, 29(12), 1173–1222.
[63] Zoppe, J. O., Ataman, N. C., Mocny, P., Wang, J., Moraes, J., & Klok, H. A. (2017). Surface-initiated controlled radical polymerization: State-of-the-art opportunities and challenges in surface and interface engineering with polymer brushes. *Chemical Reviews*, 117(3), 1105–1318.
[64] Shi, J., Fu, W., Zou, Q., Ward, N., Liao, S., Wang, J., & Zhao, R. C. (2025). Click chemistry-based nanomaterial modification for cancer targeting: A review. *ACS Applied Nano Materials*, 8(1), 10–21.
[65] Moses, J. E., & Moorhouse, A. D. (2007). The growing applications of click chemistry. *Chemical Society Reviews*, 36(8), 1249–1262.
[66] https://www.un.org/waterforlifedecade/
[67] Caturegli, L., Matteoli, S., Gaetani, M., Grossi, N., Magni, S., Minelli, A., & Volterrani, M. (2020). Effects of water stress on the spectral reflectance of bermudagrass. *Scientific Reports*, 10(1), 1–12.
[68] Schyns, J. F., Hoekstra, A. Y., Booij, M. J., Hogeboom, R. J., & Mekonnen, M. M. (2019). Limits to the world's green water resources for food, feed, fiber, timber, and bioenergy. *Proceedings of the National Academy of Sciences*, 116(11), 4893–4898.
[69] Vanham, D., Leip, A., Galli, A., Kastner, T., Bruckner, M., Uwizeye, A., & Hoekstra, A. Y. (2019). The Environmental Footprint Family to address local to planetary sustainability and deliver on the SDGs. *Science of the Total Environment*, 693, 133642.
[70] Mekonnen, M. M., & Hoekstra, A. Y. (2020). Sustainability of the blue water footprint of crops. *Advances in Water Resources*, 143, 103679.
[71] Mekonnen, M. M., & Hoekstra, A. Y. (2011). The green, blue, and grey water footprint of crops and derived crop products. *Hydrology and Earth System Sciences*, 15(5), 1577–1600.
[72] Mekonnen, M. M., & Hoekstra, A. Y. (2010). A global and high-resolution assessment of the green, blue, and grey water footprint of wheat. *Hydrology and Earth System Sciences*, 14(7), 1259–1276.
[73] Niccolucci, V., Tiezzi, E., Pulselli, F. M., & Capineri, C. (2012). Biocapacity vs. Ecological Footprint of world regions: A geopolitical interpretation. *Ecological Indicators*, 16, 23–30.
[74] Fu, W., Turner, J. C., Zhao, J., & Du, G. (2015). Ecological footprint (EF): An expanded role in calculating resource productivity (RP) using China and the G20 member countries as examples. *Ecological Indicators*, 48, 464–471.
[75] Schyns, J. F., Hoekstra, A. Y., Booij, M. J., Hogeboom, R. J., & Mekonnen, M. M. (2019). Limits to the world's green water resources for food, feed, fiber, timber, and bioenergy. *Proceedings of the National Academy of Sciences*, 116(11), 4893–4898.
[76] IPCC (2022). IPCC Sixth Assessment Report (AR6): Climate Change 2022-Impacts, Adaptation and Vulnerability, https://www.ipcc.ch/report/ar6/wg2/
[77] Ismail, M., Akhtar, K., Khan, M. I., Kamal, T., Khan, M. A., M Asiri, A., & Khan, S. B. (2019). Pollution, toxicity, and carcinogenicity of organic dyes and their catalytic bio-remediation. *Current Pharmaceutical Design*, 25(34), 3645–3663
[78] Richardson, J. B., Dancy, B. C., Horton, C. L., Lee, Y. S., Madejczyk, M. S., Xu, Z. Z., & Lewis, J. A. (2018). Exposure to toxic metals triggers unique responses from the rat gut microbiota. *Scientific Reports*, 8(1), 1–12.
[79] Singh, N. B., Nagpal, G., & Agrawal, S. (2018). Water purification by using adsorbents: A review. *Environmental Technology & Innovation*, 11, 187–240.

[80] Lu, F., & Astruc, D. (2020). Nanocatalysts and other nanomaterials for water remediation from organic pollutants. *Coordination Chemistry Reviews*, 408, 213180.
[81] Ghadimi, M., Zangenehtabar, S., & Homaeigohar, S. (2020). An overview of the water remediation potential of nanomaterials and their ecotoxicological impacts. *Water*, 12(4), 1150.
[82] Sharpe, M. (1999). Focus. Doing what comes naturally. *Journal of Environmental Monitoring*, 1(4), 55N-58N.
[83] Ziegler, P., Sree, K. S., & Appenroth, K. J. (2016). Duckweeds for water remediation and toxicity testing. *Toxicological & Environmental Chemistry*, 98(10), 1127–1154.
[84] Guerra, F. D., Attia, M. F., Whitehead, D. C., & Alexis, F. (2018). Nanotechnology for environmental remediation: Materials and applications. *Molecules*, 23(7), 1760.
[85] Huang, H. Y. (2021). Water Remediation by Natural Products Core/Shell Magnetite Nanoparticles (Doctoral dissertation, Texas A&M University-Kingsville).
[86] Shabbir, S., Faheem, M., Ali, N., Kerr, P. G., & Wu, Y. (2017). Periphyton biofilms: A novel and natural biological system for the effective removal of sulfonated azo dye methyl orange by a synergistic mechanism. *Chemosphere*, 167, 236–246.
[87] Liu, J., Wang, F., Wu, W., Wan, J., Yang, J., Xiang, S., & Wu, Y. (2018). Biosorption of high-concentration Cu (II) by periphytic biofilms and the development of a fiber periphyton bioreactor (FPBR). *Bioresource Technology*, 248, 127–134.
[88] Flanigan, D., Sandoval, R., Bridgers, K., & Hristovski, K. (2013). Bolaform amphiphile modified granular activated carbon media for removal of strong acid oxo-anions from water: Nitrate. *Journal of Environmental Chemical Engineering*, 1(4), 1188–1193.
[89] Bozbas, S. K., Ay, U., & Kayan, A. (2013). Novel inorganic-organic hybrid polymers to remove heavy metals from aqueous solution. *Desalination and Water Treatment*, 51(37–39), 7208–7215.
[90] Gyürék, L. L., & Finch, G. R. (1998). Modeling water treatment chemical disinfection kinetics. *Journal of Environmental Engineering*, 124(9), 783–793.
[91] Li, Q., & Lu, G. (2008). Controlled synthesis and photocatalytic investigation of different-shaped one-dimensional titanic acid nanomaterials. *Journal of Power Sources*, 185(1), 577–583.
[92] Myadav, S., & Jaiswar, G. (2017). Review on undoped/doped TiO_2 nanomaterial; synthesis and photocatalytic and antimicrobial activity. *Journal of the Chinese Chemical Society*, 64(1), 103–116.
[93] Liu, J., Wang, Z., Luo, Z., & Bashir, S. (2013). Effective bactericidal performance of silver-decorated titania nanocomposites. *Dalton Transactions*, 42(6), 2158–2166.
[94] Savage, N., & Diallo, M. S. (2005). Nanomaterials and water purification: Opportunities and challenges. *Journal of Nanoparticle Research*, 7(4), 331–342.
[95] Chen, P., Wang, F., Chen, Z. F., Zhang, Q., Su, Y., Shen, L., & Liu, G. (2017). Study on the photocatalytic mechanism and detoxicity of gemfibrozil by a sunlight-driven TiO2/carbon dots photocatalyst: The significant roles of reactive oxygen species. *Applied Catalysis B: Environmental*, 204, 250–259.
[96] Miklos, D. B., Remy, C., Jekel, M., Linden, K. G., Drewes, J. E., & Hübner, U. (2018). Evaluation of advanced oxidation processes for water and wastewater treatment: A critical review. *Water Research*, 139, 118–131.
[97] Dewil, R., Mantzavinos, D., Poulios, I., & Rodrigo, M. A. (2017). New perspectives for advanced oxidation processes. *Journal of Environmental Management*, 195, 93–99.
[98] Barzegar, M. H., Ghaedi, M., Avargani, V. M., Sabzehmeidani, M. M., Sadeghfar, F., & Jannesar, R. (2019). Electrochemical synthesis of Zn: ZnO/Ni_2P and efficient photocatalytic degradation of Auramine O in aqueous solution under multi-variable experimental design optimization. *Polyhedron*, 165, 1–8.
[99] Li, C., Hu, R., Lu, X., Bashir, S., & Liu, J. L. (2020). Efficiency enhancement of photocatalytic degradation of tetracycline using reduced graphene oxide-coordinated titania nanoplatelet. *Catalysis Today*, 350, 171–183.

[100] Sun, P., Lee, W. N., Zhang, R., & Huang, C. H. (2016). Degradation of DEET and caffeine under UV/chlorine and simulated sunlight/chlorine conditions. *Environmental Science & Technology*, 50(24), 13265–13273.

[101] Liu, J. L., & Bashir, S. (2015). *Advanced Nanomaterials and their Applications in Renewable Energy* (vol. 1–50). New York: Elsevier.

[102] Barrera, M. C., Escobar, J., José, A., Cortés, M. A., Viniegra, M., & Hernández, A. (2006). Effect of solvothermal treatment temperature on the properties of sol-gel ZrO_2-TiO_2 mixed oxides as HDS catalyst supports. *Catalysis Today*, 116(4), 498–504.

[103] Antonelli, D. M. (1999). Synthesis of phosphorus-free mesoporous titania via templating with amine surfactants. *Microporous and Mesoporous Materials*, 30(2–3), 315–319.

[104] Lee, G., & Cho, K. (2009). Electronic structures of zigzag graphene nanoribbons with edge hydrogenation and oxidation. *Physical Review B*, 79(16), 165440.

[105] Spyrou, K., & Rudolf, P. (2014). An introduction to graphene. *Functionalization of Graphene*, 1–20. https://onlinelibrary.wiley.com/doi/abs/10.1002/9783527672790.ch1

[106] Wang, Y., Liu, M., Liu, Y., Luo, J., Lu, X., & Sun, J. (2017). A novel mica-titania@ graphene core-shell structured antistatic composite pearlescent pigment. *Dyes and Pigments*, 136, 197–204.

[107] Lv, T., Wu, M., Guo, M., Liu, Q., & Jia, L. (2019). Self-assembly photocatalytic reduction synthesis of graphene-encapsulated $LaNiO_3$ nanoreactor with high efficiency and stability for photocatalytic water splitting to hydrogen. *Chemical Engineering Journal*, 356, 580–591.

[108] Mustapha, S., Tijani, J. O., Ndamitso, M. M., Abdulkareem, A. S., Shuaib, D. T., Amigun, A. T., & Abubakar, H. L. (2021). Facile synthesis and characterization of TiO_2 nanoparticles: X-ray peak profile analysis using Williamson-Hall and Debye-Scherrer methods. *International Nano Letters*, 11(3), 241–261.

[109] Santara, B., Giri, P. K., Imakita, K., & Fujii, M. (2014). Microscopic origin of lattice contraction and expansion in undoped rutile TiO_2 nanostructures. *Journal of Physics D: Applied Physics*, 47(21), 215302.

[110] Kamble, S. P., Sawant, S. B., Schouten, J. C., & Pangarkar, V. G. (2003). Photocatalytic and photochemical degradation of aniline using concentrated solar radiation. *Journal of Chemical Technology & Biotechnology: International Research in Process, Environmental & Clean Technology*, 78(8), 865–872.

[111] Hathway, T., & Jenks, W. S. (2008). Effects of sintering of TiO_2 particles on the mechanisms of photocatalytic degradation of organic molecules in water. *Journal of Photochemistry and Photobiology A: Chemistry*, 200(2–3), 216–224.

[112] Patterson, A. L. (1939). The Scherrer formula for X-ray particle size determination. *Physical Review*, 56(10), 978.

[113] Villarroel-Rocha, J., Barrera, D., & Sapag, K. (2014). Introducing a self-consistent test and the corresponding modification in the Barrett, Joyner, and Halenda method for pore-size determination. *Microporous and Mesoporous Materials*, 200, 68–78.

[114] Ohsaka, T., Izumi, F., & Fujiki, Y. (1978). Raman spectrum of anatase, TiO_2. *Journal of Raman Spectroscopy*, 7(6), 321–324.

[115] Minitha, C. R., & Rajendrakumar, R. T. (2013). Synthesis and characterization of reduced graphene oxide. In *Advanced Materials Research* (Vol. 678, pp. 56–60). Trans Tech Publications Ltd.

[116] Luo, Z., Vora, P. M., Mele, E. J., Johnson, A. C., & Kikkawa, J. M. (2009). Photoluminescence and bandgap modulation in graphene oxide. *Applied Physics Letters*, 94(11), 111909.

[117] Ramanan, V., Siddaiah, B., Raji, K., & Ramamurthy, P. (2018). Green synthesis of multi-functionalized, nitrogen-doped, highly fluorescent carbon dots from waste expanded polystyrene and its application in the fluorimetric detection of Au^{3+} ions in aqueous media. *Acs Sustainable Chemistry & Engineering*, 6(2), 1627–1638.

[118] Kozak, O., Sudolska, M., Pramanik, G., Cigler, P., Otyepka, M., & Zboril, R. (2016). Photoluminescent carbon nanostructures. *Chemistry of Materials*, 28(12), 4085–4128.

[119] Liang, Y. T., Vijayan, B. K., Lyandres, O., Gray, K. A., & Hersam, M. C. (2012). Effect of dimensionality on the photocatalytic behavior of carbon-titania nanosheet composites: Charge transfer at nanomaterial interfaces. *The Journal of Physical Chemistry Letters*, 3(13), 1760–1765.
[120] Duarte, H. A., Carvalho, S., Paniago, E. B., & Simas, A. M. (1999). Importance of tautomers in the chemical behavior of tetracyclines. *Journal Of Pharmaceutical Sciences*, 88(1), 111–120.
[121] Kiarii, E. M., Govender, K. K., Ndung'u, P. G., & Govender, P. P. (2017). A DFT study on the effect of supporting titania on silica, graphene, epoxy graphene, and carbon nanotubes-Interfacial properties and optical response. *Computational Condensed Matter*, 13, 615.
[122] Bashir, S., Hanumandla, P., Huang, H. Y., & Liu, J. L. (2018). Nanostructured materials for advanced energy conversion and storage devices: Safety implications at end-of-life disposal. In *Nanostructured Materials for Next-Generation Energy Storage and Conversion* (pp. 517–542). Berlin, Heidelberg: Springer.
[123] Liu, J. L., & Bashir, S. (2015). *Advanced Nanomaterials and Their Applications in Renewable Energy* (Vol. 367–405). New York: Elsevier.
[124] Elieh-Ali-Komi, D., & Hamblin, M. R. (2016). Chitin and chitosan: Production and application of versatile biomedical nanomaterials. *International Journal of Advanced Research*, 4(3), 411.
[125] Jayakumar, R., Prabaharan, M., Kumar, P. S., Nair, S. V., & Tamura, H. (2011). Biomaterials based on chitin and chitosan in wound dressing applications. *Biotechnology Advances*, 29(3), 322–337.
[126] Fernandes, J. C., Tavaria, F. K., Soares, J. C., Ramos, Ó. S., Monteiro, M. J., Pintado, M. E., & Malcata, F. X. (2008). Antimicrobial effects of chitosans and chitooligosaccharides, upon *Staphylococcus aureus* and *Escherichia coli*, in food model systems. *Food Microbiology*, 25(7), 922–928.
[127] Park, B. K., & Kim, M. M. (2010). Applications of chitin and its derivatives in biological medicine. *International Journal of Molecular Sciences*, 11(12), 5152–5164.
[128] https://gestis.dguv.de/data?name=491257&lang=en
[129] Bykova, E., Dubrovinsky, L., Dubrovinskaia, N., Bykov, M., McCammon, C., Ovsyannikov, S. V., & Prakapenka, V. (2016). The structural complexity of simple Fe_2O_3 at high pressures and temperatures. *Nature Communications*, 7(1), 1–6.
[130] Jette, E. R., & Foote, F. (1933). An X-Ray Study of the Wüstite (FeO) Solid Solutions. *The Journal of Chemical Physics*, 1(1), 29–36.
[131] Fleet, M. E. (1986). The structure of magnetite: Symmetry of cubic spinels. *Journal of Solid-State Chemistry*, 62(1), 75–82.
[132] Lavina, B., Dera, P., Kim, E., Meng, Y., Downs, R. T., Weck, P. F., & Zhao, Y. (2011). Discovery of the recoverable high-pressure iron oxide Fe_4O_5. *Proceedings of the National Academy of Sciences*, 108(42), 17281–17285.
[133] Shmakov, A. N., Kryukova, G. N., Tsybulya, S. V., Chuvilin, A. L., & Solovyeva, L. P. (1995). Vacancy ordering in γ-Fe_2O_3: Synchrotron X-ray powder diffraction and high-resolution electron microscopy studies. *Journal of Applied Crystallography*, 28(2), 141–145.
[134] Szytuła, A., Burewicz, A., Dimitrijević, Ž., Kraśnicki, S., Rżany, H., Todorović, J., & Wolski, W. (1968). Neutron diffraction studies of α-FeOOH. *Physica Status Solidi (B)*, 26(2), 429–434.
[135] Bykova, E., Dubrovinsky, L., Dubrovinskaia, N., Bykov, M., McCammon, C., Ovsyannikov, S. V., & Prakapenka, V. (2016). The structural complexity of simple Fe_2O_3 at high pressures and temperatures. *Nature Communications*, 7(1), 1–6.
[136] Post, J. E., Heaney, P. J., Von Dreele, R. B., & Hanson, J. C. (2003). Neutron and temperature-resolved synchrotron X-ray powder diffraction study of akaganéite. *American Mineralogist*, 88(5–6), 782–788.
[137] Wyckoff, R. W. G., & Wyckoff, R. W. (1963). *Crystal Structures* (Vol. 1, pp. 312). New York: Interscience Publishers.
[138] Patrat, G., De Bergevin, F., Pernet, M., & Joubert, J. C. (1983). Structure locale de δ-FeOOH. *Acta Crystallographica Section B: Structural Science*, 39(2), 165–170.

[139] Wold, A., & Dwight, K. (1993). *Solid State Chemistry: Synthesis, Structure, and Properties of Selected Oxides and Sulfides*. Springer Science & Business Media.

[140] Olsen, J. S., Cousins, C. S. G., Gerward, L., Jhans, H. T., & Sheldon, B. J. (1991). A study of the crystal structure of Fe_2O_3 in the pressure range up to 65 GPa using synchrotron radiation. *Physica Scripta*, 43(3), 327.

[141] https://materialsproject.org/materials/mp-19306/)

[142] Van Zeghbroeck, B. J. (2011). *Principles of Semiconductor Devices* (pp. 35–115). Wilmington, DE: Prentice-Hall.

[143] Gill, P. M., & von Ragué Schleyer, P. (1994). Density functional theory (DFT), Hartree–Fock (HF), and the self-consistent field. *The Journal of Chemical Physics*, 100, 5066–5075.

[144] Kohn, W., Becke, A. D., & Parr, R. G. (1996). Density functional theory of electronic structure. *The Journal of Physical Chemistry*, 100(31), 12974–12980.

[145] Cohen, A. J., Mori-Sánchez, P., & Yang, W. (2012). Challenges for density functional theory. *Chemical Reviews*, 112(1), 289–320.

[146] Perdew, J. P., & Zunger, A. (1981). Self-interaction correction to density-functional approximations for many-electron systems. *Physical Review B*, 23(10), 5048.

[147] Monkhorst, H. J. (1979). Hartree-Fock density of states for extended systems. *Physical Review B*, 20(4), 1504.

[148] Mathew, K., Zheng, C., Winston, D., Chen, C., Dozier, A., Rehr, J. J., & Persson, K. A. (2018). High-throughput computational X-ray absorption spectroscopy. *Scientific Data*, 5(1), 1–8.

[149] https://materialsproject.org/materials/mp-19306/

[150] https://materialsproject.org/materials/mp-19770/

[151] Marsman, M., Paier, J., Stroppa, A., & Kresse, G. (2008). Hybrid functionals are applied to extended systems. *Journal of Physics: Condensed Matter*, 20(6), 064201.

[152] Evans, J. (2018). *X-Ray Absorption Spectroscopy for the Chemical and Materials Sciences* (pp. 1–32). New York: John Wiley & Sons.

[153] Lamberti, C. (2004). The use of synchrotron radiation techniques in the characterization of strained semiconductor heterostructures and thin films. *Surface Science Reports*, 53(1–5), 1–197.

[154] Joseph, D., Basu, S., Jha, S. N., & Bhattacharyya, D. (2012). Chemical shifts of KX-ray absorption edges on copper in different compounds by X-ray absorption spectroscopy (XAS) with Synchrotron radiation. *Nuclear Instruments and Methods in Physics Research Section B: Beam Interactions with Materials and Atoms*, 274, 126–128.

[155] Brown, G. E., & Sturchio, N. C. (2002). An overview of synchrotron radiation applications to low-temperature geochemistry and environmental science. https://triage19.com/products/REV/REV049C01.pdf

[156] Mathew, K., Zheng, C., Winston, D., Chen, C., Dozier, A., Rehr, J. J., & Persson, K. A. (2018). High-throughput computational X-ray absorption spectroscopy. *Scientific Data*, 5(1), 1–8.

[157] Yamashita, T., & Hayes, P. (2008). Analysis of XPS spectra of Fe^{2+} and Fe^{3+} ions in oxide materials. *Applied Surface Science*, 254(8), 2441–2449.

[158] McCune, J. A., Mommer, S., Parkins, C. C., & Scherman, O. A. (2020). Design Principles for Aqueous Interactive Materials: Lessons from Small Molecules and Stimuli-Responsive Systems. *Advanced Materials*, 32(20), 1906890.

[159] Calandra, P., Caschera, D., Liveri, V. T., & Lombardo, D. (2015). How self-assembly of amphiphilic molecules can generate complexity in the nanoscale. *Colloids and Surfaces A: Physicochemical and Engineering Aspects*, 484, 164–183.

[160] Wang, Y., Song, S., Zhang, S., & Zhang, H. (2019). Stimuli-responsive nano theranostics based on lanthanide-doped upconversion nanoparticles for cancer imaging and therapy: Current advances and future challenges. *Nano Today*, 25, 38–67.

[161] Jin, H., Jiao, F., Daily, M. D., Chen, Y., Yan, F., Ding, Y. H., & Chen, C. L. (2016). Highly stable and self-repairing membrane-mimetic 2D nanomaterials assembled from lipid-like peptoids. *Nature Communications*, 7(1), 1–8.

[162] Song, H., Rioux, R. M., Hoefelmeyer, J. D., Komor, R., Niesz, K., Grass, M., Yang, P., & Somorjai, G. A. (2006). Hydrothermal growth of mesoporous SBA-15 silica in the presence of PVP-stabilized Pt nanoparticles: Synthesis, characterization, and catalytic properties. *Journal of the American Chemical Society*, 128(9), 3027–3037.

[163] Collins, S. D., Ran, N. A., Heiber, M. C., & Nguyen, T. Q. (2017). Small is powerful: Recent progress in solution-processed small-molecule solar cells. *Advanced Energy Materials*, 7(10), p.1602242.

[164] Zhang, J., Li, Y., Zhang, X., & Yang, B. (2010). Colloidal self-assembly meets nanofabrication: From two-dimensional colloidal crystals to nanostructure arrays. *Advanced Materials*, 22(38), 4249–4269.

[165] Xu, C., Lin, Y., Wang, J., Wu, L., Wei, W., Ren, J., & Qu, X. (2013). Nanoceria-Triggered Synergetic Drug Release Based on CeO_2-Capped Mesoporous Silica Host-Guest Interactions and Switchable Enzymatic Activity and Cellular Effects of CeO_2. *Advanced Healthcare Materials*, 2(12), 1591–1599.

[166] Carne, A., Carbonell, C., Imaz, I., & Maspoch, D. (2011). Nanoscale metal-organic materials. *Chemical Society Reviews*, 40(1), 291–305.

[167] Stock, N., & Biswas, S. (2012). Synthesis of metal-organic frameworks (MOFs): Routes to various MOF topologies, morphologies, and composites. *Chemical Reviews*, 112(2), 933–969.

[168] Rubio-Martinez, M., Avci-Camur, C., Thornton, A. W., Imaz, I., Maspoch, D., & Hill, M. R. (2017). New synthetic routes towards MOF production at scale. *Chemical Society Reviews*, 46(11), 3453–3480.

[169] Vallé, K., Belleville, P., Pereira, F., & Sanchez, C. (2006). Hierarchically structured transparent hybrid membranes by in situ growth of mesostructured organosilica in host polymer. *Nature Materials*, 5(2), 107–111.

[170] Kaur, P., Hupp, J. T., & Nguyen, S. T. (2011). Porous organic polymers in catalysis: Opportunities and challenges. *ACS Catalysis*, 1(7), 819–835.

[171] Hu, J., Deng, W., & Chen, D. (2017). Ceria hollow spheres as an adsorbent for efficient removal of acid dye. *ACS Sustainable Chemistry & Engineering*, 5(4), 3570–3582.

[172] Sun, J., Wang, C., Zeng, L., Xu, P., Yang, X., Chen, J., Xing, X., Jin, Q., & Yu, R. (2016). Controllable assembly of CeO_2 micro/nanospheres with adjustable size and their application in Cr (VI) adsorption. *Materials Research Bulletin*, 75, 110–114.

[173] Rego, R. M., Sriram, G., Ajeya, K. V., Jung, H. Y., Kurkuri, M. D., & Kigga, M. (2021). Cerium-based UiO-66 MOF as a multipollutant adsorbent for universal water purification. *Journal of Hazardous Materials*, 416, 125941.

[174] Gogoi, A., & Sarma, K. C. (2017). Synthesis of the novel β-cyclodextrin-supported CeO_2 nanoparticles for the catalytic degradation of methylene blue in aqueous suspension. *Materials Chemistry and Physics*, 194, 327–336.

[175] Velusamy, P., & Lakshmi, G. (2017). The enhanced photocatalytic performance of (ZnO/CeO_2)-β-CD system for the effective decolorization of Rhodamine B under UV light irradiation. *Applied Water Science*, 7(7), 4025–4036.

[176] Khan, N. A., Hasan, Z., & Jhung, S. H. (2013). Adsorptive removal of hazardous materials using metal-organic frameworks (MOFs): A review. *Journal of Hazardous Materials*, 244, 444–456.

[177] Lee, H. A., Park, E., & Lee, H. (2020). Polydopamine and Its Derivative Surface Chemistry in Material Science: A Focused Review for Studies at KAIST. *Advanced Materials*, 1907505.

[178] Long, J. R., & Yaghi, O. M. (2009). The pervasive chemistry of metal-organic frameworks. *Chemical Society Reviews*, 38(5), 1213–1214.

[179] An, J., Farha, O. K., Hupp, J. T., Pohl, E., Yeh, J. I., & Rosi, N. L. (2012). Metal-adenine vertices for the construction of an exceptionally porous metal-organic framework. *Nature Communications*, 3(1), 1–6.

[180] Koh, K., Wong-Foy, A. G., & Matzger, A. J. (2008). A crystalline mesoporous coordination copolymer with high microporosity. *Angewandte Chemie International Edition*, 47(4), 677–680.

[181] Lu, G., Li, S., Guo, Z., Farha, O. K., Hauser, B. G., Qi, X., Wang, Y., Wang, X., Han, S., Liu, X., & DuChene, J. S. (2012). Imparting functionality to a metal-organic framework material by controlled nanoparticle encapsulation. *Nature Chemistry*, 4(4), 310–316.
[182] Zhuang, W., Ma, S., Wang, X. S., Yuan, D., Li, J. R., Zhao, D., & Zhou, H. C. (2010). Introduction of cavities up to 4 nm into a hierarchically assembled metal-organic framework using an angular, tetratopic ligand. *Chemical Communications*, 46(29), 5223–5225.
[183] Eddaoudi, M., Li, H., & Yaghi, O. M. (2000). Highly porous and stable metal-organic frameworks: Structure design and sorption properties. *Journal of the American Chemical Society*, 122(7), 1391–1397.
[184] Zhao, D., Timmons, D. J., Yuan, D., & Zhou, H. C. (2011). Tuning the topology and functionality of metal-organic frameworks by ligand design. *Accounts of Chemical Research*, 44(2), 123–133.
[185] Johnsen, A. R., Wick, L. Y., & Harms, H. Principles of microbial PAH-degradation in soil. *Environmental Pollution*. 2005 Jan 1;133(1):71–84.
[186] Haritash, A. K., & Kaushik, C. P. (2009). Biodegradation aspects of polycyclic aromatic hydrocarbons (PAHs): A review. *Journal of Hazardous Materials*, 169(1–3), 1–15.
[187] An, J., Geib, S. J., & Rosi, N. L. (2009). Cation-triggered drug release from the porous zinc-adenine metal-organic framework. *Journal of the American Chemical Society*, 131(24), 8376–8377.
[188] USEPA (2004). *Cleaning up the Nation's Waste Sites: Markets and Technology Trends*. Washington, DC: US Environmental Protection Agency. EPA 542-R-04-015.
[189] Chen, Y., Cheng, J. J., & Creamer, K. S. (2008). Inhibition of anaerobic digestion process: A review. *Bioresource Technology* 99, 4044–4064.
[190] Godoy, P., Reina, R., Calderón, A., Wittich, R.-M., García-Romera, I., & Aranda, E. (2016). Exploring the potential of fungi isolated from PAH-polluted soil as a source of xenobiotic-degrading fungi. *Environmental Science and Pollution Research* 23, 20985–20996.
[191] Basak, G., Das, D., & Das, N. (2014). Dual role of acidic diacetate sophorolipid as biostabilizer for ZnO nanoparticle synthesis and biofunctionalizing agent against Salmonella enterica and Candida albicans. *Journal of Microbiology and Biotechnology*, 24(1), 87–96.
[192] Basak, G., Hazra, C., & Sen, R. (2020). Biofunctionalized nanomaterials for in situ clean-up of hydrocarbon contamination: A quantum jump in global bioremediation research. *Journal of Environmental Management* 256, 109913.
[193] Patel, V., Jain, S., & Madamwar, D. (2012). Naphthalene degradation by a bacterial consortium (DV-AL) developed from Alang-Sosiya ship-breaking yard, Gujarat, India. *Bioresource Technology* 107, 122–130.
[194] Lu, M., Zhang, Z. Z., Wang, J. X., Zhang, M., Xu, Y. X., & Wu, X. J. (2014). Interaction of heavy metals and pyrene on their fates in soil and tall fescue (Festuca arundinacea). *Environmental Science & Technology*, 48(2), 1158–1165.
[195] Gauthier, P. T., Norwood, W. P., Prepas, E. E., & Pyle, G. G. (2014). Metal-PAH mixtures in the aquatic environment: A review of co-toxic mechanisms leading to more-than-additive outcomes. *Aquatic Toxicology* 154, 253–269.
[196] Brito, E. M., De la Cruz Barrón, M., Caretta, C. A., Goñi-Urriza, M., Andrade, L. H., Cuevas-Rodríguez, G., & Guyoneaud, R. (2015). Impact of hydrocarbons, PCBs, and heavy metals on bacterial communities in Lerma River, Salamanca, Mexico: Investigation of hydrocarbon degradation potential. *Science of the Total Environment*, 521, 1–10.
[197] Zhang, Z., Guo, H., Sun, J., & Wang, H. (2020). Investigation of anaerobic phenanthrene biodegradation by a highly enriched co-culture, PheN9, with nitrate as an electron acceptor. *Journal of Hazardous Materials* 383, 121191.
[198] Zhang, D., Vahala, R., Wang, Y., & Smets, B. F. (2016). Microbes in biological processes for municipal landfill leachate treatment: Community, function, and interaction. *International Biodeterioration & Biodegradation*, 113, 88–96.

[199] Giovanella, P., Vieira, G. A. L., Ramos Otero, I. V., Pais Pellizzer, E., de Jesus Fontes, B., & Sette, L. D. (2020). Metal and organic pollutants bioremediation by extremophile microorganisms. *Journal of Hazardous Materials*, 382, 121024.
[200] Cecchin, I., Reddy, K. R., Thomé, A., Tessaro, E. F., & Schnaid, F. (2017). Nanobioremediation: Integration of nanoparticles and bioremediation for sustainable remediation of chlorinated organic contaminants in soils. *International Biodeterioration & Biodegradation* 119, 419–428.
[201] Cole, A. J., Yang, V. C., & David, A. E. (2011). Cancer theranostics: The rise of targeted magnetic nanoparticles. *Trends in Biotechnology*, 29(7), 323–332.
[202] Yang, J., & Yang, Y. W. (2020). Metal-organic framework-based cancer theranostic nanoplatforms. *View*, 1(2), e20.
[203] Wang, X. Q., Wang, W., Peng, M., & Zhang, X. Z. (2021). Free radicals for cancer theranostics. *Biomaterials*, 266, 120474.
[204] Teng, L., Song, G., Liu, Y., Han, X., Li, Z., Wang, Y., & Tan, W. (2019). Nitric oxide-activated "dual-key-one-lock" nanoprobe for in vivo molecular imaging and high-specificity cancer therapy. *Journal of the American Chemical Society*, 141(34), 13572–13581.
[205] Wang, K., Zhang, F., Wei, Y., Wei, W., Jiang, L., Liu, Z., & Liu, S. (2021). *In situ* imaging of cellular reactive oxygen species and caspase-3 activity using a multifunctional theranostic probe for cancer diagnosis and therapy. *Analytical Chemistry*, 93(22), 7870–7878.
[206] Dubey, M., Nagarkoti, S., Awasthi, D., Singh, A. K., Chandra, T., Kumaravelu, J., & Dikshit, M. (2016). Nitric oxide-mediated apoptosis of neutrophils through caspase-8 and caspase-3-dependent mechanisms. *Cell Death & Disease*, 7(9), e2348-e2348.
[207] Ball, K. A., Castello, P. R., & Poyton, R. O. (2011). Low-intensity light stimulates nitrite-dependent nitric oxide synthesis but not oxygen consumption by cytochrome c oxidase: Implications for phototherapy. *Journal of Photochemistry and Photobiology B: Biology*, 102(3), 182–191.
[208] Karu, T. I., Pyatibrat, L. V., Kolyakov, S. F., & Afanasyeva, N. I. (2005). Absorption measurements of a cell monolayer relevant to phototherapy: Reduction of cytochrome c oxidase under near IR radiation. *Journal of Photochemistry and Photobiology B: Biology*, 81(2), 98–106.
[209] Ball, K. A., Castello, P. R., & Poyton, R. O. (2011). Low-intensity light stimulates nitrite-dependent nitric oxide synthesis but not oxygen consumption by cytochrome c oxidase: Implications for phototherapy. *Journal of Photochemistry and Photobiology B: Biology*, 102(3), 182–191.
[210] To, T. L., Fadul, M. J., & Shu, X. (2014). Singlet oxygen triplet energy transfer-based imaging technology for mapping protein-protein proximity in intact cells. *Nature Communications*, 5(1), 1–9.
[211] Mahalingam, S. M., Ordaz, J. D., & Low, P. S. (2018). Targeting a photosensitizer to the mitochondrion enhances the potency of photodynamic therapy. *ACS Omega*, 3(6), 6066–6074.
[212] Zhang, C., Hong, S., Liu, M. D., Yu, W. Y., Zhang, M. K., Zhang, L., & Zhang, X. Z. (2020). pH-sensitive MOF integrated with glucose oxidase for glucose-responsive insulin delivery. *Journal of Controlled Release*, 320, 159–167.
[213] Maganti, M. (2013). *Functional metal-organic frameworks are applied in cancer diagnosis and therapy.* Texas A&M University-Kingsville.
[214] Zhou, Z., Vázquez-González, M., & Willner, I. (2021). Stimuli-responsive metal-organic framework nanoparticles for controlled drug delivery and medical applications. *Chemical Society Reviews*, 50(7), 4541–4563.
[215] Liang, J., & Liang, K. (2020). Multi-enzyme Cascade Reactions in Metal-organic Frameworks. *The Chemical Record*, 20(10), 1100–1116.
[216] Khan, N. A., Hasan, Z., & Jhung, S. H. (2013). Adsorption and removal of sulfur or nitrogen-containing compounds with metal-organic frameworks (MOFs). *Advanced Porous Materials*, 1(1), 91–102.

[217] Wang, F., Ullah, A., Fan, X., Xu, Z., Zong, R., Wang, X., & Chen, G. (2021). Delivery of nanoparticle antigens to antigen-presenting cells: From extracellular specific targeting to intracellular responsive presentation. *Journal of Controlled Release*, 333, 107–128.
[218] Mahnashi, M. H., Mahmoud, A. M., Alhazzani, K., Alanazi, A. Z., Alaseem, A. M., Algahtani, M. M., & El-Wekil, M. M. (2021). Ultrasensitive and selective molecularly imprinted electrochemical oxaliplatin sensor based on a novel nitrogen-doped carbon nanotubes/Ag@ cu MOF as a signal enhancer and reporter nanohybrid. *Microchimica Acta*, 188(4), 1–12.
[219] Wang, L., Qu, X., Zhao, Y., Weng, Y., Waterhouse, G. I., Yan, H., & Zhou, S. (2019). Exploiting single-atom iron centers in a porphyrin-like MOF for efficient cancer phototherapy. *ACS Applied Materials & Interfaces*, 11(38), 35228–35237.
[220] Zhuang, W., Yuan, D., Li, J. R., Luo, Z., Zhou, H. C., Bashir, S., & Liu, J. (2012). Highly potent bactericidal activity of porous metal-organic frameworks. *Advanced Healthcare Materials*, 1(2), 225–238.
[221] Bruck, S. D., & Mueller, E. P. (1988). Materials and biological aspects of synthetic polymers in controlled drug release systems: Problems and challenges. *Critical Reviews in Therapeutic Drug Carrier Systems*, 5(3), 171–187.
[222] He, L., Liu, Y., Lau, J., Fan, W., Li, Q., Zhang, C., & Chen, X. (2019). Recent progress in nanoscale metal-organic frameworks for drug release and cancer therapy. *Nanomedicine*, 14(10), 1343–1365.
[223] Gao, X., Cui, R., Zhang, M., & Liu, Z. (2017). Metal-organic framework nanosheets that exhibit pH-controlled drug release. *Materials Letters*, 197, 217–220.
[224] Chen, X., Tong, R., Shi, Z., Yang, B., Liu, H., Ding, S., & Fang, W. (2018). MOF nanoparticles with encapsulated autophagy inhibitor in controlled drug delivery system for antitumor. *ACS Applied Materials & Interfaces*, 10(3), 2328–2337.
[225] Jensen, S., Tan, K., Feng, L., Li, J., Zhou, H. C., & Thonhauser, T. (2020). Porous Ti-MOF-74 framework as a strong-binding nitric oxide scavenger. *Journal of the American Chemical Society*, 142(39), 16562–16568.
[226] Lu, K., He, C., Guo, N., Chan, C., Ni, K., Weichselbaum, R. R., & Lin, W. (2016). Chlorine-based nanoscale metal-organic framework systemically rejects colorectal cancers via synergistic photodynamic therapy and checkpoint blockade immunotherapy. *Journal of the American Chemical Society*, 138(38), 12502–12510.
[227] Savino, W., Mendes-da-cruz, D. A., Lepletier, A., & Dardenne, M. (2016). Hormonal control of T-cell development in health and disease. *Nature Reviews Endocrinology*, 12(2), 77–89.
[228] Tan, G., Zhong, Y., Yang, L., Jiang, Y., Liu, J., & Ren, F. (2020). A multifunctional MOF-based nanohybrid as an injectable implant platform for drug synergistic oral cancer therapy. *Chemical Engineering Journal*, 390, 124446.
[229] Nazio, F., Bordi, M., Cianfanelli, V., Locatelli, F., & Cecconi, F. (2019). Autophagy and cancer stem cells: Molecular mechanisms and therapeutic applications. *Cell Death & Differentiation*, 26(4), 690–702.
[230] Ni, W., Zhang, L., Zhang, H., Zhang, C., Jiang, K., & Cao, X. (2022). Hierarchical MOF-on-MOF Architecture for pH/GSH-Controlled Drug Delivery and Fe-Based Chemodynamic Therapy. *Inorganic Chemistry*, 61(7), 3281–3287.
[231] Malone-Povolny, M. J., Maloney, S. E., & Schoenfisch, M. H. (2019). Nitric oxide therapy for diabetic wound healing. *Advanced Healthcare Materials*, 8(12), 1801210.
[232] Negri, L. B., Martins, T. J., Alberici, L. C., Leopoldino, A. M., Hamblin, M. R., Ford, P. C., & da Silva, R. S. (2025). Synergistic effect of NO and reactive oxygen species as a new approach in PDT: Photobiological and biochemical studies involving phthalocyanine-ruthenium complexes as new photosensitizer. *Journal of Photochemistry and Photobiology A: Chemistry*, 116527.
[233] Amaroli, A., Ravera, S., Baldini, F., Benedicenti, S., Panfoli, I., & Vergani, L. (2019). Photobiomodulation with 808-nm diode laser light promotes wound healing of human endothelial

cells through increased reactive oxygen species production, stimulating mitochondrial oxidative phosphorylation. *Lasers in Medical Science*, 34(3), 495–504.
[234] Gao, X., Cui, R., Ji, G., & Liu, Z. (2018). Size and surface controllable metal-organic frameworks (MOFs) for fluorescence imaging and cancer therapy. *Nanoscale*, 10(13), 6205–6211.
[235] Wang, L., Li, J., Cheng, L., Song, Y., Zeng, P., & Wen, X. (2021). Application of hard and soft acid-base theory to uncover the destructiveness of Lewis bases to UiO-66 type metal-organic frameworks in aqueous solutions. *Journal of Materials Chemistry A*, 9(26), 14868–14876.
[236] Prasad, A., Sedlářová, M., & Pospíšil, P. (2018). Singlet oxygen imaging using fluorescent probe Singlet Oxygen Sensor Green in photosynthetic organisms. *Scientific Reports*, 8(1), 1–13.
[237] Ding, R., Li, X., Feng, Y., Liu, D., Ji, H., L., & Liu, S. (2025). Black phosphorus nanocomposites with near-infrared chemiluminescence and their application in reactive oxygen species imaging. *Talanta*, 128298.
[238] Lee, B., Hwang, J. S., & Lee, D. G. (2019). Induction of apoptosis-like death by periplanetasin-2 in Escherichia coli and contribution of SOS genes. *Applied Microbiology and Biotechnology*, 103(3), 1417–1427.
[239] Zou, Y., Park, M., Hong, S., & Lah, M. S. (2008). A designed metal-organic framework based on a metal-organic polyhedron. *Chemical Communications*, (20), 2340–2342.
[240] Deria, P., Bury, W., Hupp, J. T., & Farha, O. K. (2014). Versatile functionalization of the NU-1000 platform by solvent-assisted ligand incorporation. *Chemical Communications*, 50(16), 1965–1968.
[241] Guardia, P., Di Corato, R., Lartigue, L., Wilhelm, C., Espinosa, A., García-Hernández, M., & Pellegrino, T. (2012). Water-soluble iron oxide nanocubes with high values of specific absorption rate for cancer cell hyperthermia treatment. *ACS Nano*, 6(4), 3080–3091.
[242] Wei, Z., Maganti, M., Martinez, B., Vangara, K., Palakurthi, S., Luo, Z., & Liu, J. (2016). A biological study of metal-organic frameworks towards human ovarian cancer cell lines. *Canadian Journal of Chemistry*, 94(4), 380–385.
[243] Salgueiro, M. J., & Zubillaga, M. (2025). Theranostic nanoplatforms in nuclear medicine: Current advances, emerging trends, and perspectives for personalized oncology. *Journal of Nanotheranostics*, 6(4), 27.
[244] Ali, A., Ovais, M., Zhou, H., Rui, Y., & Chen, C. (2021). Tailoring metal-organic framework-based nanozymes for bacterial theranostics. *Biomaterials*, 275, 120951.
[245] Nasri, N., Azad, M., Mehrabi, Z., Dini, G., & Marandi, A. (2025). Metal-organic frameworks for biomedical applications: Bridging materials science and regenerative medicine. *RSC Advances*, 15(41), 34481–34509.
[246] Abbas, M., Liang, Z., Chen, M., Qu, W., Khan, S., Ashaq, M. S., & Xie, S. (2025). Toxicity Challenges and Current Advancements in Metal-Organic Frameworks (MOFs) for Biomedical Applications. *Biological Trace Element Research*, 1–17.
[247] Liu, H., Zhen, Z., Chen, F., Chen, J., & Chen, W. (2025). Advancements in iron oxide nanoparticles for multimodal imaging and tumor theranostics. *Current Medicinal Chemistry*, 32(2), 301–321.
[248] Lu, S., Zhang, Z., & Pan, Y. (2025). Assessing the Causal Relationships of Energy Poverty, Foreign Direct Investment, *and Public Health at the Global Level. Emerging Markets Finance and Trade*, 1–23.
[249] Folini, D., Friedl, A., Kübler, F., & Scheidegger, S. (2025). The climate in climate economics. *Review of Economic Studies*, 92(1), 299–338.
[250] Kondinski, A., Oyarzún-Aravena, A. M., Rihm, S. D., Bai, J., Mosbach, S., Akroyd, J., & Kraft, M. (2025). Automated Assembly Modeling of Metal-Organic Polyhedra. *European Journal of Inorganic Chemistry*, 28(26), e202500115.
[251] De Alcaraz-Fossoul, J. (2025). A preliminary study on temporal processes of latent fingermark degradation: Aged *vs* depleted impressions. *Science & Justice*, 101279.

[252] Singh, R., Singh, R., Tripathi, R. K., & Agarwal, P. (2025). Fingerprint recognition using artificial neural networks. In *Proceedings of the National Academy of Sciences, India Section A: Physical Sciences*, 1–9.

[253] Cole, S. A., & Sola, J. L. (2025). First impressions matter: Mundane obstacles to a forensic device for probabilistic reporting in fingerprint analysis. *Social Studies of Science*, 03063127251333074.

[254] Jelly, R., Patton, E. L., Lennard, C., Lewis, S. W., & Lim, K. F. (2009). The detection of latent fingermarks on porous surfaces using amino acid-sensitive reagents: A review. *Analytica Chimica Acta*, 652(1), 128–142.

[255] Patra, A., Kumari, K., Barua, A., & Pradhan, S. (2024). Advancing Visible Spectroscopy through Integrated Machine Learning and Image Processing Techniques. *Applied Sciences*, 14(11), 4527.

[256] Valenta, J. (2014). Determination of absolute quantum yields of luminescing nanomaterials over a broad spectral range: From the integrating sphere theory to the correct methodology. *Nanoscience Methods*, 3(1), 11–27.

[257] Lotey, N. K. B., Lemos, R., D'Silva, F., Deshmukh, A., Singh, N., Wankhede, S., & Kulkarni, S. (2025). Highly Fluorescent Bio-Synthesized Carbon Quantum Dots for Latent Fingerprint Detection. *Journal of Fluorescence*, 1–11.

[258] Sharma, A., Sankhla, M. S., Bhati, S. S., Agrawal, A., & Tyagi, S. (2025). Surface-specific performance of metal and metal oxide nanoparticles in latent fingerprint visualisation. *Discover Nano*, 20(1), 175.

[259] Bunescu, S., Stoica, B. A., Ciubotaru, A. D., Antohi, A. L., Bordeianu, G., & Leferman, C. E. (2025). Integrated visualization and genetic profiling of latent fingerprints via UVITEX OB and adapted fluorescence microscopy. *Forensic Science International*, 112611.

[260] Kumar, N., & Yadav, H. (2025). Application of Nanotechnology in the Development of Latent Fingerprints in the Field of Forensic Dermatoglyphics. *Nanoelectronics: Fundamentals, Advances, and Applications*, 103–128.

[261] 高尾郁也, 鏑木崇史, & 栗原陽介. (2020, August). ニューラルネットワークを用いた RGB 画像からのハイパースペクトル画像生成手法の提案. In IEICE Conferences Archives. *The Institute of Electronics, Information and Communication Engineers*. https://www.ieice.org/publications/conference-FIT-DVDs/FIT2020/data/pdf/I-040.pdf

[262] Castelló, A., Francés, F., & Verdú, F. (2013). Solving underwater crimes: Development of latent prints made on submerged objects. *Science & Justice*, 53(3), 328–331.

[263] Zahra, A., Qureshi, R., Sajjad, M., Sadak, F., Nawaz, M., Khan, H. A., & Uzair, M. (2024). Current advances in imaging spectroscopy and its state-of-the-art applications. *Expert Systems with Applications*, 238, 122172.

[264] International Fingerprint Research Group Collaboration. (2014). Guidelines for the assessment of fingermark detection techniques. *Journal of Forensic Identification*, 64(2), 174–200.

[265] Khoobehi, B., Beach, J. M., & Kawano, H. (2004). Hyperspectral imaging for measurement of oxygen saturation in the optic nerve head. *Investigative Ophthalmology & Visual Science*, 45(5), 1464–1472.

[266] Kollam, A., Jaitly, T., Ganechary, P. K., Raghottam, P., & Aseri, V. (2023). Implementing innovative strategy for visualising fingerprints on wet non-porous surfaces. *Problems of Forensic Sciences*, (133).

[267] Gomes, F. M., de Pereira, C. M. P., de Cássia Mariotti, K., Pereira, T. M., Dos Santos, N. A., & Romão, W. (2023). Study of latent fingerprints-a review. *Forensic Chemistry*, 35, 100525.

[268] Lee, H. C., & Gaensslen, R. E. (2001). Methods of latent fingerprint development. *Advances In Fingerprint Technology*, 2(105–176), 1.

[269] Laufek, F., Sejko, J., & Dušek, M. (2010). The role of silver in the crystal structure of pyrargyrite: Single crystal X-ray diffraction study. *Journal of Geosciences*, 55(2), 161–167.

[270] Kumar, A., Sahoo, S. C., Mehta, S. K., Soni, P., Sharma, V., & Kataria, R. (2022). A luminescent Zn-MOF for the detection of explosives and development of fingerprints. *Analytical Methods*, 14(7), 700–707.

[271] Jin, E., Bon, V., Das, S., Wonanke, A. D., Etter, M., Karlsen, M. A., & Kaskel, S. (2025). Engineering Photoswitching Dynamics in 3D Photochromic Metal-Organic Frameworks through a Metal-Organic Polyhedron Design. *Journal of the American Chemical Society*, 147(10), 8568–8577.
[272] Tozawa, T., Jones, J. T., Swamy, S. I., Jiang, S., Adams, D. J., Shakespeare, S., & Tang, C. (2009). Porous organic cages. *Nature Materials*, 8(12), 973–978.
[273] Vangara, K. K., Liu, J. L., & Palakurthi, S. (2013). Hyaluronic acid-decorated PLGA-PEG nanoparticles for targeted delivery of SN-38 to ovarian cancer. *Anticancer Research*, 33(6), 2425–2434.
[274] Medina-Ramirez, I., Liu, J., Gonzalez-Garcia, M., & Palakurthi, S. (2012). Application of nanometals fabricated using green synthesis in cancer diagnosis and therapy. In *Green Chemistry - Environmentally Benign Approaches* Dr. Mazaahir, K. (ed.) (pp 33–62). ISBN: 978-953-51-0334-9, Rijeka, Croatia: INTECH Publisher.
[275] Ertaş, E., Özer, H., & Alasalvar, C. (2007). A rapid HPLC method for the determination of Sudan dyes and Para Red in red chili pepper. *Food Chemistry*, 105(2), 756–760.
[276] Feng, X., Pisula, W., & Müllen, K. (2009). Large polycyclic aromatic hydrocarbons: Synthesis and discotic organization. *Pure and Applied Chemistry*, 81(12), 2203–2224.
[277] Reichardt, C. (1994). Solvatochromic dyes as solvent polarity indicators. *Chemical Reviews*, 94(8), 2319–2358.
[278] Gordon, G. W., Berry, G., Liang, X. H., Levine, B., & Herman, B. (1998). Quantitative fluorescence resonance energy transfer measurements using fluorescence microscopy. *Biophysical Journal*, 74(5), 2702–2713.
[279] Lock, N., Wu, Y., Christensen, M., Cameron, L. J., Peterson, V. K., Bridgeman, A. J., & Iversen, B. B. (2010). Elucidating negative thermal expansion in MOF-5. *The Journal of Physical Chemistry C*, 114(39), 16181–16186.
[280] Furukawa, H., Kim, J., Ockwig, N. W., O'Keeffe, M., & Yaghi, O. M. (2008). Control of vertex geometry, structure dimensionality, functionality, and pore metrics in the reticular synthesis of crystalline metal-organic frameworks and Polyhedra. *Journal of the American Chemical Society*, 130(35), 11650–11661.
[281] Fox, C. (1973). U.S. Patent No. 3,722,750. Washington, DC: U.S. Patent and Trademark Office.
[282] Boissière, C., Grosso, D., Chaumonnot, A., Nicole, L., & Sanchez, C. (2011). The aerosol route to functional nanostructured inorganic and hybrid porous materials. *Advanced Materials*, 23(5), 599–623.
[283] Han, C. C., Cheng, H. L., Lin, C. L., & Fan, K. C. (2003). Personal authentication using palmprint features. *Pattern Recognition*, 36(2), 371–381.
[284] Dos Santos, V. Z., de Cássia Mariotti, K., Santos, A. S., de Souza, M. A., da Silva, S. W., & Ferrão, M. F. (2025). Chemical changes of aged latent fingerprints by Fourier transform infrared spectrometry and machine learning. *Microchemical Journal*, 116002.
[285] Jain, R. K., Sunil, D., & Bhagavath, P. (2025). Organic fluorophores in developing latent fingerprints: An up-to-date review. *Journal of Coatings Technology and Research*, 22(1), 117–147.
[286] Mienye, I. D., Swart, T. G., Obaido, G., Jordan, M., & Ilono, P. (2025). Deep convolutional neural networks in medical image analysis: A review. *Information*, 16(3), 195.
[287] Gantz, D., Gantz, D., Walch, M., Roberts, M., & Buscaglia, J. (2014). A novel approach for latent print identification using accurate overlays to prioritize reference prints. *Forensic Science International*, 245, 162–170.
[288] Singh, R., Singh, R., Tripathi, R. K., & Agarwal, P. (2025). Fingerprint recognition using artificial neural networks. In *Proceedings of the National Academy of Sciences, India Section A: Physical Sciences* (pp. 1–9).
[289] Fang, L., & Gossard, D. C. (1995). Multidimensional curve fitting to unorganized data points by nonlinear minimization. *Computer-Aided Design*, 27(1), 48–58.

[290] Walch, M. A., Gantz, D. T., & Gantz, D. T. (2012). U.S. Patent Application No. 13/367,153-154.
[291] Grosz, S. A., & Jain, A. K. (2024, September). Genpalm: Contactless palmprint generation with diffusion models. In *2024 IEEE International Joint Conference on Biometrics (IJCB)* (pp. 1–9). IEEE.
[292] Walch, M. A. (2009). U.S. Patent Application No. 12/611,893.
[293] Schwarz, L., & Klenke, I. (2007). Enhancement of Ninhydrin-or DFO-treated latent fingerprints on thermal paper. *Journal of Forensic Sciences*, 52(3), 649–655.

Xuan Wang*, Sajid Bashir, Jingbo Louise Liu

Chapter 4
Vapor-phase synthesis of nanomaterials: structural control and functional applications

4.1 Introduction to bottom-up synthesis

Bottom-up fabrication represents a foundational principle in nanomaterials science, where complex structures are assembled from the most minor possible units, atoms, molecules, or clusters, rather than carved from bulk solids. This approach relies on wet chemistry techniques that promote self-assembly through controlled physical and chemical triggers [1]. These triggers, such as temperature gradients, pH shifts, changes in solvent polarity, and surfactant interactions, initiate nucleation and guide the growth of nanostructures into ordered monolayers or three-dimensional architectures [2]. The theoretical framework behind this process emphasizes the interplay between thermodynamic stability and kinetic control. Thermodynamic factors, such as supersaturation and interfacial energy, determine the feasibility of nucleation, while kinetic parameters, including diffusion rates and reaction time, govern the morphol-

Acknowledgments: The research efforts led by Dr Bashir and Dr Liu were supported by the American Chemical Society Petroleum Research Fund (grant 53827-UR10) federal agencies, including the Office of Naval Research, the Air Force Office of Scientific Research, the Department of Energy, the National Science Foundation, the Department of Education, the Department of Homeland Security (ORISE), and the National Institutes of Health. The Robert A. Welch Foundation (departmental grant AC-0006) is gratefully acknowledged for its support of student training. Technical assistance from Texas A&M University and Texas A&M University-Kingsville is also acknowledged with thanks. The contributions and dedication of collaborators and students are deeply appreciated and recognized as integral to the success of this work.

Authors' contributions: Dr X. Wang prepared the original manuscript for the first edition of the book. Dr J. Louise Liu provided substantial revisions that enhanced the clarity and depth of the content. Dr S. Bashir completed the final revision, performed data analysis, and obtained the necessary copyright permissions. All authors reviewed the final version and approved it for publication.

***Corresponding author: Xuan Wang**, Texas A&M Higher Education Center at McAllen, McAllen, TX, USA, e-mail: xuan.wang@science.tamu.edu
Sajid Bashir, Department of Chemistry, Texas A&M University-Kingsville, MSC161, 700 University Blvd., Kingsville, TX, USA; Texas A&M Energy Institute,1617 Research Parkway, Suite 308, College Station, TX 77843-3372, USA; Louise Nanoinnovation North American, 7534 Beau Terre, Corpus Christi, TX 78414, USA
Jingbo Louise Liu, Department of Chemistry, Texas A&M University-Kingsville, MSC161, 700 University Blvd., Kingsville, TX, USA; Texas A&M Energy Institute,1617 Research Parkway, Suite 308, College Station, TX 77843-3372, USA; Louise Nanoinnovation North American, 7534 Beau Terre, Corpus Christi, TX 78414, USA, e-mails: jingbo.liu@tamuk.edu, jingbo.liu@tamu.edu

 | https://doi.org/10.1515/9783111621111-005

ogy and uniformity of the final product [3]. This chapter will examine the principal bottom-up synthesis strategies for vapor-phase deposition. At the same time, the other three methods such as emulsion-based coprecipitation, sol–gel processing, and both hydrothermal and solvothermal methods were discussed in Chapters 1–3. Each technique possesses unique variables (as shown in Tab. 4.1) and offers advantages in controlling particle size, morphology, and crystallinity, enabling the tailored fabrication of nanomaterials for advanced functional applications. Together, these techniques embody the theoretical elegance of bottom-up design: by manipulating molecular interactions and reaction environments, researchers can engineer nanomaterials with tailored properties for applications in catalysis, electronics, energy storage, and biomedicine.

Vapor-phase deposition encompasses a suite of advanced thin-film fabrication techniques, including physical vapor deposition (PVD), chemical vapor deposition (CVD), and atomic layer deposition (ALD). Each method offers precise control over film thickness, composition, crystallinity, and microstructure. These methods rely on the transport of vapor-phase species to a substrate surface, where they condense or react to form solid films. The choice of technique depends on the desired material properties, substrate geometry, and application requirements. These form the backbone of modern nanofabrication, enabling the creation of tailored material architectures for electronics, photonics, energy storage, and biomedical devices [4]. PVD involves the physical ejection or evaporation of atoms from a solid source in a vacuum environment. These atoms travel in ballistic trajectories and condense onto the substrate. CVD uses volatile precursors that chemically react or decompose at the substrate surface to form a solid film. ALD is a self-limiting, sequential process in which precursors are pulsed one at a time, enabling atomic-scale control over film growth. Together, these vapor-phase deposition methods form the backbone of modern nanofabrication, enabling tailored material architectures for electronics, photonics, energy storage, and biomedical devices.

Microemulsion (Section 1.2.2) and coprecipitation are the bottom-up synthesis techniques that leverage nanoscale droplet environments to control the nucleation and growth of particles with exceptional precision. Microemulsions are thermodynamically stable, isotropic mixtures of oil, water, a surfactant, and, often, a cosurfactant. Unlike conventional emulsions, microemulsions form spontaneously and consist of nanoscale droplets (typically 10–100 nm) dispersed in a continuous phase. These droplets act as nanoreactors, making microemulsions ideal for synthesizing uniform nanoparticles via controlled precipitation or chemical reactions. Microemulsion-based synthesis enables precise control over nucleation and growth, yielding monodispersed nanomaterials, including metal oxides, quantum dots, and magnetic nanoparticles. At its core, the microemulsion method uses surfactants to stabilize dispersed phases, typically oil-in-water, water-in-oil, or bicontinuous systems, creating thermodynamically stable microdomains where chemical reactions can occur in confined volumes. These droplets act as nanoreactors, allowing reactants to mix and precipitate in a controlled fashion. The size, shape, and dispersion of the resulting nanoparticles are determined by the physicochemical properties of the emulsion system and the precipitation reaction kinetics [5].

Nanoemulsions are kinetically stable emulsions with droplet sizes typically ranging from 20 to 200 nm. Unlike microemulsions, they do not form spontaneously and require high-energy input, such as ultrasonication, high-pressure homogenization, or microfluidization, to achieve nanoscale dispersion. Nanoemulsions are widely used in pharmaceuticals, cosmetics, and food science due to their enhanced bioavailability and optical clarity. Nanoemulsions are particularly valuable for encapsulating hydrophobic drugs or bioactives, enabling controlled release and improved penetration across biological membranes. Nanoemulsions are kinetically stable emulsions with droplet sizes typically ranging from 20 to 200 nm. Unlike microemulsions, they do not form spontaneously and require high-energy input, such as ultrasonication, high-pressure homogenization, or microfluidization, to achieve nanoscale dispersion. Nanoemulsions are widely used in pharmaceuticals, cosmetics, and food science due to their enhanced bioavailability and optical clarity. Nanoemulsions are valuable for encapsulating hydrophobic drugs or bioactives, enabling controlled release and improved penetration across biological membranes [6]. Coprecipitation is a solution-based synthesis technique in which multiple ionic species are simultaneously precipitated from a homogeneous solution to form solid-phase nanoparticles. This method is particularly effective for producing metal oxides, hydroxides, and mixed-metal compounds with controlled stoichiometry and morphology. The process typically involves dissolving metal salts (e.g., nitrates, chlorides, and sulfates) in an aqueous or alcoholic solvent, followed by the controlled addition of a precipitating agent, such as NaOH, NH_4OH, or oxalic acid. As the solution's pH and ionic strength change, supersaturation occurs, triggering the nucleation and growth of solid particles [7].

Sol–gel synthesis (Tab. 4.1) transforms molecular precursors into porous networks via hydrolysis and condensation under tunable pH and solvent conditions. Sol–gel synthesis is a versatile bottom-up technique that transforms molecular precursors, typically metal alkoxides or salts, into solid, porous networks through a series of hydrolysis and condensation reactions. The process begins with the dissolution of precursors in a solvent, often alcohol or water, followed by controlled hydrolysis, in which metal–alkoxide bonds (M–OR) are replaced with hydroxyl groups (M–OH). These hydroxylated species then undergo condensation reactions, forming M–O–M linkages and gradually evolving into a three-dimensional oxide network. The term "sol" refers to the colloidal suspension of precursor particles, while "gel" denotes the interconnected solid matrix formed as the reaction progresses. Its ability to produce uniform, high-surface-area materials with controlled porosity and composition makes sol–gel synthesis indispensable in nanotechnology, photonics, and biomedical engineering. This transformation is highly sensitive to reaction conditions, making sol–gel synthesis a tunable platform for tailoring material properties. Its ability to produce uniform, high-surface-area materials with controlled porosity and composition makes sol–gel synthesis indispensable in nanotechnology, photonics, and biomedical engineering [8].

Hydrothermal and solvothermal methods (Tab. 4.1) exploit high-pressure, high-temperature environments to crystallize materials from solution, enabling phase-selective growth and compositional control [9]. Hydrothermal and solvothermal synthesis methods have played a foundational role in materials chemistry since their industrial use in the nineteenth-century mining sector, particularly for crystal growth and mineral extraction. These techniques are defined as solution-based chemical reactions conducted in sealed, high-pressure vessels, commonly autoclaves, where the temperature and pressure exceed the solvent's boiling point. Both methods offer precise control over nucleation, crystal growth, and phase purity, making them indispensable in nanomaterials research, energy storage, catalysis, and optoelectronics. The term "hydrothermal" applies when water is used as the solvent, while "solvothermal" refers to reactions carried out in organic solvents such as alcohols, amines, or ketones. Hydrothermal synthesis involves aqueous-phase reactions in a sealed vessel, typically at temperatures between 100 and 250 °C and autogenous pressures ranging from 1 to 10 MPa. The solvothermal approach is widely used for the growth of oxide-based nanomaterials, zeolites, and piezoelectric crystals [10]. This synthesis method extends the hydrothermal concept to nonaqueous solvents, enabling the formation of materials that are unstable or insoluble in water. It is especially effective for producing metal chalcogenides, hybrid nanostructures, and coordination polymers [11] (Table 4.1).

Tab. 4.1: Summary of the technical variables and materials provided using different bottom-up synthesis approaches.

Technical parameters	**Values**	**Variants and materials**
Method: **Physical vapor deposition (PVD)**		
Source temperature	500–3,000 °C	*Variants*: Thermal evaporation, electron beam evaporation, and magnetron sputtering *Materials*: Metal films, hard coatings (TiN and CrN), and optical layers
Chamber pressure	10^{-6}–10^{-8} torr	
Deposition rate	0.1–10 nm/s	
Substrate bias voltage	−50 to −300 V	
Target–substrate distance and angle vary		
Method: **Chemical vapor deposition (CVD)**		
Precursor flow rate	10–500 sccm	*Variants*: Plasma-enhanced CVD (PECVD) and metal–organic CVD (MOCVD) *Materials*: Semiconductor layers (Si, SiO_2, and GaN), dielectric coatings, and barrier films
Reactor pressure	0.1–10 torr for LPCVD; atmospheric for APCVD	
Substrate temperature	300–1,000 °C	
Carrier gas composition	H_2, N_2, Ar, and He	
Reaction kinetics and residence time vary		

Tab. 4.1 (continued)

Technical parameters	**Values**	**Variants and materials**
Method: Atomic layer deposition (ALD)		
Pulse and purge durations	0.5–5 s	*Variants*: Sub-nanometer thickness control and conformality on 3D structures *Materials*: Gate oxides, passivation layers, nanolaminates, and high-k dielectrics
Number of cycles	Hundreds to thousands	
Reactor temperature	100–300 °C	
Precursor reactivity and saturation behavior vary		
Method: **Microemulsion**		
Surfactant type and concentration	Nonionic: cation or anionic surfactant	Metal oxides (e.g., Fe_3O_4 and ZnO) Quantum dots (e.g., CdSe and PbS) Magnetic nanoparticles Drug delivery carriers and biosensors Electro- and photochemical catalysts
Oil-to-water ratio	1:9 to 9:1	
Cosurfactant	Short-chain alcohols	
Reagent addition rate	Dropwise, continuous, and rapid addition	
Reaction temperature	25–80 °C	
pH and ionic strength	Buffering solution	
Droplet size distribution	1–100 nm	
Method: **Nanoemulsion**		
Droplet size distribution	20–200 nm	*Applications*: Drug delivery, vaccine adjuvants, nutraceuticals, and topical formulations *Advantages*: Improved solubility, rapid absorption, and transparent appearance
Formation requirement	Mechanical energy	
Surfactants	Tween 80, lecithin, and PEG	
Stability	Kinetics are prone to Ostwald ripening	
Method: **Coprecipitation**		
Precursor concentration	Various	Magnetic nanoparticles (e.g., Fe_3O_4 and $CoFe_2O_4$) Catalysts (e.g., mixed oxides of Cu, Zn, and Al) Drug delivery carriers Adsorbents for water remediation
pH control	Typically, 8–12	
Temperature	25–80 °C	

Tab. 4.1 (continued)

Technical parameters	Values	Variants and materials
Stirring rate	Uniform mixing	
Addition rate	Dropwise, controlled rate	
Aging time	Controlled particle growth	
Posttreatment	Washing, drying, and heating	
Method: **Sol–gel**		
Precursor type	Tetraethyl orthosilicate and metal alkoxide	*Applications*: Porous ceramics and aerogels Optical coatings and antireflective films Catalysts and adsorbents Bioactive glasses and drug delivery systems
Solvent system	Alcohols	
Water-to-precursor ratio	1:10	
pH control	2–4 or 8–10	
Temperature	25–80 °C	
Aging time	Post-gelation varying	
Drying and calcination	Varying	
Method: **Hydrothermal synthesis**		
Solvent	Water (H_2O)	*Applications*: Metal oxide (TiO_2 and ZnO) Ceramics: $BaTiO_3$ Crystals: Quartz crystals Metal–organic frameworks (MOFs)
Temperature	100–300 °C	
Pressure	1–10 MPa (autoclave)	
Reaction time	6–48 h	
pH control	Acid or base	
Mineralizers	NaOH, KOH, or HCl	
Seeding	Various: Guide nucleation	
Method: **Solvothermal synthesis**		
Solvent	Organic liquid	*Applications*: Metal sulfide (CdS and PbS) MOFs Perovskites Metal oxides Nanocomposites
Temperature	120–300 °C	
Pressure	2–15 MPa	
Reaction time	6–72 h	

Tab. 4.1 (continued)

Technical parameters	Values	Variants and materials
Solvent polarity	Affects solubility and reaction kinetics	
Ligand/additives	Control morphology	
Atmosphere	Inert gases: H_2, N_2, and Ar	

4.2 Vapor-phase deposition

4.2.1 Overview

This chapter provides a detailed discussion of vapor-phase deposition, a versatile family of thin-film fabrication techniques that coat substrates by transporting material in the vapor phase and condensing it into a solid layer [8]. Vapor deposition is a high-temperature thin-film technique that "grows" glass one molecular layer at a time, instead of cooling and solidifying a bulk melt. The roots of vapor deposition trace back to prehistoric times, when soot from open flames deposited carbon films on cave walls, a rudimentary form of CVD noted by MIT's Sreenivasan and Karen Gleason [14]. A historical overview of vapor-phase deposition illustrates the evolution of its processing techniques, as shown in Fig. 4.1. Ancient artisans later harnessed similar principles, applying gold leaf to copper and bronze using vapors from heated metal salts to create decorative thin films. The seventeenth century ushered in vacuum science with primitive pumps [10], setting the stage for nineteenth-century sputter deposition [15]. Early researchers observed that electric discharges striking metal surfaces ejected atoms onto nearby substrates, establishing PVD techniques by the 1880s [16]. By the mid-1800s, sputtering dominated optical coating processes, and foundational measurements of sputter yields appeared by the 1930s [17]. Industrial demands in the mid-twentieth century drove the refinement of both PVD and CVD [18]. Thermal evaporation and electron beam (e-beam) PVD achieved high-purity metal films under ultra-high vacuum, while magnetron sputtering (1938 patent) enhanced deposition rates and film density [19]. In parallel, low-pressure CVD (LPCVD) [20] and atmospheric pressure CVD (APCVD) emerged for the deposition of silicon and silicon dioxide films in microelectronics [21]. The late 1960s saw the birth of ALD, offering atomic-scale thickness control via sequential precursor pulses [22]. Today, hybrid systems blend PVD's material versatility with CVD's conformality, and advanced methods like plasma-enhanced CVD (PECVD) [23] and high-power impulse magnetron sputtering (HiPIMS) push film performance to new heights [24]. Across optics, semiconductors, and energy devices, the evolution of vapor deposition, from cave-wall soot to atomic-layer precision, remains central to modern nanotechnology [25].

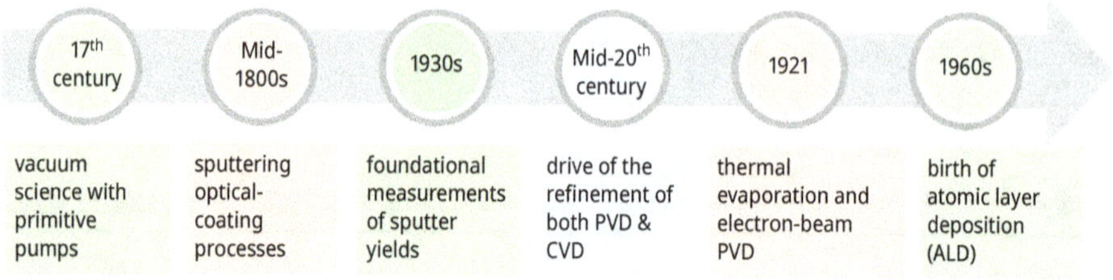

Fig. 4.1: A historical overview of vapor-phase deposition illustrates the evolution of its processing techniques.

Corning has applied vapor deposition technologies for many decades, most prominently through the outside vapor deposition (OVD) method that established its global leadership in optical fiber, and through "direct-to-glass" techniques used in specialty telescope mirrors and spacecraft windows [26]. These processes highlight Corning's expertise in materials engineering, precision process control, equipment innovation, and automation. In OVD, silicon-based precursors mixed with dopants are vaporized and transported by an inert carrier gas to a high-temperature burner. Combustion transforms the vapors into a fine aerosol of oxidized glass particles suspended in residual gases [27]. This stream is directed onto a rotating ceramic mandrel, where successive deposits build an ultra-pure preform. Once deposition is complete, the mandrel is removed, and the porous preform undergoes high-temperature sintering, consolidating the soot into a transparent, uniform tube [28]. The finished preform is then drawn under carefully regulated tension and heat into kilometers of optical fiber. For significant optical components, the direct-to-glass approach replaces the mandrel with a mold or substrate, enabling the gradual formation of monolithic glass structures through continuous layering over extended periods [29].

By fine-tuning precursor chemistry, flame characteristics, and deposition parameters, engineers achieve exceptional control over oxygen content, impurity levels, and dopant distribution, critical for minimizing light attenuation and ensuring thermal stability. The hallmark of Corning's vapor deposition technology is its ability to produce glass of unrivaled purity and uniformity at an industrial scale [30]. Tight dopant control delivers precise refractive index profiles for specialty fibers, while the inherent low-expansion properties of pure silica glass guarantee dimensional stability in extreme environments, from deep-space telescopes to high-speed data networks [31]. These methods bifurcate into PVD and CVD, distinguished by how the vapor species are generated and delivered to the substrate.

To restate, vapor-phase deposition, as a core technology, enables the deterministic, atomically precise construction of functional thin films and nanomaterials. By delivering reactive or inert species in the vapor state to engineered surfaces, it couples surface thermodynamics with controllable kinetics to build layers with specified thickness, composition, crystallinity, and defect density. The process space is defined by substrate temperature, total and partial pressures, precursor chemistry, carrier gas composition, plasma power and frequency (when used), substrate bias, and, in pulsed modes, timing sequences, all of which steer nucleation, growth mode, and microstructure across flat and high-aspect-ratio geometries. This tunability allows properties such as conductivity, bandgap, refractive index, barrier performance, and mechanical hardness to be dialed in for target applications spanning microelectronics, photonics, energy storage, sensors, and biomedical devices. Within this framework, three complementary methods PVD (Section 4.3), CVD (Section 4.4), and ALD (briefly discussed as an extension of PVD and CVD) provide distinct advantages, emphasizing their governing variables, growth mechanisms, and application-driven trade-offs. PVD employs evaporation or sputtering to deliver high-purity, largely line-of-sight fluxes that favor dense metallic and ceramic films with excellent adhesion and controlled stress states. CVD uses volatile precursors that react or decompose at the substrate, yielding conformal coverage and precise stoichiometry over complex topographies, with process windows set by gas flows, residence time, and surface reaction energetics. ALD extends chemical control to the limit of self-limiting reactions, achieving sub-nanometer thickness precision and unrivaled 3D uniformity through sequential precursor dosing and purge cycles.

In PVD, source materials, such as metals, ceramics, or polymers, are ablated into the gas phase via purely physical processes [32]. Thermal evaporation heats the source in a high-vacuum (10^{-6}–10^{-8} torr) environment using resistive filaments or e-beam guns, vaporizing atoms that condense into films with deposition rates of 0.1–10 nm/s and purities exceeding 99.99% [33]. Sputter deposition employs an inert gas plasma (typically Ar^+) to bombard a solid target, ejecting atoms that deposit onto substrates at pressures of 0.5–10 mtorr [34]. Magnetron sputtering enhances ionization efficiency and film density by confining electrons with magnetic fields. In contrast, reactive sputtering introduces controlled flows of O_2 or N_2 to form oxides, nitrides, and other compound films in a single step [35]. By contrast, CVD synthesizes films via surface chemistry: volatile precursors decompose, react, or adsorb on a heated substrate (200–1,000 °C) under pressures ranging from sub-torr (LPCVD) to ambient (APCVD) [36]. Metal–organic CVD (MOCVD) uses organometallic compounds to deposit III–V semiconductors for LEDs and lasers, with precursor flow rates, reactor temperature, and carrier gas composition dictating film stoichiometry and crystallinity. PECVD utilizes radio frequency (RF) or microwave plasmas to generate reactive radicals, lowering deposition temperatures to 50–400 °C and enabling conformal coatings on temperature-sensitive polymers and microelectronic devices. ALD extends CVD's precision by alternating precursor pulses and inert gas purges, yielding sub-ångström control over thickness and unrivaled uniformity on complex three-dimensional architectures [37]. Recent literature surveys highlight hybrid PVD/CVD configurations that combine high-purity physical ablation with in situ chemical reactions, as well as

HiPIMS for ultra-dense coatings with tailored microstructures [38]. Across microelectronics, optics, energy storage, and protective surfaces, vapor-phase deposition remains indispensable: by mastering precursor chemistry, plasma dynamics, and thermo-kinetic balances, researchers can engineer films with precise composition, morphology, and functionality, cornerstones of next-generation nanotechnology [39].

4.3 Physical vapor deposition (PVD)

4.3.1 Overview

PVD is a vacuum-based thin-film process in which a solid material is physically converted to vapor (by evaporation or sputtering) and transported to a substrate where it condenses into a coating. By tuning pressure, geometry, energy (e.g., substrate bias), and deposition rate, PVD delivers high-purity, controllable films with engineered microstructure and interfaces. The basic PVD processes (Fig. 4.2A) involve condensing vapor-phase species onto a surface to form a nanoscale thin film under vacuum, producing various materials (Fig. 4.2B) used in different fields. The conventional PVD processes are further categorized into sputtering and evaporation, encompassing a suite of vacuum-based techniques that condense vaporized species onto substrates to form nanoscale thin films. In thermal evaporation, a solid source, such as aluminum, gold, or organic dyes, is resistively or e-beam heated in a high-vacuum chamber (10^{-6}–10^{-8} torr) until it vaporizes. The volatile component (evaporant) flux travels ballistically, and deposition rates, governed by the Langmuir–Knudsen relation, range from 0.01 to 10 nm/s, with film uniformity dictated by the source–substrate geometry and the chamber pressure. Evaporation produces high-purity, line-of-sight films, but struggles on three-dimensional or recessed features due to its directional nature. Advanced PVD variants, including HiPIMS, pulsed-direct current (DC) sputtering, and hybrid PVD/CVD systems, further push film performance by generating highly ionized plasmas, achieving ultra-dense microstructures, and providing conformal coverage on complex geometries. Across industries, from wear-resistant cutting-tool coatings that halve friction to transparent conductive zinc oxide (ZnO):Al films in displays, PVD's flexibility in material choice, film thickness control, and deposition environment remains indispensable for next-generation nanotechnology [40].

4.3.2 Synthesis parameters

The PVD process vaporizes a solid source (evaporation or sputtering) in high vacuum and condenses the vapor onto a substrate. Film properties emerge from the interplay of energetic flux, surface mobility, and gas-phase scattering. The synthesis parameters

A

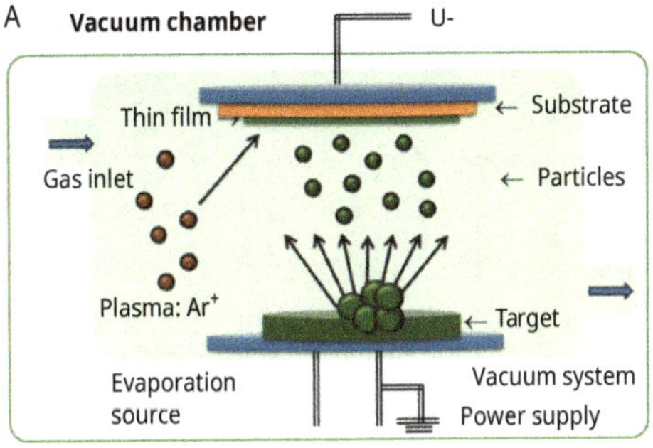

B

Metals and Alloys:	A1, Au, Ag, Cu, Ti, Pt
	NiCr, Cu-Mo composites
Ceramic and Compound Coatings	
Transition-metal nitrides	TiN, CrN, and ZrN
Metal oxides	$A1_2O_3$, SiO_2, ITO, ZnO:A1
Diamond-like carbon	Carbon-based film
Semiconductors and Organics	
Elemental semiconductors	Si, Ge, GaAs, InP
Organic semiconductors	OLEDs (e.g., A1q3)

C PVD synthesis parameters and their influences on film growth and quality

A. Source temperature:	***B. Chamber pressure:***	***C. Depositrate:***	***D. Substrate bias voltage:***	***E. Target-substrate distance and angle:***
1) elevate evaporation rate or sputter yield 2) boost deposition rate	1) control on transport, plasma behavior 2) favor interfaces and high-purity films	1) Control film-forming species 2) Scale target power, ion current, & sputter yield	1) tune ion energy 2) control nucleation, densification, stress, texture, & adhesion	1) tune ion energy 2) control nucleation, densification, stress, texture, & adhesion

Fig. 4.2: A. Representation of physical vapor deposition (PVD). Reproduced from [5]. Copyright © Sigma Aldrich 2018. **B.** Materials produced using PVD, and **C.** PVD synthesis parameters and their influences on film growth and quality.

such as **A.** source temperature, **B.** chamber pressure, **C.** deposition rate, **D.** substrate bias voltage, and **E.** target–substrate distance and angle are primary levers, and their influences are tightly coupled as shown in Fig. 4.2C.

A. **Source temperature**: Raising the source temperature within 500–3,000 °C increases vapor flux, which elevates the evaporation rate or sputter yield and, in turn, boosts the deposition rate to achieve thicker films in less time. Higher temperatures also raise the substrate temperature by radiation or deliberate heating, enhancing adatom mobility and surface diffusion; this typically produces larger grains, fewer defects, and improved crystallinity. At elevated temperatures, additional phases can become accessible, including metastable or high-temperature structures, but excessive heating can lead to interdiffusion at interfaces, dewetting of metals on oxides, and surface roughening. For multicomponent targets or sources, differing vapor pressures among constituents can drive composition drift at higher temperatures, leading to off-stoichiometry; using co-evaporation with independently controlled sources or separate sputter targets helps maintain the desired composition. The influence of source temperature and vapor flux on deposition rate, adatom mobility, phase formation, and stoichiometry control in multicomponent PVD processes was demonstrated in Fig. 4.3 and discussed below [41].

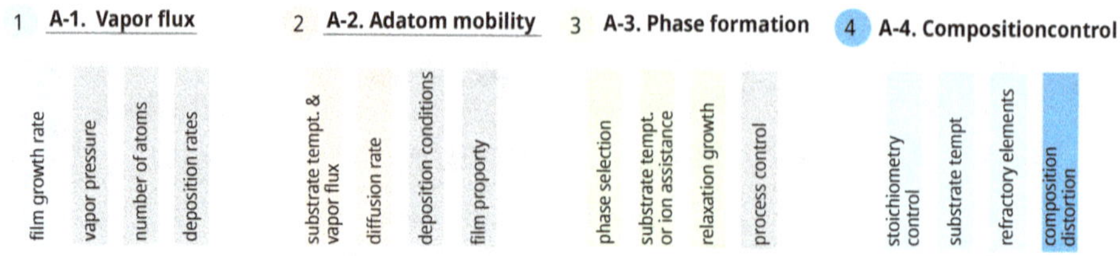

Fig. 4.3: Influence of source temperature on vapor flux, deposition rate, adatom mobility, phase formation, and stoichiometry control in multicomponent physical vapor deposition processes.

A-1. Vapor flux: In PVD, it is the primary driver of film growth rate and scales strongly with temperature and ion/target power. In evaporation, raising the source temperature exponentially increases the material's equilibrium vapor pressure, boosting the number of atoms leaving the source per unit time and resulting in higher deposition rates at the substrate. In sputtering, increasing target power elevates ion current and ion energy, raising sputter yield and, therefore, the flux of ejected atoms. Greater flux shortens run times and enables thicker films within practical production windows, but it also introduces trade-offs. Excessive flux can outpace adatom surface diffusion, trapping defects, elevating intrinsic stress, and roughening interfaces. Effective control couples flux with chamber pressure, source-to-substrate distance, and substrate temperature to maintain desired adatom mobility. Closed-loop rate control using quartz crystal microbalances (QCMs), stable power supplies, and thermal shielding keeps flux within tight tolerances. Process recipes often ramp flux gradually to nucleate uniformly, then increase it to efficiently accumulate thickness without sacrificing microstructure and reliability [42].

A-2. **Adatom mobility**: It governs how arriving species rearrange on the surface before being buried and is tightly coupled to both substrate temperature and vapor flux in PVD. Elevated substrate heating, whether from radiative load of a hot source or intentional wafer heating, increases surface diffusion coefficients, lengthens the adatom diffusion length, and enables atoms to find low-energy lattice sites. This promotes coalescence of nuclei into larger grains, reduces point and line defect densities, sharpens texture, and generally improves crystallinity. However, mobility is not determined solely by temperature; the ratio of arrival rate to diffusion rate is decisive. At excessively high vapor flux, adatoms are buried before equilibrating, yielding fine columnar morphologies, higher intrinsic stress, and rougher interfaces. Moderate flux paired with adequate temperature (and, where appropriate, gentle ion assistance) optimizes the diffusion-to-deposition balance, densifying films without inducing damage. Practical controls include ramping the temperature during nucleation, throttling the rate as layers approach closure, and operating at low pressures to increase the mean free path while avoiding re-sputtering. When tuned correctly, enhanced adatom mobility lowers defect formation energies, stabilizes desired phases, and produces films with superior electrical, optical, and mechanical performance [43].

A-3. Phase formation: Vapor flux is a kinetic lever that steers which phases nucleate and persist during PVD. High flux drives rapid supersaturation at the growth front, shortening surface-diffusion times and favoring kinetic trapping of metastable or high-temperature phases that would be inaccessible at equilibrium. When paired with adequate substrate temperature or mild ion assistance, this can stabilize textured phases, enforce preferred orientations, and lock in nonequilibrium compositions in alloys and nitrides. However, excessive flux relative to adatom mobility shifts growth into a bury-before-relaxation regime. The consequences include interface reactions and interdiffusion (especially in multilayers with reactive couples), dewetting of metals on oxides, and roughening via shadowing-driven columnar growth. Elevated flux can also amplify re-sputtering and preferential sputter loss of volatile components, pushing films out of stoichiometry. Practical control balances flux, temperature, and ion energy. Use moderate flux during nucleation for wetting, then step up to build thickness; raise the substrate temperature or apply a gentle bias to extend the diffusion length without inducing damage; insert wetting/adhesion layers (e.g., Ti and Cr) to suppress dewetting; and employ graded or diffusion-barrier interlayers to mitigate intermixing. For multicomponent films, decouple sources and close the loop with in situ rate/thickness monitoring to maintain phase stability and interface planarity. It enables metastable or high-temperature phases; too high can induce interdiffusion, metal dewetting on oxides, or roughening [44].

A-4. Composition control: Achieving target stoichiometry in multicomponent films is fundamentally constrained by the disparate vapor pressures, sputter yields, and sticking coefficients of the constituent elements. At elevated temperatures, components with higher equilibrium vapor pressures (or lower cohesive energies) desorb or evaporate preferentially, driving composition drift and phase segregation. In evaporation, this manifests as enrichment of refractory species in the film and depletion of volatile constituents; in sputtering, preferential re-sputtering and angle-dependent yields can further skew ratios, especially for light or reactive elements. Robust control strategies decouple sources and close the flux loop [45]:

1) Co-evaporation with independently heated crucibles or effusion cells allows real-time rate matching; QCMs near each source maintain setpoints and compensate for drift.
2) Co-sputtering with separate targets provides orthogonal power control; adjusting DC/RF power (and, in alloys, target composition) tunes elemental arrival rates without overheating.
3) Reactive processes demand tight control of partial pressures; feedback-controlled mass flow and optical emission or plasma probes prevent target "poisoning" that abruptly changes yields.
4) Substrate temperature and bias are tuned to equalize sticking probabilities and suppress re-evaporation of volatile species. At the same time, low total pressure preserves line-of-sight transport and reduces scattering-induced composition gradients.

5) For steeply mismatched vapor pressures, use multilayer or nanolaminate stacking followed by low-temperature interdiffusion to reach the desired average stoichiometry without overshooting during growth.
6) Geometry and motion matter: off-axis or confocal arrangements and planetary rotation average angular yield asymmetries, improving across-wafer uniformity.
7) Post-deposition anneals under controlled atmospheres (inert, reducing, or reactive) can rectify slight off-stoichiometry, activate desired phases, and homogenize dopant distributions.

When precision is paramount, in situ metrology, QCM, spectroscopic ellipsometry, and reflection high-energy electron diffraction (RHEED) for epitaxy are required, with recipe logic that adjusts source power, shutter timing, or gas flows on the fly. This system's approach mitigates differential vapor pressure effects, stabilizes phase formation, and delivers compositionally accurate films across batches and toolsets. For multicomponent sources, differential vapor pressures can cause off-stoichiometry at high temperatures; co-evaporation or separate targets mitigate this [46].

B. Chamber pressure: Chamber pressure is a first-order control on transport, plasma behavior, and film growth in PVD, because it sets the mean free path, collision frequency, and impurity incorporation. Tuning pressure shifts the process between ballistic, low-scatter regimes that favor sharp interfaces and high-purity films, and collisional regimes that enhance angular averaging and step coverage. Chamber pressure is a primary knob in PVD because it governs how atoms travel, how plasmas behave, and how films grow. By setting the gas density, pressure determines the mean free path, collision frequency, and contamination risk. Lower pressures create a ballistic, low-scattering environment that produces sharp interfaces and high purity. Higher pressures increase collisions, broadening the angular spread of arriving atoms and improving coverage on complex shapes, but may reduce density and purity. Chamber pressure governs atomic transport, plasma behavior, and film growth, shifting PVD between ballistic purity and collisional conformity across five process dimensions (Fig. 4.4). Five perspectives will be discussed in detail: **B-1**. transport and purity; **B-2**. plasma and sputter dynamics; **B-3**. microstructure and stress; **B-4**. step coverage and uniformity; and **B-5**. rate, composition, and defects [47].

B-1. Transport and purity: Lower chamber pressure increases mean free path, keeping ballistic trajectories and interfaces sharp. At 10^{-6}–10^{-8} torr, the mean free path typically exceeds tens of meters, far larger than source–substrate distances (10–30 cm), minimizing gas-phase scattering and preserving line-of-sight deposition. In contrast, at 0.5–10 mtorr (common for sputtering), the mean free path shrinks to centimeters, broadening the angular distribution and reducing the abruptness of the interface. Control levers include total pressure, working-gas flow (e.g., Ar 5–50 sccm), target–substrate spacing, and temperature, which influences desorption and re-evaporation. Purity improves as the partial pressures of residual gas drop. A clean base pressure $\leq 10^{-7}$ torr limits O_2, H_2O, N_2, and hydrocarbon uptake, lowering resistivity

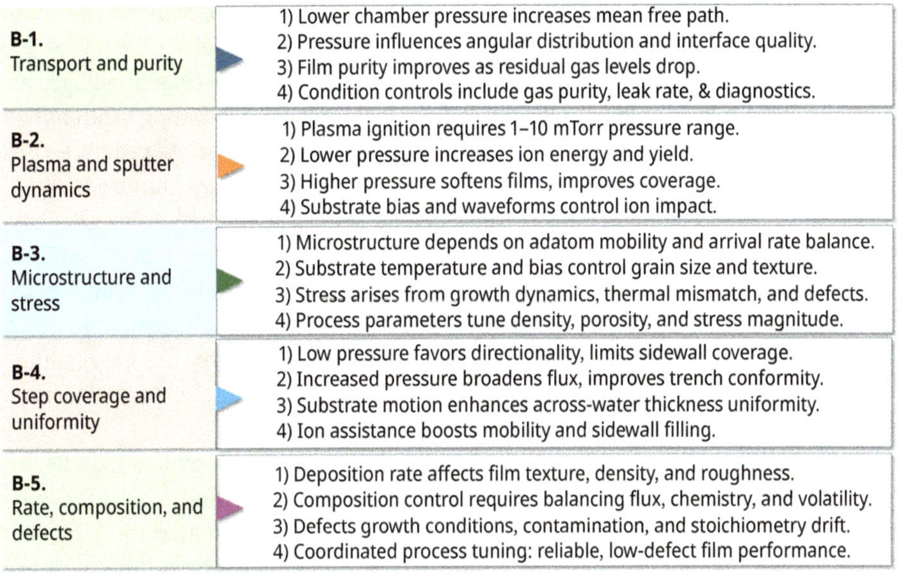

Fig. 4.4: Chamber pressure governs atomic transport, plasma behavior, and film growth, shifting PVD between ballistic purity and collisional conformity across five process dimensions.

and optical loss. Practical controls: ultra-high-purity gases (≥99.999%, O_2/H_2O <1 ppm), leak-up rates $<1 \times 10^{-9}$ torr · L/s, chamber bake-outs (100–200 °C, 8–24 h) to reduce outgassing, cryotrapping or sorption trapping to suppress H_2O/hydrocarbons, and substrate prebakes (150–300 °C) or plasma cleans to desorb adventitious films. Continuous residual gas analysis and rate/thickness metrology help correlate impurity spikes with property drift and adjust pressure/flow setpoints in real time [48].

B-2. Plasma and sputter dynamics hinge on pressure and power, affecting stable ignition. Typically, stable ignition requires 1^{-10} mtorr with sufficient target power and magnetic confinement, while excessive pressure increases charge exchange, cools ions, and raises arc risk. At lower pressures, higher sheath voltages deliver more energetic ions and a higher sputter yield, densifying films. In contrast, higher pressures reduce ion energy, soften films, and broaden the angular flux, thereby improving step coverage. Substrate bias and pulsed waveforms fine-tune this trade-off.

Ignition and stability: Reliable DC/RF magnetron ignition typically occurs within 1^{-10} mtorr Ar, where electron-neutral collision frequency supports ionization without over-thermalizing electrons. Below ≈0.5 mtorr, mean free paths become too long to sustain avalanche ionization; above $\approx 10^{-20}$ mtorr, charge-exchange collisions proliferate, cooling ions and increasing arc probability. Stable operation balances gas flow (e.g., Ar 5–50 sccm), target power density (1–20 W/cm^2 DC; 1–10 W/cm^2 RF), and magnetic confinement to maintain a steady electron energy distribution. In reactive sputtering, tight control of reactive partial pressure (O_2/N_2 a few percent of the total) avoids target poisoning and hysteresis that destabilize the discharge.

Ion energy and yield: At lower pressure (≈0.5^{-2} mtorr), larger sheath voltages accelerate ions to higher energies (often 100–500 eV at the target), boosting sputter yield and increasing film densification through energetic bombardment. As pressure rises (≈5^{-10} mtorr), frequent collisions reduce ion energy to tens to low hundreds of eV, lowering yield and softening films but broadening angular arrival for better step coverage. Substrate bias (–50 to –300 V), pulsed-DC/HiPIMS waveforms, and RF self-bias tune the ion energy distribution at the substrate, enabling a trade-off between densification/adhesion and damage/stress. Adjusting total pressure, gas composition (e.g., Ar/He/Ne mixtures), and power waveforms provides fine control over ion energy, flux, and the metal-ion fraction, which ultimately determine growth rate, microstructure, and stress [49].

B-3. Microstructure and stress: Microstructure and stress evolve together in PVD, within the same knobs that shape grains, texture, and porosity, and set the magnitude and sign of residual stress. Managing this coupling is essential for achieving density, adhesion, and reliability without warping wafers or cracking films, further tuning the growth regimes and grain structure. Microstructure arises from the balance between adatom mobility and the arrival rate of depositing species. When the substrate temperature is low or the flux is high, surface diffusion is limited, nucleation coalescence is less efficient, and the film develops fine, shadowed columns with a higher void fraction. Raising the temperature or introducing gentle ion assistance extend diffusion lengths, promote coalescence, and yield larger, more tightly packed grains. The Thornton framework captures these regimes qualitatively: in Zone I (low temperature and higher pressure), films form porous, tapered columns with weak interfaces; in Zone T (moderate temperature or ion energy), they transition to denser fibrous or columnar grains with improved adhesion; and in Zone II (high temperature), growth produces equiaxed grains with strong texture and high density. The texture of the final products can be controlled by adjusting the substrate temperature, bias, and pressure, which steer preferred orientations (e.g., (111) versus (200) in fcc nitrides), in turn setting hardness, conductivity, and optical response. Residual stress develops from coupled growth, thermal, and interfacial effects. Intrinsic compressive stress often arises from energetic ion "peening" at low pressure, where near-surface bombardment densifies the film. Tensile stress typically accompanies island coalescence and grain-boundary formation during the early stages or under higher-pressure, lower-energy growth [50].

The defect can be incorporated through trapped argon, vacancies, and voids, distorting the lattice, elevating brittleness, and increasing resistivity. Superimposed on these are thermal stresses, which arise during cooldown as the thermal expansion mismatch between the film and substrate introduces compressive or tensile components. Interfacial chemistry further sets the initial stress state through wetting, intermixing, and thin adhesion layers, such as Ti or Cr, which can suppress delamination and distribute strain parameter levers and their influence, with the process parameters providing direct control over microstructure and stress. Lower pressure increases ion energy and densification, tending toward compressive stress, whereas higher pressure broadens angular flux, lowers energy, and favors more tensile, porous films. Applying a substrate bias in the range of −50 to −300 V increases the near-surface ion energy, enhancing adhesion and density. Still, excessive bias can lead to over-compressive stress and re-sputtering. Higher substrate temperatures promote defect annihilation and grain growth, shifting films toward dense, low-stress, or mildly tensile states, while insufficient temperatures trap defects and gas. Slower deposition rates allow for relaxation and smoother interfaces; very high rates, on the other hand, trap defects and amplify roughness and tensile coalescence stress. Finally, reactive gas fractions (O_2 and N_2) modify target state and ion chemistry, altering energy accommodation, composition, and the resulting stress profile. Diagnostics and control strategies can be achieved through real-time monitoring and recipe design, ensuring the structure and stress remain within specification [51].

Wafer curvature measurements track stress evolution, QCMs stabilize the rate, optical emission probes plasma state, and RHEED or ellipsometry follow structure and thickness for feedback control. Practical recipes often nucleate at moderate energy for wetting, then raise mobility through temperature or mild bias to densify. Pulsed power or pulsed bias mitigates over-peening, and stepwise pressure changes improve conformity without sacrificing density. Architecture choices, including the adhesion or wetting layers, diffusion barriers, and graded stacks, can further distribute the strain effect [52].

B-4. Step coverage and uniformity: Step coverage and across-substrate uniformity in PVD emerge from the interplay of transport geometry, gas-phase scattering, and surface mobility. Directionality dominates at low pressure with long mean free paths, so features facing the source accrue more flux while sidewalls and re-entrant corners are shadowed. Introducing controlled scattering, mechanical motion, or ion assistance redistributes the arriving species, improving conformity without sacrificing film integrity.

Governing factors and levers: Pressure sets the mean free path and, therefore, the angular spread of arriving species. At lower pressure (UHV evaporation or ~0.5–2 mtorr sputtering), transport remains largely line-of-sight, maximizing flux to center-facing surfaces but compromising sidewall coverage. Modestly increasing pressure by ~1–3 mtorr broadens the angular distribution and improves trench and via conformity, though at the cost of density and rate; pairing this adjustment with a mild substrate bias helps recover densification [53].

Target–substrate distance and emission geometry also shape uniformity. Larger throws (e.g., 15–30 cm) and cosine-like emission average incident angles and smooth radial thickness roll-off, whereas short throws boost rate at the expense of uniformity. Collimators or off-axis configurations can filter high-angle flux to enable clean lift-off, while profile masks or baffling can correct radial nonuniformities. Substrate motion averages geometric effects. Continuous rotation or planetary motion improves across-substrate uniformity and mitigates facet-dependent growth in 3D topographies; when using oblique or off-axis incidence, synchronized rotation counters shadowing and aligns columnar tilt. Angular incidence and ion assistance provide additional control. Near-normal incidence favors dense, fine-grained films on open surfaces, while intentional oblique incidence can engineer anisotropy but typically worsens step coverage unless combined with rotation. Gentle ion assistance via a substrate bias of –50 to –150 V or ionized PVD/HiPIMS enhances adatom mobility and re-emission, enabling self-sputter sidewall filling and higher conformality. Deposition rate and temperature determine whether adatoms can redistribute before deposition. Slower rates and moderate substrate heating improve wetting, coalescence, and corner rounding, while excessively high rates or low temperatures trap roughness and widen top-to-sidewall thickness disparities. Feature scale and aspect ratio magnify geometric shadowing as dimensions shrink and depths increase. Practical mitigations include multistep recipes that nucleate at a low rate before a higher-rate fill, alternating pressure setpoints to balance directionality and scattering, and ultrathin wetting or adhesion layers to seed uniform sidewall growth.

Practical guidance: For planar optics and wafers, use low pressure, a larger throw distance, and substrate rotation to achieve tight thickness tolerances, adding profile masks only when necessary. For high-aspect features, raise pressure slightly, introduce ionized flux with modest substrate bias, and slow initial nucleation to establish continuous sidewalls before ramping rate; verify progress with in situ rate/thickness control and post-process mapping, then iteratively tune spacing, pressure, and motion to lock in target uniformity and conformity [54].

B-5. Rate, composition, and defects: Deposition rate, film composition, and defect landscape are tightly coupled in PVD; tuning one almost always perturbs the others. Robust processes deliberately balance flux, energy, and chemistry to achieve target stoichiometry at acceptable throughput while suppressing defect formation and variability.

Deposition rate: Deposition rate is governed by flux from the source and by gas-phase transport. In evaporation, the rate rises exponentially with source temperature, whereas in sputtering it scales with target power and ion current. Lower chamber pressure reduces scattering and increases the net arrival rate at the substrate; higher pressure broadens the angular spread but can diminish the effective rate. Microstructurally, low rates enhance surface diffusion, sharpen interfaces, strengthen texture, and reduce roughness and intermixing, while high rates risk bury-before-relaxation, leading to voids, tensile coalescence stress, rougher surfaces, and columnar porosity. Across-wafer uniformity is set by geometry and motion, as target–substrate spacing, incidence angle, and substrate rotation determine radial and lateral rate distribution [55].

Composition: Film stoichiometry reflects the interplay of source physics and surface chemistry. In evaporation, differential vapor pressures bias incorporation toward refractory species while volatile elements under-incorporate or re-evaporate. In sputtering, preferential sputter yields and re-sputtering at the substrate skew elemental ratios, and gas-phase scattering can differentially deflect light versus heavy species. Reactive processes hinge on partial pressures: O_2, N_2, or CH_4 flows control compound formation, target poisoning, and hysteresis, demanding tight partial-pressure regulation. Substrate temperature and bias modify sticking coefficients and desorption rates, especially for volatile constituents, thereby shifting the balance of incorporation. Mitigation relies on decoupling sources, co-evaporation or co-sputtering with independent rate or power control, and on process windows that favor line-of-sight stoichiometry at lower total pressure, modest bias to suppress re-evaporation, and nanolaminate stacking followed by low-temperature interdiffusion to reach target composition [56].

Defects: Defect formation arises from both growth dynamics and the environment. Nodules and particles arise from dust, target flakes, or arc droplets and overgrow into protrusions that shadow subsequent layers. Voids and pinholes form when adatom mobility is limited at low temperature or at high rates, or when shadowing dominates in high-aspect features. Gas entrapment increases with pressure; inert species such as argon can raise brittleness and resistivity. Compositional inhomogeneity, including off-stoichiometry grains, phase separation, or interdiffusion at interfaces, degrades functional properties. Stress-linked failures follow similar logic: excessive tensile stress from coalescence or thermal mismatch promotes cracking, while over-compressive peening stress or weak interfaces drive buckling and delamination.

Interdependencies: Rate, composition, and defects are mutually constraining. Higher growth rates can freeze nonequilibrium stoichiometry, whereas slower rates permit equilibration of volatile species but may encourage re-evaporation at elevated temperature. Faster growth tends to increase roughness and voiding, while slower, high-mobility growth smooths films yet can enhance intermixing at hot interfaces. Off-stoichiometry phases and impurities serve as defect nucleation sites, elevating pinhole density and weakening grain boundaries [57].

Control strategies: Stable processes align setpoints, chemistry, and cleanliness with real-time metrology. In sputtering, operating at roughly 0.5–2 mtorr yields dense films, whereas evaporation benefits from lower pressure for greater purity and sharper interfaces. A substrate bias in the –50 to –150 V range typically densifies films without incurring severe compressive stress or re-sputter. Pair moderate substrate temperature with moderate rate, ramping from slow nucleation to faster fill to balance relaxation and throughput. Maintain reactive gas flows with feedback to avoid target-poisoning hysteresis, and manage sources with stable power, precise shutter timing, and independent guns or cells to hold stoichiometry. Reduce particulates through prebake, plasma cleaning, high-purity gases, and shield maintenance, and shape profiles with larger throw, rotation, or planetary motion, and appropriate baffling. Close the loop with QCMs for rate, residual gas analysis for contaminants, in situ curvature for stress, and post-deposition mapping for thickness and defect metrics.

Coordinated in this way, the levers deliver compositionally accurate, low-defect films at practical deposition rates with reproducible device-level performance [58].

C. Deposition rate: Deposition rate is the net arrival of film-forming species at the substrate per unit time, set by source output and gas-phase transport. In evaporation, the rate rises exponentially with source temperature as vapor pressure increases; in sputtering, it scales with target power, ion current, and sputter yield, moderated by working pressure and magnetron efficiency. Pressure governs how much of that flux reaches the substrate: low pressure preserves ballistic transport and maximizes effective rate, while higher pressure increases scattering, reducing net arrival but broadening angular distribution. Deposition rate governs film thickness, microstructure, stress, and composition; optimized through temperature, pressure, bias, and in situ metrology for reliable performance (as shown in Fig. 4.5). The corresponding four key points will be discussed in this section: **C-1**. Microstructure and interfaces, **C-2**. stress and density, **C-3**. composition and stoichiometry, and **C-4**. uniformity and step coverage stoichiometry. The synthesis process can be set up at low pressure to maximize the effective arrival rate and purity, and then adjusted to source power or temperature to achieve the target nm/s while preserving surface mobility. As a sequence, a low-rate nucleation step will be implemented to improve wetting and interface sharpness, followed by a controlled ramp to a higher rate for efficient thickness accumulation. Balance energy by pairing a moderate rate with mild substrate heating or a slight negative bias to enhance diffusion and densification without inducing over-peening damage. Stabilization of growth with in situ metrology will then be achieved using QCM or ellipsometry to control the rate, perform post-deposition thickness mapping to refine geometry and motion settings, and track residual gas to link rate drifts to contamination events. When calibrated in this manner, the deposition rate becomes a precise lever for co-optimizing film density, stress state, composition, interface quality, and overall throughput [60].

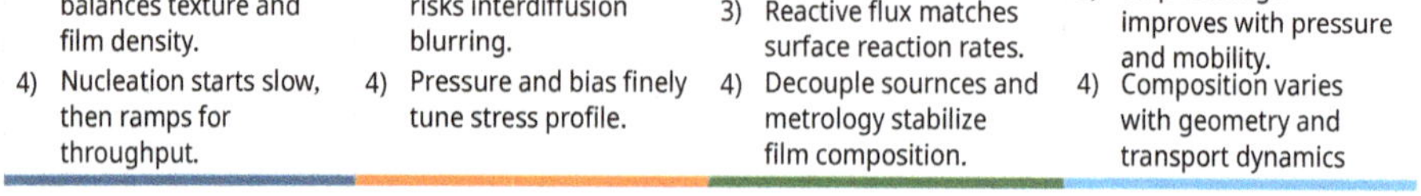

Fig. 4.5: Deposition rate governs film thickness, microstructure, stress, and composition; optimized through temperature, pressure, bias, and in situ metrology for reliable performance.

C-1. **Impact on microstructure and interfaces**: Deposition rate sets the tempo of growth, dictating how long adatoms can diffuse before being buried. When the rate is low relative to surface mobility, atoms sample more lattice sites and step edges, thereby smoothing terraces, sharpening interfaces, and promoting larger grains with stronger crystallographic texture. The resulting films are denser, with fewer voids and lower roughness; multilayers benefit from abrupt, well-registered interfaces and tighter thickness control, improving electronic, optical, and barrier performance. As the rate increases, arrivals outpace relaxations. Nuclei coalesce under kinetic constraint, producing finer, columnar grains with higher void fraction and rougher topography. This "bury-before-relaxation" regime increases the tensile coalescence stress and amplifies shadowing, especially at lower temperatures or higher pressures, where mobility is limited. In heterostructures, fast growth can blur interfaces through roughness-induced intermixing or by trapping nonequilibrium configurations; if substrate temperature is high, volatile species may re-evaporate, skewing composition and further degrading interface planarity. Practical recipes often use low nucleation to establish wetting and sharpness, then ramp to a moderate rate that maintains adequate diffusion lengths, balancing throughput with smoothness, density, and interface integrity [61].

C-2. **Impact on stress and density**: Moderate deposition rates, coupled with adequate adatom mobility achieved through substrate heating or gentle ion assistance, promote densification mechanisms such as surface diffusion and subplantation, yielding compact microstructures with controlled, often mildly compressive, residual stress. In this regime, atoms have sufficient time and energy to occupy low-energy lattice sites, annihilate point defects, and coalesce grain boundaries, thereby lowering the void fraction and improving mechanical integrity, electrical conductivity, and barrier performance. When the rate is pushed high relative to mobility, atoms are buried before relaxing, thereby trapping vacancies and inert gas and elevating the tensile coalescence stress. The resulting columnar morphology carries higher porosity and roughness, and the film is more prone to cracking, electromigration, and moisture ingress. Conversely, very low rates at elevated temperature can over-activate interdiffusion across interfaces, relaxing stress but blurring heterointerfaces and shifting composition. Practical recipes, therefore, decouple early-stage nucleation from bulk fill: a slow, moderately energetic nucleation step establishes wetting and continuous coverage with low tensile stress, followed by a higher-rate fill that maintains sufficient mobility (via modest heating or –50 to –150 V bias) to preserve density without inducing over-compressive "peening." Fine control of pressure further tunes the stress sign. Lower pressure increases ion energy and compressive tendencies, while slightly higher pressure softens bombardment and reduces compressive load, allowing stress-targeting without sacrificing film density or interface quality [62].

C-3. **Impact on composition and stoichiometry**: In multicomponent PVD, the deposition rate is influenced by volatility, sticking coefficients, and surface residence time, thereby determining the final composition and stoichiometry. When the rate is high relative to adatom mobility, nonequilibrium configurations can be "frozen in," enabling metastable alloys and dopant supersaturation, but volatile species (e.g., Zn, Pb, S, and Se) may under-incorporate or re-evaporate if the substrate is warm enough to lower their residence time. Slower growth extends equilibration time at the surface, improving stoichiometric fidelity and layer-to-layer reproducibility; however, if the temperature is too high or the bias is too low, light/volatile elements can desorb between arrivals, shifting the film off-target. In reactive PVD, the kinetic budget tightens further: a high metal flux can outpace surface reactions with $O_2/N_2/CH_4$, leading to sub-stoichiometric compounds or phase mixtures unless reactive partial pressures, residence time, or plasma reactivity are increased. Practical control aligns rate with chemistry and energy. Decouple sources (co-evaporation or co-sputtering) to tune elemental fluxes independently; hold total pressure low enough to limit differential scattering of light versus heavy species; and use modest substrate bias and temperature to raise sticking while suppressing re-evaporation. For severe volatility mismatches, employ nanolaminate stacking followed by low-temperature interdiffusion to reach the target average composition without overshooting during growth. Closed-loop metrology, rate monitors at each source, in situ optical/ellipsometric thickness, and partial-pressure control in reactive flows keep stoichiometry within specification as rate and process conditions evolve [63].

C-4. **Impact on uniformity and step coverage stoichiometry**: Deposition rate modulates how geometric nonuniformities average out across the substrate. Lower rates, when combined with substrate rotation and larger target–substrate distances, give atoms more time to redistribute and smooth out radial flux gradients, improving center-to-edge thickness uniformity. At higher rates, the process becomes more directional: flux is dominated by line-of-sight trajectories, so center-facing regions thicken faster, accentuating radial nonuniformity and making it harder to achieve uniform films without aggressive motion or masking. In high-aspect-ratio features, the same kinetics govern step coverage. Slower rates and slightly higher working pressure broaden the angular distribution and extend surface residence time, which promotes sidewall nucleation, corner rounding, and more continuous films down into trenches and vias. High-speed rates, especially at low pressure, intensify shadowing and top-surface dominance, leaving sidewalls thinner or discontinuous. Because composition couples to transport, these geometric effects can introduce stoichiometric gradients: light or volatile species scatter and re-evaporate differently than heavy constituents, so directional, high-rate growth tends to enrich center-facing areas or top corners, while sidewalls may become off-stoichiometric. Mitigation pairs moderate rates with controlled pressure, rotation, or planetary motion, and, when needed, ion-assistance to enhance sidewall incorporation and maintain uniform thickness and stoichiometry throughout complex topographies [64].

D. Substrate bias voltage: Substrate bias voltage plays a critical role in PVD by controlling the energy of ions arriving at the substrate surface. This energy directly influences nucleation behavior, film densification, crystallographic texture, adhesion strength, and residual stress development. When properly applied, bias voltage serves as a precise, low-temperature tuning mechanism that enhances film quality without degrading interface sharpness or altering composition. Substrate bias voltage tunes ion energy to control nucleation, density, stress, and texture, enabling low-temperature optimization of film structure and interface quality (as shown in Fig. 4.6). The detailed discussion is organized from four key perspectives: **D-1**. Mechanisms of ion acceleration and energy transfer, **D-2**. typical ranges and coupling variables, **D-3**. microstructure and property impacts, and **D-4**. integration with step coverage and interfaces. By adjusting bias voltage within controlled limits, typically between –50 and –300 V, practitioners can improve conformity, suppress defect formation, and tailor stress profiles. This enables the deposition of dense, adherent films with optimized performance across electronic, optical, and barrier applications [65].

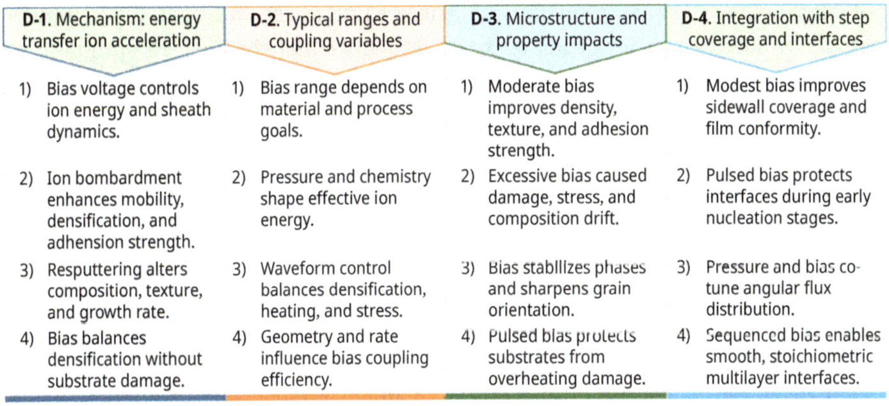

Fig. 4.6: Substrate bias voltage tunes ion energy to control nucleation, density, stress, and texture, enabling low-temperature optimization of film structure and interface quality.

D-1. Mechanisms: Substrate bias voltage shapes the plasma–surface interaction by forming an ion-accelerating sheath at the substrate. Applying a negative bias lowers the substrate potential relative to the plasma, so positive ions are accelerated across the sheath with an energy that approximates the magnitude of the bias (reduced by collisional losses at higher pressure). The result is a tunable ion energy distribution at the surface, controlled by the bias amplitude, waveform, duty cycle, and frequency (for RF/self-bias), and by ambient pressure (which sets the sheath thickness and collisionality). Ion bombardment modifies growth through several coupled pathways. First, momentum transfer enhances adatom mobility: shallow implantation and near-surface "knock-on" events help atoms overcome local barriers and reach low-e nergy sites before burial, increasing packing density and continuity at lower substrate temperatures. Second, subplantation and peening compact the near-surface region, closing voids and driving densification and hardness, while simultaneously introducing compressive stress. Third, preferential resputtering can remove weakly bound or volatile species, altering surface composition, local growth rate, and texture; careful tuning prevents stoichiometric drift [66]. Fourth, ion-stimulated surface reactions and defect creation increase the number of adsorption/incorporation sites, assisting compound formation in reactive PVD and promoting nucleation on challenging surfaces or sidewalls. Bias also enables in situ surface conditioning. Pre-deposition ion cleaning (at higher bias or with reactive ions) removes native oxides and adventitious contamination, improving wetting and adhesion of the first monolayers. During growth, a modest, continuous, or pulsed bias maintains a controlled ion-to-neutral flux ratio, balancing densification with damage [67]. Waveform engineering, DC, RF self-bias for insulating substrates, or pulsed DC/HiPIMS, shapes the ion energy distribution, peak metal-ion fraction, and instantaneous power density, allowing densification with lower average heating. Because sheath dynamics depend on pressure, the same applied bias delivers higher effective ion energies at lower pressure (collisionless sheaths) and lower energies at higher pressure (collisional sheaths), so optimal bias must be co-tuned with gas pressure and composition. On insulating or partially insulating films, charge accumulation can distort local fields and broaden the energy/angle distribution of arriving ions; RF coupling or conductive seed layers mitigate charging and enable stable biasing. Finally, angular effects matter in ion bombardment, which increases re-emission and angular redistribution of adatoms, reducing geometric shadowing and aiding step coverage. Still, excessive bias can roughen interfaces, enhance intermixing at heterojunctions, or induce defect states. The practical mechanism "sweet spot" is a regime where ion energy is high enough to activate mobility and subplantation for densification and adhesion, yet low enough to avoid significant resputter, over-compressive stress, or damage to sensitive chemistries and substrates [68].

D-2. Typical ranges and coupling variables: Substrate bias typically spans a few regimes that map to material class and process goals. For metals and conductive alloys, modest biases in the –50 to –150 V range generally achieve densification and improved adhesion without excessive resputtering. For hard coatings (nitrides, carbides, borides, and some oxides), higher biases up to roughly –300 V are common to drive subplantation and hardness, while carefully monitoring compressive stress. Beyond about –300 to –500 V, damage, resputter-induced composition shifts, and over-compressive stress become likely unless pulse shaping and low pressure are used to limit ion energy at the surface. Effective ion energy at a given bias depends strongly on pressure and plasma chemistry. Lower total pressure thickens the sheath and reduces collisions, so ions traverse with energies closer to the applied bias; higher pressure increases collisionality, reducing ion energy and broadening arrival angles [69]. Gas species and mixtures (e.g., Ar/Ne/He or reactive O_2/N_2) alter ion masses and charge-exchange rates, thereby shifting energy-transfer and resputter thresholds. Waveform control is a second major lever. DC bias is straightforward on conductive substrates; RF bias (e.g., 13.56 MHz) generates a self-bias on insulators and partially resistive films; pulsed DC and HiPIMS modulate peak ion energy, duty cycle, and the metal-ion fraction to balance densification against heating and stress. Duty cycles of 5–50% with micro- to millisecond pulses often reduce average damage while preserving high instantaneous ion energy for compaction and surface activation. Substrate electrical properties and grounding define how bias couples to the wafer. Conductive wafers track the set voltage; insulating substrates float unless they are given a conductive seed, backside metallization, RF coupling, or biased fixtures/grids to stabilize the sheath [70]. Charging on dielectrics can distort local fields, so RF or pulsed schemes are preferred to avoid arcing and nonuniform bombardment. Temperature limits and thermal budget shape the usable bias window. At low substrate temperatures, bias substitutes for thermal mobility, enabling dense films without exceeding thermal constraints on polymers, glasses, or back-end semiconductor stacks. At elevated temperatures, the same bias can overshoot into peening, so lowering bias or pressure, or switching to pulsed bias, helps avoid over-compressive stress and intermixing at interfaces. Film chemistry and volatility further constrain bias. Multicomponent and reactive films are susceptible to preferential resputtering of light or volatile species (B, C, N, O, S, and Zn); keeping the bias at the lower end of the effective window, lowering pressure, or increasing the neutral flux (rate) mitigates composition drift [71]. Conversely, to clean and activate surfaces before nucleation, a short, higher-bias preclean step followed by reduced bias growth improves adhesion without long-term damage. Finally, tool geometry and rate couple to bias efficacy. Larger target–substrate distances and lower rates accentuate the benefits of bias by extending adatom residence time and enabling ion-assisted diffusion. At very high rates or short throws, bias may need to be increased or pulsed to maintain density, while compensating elsewhere (pressure, power) to prevent resputter and stress buildup. In practice, optimal bias is co-tuned with pressure, rate, gas composition, and temperature to achieve the minimum ion energy required to meet density and adhesion targets while maintaining stable stress and preserving stoichiometry [72].

D-3. Microstructure and property impacts: Substrate bias shapes film formation at the atomic scale, converting a large thermal growth process into one driven by controlled ion energy at the surface. With a modest negative bias, ions transfer momentum to adatoms, extend diffusion lengths, and compact near-surface layers; at excessive bias, the same bombardment introduces damage, resputters volatile species, and drives over-compressive stress. In practice, the best results come from a bias "just sufficient" to achieve densification and adhesion, often with pulsed or RF waveforms, co-tuned with pressure, temperature, and rate to lock in the desired microstructure and properties without sacrificing interface integrity or composition.

Density and porosity: Mild to moderate bias (≈–50 to –150 V for metals; up to ≈–300 V for hard coatings) closes voids and columns, yielding high packing density, smoother topography, and better hermeticity. Over-biasing can roughen surfaces via resputtering and defect creation.

Grain size and texture: Bias-assisted mobility promotes coalescence and texture sharpening (e.g., shifting fcc films between (111) and (200) orientations). Moderate bias refines grains without excessive defect density; very high bias can drive grain refinement to the point of high dislocation density or amorphization in sensitive systems [73].

Phase formation: Nonequilibrium ion energy can stabilize metastable phases and tailor stress-mediated phase selection (useful in nitrides, carbides, and alloys). Excess energy can lead to phase segregation or intermixing at heterointerfaces.

Adhesion and interfaces: Pre-bias ion cleaning removes native oxides and hydrocarbons, improving wetting. Biased nucleation enhances interfacial bonding and reduces incubation time for continuous films, but high bias during critical interfaces can erode layers or broaden junctions.

Stress state: Ion peening induces compressive stress, improving toughness and adhesion at moderate levels. If bias is too high or pressure is too low (collisionless sheath), compressive stress can exceed adhesion limits, causing buckling or warpage; at too low a bias, films remain tensile and porous.

Composition and stoichiometry: Under bias, preferential resputtering disproportionately removes light/volatile elements (B, C, N, O, S, and Zn), skewing composition in multicomponent and reactive films [74]. Keeping bias at the low end of the effective window, lowering pressure, or increasing neutral flux mitigates drift.

Electrical and optical properties: Higher density and lower impurity content reduce resistivity and optical loss and sharpen interference fringes in optical stacks. Over-bias can introduce defect states, increase scattering, and degrade conductivity/transparency.

Mechanical and environmental performance: Densified, textured films exhibit higher hardness and wear resistance, and improved corrosion resistance due to reduced pathways for moisture and ions. Over-compressive stress or bias-induced damage undermines these gains.

Thermal budget and substrate sensitivity: Bias substitutes for heat, enabling dense films on temperature-limited substrates (polymers and back end-of-line stacks). Use pulsed or RF bias to prevent charging on dielectrics and to limit average energy, avoiding arcing and local overheating.

Coupling with pressure and rate: Lower pressure amplifies the effective ion energy at a given bias (more compressive tendency), while higher pressure collisionally cools ions, softening bombardment and aiding conformity. At high deposition rates, bias may need to be increased or pulsed to retain density; at low rates and elevated temperature, reduce bias to avoid over-peening and intermixing [75].

D-4. Integration with step coverage and interfaces: Substrate bias integrates with geometry, pressure, and rate to turn directional PVD flux into conformal, well-bonded films. A modest negative bias accelerates ions toward feature sidewalls, increasing near-surface mobility, re-emission, and shallow subplantation; this helps nucleate continuous films along trench/via walls, reduces overhang, and mitigates shadowing without requiring excessive pressure that would compromise density. Because ion energy at the sidewall is inherently lower than at the field, bias effectively "levels the playing field," improving sidewall coalescence and closing micro-voids while keeping top-surface growth in check. Interface quality depends on when and how strongly bias is applied. Before deposition, a short, high-bias ion clean removes native oxides and adventitious carbon, thereby improving wetting and chemical bonding [76]. During the first nanometers of nucleation at critical heterointerfaces, reduce or pulse the bias to avoid resputtering sensitive layers, intermixing, or roughening; after a continuous film forms, increase bias modestly to densify the bulk. In multilayers, repeat this low bias "interface window" at each boundary to preserve abruptness and minimize interdiffusion, then resume the densification bias for thickness build. Bias must be co-tuned with pressure and angle to balance conformity and damage. Slightly higher working pressure broadens angular distributions, improving sidewall reach, while bias restores density lost to scattering. At very low pressure, the same bias delivers higher ion energy (a collisionless sheath), so reduce the setpoint or use pulsed duty cycles to prevent over-compressive peening and resputtering, especially in narrow features. Rotation or planetary motion remains essential; bias complements motion by activating sidewall incorporation each time a facet faces the plasma, improving uniformity across complex topographies. Composition and stoichiometry can drift under excessive bias due to preferential resputtering of light or volatile species on sidewalls. Keep bias at the low end of the effective window for multicomponent or reactive films and compensate with slightly higher neutral flux or temperature to maintain incorporation without roughening interfaces [77]. For extreme aspect ratios, sequence the recipe: initiate with low rate, modest pressure, and low bias to seed continuous sidewalls; alternate short high-bias bursts to activate re-emission at pinch points; then transition to a moderate bias and rate for uniform fill. This choreography yields continuous, stoichiometrically faithful sidewalls, smooth and abrupt interfaces, and field regions free of overhang, all while preserving density and manageable residual stress [78].

E. Target–substrate distance and angle: Target–substrate distance and incident angle are core geometric levers that determine thickness uniformity, microstructure, and step coverage in PVD. They set the directionality of the arriving flux, how much of the wafer "sees" the source at any instant, and how effectively atoms reach and coat sidewalls and recessed features without sacrificing density or interface sharpness. Calibrate spacing to balance rate and uniformity, then establish substrate motion (continuous rotation or planetary) to average angular flux. Begin deposition at near-normal incidence to promote wetting and abrupt interfaces; introduce controlled tilt or off-axis passes only after a continuous film is established to tailor texture or anisotropy. Maintain low total pressure to extend the mean free path and add slight pressure only when step coverage demands it, paired with modest bias to recover density. Use fixture baffling or profile masks to correct residual radial roll-off and validate choices with thickness maps and cross-sectional imaging to iteratively converge on the desired combination of uniformity, conformity, and microstructural quality. The substrate bias voltage tunes ion energy to control nucleation, density, stress, and texture, enabling low-temperature optimization of film structure and interface quality (as shown in Fig. 4.7). Four perspectives, **E-1**. Distance: length and transport balance, **E-2**. angle: incidence direction and morphological control, **E-3**. step coverage and uniformity across features and wafers, and **E-4**. coupling with pressure, rate, and bias will be examined in this section [79].

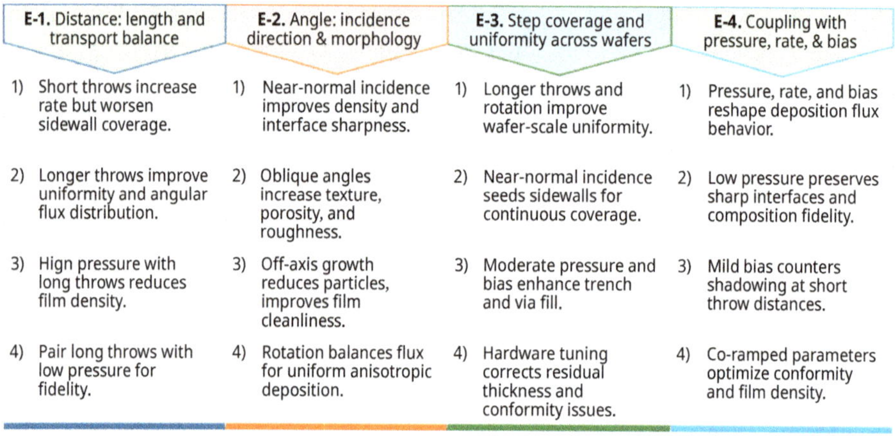

Fig. 4.7: Substrate bias voltage tunes ion energy to control nucleation, density, stress, and texture, enabling low-temperature optimization of film structure and interface quality.

E-1. Distance: throw length and transport balance: Short target–substrate distances intensify flux density and raise deposition rate, but they also magnify geometric shadowing and center-weighted thickness profiles, which reinforce columnar growth and undermine coverage in high-aspect-ratio features [80]. The near-field angular distribution remains narrow at short throws, so facets facing the source thicken rapidly while sidewalls lag; unless rotation or masking is aggressive, this yields rougher tops, thinner sidewalls, and greater micro-void content. In sputtering, very short throws can further elevate energetic neutral bombardment at the wafer center, compounding roughness and stress nonuniformity. Increasing throw distance broadens the effective incidence distribution at the wafer, averaging out azimuthal variations and smoothing radial thickness roll-off. With adequate vacuum – low mtorr for sputtering or high/ultra-high vacuum for evaporation – the longer path still supports largely ballistic transport, preserving interface sharpness and limiting gas-phase reactions that would blur composition. Practical ranges of roughly 10–30 cm in magnetron sputtering balance rate, uniformity, and step coverage; at the longer end, coupling with continuous rotation or planetary motion further homogenizes flux without sacrificing density [81]. If chamber pressure is not minimized, an increased path length increases ion–neutral and neutral–neutral scattering, broadens angular distributions, lowers the effective arrival rate, and potentially dilutes film density by softening the energy of arriving species. In multicomponent films, differential light scattering by heavy atoms can skew stoichiometry and shift phase formation, especially at elevated pressures or long throw distances. To counter these effects, pair longer throws with lower total pressure to extend mean free path, and, when conformity requires a slight pressure increase, recover density and composition fidelity with modest substrate bias and a moderate deposition rate [82].

E-2. Angle: incidence direction and morphological control: Incident angle dictates how adatoms land, migrate, and ultimately assemble, making it a first-order control on density, texture, and anisotropy. Near-normal incidence delivers flux with minimal lateral shadowing, so adatoms efficiently fill micro-facets and step edges, producing compact, fine-grained films with low porosity, strong adhesion, and crisp, abrupt interfaces. As the angle becomes more oblique, line-of-sight effects dominate taller surface features that self-shadow their neighbors, nucleation becomes facet-selective, and columns tilt toward the source, strengthening anisotropic texture while increasing void fraction and surface roughness. These oblique-angle regimes are powerful when intentionally harnessed; glancing-angle or off-axis growth can align nanocolumns, tune birefringent optical responses, or engineer graded porosity, yet they typically degrade step coverage and can weaken mechanical integrity if not balanced by motion or ion assistance. Off-axis geometries offer a practical advantage by diverting macroparticles and droplets (notably in arc sources) away from the substrate, trading some directionality for cleaner films; however, they benefit from synchronized substrate rotation or planetary motion to average the azimuthal flux and restore across-wafer uniformity. In practice, recipes often begin at near-normal incidence to secure wetting and continuous nucleation, then introduce controlled tilt or off-axis passes, co-tuned with pressure, rate, and mild bias, to tailor anisotropy without sacrificing density, interface sharpness, or sidewall continuity [83].

E-3. Step coverage and uniformity across features and wafers: Achieving uniform films over large substrates relies on geometries that average incident angles, typically targeting target–substrate distances of roughly 10–30 cm in magnetron sputtering, paired with continuous rotation or planetary motion to smooth azimuthal variation. Longer throws broaden the effective angular distribution across the wafer and reduce center-to-edge thickness roll-off. At the same time, motion continually reorients local facets to the source, converting directional flux into a more isotropic, wafer-scale average. High-aspect-ratio features require a different balance of directionality, scattering, and surface mobility. Starting nucleation at near-normal incidence establishes continuous seeds on sidewalls and corners, minimizing overhangs at feature mouths. After continuity is secured, a slight increase in working pressure widens the angular spread, and synchronized rotation exposes sidewalls repeatedly to replenished flux, improving trench and via fill without resorting to excessive pressure that would erode density. A carefully chosen throw maintains sufficient ballistic flux to preserve interface sharpness, while modest substrate bias restores densification and mitigates porosity that higher pressure would otherwise introduce. Rate and energy must be co-tuned to prevent bury-before-relaxation on sidewalls. Slower initial rates give adatoms time to migrate around corners and bridge micro-facets; once a continuous film forms, ramping to a moderate rate maintains throughput without reopening void pathways. Gentle ion assistance (e.g., small negative bias) increases re-emission and shallow sub-plantation, promoting sidewall coalescence and reducing micro-voids, yet avoids resputtering that could thin corners or disturb stoichiometry. Residual nonuniformities are best addressed with hardware and feedback rather than brute-force process changes. Profile masks or fixture baffling correct radial thickness slope; confocal gun arrangements or adjusted gun tilts counter-center bias; and planetary motion averages tool asymmetries. Close the loop with in situ rate control and post-process thickness mapping, iterating through distance, pressure, bias, and motion until both across-wafer uniformity and within-feature conformity meet specification without sacrificing density, interface sharpness, or composition fidelity [84].

E-4. Coupling with pressure, rate, and bias: Geometry cannot be tuned in isolation; pressure, rate, and substrate bias reshape the same flux that distance and angle define. Low pressure extends the mean free path at longer throws, preserving ballistic transport, interface sharpness, and composition fidelity, whereas any necessary pressure increase to aid conformity should be paired with a gentle negative bias and a moderate rate to restore densification and suppress porosity without driving excessive compressive stress. At short throws and high rates, directional flux and rapid burial amplify shadowing and roughness; mild continuous or pulsed bias activates near-surface mobility to counter bury-before-relaxation, while continuous rotation or planetary motion averages center-thick profiles. Conversely, at long throws and low rates where angular averaging is strong and diffusion times are long, bias should be reduced near critical interfaces to avoid over-peening, resputter, and intermixing; after interface closure, bias can be increased modestly to compact the bulk. Practical recipes co-ramp these levers: set the lowest pressure consistent with step coverage, sequence rate from slow nucleation to moderate fill, and bias only as high as needed for density and adhesion, adjusting with motion to lock in both across-wafer uniformity and within-feature conformity [85].

In summary, the PVD is, at its core, controlled transport: the researchers can set the chemistry, energy, geometry, and time so that atoms land at their desired locations. Its power comes from the independence of its primary levers, pressure (mean free path and scattering), plasma/bias (ion energy and flux), rate (kinetics vs. relaxation), and geometry (distance and angle), which can be co-tuned to sculpt microstructure, interfaces, and device-scale uniformity without resorting to high thermal budgets. PVD's strengths are repeatability, purity, and design flexibility. At low pressure and with disciplined gas control, you achieve sharp interfaces and low impurity content; with modest substrate bias and temperature, you densify and texture films at wafer-safe thermal loads; with recipe sequencing (slow nucleation and faster fill) and motion, you lock in conformity and control thickness across the wafer. These knobs extend naturally to complex stacks and reactive chemistries: partial-pressure control stabilizes compounds, while waveform engineering and bias windows protect stoichiometry in multicomponent and volatility-mismatched systems. Limits are absolute but tractable. Excess bias drives compressive stress and re-sputtering; high rates with insufficient mobility trap voids and elevate tensile stress; long throws at inadequate vacuum invite scattering and composition drift. Mitigations are procedural: co-tune pressure with bias, pair rate with mobility, synchronize geometry with motion, and validate with in situ metrology (rate, plasma, and stress) and post-maps (thickness, defects, and composition). The net is pragmatic: PVD offers a precise, low-temperature route to engineer density, stress, texture, composition, and conformity in tandem. When these levers are calibrated to the device's functional targets, electrical, optical, mechanical, or barrier performance, PVD translates atomic control into reliable, scalable films from pilot to production.

F-1. Variants and material coverage of PVD: PVD includes sputtering, thermal, and e-beam evaporation, each offering distinct energy delivery and material compatibility for tailored thin-film growth. Thermal evaporation is suitable for metals and chalcogenides, while e-beam enables high-purity coatings from refractory sources with enhanced adhesion via ion-assisted deposition. Magnetron sputtering excels at producing dense, conformal films across metals, ceramics, and compounds, with bias and pressure tuning that control microstructure and step coverage precision [86].

I. PVD method spans complementary variants using these approaches: sputtering deposition, thermal evaporation, e-beam evaporation, and magnetron sputtering, each defined by how source atoms are liberated and how much energy arrives at the substrate. Thermal evaporation uses resistive heating to vaporize materials with moderate melting points, delivering highly directional, high-rate flux ideal for metals, soft solders, and some chalcogenides when interface sharpness and low contamination are paramount. E-beam evaporation concentrates energy into refractory sources, enabling high-purity deposition of high-melting-point metals and complex optical materials; paired with ion assistance, it increases packing density and adhesion while maintaining the clean, line-of-sight character preferred for precision multilayers. Magnetron sputtering is a plasma-driven process that ejects atoms from a target via ion bombardment, producing dense, adherent films with tunable stress, texture, and composition at comparatively low substrate temperatures [87].

II. The PVD procedure draws close attention due to its excellence for metal and alloys, ceramic and compound coatings, hard coatings such as TiN and CrN, conductive and dielectric layers, and compound films formed via reactive sputtering (e.g., TiN from Ti in N_2, oxides in O_2), where bias and pressure tailor microstructure and step coverage. Across these variants, the accessible material set is broad: metals and alloys; hard nitrides and carbides (TiN, CrN, AlTiN, and TiC); and optical layers, including oxides (SiO_2, Al_2O_3, HfO_2, and Ta_2O_5), nitrides (SiN_x), and selected fluorides

(MgF_2 and CaF_2). Reactive gases extend composition control in both e-beam and sputter systems, while substrate motion, low pressure, and optional bias unify the approach: sharp interfaces from evaporation, high density and conformity from sputtering, and ion-assisted densification where optical or mechanical performance demands it [88].

F-I PVD approach: PVD encompasses vacuum processes that physically eject or evaporate atoms from a source and condense them as controlled thin films. Four methods will be examined: sputtering deposition, thermal evaporation, e-beam evaporation, and magnetron sputtering. **F-I1. Sputter deposition** ejects atoms from a solid target by ion bombardment in a plasma, producing dense, adherent films with good step coverage at relatively low substrate temperatures and enabling reactive formation of compounds (e.g., oxides and nitrides). **F-I2. Thermal evaporation** resistively heats a source to generate a molecular beam with highly directional, line-of-sight transport, yielding high-purity, sharp-interface metal and optical films but with limited conformality and challenges for very high-melting materials. **F-II1. E-beam evaporation** focuses a high-energy e-beam onto the source to evaporate refractory materials cleanly at high rates, supporting complex optical stacks and metals, often paired with optional ion assistance to boost density and adhesion. **F-II2. Magnetron sputtering** is a sputtering variant that uses magnetic fields to confine electrons and intensify the plasma at the target, enabling higher deposition rates, lower operating pressures, improved film density, and control of stress/texture, as well as efficient reactive sputtering. It can source and condense them as controlled thin films [89].

F-I1. Sputter deposition: Sputter deposition (Fig. 4.8A) is a core PVD method in which a glow-discharge plasma (typically Ar at 0.5–10 mtorr) accelerates ions toward a solid target, ejecting atoms that condense on a cooler substrate to form dense, adherent thin films. Ion energy and flux are set by target power density (about 1–20 W/cm^2), discharge mode (DC, RF, pulsed DC, or HiPIMS), working pressure, and any applied substrate bias (commonly –50 to –300 V), while target–substrate distance and substrate temperature (room temperature to roughly 400 °C) govern transport, step coverage, and microstructure. DC sputtering efficiently deposits a variety of materials (Fig. 4.8B) with unique applications (Fig. 4.8C). These classes are accessible via direct sputtering from compound targets or by reactive sputtering (introducing O_2, N_2, CH_4, etc.) from metallic targets to synthesize the desired stoichiometry in situ. Composition, phase, and properties are tuned by gas partial pressures, substrate bias, pressure, and temperature to meet electrical, optical, magnetic, and mechanical specifications [90]. Pure metals and alloys: Common metals (Al, Cu, Ti, W, Mo, Ni, Cr, Pt, Au, and Ag) and engineered alloys (NiFe, CoPt, AlCu, and TiAl) for interconnects, contacts, seed/barrier stacks, reflectors, and catalytic layers. Nitrides: Transition metal nitrides (TiN, CrN, ZrN, HfN, TaN, and AlTiN) and III-nitrides (AlN) for hard/wear coatings, diffusion barriers, plasmonic/optical layers, and piezoelectric films. Carbides and carbon-based films: Metal carbides (TiC, WC, and SiC) and amorphous carbon variants (DLC, a-C: H, ta-C) for ultra-hard, low-friction coatings, microelectromechanical systems (MEMS) tribology, and barrier applications. Oxides: Binary oxides (SiO_2, Al_2O_3, HfO_2, Ta_2O_5, TiO_2, and ZrO_2) and complex oxides (indium tin oxide (ITO), aluminum-doped zinc oxide (AZO), ZnO, SnO_2, and CeO_2) for dielectrics, optical interference stacks, transparent conductors, and barrier layers [91]. Oxynitrides and oxycarbides: Tailored compounds (SiO_xN_y, TiO_xN_y, and AlO_xN_y) with tunable refractive index, bandgap, and barrier properties for optical coatings and passivation. Borides and silicides: Borides (TiB_2, ZrB_2, and CrB_2) for hardness and conductivity; silicides (NiSi, $CoSi_2$, $TiSi_2$, and $MoSi_2$) for low-resistance contacts and diffusion barriers in microelectronics. Chalcogenides: Sulfides/selenides/tellurides (MoS_2, WS_2, ZnS, ZnSe, and $Ge_2Sb_2Te_5$) for solid lubricants, IR optics, and phase-change memory/media. Fluorides: MgF_2, CaF_2, LaF_3 for low-index optical coatings, AR stacks, and UV optics. Semiconductors: Elemental and compound semiconductors (Si, Ge, Ga_2O_3, ZnO, Cu_2O, In_2O_3; Cu(In, Ga)Se_2 layers through reactive/sequential sputter) for thin-film electronics, sensors, and photovoltaics. Magnetic materials and multilayers: Soft and hard magnetics (NiFe, CoFe, Co/Pt, Co/Pd, and FeCoB) and spintronic stacks (FM/Ta, Ru, IrMn, and Pt) for MRAM (Magnetoresistive Random Access Memory), read heads, and magnetic sensors. Superconductors: Nb, NbN, NbTiN, MgB_2 thin films for superconducting electronics, detectors, and resonators. Cermets and nanocomposites: Metal-in-dielectric or metal-in-ceramic systems (e.g., Ag: SiO_2, TiN: Si_3N_4) for tunable plasmonics, resistive switching, and wear-resistant composites [92]. Barrier and adhesion layers: Refractory nitrides/carbides (TiN, TaN, and WN) and adhesion promoters (Ti and Cr) for Cu/Al interconnects and multilayer stacks. Transparent conductors: ITO, AZO, fluorine-doped tin oxide (FTO) for displays, PV, and optoelectronics. Dielectrics for microelectronics: High-k and low-k films (HfO_2, Al_2O_3, SiN_x, and SiO_2) for gates, capacitors, and passivation. Catalytic and energy materials: Pt, Pd, Ni, NiFe, and mixed oxides (NiO and Co_3O_4) for electrocatalysis, gas sensors, and battery interfaces. Specialty coatings: Antireflection/HR stacks (SiO_2/TiO_2/HfO_2/Ta_2O_5), IR reflectors (Ag-based multilayers), decorative/protective nitrides (ZrN and TiAlN), and low-emissivity glass coatings (Ag with oxide/nitride barriers) [93].

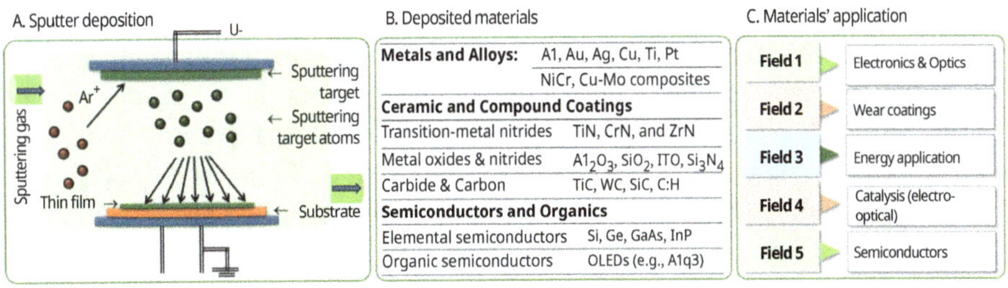

Fig. 4.8: Demonstration of sputter deposition: **A**. sputter deposition uses plasma-driven ion bombardment for dense films; **B**. deposited materials span metals, nitrides, oxides, and semiconductors; and **C**. materials' applications.

Advanced variants expand capability and film quality. HiPIMS generates a highly ionized metal flux that enhances densification, adhesion, and phase control at comparatively low thermal budgets, while pulsed-DC schemes moderate average energy to limit damage on temperature-sensitive stacks. Typical deposition rates range from ~0.1 to 2 nm/s depending on the target material, geometry, and power, and can be stabilized with in situ rate monitors and plasma diagnostics. Microstructure and stress respond sensitively to the balance of rate and mobility: lower pressure and modest bias improve packing density and reduce void fraction; slightly higher pressure widens angular spread for better sidewall reach, with bias restoring densification to avoid porosity. Composition control in multicomponent and reactive systems hinges on minimizing differential scattering and resputtering of light or volatile species, aided by low pressure, moderate bias, and precise gas-flow feedback [94].

A widely used configuration is the planar diode glow discharge, in which the target serves as the cathode, facing an anode, and the substrate is inside a vacuum chamber. Applying a high negative voltage ignites the plasma, and sputtered atoms leave the target with a broad angular distribution, a fraction of which arrive at the substrate to form the film. Deposition rate is tuned by target power, erosion area, pressure, and throw distance, while rotation or planetary motion averages geometric flux to improve across-wafer uniformity. The planar diode configuration accommodates a wide range of material systems, including elemental metals, engineered alloys, wear-resistant compounds such as TiN, CrN, and ZrN, as well as optical and magnetic thin films. Its adaptable geometry enables deposition of oxides, nitrides, and barrier layers for use in electronics, photonics, and protective coatings. However, practical constraints remain: each cathode covers a limited area, and mechanical motion is often required to counteract radial flux gradients. Additionally, heat-sensitive substrates may restrict power input due to thermal stress from ion bombardment. By carefully adjusting operating parameters such as pressure, waveform, bias voltage, deposition rate, and spatial layout, this method can yield dense, low-defect films with tailored stress profiles and crystallographic texture, suitable for high-throughput manufacturing [95].

F-I2. Thermal evaporation: Thermal evaporation resistively heats a source (wire or boat, crucible, or filament-coated slug) under high vacuum until it sublimates or boils, creating a molecular beam that travels line of sight to the substrate and condenses (Fig. 4.9A). Because transport is largely ballistic, films achieve high purity and sharp interfaces at low substrate temperatures, although conformality is limited without rotation, a tailored fixture, or ion assistance. Process integrity starts with vacuum and cleanliness: base pressure is typically 1×10^{-6} to 1×10^{-8} torr, with process pressure $\leq 1\times10^{-5}$ torr, maintained by a turbomolecular pump with backing and, optionally, augmented by cryo or titanium sublimation pumping for ultra-clean optical stacks. Outgassing control relies on prebakes, load-locks to isolate the main chamber, and source-degas runs to stabilize emissions before production [96]. Heating configuration and crucible choices dictate stability and chemistry. Sources include W/Mo boats, baskets, and crucibles; filaments wrapped with the charge; and crucibles lined with Al_2O_3, BN, or graphite to prevent reactions; power ranges from tens to hundreds of watts for low-melting metals and up to kilowatts for higher-melting alloys. Charge size and geometry set rate stability and spitting risk, which are mitigated by slow ramps through phase transitions and by using pellets or shot rather than powders. Deposition control is set by rate, temperature, geometry, and motion. Typical rates range from ~0.1 to 50 nm/s (often 0.5–5 nm/s for metals) measured by a QCM with geometry-specific tooling factors; substrates are held near ambient up to ~300 °C (higher with radiant heaters if needed). Throw distances of ~20–80 cm balance uniformity and throughput, with shorter throws increasing rate but accentuating angular nonuniformity; continuous rotation or planetary motion improves thickness uniformity, while shutters define steps and multilayers [97]. Optional assists include a low-pressure backfill to improve wetting or mild ion/e-beam assistance to increase density when purity trade-offs are acceptable. Crucible and liner compatibility is essential to avoid reactions and contamination. BN or graphite liners pair well with Al; Al_2O_3 liners suit Au and Ag; reactive halides should avoid Al_2O_3. Managing spatter through controlled thermal ramps and appropriate charge morphology reduces particle formation and improves interface quality [98].

A wide material palette (Fig. 4.9B) is accessible, classified by chemistry and application (Fig. 4.9C). Pure metals and simple alloys (Al, Ag, Au, Cu, Cr, Ti, Ni, Pt, and Pd); low-temperature solders (In, Sn, and Pb); and alloys such as NiCr, TiW, and AlCu serve as adhesion and seed layers, mirrors and reflectors, electrodes and interconnects, IR/UV optics, and decorative finishes. Fluorides and other low-index dielectrics, such as MgF_2, LaF_3, and LiF, enable antireflection coatings, UV optics, and laser cavities. Oxides from evaporants or oxide granules build multilayer interference stacks, passivation, and moisture/oxygen barriers. Chalcogenides and phase-change materials, such as ZnS, ZnSe, Ge–Sb–Te, and Se-based layers, serve IR windows, dielectric spacers, optical/data storage, and metasurfaces. Organic small molecules, Alq_3, NPB, rubrene, C_{60}, and doped host–guest mixtures are deposited in dedicated organic evaporators for organic light-emitting diode (OLED) emissive/transport layers, OPV (Organic Photovoltaic), and sensors [99]. Reactive or assisted growth can convert metals into oxides or nitrides in situ, e.g., $Al \rightarrow Al_2O_3$ $Ti \rightarrow TiO_2$, $SiO \rightarrow SiO_2$, by introducing oxygen or nitrogen, or by post-oxidation. Thermal evaporation's strengths are ultra-clean films, high rates, low substrate temperature, and sharp interfaces, with limitations in intrinsic step coverage, handling of very high-melting or decomposing materials, spitting through phase changes, and line-of-sight roll-off on large wafers. Quality is maintained by low base pressure, thorough source degassing, carefully tuned rate ramps, judicious use of rotation and masks, QCM-based thickness control with cross-wafer mapping, and only minimal ion assistance when densification is needed [100].

F-II1. E-beam evaporation: E-beam evaporation concentrates a high-energy e-beam onto a solid charge inside high vacuum, locally melting and vaporizing even refractory materials; the resulting vapor travels largely line of sight to the substrate,

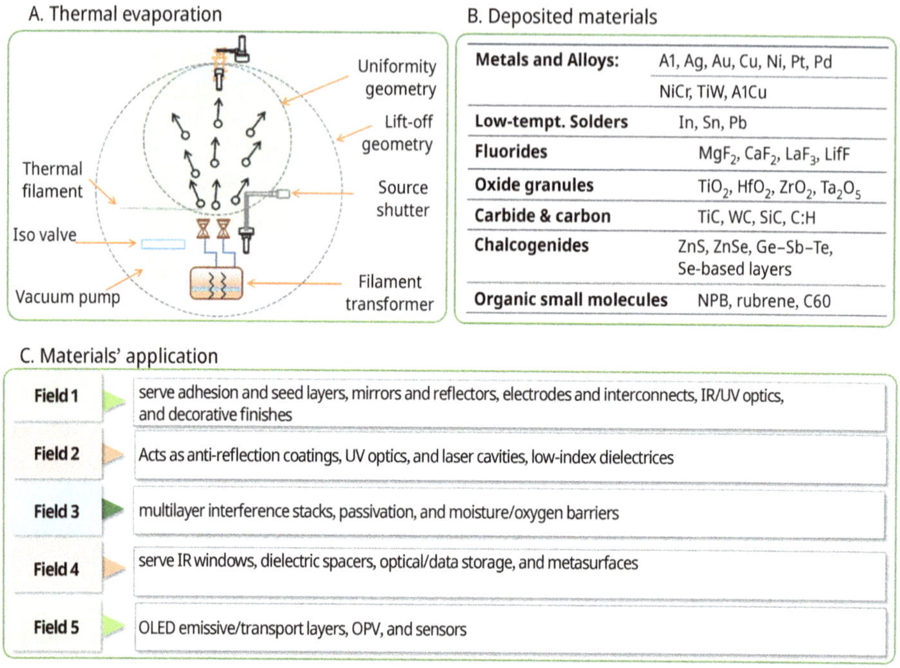

Fig. 4.9: Demonstration of thermal evaporation: **A**. thermal evaporation heats a source under vacuum to produce line-of-sight films; **B**. representative deposited materials; and **C**. five key application areas.

enabling high-rate, high-purity deposition with sharp interfaces, while optional low-energy ion assistance can boost density, adhesion, and stress control (Fig. 4.10A). Stable operation relies on clean vacuum (base pressure on the order of 10^{-6}–10^{-8} torr and process pressure $\leq 10^{-5}$ torr), rigorous preconditioning of the chamber and charges (bakeouts, load-lock transfers, and pre-melts/degassing), and adequate pumping capacity using turbomolecular stages, often augmented by cryogenic or titanium-sublimation elements for ultra-low contamination loads [101]. The electron gun typically operates at 5–30 kV with kilowatt-scale beam power, and electromagnetic beam steering rasters the melt pool to even out erosion and stabilize the rate; emission is provided by thermionic filaments or field-assisted emitters, with shielding to prevent filament by-products from reaching the substrate. Source hardware includes single- or multi-pocket water-cooled copper hearths fitted with chemically compatible liners such as graphite, BN, metal oxides (Al_2O_3, ZrO_2, or Y_2O_3); charge geometry and pellet size are chosen to minimize spitting during phase transitions, and continuous-feed ingots or rotating/tilting hearths support long runs and uniform composition. Deposition control centers on rate, temperature, and geometry: thin-film rates commonly span ~0.1–50 nm/s measured by QCM with calibrated tooling factors; substrates are held near ambient up to roughly 300–400 °C with optional radiant heating; throw distances of ~30–100 cm and continuous rotation or planetary motion are used to balance throughput and across-substrate uniformity; and shutters define sharp layer transitions in complex stacks. Ion assistance with Ar, O_2, or N_2 beams at tens to a few hundred electronvolts can densify films, tailor texture, and help lock stoichiometry during reactive sequences, while controlled backfilling enables in situ formation of oxides or nitrides without unduly shortening mean free path or destabilizing the beam [102].

The palette of materials and their application spans pure and alloyed metals (Al, Ag, Au, Cu, Cr, Ti, Ni, Pt, Pd, NiCr, and TiW) for mirrors, electrodes, seed and adhesion layers, decorative and barrier coatings; high-melting metals (W, Mo, Ta, and Nb) and compatible silicides/borides for diffusion barriers, high-temperature reflectors, X-ray and SEM coatings, and wear-resistant layers; optical oxides (as mentioned above) for precision AR/HR stacks, laser mirrors, waveguides, and passivation; fluorides for broadband AR and UV optics; chalcogenides and phase-change materials (ZnS, ZnSe, Ge-Sb-Te, As_2S_3, and Se-based glasses) for IR windows, dielectric spacers, optical data storage, and metasurfaces; and in dedicated organic toolsets, small-molecule organics for OLED emissive/transport layers, OPV, and sensors. The principal advantages are very high deposition rates over large areas, low contamination, access to refractory materials, precise thickness control via shuttering and QCM, and the option to achieve high density at comparatively low substrate temperatures when assisted; practical limits include line-of-sight conformality, spitting risks during melting/boiling, and differential vapor pressures that complicate multicomponent stoichiometry, all of which are mitigated by careful liner selection, pre-melting, controlled ramps, appropriate throw and motion, judicious ion assistance, and routine verification of rate, thickness uniformity, and optical absorption/scatter [103].

F-II2. Magnetron sputtering: Magnetron sputtering confines electrons with magnetic fields near a negatively biased target, intensifying the plasma at the target surface and boosting ionization so more ions strike the target and eject atoms efficiently (experimental procedure shown in Fig. 4.10B). The higher ion flux at lower operating pressures increases deposition rates. It enables the formation of compact films with tunable stress and texture, while also stabilizing reactive sputtering to form compounds in situ. Process windows typically use argon at about 0.2–5 mtorr, with optional O_2, N_2, CH_4, or H_2 to synthesize oxides, nitrides, carbides, or hydrides under partial-pressure control to avoid hysteresis and target poisoning [104]. Power delivery is matched to the material and goal: DC magnetron supports conductive targets for high-rate metals; RF at 13.56 MHz enables insulating targets for dielectrics and semiconductors; pulsed DC mitigates arcing in reactive modes; and HiPIMS delivers short, high-power pulses that produce a highly ionized metal flux for exceptional density, adhesion, and refined microstructure. Magnetic topology is tuned for application needs: balanced magnetrons confine plasma tightly to the target, while unbalanced designs increase ion flux to the substrate to boost adhesion; planar cathodes serve most wafer and optics tools, rotary (cylindrical) cathodes provide long-length uniformity and high utilization for architectural glass or web lines, and S-gun or remote geometries isolate plasma for thermally sensitive films. Substrate conditions are co-tuned with the plasma: a modest negative bias around −30 to −300 V enhances mobility and densification with pulsed bias limiting stress and resputtering, and substrate temperatures from ambient to roughly 400 °C balance densification against intermixing at interfaces [105]. Throughput and uniformity are set by power density, target–substrate distance of roughly 10–30 cm, rotation or planetary motion, and confocal multi-cathode layouts, yielding rates on the order of 0.1–10 nm/s depending on material and geometry. The mechanistic advantages stem from magnetic confinement increasing electron path length, which sustains ionization at low pressure and reduces gas-phase scattering, thereby keeping interfaces sharp and composition faithful. Concentrating energy at the target rather than the substrate improves sputter yield, with lower substrate heating than in diode sputtering, and combining pulsed power with feedback-controlled gas flows stabilizes metal and compound modes for repeatable stoichiometry [106].

The resulting materials span metals and alloys such as Cu, Al, Ti, W, Mo, Ni, Cr, Pt, Au, NiFe, TiW, and AlCu for interconnect seed and barrier stacks, reflective layers, diffusion barriers, catalytic electrodes, and large-area low-emissivity coatings where dielectric barriers protect Ag. Hard coatings and refractory compounds, including TiN, CrN, ZrN, AlTiN, TiC, WC, and TaN, extend tool life, protect aerospace valves and turbine components, and enable durable decorative finishes with color tuned by nitride stoichiometry. Dielectrics and optical layers such as metal oxides and nitrides are used to build AR/HR filters, low-loss laser mirrors, waveguides, and robust passivation or moisture barriers for displays and photovoltaic modules. Transparent conductors, including ITO, AZO, FTO, and ZnO variants, are used in displays, touch sensors, thin-film photovoltaics, and low-E glass, where large-area, uniform sheet resistance and transparency are essential. Magnetic and spintronic stacks based on NiFe, CoFe, Co/Pt, Co/Pd, IrMn, and Ru spacers enable MRAM, GMR/TMR (Giant Magnetoresistance /Tunnel Magnetoresistance) heads, magnetic sensors, and microwave devices by tailoring stress and texture to tune anisotropy and switching fields [107]. Solid lubricants such as MoS_2 and WS_2 provide ultra-low friction for space mechanisms and vacuum tribology, while ZnS and ZnSe supply dense, low-scatter IR optics and beamsplitters; carbon-based films such as DLC (Diamond-Like Carbon), a-C: H (Amorphous Carbon:Hydrogen), and ta-C (Tetrahedral Amorphous Carbon) reduce wear and friction in automotive and biomedical components and serve as protective overcoats and barriers. In practice, co-tuning pressure and bias preserves density and interface sharpness while avoiding over-compressive stress. Reactive modes are stabilized with partial-pressure feedback and pulsed DC (Direct Current) or HiPIMS (High Power Impulse Magnetron Sputtering), and uniformity is engineered through cathode geometry, throw distance, and synchronized motion with confocal guns for complex stacks. Production recipes are validated by mapping thickness, sheet resistance, stress, and optical constants, then iterating power, pressure, and gas flows to lock performance from pilot to high-volume manufacturing [108].

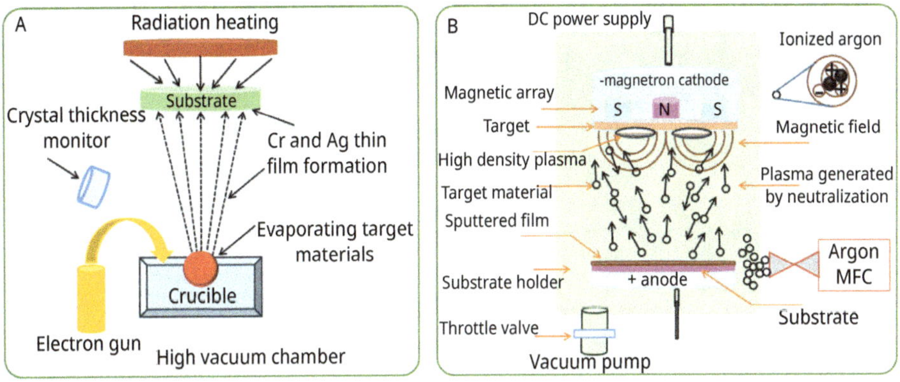

Fig. 4.10: Schematic representation of two physical vapor deposition (PVD) techniques: (A) electron-beam evaporation and (B) magnetron sputtering.

4.4 Chemical vapor deposition

CVD is a versatile gas-phase thin-film fabrication technique widely employed to produce high-purity, high-performance coatings for electronic, optical, mechanical, and protective applications. In contrast to PVD, which relies on physical transport of material, CVD introduces one or more volatile precursors into a controlled reaction chamber, where they undergo thermal or energy-assisted decomposition and/or chemical reaction at the heated substrate surface. The process is driven by multidirectional molecular diffusion, enabling conformal coverage even on complex geometries. Typical precursors for CVD processes include oxygen, halides (H_2SiCl_2, $HSiCl_3$, $TiCl_4$, and WF_6), hydrides (SiH_4, GeH_4, $AlH_3(NMe_3)_2$, and NH_3), organometallics ($AlMe_3$, $Ti(CH_2{}^tBu)_4$), metal alkoxides (TEOS, tetracid dimethylaminotitanium (TDMAT), $Ti(O^iPr)_4$), metal dialkylamides ($Ti(NMe_2)_4$), metal diketonates ($Cu(acac)_2$), and metal carbonyls ($Ni(CO)_4$). Growth rate depends on surface chemistry, substrate temperature, and precursor concentration. Most CVD reactions are endothermic and require energy supplied via resistive, radiant, or RF heating. Based on the activation method, CVD is classified into thermal, plasma-assisted, photo-assisted, and gas-assisted variants, with plasma and photo-assisted approaches enabling reduced substrate temperatures. Optimal precursors exhibit adequate volatility, thermal stability, high chemical purity, and cost-effectiveness [109]. This section classifies CVD techniques, the resulting deposited materials, and their respective applications (Fig. 4.11).

In summary, CVD is a versatile gas-phase thin-film fabrication technique in which volatile precursors are delivered into a controlled reaction chamber, where they thermally decompose or chemically react at the substrate surface to form a solid layer (Fig. 4.11A). This process produces high-purity, dense, conformal coatings with precise thickness control, making it essential for advanced microelectronics, optical compo-

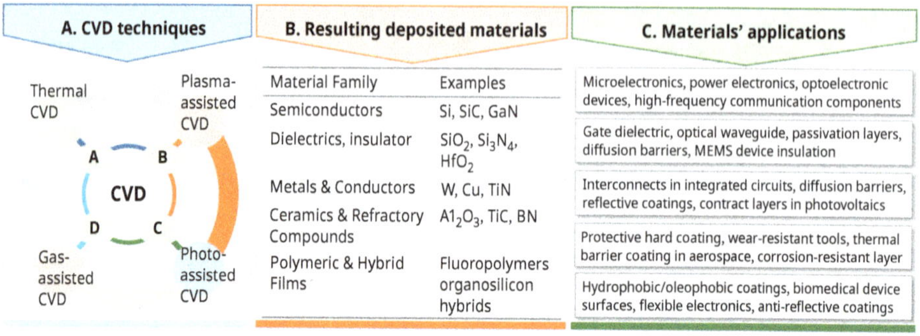

Fig. 4.11: Demonstration classifies CVD techniques, the resulting deposited materials, and their respective applications.

nents, protective barriers, and energy conversion or storage devices. Its multidirectional molecular transport enables uniform coverage over complex geometries, while process parameters can be tuned to tailor microstructure, composition, and functional performance. CVD encompasses four principal variants distinguished by activation method. Thermal CVD (**A**) relies on substrate heating to activate the reaction. Plasma-assisted CVD (PACVD) (**B**) uses energetic plasma species to enable lower-temperature growth. Photo-assisted CVD (**C**) employs ultraviolet (UV) or laser photons to initiate reactions with minimal thermal load. Gas-assisted CVD (GACVD) (**D**) introduces reactive gases to promote targeted chemistries and improve stoichiometry. Each method offers unique advantages, constraints, and material compatibilities, allowing CVD to address a broad spectrum of industrial and research applications.

A. Thermal CVD: Thermal CVD relies solely on substrate heating to activate precursor decomposition and surface reactions. It produces dense, high-purity films with excellent crystallinity, precise thickness control, and uniformity across large substrates. This method is ideal for applications requiring superior structural integrity and reproducibility, such as semiconductor dielectrics and epitaxial layers. However, its high thermal budget limits use with temperature-sensitive substrates and demands thermally stable precursors with suitable volatility and decomposition characteristics, making precursor selection and substrate compatibility critical to process success (Fig. 4.11B).

B. PACVD: PACVD uses an electrical discharge to generate reactive ions, radicals, and excited species, enabling film growth at significantly lower substrate temperatures than thermal CVD. This makes it compatible with polymers, glasses, and other heat-sensitive materials while maintaining good film density and adhesion. It also allows the deposition of materials with high activation energy requirements. However, plasma exposure can cause ion-induced surface damage, alter near-surface chemistry, and introduce stress [110]. The equipment is more complex and costly, requiring careful control of plasma parameters (Fig. 4.11C).

C. Photo-CVD: Photo-CVD employs UV or laser photons to initiate precursor dissociation and surface reactions without relying heavily on thermal energy. This enables deposition on highly temperature-sensitive substrates, including flexible polymers and hybrid materials, with minimal thermal damage. It offers precise control over reaction pathways via wavelength and intensity tuning. Limitations include slower deposition rates, reduced suitability for thick films, and challenges in coating complex geometries due to limited photon penetration depth, making it best suited for specialized, high-precision coatings.

D. Gas-assisted CVD (GACVD): The GACVD introduces reactive assisting gases (such as O_2, H_2, NH_3, or halogens) alongside primary precursors to promote targeted chemical reactions during film growth. This approach enables lower-temperature processing, improved stoichiometry, and enhanced incorporation of specific elements, making it valuable for oxides, nitrides, carbides, and tailored functional coatings. Precise control of gas composition, flow rate, and partial pressure is essential, as assisting gases can trigger unwanted side reactions or alter film composition unpredictably [111]. It offers high flexibility but demands rigorous process optimization. Table 4.2 presents a side-by-side comparison of the four CVD variants, aligning their core parameters, operating conditions, and suitability for various materials and applications.

Tab. 4.2: Comparison of four CVD variants to prepare materials and their applications.

Parameters	Thermal CVD	Plasma-assisted CVD (PACVD)	Photo-assisted CVD (photo-CVD)	Gas-assisted CVD (GACVD)
Primary activation method	A high substrate temperature provides the activation energy for precursor decomposition	Plasma generates energetic ions/radicals to activate reactions at lower temperatures	UV or laser photons break precursor bonds, initiating surface reactions	Reactive assisting gas (e.g., O_2, H_2O, and NH_3) promotes specific chemical reactions
Substrate temperature	300–1,200 °C	100–500 °C	Ambient to 400 °C	Varies widely; often lower than pure thermal CVD due to gas reactivity
Energy source	Thermal heating (resistance, radiant, or induction)	Electrical energy for plasma generation (RF, microwave, and DC)	Ultraviolet or laser radiation	Thermal plus chemical activation from assisting gas
Deposition rate	~0.1–10 μm/h	~0.1–5 μm/h	~0.01–2 μm/h	~0.05–5 μm/h (process-dependent)
Material compatibility	Thermally stable substrates (Si, ceramics, and refractory metals)	Temperature-sensitive substrates (polymers, certain metals, and glass)	Highly temperature-sensitive substrates (plastics and flexible electronics)	Broad range; tailored for oxides, nitrides, carbides, and specialty films

Tab. 4.2 (continued)

Parameters	Thermal CVD	Plasma-assisted CVD (PACVD)	Photo-assisted CVD (photo-CVD)	Gas-assisted CVD (GACVD)
Advantages	Film density, crystallinity, and low contamination	Low thermal budget, good step coverage, and wide material range	Minimal thermal damage and selective area deposition possible	Enables specific chemistries, improved stoichiometry, and lower temperature growth
Limitations	Not suitable for heat-sensitive materials; high energy cost	Possible ion damage; more complex equipment	Slower rates; limited photon penetration	Requires precise gas control; may introduce unwanted reactions
Typical applications	Semiconductors, dielectrics, metals in microelectronics, photovoltaics, and MEMS	Optical coatings, barrier layers, biomedical devices, and semiconductor passivation	Optoelectronic materials, transparent conductive films, and flexible electronics	Oxide/nitride coatings, catalytic layers, protective films, and energy storage materials

4.4.1 Thermal chemical vapor deposition (thermal CVD)

Thermal CVD is a vapor-phase thin-film fabrication process in which elevated substrate temperatures provide the activation energy necessary for precursor decomposition and surface reactions (Fig. 4.12A). Thermal CVD employs elevated substrate temperatures to initiate and sustain precursor decomposition, enabling uniform thin-film growth through controlled surface reactions within a precisely regulated reaction chamber. The process is governed by thermally activated kinetics, in which increasing temperature accelerates reaction rates, enabling fine control over film thickness, crystallinity, grain size, and overall microstructural quality. Elevated temperatures facilitate complete precursor decomposition, thereby reducing residual contaminants and enhancing film purity, density, and adhesion. This high thermal budget, however, can restrict its use with temperature-sensitive substrates or device architectures that may suffer thermal degradation or interdiffusion. Thermal CVD is a mature, industry-proven technique widely used for depositing semiconductors, dielectric layers, and metallic coatings for microelectronics, photovoltaic devices, optical components, and protective barriers. Its ability to deliver conformal, high-quality films with reproducible properties across large-area substrates makes it indispensable for applications that demand structural integrity, chemical stability, and precise dimensional control, if substrate and precursor compatibility are carefully matched to the process conditions [112].

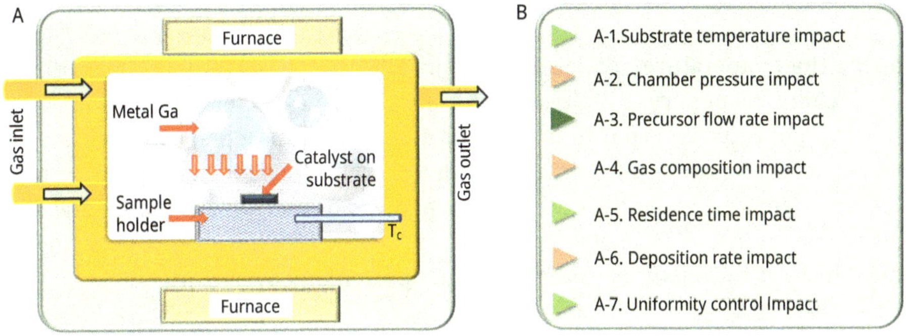

Fig. 4.12: Demonstration of thermal CVD techniques: A. schematic diagram of thermal CVD and B. summary of synthesis parameters.

Thermal CVD process parameters such as substrate temperature, precursor flow rate, and chamber pressure are carefully optimized to achieve the desired film uniformity, composition, and functional performance (Fig. 4.12B and Tab. 4.3). These parameters and their impact will be discussed in the following sections. In a controlled reaction chamber, volatile precursors are introduced, often via a carrier gas (nitrogen, argon, or hydrogen), and transported to the heated substrate. Substrate temperatures typically range from 300 °C to 1,200 °C, depending on the material system, with higher temperatures enhancing film density, crystallinity, and adhesion by promoting complete precursor decomposition. Chamber pressures can range from APCVD (Atmospheric Pressure Chemical Vapor Deposition) to LPCVD (Low Pressure Chemical Vapor Deposition), from 10^{-3} to 10^{2} torr, where reduced pressures improve step coverage and minimize unwanted gas-phase nucleation. The process follows thermally activated kinetics, in which reaction rates increase exponentially with temperature, enabling precise control over film thickness (nanometers to several micrometers), grain size, and microstructural properties. Deposition rates typically range from 0.1 to 10 µm/h and can be adjusted by varying precursor flow rates, gas composition, and residence time. Uniformity across large substrates is achieved through optimized gas distribution, substrate rotation, and careful thermal profile management. Thermal CVD is widely used to deposit semiconductors (Si, SiC, and GaN), dielectrics (SiO_2, Si_3N_4, and HfO_2), and metals (W and TiN) across applications such as microelectronics, photovoltaics, MEMS, and protective coatings. While its high thermal budget ensures high-purity, high-density films with excellent adhesion, it also limits compatibility with temperature-sensitive substrates, polymers, and specific prefabricated device layers. Consequently, precursor selection is critical, requiring adequate volatility, high thermal stability, and clean decomposition pathways to achieve the desired film quality and performance [113] (Fig. 4.12B).

A-1. Substrate temperature impact: Substrate temperature is the primary lever for controlling surface reaction rates in thermal CVD, which follows Arrhenius behavior (eq. 4.1):

$$k = Ae^{E_a/RT} \tag{4.1}$$

Raising the temperature accelerates precursor decomposition and surface diffusion, yielding higher crystallinity, larger grain size, and greater film density while reducing impurity incorporation from partially decomposed species. Excessive temperatures can cause interdiffusion with underlying layers, modify dopant profiles, and damage temperature-sensitive substrates or metallization. Practical setpoints span roughly 300–1,200 °C depending on the material system, with lower limits dictated by kinetic requirements and upper limits by substrate/device thermal budgets [114].

A-2. Chamber pressure impact: Chamber pressure shapes transport by setting the mean free path and boundary-layer thickness at the substrate. LPCVD reduces gas-phase collisions, suppresses homogeneous nucleation and particle formation, and improves conformality in high-aspect-ratio features; it also shifts control toward surface kinetics. Higher pressures increase reactant partial pressures and can raise deposition rates, but they also increase the risk of gas-phase reactions and nonuniformity across the substrate. Typical ranges include ~10^{-3} to 10^2 torr for LPCVD and ~760 torr for atmospheric systems, selected to balance kinetics, transport, and cleanliness [115].

A-3. Precursor flow rate impact: Precursor flow rate determines reactant partial pressures and surface supersaturation, directly influencing growth rate and film stoichiometry. Insufficient flow starves the surface, leading to thickness nonuniformity and composition drift, especially on large wafers or in batch loads. Excessive flow promotes gas-phase reactions and aerosol formation, increasing particle contamination and reducing step coverage, particularly under higher pressures. Optimal flow is established alongside pressure and temperature to maintain laminar transport, efficient utilization, and a stable, surface-reaction-limited regime [116].

A-4. Gas composition impact: Carrier and co-reactant gases determine reaction pathways, heat and mass transport, and final film stoichiometry. Inert carriers such as nitrogen or argon moderate dilution and transport, whereas hydrogen can serve as a reductant and passivating agent, improving purity and surface termination. Reactive co-reactants (e.g., oxygen, water, ammonia, or silane partners) fix anion/cation ratios, influence crystallographic phase selection, and tune defect density. Additives like halogen-containing species can enable in situ etching, suppressing parasitic deposition and improving conformality on complex topographies [117].

A-5. Residence time impact: Residence time, governed by chamber volume and total flow, controls the extent of precursor conversion and by-product removal. Longer residence times generally increase utilization efficiency and growth rate until secondary gas-phase reactions emerge and degrade film quality. Shorter residence times minimize homogeneous reactions and particle formation but may underutilize precursors and create center-to-edge thickness gradients.

Engineering a narrow residence time distribution through reactor geometry (e.g., showerhead plenums and hot-wall vs. cold-wall designs) supports uniformity at scale [118].

A-6. Deposition rate impact: Deposition rate reflects the balance between surface kinetics and mass transport: at lower temperatures, it is reaction-limited and increases steeply with temperature, while at higher temperatures or partial pressures, it becomes transport-limited. Excessive rates can outpace adatom diffusion, producing rougher surfaces, smaller grains, higher intrinsic stress, and elevated hydrogen or impurity incorporation. Slower rates promote smoother morphology, improved crystallinity, and better step coverage, albeit at the expense of throughput. Practical target rates of 0.1–10 μm/h are tuned by temperature, pressure, and precursor concentration to meet device specifications [119].

A-7. Uniformity control impact: Film uniformity directly affects electrical, optical, and mechanical performance, making thermal management and flow distribution critical design priorities. Multizone heaters, calibrated susceptor emissivity, and wafer rotation minimize thermal gradients that otherwise drive thickness and property variations. Gas-delivery strategies, such as showerhead distribution, edge-flow compensation, and carefully chosen dilutions, equalize reactant availability across the substrate and within features. When combined with appropriate pressure and flow regimes, these measures deliver the conformality and across-wafer uniformity required for advanced device manufacturing [120].

4.4.2 Plasma-assisted chemical vapor deposition (PACVD)

PACVD uses plasma to generate reactive species from gaseous precursors, enabling film deposition at significantly lower substrate temperatures than in conventional thermal CVD processes (Fig. 4.13A). The plasma provides energetic ions and radicals, enhancing reaction rates and enabling deposition on temperature-sensitive materials without compromising film quality or adhesion [121]. By adjusting plasma power, frequency, and gas composition, researchers can tailor the properties of films, including hardness, refractive index, electrical conductivity, and chemical resistance. PACVD is extensively applied in optical coatings, barrier layers, biomedical devices, and semiconductor passivation, where low thermal budgets are essential. The technique offers high deposition rates, excellent step coverage, and the ability to deposit films with complex chemistries, including inorganic, organic, and hybrid materials. PACVD uses an electrical discharge to generate a reactive plasma from gaseous precursors, enabling high-quality thin-film growth at substrate temperatures as low as 100 °C [122]. Energetic ions, radicals, and photons in the plasma enhance surface reaction rates, film densification, and adhesion on heat-sensitive substrates without compromising chemical resistance [123]. Critical process parameters and their impacts are summarized in Fig. 4.13B and Tab. 4.3, and discussed in detail below.

Tab. 4.3: Comparison of the thermal CVD technique's seven primary parameters for exceptional material tunability.

Parameters	Definition in thermal CVD	Primary impact on process	Effect on final film properties
A-1. Substrate temperature	Controlled heating of the substrate to activate precursor decomposition and surface reactions	Governs reaction kinetics, adatom mobility, and crystallization rate	High temperatures improve crystallinity, density, and adhesion; excessive heat may cause interdiffusion, stress, or damage to T-sensitive substrates
A-2. Chamber pressure	Total operating pressure inside the CVD reactor during deposition	Influences mean free path, collision frequency, and transport regime (ballistic vs. diffusive)	Low pressure yields sharp interfaces and high purity; higher pressure improves conformality but may reduce density or increase defect incorporation
A-3. Precursor flow rate	Volumetric delivery rate of precursor gases into the reactor	Controls precursor availability, reaction rate, and by-product removal	Proper flow ensures uniform thickness and composition; excessive flow can cause gas-phase nucleation
A-4. Gas composition	Chemical identity and ratio of precursor and carrier/ reactive gases	Determines dominant chemical pathways and intermediate species formation	Directly affects stoichiometry, phase purity, and functional properties such as conductivity, refractive index, or hardness
A-5. Residence time	Average time precursor molecules are in the reaction zone before exhaust	Balances surface reaction completion with by-product removal	Long residence improves incorporation and conformality; excessive time causes secondary reactions or contamination
A-6. Deposition rate	The rate at which film material accumulates on the substrate surface	Controlled by temperature, precursor supply, and reaction kinetics	High rates increase throughput but reduce density or smoothness; low rates improve structural control and interface sharpness
A-7. Uniformity control	Strategies to ensure consistent film thickness and composition across the substrate	Achieved via gas flow design, substrate rotation, and temperature uniformity	Enhances across-wafer consistency, reduces defects, and ensures reproducible electrical, optical, or mechanical performance

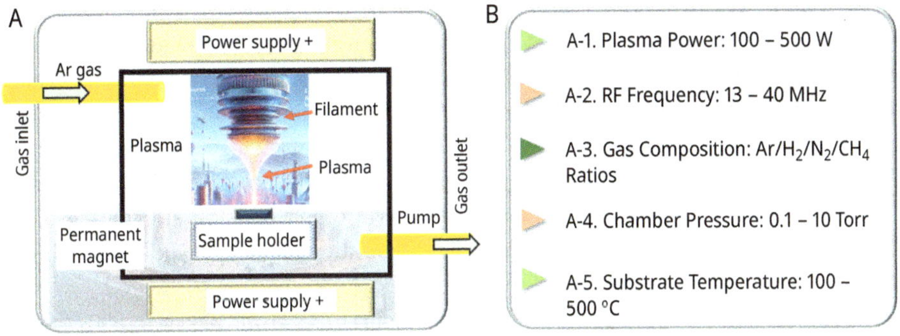

Fig. 4.13: Demonstration of plasma-assisted CVD techniques: A. schematic diagram of PACVD and B. summary of synthesis parameters.

B-1. Plasma power: Plasma power settings between 100 W and 500 W determine the ion and radical density in the discharge region, directly influencing deposition kinetics and film microstructure. At lower power, limited reactive species flux produces slower growth rates, lower film density, and higher porosity; at higher power, increased radical flux and ion bombardment accelerate deposition while promoting densification. However, excessive plasma power can overheat the substrate, induce ion damage, and generate elevated residual stress in the film, compromising integrity and adhesion. Optimizing power levels is therefore critical to balance throughput, film quality, and substrate compatibility [124].

B-2. RF: RF, typically between 13.56 MHz and 40 MHz, governs the temporal dynamics of the plasma sheath and the energy distribution of ions bombarding the substrate. Higher frequencies shorten the sheath oscillation period, yielding more uniform plasma densities and enhanced conformal coverage over complex topographies. Conversely, lower frequencies increase average ion bombardment energy, improving precursor dissociation and surface activation but also raising substrate heating and potential damage. By fine-tuning the RF, engineers can selectively modulate film density, adhesion, refractive index, and electrical conductivity without exceeding the thermal or damage thresholds of sensitive substrates [125].

B-3. Gas composition ratio: The mixture of inert carriers (argon and nitrogen) and reactive gases (hydrogen, methane, ammonia) defines both the chemical pathways within the plasma and the resulting film stoichiometry. Hydrogen serves as a reducing agent, passivating dangling bonds and lowering defect density, while argon stabilizes plasma density and enhances sputtering of loosely bound species. Carbon- or nitrogen-containing precursors adjust mechanical hardness, refractive index, and chemical resistance. Precise gas ratio control is essential for tailoring film microstructure, growth rate, and process stability, ensuring the repeatable production of coatings ranging from hard protective layers to optically transparent films [126].

B-4. Chamber pressure: Operating pressure between 0.1 torr and 10 torr sets the mean free path for reactive species and shapes boundary-layer thickness at the substrate surface. Low pressures extend thee mean free paths, minimizing gas-phase collisions, suppressing homogeneous nucleation, and improving conformality in high-aspect-ratio features. Higher pressures elevate the partial pressures of reactive species, boosting growth rates but increasing the risk of gas-phase reactions, particle formation, and nonuniform deposition. Careful pressure tuning enables control over film thickness uniformity, surface morphology, and deposition efficiency while maintaining plasma stability and minimizing defect densities [127].

B-5. Substrate temperature: Substrate temperatures from 100 °C to 500 °C govern adatom mobility, nucleation behavior, and overall film morphology during PACVD. Elevated temperatures enhance surface diffusion, resulting in improved crystallinity, larger grain structures, and stronger adhesion by enabling adatoms to find energetically favorable sites. However, excessive heating can deform or damage heat-sensitive substrates and alter underlying device architectures. Conversely, lower temperatures preserve delicate materials but may yield amorphous or porous films due to limited adatom mobility. Precise thermal management using multizone heating and real-time monitoring is vital to balance crystallinity, uniformity, and substrate compatibility in advanced applications [128].

4.4.3 Photo-assisted chemical vapor deposition (photo-CVD)

Photo-induced CVD (photo-CVD, Tab. 4.4, Fig. 4.14A) is an advanced thin-film fabrication technique that uses UV or laser radiation to initiate and control chemical reactions directly at the substrate surface. Unlike conventional thermal CVD, which relies on elevated substrate temperatures to activate precursor molecules, Photo-CVD uses photon energy to drive the process. This fundamental difference allows deposition to occur at much lower substrate temperatures, making it particularly well-suited for temperature-sensitive materials such as polymers, flexible electronic components, and hybrid organic–inorganic systems that would otherwise degrade or deform under high thermal loads [129]. In Photo-CVD, photons interact with gaseous precursor molecules in the vapor phase, breaking specific chemical bonds and generating reactive intermediates. These intermediates then adsorb onto the substrate surface, where they undergo further reactions to form the desired thin film. Because the activation energy is supplied by light rather than heat, the process decouples thermal energy from chemical activation (Fig. 4.14B). This decoupling is a key advantage, as it enables precise control over reaction pathways, minimizes thermal damage, and expands the range of compatible substrates. The photo-CVD is particularly valuable for producing high-performance films in applications where both material quality and substrate integrity are critical. It is widely used for optoelectronic materials, transparent conductive films, dielectric layers, and specialized barrier coatings. The ability to deposit on flexible or low-melting substrates opens opportunities in flexible dis-

Tab. 4.4: A comparison of the five primary parameters of the plasma-assisted CVD technique for exceptional material tunability.

Parameters	Definition in thermal CVD	Primary impact on process	Effect on final film properties
B-1. Plasma power	Electrical power is applied to sustain the plasma discharge, typically 100–500 W	Controls ion and radical densities, influencing deposition kinetics, surface activation, and ion-bombardment intensity	Low power yields slower growth, lower density, and higher porosity; high power promotes densification and faster growth but may cause ion damage, overheating, and residual stress
B-2. Radio frequency	Radio frequency excitation, 13–40 MHz, is used to generate and sustain the plasma	Governs plasma sheath dynamics and ion energy distribution at the substrate	High frequencies improve plasma uniformity and conformal coverage; low frequencies enhance precursor dissociation and adhesion, but increase heating and the risk of substrate damage
B-3. Gas composition ratio	Proportion of inert carrier gases (Ar and N_2) to reactive gases (H_2, CH_4, and NH_3)	Defines plasma chemistry, reaction pathways, and the generation of reactive species	Tailors stoichiometry, hardness, refractive index, and defect density; precise ratios ensure stable growth and properties
B-4. Chamber pressure	The operating pressure during deposition is typically 0.1–10 torr	Sets mean free path, collision frequency, and boundary-layer thickness	Low pressure improves conformality and purity; high pressure increases growth rate but may cause particle formation, nonuniformity, and reduced density
B-5. Substrate temperature	Controlled heating of the substrate, ranging from 100 to 500 °C	Influences adatom mobility, nucleation, and crystallization	High temperatures improve crystallinity, grain size, and adhesion; low *T* protects sensitive substrates but may yield amorphous or porous films

plays, wearable electronics, and advanced packaging. Additionally, the photon-driven nature of the process allows spatially selective deposition by patterning the light source, enabling direct-write fabrication of micro- and nanoscale features without complex lithography steps [130]. Photo-CVD represents a powerful, low-temperature alternative to traditional CVD methods, offering precise control over chemical activation and film properties by manipulating light parameters. Decoupling thermal and chemical processes enables the deposition of high-quality films on a broad range of substrates, including thermally fragile ones. Its flexibility, precision, and compatibility with diverse materials make it an indispensable tool in modern thin-film technology, particularly in fields demanding both functional performance and substrate preservation [131].

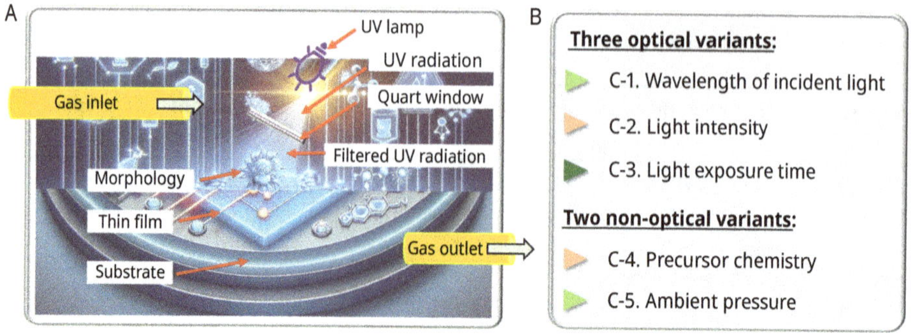

Fig. 4.14: Demonstration of photo-assisted CVD techniques: A. schematic diagram of photo-CVD and B. summary of synthesis parameters.

The technique offers tunability through five key parameters (Fig. 4.14 and Tab. 4.5): *C-1. Wavelength of incident light, C-2. light intensity, C-3. exposure time, C-4. precursor chemistry*, and *C-5. ambient pressure*. The wavelength of incident light determines photon energy and, consequently, which molecular bonds in the precursor can be selectively broken. Deep UV radiation can target high-energy bonds, enabling the activation of complex or refractory precursors that are difficult to decompose thermally. Light intensity governs the rate of photodissociation, influencing the concentration of reactive species in the reaction zone. While higher intensities can accelerate deposition rates and increase throughput, they must be balanced to avoid unwanted side reactions or substrate damage [132]. Exposure duration controls the total energy delivered, affecting film thickness, uniformity, and microstructural evolution. Shorter wavelengths combined with moderate intensities can promote selective activation while maintaining smooth, dense films. In contrast, higher intensities with longer exposures may be used for rapid deposition if conditions are optimized to prevent roughness or defects. Precursor chemistry strongly influences the types of reactive intermediates formed and their surface reactivity; volatile, photoreactive precursors are preferred for efficient deposition, while complex organometallics may require tailored light sources and ambient conditions. Ambient pressure affects the transport of reactive species and their residence time on the substrate. Lower pressures favor ballistic transport and sharp interfaces, while higher pressures can improve conformality but may introduce scattering and reduce film purity. By carefully tuning wavelength, intensity, exposure time, precursor chemistry, and pressure, researchers can precisely control film morphology, density, adhesion, and functional performance. Photo-CVD is particularly valuable for producing optoelectronic materials, transparent conductive films, and specialized coatings where thermal damage must be avoided. The method supports fabrication of high-purity, low-defect films with tailored optical, electrical, and mechanical properties, and is compatible with in situ monitoring for real-time process optimization [133].

C-1. Wavelength of the incident light: In PACVD, the wavelength of the incident light is a primary synthesis parameter because it directly determines the photon energy delivered to the precursor molecules. For effective activation, this photon energy must match or exceed the bond dissociation energy of specific chemical bonds in the precursor. When the wavelength is optimally selected, targeted bonds are cleaved, producing reactive intermediates that drive controlled surface reactions. This selectivity governs which species are incorporated into the growing film, influencing its stoichiometry, phase composition, and crystallographic structure. Shorter wavelengths, such as deep UV, provide higher photon energies capable of breaking stronger bonds, enabling the activation of refractory or complex precursors that are otherwise difficult to decompose thermally. Conversely, longer wavelengths may limit activation to weaker bonds, potentially restricting reaction pathways and altering film chemistry. The chosen wavelength also interacts with other synthesis parameters, such as light intensity, exposure time, precursor chemistry, and ambient pressure, to control growth kinetics, defect density, and interface sharpness, ultimately determining the film's density, morphology, and functional performance [134].

C-2. Light intensity: In photo-CVD, it is a critical synthesis parameter because it directly governs the generation rate of reactive species via photodissociation of precursor molecules. Higher intensities increase photon flux, leading to a greater concentration of radicals and ions in the reaction zone and accelerating deposition rates, potentially improving throughput. This enhanced reactivity can also promote more uniform nucleation across the substrate, aiding in thickness uniformity when combined with appropriate substrate motion and precursor delivery. However, excessive intensity can oversaturate the surface with reactive species, leading to uncontrolled secondary reactions, defect incorporation, or localized heating that damages temperature-sensitive substrates. The optimal intensity must be co-tuned with other parameters, wavelength (to ensure photon energy matches bond dissociation thresholds), exposure time (to control total energy dose), precursor chemistry (to match photoreactivity and volatility), and ambient pressure (which influences species transport and collision frequency). Proper calibration balances reactivity with control, enabling dense, smooth films with precise stoichiometry while avoiding roughness, stress buildup, or compositional drift, ultimately ensuring high-performance coatings tailored to the application [135].

C-3. Exposure time: In photo-CVD, it is a critical synthesis parameter that directly determines the total photon dose delivered to the substrate, thereby influencing both film thickness and microstructural evolution. Longer exposure durations allow more reactive species to interact with the surface, enabling the growth of thicker films and, when well-controlled, improving coverage uniformity. However, excessive exposure without proper adjustment of precursor flow rate can lead to precursor depletion, overgrowth, or surface roughening [136]. The impact of exposure time is closely coupled with other variables: wavelength must be matched to the bond dissociation energy to ensure efficient activation throughout the deposition period; light intensity sets the instantaneous generation rate of reactive species, so higher intensities may require shorter exposures to avoid overheating or defect formation; precursor chemistry dictates how quickly intermediates are consumed or desorb, affecting optimal exposure length; and ambient pressure influences species transport and residence time, modifying how exposure translates into growth. Properly tuned, exposure time balances throughput with interface sharpness, enabling dense, smooth films with controlled morphology and precise thickness for high-performance applications [137].

C-4. Precursor chemistry: In photo-CVD, precursor chemistry is a decisive synthesis parameter because it governs the nature, yield, and reactivity of the intermediates generated upon photon activation. The molecular structure, volatility, and photoreactivity of the precursor determine how efficiently it absorbs incident photons at different wavelengths and how readily it undergoes bond cleavage to form reactive species. Volatile, photoreactive precursors enable high deposition rates and uniform coverage. At the same time, complex organometallics or refractory compounds may require tailored wavelengths, higher light intensity, or modified exposure times to achieve sufficient activation. Precursor chemistry also interacts with ambient pressure, which influences the transport and residence time of intermediates, affecting their likelihood of reaching and reacting at the substrate surface. The choice of precursors directly affects film stoichiometry, phase purity, and defect density: highly pure, well-matched precursors yield dense, compositionally accurate films, whereas poorly matched precursors can lead to incomplete decomposition, contamination, or unwanted phases. By aligning precursor properties with the other synthesis parameters, photo-CVD can deliver films with controlled microstructure, sharp interfaces, and tailored functional performance for optical, electronic, or barrier applications [138].

C-5. Ambient pressure: In photo-CVD, ambient pressure is a critical synthesis parameter because it dictates the mean free path, collision frequency, and residence time of reactive species, thereby shaping both transport dynamics and surface reaction behavior. At lower pressures, extended mean free paths enable ballistic transport, allowing photogenerated intermediates to retain higher kinetic energy upon reaching the substrate. This promotes sharp interfaces, high film purity, and minimal scattering losses, but can reduce conformality in high-aspect-ratio structures due to limited angular spread. In contrast, higher pressures increase gas-phase collisions, broadening the angular distribution of arriving species and improving sidewall coverage. However, this also reduces their kinetic energy, which can lower film density, introduce impurities, or increase defect incorporation. Pressure effects are strongly coupled with other synthesis parameters: wavelength and light intensity determine species generation rates; exposure time influences total surface interaction; and precursor chemistry affects volatility and reactivity under varying pressures. Optimizing ambient pressure in concert with these variables enables precise control over conformality, density, stoichiometry, and defect levels, ensuring films meet both structural and functional performance requirements [139].

Tab. 4.5: Comparison of the photo-CVD technique's five primary parameters for exceptional material tunability.

Parameters	Definition in photo-CVD	Primary impact on process	Effect on final film properties
Wavelength (λ) of incident light	The photon energy determined by the light's λ must match or exceed the bond dissociation energy of the precursor molecules	Selectively breaks specific chemical bonds, controlling which reactive intermediates are generated	Governs film composition, phase structure, and stoichiometric accuracy; shorter wavelengths enable activation of refractory precursors and sharper interfaces
Light intensity	The radiant power delivered per unit area to the reaction zone	Controls the rate of photodissociation, throughout, and reactive species generation	Higher intensity increases deposition rate; excessive intensity risks substrate damage, roughness, or side reactions
Light exposure time	The total duration the substrate is illuminated during deposition	Determines total photon dose and cumulative energy delivered to the surface	Longer exposures increase film thickness and can improve uniformity; overexposure may cause overgrowth, roughness, or defect formation
Precursor chemistry	The molecular composition, volatility, and photo-reactivity of the source gases	Influences the type and reactivity of intermediates formed upon photon activation	Affects stoichiometry, phase purity, and microstructure; volatile, photoreactive precursors yield smoother, denser films
Ambient pressure	The total pressure in the deposition chamber during processing	Modulates transport dynamics, mean free path, and surface residence time of reactive species	Low pressure favors ballistic transport and sharp interfaces; higher pressure improves conformality

4.4.4 Gas-assisted chemical vapor deposition (GACVD)

GACVD (Fig. 4.15A) is a versatile thin-film fabrication technique in which one or more reactive "assisting" gases are introduced alongside the primary precursor feed to actively influence the chemical reactions occurring during film growth. Unlike conventional CVD, where the precursor decomposition pathway is primarily dictated by temperature and inherent chemistry, GACVD leverages the chemical reactivity of the assisting gas to modify reaction kinetics, alter intermediate species, and enable film formation under milder thermal conditions. This makes it particularly valuable for depositing high-quality films on temperature-sensitive substrates, improving stoichiometric control, and enhancing the incorporation of specific elements into the growing layer [140]. The method's flexibility stems from the ability to tailor the assisting gas chemistry to the target material system. GACVD's adaptability makes it a powerful tool for engineering films with application-specific properties. In catalysis, it enables active oxide or nitride surfaces with controlled porosity. In energy storage, it supports conformal coatings on battery electrodes to enhance stability. For protective coatings, it delivers dense, adherent barrier layers with tailored hardness and corrosion resistance (Fig. 4.15B). In advanced microelectronics, GACVD allows precise dielectric, conductive, or semiconducting layers with minimal thermal budget, supporting next-generation device architectures. By strategically selecting the assisting gas and optimizing its delivery parameters, GACVD provides a finely tunable platform for achieving targeted film compositions, microstructures, and functional performance that are difficult to reach with conventional CVD alone [141].

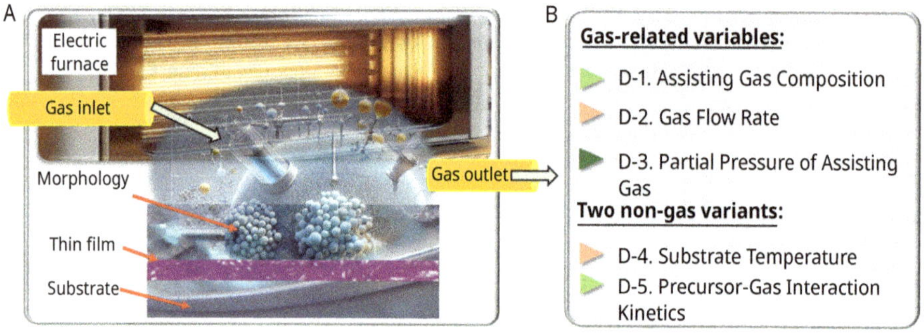

Fig. 4.15: Demonstration of gas-assisted CVD techniques: A. schematic diagram of GACVD and B. summary of synthesis parameters.

Five key technical synthesis variables, *D-1. assisting gas composition, D-2. gas flow rate, D-3. partial pressure of assisting gas, D-4. substrate temperature, D-5. precursor–gas interaction kinetics* govern GACVD and, together, determine both the process dynamics and the properties of the resulting films, as illustrated in Fig. 4.15B and Tab. 4.6. The first is assisting gas composition, whose chemical identity of the assisting gas, such as O_2, H_2, NH_3, H_2O, or halogens, defines the dominant reaction pathways, oxidation, reduction, nitridation, or etching, and directly influences film composition, phase formation, and functional performance. The second variable is the gas flow rate, which regulates the supply of reactive species and the removal of by-products. Higher flow rates can accelerate reaction kinetics and improve thickness uniformity, whereas excessive flow may dilute the precursor concentration or trigger undesirable gas-phase reactions. The third variable is the partial pressure of the assisting gas, which balances surface-limited and gas-phase reaction regimes [142]. Elevated partial pressures enhance the incorporation of the assisting element but may increase scattering or particle formation, while lower pressures preserve interface sharpness but can reduce conformality. The fourth variable is substrate temperature, which works synergistically with the assisting gas to control reaction activation, crystallinity, and defect density. Lower temperatures protect heat-sensitive substrates, while higher temperatures promote grain growth and phase stability. The fifth variable is precursor–gas interaction kinetics. The chemical compatibility and reaction rate between the primary precursor and the assisting gas govern intermediate formation, incorporation efficiency, and the likelihood of side reactions, ultimately determining stoichiometry, microstructure, and defect levels. Collectively, these five core synthesis variables assist with gas composition, gas flow rate, partial pressure, substrate temperature, and precursor–gas interaction kinetics in controlling reaction pathways and growth dynamics [143]. The assisting gas chemistry and delivery conditions dictate the type and rate of surface reactions, while temperature and precursor compatibility influence crystallinity, stoichiometry, and defect density. By precisely tuning these parameters in concert, GACVD can produce films with tailored composition, microstructure, and functional properties for diverse advanced applications [144].

D-1. Assisting gas composition: In GACVD, the assisting gas composition is a primary determinant of the chemical pathways that dominate film growth. The identity of the gas, whether oxygen, hydrogen, ammonia, water vapor, or halogens, dictates whether oxidation, reduction, nitridation, hydrolysis, or etching reactions occur at the substrate surface. Oxygen-based assists promote oxide formation with improved stoichiometry and dielectric performance, while hydrogen assists can reduce oxides, remove ligands, and enhance grain growth in metals and semiconductors. Ammonia provides nitrogen for nitride films, enabling hardness, thermal stability, and tailored electrical properties. Water vapor assists in hydroxylation and oxide growth, improving adhesion and conformality, particularly in porous or high-aspect-ratio structures. Halogen gases can activate surfaces or enable selective area deposition through in situ etching. The chosen composition must be compatible with the primary precursor chemistry to avoid unwanted side reactions or contamination. Its selection directly impacts film phase, crystallinity, defect density, and functional performance, making it a critical lever for tailoring properties to meet application-specific requirements in electronics, optics, catalysis, and protective coatings [145].

D-2. Gas flow rate: The assisting gas flow rate in GACVD governs the delivery of reactive species to the substrate and the removal of gaseous by-products from the reaction zone. An optimal flow rate ensures a steady supply of active intermediates while preventing precursor depletion or stagnation, which could lead to non-uniform growth. Higher flow rates can accelerate reaction kinetics, improve across-wafer uniformity, and enhance the replenishment of reactive species in recessed features. However, excessively high flow can dilute precursor concentration, shift reaction equilibria, or promote unwanted homogeneous gas-phase reactions that generate particles. Conversely, too low a flow rate may limit the availability of reactive species, slow deposition rates, and cause compositional gradients. The ideal flow rate depends on the composition of the assisting gas, precursor volatility, reactor geometry, and substrate temperature. It must be balanced with partial pressure to maintain the desired surface-limited reaction regime. Properly tuned, gas flow rate contributes to consistent film thickness, controlled microstructure, and reproducible material properties across diverse GACVD applications [146].

D-3. Partial pressure of assisting gas: Partial pressure of the assisting gas is a decisive parameter in GACVD because it modulates the balance between surface-limited and gas-phase reaction pathways. At higher partial pressures, the concentration of the assisting species increases, enhancing the incorporation of the desired element, such as oxygen in oxides or nitrogen in nitrides, and accelerating reaction rates [147]. This can improve conformality in complex geometries by broadening the angular distribution of reactive species. However, elevated partial pressures also increase collision frequency, which can scatter species, reduce their kinetic energy, and promote gas-phase nucleation, leading to particle contamination or reduced density. Lower partial pressures favor ballistic transport, preserving interface sharpness and film purity, but may limit step coverage in high-aspect-ratio features. The optimal partial pressure depends on the assisting gas chemistry, precursor reactivity, substrate temperature, and desired film properties. Careful control ensures the right trade-off between conformality, density, and stoichiometric precision, enabling GACVD to produce films with both structural integrity and functional performance tailored to the application [148].

D-4. Substrate temperature: Substrate temperature in GACVD plays a central role in determining reaction activation, film crystallinity, and defect density. It works in synergy with the assisting gas to control the energy available for surface reactions and adatom mobility. Lower substrate temperatures are advantageous for heat-sensitive substrates such as polymers or certain device layers, enabling deposition without thermal degradation. However, low temperatures can limit surface diffusion, potentially leading to amorphous or porous films unless compensated by reactive assisting gases or plasma activation. Higher substrate temperatures promote grain growth, phase stability, and densification, often improving electrical, optical, and mechanical properties. Yet, excessive heating can cause interdiffusion, stress buildup, or damage to underlying layers. The optimal temperature is determined by the precursor chemistry, assisting gas composition, and desired microstructure. It must also be balanced with partial pressure and flow rate to maintain controlled reaction kinetics. Proper temperature control ensures films meet stringent requirements for crystallinity, adhesion, and functional performance in advanced applications [149].

D-5. Precursor–gas interaction kinetics: Precursor–gas interaction kinetics define how efficiently the primary precursor reacts with the assisting gas to form the desired film. This parameter encompasses the chemical compatibility between the two species, the activation energy of their reaction, and the rate at which reactive intermediates are generated. Fast, well-matched interactions can yield high deposition rates, precise stoichiometry, and dense microstructures. Poorly matched chemistries may result in incomplete reactions, contamination, or the formation of unwanted secondary phases. The kinetics are influenced by assisting gas composition, wavelength, or energy source (in photo- or plasma-assisted variants), substrate temperature, and ambient pressure. For example, highly reactive oxygen can rapidly oxidize metal precursors, while ammonia requires sufficient activation to incorporate nitrogen effectively. The interaction rate also affects by-product formation and removal, impacting film purity and interface quality. Optimizing precursor–gas kinetics in concert with other synthesis variables allows GACVD to deliver films with tailored composition, microstructure, and functional properties, meeting the performance demands of electronics, optics, catalysis, and protective coating applications [150].

Tab. 4.6: The condensed comparative table for the five GACVD synthesis variables is organized into the four requested columns for clarity and publication-ready use.

Parameters	Definition in GA-CVD	Primary impact on process	Effect on final film properties
D-1. Assisting gas composition	The chemical identity of the assisting gas (O_2, H_2, NH_3, H_2O, and halogens) introduced alongside the primary precursor	Determines dominant reaction pathways, such as oxidation, reduction, nitridation, hydrolysis, or etching, and influences the formation of intermediate species	Controls film composition, phase structure, and functional attributes, including dielectric constant, hardness, optical transparency, catalytic activity, and corrosion resistance
D-2. Assisting gas flow rate	The volumetric delivery rate of the assisting gas into the reaction zone	Regulates the supply of reactive species and removal of by-products; influences reaction kinetics and uniformity	Affects thickness uniformity, deposition rate, and defect density; excessive flow may dilute precursors or cause particle contamination
D-3. Partial pressure of assisting gas	The proportion of total chamber pressure contributed by the assisting gas	Balances surface-limited versus gas-phase reactions; modifies collision frequency and angular distribution of species	Higher pressures improve conformality but may reduce density and purity; lower pressures preserve interface sharpness but limit step coverage
D-4. Substrate temperature	The controlled temperature of the substrate during deposition	Works with assisting gas chemistry to activate reactions, enhance adatom mobility, and influence crystallization	Determines crystallinity, grain size, stress state, and adhesion; low temperature protects sensitive substrates, high T improves phase stability and density
D-5. Precursor–gas interaction kinetics	The reaction rate and compatibility between the primary precursor and the assisting gas are essential	Governs intermediate formation, incorporation efficiency, and likelihood of side reactions	Controls stoichiometry, microstructure, and defects; optimized kinetics produce dense, precise, high-performance films

4.5 CVD-derived nanomaterials and structural tunability

A central challenge in contemporary nanotechnology is the scalable synthesis of nanomaterials with controlled structure, composition, and functional properties. Achieving large-scale production while maintaining nanoscale precision is essential for translating laboratory advances into industrial applications. Among the available fabrication techniques, CVD has emerged as one of the most widely adopted and versatile methods for producing high-quality nanostructures. Its ability to deliver uniform, high-purity materials with tunable morphology makes it particularly suited for the synthesis of graphene, carbon nanotubes (CNTs), nanofibers, fullerenes, and other nanoscale polymers [151]. In CVD, volatile precursors are introduced into a controlled reaction environment, where they decompose or react at a substrate surface to form the desired nanomaterial. By adjusting parameters such as temperature, pressure, gas composition, and flow rate, researchers can tailor the size, crystallinity, and alignment of the resulting structures [152]. The following sections present representative examples of CVD-based synthesis routes for carbon-based materials. Five examples, graphene, CNTs, carbon nanofibers (CNFs), and fullerenes, highlighting their process conditions, growth mechanisms, and application potential, will be discussed in this section. ***A. Graphene***: Graphene, a single atomic layer of graphite, is a novel two-dimensional building block with exceptional mechanical, thermal, and electrical properties. It is widely studied for applications in energy harvesting, storage, strain sensing, and the steel industry. ***B. CNTs***: CNTs are a carbon allotrope formed by seamlessly rolling graphene sheets into cylindrical structures. These cylinders are held together by van der Waals interactions, giving CNTs remarkable mechanical, electrical, and thermal properties. ***C. CNFs***: CNFs are cylindrical structures formed by stacking single or double graphene layers into cones, cups, or plates [153]. They are carbon allotropes with diameters typically ranging from 50 to 200 nm. ***D. Fullerene***: Fullerene, an allotrope of carbon, forms a closed or partially closed cage linked by carbon–carbon single and double bonds. Its structure consists of fused pentagonal and hexagonal rings, forming a stable, symmetrical molecular framework. The subsequent sections provide a detailed examination of these four carbon allotropes, with emphasis on their structural characteristics and fundamental physicochemical properties [154].

A. Graphene: Graphene, a single atomic layer of graphite, is a novel two-dimensional building block for numerous nanomaterials, distinguished by exceptional mechanical strength, thermal conductivity, and electrical performance. It has attracted significant interest for applications in energy harvesting and storage, strain sensing, and steel manufacturing [155]. The first successful isolation of a single graphene layer was reported in 2004, achieved by pressing cleaved graphite onto an oxidized SiO_2 surface [156]. Following this breakthrough, research efforts focused on improving the quality of graphene sheets using techniques such as electrostatic-voltage-assisted exfoliation and solution-based diffusion [157]. At present, large-scale synthesis of high-quality graphene is predominantly accomplished via CVD using two principal routes [158]. The first involves depositing a carbon-containing gas onto the surface of a metal catalyst. For instance, preheating a H_2–CH_4 gas mixture to 1,000 °C facilitates the growth of graphene layers on a nickel surface before saturation and subsequent graphite formation [159]. The second route relies on carbon segregation at the surface of a metastable carbon–metal solid solution, demonstrated with transition metals such as Ni, Pt, Re, Pd, Ru, and Ir [160]. Both approaches enable controlled graphene growth, with process parameters tailored to optimize layer uniformity, crystallinity, and scalability for advanced applications. During annealing and cooling, carbon impurities segregate and diffuse to the surface of the bulk metal [161]. In this process, the solubility of carbon in the metal–carbon solution, along with temperature and other growth parameters [162], serves as a key set of controllable factors governing the morphology and thickness of the resulting graphene films. Li et al. [163] demonstrated the growth of large-area, high-quality, uniform graphene films by precisely regulating the carbon content in a copper-carbon solution to control the film growth rate. Through this approach, they successfully synthesized centimeter-scale single-layer graphene on copper foil. Building on this work, subsequent efforts exploited the inherent flexibility of copper foil to enable high-speed roll-to-roll transfer of graphene, achieving continuous sheets up to 30 inches in length [163]. These advances highlight the importance of optimizing process parameters to scale graphene production while maintaining structural integrity, uniformity, and performance for industrial and technological applications.

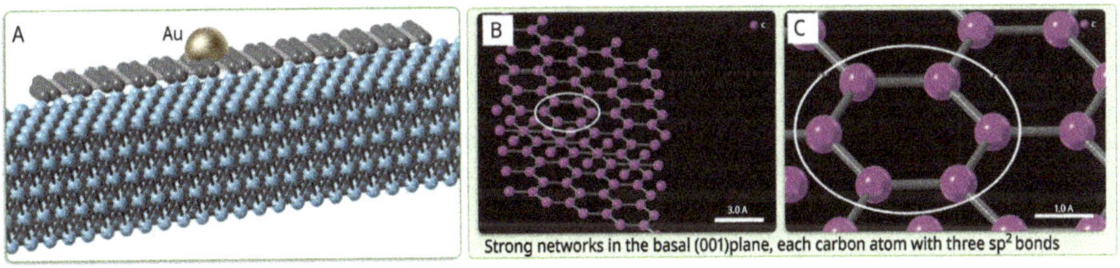

Fig. 4.16: The schematic diagram of Ni-supported graphene: A. the Au single atom on top acts as a catalyst; B. the ball-and-stick model shows each carbon atom with three sp^2 bonds, forming strong networks in the basal (001) plane; C. the enlarged hexagonal unit cell. Crystal structure data taken from Wyckoff [165].

In addition to copper, nickel, another relatively inexpensive polycrystalline substrate, has attracted considerable attention for large-scale graphene film deposition [165]. Early studies demonstrated the synthesis of 1–12 graphene layers on nickel films via ambient-pressure CVD [166]. Another investigation reported the formation of single- and few-layer graphene over regions spanning only several tens of micrometers on nickel surfaces [167], attributed to rapid carbon segregation from nickel carbide during cooling and the influence of heterogeneous grain boundaries. Carbon isotope labeling studies have revealed two distinct, metal-dependent growth mechanisms for graphene on metallic substrates [168]. For nickel, carbon-containing gaseous precursors decompose at elevated temperatures, and the resulting carbon atoms dissolve into the catalyst, mix to near saturation, and subsequently segregate to the surface during cooling. In contrast, copper exhibits low carbon solubility at high temperatures, eliminating the dissolution step; carbon atoms instead segregate and grow directly on the surface [169]. On copper, monolayer graphene formation is self-limiting due to surface blockage by the initial layer [170]. In nickel, carbon solubility is temperature-dependent, and precipitation during cooling governs layer thickness [171]. Precise cooling control is therefore critical for achieving single-layer films. Optimal CVD growth occurs near 550 °C; temperatures above 600 °C promote excessive carbon dissolution, while those below 500 °C favor competing surface carbide formation, hindering graphene growth [172]. Leveraging the contrasting carbon solubilities of copper and nickel, Cu–Ni binary alloys have been explored for CVD synthesis [173]. Increasing nickel content enhances graphene layer thickness and uniformity, and careful adjustment of alloy composition has yielded over 95% single-layer and 91% bilayer graphene films [174].

B. ***CNTs***: CNTs are a unique carbon allotrope formed by seamlessly rolling graphene sheets into cylindrical nanostructures stabilized by van der Waals forces. They exhibit extraordinary tensile strength, high Young's modulus, exceptional thermal conductivity (~3,000 W/m K), and remarkable electrical properties, ranging from metallic to semiconducting depending on chirality. CNTs are classified as single-walled (SWCNTs) or multi-walled (MWCNTs), with diameters from sub-nanometer to ~100 nm and lengths up to several centimeters. Their high aspect ratio, chemical stability, and tunable surface chemistry make them well-suited for applications in nanoelectronics, composite reinforcement, energy storage, sensors, and field-emission devices [175]. Based on wall count, CNTs are classified as SWCNTs (Fig, 4.17.1A-1 and A-2) and MWCNTs (Fig. 4.17.1B-1 and B-2). SWCNTs consist of a single seamless graphene cylinder with diameters typically ranging from 0.4 to 3 nm and lengths extending up to several centimeters. They are defined by their chiral vector (n,m), which governs electronic band structure and thus determines whether they exhibit metallic or semiconducting behavior. Additional key parameters include aspect ratio, defect density, and wall smoothness, all of which influence electrical conductivity, carrier mobility, optical absorption, thermal conductivity, and mechanical strength. MWCNTs consist of multiple concentric graphene cylinders, with adjacent walls separated by an interlayer spacing of approximately 0.34 nm, analogous to turbostratic graphite [176]. They are further characterized by the number of walls, which can range from a few to several dozen, overall outer diameter (typically 2–100 nm), and degree of structural alignment along the tube axis.

Additional key parameters include wall integrity, defect density, and interwall coupling, all of which influence their mechanical strength (tensile strength >60 GPa), electrical conductivity, thermal conductivity (>3,000 W/m K), and chemical stability. The combination of high aspect ratio, tunable electronic properties, and exceptional load-bearing capacity makes MWCNTs suitable for applications in structural composites, nanoelectronics interconnects, thermal interface materials, and energy storage devices [177]. SWCNTs can range in length from a few nanometers to nearly 20 cm [178] and typically exhibit diameters between 0.4 and 3 nm [179], offering exceptionally high aspect ratios that influence mechanical flexibility, electrical transport, and thermal conduction. MWCNTs consist of multiple concentric graphene cylinders and can reach outer diameters of up to ~100 nm [180], with an interlayer spacing of approximately 0.34 nm analogous to turbostratic graphite. CNTs are further classified by their chiral vector zigzag, armchair, or chiral defined by the (*n*,*m*) indices of the graphene lattice roll-up, which dictate electronic band structure, thereby determining whether the nanotube exhibits metallic or semiconducting behavior, as well as influencing optical absorption, carrier mobility, and anisotropic mechanical properties [181], as shown in Fig. 4.17.1a.

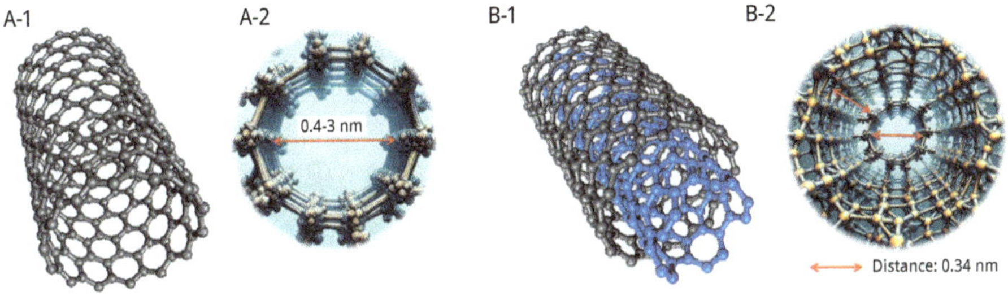

Fig. 4.17.1: Structural representation of carbon nanotube: A. single-walled (A-1 and A-2) nanotubes, and B. multi-walled (B-1 and B-2) carbon nanotubes. Images of A-1 and B-1 were reproduced with permission from [182]. Copyright @ American Chemical Society 2009. Images of A-2 and B-2 are created using Microsoft Designer (Sept 2025), a licensed product.

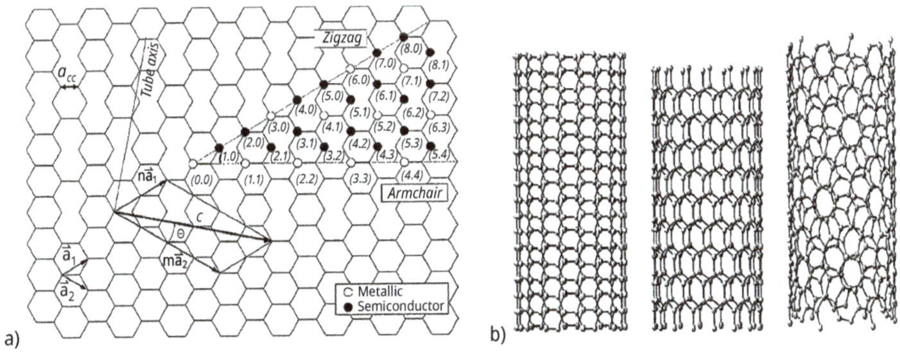

Fig. 4.17.2: Use the chair vector (a) to obtain the three configurations (b) of zigzag (left), armchair (middle), and chair (right) nanotubes. Reproduced from ref. [183]. Copyright @ American Chemical Society 2007.

In addition to laser ablation and electric arc discharge, CVD synthesis of CNTs has been established as an effective method for large-scale production [184]. The high yield of CNTs requires a relatively low temperature (500–1,000 °C). The CNT morphology and structure can be controlled by varying the alignment angle, the number of walls, the diameter, and the length [185]. A wide range of catalysts, including Fe, Mo, Co, Ni, and other metal-containing species, has been studied in the CVD synthesis of CNTs [186]. The most common CVD methods are thermal CVD, plasma-assisted CVD, aerosol-assisted CVD, and catalytic pyrolysis of hydration. Early studies developed floating-catalyst CVD for the continuous production of CNTs. This method simultaneously introduced the catalyst precursors and the carbon feedstock gas into the reactor using a carrier [187]. In this design, continuous production of CNT was reported by the in situ decomposition of ferrocene of the fluidized Fe/MgO, which facilitated the discharging of CNTs with the formation of small iron nanoparticles [188]. Another study also reported the utilization of small cobalt nanoparticles as catalysts to produce highly pure SWCNTs [189]. In this study, a small oxygen flow was introduced into the reactor to remove unreacted catalyst particles, yielding high-purity SWCNTs with few severe defects. Lehman and other researchers grew MWCNTs on a pyroelectric detector using hot-wire CVD with a nickel film as the catalyst [190] Fig. 4.17.1b. The reactor is enclosed in a clamshell furnace to produce CNTs with good resistance to higher temperatures.

Multiple groups have also reported aerosol-assisted CVD synthesis of CNTs due to the effective dissolving of liquid hydrocarbons into the metal catalysts [191]. Aerosol-assisted CVD produces free-standing, vertically aligned, MWCNTs using benzene derivatives as precursors and metallocenes as catalysts, which can finely tune CNT morphology and properties. Similar efforts have been made by injecting a ferrocene solution into toluene to control the catalyst-to-carbon ratio [192]. In this study, researchers found that the length of MWCNTs increased, but the growth rate decreased with the injection time. A detailed investigation was conducted into the synthesis of MWCNTs from various carbon sources using cobaltocene and nickelocene as catalysts [193–196]. The combination of ferrocene and nickelocene yielded the highest CNT yield. In addition, Hayashi and coworkers [197] further improved the floating reactant methods using zeolite particles as the floating catalyst to synthesize the smallest free-standing SWCNTs. Those CNTs were so small, with a diameter of ~0.43 nm, that they could be treated as a 1D material.

C. ***CNFs***: CNFs are cylindrical nanostructures, exemplified by the morphology depicted in Fig. 4.18A. The CNFs are synthesized by stacking one or two graphene layers into distinct morphologies, conical, cup-stacked, herringbone, fishbone, "gil-step" (variant fishbone), or platelet, each defined by the orientation of graphene planes relative to the fiber axis and the resulting structural geometry (Fig. 4.18B–H and Tab. 4.7), each defined by the orientation of graphene planes relative to the fiber axis [195]. They represent a distinct carbon allotrope with diameters typically ranging from 50 nm to 200 nm and lengths extending from micrometers to millimeters, depending on the synthesis conditions. Compared with CNTs, CNFs possess a lower density (~1.3–1.6 g/cm^3), higher compressive strength (>2 GPa), and greater defect density, which increases active edge sites and enhances interfacial bonding in composite matrices. Their elevated surface roughness, tunable functional groups, and defect-induced reactivity improve load transfer efficiency, making CNFs advantageous for reinforcing polymer, metal, and ceramic composites in structural and multifunctional applications [196]. They are also more cost-effective to produce, require simpler synthesis protocols, and are more readily available, making them attractive for large-scale applications. CNFs possess tensile strengths up to ~7 GPa, Young's modulus in the range of 100–600 GPa, electrical conductivity up to 10^3 S/cm, and thermal conductivity exceeding 200 W/m K. Their tunable surface chemistry enables functionalization for targeted applications, and they have attracted significant interest in high-performance polymer and metal–matrix composites, energy storage electrodes, thermal interface materials, filtration membranes, and catalytic supports [197].

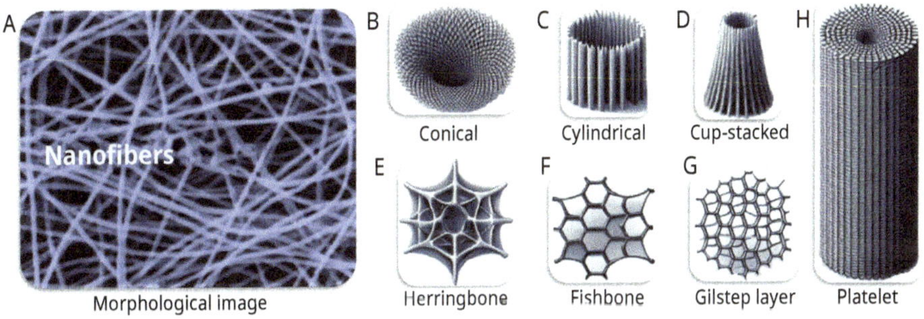

Fig. 4.18: Demonstration of carbon nanofibers (CNFs): A. image of nanofibers and B. schematic diagrams of CNF shapes.

Catalytic CVD is one of the most versatile and scalable methods for synthesizing CNFs, relying on the catalytic decomposition of small hydrocarbon feedstocks, such as methane, carbon monoxide, ethylene, ethane, propane, acetylene, or benzene, over transition metal catalysts, including Fe, Co, Ni, or their alloys [198]. Similar to CNT growth, the CVD process offers precise control over CNF diameter, length, alignment, and morphology by carefully adjusting process variables such as reaction temperature, hydrocarbon-to-carrier gas ratio, total gas flow rate, reactor pressure, and catalyst particle size distribution. Carbon conversion from gaseous precursors typically occurs within the temperature range of 700–1,200 K [199], where hydrocarbon molecules undergo catalytic decomposition to release carbon atoms. These atoms either dissolve into the bulk of the metal catalyst particles or react to form metastable metal carbides, depending on the catalyst composition and temperature. Upon supersaturation, carbon precipitates from the catalyst surface in the form of graphitic layers, which assemble into the

Tab. 4.7: Morphological variants of carbon nanofibers (CNFs) are formed by stacking one or two graphene layers in distinct orientations relative to the fiber axis.

Label	Morphology	Technical description
B	Conical	Graphene layers are inclined toward the fiber axis, forming a continuous cone-like taper with high axial alignment and gradual diameter reduction
C	Cylindrical	Graphene layers parallel to the fiber axis, producing a smooth, tube-like structure with minimal taper and high structural uniformity
D	Cup-stacked	Sequential truncated conical graphene segments ("cups") stacked end to end, with open interiors that narrow toward one end, provide high surface area
E	Herringbone	Graphene layers angled alternately to the fiber axis, creating a chevron-like pattern, offer increased defect density and active edge sites
F	Fishbone (open)	Graphene layers are inclined from the core toward the surface, producing an open skeletal appearance with exposed edge planes
G	Fishbone (solid)/ "gilstep"	Like fishbone but with denser packing of graphene layers, resulting in a more solid cone-like form; "gilstep" refers to stepped, tightly packed layer transitions
H	Platelet	Graphene layers stacked perpendicular to the fiber axis, forming segmented, plate-like cross sections with high edge exposure

characteristic CNF morphologies. The resulting CNF structure is determined not only by macroscopic growth parameters but also by nanoscale catalyst characteristics, including crystallographic orientation, particle size, surface energy, and shape anisotropy. For example, spherical catalyst particles tend to promote cylindrical or conical CNFs, whereas faceted particles can induce herringbone or platelet morphologies [200]. The angle of graphene layer stacking relative to the fiber axis governs defect density, surface area, and the proportion of exposed edge sites, which in turn influence mechanical reinforcement capability, electrical conductivity, and chemical reactivity. Higher growth temperatures generally enhance graphitization, improving thermal conductivity (>200 W/m K) and electrical conductivity (up to 10^3 S/cm). In comparison, lower temperatures can increase defect density, which is beneficial for catalytic or adsorption applications. By tailoring CVD parameters, porous CNFs can be engineered to achieve specific combinations of tensile strength (up to ~7 GPa), Young's modulus (100–600 GPa), and surface functionality to meet the requirements of targeted applications. These include structural reinforcement in polymer and metal–matrix composites, conductive networks in energy storage electrodes, thermally conductive fillers in heat management systems, and high-surface-area supports for heterogeneous catalysis. Thus, CVD remains a critical synthesis route for producing CNFs with tunable properties and morphologies optimized for both performance and scalability [201].

CNFs are synthesized via catalytic CVD, in which hydrocarbon feedstocks decompose on heated transition metal catalysts such as Fe, Co, or Ni, as mentioned above. The CNF growth process comprises eight distinct stages, each contributing to the structural evolution of CNFs: ***1. Hydrocarbon introduction***, ***2. catalyst activation***, ***3. hydrocarbon decomposition***, ***4. carbon dissolution/carbide formation***, 5. ***supersaturation***, 6. ***graphitic layer precipitation***, 7. ***fiber elongation***, and 8. ***termination***. During decomposition, carbon atoms are released and either dissolve into the catalyst or form metastable metal carbides; upon reaching supersaturation, these atoms precipitate as graphitic layers. The resulting fiber morphology, conical, cup-stacked, herringbone, fishbone, or platelet, is governed by catalyst geometry, reaction temperature, and gas-phase composition. Continuous carbon supply drives axial elongation, while growth terminates upon feedstock exhaustion, catalyst encapsulation, or deactivation, thereby preserving the final CNF structure and properties [202]. A detailed stepwise explanation of each stage in the catalytic growth mechanism is presented in Tab. 4.8 and Fig. 4.19 and elaborated in the following section. In short, the CNFs exhibit tunable thermal conductivity (>200 W/m K), electrical conductivity (up to 10^3 S/cm), and mechanical strength (up to ~7 GPa) depending on the CVD growth conditions (Fig. 4.19A). Their adjustable defect density and surface functionality make them ideal for applications in structural composites, energy storage, thermal management, and catalysis. Key process variables influence the catalytic CVD growth of CNFs [203]. Each parameter, ranging from catalyst type and particle size to temperature and gas composition, affects the resulting CNF morphology, structural integrity, and functional properties, including conductivity, defect density, and mechanical strength (Fig. 4.19B) [204].

Step 1. Hydrocarbon introduction: A controlled mixture of hydrocarbon feedstock gases, such as methane (CH_4), acetylene (C_2H_2), or ethylene (C_2H_4), is introduced into a heated reaction chamber along with inert or reducing carrier gases, such as argon, nitrogen, or hydrogen. The carrier gas regulates partial pressure, controls residence time, and ensures uniform dispersion of reactive species. This mixture flows over the catalyst bed, initiating the chemical environment required for CNF formation. Proper gas composition and flow dynamics are critical for consistent carbon delivery and morphology control [205].

Step 2. Catalyst activation: Transition metal nanoparticles (Fe, Co, Ni, or their alloys), typically supported on substrates such as SiO_2 or Al_2O_3, are thermally activated by heating to 700–1,200 K. This temperature range enhances surface mobility, increases catalytic reactivity, and facilitates carbon solubility. The metal particles restructure slightly under thermal stress, exposing active sites for hydrocarbon adsorption. Catalyst activation is essential for initiating decomposition and controlling the fiber nucleation step [206].

Step 3. Hydrocarbon decomposition: Hydrocarbon molecules adsorb onto the surface of the activated catalyst particles, where catalytic dissociation cleaves C–H bonds. This reaction releases carbon atoms and hydrogen gas, with the decomposition rate governed by feedstock type, temperature, and catalyst composition. Efficient decomposition

Tab. 4.8: Key process variables influencing the catalytic chemical vapor deposition (CVD) growth of carbon nanofibers (CNFs).

Variables	Range/example	Influence on CNF properties
Catalyst type	Fe, Co, Ni, Fe–Co, and Fe–Ni	Determines carbon solubility, carbide formation tendency, and growth mode
Particle size	5–200 nm	Controls CNF diameter: smaller particles yield finer fibers
Reaction T	700–1,200 K	Higher $T \rightarrow$ greater graphitization, conductivity; lower $T \rightarrow$ higher defect density, more active sites
Feedstock gas	CH_4, C_2H_4, C_3H_8, and C_6H_6	Affects decomposition rate, carbon yield, and morphology
Carrier gas	Ar, N_2, and H_2	Dilutes feedstock, controls partial pressure, and influences defect density
Gas flow rate	10–500 sccm	Impacts residence time, carbon supply rate, and fiber uniformity
Pressure	1 torr to 1 atm	Alters diffusion kinetics and carbon precipitation rate
Growth time	Seconds to hours	Controls fiber length and degree of alignment
Support	SiO_2, Al_2O_3, and MgO	Affects catalyst dispersion, stability, and morphology control
Cooling rate	Rapid versus slow	Influences precipitation dynamics, defect density, and wall orientation

ensures a steady carbon supply for subsequent growth stages. The nature of the hydrocarbon influences carbon yield and defect density in the final CNF structure [207].

Step 4. Carbon dissolution | carbide formation: Released carbon atoms either dissolve into the bulk of the catalyst or react to form metastable metal carbides, depending on the metal's carbon solubility and the reaction temperature. This internal carbon accumulation sets the stage for nucleation and growth. The balance between dissolution and carbide formation affects precipitation dynamics and fiber morphology. Catalyst composition and thermal stability are key to this phase [208].

Step 5. Supersaturation: As the carbon concentration in the catalyst increases, it reaches a threshold, triggering precipitation. Carbon atoms begin to segregate at specific crystallographic facets, initiating the formation of graphitic layers. Supersaturation dynamics influence growth rate, wall orientation, and structural uniformity. Controlled saturation ensures consistent fiber morphology and minimizes unwanted amorphous carbon deposition [209].

Step 6. Graphitic layer precipitation: Carbon atoms precipitate from the catalyst surface as sp^2-bonded graphitic layers, forming the structural backbone of CNFs. The stacking orientation, conical, cup-stacked, herringbone, fishbone, or platelet, is determined by catalyst geometry, surface energy, and local growth kinetics. These morphologies influence mechanical strength, electrical conductivity, and surface area. Layer arrangement also affects defect density and functionalization potential [210].

Step 7. Fiber elongation: Continuous carbon supply sustains axial fiber growth, with elongation driven by ongoing precipitation at the catalyst interface. The growth rate and final morphology are dictated by catalyst size, shape, temperature, and gas composition. Stable catalyst activity ensures long, uniform fibers with consistent structural properties. The elongation phase determines fiber length, alignment, and suitability for application [211].

Step 8. Termination: Growth terminates when hydrocarbon feedstock is removed, catalyst activity declines, or encapsulation blocks active sites. Catalyst deactivation may result from sintering, poisoning, or carbon overcoating. The final CNF structure is locked in at this stage, preserving its morphology and properties. Controlled termination is essential to prevent structural defects and ensure reproducibility [212].

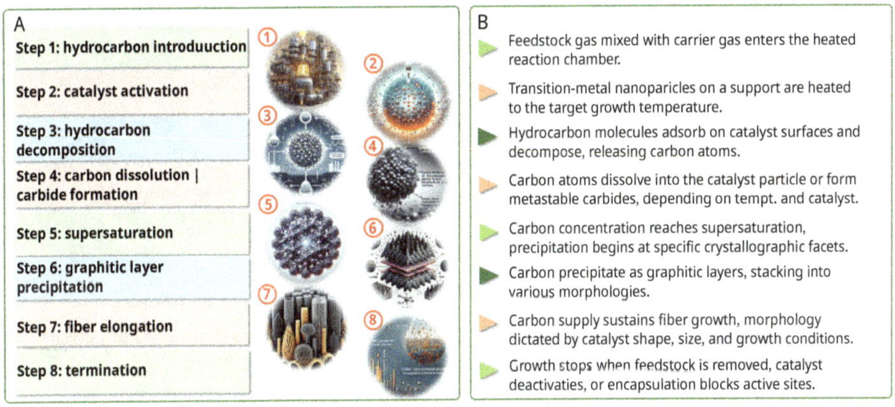

Fig. 4.19: Proposed catalytic CVD growth mechanism of carbon nanofibers (CNFs): **A**. flow diagram of catalytic CVD growth mechanism of CNFs to produce CNFs and **B**. stepwise explanation of each stage in the catalytic growth mechanism.

Fullerene: Fullerene, a distinct allotrope of carbon, comprises closed or partially closed cage-like structures formed by carbon–carbon single and double bonds arranged in fused pentagonal and hexagonal rings [213]. Their geometry, ranging from spherical (C_{60}) to ellipsoidal (C_{70}) and tubular forms, enables tunable electronic, photodynamic, and encapsulation properties [214]. Their geometries of fullerene, ranging from spherical (e.g., C_{60}) to ellipsoidal (e.g., C_{70}) and extended forms such as tubular or bowl-shaped, exhibit high symmetry and curvature, which impart unique electronic, mechanical, and photophysical properties (Tab. 4.9) [215]. Eight representative fullerene formulations (Fig. 4.20A–**H**) are depicted to highlight structural variation across molecular sizes and symmetries: C_{60}, C_{70}, C_{74}, C_{80}, C_{108}, C_{180}, C_{240}, and C_{540} [216]. These delocalized π-electron systems confer exceptional electron affinity, redox versatility, and photostability, making fullerenes attractive for applications in organic electronics, catalysis, and biomedicine. Structurally, fullerenes offer tunable cavity volumes, hydrophobic interiors, and modifiable surfaces, enabling encapsulation of small molecules, ions, or therapeutic agents. Their low toxicity, high biocompatibility, and membrane permeability support their use in drug delivery, gene transfection, and targeted imaging [217]. Functionalized derivatives further enhance solubility and binding specificity, facilitating roles in photodynamic therapy, antioxidant formulations, and biosensing platforms. Key photophysical attributes, such as singlet oxygen generation, long triplet lifetimes, and near-infrared absorption, enable their deployment as photosensitizers in cancer treatment and antimicrobial applications [218].

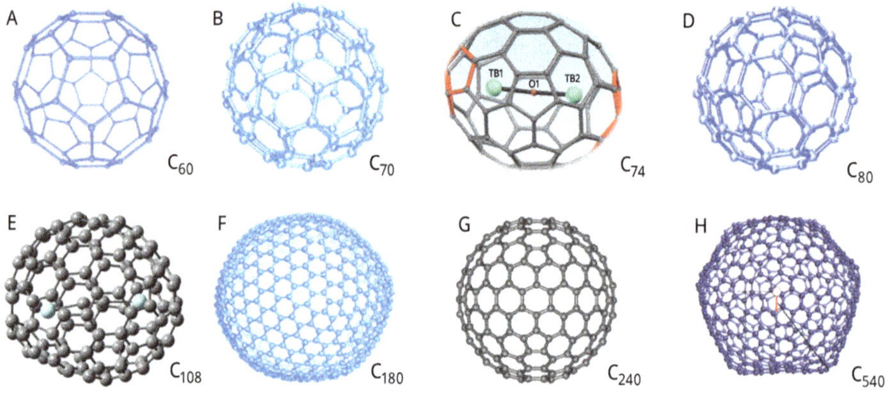

Fig. 4.20: Illustration of fullerene structures, a unique carbon allotrope composed of closed or partially closed cage-like frameworks formed by fused pentagonal and hexagonal rings. These cages are stabilized by carbon–carbon single and double bonds, enabling diverse geometries and electronic properties. Eight representative fullerene formulations are depicted to highlight structural variation across molecular sizes and symmetries: **A.** C_{60}, **B.** C_{70}, **C.** C_{74}, **D.** C_{80}, **E.** C_{108}, **F.** C_{180}, **G.** C_{240}, and **H.** C_{540}.

***A. C_{60} fullerene (buckminsterfullerene)*:** C_{60} is a truncated icosahedron composed of 12 pentagons and 20 hexagons, forming a perfectly spherical cage with I_h symmetry, the highest possible for a molecule (Fig. 4.20A). Each carbon atom is sp^2-hybridized, forming three σ bonds and one delocalized π bond, contributing to a conjugated electron system across the surface. The molecule contains 90 covalent bonds: 60 single and 30 double, though resonance delocalization blurs this distinction. Its π-electron cloud imparts high electron affinity and redox activity. Intermolecular forces are dominated by van der Waals interactions, enabling solid-state packing into face-centered cubic lattices. C_{60}'s curvature induces strain, but its symmetry stabilizes the structure. The molecule behaves as a soft electron acceptor and exhibits photophysical properties, including the generation of singlet oxygen. Its low polarizability and hydrophobic surface make it suitable for encapsulating small molecules. C_{60}'s unique topology supports applications in photovoltaics, drug delivery, and superconductivity.

Its high symmetry also allows for extensive functionalization without disrupting the core geometry, making it a versatile nanomaterial [219].

B. C_{70} fullerene: C_{70} consists of 25 hexagons and 12 pentagons arranged in an elongated ellipsoidal cage, resembling a rugby ball. It exhibits D_{5h} symmetry (Fig. 4.20B), lower than C_{60}'s I_h, resulting in anisotropic electronic and vibrational properties. Each carbon atom remains sp^2-hybridized, forming three σ bonds and one delocalized π bond, but the curvature of the molecule varies, influencing bond lengths and electron density. The π-system extends along the longitudinal axis, enhancing redox behavior and charge mobility. C_{70} contains 105 covalent bonds, with alternating single and double character due to resonance. Van der Waals interactions dominate intermolecular forces, though their elongated shape affects packing efficiency in solid-state lattices. The molecule's larger surface area and reduced symmetry allow for more diverse functionalization sites. C_{70}'s photophysical properties, such as longer triplet lifetimes and broader absorption spectra, make it suitable for organic electronics, biosensing, and photodynamic therapy. Its enhanced electron affinity and structural anisotropy support applications in photovoltaics, molecular semiconductors, and targeted drug delivery systems [220].

C. C_{74} fullerene: C_{74} is a higher fullerene with an asymmetric cage composed of 12 pentagons and 31 hexagons, resulting in lower symmetry (D_3 or C_{2V} depending on isomer). Its geometry is elongated and slightly distorted, producing uneven curvature and localized strain (Fig. 4.20C). Carbon atoms are sp^2-hybridized, forming three σ-bonds and one π-bond, but the π-electron delocalization is less uniform than in C_{60} or C_{70}. The molecule contains 111 covalent bonds, with variable bond lengths due to curvature gradients. The reduced symmetry and increased surface area enhance its reactivity and functionalization potential [221]. Intermolecular forces remain van der Waals in nature, but packing behavior is influenced by its irregular shape. C_{74}'s electronic structure supports multiple redox states, making it useful in charge transfer complexes and supramolecular assemblies. Its larger cavity allows encapsulation of small atoms or ions, and its photophysical properties enable the generation of singlet oxygen. C_{74} is studied for applications in catalysis, molecular electronics, and as a scaffold for endohedral metallofullerenes in quantum and magnetic devices [222].

D. C_{80} fullerene: C_{80} is a giant fullerene composed of 12 pentagons and 38 hexagons, forming a nearly spherical cage with multiple isomeric forms, including I_h and D_{5h} symmetries (Fig. 4.20D). Its structure accommodates more delocalized π-electrons, enhancing redox versatility and electron mobility. Each carbon atom is sp^2-hybridized, forming three σ bonds and contributing to a conjugated π system on the surface. C_{80} contains 120 covalent bonds, with alternating single and double character influenced by curvature and symmetry. The molecule's increased diameter (~8.2 Å) allows for encapsulation of metal atoms or clusters, forming endohedral fullerenes with tailored magnetic and electronic properties. Intermolecular forces are primarily van der Waals, but functionalization can introduce dipolar interactions. C_{80}'s photophysical behavior includes strong near-infrared absorption and efficient triplet state formation, supporting applications in photodynamic therapy and nonlinear optics. Its structural stability and electron-accepting capacity make it suitable for use in organic solar cells, molecular electronics, and as a host for lanthanide-based imaging agents and spintronic materials [223].

E. C_{108} fullerene: C_{108} is a large fullerene composed of 12 pentagons and 51 hexagons, forming an elongated cage with low symmetry (often C_{1V} or C_{2V} depending on isomer). Its geometry deviates significantly from spherical, introducing curvature gradients and localized strain (Fig. 4.20E). Carbon atoms remain sp^2-hybridized, forming three σ bonds and contributing to a complex π-electron system with partial delocalization. The molecule contains 162 covalent bonds, with bond lengths varying due to structural distortion. C_{108}'s expanded cavity supports encapsulation of metal clusters, enabling formation of endohedral fullerenes with tunable magnetic and electronic properties. Van der Waals interactions dominate intermolecular forces, but functionalization can introduce hydrogen-bonding or dipolar interactions. Its low symmetry enhances chemical reactivity and allows for regioselective functionalization. C_{108} exhibits a broad absorption spectrum and a high electron affinity, making it suitable for molecular electronics, catalysis, and quantum computing platforms. Its structural complexity and electronic richness support advanced applications in supramolecular chemistry and as a scaffold for multifunctional nanomaterials [224].

F. C_{180} fullerene: C_{180} consists of 12 pentagons and 68 hexagons, forming a large, nearly spherical cage with moderate symmetry (often D_3 or C_3v depending on the isomer). Its diameter exceeds 10 Å, providing substantial internal volume for encapsulation of atoms or small molecules (Fig. 4.20F). Each carbon atom is sp^2-hybridized, forming three σ bonds and contributing to a delocalized π system across the surface. The molecule contains 270 covalent bonds, with alternating single and double character influenced by curvature and resonance. C_{180}'s π-electron cloud supports high electron mobility and redox flexibility. Intermolecular forces are primarily van der Waals, though functionalization can introduce dipole–dipole interactions [225]. Its large surface area and curvature gradients enable a wide range of chemical modifications. C_{180} is studied for use in energy storage, molecular electronics, and as a host for endohedral complexes. Its photophysical properties include broad-spectrum absorption and efficient triplet state formation. The molecule's size and electronic richness make it a candidate for advanced optoelectronic devices, catalytic platforms, and biomedical imaging systems [226].

G. C_{240} fullerene: C_{240} is a giant fullerene composed of 12 pentagons and 88 hexagons, forming a highly symmetric, nearly spherical cage with Ih or D_5h symmetry depending on the isomer. Its diameter (~12 Å) allows for encapsulation of large metal clusters or small molecules (Fig. 4.20G). Each carbon atom is sp^2-hybridized, forming three σ bonds and contributing to a fully conjugated π-electron system. The molecule contains 360 covalent bonds, with resonance stabilization provided by alternating single and double bonds. C_{240}'s extensive π-system supports high electron mobility, redox activity, and photostability. Van der Waals interactions dominate intermolecular forces, but functionalization can introduce dipole–dipole or hydrogen-bonding effects. Its large surface area and high symmetry enable uniform functionalization and predictable packing behavior. C_{240} is explored for use in organic photovoltaics, nonlinear optics, and as a scaffold for endohedral metallofullerenes. Its structural robustness and electronic versatility make it suitable for quantum materials, molecular electronics, and targeted drug-delivery systems that require high payload capacity [227].

H. C_{540} fullerene: C_{540} is an ultra-large fullerene composed of 12 pentagons and 168 hexagons, forming a multi-shell icosahedral structure with near-perfect spherical geometry and Ih symmetry (Fig. 4.20H). Its diameter exceeds 16 Å, offering exceptional internal volume for encapsulating atoms, clusters, or functional payloads. Each carbon atom is sp^2-hybridized, forming three σ bonds and contributing to a vast, delocalized π-electron network. The molecule contains 810 covalent bonds, stabilized by resonance and curvature-induced strain distribution. C_{540}'s π-system supports high electron mobility, multiple redox states, and topologically protected electronic modes. Intermolecular forces are primarily van der Waals, though functionalization can introduce polar interactions. Its symmetry and size enable uniform surface modification and predictable supramolecular assembly. C_{540} is investigated for use in quantum computing, topological metamaterials, and as a host for multifunctional endohedral complexes. Its photophysical properties include broad-spectrum absorption and efficient energy transfer. The molecule's scale and electronic richness make it a frontier platform for advanced nanotechnology, molecular electronics, and high-capacity biomedical carriers [228].

The structural diversity of fullerenes illustrated in Fig. 4.20 and summarized in Tab. 4.9 collectively reveals how variations in cage geometry, symmetry, and bonding topology directly influence their electronic behavior and functional adaptability. These molecular architectures, defined by fused pentagonal and hexagonal rings, support delocalized π-systems and curvature-induced strain, which underpin their redox activity, photostability, and encapsulation capacity. As shown in Fig. 4.21A, fullerenes exhibit tunable mechanical and electronic properties, including high tensile strength, thermal resilience, and charge mobility [229]. These attributes enable their integration into advanced technologies, such as organic photovoltaics, field-effect transistors, and molecular electronics (Fig. 4.21B), where structural precision translates into performance optimization [230]. A distinctive feature of fullerenes is their capacity to undergo repeated multi-electron uptake without compromising the integrity of the carbon cage, a property that positions them as highly effective charge storage components in lithium-ion batteries, electrochemical capacitors, and a range of nonlinear optical systems. Their electron-rich surfaces also make functionalized fullerenes valuable catalytic platforms, where they participate in oxidation and hydrogenation pathways by stabilizing transient reactive species. This combination of redox flexibility, surface reactivity, and structural robustness continues to stimulate the development of fullerene-enhanced nanocomposites, biohybrid materials, and analytical technologies [231]. Consequently, fullerenes occupy a foundational role in modern nanotechnology and biomedical engineering, offering a modular molecular scaffold for precision functionalization, selective cargo transport, and seamless integration into emerging device architectures [232].

Tab. 4.9: Summary of symmetry, bonding characteristics, and applications of fullerene.

Type	Geometry and symmetry	Bonding and π-system	Key applications
C_{60}	Spherical, I_h symmetry	90 covalent bonds; even π-delocalization	Drug delivery, photovoltaics, and superconductors
C_{70}	Ellipsoidal, D_5h symmetry	105 bonds; anisotropic π-system	Sensing, photodynamic therapy, and electronics
C_{74}	Asymmetric, D_3/C_2v	111 bonds; uneven π-distribution	Catalysis, supramolecular hosts, and quantum devices
C_{80}	Near-spherical, Ih/D_5h	120 bonds; extended π-network	Imaging agents, solar cells, and nonlinear optics
C_{108}	Elongated, C_1/C_2v	162 bonds; partial π-delocalization	Quantum computing, catalysis, and molecular electronics
C_{180}	Large spherical, D_3/C_3v	270 bonds; high electron mobility	Energy storage, optoelectronics, and imaging
C_{240}	Giant spherical, Ih/D_5h	360 bonds; robust π-conjugation	Organic electronics, drug carriers, and quantum materials
C_{540}	Multi-shell spherical, Ih	810 bonds; topologically protected π-states	Metamaterials, spintronics, and high-capacity nanocarriers

A. C_{60} fullerene: C_{60} exhibits high symmetry (I_h), a molecular diameter of ~7. Å, and a molecular weight of 720 g/mol. It has a HOMO–LUMO gap of ~1.7 eV and supports reversible redox transitions up to $C_{60}{}^{6-}$. Its photostability, singlet-oxygen generation efficiency, and low toxicity make it ideal for biomedical applications. C_{60} exhibits a high electron affinity (~2.6 eV) and serves as an efficient electron acceptor in organic solar cells. It has a melting point above 600 °C and forms face-centered cubic crystals under ambient conditions. Applications include drug delivery (via endohedral encapsulation), photodynamic therapy, and radical scavenging [233]. In electronics, it is used in organic field-effect transistors (OFETs), superconductors (e.g., alkali-doped C_{60}), and nonlinear optical devices. C_{60} also serves as a molecular probe in biosensing and imaging. Its ability to form stable adducts and derivatives supports functionalization for targeted therapies. C_{60} remains a benchmark material in nanomedicine, photovoltaics, and supramolecular chemistry due to its tunable electronic properties and scalable synthesis [234].

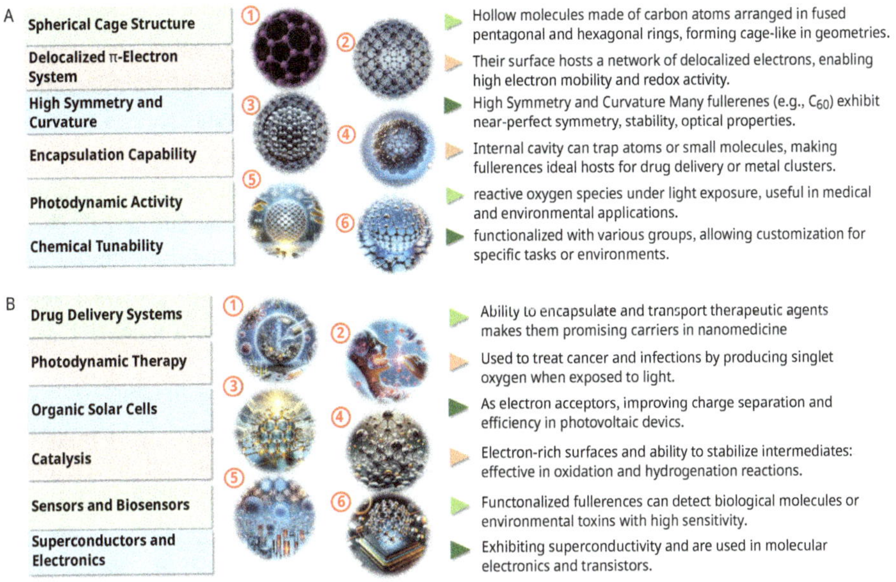

Fig. 4.21: Fullerene's properties and applications: **A.** structural and electronic features, including tunable π-electron systems, photodynamic activity, and encapsulation capacity; and **B.** representative applications spanning drug delivery, photodynamic therapy, organic photovoltaics, catalysis, biosensing, and molecular electronics.

B. C_{70} fullerene: C_{70} has D_5h symmetry, a molecular diameter of ~7.8 Å, and a molecular weight of 840 g/mol. Its HOMO–LUMO gap (~1.6 eV) is slightly narrower than that of C_{70}, thereby enhancing charge mobility and redox behavior. C_{70} absorbs more strongly in the visible and near-infrared regions and has extended triplet lifetimes (~100 μs), making it suitable for photodynamic therapy and imaging. It supports multiple redox states and forms monoclinic crystals. C_{70}'s anisotropic geometry improves packing in organic solar cells, enhancing charge separation and device efficiency. It is used in OFETs, biosensors, and antioxidant formulations. In biomedical applications, C_{70} derivatives show improved solubility and selective binding for targeted delivery. Its higher surface area supports functionalization for catalytic and supramolecular systems. C_{70} is also employed in hybrid nanocomposites and energy storage devices. Its photophysical properties enable applications in nonlinear optics and light-harvesting systems. C_{70}'s enhanced reactivity and broader absorption spectrum make it a versatile platform for optoelectronics, nanomedicine, and molecular electronics [235].

C. C_{74} fullerene: C_{74} has lower symmetry (D_3 or C_{2V}), a molecular diameter of ~8.0 Å, and a molecular weight of 888 g/mol. Its HOMO–LUMO gap (~1.5 eV) and irregular geometry lead to localized electronic states and enhanced chemical reactivity. C_{74} supports multiple redox transitions and exhibits broad-spectrum absorption, including near-infrared regions. It is suitable for supramolecular assembly and host–guest chemistry due to its larger cavity and asymmetric surface. C_{74} derivatives are used in catalysis, especially in oxidation reactions, where electron-rich sites stabilize intermediates. In biomedical applications, C_{74} supports drug encapsulation and targeted delivery, particularly for hydrophobic payloads. Its photodynamic properties enable the generation of singlet oxygen for cancer therapy and antimicrobial applications. C_{74} is also explored in quantum materials and spintronic devices due to its orbital flexibility. It forms stable adducts with transition metals and lanthanides, supporting magnetic and electronic tuning. C_{74}'s low symmetry and high reactivity make it ideal for designing multifunctional nanomaterials, hybrid composites, and molecular sensors [236].

D. C_{80} fullerene: C_{80} features multiple isomers (I_h, D_{5h}), a molecular diameter of ~8.2 Å, and a molecular weight of 960 g/mol. Its HOMO–LUMO gap (~1.4 eV) supports enhanced electron transport and redox cycling. C_{80} is widely used in endohedral metallofullerenes (e.g., $M_3N@C_{80}$), enabling applications in magnetism, electronics, and imaging. It exhibits strong near-infrared absorption and efficient triplet-state formation, making it ideal for photodynamic therapy and nonlinear optics. C_{80} derivatives show high solubility and stability, supporting use in organic solar cells and OFETs. Its large cavity allows encapsulation of metal clusters, making it suitable for MRI contrast agents and spintronic platforms. C_{80} is also employed in catalytic applications, particularly for hydrogenation and electron-transfer reactions [237]. It forms ordered thin films and nanocomposites for optoelectronic devices. Biomedical applications include targeted drug delivery and biosensing, supported by its low toxicity and membrane permeability. C_{80}'s electronic versatility and structural robustness make it a key material in advanced energy systems, quantum electronics, and multifunctional nanocarriers [238].

E. C_{108} fullerene: C_{108} has a molecular diameter of ~9. Å and a molecular weight of 1,296 g/mol. Its HOMO–LUMO gap (~1.3 eV) and low symmetry support orbital flexibility and enhanced reactivity. C_{108} is often used in endohedral metallofullerenes, where its expanded cavity accommodates metal clusters to tune magnetic and electronic properties. It exhibits broad-spectrum absorption and high electron affinity, making it suitable for optoelectronic and quantum applications. C_{108} derivatives are explored in spintronic devices, supramolecular assemblies, and charge transfer complexes. In catalysis, its electron-rich surface stabilizes reactive intermediates, especially in redox and hydrogenation reactions. Biomedical applications include targeted drug delivery and biosensing, supported by its encapsulation capacity and functionalization potential. C_{108} is also used in molecular electronics, where its orbital diversity enhances device performance. Its structural complexity enables regioselective modification, enabling the synthesis of multifunctional nanomaterials. C_{108}'s combination of size, electronic richness, and reactivity makes it a promising candidate for hybrid composites, quantum computing platforms, and advanced energy storage systems [239].

F. C_{180} fullerene: C_{180} has a molecular diameter of ~10.5 Å and a molecular weight of 2,160 g/mol. Its HOMO–LUMO gap (~1.2 eV) and high electron mobility support applications in energy conversion and storage. C_{180}'s large surface area and curvature gradients enable extensive functionalization and encapsulation of atoms or small molecules. It is used in organic solar cells, supercapacitors, and lithium-ion batteries, where its redox stability enhances charge retention. C_{180} derivatives are also explored in biosensing and imaging, supported by their photodynamic activity and near-infrared absorption. In catalysis, C_{180} stabilizes transition states and facilitates electron transfer in oxidation reactions. Its low toxicity and membrane permeability support biomedical applications, including drug delivery and gene transfection. C_{180} is also used in molecular electronics and nonlinear optical devices. Its ability to form ordered nanostructures makes it suitable for supramolecular engineering and the fabrication of hybrid materials. C_{180}'s combination of scale, electronic versatility, and structural adaptability positions it as a multifunctional platform for advanced nanotechnology and sustainable energy systems [240].

G. C_{240} fullerene: C_{240} has a molecular diameter of ~12.0 Å and a molecular weight of 2,880 g/mol. Its HOMO–LUMO gap (~1.1 eV) and extensive π-conjugation support high electron mobility and redox cycling. C_{240} is widely used in organic electronics, including photovoltaics and field-effect transistors, where its large surface area enhances charge transport. Its cavity accommodates larger metal clusters, enabling the formation of endohedral complexes for magnetic and imaging applications. C_{240} derivatives exhibit strong near-infrared absorption and efficient triplet-state formation, making them ideal for photodynamic therapy and biosensing [241]. In catalysis, C_{240} supports multielectron transfer and stabilizes reactive intermediates. It is also used in supramolecular chemistry and nanocomposites, where its symmetry enables predictable assembly. Biomedical applications include high-capacity drug delivery and targeted imaging, supported by its functionalization potential and biocompatibility. C_{240}'s structural robustness and electronic richness make it suitable for quantum materials, hybrid biomaterials, and multifunctional device platforms. Its scalability and tunable properties position it as a key material in next-generation optoelectronics and molecular engineering [242].

H. C_{540} fullerene: C_{540} has a molecular diameter of ~16.4 Å and a molecular weight of 6,480 g/mol. Its HOMO–LUMO gap (~1.0 eV) and multi-shell π-conjugation support topologically protected electronic states and high charge mobility. C_{540} is used in metamaterials and quantum computing platforms, where its symmetry and orbital diversity enable advanced electronic behavior. It supports encapsulation of large clusters or payloads, making it ideal for high-capacity drug delivery and molecular storage. C_{540} derivatives exhibit broad-spectrum absorption and efficient energy transfer, suitable for nonlinear optics and photodynamic therapy. In catalysis, its electron-rich surface facilitates multistep reactions and stabilizes transition states. C_{540} is also employed in spintronic devices and superconducting materials. Its ability to form ordered lattices enables supramolecular engineering and the fabrication of hybrid composites. Biomedical applications include imaging, biosensing, and targeted therapeutics, enabled by its functionalization flexibility and biocompatibility. C_{540}'s scale, electronic richness, and structural precision make it a frontier material for multifunctional nanotechnology, advanced electronics, and topological quantum systems [243].

4.5.5 Challenges and opportunities in fullerene synthesis via CVD

The commercialization of fullerenes for biomedical applications remains constrained by the lack of scalable, high-yield synthesis routes that produce structurally uniform, functionally tunable materials. Current efforts focus on optimizing production throughput, cost-efficiency, and purity key metrics for clinical translation. Traditional methods such as laser ablation and arc discharge require temperatures exceeding 3,000 K and inert gas environments, resulting in low yield-to-energy ratios and limited control over fullerene size distribution. In contrast, CVD offers a more scalable alternative, operating at temperatures between 700 and 120 °C and enabling continuous growth under controlled gas-phase conditions. Notably, Chow et al. [244] reported the incidental formation of fullerenes as by-products during hot filament CVD synthesis of diamond films. Kleckley et al. [245] later confirmed this co-production, proposing that fullerene intermediates may act as nucleation templates for diamond crystallization. Their microwave-assisted CVD experiments revealed transient corannulene formation, an aromatic bowl-shaped molecule considered a key precursor in fullerene assembly [246]. Beyond fullerenes, CVD has enabled the synthesis of other carbon nanostructures, including graphene (via CH_4/H_2 flows on Cu substrates), CNTs (using Fe or Co catalysts), and CNFs (via Ni-catalyzed hydrocarbon decomposition). Emerging targets include carbon nano-onions, ultrathin nanosheets, and high-aspect-ratio nanowires, each requiring precise control over precursor chemistry, plasma parameters, and substrate temperature [247]. Despite these advances, the mass production of defect-controlled, application-grade nanomaterials remains a central challenge. Yield optimization, reactor scalability, and post-synthesis purification protocols must be addressed to unlock commercial viability across sectors ranging from drug delivery and imaging to energy storage and nanoelectronics [248].

4.6 Summary and outlook

Vapor deposition encompasses a family of thin-film fabrication techniques in which materials are deposited onto substrates from a vapor phase. These methods are essential for producing high-performance coatings in microelectronics, optics, energy systems, and advanced manufacturing. The two primary categories, PVD and CVD, differ fundamentally in their mechanisms and material versatility. PVD is a line-of-sight process that involves the physical transformation of a solid source material into vapor, followed by condensation onto a substrate. Standard PVD techniques include thermal evaporation and sputtering, typically performed under vacuum. The process requires high temperatures or energetic particle bombardment to vaporize the source material. PVD is ideal for depositing metals (e.g., Al, Cu, and Ti), metal nitrides (e.g., TiN and CrN), and oxides (e.g., Al_2O_3 and ITO) with excellent hardness, adhesion, and corrosion resistance. However, due to its directional nature, PVD struggles with coating complex geometries and high-aspect-ratio features. CVD, by contrast, relies on chemical reactions between gaseous precursors at the substrate surface to form solid films. These reactions can be thermally activated (LPCVD), plasma-assisted (PECVD), or MOCVD, depending on the application. CVD offers superior conformality and uniformity, making it suitable for coating intricate structures and porous materials. It enables the deposition of a wide range of compounds, including semiconductors (e.g., Si and Si_3N_4), ceramics (e.g., SiC and TiC), and carbon-based materials (e.g., graphene and CNTs). CVD is widely used in semiconductor fabrication, diffusion barriers, solar cells, and MEMS devices. While PVD is favored for high-purity metallic coatings and decorative finishes, CVD excels in producing chemically complex, conformal layers for functional applications. Both techniques continue to evolve, with hybrid approaches and low-temperature variants expanding their reach into flexible electronics, biomedical devices, and nanostructured materials. Understanding their distinct principles and material capabilities is essential for selecting the appropriate deposition method in advanced thin-film engineering.

As the demand for advanced thin films and coatings continues to grow across industries, future research in vapor deposition, particularly PVD and CVD, is increasingly focused on precision engineering, material diversification, and application-specific optimization. These techniques are evolving beyond traditional boundaries to support emerging technologies in flexible electronics, energy conversion, biomedical devices, and quantum systems. One central area of advancement involves expanding the library of materials that can be deposited. Researchers are developing high-entropy alloys and nanocomposites for wear-resistant and multifunctional coatings. Two-dimensional materials such as molybdenum disulfide (MoS_2), tungsten disulfide (WS_2), and graphene are being synthesized via low-temperature CVD for use in nanoelectronics and sensors. Functional oxides, including ZnO, titanium dioxide (TiO_2), and vanadium dioxide (VO_2), are being tailored to exhibit tunable optical, electrical, and catalytic properties. In the biomedical field, PVD enables the deposition of bioactive and antimicrobial coatings, including silver, copper, and hybrid layers, for medical implants.

Process innovation is also driving the evolution of vapor deposition. Techniques such as field-enhanced CVD, plasma-assisted PVD, and atomic layer hybridization are enabling lower-temperature deposition on polymer substrates and flexible electronics. These methods offer enhanced control over grain size, texture, and phase composition through the use of electric, magnetic, and plasma fields. Additionally, the integration of artificial intelligence and Internet of Things technologies is facilitating real-time feedback and adaptive control for intelligent manufacturing environments. Deposition strategies are becoming increasingly tailored to specific applications. In energy systems, thin-film electrocatalysts are being developed for hydrogen production and solid-state batteries. Optoelectronic devices benefit from transparent conductive layers and quantum dot interfaces. Aerospace and automotive sectors require corrosion-resistant, lightweight coatings with high thermal stability. Microelectronics demand ultra-thin barrier layers and interconnects fabricated with atomic precision. Finally, sustainability and scalability are central to future development. Researchers are working to reduce hazardous by-products in CVD, recycle target materials in PVD, and transition to atmospheric pressure systems for roll-to-roll manufacturing. These efforts aim to make vapor deposition more environmentally responsible and industrially viable. Collectively, these innovations are transforming vapor deposition into a platform for the design of functional materials, enabling the next generation of high-performance technologies across disciplines.

References

[1] Bakshi, M. S. (2016). How surfactants control crystal growth of nanomaterials. *Crystal Growth & Design*, 16(2), 1104–1133.

[2] Tripathi, S. K., Kaur, R., Kaur, H., Rani, M., Kaur, J., & Kaur, H. (2015, May). Fabrication and electrical characterization of memristor with TiO_2 as an active layer. In (vol. *In AIP Conference Proceedings* (Vol. 1661, No. 1, pp. 110027). AIP Publishing LLC.

[3] Malik, M. A., Wani, M. Y., & Hashim, M. A. (2012). Microemulsion method: A novel route to synthesize organic and inorganic nanomaterials: 1st Nano Update. *Arabian Journal of Chemistry*, 5(4), 397–417.

[4] Wahl, G., Davies, P. B., Bunshah, R. F., Joyce, B. A., Bain, C. D., Wegner, G., & Edler, K. J. (2000). Thin Films. *Ullmann's Encyclopedia of Industrial Chemistry*, 1–75.

[5] Bunton, C. A. (1979). Reaction kinetics in aqueous surfactant solutions. *Catalysis Reviews Science and Engineering*, 20(1), 1–56.

[6] Shaker, D. S., Ishak, R. A., Ghoneim, A., & Elhuoni, M. A. (2019). Nanoemulsion: A review on mechanisms for the transdermal delivery of hydrophobic and hydrophilic drugs. *Scientia Pharmaceutica*, 87(3), 17.

[7] Thanh, N. T., Maclean, N., & Mahiddine, S. (2014). Mechanisms of nucleation and growth of nanoparticles in solution. *Chemical Reviews*, 114(15), 7610–7630.

[8] Guo, X., Zhang, Q., Ding, X., Shen, Q., Wu, C., Zhang, L., & Yang, H. (2016). Synthesis and application of several sol-gel-derived materials via sol-gel process combining with other technologies: A review. *Journal of Sol-Gel Science and Technology*, 79(2), 328–358.

[9] Lei, Q., Guo, J., Noureddine, A., Wang, A., Wuttke, S., Brinker, C. J., & Zhu, W. (2020). Sol-gel-based advanced porous silica materials for biomedical applications. *Advanced Functional Materials*, 30(41), 1909539.

[10] Yun, Q., Ge, Y., Shi, Z., Liu, J., Wang, X., Zhang, A., & Zhang, H. (2023). Recent progress on phase engineering of nanomaterials. *Chemical Reviews*, 123(23), 13489–13692.

[11] Ossai, C. I., & Raghavan, N. (2018). Nanostructure and nanomaterial characterization, growth mechanisms, and applications. *Nanotechnology Reviews*, 7(2), 209–231.

[12] He, C., Liu, D., & Lin, W. (2015). Nanomedicine applications of hybrid nanomaterials built from metal-ligand coordination bonds: Nanoscale metal-organic frameworks and nanoscale coordination polymers. *Chemical Reviews*, 115(19), 11079–11108.

[13] Sundgren, J. E., & Hentzell, H. G. (1986). A review of the present state of art in hard coatings grown from the vapor phase. *Journal of Vacuum Science & Technology A: Vacuum, Surfaces, and Films*, 4(5), 2259–2279.

[14] Sreenivasan, R., & Gleason, K. K. (2009). Overview of strategies for the CVD of organic films and functional polymer layers. *Chemical Vapor Deposition*, 15(4-6), 77–90.

[15] Kusano, E. (2019). Structure-zone modeling of sputter-deposited thin films: A brief review. *Applied Science and Convergence Technology*, 28(6), 179–185.

[16] Shahidi, S., Moazzenchi, B., & Ghoranneviss, M. (2015). A review-application of physical vapor deposition (PVD) and related methods in the textile industry. *The European Physical Journal Applied Physics*, 71(3), 31302.

[17] Sigmund, P. (1969). Theory of sputtering. I. Sputtering yield of amorphous and polycrystalline targets. *Physical Review*, 184(2), 383.

[18] Stüber, M., Leiste, H., Ulrich, S., & Skokan, A. (1999). Development of tailored coating concepts for CVD and PVD deposition of multifunctional coatings-a review. *International Journal of Materials Research*, 90(10), 774–779.

[19] Scott, H. J. (1938). U.S. Patent No. 2,122,495. Washington, DC: U.S. Patent and Trademark Office.

[20] Langmuir, I. (1913). Chemical reactions at very low pressures. II. The chemical clean-up of nitrogen in a tungsten lamp. *Journal of the American Chemical Society*, 35(8), 931–945.

[21] Bhatt, V., & Chandra, S. (2007). Silicon dioxide films by RF sputtering for microelectronic and MEMS applications. *Journal of Micromechanics and Microengineering*, 17(5), 1066.

[22] Kessels, E., Devi, A., Park, J. S., Ritala, M., Yanguas-Gil, A., & Wiemer, C. (2025). Atomic layer deposition. *Nature Reviews Methods Primers*, 5(1), 66.

[23] Hamedani, Y., Macha, P., Bunning, T. J., Naik, R. R., & Vasudev, M. C. (2016). Plasma-enhanced chemical vapor deposition: Where we are and the outlook for the future. *Chemical Vapor Deposition-Recent Advances and Applications in Optical, Solar Cells and Solid State Devices. IntechOpen*. doi: 10.5772/64654.

[24] Anders, A., & Shohet, J. L. (2017). High-power impulse magnetron sputtering (HiPIMS). In *Encyclopedia of Plasma Technology* Shohet, J. L. ed. (pp. 588).

[25] Ritala, M., & Leskelä, M. (1999). Atomic layer epitaxy-a valuable tool for nanotechnology? *Nanotechnology*, 10(1), 19.

[26] Graham, G. M., & Alam, M. K. (1991). Experimental study of the outside vapor deposition process. *Aerosol Science and Technology*, 15(1), 69–76.

[27] Gordon, R. (1997). Chemical vapor deposition of coatings on glass. *Journal of Non-Crystalline Solids*, 218, 81–91.

[28] Kellerer, H., Müller, A., Bauer, H. J., & Wittig, S. (1996). Soot formation in a shock tube under elevated pressure conditions. *Combustion Science and Technology*, 113(1), 67–80.

[29] Dragic, P., Hawkins, T., Foy, P., Morris, S., & Ballato, J. (2012). Sapphire-derived all-glass optical fibres. *Nature Photonics*, 6(9), 627–633.

[30] Symoens, E., Van Coile, R., & Belis, J. (2020). Behaviour of monolithic and layered glass elements subjected to elevated temperatures: State of the art. In *CGC 7, Challenging Glass 7, Conference on Architectural and Structural Applications of Glass* (Vol. 7). TU Delft Open. https://doi.org/10.7480/cgc.7.4489.

[31] Butt, M. A. (2024, September). A Comprehensive Exploration of Contemporary Photonic Devices in Space Exploration: A Review. *Photonics*, 11(9), 873. MDPI.

[32] Helmersson, U., Lattemann, M., Bohlmark, J., Ehiasarian, A. P., & Gudmundsson, J. T. (2006). Ionized physical vapor deposition (IPVD): A review of technology and applications. *Thin Solid Films*, 513(1-2).

[33] Hamberg, I., & Granqvist, C. G. (1986). Evaporated Sn-doped In_2O_3 films: Basic optical properties and applications to energy-efficient windows. *Journal of Applied Physics*, 60(11), R123–R160.

[34] Raman, P., Shchelkanov, I. A., McLain, J., & Ruzic, D. N. (2015). High power pulsed magnetron sputtering: A method to increase deposition rate. *Journal of Vacuum Science & Technology A*, 33(3), 031304.

[35] LeClair, P., Berera, G. P., & Moodera, J. S. (2000). Titanium nitride thin films obtained by a modified physical vapor deposition process. *Thin Solid Films*, 376(1-2), 9–15.

[36] Awan, T. I., Afsheen, S., & Kausar, S. (2025). Chemical vapor deposition technique. In *Thin Film Deposition Techniques: Thin Film Deposition Techniques and Its Applications in Different Fields* (pp. 65–96). Singapore: Springer Nature Singapore.

[37] Le Trinh, N., Kim, W., Lee, M., Gu, B., Thi, S. V., Liu, J., & Lee, H. B. R. (2025). Overcoming the Critical Thickness Limit. *Interfacial Control of Crystallization Pathways in Atomic-Scale Dielectric Thin Films*. doi: 10.26434/chemrxiv-2025-5x13x.

[38] Borowski, P., & Myśliwiec, J. (2025). Recent advances in magnetron sputtering: From fundamentals to industrial applications. *Coatings*, 15(8), 922.

[39] Das, H. S., Basak, A., & Maity, S. (2025). Materials science and nanotechnology. In *Innovations in Energy Efficient Construction Through Sustainable Materials* (pp. 175–206). IGI Global.

[40] Ahmad, W., Nawaz, M. Z., Kazmi, J., Younis, U., Illarionov, Y., & Wang, Z. (2025). Two-dimensional indium chalcogenides: from fundamental properties to next-generation optoelectronic devices. *Advanced Optical Materials*, 13(24), e01067.

[41] Gudmundsson, J. T., Anders, A., & Von Keudell, A. (2022). Foundations of physical vapor deposition with plasma assistance. *Plasma Sources Science and Technology*, 31(8), 083001.

[42] Musálek, T., Liška, P., Morsa, A., Arregi, J. A., Kratochvíl, M., Sergeev, D., & Kolíbal, M. (2024). Achieving Different Stoichiometries and Morphologies in Vapor Phase Deposition of Inorganic Halide Perovskites: Single or Dual Precursor Sources? *arXiv Preprint arXiv:2409.18294*.

[43] Qie, H., Liu, J., Sun, Q., Gao, H., Sun, X., Zhou, Y., & Yang, H. (2022). Selective area epitaxy of degenerate n-GaN for HEMT ohmic contact by MOCVD. *Applied Physics Letters*, 121(21), 212106.

[44] Huang, R., Yang, W., Fan, X., Lai, C., Mao, J., Liu, M., & Zhang, X. (2023). Insights into the primitives and sequence deposition mechanism of 7YSZ TBCs, respectively, by EB-PVD and PS-PVD. *Surface and Coatings Technology*, 468, 129673.

[45] Sierra-Castillo, A., Haye, E., Acosta, S., Arenal, R., Bittencourt, C., & Colomer, J. F. (2021). Atmospheric pressure chemical vapor deposition growth of vertically aligned SnS_2 and $SnSe_2$ nanosheets. *RSC Advances*, 11(58), 36483–36493.

[46] Dirko, V. V., Lozovoy, K. A., Kokhanenko, A. P., & Voitsekhovskii, A. V. (2021). High-resolution RHEED analysis of dynamics of low-temperature superstructure transitions in Ge/Si (001) epitaxial system. *Nanotechnology*, 33(11), 115603.

[47] Carazzetti, P., Drechsel, C., Rettenmeier, R., Weichart, J., Viehweger, K., & Strolz, E. (2024, September)). Study of cross-contamination in multi-chamber PVD systems used for high-throughput seed layer deposition. In *2024 IEEE 10th Electronics System-Integration Technology Conference (ESTC)* (pp. 1–9). IEEE.

[48] Grigoriev, S., Vereschaka, A., Zelenkov, V., Sitnikov, N., Bublikov, J., Milovich, F., & Sotova, C. (2022). Investigation of the influence of the features of the deposition process on the structural features of microparticles in PVD coatings. *Vacuum*, 202, 111144.

[49] Patel, R., Van Oijen, J., Dam, N., & Nijdam, S. (2023). Low-temperature filamentary plasma for ignition-stabilized combustion. *Combustion and Flame*, 247, 112501.

[50] Quillin, K., Yeom, H., Pu, X., Frazer, D., Sasidhar, K. N., & Sridharan, K. (2023). Effects of elevated temperature exposure on the residual stress state and microstructure of PVD Cr coatings on SiC investigated via in situ X-ray diffraction and transmission electron microscopy. *Materials Science and Engineering: A*, 879, 145273.

[51] Zhang, Y., Wang, Q., Ramachandran, C. S., Guo, P., & Wang, A. (2022). Microstructure and performance of high-velocity oxygen-fuel coupled physical vapor deposition (HVOF-PVD) duplex protective coatings: A review. *Coatings*, 12(10), 1395.
[52] Xavier, F. B., Patrick, C., & Ewald, S. (2021, June). Investigation of low stress and low temperature SiN and SiCN PVD films for advanced packaging applications. In *2021, IEEE 71st Electronic Components and Technology Conference (ECTC)* (pp. 2111–2117). IEEE.
[53] Sai, L. K., Swain, K. K., & Pradhan, S. K. (2023). Thin Film Fabrication Techniques. *Metal Oxide Nanocomposite Thin Films for Optoelectronic Device Applications*, Elsevier, 155–177.
[54] Butt, M. A. (2022). Thin-film coating methods: A successful marriage of high-quality and cost-effectiveness: A brief exploration. *Coatings*, 12(8), 1115.
[55] Gupta, G., Tyagi, R. K., Rajput, S. K., Saxena, P., Vashisth, A., & Mehndiratta, S. (2021). PVD-based thin film deposition methods and characterization/property of different compositional coatings: A critical analysis. In *Materials Today: Proceedings* (Vol. 38, pp. 259–264).
[56] Panjan, P., Drnovšek, A., Mahne, N., Čekada, M., & Panjan, M. (2021). Surface topography of PVD hard coatings. *Coatings*, 11, 1387.
[57] Baptista, A., Silva, F., Porteiro, J., Míguez, J., & Pinto, G. (2018). Sputtering physical vapour deposition (PVD) coatings: A critical review on process improvement and market trend demands. *Coatings*, 8(11), 402.
[58] Chatterjee, S., Abadie, T., Wang, M., Matar, O. K., & Ruoff, R. S. (2024). Repeatability and Reproducibility in the Chemical Vapor Deposition of 2D Films: A Physics-Driven Exploration of the Reactor Black Box. *Chemistry of Materials*, 36(3), 1290–1298.
[59] Mattox, D. M. (2002). Physical vapor deposition (PVD) processes. *Metal Finishing*, 100, 394–408.
[60] Broszeit, E., Friedrich, C., & Berg, G. (1999). Deposition, properties, and applications of PVD Cr_xN coatings. *Surface and Coatings Technology*, 115(1), 9–16.
[61] Wang, X., Pei, Y., & Ma, Y. (2013). The effect of microstructure at interface between coating and substrate on damping capacity of coating systems. *Applied Surface Science*, 282, 60–66.
[62] Oettel, H., & Wiedemann, R. (1995). Residual stresses in PVD hard coatings. *Surface and Coatings Technology*, 76, 265–273.
[63] Mitterery, C., Mayrhofen, P. H., Kelesoglu, E., Wiedemann, R., & Oettel, H. (1999). Internal growth parameters – a suitable basis for comparison of PVD coatings. *International Journal of Materials Research*, 90(8), 602–607.
[64] Coronell, D. G., Egan, E. W., Hamilton, G., Jain, A., Venkatraman, R., & Weitzman, B. (1998). Monte Carlo simulations of sputter deposition and step coverage of thin films. *Thin Solid Films*, 333(1-2), 77–81.
[65] Yate, L., Martínez-de-olcoz, L., Esteve, J., & Lousa, A. (2009). Control of the bias voltage in DC PVD processes on insulator substrates. *Vacuum*, 83(10), 1287–1290.
[66] Boing, D., De Oliveira, A. J., & Schroeter, R. B. (2020). Evaluation of wear mechanisms of PVD and CVD coatings deposited on cemented carbide substrates applied to hard turning. *The International Journal of Advanced Manufacturing Technology*, 106(11), 5441–5451.
[67] Inspektor, A., & Salvador, P. A. (2014). Architecture of PVD coatings for metalcutting applications: A review. *Surface and Coatings Technology*, 257, 138–153.
[68] Sohn, Y. H., Kim, J. H., Jordan, E. H., & Gell, M. (2001). Thermal cycling of EB-PVD/MCrAlY thermal barrier coatings: I. Microstructural development and spallation mechanisms. *Surface and Coatings Technology*, 146, 70–78.
[69] Rickerby, D. S., & Burnett, P. J. (1988). Correlation of process and system parameters with structure and properties of physically vapour-deposited hard coatings. *Thin Solid Films*, 157(2), 195–222.
[70] Hovsepian, P. E., & Ehiasarian, A. P. (2019). Six strategies to produce application tailored nanoscale multilayer structured PVD coatings by conventional and High Power Impulse Magnetron Sputtering (HIPIMS). *Thin Solid Films*, 688, 137409.

[71] Altaf, S. F., Rahman, A., & Wani, M. F. (2025). Nanostructured Thermal Barrier Coatings: Enhancing Performance for High-Temperature Applications: A Comprehensive Review. *Surfaces and Interfaces*, 107033.

[72] Castro, J. D., Sánchez-López, J. C., Rojas, T. C., Escobar-Galindo, R., & Carvalho, S. (2025). Diffusion mechanisms and corrosion resistance of nanostructured ZrN-Cu coating obtained by hybrid HiPIMS-DCMS. *Applied Surface Science*, 682, 161635.

[73] Ichou, H., Arrousse, N., Berdimurodov, E., & Aliev, N. (2024). Exploring the advancements in physical vapor deposition coating: A review. *Journal of Bio- and Tribo-Corrosion*, 10(1), 3.

[74] Vereschaka, A., Milovich, F., Andreev, N., Seleznev, A., Alexandrov, I., Muranov, A., & Tatarkanov, A. (2023). Comparison of properties of ZrHf-(Zr, Hf) N-(Zr, Hf, Cr, Mo, Al) N and Ti-TiN-(Ti, Cr, Al). *N Nanostructured Multilayer Coatings and Cutting Properties of Tools with These Coatings During Turning of Nickel Alloy, Journal of Manufacturing Processes*, 88, 184–201.

[75] Yang, C. C., Edelstein, D., Clevenger, L., Cowley, A., Gill, J., Chanda, K., & Stamper, A. (2005, June). Extendibility of PVD barrier/seed for BEOL Cu metallization. In *Proceedings of the IEEE 2005 International Interconnect Technology Conference 2005* (pp. 135–137). IEEE.

[76] Ionascu, G., Manea, E., Cernica, I., & Moraru, E. (2022, August). Theoretical and experimental research on step coverage optimization for integrated microstructures of thin films. In *International Conference on Reliable Systems Engineering* (pp. 185–202). Cham: Springer International Publishing.

[77] Chun, S. Y., & Jeong, P. G. (2016). Microstructure, crystal structure, and mechanical properties of VN coatings using asymmetric bipolar pulsed DC sputtering. *Journal of the Korean Institute of Surface Engineering*, 49(5), 461–466.

[78] Teixeira, V. (2002). Residual stress and cracking in thin PVD coatings. *Vacuum*, 64(3-4), 393–399.

[79] Grant-Jacob, J. A., Prentice, J. J., Beecher, S. J., Shepherd, D. P., Eason, R. W., & Mackenzie, J. I. (2018). Particulate reduction in ternary-compound film growth via pulsed laser deposition from segmented binary-targets. *Materials Research Express*, 5(3), 036402.

[80] Wang, B., Fu, X., Song, S., Chu, H. O., Gibson, D., Li, C., & Wu, Z. (2018). Simulation and optimization of film thickness uniformity in physical vapor deposition. *Coatings*, 8(9), 325.

[81] Abegunde, O. O., Akinlabi, E. T., Oladijo, O. P., Akinlabi, S., & Ude, A. U. (2019). Overview of thin film deposition techniques. *AIMS Materials Science*, 6(2), 174–199.

[82] Westwood, W. D. (1989). Physical vapor deposition. In *Microelectronic Materials and Processes* (pp. 133–201). Dordrecht: Springer Netherlands.

[83] Gruner, C., Liedtke, S., Bauer, J., Mayr, S. G., & Rauschenbach, B. (2018). Morphology of thin films formed by oblique physical vapor deposition. *ACS Applied Nano Materials*, 1(3), 1370–1376.

[84] Zhu, Y., & Wu, X. (2023). Heterostructured materials. *Progress in Materials Science*, 131, 101019.

[85] Kozuch, J., Ataka, K., & Heberle, J. (2023). Surface-enhanced infrared absorption spectroscopy. *Nature Reviews Methods Primers*, 3(1), 70.

[86] Bhushan, B., & Jung, Y. C. (2011). Natural and biomimetic artificial surfaces for superhydrophobicity, self-cleaning, low adhesion, and drag reduction. *Progress in Materials Science*, 56(1), 1–108.

[87] Zhu, Y., Ameyama, K., Anderson, P. M., Beyerlein, I. J., Gao, H., Kim, H. S., & Wu, X. (2021). Heterostructured materials: Superior properties from hetero-zone interaction. *Materials Research Letters*, 9(1), 1–31.

[88] Anders, A. (2017). Tutorial: Reactive high power impulse magnetron sputtering (R-HiPIMS). *Journal of Applied Physics*, 121(17).

[89] Rossnagel, S. M. (1998). Directional and ionized physical vapor deposition for microelectronics applications. *Journal of Vacuum Science & Technology B: Microelectronics and Nanometer Structures Processing, Measurement, and Phenomena*, 16(5), 2585–2608.

[90] Sankaran, K. J., Kunuku, S., Leou, K. C., Tai, N. H., & Lin, I. N. (2014). Enhancement of the electron field emission properties of ultrananocrystalline diamond films via hydrogen post-treatment. *ACS Applied Materials & Interfaces*, 6(16), 14543–14551.

[91] Svetlizky, D., Das, M., Zheng, B., Vyatskikh, A. L., Bose, S., Bandyopadhyay, A., & Eliaz, N. (2021). Directed energy deposition (DED) additive manufacturing: Physical characteristics, defects, challenges, and applications. *Materials Today*, 49, 271–295.
[92] Oke, J. A., & Jen, T. C. (2022). Atomic layer deposition and other thin film deposition techniques: From principles to film properties. Journal of Materials Research andTechnology. 21, 2481–2514, Granqvist C. G. (2007). Transparent conductors as solar energy materials: A panoramic review. *Solar Energy Materials and Solar Cells*, 91(17), 1529-1598.
[93] Berg, S., & Nyberg, T. (2005). Fundamental understanding and modeling of reactive sputtering processes. *Thin Solid Films*, 476(2), 215–230.
[94] Raoux, S. (2009). Phase change materials. *Annual Review of Materials Research*, 39(1), 25–48.
[95] Greene, J. E. (2017). Tracing the recorded history of thin-film sputter deposition: From the 1800s to 2017. *Journal of Vacuum Science & Technology A*, 35(5).
[96] Pashley, R. M., & Israelachvili, J. N. (1984). Molecular layering of water in thin films between mica surfaces and its relation to hydration forces. *Journal of Colloid and Interface Science*, 101(2), 511–523.
[97] Brune, H. (2014). Epitaxial growth of thin films. *Surface and Interface Science*, 4, 421–477.
[98] Forrest, S. R. (2004). The path to ubiquitous and low-cost organic electronic appliances on plastic. *Nature*, 428(6986), 911–918.
[99] Dahotre, N. B., Kadolkar, P., & Shah, S. (2001). Refractory ceramic coatings: Processes, systems, and wettability/adhesion. Surface and Interface Analysis: An *International Journal devoted to the development and application of techniques for the analysis of surfaces, interfaces and thin films*. 31, 7, 659–672.
[100] Wang, Z., & Zhang, Z. (2016). Electron beam evaporation deposition. *Advanced Nano Deposition Methods*, 33–58.
[101] Petrov, I., Adibi, F., Greene, J. E., Hultman, L., & Sundgren, J. E. (1993). Average energy deposited per atom: A universal parameter for describing ion-assisted film growth? *Applied Physics Letters*, 63(1), 36–38.
[102] Alivisatos, A. P., Cohen, M. L., Fréchet, J. M., Head-Gordon, M., Long, J. R., Louie, S. G., & Zettl, A. (2006). *A synergistic approach to the development of new classes of hydrogen storage materials* . Arlington, Virginia: Crystal Gateway Marriott Hotel, 50. osti.gov/servlets/purl/1275515#page=66.
[103] Gudmundsson, J. T. (2020). Physics and technology of magnetron sputtering discharges. *Plasma Sources Science and Technology*, 29(11), 113001.
[104] Musil, J. (2000). Hard and superhard nanocomposite coatings. *Surface and Coatings Technology*, 125 (1-3), 322–330.
[105] Hopwood, J. (1998). Ionized physical vapor deposition of integrated circuit interconnects. *Physics of Plasmas*, 5(5), 1624–1631.
[106] Ikeda, S., Miura, K., Yamamoto, H., Mizunuma, K., Gan, H. D., Endo, M., & Ohno, H. (2010). A perpendicular-anisotropy CoFeB-MgO magnetic tunnel junction. *Nature Materials*, 9, 721–724.
[107] Erdemir, A. (2005). Review of engineered tribological interfaces for improved boundary lubrication. *Tribology International*, 38(3), 249–256.
[108] Marichy, C., Bechelany, M., & Pinna, N. (2012). Atomic layer deposition of nanostructured materials for energy and environmental applications. *Advanced Materials*, 24(8), 1017–1032.
[109] Vallée, C., Bonvalot, M., Belahcen, S., Yeghoyan, T., Jaffal, M., Vallat, R., & Granier, A. (2020). Plasma deposition: Impact of ions in plasma enhanced chemical vapor deposition, plasma enhanced atomic layer deposition, and applications to area selective deposition. *Journal of Vacuum Science & Technology A*, 38(3), 033007.
[110] Roy, A., Wang, S., & Komvopoulos, K. (2023). A review of plasma-assisted deposition methods for amorphous carbon thin and ultrathin films with a focus on the cathodic vacuum arc technique. *Journal of Materials Research*, 38(3), 586–616.

[111] Pierson, H. O. (1999). Handbook of chemical vapor deposition: Principles. In *Technology and Applications* (pp. 36–67). William Andrew: ISBN 978-0-8155-1432-9.
[112] Corbella, C., Aijaz, A., Kubart, T., Lin, L., Portal, S., & Keidar, M. (2025). Pulsed plasma vapour deposition of carbon materials: Advantages and challenges. *Carbon*, 232, 119772.
[113] Binner, J., Porter, M., Baker, B., Zou, J., Venkatachalam, V., Diaz, V. R., & Murthy, T. S. R. C. (2020). Selection, processing, properties, and applications of ultra-high temperature ceramic matrix composites, UHTCMCs: A review. *International Materials Reviews*, 65(7), 389–444.
[114] Van Mol, A. M. B., Chae, Y., McDaniel, A. H., & Allendorf, M. D. (2006). Chemical vapor deposition of tin oxide: Fundamentals and applications. *Thin Solid Films*, 502(1-2), 72–78.
[115] Kern, W. (1990). The evolution of silicon wafer cleaning technology. *Journal of the Electrochemical Society*, 137(6), 1887.
[116] Coltrin, M. E., Kee, R. J., & Miller, J. A. (1984). A mathematical model of the coupled fluid mechanics and chemical kinetics in a chemical vapor deposition reactor. *Journal of the Electrochemical Society*, 131(2), 425.
[117] Carlsson, J. O., & Martin, P. M. (2010). Chemical vapor deposition. In *Handbook of Deposition Technologies for Films and Coatings* (pp. 314–363). William Andrew Publishing.
[118] Kern, W. (1970). Cleaning solution based on hydrogen peroxide for use in silicon semiconductor technology. *RCA Review*, 31(2), 187–206.
[119] Lieberman, M. A., & Lichtenberg, A. J. (1994). Principles of plasma discharges and materials processing. *MRS Bulletin*, 30(12), 899–901.
[120] Jhansirani, K., Dubey, R. S., More, M. A., & Singh, S. (2016). Deposition of silicon nitride films using chemical vapor deposition for photovoltaic applications. *Results in Physics*, 6, 1059–1063.
[121] Gumpher, J., Bather, W. A., & Wedel, D. (2003). LPCVD silicon nitride uniformity improvement using adaptive real-time temperature control. *IEEE Transactions on Semiconductor Manufacturing*, 16(1), 26–35.
[122] Ojha, S. M. (1982). Plasma-enhanced chemical vapor deposition of thin films. In *Physics of Thin Films* (Vol. 12, pp. 237–296). Elsevier.
[123] Batey, J., & Tierney, E. (1986). Low-temperature deposition of high-quality silicon dioxide by plasma-enhanced chemical vapor deposition. *Journal of Applied Physics*, 60(9), 3136–3145.
[124] Chen, N., Kim, D. H., Kovacik, P., Sojoudi, H., Wang, M., & Gleason, K. K. (2016). Polymer thin films and surface modification by chemical vapor deposition: Recent progress. *Annual Review of Chemical and Biomolecular Engineering*, 7(1), 373–393.
[125] Snyders, R., Hegemann, D., Thiry, D., Zabeida, O., Klemberg-Sapieha, J., & Martinu, L. (2023). Foundations of plasma enhanced chemical vapor deposition of functional coatings. *Plasma Sources Science and Technology*, 32(7), 074001.
[126] Piallat, F., Vallée, C., Gassilloud, R., Michallon, P., Pélissier, B., & Caubet, P. (2014). PECVD RF versus dual frequency: An investigation of plasma influence on metal-organic precursors' decomposition and material characteristics. *Journal of Physics D: Applied Physics*, 47(18), 185201.
[127] Iliescu, C. (2021). A comprehensive review on thin film depositions on PECVD reactors. *Annals of the Academy of Romanian Scientists Series on Science and Technology of Information*, 14(1–2), 12.
[128] Lee, J., Jang, J., & Hong, S. J. (2025). Understanding the chamber wall-deposited thin film of plasma deposition equipment for the efficiency of in situ dry-cleaning. *Coatings*, 15(5), 563.
[129] Hariharan, V., Vanjaria, J., Arjunan, A. C., Tompa, G. S., & Yu, H. (2021). Effect of chamber conditions and substrate type on PECVD of SiGeSn films. *Crystal Structure Theory and Applications*, 10(3), 39–56.
[130] Singh, B., Fischer, T., & Mathur, S. (2025). Field-enhanced chemical vapor deposition: New perspectives for thin film growth. *Journal of Materials Chemistry A*, 13, 20104–20142.
[131] Irvine, S. J. C. (2002). Photochemical vapour deposition of thin films. In *Chemical Physics of Thin Film Deposition Processes for Micro-and Nano-Technologies* (pp. 199–222). Dordrecht: Springer Netherlands.

[132] Shen, Y., Qiu, M., Huang, X., Du, W., He, X., Luo, Y., & Zhao, J. (2024). Photocatalytic surface initiation for area-selective chemical vapor deposition of polymer thin film. *ACS Materials Letters*, 6(9), 4058–4065.
[133] Prasad, R. D., Teli, B., Prasad, R. S., Prasad, R. B., Prasad, S. R., Sinha, P., & Guo, Z. (2024). A review on thin film technology and nanomaterial characterization techniques. *ES Materials & Manufacturing*, 25, 1198.
[134] Lisha, F., Fan, L., Guolong, W., Kovalenko, V. S., & Jianhua, Y. (2025). Research progress of laser-assisted chemical vapor deposition. *Opto-Electronic Engineering*, 49(2), 210333–1.
[135] Hanabusa, M. (1987). Photoinduced deposition of thin films. *Materials Science Reports*, 2((2)), 51–97.
[136] Braichotte, D., Garrido, C., & Van den Bergh, H. (1990). The photolytic laser chemical vapor deposition rate of platinum, its dependence on wavelength, precursor vapor pressure, light intensity, and laser beam diameter. *Applied Surface Science*, 46(1–4), 9–18.
[137] Eres, D., Lowndes, D. H., Geohegan, D. B., & Mashburn, D. N. (1987). Laser photochemical growth of amorphous silicon at low temperatures and comparison with thermal chemical vapor deposition. *MRS Online Proceedings Library*, 101(1), 355–360.
[138] Fowler, B., Lian, S., Krishnan, S., Jung, L., Li, C., Samara, D., & Banerjee, S. (1992). Epitaxial silicon growth conditions and kinetics in low-temperature ArF excimer laser photochemical-vapor deposition from disilane. *Journal of Applied Physics*, 72(3), 1137–1148.
[139] Blair, S. L., Chu, C. W., Dammel, R. R., & Hill, R. H. (1997, July). Use of complex coordination chemistry for the deposition of inorganic materials: Spin on metals and photoresist-free lithography. In *Advances in Resist Technology and Processing XIV* (Vol. 3049, pp. 829–837). SPIE.
[140] Shtein, M., Gossenberger, H. F., Benziger, J. B., & Forrest, S. R. (2001). Material transport regimes and mechanisms for growth of molecular organic thin films using low-pressure organic vapor phase deposition. *Journal of Applied Physics*, 89(2), 1470–1476.
[141] Sun, L., Yuan, G., Gao, L., Yang, J., Chhowalla, M., Gharahcheshmeh, M. H., & Liu, Z. (2021). Chemical vapour deposition. *Nature Reviews Methods Primers*, 1(1), 5.
[142] Creighton, J. R., & Ho, P. (2001). Introduction to chemical vapor deposition (CVD). *ASM International*, 407.
[143] Teyssandier, F., & Wang, Y. B. (1995). Gas phase reactivity in chemical vapor deposition. *Surface and Coatings Technology*, 76, 303–310.
[144] Choy, K. L. (2003). Chemical vapour deposition of coatings. *Progress in Materials Science*, 48(2), 57–170.
[145] Majidi, S., Zeinali Sehrig, F., Farkhani, S. M., Soleymani Goloujeh, M., & Akbarzadeh, A. (2016). Current methods for synthesis of magnetic nanoparticles. *Artificial Cells, Nanomedicine, and Biotechnology*, 44(2), 722–734.
[146] Nguyen, S. V. (1999). High-density plasma chemical vapor deposition of silicon-based dielectric films for integrated circuits. *IBM Journal of Research and Development*, 43(1.2), 109–126.
[147] Creighton, J. R., & Parmeter, J. E. (1993). Metal CVD for microelectronic applications: An examination of surface chemistry and kinetics. *Critical Reviews in Solid State and Material Sciences*, 18(2), 175–237.
[148] Ban, V. S., & Gilbert, S. L. (1975). The chemistry and transport phenomena of chemical vapor deposition of silicon from $SiCl_4$. *Journal of Crystal Growth*, 31, 284–289.
[149] Perrin, J., Leroy, O., & Bordage, M. C. (1996). Cross-sections, rate constants, and transport coefficients in silane plasma chemistry. *Contributions to Plasma Physics*, 36(1), 3–49.
[150] Komiyama, H., Shimogaki, Y., & Egashira, Y. (1999). Chemical reaction engineering in the design of CVD reactors. *Chemical Engineering Science*, 54(13-14), 1941–1957.
[151] Jones, A. C., & Hitchman, M. L. (2009). Overview of chemical vapour deposition. Chemical vapour deposition: Precursors, processes and applications. *Royal Society of Chemistry (RSC)*, 1, 1–36.
[152] Wang, Y. F., & Pollard, R. (1995). An approach for modeling surface reaction kinetics in chemical vapor deposition processes. *Journal of the Electrochemical Society*, 142(5), 1712.

[153] Manawi, Y. M., Ihsanullah, Samara, A., Al-Ansari, T., & Atieh, M. A. (2018). A review of carbon nanomaterials' synthesis via the chemical vapor deposition (CVD) method. *Materials*, 11(5), 822.
[154] Dresselhaus, M. S., & Terrones, M. (2013). Carbon-based nanomaterials from a historical perspective. *Proceedings of the IEEE*, 101(7), 1522–1535.
[155] Kroto, H. W., Heath, J. R., O'Brien, S. C., Curl, R. F., & Smalley, R. E. (1985). C60: Buckminsterfullerene. *Nature*, 318(6042), 162–163.
[156] Geim, A. K., & Novoselov, K. S. (2007). The rise of graphene. *Nature Materials*, 6(3), 183–191.
[157] Novoselov, K. S., Geim, A. K., Morozov, S. V., Jiang, D. E., Zhang, Y., Dubonos, S. V., & Firsov, A. A. (2004). Electric field effect in atomically thin carbon films. *Science*, 306(5696), 666–669.
[158] Hernandez, Y., Nicolosi, V., Lotya, M., Blighe, F. M., Sun, Z., De, S., & Coleman, J. N. (2008). High-yield production of graphene by liquid-phase exfoliation of graphite. *Nature Nanotechnology*, 3(9), 563–568.
[159] Mattevi, C., Kim, H., & Chhowalla, M. (2011). A review of chemical vapour deposition of graphene on copper. *Journal of Materials Chemistry*, 21(10), 3324–3334.
[160] Khan, Y., Sadia, H., Ali Shah, S. Z., Khan, M. N., Shah, A. A., Ullah, N., & Khan, M. I. (2022). Classification, synthetic, and characterization approaches to nanoparticles, and their applications in various fields of nanotechnology: A review. *Catalysts*, 12(11), 1386.
[161] Sutter, P. W., Flege, J. I., & Sutter, E. A. (2008). Epitaxial graphene on ruthenium. *Nature Materials*, 7(5), 406–411.
[162] McCarty, K. F., Feibelman, P. J., Loginova, E., & Bartelt, N. C. (2009). Kinetics and thermodynamics of carbon segregation and graphene growth on Ru (0 0 0 1). *Carbon*, 47(7), 1806–1813.
[163] Li, X., Cai, W., An, J., Kim, S., Nah, J., Yang, D., & Ruoff, R. S. (2009). Large-area synthesis of high-quality and uniform graphene films on copper foils. *Science*, 324(5932), 1312–1314.
[164] Avishan, N., Hussain, N., & Nosheen, F. (2022). Large-scale graphene production and transfer for industrial applications. *Materinnov*, 2, 15.
[165] Wyckoff, R. W. G., & Wyckoff, R. W. (1963). *Crystal Structures* (Vol. 1, pp. 26–27). New York: Interscience Publishers.
[166] Reina, A., Jia, X., Ho, J., Nezich, D., Son, H., Bulovic, V., & Kong, J. (2009). Large area, few-layer graphene films on arbitrary substrates by chemical vapor deposition. *Nano Letters*, 9(1), 30–35.
[167] Kim, K. S., Zhao, Y., Jang, H., Lee, S. Y., Kim, J. M., Kim, K. S., & Hong, B. H. (2009). Large-scale pattern growth of graphene films for stretchable transparent electrodes. *Nature*, 457(7230), 706–710.
[168] Obraztsov, A. N. (2009). Making graphene on a large scale. *Nature Nanotechnology*, 4(4), 212–213.
[169] Li, X., Magnuson, C. W., Venugopal, A., Tromp, R. M., Hannon, J. B., Vogel, E. M., & Ruoff, R. S. (2011). Large-area graphene single crystals grown by low-pressure chemical vapor deposition of methane on copper. *Journal of the American Chemical Society*, 133(9), 2816–2819.
[170] Li, X., Zhu, Y., Cai, W., Borysiak, M., Han, B., Chen, D., & Ruoff, R. S. (2009). Transfer of large-area graphene films for high-performance transparent conductive electrodes. *Nano Letters*, 9(12), 4359–4363.
[171] Bae, S., Kim, H., Lee, Y., Xu, X., Park, J. S., Zheng, Y., & Iijima, S. (2010). Roll-to-roll production of 30-inch graphene films for transparent electrodes. *Nature Nanotechnology*, 5(8), 574–578.
[172] Sutter, P., Sadowski, J. T., & Sutter, E. A. (2010). Chemistry under cover: Tuning metal–graphene interaction by reactive intercalation. *Journal of the American Chemical Society*, 132(23), 8175–8179.
[173] Zhang, Y. I., Zhang, L., & Zhou, C. (2013). Review of chemical vapor deposition of graphene and related applications. *Accounts of Chemical Research*, 46(10), 2329–2339.
[174] Zhang, Y., Zhang, L., Kim, P., Ge, M., Li, Z., & Zhou, C. (2012). Vapor trapping growth of single-crystalline graphene flowers: Synthesis, morphology, and electronic properties. *Nano Letters*, 12(6), 2810–2816.
[175] Hao, Y., Bharathi, M. S., Wang, L., Liu, Y., Chen, H., Nie, S., & Ruoff, R. S. (2013). The role of surface oxygen in the growth of large single-crystal graphene on copper. *Science*, 342(6159), 720–723.

[176] Dresselhaus, M. S., & Avouris, P. (2001). Introduction to carbon materials research. In *Carbon Nanotubes: Synthesis, Structure, Properties, and Applications* (pp. 1–9). Berlin, Heidelberg: Springer Berlin Heidelberg.
[177] Iijima, S. (1991). Helical microtubules of graphitic carbon. *Nature*, 354(6348), 56–58.
[178] Yu, M. F., Files, B. S., Arepalli, S., & Ruoff, R. S. (2000). Tensile loading of ropes of single wall carbon nanotubes and their mechanical properties. *Physical Review Letters*, 84(24), 5552.
[179] Zhang, Y., Chang, A., Cao, J., Wang, Q., Kim, W., Li, Y., & Dai, H. (2001). Electric-field-directed growth of aligned single-walled carbon nanotubes. *Applied Physics Letters*, 79(19), 3155–3157.
[180] Saito, R., Dresselhaus, G., Dresselhaus, M. S., & Dresselhaus, G. (1998). *Synthesis of Carbon Nanotubes In Physical properties of carbon nanotubes* (Vol. 203, pp. 73–93). London: Imperial College Press. doi: https://doi.org/10.1142/p080.
[181] Ebbesen, T. W., & Ajayan, P. M. (1992). Large-scale synthesis of carbon nanotubes. *Nature*, 358(6383), 220–222.
[182] Hamada, N., Sawada, S. I., & Oshiyama, A. (1992). New one-dimensional conductors: Graphitic microtubules. *Physical Review Letters*, 68(10), 1579.
[183] Dresselhaus, M. S., Dresselhaus, G., & Saito, R. (1995). Physics of carbon nanotubes. *Carbon*, 33(7), 883–891.
[184] Journet, C., & Bernier, P. (1998). Production of carbon nanotubes. *Applied Physics A: Materials Science & Processing*, 67(1).
[185] Kumar, M., & Ando, Y. (2010). Chemical vapor deposition of carbon nanotubes: A review on growth mechanism and mass production. *Journal of Nanoscience and Nanotechnology*, 10(6), 3739–3758.
[186] Lim, Y. D., Avramchuck, A. V., Grapov, D., Tan, C. W., Tay, B. K., Aditya, S., & Labunov, V. (2017). Enhanced carbon nanotubes growth using nickel/ferrocene-hybridized catalyst. *ACS Omega*, 2(9), 6063–6071.
[187] Chauhan, D., Pujari, A., Zhang, G., Dasgupta, K., Shanov, V. N., & Schulz, M. J. (2022). Effect of a metallocene catalyst mixture on CNT yield using the FC-CVD process. *Catalysts*, 12(3), 287.
[188] Moon, S. Y., Jeon, S. Y., Lee, S. H., Lee, A., & Kim, S. M. (2022). High purity single wall carbon nanotube by oxygen-containing functional group of ferrocene-derived catalyst precursor by floating catalyst chemical vapor deposition. *Nanomaterials*, 12(5), 863.
[189] Zhou, O., Shimoda, H., Gao, B., Oh, S., Fleming, L., & Yue, G. (2002). Materials science of carbon nanotubes: Fabrication, integration, and properties of macroscopic structures of carbon nanotubes. *Accounts of Chemical Research*, 35(12), 1045–1053.
[190] Lehman, J. H., Terrones, M., Mansfield, E., Hurst, K. E., & Meunier, V. (2011). Evaluating the characteristics of multiwall carbon nanotubes. *Carbon*, 49(8), 2581–2602.
[191] Maruyama, S., Einarsson, E., Murakami, Y., & Edamura, T. (2005). Growth process of vertically aligned single-walled carbon nanotubes. *Chemical Physics Letters*, 403(4-6), 320–323.
[192] Singh, C., Shaffer, M. S., & Windle, A. H. (2003). Production of controlled architectures of aligned carbon nanotubes by an injection chemical vapour deposition method. *Carbon*, 41(2), 359–368.
[193] Jiang, Q., Wu, Y., Wang, F., Zhu, P., Li, R., Zhao, Y., & Zhang, R. (2024). Floating bimetallic catalysts for growing 30 cm-long carbon nanotube arrays with high yields and uniformity. *Advanced Materials*, 36(32), 2402257.
[194] Yang, Z., Xia, Y., & Mokaya, R. (2005). Aligned N-doped carbon nanotube bundles prepared via CVD using zeolite substrates. *Chemistry of Materials*, 17(17), 4502–4508.
[195] Rodriguez, N. M. (1993). A review of catalytically grown carbon nanofibers. *Journal of Materials Research*, 8(12), 3233–3250.
[196] Rodriguez, N. M., Chambers, A., & Baker, R. T. K. (1995). Catalytic engineering of carbon nanostructures. *Langmuir*, 11(10), 3862–3866.
[197] Hayashi, T., Kim, Y. A., Matoba, T., Esaka, M., Nishimura, K., Tsukada, T., & Dresselhaus, M. S. (2003). Smallest freestanding single-walled carbon nanotube. *Nano Letters*, 3(7), 887–889.

[198] Yoon, S. H., Lim, S., Hong, S. H., Qiao, W., Whitehurst, D. D., Mochida, I., & Yokogawa, K. (2005). A conceptual model for the structure of catalytically grown carbon nano-fibers. *Carbon*, 43(9), 1828–1838.
[199] Baker, R. T. K. (1989). Catalytic growth of carbon filaments. *Carbon*, 27(3), 315–323.
[200] Hofmann, S., Ducati, C., Robertson, J., & Kleinsorge, B. (2003). Low-temperature growth of carbon nanotubes by plasma-enhanced chemical vapor deposition. *Applied Physics Letters*, 83(1), 135–137.
[201] Ruiz-Cornejo, J. C., Sebastián, D., & Lázaro, M. J. (2020). Synthesis and applications of carbon nanofibers: A review. *Reviews in Chemical Engineering*, 36(4), 493–511.
[202] Helveg, S., López-Cartes, C., Sehested, J., Hansen, P. L., Clausen, B. S., Rostrup-Nielsen, J. R., & Nørskov, J. K. (2004). Atomic-scale imaging of carbon nanofibre growth. *Nature*, 427(6973), 426–429.
[203] Sevilla, M., & Fuertes, A. B. (2006). Catalytic graphitization of templated mesoporous carbons. *Carbon*, 44(3), 468–474.
[204] Serp, P., Corrias, M., & Kalck, P. (2003). Carbon nanotubes and nanofibers in catalysis. *Applied Catalysis A: General*, 253(2), 337–358.
[205] Tibbetts, G. G., Bernardo, C. A., Gorkiewicz, D. W., & Alig, R. L. (1994). Role of sulfur in the production of carbon fibers in the vapor phase. *Carbon*, 32(4), 569–576.
[206] Baker, R. T. K., Barber, M. A., Harris, P. S., Feates, F. S., & Waite, R. J. (1972). Nucleation and growth of carbon deposits from the nickel catalyzed decomposition of acetylene. *Journal of Catalysis*, 26(1), 51–62.
[207] Bartholomew, C. H. (2001). Mechanisms of catalyst deactivation. *Applied Catalysis A: General*, 212(1-2), 17–60.
[208] Rostrup-Nielsen, J. R. (1973). Activity of nickel catalysts for steam reforming of hydrocarbons. *Journal of Catalysis*, 31(2), 173–199.
[209] Tibbetts, G. G., Lake, M. L., Strong, K. L., & Rice, B. P. (2007). A review of the fabrication and properties of vapor-grown carbon nanofiber/polymer composites. *Composites Science and Technology*, 67(7-8), 1709–1718.
[210] De Jong, K. P., & Geus, J. W. (2000). Carbon nanofibers: Catalytic synthesis and applications. *Catalysis Reviews*, 42(4), 481–510.
[211] Su, D. S., Perathoner, S., & Centi, G. (2013). Nanocarbons for the development of advanced catalysts. *Chemical Reviews*, 113(8), 5782–5816.
[212] Mattos, L. V., Jacobs, G., Davis, B. H., & Noronha, F. B. (2012). Production of hydrogen from ethanol: Review of reaction mechanism and catalyst deactivation. *Chemical Reviews*, 112(7), 4094–4123.
[213] Krätschmer, W., Lamb, L. D., Fostiropoulos, K. H. D. R., & Huffman, D. R. (1990). Solid C60: A new form of carbon. *Nature*, 347(6291), 354–358.
[214] Foote, C. S. (2005). Photophysical and photochemical properties of fullerenes. In *Electron Transfer I* (pp. 347–363). Berlin, Heidelberg: Springer.
[215] Kroto, H. W., Allaf, A. W., & Balm, S. P. (1991). C60: Buckminsterfullerene. *Chemical Reviews*, 91(6), 1213–1235.
[216] Ajie, H., Alvarez, M. M., Anz, S. J., Beck, R. D., Diederich, F., Fostiropoulos, K., & Rubin, Y. (1990). Characterization of the soluble all-carbon molecules C_{60} and C_{70}. *Journal of Physical Chemistry*, 94(24), 8630–8633.
[217] Bratovcic, A. (2023, May). Biomedical application of nanocomposites based on fullerenes C60. In *International Conference "New Technologies, Development and Applications"* (pp. 107–117). Cham: Springer Nature Switzerland.
[218] Arbogast, J. W., & Foote, C. S. (1991). Photophysical properties of C_{70}. *Journal of the American Chemical Society*, 113(23), 8886–8889.
[219] Sherigara, B. S., Kutner, W., & D'Souza, F. (2003). Electrocatalytic properties and sensor applications of fullerenes and carbon nanotubes. *Electroanalysis: An International Journal Devoted to Fundamental and Practical Aspects of Electroanalysis*, 15(9), 753–772.

[220] Achiba, Y., Kikuchi, K., Aihara, Y., Wakabayashi, T., Miyake, Y., & Kainosho, M. (1994). Higher fullerenes: Structure and properties. *MRS Online Proceedings Library*, 359(1), 3–9.

[221] Schwerdtfeger, P., Wirz, L. N., & Avery, J. (2015). The topology of fullerenes. *Wiley Interdisciplinary Reviews: Computational Molecular Science*, 5(1), 96–145.

[222] Shinohara, H. (2000). Endohedral metallofullerenes. *Reports on Progress in Physics*, 63(6), 843.

[223] Stevenson, S., Rice, G., Glass, T., Harich, K., Cromer, F., Jordan, M. R., & Dorn, A. H. (1999). Small-bandgap endohedral metallofullerenes in high yield and purity. *Nature*, 401(6748), 55–57; and Erratum: Small-bandgap endohedral metallofullerenes in high yield and purity. *Nature*, 402(6764), 898-898.

[224] Thilgen, C., & Diederich, F. (2006). Structural aspects of fullerene chemistry: A journey through fullerene chirality. *Chemical Reviews*, 106(12), 5049–5135.

[225] Wang, X. Q., Wang, C. Z., Zhang, B. L., & Ho, K. M. (1993). Structural and electronic properties of large fullerenes. *Zeitschrift Für Physik D: Atoms, Molecules and Clusters*, 26(Suppl. 1), 264–266.

[226] Thilgen, C., Herrmann, A., & Diederich, F. (1997). The covalent chemistry of higher fullerenes: C_{70} and beyond. *Angewandte Chemie International Edition in English*, 36(21), 2268–2280.

[227] Balasubramanian, K. (2020). Combinatorics of edge symmetry: Chiral and achiral edge colorings of icosahedral giant fullerenes: C_{80}, C_{180}, and C_{240}. *Symmetry*, 12(8), 1308.

[228] Curl, R. F., & Smalley, R. E. (1991). Fullerenes. *Scientific American*, 265(4), 54–63.

[229] Ruoff, R. S., Tse, D. S., Malhotra, R., & Lorents, D. C. (1993). Solubility of fullerene (C60) in a variety of solvents. *The Journal of Physical Chemistry*, 97(13), 3379–3383.

[230] Liu, J., Qiu, L., & Shao, S. (2021). Emerging electronic applications of fullerene derivatives: An era beyond OPV. *Journal of Materials Chemistry C*, 9(45), 16143–16163.

[231] Prato, M., Kostarelos, K., & Bianco, A. (2008). Functionalized carbon nanotubes in drug design and discovery. *Accounts of Chemical Research*, 41(1), 60–68.

[232] Castro, E., García, A. H., Zavala, G., & Echegoyen, L. (2017). Fullerenes in biology and medicine. *Journal of Materials Chemistry B*, 5(32), 6523–6535.

[233] Partha, R., & Conyers, J. L. (2009). Biomedical applications of functionalized fullerene-based nanomaterials. *International Journal of Nanomedicine*, 261–275.

[234] Brabec, C. J., Sariciftci, N. S., & Hummelen, J. C. (2001). Plastic solar cells. *Advanced Functional Materials*, 11(1), 15–26.

[235] Murata, Y., Murata, M., & Komatsu, K. (2003). 100% encapsulation of a hydrogen molecule into an open-cage fullerene derivative and gas-phase generation of H_2@ C_{60}. *Journal of the American Chemical Society*, 125(24), 7152–7153.

[236] Liu, H., Wang, J., Chanana, A., & Vardeny, Z. V. (2019). Studies of spin transport in fullerene films. *Journal of Applied Physics*, 125(14), 142908.

[237] Dunsch, L., & Yang, S. (2007). Metal nitride cluster fullerenes: Their current state and future prospects. *Small*, 3(8), 1298–1320.

[238] Loboda, O., Zalesny, R., Avramopoulos, A., Luis, J. M., Kirtman, B., Tagmatarchis, N., & Papadopoulos, M. G. (2009). Linear and nonlinear optical properties of [60] fullerene derivatives. *The Journal of Physical Chemistry A*, 113(6), 1159–1170.

[239] Fowler, P. W., Rogers, K. M., Fajtlowicz, S., Hansen, P., & Caporossi, G. (2001). Facts and conjectures about fullerene graphs: Leapfrog, cylinder, and Ramanujan fullerenes. In *Algebraic Combinatorics and Applications: Proceedings of the Euroconference, Algebraic Combinatorics and Applications (ALCOMA), held in Gößweinstein Germany, September 12-19, 1999* (pp. 134–146). Berlin, Heidelberg:: Springer, Berlin Heidelberg.

[240] Guldi, D. M., & Asmus, K. D. (1997). Photophysical properties of mono-and multiply-functionalized fullerene derivatives. *The Journal of Physical Chemistry A*, 101(8), 1472–1481.

[241] Hirahara, K., Suenaga, K., Bandow, S., Kato, H., Okazaki, T., Shinohara, H., & Iijima, S. (2000). One-dimensional metallofullerene crystal generated inside single-walled carbon nanotubes. *Physical Review Letters*, 85(25), 5384.

[242] Khamatgalimov, A. R., & Kovalenko, V. I. (2021). Substructural approach for assessing the stability of higher fullerenes. *International Journal of Molecular Sciences*, 22(7), 3760.

[243] Bandosz, T. J., Biggs, M. J., Gubbins, K. E., Hattori, Y., Uyama, T., Kaneko, K., & Thomson, K. T. (2003). Molecular models of porous carbons. In *Chemistry & Physics of Carbon* (pp. 61–81). CRC Press.

[244] Chow, L., Wang, H., Kleckley, S., Daly, T. K., & Buseck, P. R. (1995). Fullerene formation during production of chemical vapor deposited diamond. *Applied Physics Letters*, 66(4), 430–432.

[245] Kleckley, S., Wang, H., Oladeji, I., Chow, L., Daly, T. K., Buseck, P. R., & Marshall, A. (1998). Fullerenes and polymers produced by the chemical vapor deposition method. *Synthesis and Characterization of Advanced Materials, Chapter 6*, 51–60.

[246] Scott, L. T. (2010). Polycyclic aromatic hydrocarbon bowls, baskets, balls, and tubes: Challenging targets for chemical synthesis. *Polycyclic Aromatic Compounds*, 30(5), 247–259.

[247] Shoukat, R., & Khan, M. I. (2022). Carbon nanotubes/nanofibers (CNTs/CNFs): A review on state of the art synthesis methods. *Microsystem Technologies*, 28(4), 885–901.

[248] Chatterjee, S., Abadie, T., Wang, M., Matar, O. K., & Ruoff, R. S. (2024). Repeatability and reproducibility in the chemical vapor deposition of 2D films: A physics-driven exploration of the reactor black box. *Chemistry of Materials*, 36(3), 1290–1298.

Xuan Wang*, Sajid Bashir, Jingbo Louise Liu

Chapter 5
Atomic layer deposition of nanomaterials: fabrication, structure, and property evaluation

5.1 Introduction: atomic layer deposition

Atomic layer deposition (ALD) is a thin-film fabrication technique that enables precise control over material thickness at the atomic scale. It works by alternating self-limiting chemical reactions between gaseous precursors and a substrate surface, allowing for layer-by-layer growth. ALD is widely used in semiconductor manufacturing, energy storage devices, and biomedical coatings due to its uniformity, conformality, and ability to coat complex 3D structures. Typical deposition temperatures range from 10 °C to 300 °C, depending on the precursor chemistry and substrate compatibility. Its precision makes ALD ideal for producing ultrathin dielectric layers, barrier coatings, and nanoscale heterostructures with atomic-level accuracy. ALD offers precise, atomic-scale control over ultrathin film growth, enabling uniform and conformal coatings on complex, high-aspect-ratio structures. The self-limiting surface reactions of ALD ensure reproducible thickness and excellent film quality across diverse substrates, positioning ALD as a foundational technique for next-generation electronic, energy, and biomedical

Acknowledgments: The research efforts led by Dr Bashir and Dr Liu were supported by the American Chemical Society Petroleum Research Fund (Grant 53827-UR10) federal agencies, including the Office of Naval Research, the Air Force Office of Scientific Research, the Department of Energy, the National Science Foundation, the Department of Education, the Department of Homeland Security (ORISE), and the National Institutes of Health. The Robert A. Welch Foundation (Departmental Grant AC-0006) is gratefully acknowledged for its support to student training. Technical assistance from Texas A&M University and Texas A&M University-Kingsville is also acknowledged with thanks. The contributions and dedication of collaborators and students are deeply appreciated and recognized as integral to the success of this work.

Authors' contributions: Dr X. Wang prepared the original manuscript for the first edition of the book. Dr J. Louise Liu provided substantial revisions that enhanced the clarity and depth of the content. Dr S. Bashir completed the final revision, performed data analysis, and obtained the necessary copyright permissions. All authors reviewed the final version and approved it for publication.

***Corresponding author: Xuan Wang,** Texas A&M Higher Education Center at McAllen, McAllen, TX, USA, e-mail: xuan.wang@science.tamu.edu
Sajid Bashir, Jingbo Louise Liu: The Department of Chemistry, Texas A&M University-Kingsville, MSC161, 700 University Blvd., Kingsville, TX 78363, USA; Texas A&M Energy Institute,1617 Research Parkway, Suite 308, College Station, TX 77843-3372, USA; Louise Nanoinnovation North American, 7534 Beau Terre, Corpus Christi, TX 78414, USA; e-mails: jingbo.liu@tamuk.edu, jingbo.liu@tamu.edu

 | https://doi.org/10.1515/9783111621111-006

technologies. Metal oxides (e.g., Al_2O_3) remain the model ALD system, with typical growth rates of 1.1–1.2 Å per cycle that exemplify the precision achievable through sequential, self-terminating reactions. Both thermal ALD and plasma- or radical-enhanced variants expand precursor flexibility and support lower temperature processing, broadening compatibility with temperature-sensitive materials. Reactor architecture, metal-film chemistries, and substrate-specific nucleation behavior, each of which governs the quality of the final films on substrates such as H-terminated Si, oxides, carbon nanotubes, and graphene, play central roles in determining and tuning the properties of ALD-derived materials. Beyond semiconductors, ALD now enables coatings on polymers, biological templates, and nanostructured architectures, underscoring its rapidly growing technological impact [1].

This chapter is organized into four parts: Section 5.1 provides an overview of ALD; Section 5.2 outlines its governing principles and key synthesis parameters; Section 5.3 compares various deposition approaches; and Section 5.4 highlights ALD-derived materials and their functional applications. Section 5.1 introduces ALD as a vapor-phase, self-limiting thin-film technique capable of delivering atomic-scale control over thickness, composition, and conformality. The discussion highlights ALD's distinction from conventional chemical vapor deposition, focusing on its sequential surface reactions, exceptional uniformity on complex three-dimensional (3D) substrates, and compatibility with a wide range of materials systems. Section 5.2 provides an in-depth analysis of the governing principles and key synthesis parameters that define ALD performance. Critical variables such as precursor pulse time, purge duration, deposition temperature, precursor reactivity, surface activation, and cycle count are examined in relation to nucleation behavior, growth per cycle (GPC), film purity, and conformality. This section establishes how precise control over these parameters enables deterministic tuning of film thickness, crystallinity, interface quality, and defect density, particularly on high aspect ratio and chemically heterogeneous surfaces. Section 5.3 compares a range of ALD deposition approaches, including thermal ALD, plasma-enhanced ALD, radical-enhanced ALD, spatial ALD, roll-to-roll ALD, and molecular layer deposition. Each method is evaluated in terms of reaction mechanisms, temperature windows, throughput, and material compatibility, illustrating how process innovations expand ALD's applicability to temperature-sensitive substrates, flexible electronics, hybrid organic–inorganic materials, and large-area manufacturing. Finally, Section 5.4 highlights ALD-derived materials and their functional applications, spanning metal oxides, transition-metal dichalcogenides, nanomembranes, and metal–organic frameworks. The section emphasizes structure–property relationships and demonstrates how ALD enables advanced materials for energy storage and conversion, electronic and optoelectronic devices, catalysis, sensing, and biomedical technologies. Collectively, the chapter positions ALD as a foundational tool for next-generation nanomaterial engineering and device integration.

5.2 Overview of atomic layer deposition

ALD is a vapor-phase thin-film technique derived from chemical vapor deposition (CVD), distinguished by its use of self-limiting surface reactions to achieve atomic-scale precision. Unlike conventional CVD, which relies on simultaneous precursor reactions in the gas phase, ALD introduces gaseous precursors sequentially into the reaction chamber, allowing each to adsorb onto the substrate surface until saturation is reached [2]. Once a monolayer is formed, the reaction halts automatically, ensuring uniform coverage and eliminating the need for precise precursor dosing. A purge step, typically using an inert noble gas or vacuum evacuation, is then employed to remove excess reactants and by-products before introducing the next precursor. This cycle is repeated to build up the desired film thickness with angstrom-level control. Importantly, ALD precursors do not need to be highly reactive, as they react independently with the surface rather than with each other [2]. To ensure high film purity and conformality, the pulse duration must be sufficient to thoroughly saturate the surface, and the purge time must be long enough to eliminate residual contaminants.

Historically, ALD has been employed to produce dense, uniform films of metal oxides, nitrides, sulfides, and elemental materials, reflecting its origins as a versatile gas–solid reaction technique capable of forming conformal inorganic layers with nanometer-scale precision. As summarized by Puurunen [3], these processes rely on sequential, self-terminating surface reactions that enable controlled growth across a wide range of chemistries and substrate types. More recent advances have extended ALD's utility to post-synthetic surface modification of transition metal dichalcogenides (TMDCs) and metal–organic frameworks (MOFs), enabling tailored surface chemistry and enhanced functionality [4]. These developments have broadened ALD's relevance in fields such as catalysis, energy storage, and molecular electronics. To address limitations in deposition rate and thermal sensitivity, several innovative ALD variants have emerged. Lift-off ALD, introduced by Biercuk et al. [4], enables low-temperature processing by applying a resist pattern before film growth; the resist and overlying material are removed post deposition, preserving substrate integrity. Spatial ALD eliminates purge steps by physically separating half-reactions using modular reactor designs, such as rotating cylinders or slit-based gas sources, allowing continuous flow and high-throughput processing [5]. Radical-enhanced ALD further accelerates reaction kinetics by introducing plasma or radical species, improving film quality at reduced temperatures. Together, these innovations reinforce ALD's position as a cornerstone technique in nanoscale fabrication, offering unmatched control over film thickness, composition, and interface engineering across a wide range of materials and applications. Although ALD is inherently slower than CVD due to its cyclic nature, it offers superior control over film uniformity, step coverage, and thickness reproducibility, especially on high aspect ratio of 3D nanostructures. However, its surface-dependent growth mechanism can result in low deposition rates, particularly on inert or poorly activated substrates. ALD is widely used in semiconductor fabrication, energy storage devices, and biomedical coatings, where precision and conformality are critical. Its ability to deposit ultrathin layers of oxides, nitrides, and metals with sub-nanometer accuracy makes it indispensable for gate dielectrics, diffusion barriers, and surface passivation layers. As research advances, efforts continue to optimize precursor chemistry, reactor design, and cycle efficiency to expand ALD's applicability across emerging fields such as flexible electronics, quantum devices, and nanoscale catalysis [6].

5.3 Principles and synthesis variables

ALD operates on the principle of sequential, self-limiting surface reactions to fabricate thin films with atomic-scale precision. Unlike conventional CVD, which relies on simultaneous gas-phase reactions, ALD introduces gaseous precursors into the reaction chamber in an alternating sequence (Fig. 5.1A). Each precursor reacts only with available surface sites, forming a saturated monolayer and halting further reaction. This self-limiting behavior ensures exceptional conformality and uniform film growth, even on complex geometries and high-aspect-ratio structures [7]. The ALD cycle consists of four discrete steps: ***A. Precursor A pulse***, ***B. purge step***, ***C. precursor B pulse***, and ***D. second purge***. Each cycle deposits a sub-monolayer, and the number of cycles precisely controls film thickness. ALD typically operates at temperatures between 100 and 300 °C, depending on the precursor chemistry and substrate compatibility. Significantly, precursors in ALD do not decompose in the gas phase but react independently on the surface, eliminating the need for highly reactive species required in CVD [8]. Because growth terminates upon saturation, precursor dosing does not require fine control, though pulse and purge durations must be optimized to ensure complete surface coverage and impurity removal. This stepwise mechanism results in longer process times, compared to CVD, and yields that are highly surface-dependent, often leading to lower deposition rates. The following section provides a detailed discussion of the four sequential steps involved in the ALD process, each conducted within a precisely controlled reactor chamber.

A. Precursor A pulse – chemisorption of the first precursor: The ALD cycle begins with the introduction of precursor A into the reactor chamber under controlled flow conditions. This precursor, typically a volatile metal-organic compound such as trimethylaluminum (TMA) or titanium tetrachloride ($TiCl_4$), is delivered at a flow rate of 10–100 sccm and with pulse duration of 0.5–5 s, depending on surface reactivity and chamber volume. Upon contacting the substrate, the precursor undergoes chemisorption, forming a monolayer via ligand exchange or surface coordination. The reaction is self-limiting: once all reactive surface sites are occupied, further adsorption ceases, ensuring uniform coverage. Substrate temperature (T_{sub}) is typically maintained between 100 and 30 °C to promote surface mobility without thermal decomposition. The surface reaction rate (k_{surf}) and sticking coefficient (α) govern the saturation kinetics. No gas-phase reaction occurs during this step, preserving film purity. The success of this phase depends on precursor volatility, molecular size, and steric accessibility. This step sets the foundation for atomic-scale control and conformal film growth across complex geometries [9].

B. Purge step – removal of excess precursor and by-products: Following the precursor, a pulse is initiated to remove unreacted precursor molecules and volatile reaction by-products from the reactor chamber. This is achieved by introducing an inert carrier gas, typically nitrogen (N_2) or argon (Ar), at a flow rate of 100–500 sccm, depending on the reactor volume and precursor volatility. The purge duration (t_{purge}) typically ranges from 2 to 10 s, optimized to ensure complete evacuation without disrupting surface-bound species. Reactor pressure ($P_{reactor}$) is maintained between 0.1 and 10 torr to facilitate laminar flow and prevent back-diffusion. The purge gas displaces residual molecules via convective transport and molecular diffusion, governed by the Reynolds number (Re) and the Knudsen number (K_n). Efficient purging is critical to prevent gas-phase reactions during subsequent precursor pulses, which could compromise film uniformity and stoichiometry. Advanced systems may use vacuum-assisted purging or pulsed pressure modulation to enhance throughput. This step ensures chemical isolation between half-reactions, preserving ALD's self-limiting nature and enabling high-purity, defect-free film growth [10].

C. Precursor B pulse – surface reaction to form the desired material: Once the chamber is purged, precursor B is introduced to react with the surface-adsorbed layer from precursor A. Common co-reactants include H_2O, O_3, NH_3, or plasma-activated species, depending on the target material (e.g., Al_2O_3, TiN, and ZnO). The pulse duration (t_B) typically ranges from 0.5 to 5 s, with flow rates between 10 and 100 sccm. The reaction proceeds via ligand exchange, oxidation, or nitridation, forming a stable monolayer of the desired compound. Surface reaction kinetics is governed by activation energy (E_a), temperature (T_{sub}), and precursor partial pressure (P_B). No gas-phase reaction occurs, preserving atomic-scale control. The GPC is typically 0.1–0.3 nm, depending on precursor chemistry and surface saturation. This step completes one full ALD cycle and determines film composition, crystallinity, and interface quality. The reaction must be thermodynamically favorable and kinetically accessible under ALD conditions. Precursor B's reactivity, steric profile, and volatility are critical for achieving uniform, conformal deposition across planar and 3D substrates (Fig. 5.1B) [11].

D. Second purge – clearing residual gases before the next cycle: The final step in the ALD cycle is the second purge, which removes excess precursor B and any volatile by-products generated during the surface reaction. This step is essential to reset the reactor environment and prevent cross-contamination in subsequent cycles. Inert gases such as N_2 or Ar are introduced at flow rates of 100–500 sccm, with purge durations (t_{purge2}) typically ranging from 2 to 10 s. Reactor pressure ($P_{reactor}$) is maintained between 0.1 and 10 torr to ensure efficient gas exchange. The purge mechanism relies on convective displacement and molecular diffusion, with flow dynamics influenced by chamber geometry and precursor volatility. Advanced systems may employ vacuum-assisted purging or spatial separation to accelerate throughput. The effectiveness of this step directly impacts film purity, interface sharpness, and cycle reproducibility. Residual gas concentration (C_{res}) must be minimized to below ppm levels to avoid parasitic reactions. Once complete, the reactor is ready for the next cycle, enabling precise layer-by-layer growth of ultrathin films with atomic-level control [12].

The quality and precision of ALD films are governed by six interdependent synthesis parameters (Fig. 5.1B): (1) precursor pulse time (t_{pulse}), (2) purge time (t_{purge}), (3) deposition temperature (T_{sub}), (4) precursor reactivity (k_{surf}), (5) surface activation (A_{surf}), and (6) cycle count (N_{cycles}). The t_{pulse} must be long enough to achieve full surface saturation; typical values range from 0.5 to 5 s, depending on the precursor's volatility and the substrate's reactivity. Purge time (t_{purge}), often 2–10 s, ensures complete removal of excess precursor and by-products using inert gases (e.g., N_2 at 100–500 sccm), preventing parasitic gas-phase reactions. Deposition temperature (T_{sub}), typically between 100 and 300 °C, must be optimized to balance surface reaction kinetics without decomposing precursors or damaging temperature-sensitive substrates. Precursor reactivity (k_{surf}) affects growth efficiency; ALD tolerates lower reactivity than CVD because of its sequential, surface-limited reactions. Surface activation, influenced by substrate functional groups and surface energy (γ_{surf}), determines nucleation density and conformality. Finally, cycle count (N_{cycles}) directly controls film thickness, with GPC typically 0.1–0.3 nm. Precise control of these variables enables atomic-scale uniformity, high-aspect-ratio coverage, and tailored material properties across diverse applications. Figure 5.1C outlines the individual impact of key ALD synthesis parameters, such as precursor pulse time, purge duration, deposition temperature, and surface activation, and a proposed six-step mechanism, including nucleation dynamics, GPC, film conformality, and overall process efficiency [13].

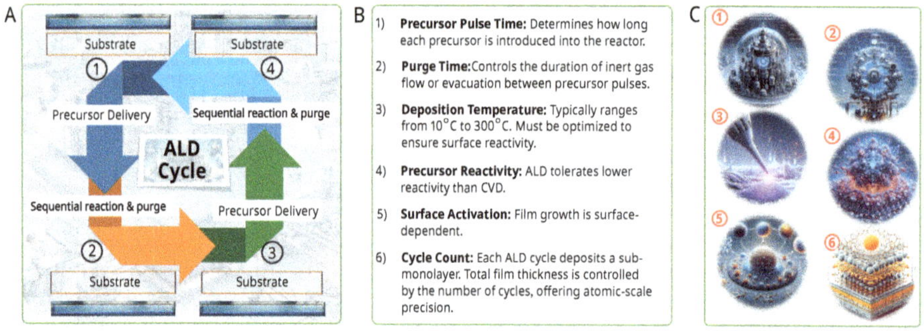

Fig. 5.1: Illustration of the atomic layer deposition (ALD) method, highlighting its core operational framework. **A**. The four-step ALD cycle, sequential precursor pulses, and purge phases, used to achieve self-limiting, layer-by-layer film growth. **B**. Summarizes key synthesis variables, including precursor pulse time, purge duration, deposition temperature, precursor reactivity, surface activation, and cycle count, each critical to process fidelity and film quality. **C**. The proposed influence of each parameter on nucleation behavior, growth per cycle (GPC), conformality, and overall deposition efficiency.

1) Precursor pulse time (t_{pulse}): The t_{pulse} is a critical parameter in ALD, directly influencing surface saturation, film uniformity, and growth efficiency. It defines the duration during which a gaseous precursor is introduced into the reactor chamber, typically 0.5 to 5 s, depending on precursor volatility, molecular size, and substrate reactivity. The goal is to ensure complete chemisorption of the precursor onto all available reactive sites on the substrate surface. Insufficient pulse time can result in incomplete coverage, leading to nonuniform films, reduced GPC, and compromised conformality, especially on high-aspect-ratio structures. Conversely, excessively long pulses may waste precursor material and reduce throughput, without improving film quality. Optimization of pulse time depends on reactor volume, flow rate (10–100 sccm), and surface kinetics, including the sticking coefficient (α) and surface reaction rate constant (k_{surf}). In advanced ALD systems, real-time monitoring techniques such as quartz crystal microbalance (QCM) or in situ ellipsometry are used to fine-tune pulse durations. Proper control of precursor pulse time is essential for achieving atomic-scale precision and reproducible film properties [14].

2) Purge time (t_{purge}): Precursor pulse time is a critical parameter in ALD, directly influencing surface saturation, film uniformity, and growth efficiency. It defines the duration during which a gaseous precursor is introduced into the reactor chamber, typically 0.5 to 5 s, depending on precursor volatility, molecular size, and substrate reactivity [15]. The goal is to ensure complete chemisorption of the precursor onto all available reactive sites on the substrate surface. Insufficient pulse time can result in incomplete coverage, leading to nonuniform films, reduced GPC, and compromised conformality, especially on high-aspect-ratio structures. Conversely, excessively long pulses may waste precursor material and reduce throughput without improving film quality. Optimization of pulse time depends on reactor volume, flow rate (10–100 sccm), and surface kinetics, including the sticking coefficient (α) and surface reaction rate constant (k_{surf}). In advanced ALD systems, real-time monitoring techniques such as QCM or in situ ellipsometry are used to fine-tune pulse durations. Proper control of precursor pulse time is essential for achieving atomic-scale precision and reproducible film properties [16].

3) Deposition temperature (T_{sub}): T_{sub} is a critical parameter in ALD, directly influencing surface reaction kinetics, precursor stability, and film quality. Typically ranging from 1 °C to 300 °C, the optimal temperature must balance sufficient thermal energy to drive chemisorption and surface reactions without causing precursor decomposition or substrate degradation. At lower temperatures (<100 °C), reaction rates may be limited, resulting in incomplete surface coverage and reduced GPC. At higher temperatures (>300 °C), volatile precursors may decompose prematurely, leading to nonuniform films and contamination. The activation energy (E_a) of the surface reaction, precursor volatility, and substrate thermal budget all guide the selection of temperature. For temperature-sensitive substrates such as polymers or flexible electronics, low-temperature ALD variants, e.g., plasma-enhanced or radical-assisted ALD, are often employed. Temperature also affects film crystallinity, density, and impurity incorporation. Precise control of deposition temperature ensures reproducibility, conformality, and stoichiometric accuracy across diverse materials systems. Monitoring substrate temperature (T_{sub}) and maintaining thermal uniformity within ± 1 °C are essential for achieving atomic-scale precision in ALD processes [17].

4) Precursor reactivity (k_{surf}): k_{surf} is a defining parameter in ALD, influencing surface reaction efficiency, film uniformity, and process selectivity. Unlike CVD, which requires highly reactive precursors for simultaneous gas-phase reactions, ALD tolerates lower reactivity due to its sequential, surface-limited mechanism. In ALD, each precursor interacts independently with the substrate, allowing reactions to proceed via chemisorption rather than decomposition. This enables the use of thermally stable, less reactive compounds such as metal alkyls, halides, and amidinates. Surface reaction rate constant (k_{surf}), activation energy (E_a), and sticking coefficient (α) govern the kinetics of precursor adsorption and conversion. Low reactivity can be compensated for by longer pulse durations or elevated substrate temperature (T_{sub}), typically between 100 and 300 °C. However, excessively inert precursors may result in incomplete coverage or reduced GPC [18]. Selecting precursors with appropriate ligand chemistry, steric accessibility, and volatility is essential for achieving self-limiting behavior. Optimizing precursor reactivity ensures conformal film growth, high purity, and reproducibility across diverse material systems and substrate geometries [19].

5) Surface activation (A_{surf}): A_{surf} is a critical determinant of film nucleation, growth kinetics, and uniformity in ALD. Because ALD relies on self-limiting surface reactions, the chemical nature of the substrate, including its functional groups, surface energy (γ_{surf}), and defect density, directly influences precursor adsorption and reaction efficiency. Substrates with hydroxyl (–OH), amine ($–NH_2$), or carboxyl (–COOH) groups typically promote strong chemisorption and rapid nucleation. In contrast, inert or passivated surfaces may require pre-treatment, such as plasma activation, ozone exposure, or seed-layer deposition, to enhance reactivity. The initial nucleation delay (t_{nuc}) and GPC are sensitive to surface termination and steric accessibility. Poor surface activation can lead to island growth, nonuniform coverage, and reduced conformality, especially on high-aspect-ratio structures. Surface activation also affects film crystallinity, interface sharpness, and adhesion. For advanced applications, tailoring surface chemistry enables selective area deposition (SAD) and interface engineering. Optimizing surface activation ensures reproducible ALD performance across diverse substrates, including oxides, polymers, 2D materials, and MOFs [20].

6) Cycle count (N_{cycles}). N_{cycles} is a fundamental parameter in ALD, directly governing the total film thickness and enabling atomic-scale precision. Each ALD cycle typically deposits a sub-monolayer of material, with a GPC of 0.1–0.3 nm, depending on precursor chemistry, surface saturation, and process conditions. The total thickness (t_{total}) is calculated as the product of GPC and Ncycles, enabling deterministic control over nanoscale film architecture. Accurate cycle control is essential for achieving target specifications in applications such as gate dielectrics, diffusion barriers, and multilayer heterostructures. In high-throughput systems, cycle timing must be optimized to balance deposition rate and film quality, with typical cycle durations ranging from 5 to 20 s. Variability in surface activation or precursor delivery can affect GPC, so in situ monitoring tools like QCM or spectroscopic ellipsometry are often used to validate cycle performance. Ultimately, cycle count provides a scalable and reproducible framework for tailoring film thickness, interface sharpness, and functional layering in advanced nanofabrication [21].

5.4 Various deposition approaches

ALD encompasses several specialized methods, each tailored to overcome limitations of conventional thermal ALD and expand its applicability across diverse substrates and materials. ALD has evolved into a suite of specialized methods designed to overcome the limitations of traditional thermal ALD and broaden its applicability across diverse substrates and material systems. In this section, six methods will be briefly discussed to showcase their key technical synthesis variables, mechanism, and advantages (Tab. 5.1). ***A. Thermal ALD***, the foundational technique, relies solely on heat to drive surface reactions and is ideal for depositing oxides, nitrides, and sulfides on thermally stable substrates, typically within the 100–300 °C range [22]. ***B. Plasma-enhanced ALD*** (PEALD) introduces plasma to activate reactions at lower temperatures, improving film density and enabling deposition on flexible electronics and high-k dielectrics. ***C. Radical-enhanced ALD*** uses reactive species like $O^{\bullet}$ or $N^{\bullet}$ to achieve similar low-temperature benefits with reduced substrate damage, making it suitable for organic and sensitive surfaces. ***D. Spatial ALD*** separates precursor exposure spatially, allowing continuous, high-throughput processing in applications such as photovoltaics and display technologies. ***F. Roll-to-roll ALD*** adapts this principle for flexible substrates in web formats, ideal for packaging and wearable devices. ***G. Molecular layer deposition (MLD)*** extends ALD to hybrid organic–inorganic films using bifunctional precursors, enabling precise coatings on polymers and MOFs [23].

A. Thermal ALD: Thermal ALD is a vapor-phase technique that enables precise, layer-by-layer growth of thin films through sequential, self-limiting surface reactions. It operates exclusively by thermal activation, typically at 100–300 °C, making it suitable for substrates with high thermal stability. Each ALD cycle involves alternating pulses of gaseous precursors, such as TMA and H_2O for Al_2O_3, separated by inert gas purges to prevent gas-phase reactions. The surface reactions are kinetically controlled, ensuring uniform film growth with sub-nanometer resolution across complex geometries. Thermal ALD is widely used to deposit oxides (Al_2O_3 and HfO_2), nitrides (TiN and AlN), and sulfides (ZnS), offering excellent conformality, thickness control, and material purity. Its reaction-limited regime supports predictable growth rates, typically 0.5–2.0 Å per cycle. However, its elevated temperature requirements limit

Tab. 5.1: Comparative summary of atomic layer deposition (ALD) methods: techniques, synthesis variables, mechanisms, and material examples.

Method	Synthesis variables	Mechanism	Representative materials	Advantages
Thermal ALD	*Temperature*: 100–300 °C; *Precursors*: metal alkyls, halides, amides *Substrate*: thermally stable (Si, Al_2O_3, TiN)	Surface-limited reactions, driven by thermal energy; alternating precursor pulses form monolayers	Al_2O_3 (TMA + H_2O), TiO_2 ($TiCl_4$ + H_2O), ZnO (DEZ + H_2O), HfO_2 ($HfCl_4$ + H_2O)	High conformality; excellent thickness control; ideal for oxides and nitrides on stable substrates
Plasma-enhanced ALD	*Temperature*: 25–150 °C; *Plasma gases*: O_2, N_2, H_2, Ar; *Substrate*: flexible, low-temp tolerant (PET, polymers)	Plasma generates reactive species (radicals, ions) to enhance surface reactions at lower temperatures	SiO_2 (SiH_4 + O_2 plasma), AlN (TMA + N_2 plasma), TiO_2 (TTIP + O_2 plasma), Ta_2O_5 ($TaCl_5$ + O_2 plasma)	Enables low-temp deposition; improved film density; suitable for flexible electronics and high-k dielectrics
Radical-enhanced ALD	*Temperature*: 25–100 °C *Radical sources*: ozone, UV-activated O_2 or N_2 *Substrate*: organic, polymeric, MOFs	Reactive radicals ($O^\bullet$, $N^\bullet$) initiate surface reactions without ion bombardment or plasma damage	Al_2O_3 (TMA + ozone), ZnO (DEZ + ozone), organic–inorganic hybrids (alucone)	Gentle on sensitive substrates; avoids plasma damage; compatible with organics and MOFs
Spatial ALD	No purge steps; Precursor zones are separated spatially; *Substrate motion*: linear or rotary	Al_2O_3 (TMA + H_2O), ZnO (DEZ + H_2O), SnO_2 ($SnCl_4$ + H_2O)	Continuous substrate movement through alternating precursor zones; reactions occur in real time	High throughput; scalable for large-area coatings; ideal for photovoltaics and display panels
Roll-to-roll ALD	*Substrate*: flexible web (PET, PI); *Ambient control*: inert atmosphere; *Precursor delivery*: slot-die or vapor	Extension of spatial ALD to moving flexible substrates; sequential exposure during continuous motion	Al_2O_3 barrier layers on PET, ZnO for transparent electrodes, TiO_2 for UV protection	Scalable for packaging, wearables, and flexible electronics; compatible with industrial formats
Molecular-layer deposition	*Temperature*: 80–200 °C; Bifunctional organic precursors; *Substrate*: polymers, MOFs, oxides;	Alternating organic/inorganic reactions form hybrid layers with tailored functionality	Alucone (TMA + ethylene glycol), titanicone ($TiCl_4$ + glycerol), hybrid MOF coatings	Enables hybrid films; precise control over organic–inorganic interfaces; functionalized surfaces

Note: **TMA**, trimethylaluminum; **DEZ**, diethylzinc; **TTIP**, titanium isopropoxide.

compatibility with temperature-sensitive substrates. Despite this, thermal ALD remains a cornerstone method for high-performance coatings in semiconductor fabrication, energy storage, and advanced surface engineering [24].

A.1. Working principle and reaction mechanism: The thermal ALD process operates through a well-defined cyclic sequence of surface-mediated reactions, enabling the deposition of ultrathin films with atomic-scale precision. The process is fundamentally governed by two self-limiting half-reactions between gaseous precursors and reactive surface sites, executed sequentially to ensure controlled layer growth. In a typical cycle, Precursor A, commonly a volatile metal-organic compound such as TMA for aluminum oxide, is introduced into the reaction chamber. Upon exposure, the precursor chemisorbs onto the substrate surface, selectively reacting with available functional groups (e.g., hydroxyls) to form a saturated monolayer. This chemisorption step is inherently self-limiting: once all reactive sites are occupied, further adsorption is kinetically inhibited, preventing uncontrolled film growth. Following the adsorption phase, an inert gas purge (typically N_2 or Ar) is applied to evacuate excess precursor and volatile reaction byproducts. Subsequently, Precursor B, often a counter-reactant such as H_2O, O_3, or NH_3, is pulsed into the chamber. This second precursor reacts with the surface-bound species to complete the formation of the desired compound (e.g., Al_2O_3) and regenerate the reactive surface groups, thereby preparing the substrate for the next cycle. The entire sequence is repeated until the target film thickness is achieved [25]. These reactions are thermally activated, typically occurring within a temperature window of 100–300 °C, selected to optimize surface reaction kinetics while avoiding precursor decomposition or substrate degradation. Unlike plasma-enhanced or photonic ALD variants, thermal ALD does not rely on external energy sources such as ionized gases or UV radiation. This absence of energetic activation preserves substrate integrity, minimizes damage, and simplifies reactor design, making thermal ALD particularly suitable for high-purity, conformal coatings on thermally robust substrates. The self-limiting nature of each half-reaction is central to the precision of thermal ALD. It ensures that film growth proceeds layer by layer, with each cycle contributing a well-defined increment to the overall thickness. Typical GPC values range from 0.5 to 2.0 Å, depending on the precursor pair, surface saturation efficiency, and reactor conditions [26].

A.2. Kinetic modeling and growth rate equations: The growth rate in thermal ALD is governed primarily by surface reaction kinetics rather than gas-phase transport, because each half-reaction proceeds only through self-limiting chemisorption on the available reactive sites. A simplified kinetic description often assumes Langmuir-type adsorption behavior, in which the fractional surface coverage increases monotonically with precursor exposure. This kinetic model assumes Langmuir-type adsorption behavior (eq. 5.1):

$$\theta_t = 1 - e^{\quad k_{\text{ads}} P t} \tag{5.1}$$

where θ_t is the fractional surface coverage at time t, k_{ads} is the adsorption rate constant, P is the precursor partial pressure, and t is the pulse duration.

The GPC is proportional to the saturated surface coverage (eq. 5.2):

$$\mathrm{GPC} = \theta_{max} \times d_{ML} \tag{5.2}$$

where $\theta_{max} \approx 1$ for ideal saturation and θ_{max} is the thickness of a monolayer (typically 0.1–0.2 nm).

In practice, deviations from ideal behavior arise due to steric hindrance, incomplete reactions, or precursor decomposition. These effects are captured in more advanced models using **reaction-limited** or **transport-limited** regimes, often described as follows (eq. 5.3):

$$r = k \times C_s \tag{5.3}$$

where r is the reaction rate; k is the surface reaction rate constant; and C_s the surface concentration of the reactive sites [27].

A.3. Synthesis variables: Thermal ALD synthesis is governed by tightly controlled process variables that directly influence film quality, growth rate, and material purity. Four factors will be discussed: ***A.3A. Temperature***, ***A.3B. precursor chemistry***, ***A.3C. pulse/purge timing***, and ***A.3D. substrate compatibility***. In general, the temperature is typically maintained between 100 and 300 °C, which must be high enough to activate surface reactions, yet low enough to prevent precursor decomposition or substrate degradation. Precursor chemistry is equally critical: metal alkyls such as TMA enable rapid, self-limiting reactions for Al_2O_3; metal halides such as $TiCl_4$ offer high volatility and reactivity for TiO_2; and metal amides such as TDMAH provide clean decomposition pathways for HfO_2 [28]. Pulse and purge timing must be precisely tuned; short pulses risk incomplete surface saturation, while insufficient purging can lead to gas-phase reactions and CVD-like growth. Inert gases such as N_2 or Ar are used to evacuate residual reactants and byproducts between steps. Substrate compatibility is essential; only thermally stable materials like Si, TiN, Al_2O_3, and sapphire can withstand repeated thermal cycling without structural or chemical degradation. Each variable must be optimized in concert to ensure uniform, conformal film growth with atomic-scale precision across complex geometries. Four key synthesis variables and their impacts are summarized in Fig. 5.2A–D, with a detailed discussion provided in the following sections [29].

A.3A. Temperature: Temperature plays a pivotal role in determining film quality, growth kinetics, and precursor stability in thermal ALD. Operating typically within the 100–300 °C range, the temperature must be carefully optimized to activate surface reactions without inducing thermal decomposition of precursors or damaging the substrate. At lower temperatures (<150 °C), incomplete chemisorption and reduced reaction rates can lead to sub-monolayer coverage, poor film uniformity, and increased impurity incorporation. Conversely, excessively high temperatures (>300 °C) may cause precursor decomposition, parasitic CVD, and loss of self-limiting behavior,

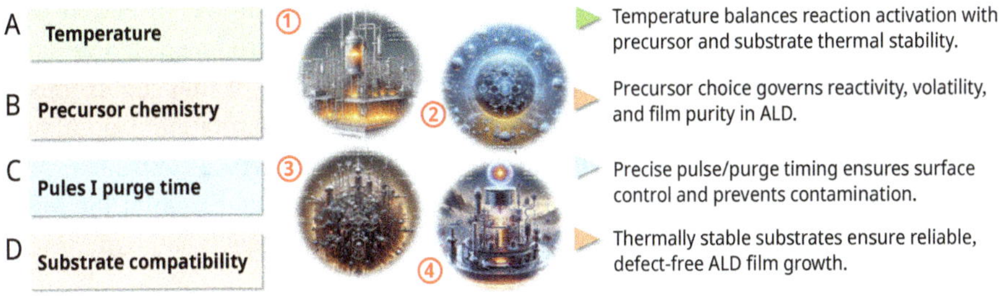

Fig. 5.2: Influence of synthesis variables on the final products derived using thermal atomic layer deposition: **A**. Temperature, **B**. precursor chemistry, **C**. pulse/purge timing, and **D**. substrate compatibility.

compromising thickness control and conformality. The ideal temperature window varies by precursor chemistry: for example, TMA and H_2O for Al_2O_3 exhibit optimal reactivity near 200 °C, while metal halides like $TiCl_4$ require slightly higher activation thresholds. Substrate thermal stability also constrains process temperature, limiting compatibility with polymers and low-melting materials. Ultimately, precise thermal control ensures reproducible GPC, high film purity, and defect-free interfaces, making temperature one of the most critical synthesis variables in thermal ALD [30].

A.3B. Precursor chemistry: Precursor chemistry critically influences film quality, growth kinetics, and impurity levels in thermal ALD. Metal alkyls, such as TMA, are widely used for Al_2O_3 due to their high volatility and rapid surface reactivity with H_2O, yielding dense, conformal films with minimal carbon contamination [31]. TMA's self-limiting behavior ensures excellent thickness control and uniformity across high-aspect-ratio structures. Metal halides such as titanium tetrachloride ($TiCl_4$) are preferred for TiO_2 deposition, owing to their strong reactivity with water and high thermal stability. However, chloride byproducts may require extended purge cycles to prevent residual contamination, impacting throughput. Metal amides, such as tetrakis(dimethylamido)hafnium (TDMAH), are commonly employed for HfO_2 due to their clean decomposition pathways and compatibility with lower temperatures. TDMAH reacts efficiently with oxidants such as H_2O or ozone, yielding high-k dielectric films with low impurity levels. Each precursor class presents trade-offs in volatility, ligand stability, and reaction kinetics, which must be carefully matched to substrate conditions and desired film properties to ensure reproducible, high-performance ALD coatings [32].

A.3C. Pulse/purge timing: Pulse and purge timing are critical parameters in thermal ALD, directly influencing film uniformity, purity, and growth reproducibility. During each ALD cycle, precursor pulses must be long enough to achieve full surface saturation without inducing gas-phase reactions. Insufficient pulse duration can result in incomplete chemisorption, leading to nonuniform coverage and reduced GPC. Conversely, excessive pulse times may waste precursor and increase cycle time without improving film quality. The subsequent purge step, typically using inert gases such as nitrogen or argon, is equally vital. It must effectively remove unreacted precursor and volatile byproducts before the next reactant is introduced. Inadequate purging can cause precursor overlaps, triggering CVD-like behavior that compromises the self-limiting nature of ALD and introduces defects or contamination. Optimizing pulse/purge timing ensures that each half-reaction remains surface-controlled, preserving atomic-scale precision and conformality. This balance is essential for high-aspect ratio structures and multilayer stacks, where even minor deviations can propagate into significant performance losses [33].

A.3D. Substrate compatibility: Substrate compatibility is a critical determinant of film quality, process reliability, and application scope in thermal ALD. Because thermal ALD relies solely on heat to activate surface reactions, substrates must withstand repeated exposure to elevated temperatures, typically in the range of 100–300 °C, without undergoing chemical degradation, phase transitions, or structural distortion. Thermally stable substrates such as silicon (Si), titanium nitride (TiN), aluminum oxide (Al_2O_3), and sapphire are commonly used due to their inertness and mechanical robustness under ALD conditions. These substrates support uniform nucleation and facilitate consistent GPC, yielding conformal, defect-free films. In contrast, temperature-sensitive materials such as polymers, organic layers, and low-melting-point metals may deform, outgas, or react chemically with precursors, leading to poor adhesion, incomplete coverage, or contamination. Additionally, substrate surface chemistry, such as hydroxyl termination or the presence of native oxide, affects precursor adsorption and reaction kinetics. Therefore, careful substrate selection and pre-treatment are essential to ensure reproducible film growth, especially in applications requiring high-aspect-ratio coverage, dielectric integrity, or multilayer stack integration [34].

A.4. Applications: Thermal ALD is widely employed across advanced manufacturing sectors due to its ability to produce conformal, high-purity films with atomic-scale thickness control. In microelectronics, thermal ALD is integral to gate dielectric formation, spacer layers, and diffusion barriers, with materials such as Al_2O_3, HfO_2, and TiN supporting device scaling and reliability. In energy systems, it enables the fabrication of protective coatings and ion-conductive layers in batteries, fuel cells, and photovoltaic devices, where uniform coverage over porous or high-aspect-ratio structures is essential. Optical applications benefit from ALD-grown TiO_2 and ZnS films used in antireflective coatings, waveguides, and photonic structures. Additionally, ALD is increasingly applied in catalysis and surface passivation, where precise control of active-site exposure enhances performance. Its compatibility with complex geometries and multilayer stacks makes it ideal for 3D architecture and nanoscale interfaces. However, its reliance on elevated temperatures limits its usage on polymeric or biologically sensitive substrates, prompting the development of low-temperature ALD variants for broader applicability [35].

B. Plasma-enhanced ALD (PEALD): PEALD is a variant of conventional ALD that incorporates plasma activation to enhance surface reactivity and enable film growth at reduced temperatures. Unlike thermal ALD, which relies solely on heat to drive surface reactions, PEALD utilizes plasma-generated radicals, ions, and excited species to facilitate precursor decomposition and surface functionalization. This approach is particularly advantageous for temperature-sensitive substrates such as polymers and low-melting-point metals, and for materials requiring high film density or improved stoichiometry. PEALD typically operates at temperatures between 25 and 150 °C, expanding its applicability to flexible electronics, organic devices, and advanced packaging. Common plasma gases include O_2, N_2, H_2, and Ar, selected based on the desired film composition and reactivity. Materials deposited via PEALD include oxides (e.g., SiO_2, Al_2O_3, and TiO_2), nitrides (e.g., AlN and TiN), and metals. The technique offers superior conformality, reduced impurity levels, and enhanced control over film properties. However, plasma exposure may introduce substrate damage or require complex reactor designs, necessitating careful process optimization [36].

B.1. Reaction mechanism and kinetics of PEALD: PEALD preserves the cyclic, self-limiting architecture of ALD while replacing the thermal co-reactant with plasma-generated species to unlock low-temperature, highly reactive surface chemistry. A typical cycle comprises: (1) precursor pulse, during which chemisorption saturates reactive surface sites; (2) purge; (3) plasma exposure, where radicals (e.g., O, N, and H), metastables, and, depending on configuration, low-energy ions complete ligand removal and lattice formation; and **(4)** purge. Remote or downstream plasma configurations minimize ion bombardment, while direct plasma modes exploit controlled ion assistance to tailor density and phase. Plasma variables (power, duty cycle, pressure, gas mixture, and substrate bias) determine radical flux, ion energy, and the balance between reaction acceleration and potential damage. Three kinetic models and two pathways are discussed as follows [37]:

B.1.A. Precursor half-reaction: In PEALD, the precursor half-reaction follows Langmuir adsorption kinetics (eq. (5.4)), where molecules chemisorb onto reactive surface sites until saturation is reached. The adsorption rate depends on precursor flux, sticking coefficient, and available surface coverage, ensuring self-limiting behavior. This controlled chemisorption forms a uniform monolayer, critical for achieving atomic-scale precision and maintaining conformality across complex substrate geometries (eq. 5.4):

$$\frac{d\theta}{dt} = s_A \times J_A(1-\theta) - k_{\mathrm{des}} \times \theta \tag{5.4}$$

where θ is the surface coverage, s_A is the effective sticking coefficient, J_A is the molecular flux, and k_{des} is desorption. Saturation yields the self-limiting condition $\theta \rightarrow \theta_{\mathrm{sat}}$.

B.1.B. Plasma half-reaction: In PEALD, the plasma half-reaction involves reactive species, such as O, N, or H*, interacting with surface-bound ligands via a radical-driven Eley-Rideal or Langmuir–Hinshelwood mechanism (eq. (5.5)). These radicals facilitate ligand abstraction and bond formation, enabling low-temperature synthesis of dense, stoichiometric films. Reaction efficiency depends on radical flux, surface coverage, and plasma parameters, which must be carefully tuned to preserve self-limiting behavior and avoid substrate damage:

$$\frac{d\theta^*}{dt} = s - k_{ER} \times \tau_{rad} \times \theta^* - k_{LH} \times \theta_{rad} \times \theta^* \quad (5.5)$$

where θ^* is the ligand-terminated coverage, τ_{rad} is the radical flux, θ_{rad} is the adsorbed-radical coverage, and k_{ER} and k_{LH} are the Eley-Rideal and Langmuir–Hinshelwood rate constants, respectively. The net GPC follows eq. (5.6):

$$\text{GPS} = \theta_{sat} \times d_{ML} \times \eta_{rem} \quad (5.6)$$

where d_{ML} is the monolayer equivalent and η_{rem} is the ligand removal efficiency.

B.1.C. Aspect ratio/conformality (diffusion reaction balance): Aspect ratio and conformality in PEALD are governed by the balance between precursor diffusion and surface reaction kinetics. The Damköhler number (D_a) quantifies this interplay, where low D_a values favor uniform coverage in deep or narrow features (eqs. (5.7) and (5.8)). Optimizing exposure time, precursor reactivity, and plasma conditions ensures consistent film thickness across complex geometries, which is essential for high-performance devices with 3D architectures or high-aspect-ratio structures [38]:

$$D_a = \frac{k_s \times L}{D_{eff}} \quad (5.7)$$

$$\phi = \frac{1}{1 + D_a} \quad (5.8)$$

where D_a is the Damköhler number, k_s, is the surface reaction velocity, L, is the feature length, D_{eff} is the effective (Knudsen) diffusivity, and ϕ is the conformality proxy. Lower D_a via moderated reactivity or increased exposure, step coverage improves (eq. 5.8).

B.1.D. Ion-assisted pathways: Ion-assisted pathways in PEALD leverage low-energy ion bombardment to enhance surface reactions, densify films, and promote phase control. Ion flux and energy must be carefully tuned to avoid substrate damage, while enabling ligand removal and lattice restructuring. These pathways complement radical-driven mechanisms, offering additional activation routes that improve film quality, especially in applications requiring high density, crystallinity, or tailored interfacial properties. Ion flux increases with the energy-dependent reaction or etch yield (eq. (5.9)), meaning higher ion energy can enhance surface activation. However, to maintain the precise, layer-by-layer control of PEALD, ion energy must remain below damage thresholds. This ensures that the process stays self-limiting and avoids unwanted substrate erosion or film defects [39]:

$$\gamma_{ion} \propto \tau_{ion} \times Y(E_{ion}) \quad (5.9)$$

where γ_{ion} Is the ion flux and $Y(E_{ion})$ is the energy-dependent reaction/etch yield, keeping E_{ion} below damage thresholds, preserving self-limiting behavior (eq. 5.9).

B.1.E. Surface reaction pathways: In the PEALD, ligand elimination is primarily driven by plasma-generated radicals that facilitate the removal of surface-bound organic or inorganic groups. For example, oxygen radicals (O) can abstract methyl ligands ($-CH_3$), converting them into volatile byproducts such as CO, CO_2, or H_2O, while hydrogen radicals (H) promote hydrogenolysis of halide (–Cl) or amide ($-NR_2$) ligands. These reactions result in the formation of stable metal–oxygen (M–O) or metal–nitrogen (M–N) bonds, completing the desired film structure. Following ligand removal, the surface undergoes regeneration, wherein reactive terminations such as hydroxyl (–OH) or amine ($-NH_x$) groups are re-established. These functional groups are essential for initiating the next precursor adsorption step, thereby preserving the self-limiting nature of the ALD cycle and ensuring consistent GPC. Plasma-specific effects further influence film properties by enhancing densification and improving stoichiometric control at reduced temperatures. However, excessive exposure to energetic ions or metastable species may induce sub-surface activation, defect formation, or unintended etching, necessitating careful optimization of plasma parameters to maintain film integrity and process reproducibility [40].

Aluminum oxide (Al_2O_3) formation (eq. (5.10)) occurs by ligand elimination via oxygen radical ($O^•$) using precursor TMA ($Al(CH_3)_3$): Surface-bound TMA reacts with oxygen radicals, abstracting methyl ligands and forming Al-O bonds. The methyl groups are oxidized into volatile species such as CO, CO_2, and H_2O, which are purged from the chamber:

$$Al(CH_3)_3{}^* + 3O^* \rightarrow Al_2O_3 + 3CH_3{}^\bullet \rightarrow CO/CO + H_2O \qquad (5.10)$$

Hydrogenolysis of chloride ligands leads to titanium nitride (TiN, eq. (5.11)). Titanium tetrachloride ($TiCl_4$) is a common precursor for TiN synthesis. During the plasma half-reaction, hydrogen radicals (H^*) selectively cleave Ti–Cl bonds through hydrogenolysis, producing volatile hydrogen chloride (HCl) and exposing reactive titanium sites. Subsequent exposure to nitrogen plasma introduces active nitrogen species that react with the titanium surface, forming robust Ti–N bonds and completing the deposition cycle. This sequence highlights the role of plasma-generated radicals in enabling efficient ligand removal and precise bond formation, facilitating low-temperature growth of high-purity, conformal nitride films (eq. 5.11) [41]:

$$TiCl_4{}^* + 4H^* \rightarrow TiN + 4HCl\,(g) \qquad (5.11)$$

B.2. Synthesis variables: Synthesis variables in PEALD play a pivotal role in determining film quality, growth rate, and process reproducibility. Key parameters include ***B.2.A precursor chemistry***, ***B.2.B plasma composition***, ***B.2.C plasma power***, ***B.2.D pressure***, ***B.2.E substrate temperature***, and ***B.2.F timing of pulse and purge*** steps. Precursor selection governs volatility, reactivity, and ligand removal efficiency, while plasma gas type (e.g., O_2, N_2, H_2) influences the nature of reactive species and final film stoichiometry. Plasma power and exposure duration affect radical density and ion energy, which must be carefully balanced to enhance surface reactions without damaging the substrate. Chamber pressure controls mean free path and species transport, impacting conformality in high-aspect-ratio structures. Substrate temperature, typically lower than in thermal ALD, must still support sufficient surface mobility and reaction kinetics. Precise timing of precursor pulses and plasma exposure ensures self-limiting behavior and avoids gas-phase reactions. These variables must be optimized in concert to achieve uniform, high-purity films with tailored properties across diverse materials and geometries [42].

B.2.A Precursor chemistry: Precursor chemistry is a foundational element in the PEALD process, directly influencing film composition, growth kinetics, and overall deposition quality. Ideal precursors must exhibit high volatility, thermal stability, and selective reactivity with plasma-generated species. Metal alkyls, such as TMA, are widely used for oxide deposition due to their rapid, self-limiting surface reactions and compatibility with oxygen plasma. Metal halides such as titanium tetrachloride ($TiCl_4$, eq. (5.11)) exhibit high volatility and strong ligand reactivity, making them suitable for nitride and oxide films when used with hydrogen or nitrogen plasma. Metal amides, such as tetrakis(dimethylamido)hafnium (TDMAH), provide clean decomposition pathways and are favored for high-k dielectric materials. The choice of precursors also affects ligand elimination efficiency, byproduct volatility, and surface regeneration, all of which are critical for maintaining conformality and atomic-scale control. Additionally, precursor–plasma interactions must be carefully tuned to avoid unwanted gas-phase reactions or substrate damage, ensuring reproducible, high-purity film growth across diverse applications and temperature-sensitive substrates [43].

B.2.B Plasma composition: Plasma composition is a critical variable in PEALD, directly influencing film stoichiometry, density, and reaction kinetics. The choice of plasma gas, such as oxygen (O_2), nitrogen (N_2), hydrogen (H_2), ammonia (NH_3), or argon (Ar), determines the type and concentration of reactive species generated, including radicals, ions, and metastables. Oxygen plasma is commonly used for oxide deposition, providing strong oxidizing radicals (O^*) that facilitate ligand removal and metal–oxygen bond formation. Nitrogen and ammonia plasmas are employed for nitride synthesis by introducing active nitrogen species that form metal–nitrogen bonds. Hydrogen plasma supports halide ligand elimination and surface passivation. Inert gases such as argon are often used to dilute reactive gases or to enhance physical activation through ion bombardment. Plasma parameters such as power, pressure, and gas flow rates must be carefully tuned to balance radical density with ion energy, minimizing substrate damage while maximizing reaction efficiency. The plasma composition affects surface functionalization and site regeneration, which are essential for maintaining self-limiting behavior and achieving uniform, high-purity films across complex geometries and temperature-sensitive substrates [44].

B.2.C Plasma power: Plasma power is a key synthesis variable in PEALD, directly influencing the density, energy, and composition of reactive species that drive surface reactions. Higher power levels increase the generation of radicals and ions, enhancing ligand removal efficiency and enabling low-temperature film growth with improved stoichiometry and density. However, excessive power can lead to substrate damage, ion-induced defects, and loss of self-limiting behavior, especially in sensitive materials or nanoscale features. The optimal power setting depends on the precursor chemistry, desired film properties, and plasma configuration (e.g., remote vs. direct). Pulsed plasma modes are often employed to balance reactivity with thermal and physical control, reducing cumulative damage while maintaining process efficiency. Power also affects the electron temperature and plasma potential, which in turn influence ion energy and surface activation pathways. Careful tuning of plasma power, in conjunction with gas composition and pressure, ensures reproducible, high-quality film deposition across diverse substrates and geometries. Ultimately, plasma power must be optimized to achieve precise control over reaction kinetics, conformality, and material integrity [45].

B.2.D Pressure: Operating pressure in PEALD significantly influences plasma behavior, species transport, and film conformality. Lower pressures (typically 1–10 mtorr) increase the mean free path of reactive species, enhancing radical penetration into high-aspect-ratio features and reducing gas-phase collisions. This favors surface-limited reactions and uniform coverage. Conversely, higher pressures (10–100 mtorr) promote greater plasma density and radical generation but may reduce conformality due to increased scattering and recombination. Pressure also affects the ion energy distribution and residence time, thereby impacting substrate interactions and potential damage. In remote plasma configurations, pressure tuning helps isolate radical-driven chemistry from ion bombardment, preserving film integrity. Additionally, pressure interacts with gas flow rates and precursor volatility, influencing purge efficiency and cycle reproducibility. Optimal pressure settings depend on precursor type, plasma composition, and desired film properties, requiring careful calibration to balance reactivity, uniformity, and substrate compatibility [46]. Ultimately, precise pressure control is essential for achieving self-limiting growth, minimizing defects, and ensuring high-quality deposition across diverse materials and geometries [47].

B.2.E Substrate temperature: Substrate temperature is a critical parameter in PEALD, influencing surface reaction kinetics, film crystallinity, and overall process stability. Unlike thermal ALD, PEALD enables deposition at significantly lower temperatures, typically between 25 °C and 150 °C, due to the enhanced reactivity of plasma-generated species. This low-temperature capability is essential for processing temperature-sensitive substrates such as polymers, organic electronics, and flexible devices. However, the substrate temperature must remain sufficiently high to support precursor adsorption, surface mobility, and complete ligand removal. Inadequate temperatures may lead to incomplete reactions, poor film quality, or residual contamination. Conversely, elevated temperatures can enhance densification and reduce impurities but may compromise substrate integrity or induce unwanted reactions. Temperature also affects the regeneration of reactive surface sites, which is vital for maintaining self-limiting growth behavior. Optimal temperature settings depend on precursor chemistry, plasma composition, and desired film properties, requiring careful calibration to balance reactivity, conformality, and thermal compatibility. Ultimately, precise temperature control ensures reproducible, high-quality deposition across diverse materials and device architectures [48].

B.2.F Timing of pulse and purge: Precise timing of precursor pulse and purge steps is essential in PEALD to maintain self-limiting growth, prevent gas-phase reactions, and ensure film uniformity. During the precursor pulse, sufficient exposure time allows complete surface saturation without over-adsorption or condensation. The subsequent purge must fully evacuate excess precursor and byproducts before plasma activation to avoid undesired reactions or particle formation. Plasma exposure timing determines the interaction of radicals with surface-bound species, thereby influencing ligand removal efficiency and film quality. Overexposure may cause substrate damage or suboptimal film growth, while underexposure risks incomplete reactions. Post-plasma purge is equally critical for removing long-lived reactive species and volatile byproducts, thereby preserving cycle reproducibility. Timing must be optimized based on precursor volatility, plasma composition, chamber geometry, and target material properties. Advanced timing schemes, such as pulsed plasma or extended purge protocols, are often employed to balance reactivity with substrate protection. Ultimately, synchronized control of pulse and purge durations ensures high-purity, conformal films with atomic-scale precision across diverse substrates and device architectures [49].

B.3. Applications and diagnosis: PEALD has become a cornerstone thin-film technology across sectors demanding atomic-scale precision, low-temperature processing, and high material quality. Its ability to deposit conformal, stoichiometric films on complex geometries makes it indispensable in semiconductor fabrication, flexible electronics, energy systems, and advanced packaging [50]. In semiconductor manufacturing, PEALD is widely used for gate dielectrics (e.g., Al_2O_3 and HfO_2), barrier layers (e.g., TiN), and spacer films, where plasma activation ensures high purity and density while preserving device integrity under reduced thermal budgets. Its conformality supports intricate 3D architectures such as FinFETs and high-aspect-ratio trenches in DRAM and NAND flash. In flexible and organic electronics, PEALD enables encapsulation and barrier coatings on temperature-sensitive substrates like polymers using low-energy plasma and tailored precursor chemistry to maintain film uniformity without degradation. Energy applications benefit from PEALD's precision in electrode coatings, solid electrolytes, and passivation layers for batteries, fuel cells, and photovoltaics. The method also supports MEMS/NEMS, biosensors, and optical coatings, where surface functionalization and reproducibility are critical [51].

To achieve such versatility, plasma diagnostics play a vital role in tuning PEALD processes for optimal performance. Techniques such as Optical Emission Spectroscopy (OES) provide real-time monitoring of excited-state species, supporting endpoint detection and plasma stability. Actinometry enhances OES by quantifying the densities of reactive atoms under varying synthesis conditions. Langmuir probes offer insight into electron temperature and plasma density, informing ion flux and energy distribution, though they require careful application near sensitive substrates. Quadrupole mass spectrometry (QMS) detects volatile byproducts such as CH_4 and HCl to assess ligand removal and purge efficiency. Ground-state radicals such as O, H, and N are probed using absorption spectroscopy methods, including UV, VUV, and Molecular Beam Mass Spectrometry (MBMS). In situ ellipsometry and QCM track GPC, nucleation delays, and densification, while in situ FTIR, via RAIRS or ATR, monitors surface functional groups and confirms site regeneration. Together, these diagnostics enable reproducible, self-limiting PEALD processes with controlled stoichiometry, conformality, and minimal plasma-induced damage, supporting the method's broad applicability across emerging device platforms [52].

C. Radical-enhanced ALD (REALD): REALD is a surface-controlled thin-film technique that utilizes neutral radicals, rather than ionized plasma species, to drive chemical reactions during the ALD cycle. Like PEALD, REALD maintains the self-limiting, cyclic nature of ALD. Still, it eliminates ion bombardment and electrical bias, making it particularly suitable for temperature-sensitive or damage-prone substrates. Reactive radicals such as O, N, or H are typically generated through remote plasma sources or thermal cracking units and introduced into the reaction chamber to facilitate ligand abstraction and bond formation. REALD is especially advantageous in applications requiring ultra-low thermal budgets, such as flexible electronics, organic semiconductors, and bio-integrated devices. It enables the deposition of high-quality oxides, nitrides, and hybrid materials with excellent conformality and reduced impurity levels. The absence of charged species minimizes substrate damage and defect formation, while preserving atomic-scale precision. Diagnostic tools such as UV absorption spectroscopy, in situ FTIR, ellipsometry, and QCM are used to monitor radical flux, surface reactions, and growth kinetics. REALD complements PEALD by offering a gentler, yet equally precise, pathway for advanced material synthesis [53].

C.1. Working principle and reaction mechanism of REALD: REALD is a variant of conventional ALD that employs neutral radicals rather than ionized plasma species to activate surface reactions and enable film growth at reduced temperatures. Unlike thermal ALD, which relies solely on thermal energy, and PEALD, which introduces ions and metastables, REALD utilizes chemically reactive radicals (e.g., $O^{\bullet}$, $N^{\bullet}$, $H^{\bullet}$) generated remotely or via thermal dissociation. These radicals facilitate ligand abstraction and bond formation without introducing ion bombardment, making REALD particularly suitable for ultra-sensitive substrates such as polymers, organic semiconductors, and low-melting-point metals. REALD typically operates at temperatures between 25 and 150 °C, offering compatibility with flexible electronics, bio-integrated devices, and low-temperature packaging. Familiar radical sources include remote O_2, NH_3, H_2, and N_2 plasmas, as well as catalytic cracking units. Materials deposited via REALD include oxides (e.g., Al_2O_3 and SiO_2), nitrides (e.g., AlN and TiN), and hybrid organic–inorganic films. The technique offers excellent conformality, reduced risk of damage, and precise stoichiometric control. However, radical flux and surface reactivity must be carefully tuned to maintain self-limiting behavior and avoid incomplete reactions. This distinguishes REALD from PEALD, which introduces both radicals and energetic ions, often requiring careful mitigation of ion-induced effects. REALD offers a gentler, highly selective alternative to PEALD, especially for applications requiring pristine interfaces and low thermal budgets. While PEALD provides additional activation via ions, REALD achieves comparable film quality through radical chemistry alone, making it ideal for next-generation flexible, bio-integrated, and low-damage device platforms [54]. The reaction mechanisms of the REALD and PEALD methods are summarized in Tab. 5.2.

Tab. 5.2: Summary of the reaction mechanism of both REALD and PEALD methods.

Features	Radial-enhanced ALD	Plasma-enhanced ALD
Activation species	Neutral radicals ($O^{\bullet}$, $H^{\bullet}$, $N^{\bullet}$)	Radicals + ions + metastables
Ion bombardment	Absent (no substrate damage)	Present (can cause damage or densification)
Radical source	Remote plasma or thermal cracking	Direct or remote plasma
Substrate compatibility	Excellent for sensitive materials (e.g., polymers, organics)	Requires tuning to avoid ion damage
Temperature range	25–150 °C	25–150 °C
Reaction mechanism	Eley–Rideal/Langmuir–Hinshelwood (radical-driven)	Exact mechanisms, but ion-enhanced
Film types	Metals (Cu, Ag), oxides (Al_2O_3, TiO_2), nitrides (TiN)	Oxides, nitrides, metals
Conformality	High, especially in high-aspect-ratio features	High, but ion effects may reduce uniformity
Process complexity	Simpler reactor design (no biasing or ion control)	Requires control of plasma power, bias, and ion energy
Surface regeneration	Via radical re-termination (e.g., –OH and $-NH_x$)	Via plasma-induced functionalization
Risk of subsurface activation	Minimal	Possible with high ion energy

REALD offers a gentler, highly selective alternative to PEALD, especially for applications requiring pristine interfaces and low thermal budgets. The surface reaction mechanisms in REALD are also governed by either Langmuir–Hinshelwood (eq. (5.4)) or Eley–Rideal (eq. (5.5)) pathways, depending on the interaction between gas-phase radicals and surface-bound precursor ligands. Its key mechanistic features of REALD include ***C.1.A. neutral radical activation, C.1.B. self-limiting surface chemistry, C.1.C. radical-driven reaction pathways, C.1.D. damage-free deposition, and C.1.***

E. versatile material compatibility. Radical generation: Radicals are typically produced via remote plasma sources or thermal dissociation. Unlike PEALD, REALD avoids ion bombardment and relies solely on chemically reactive neutral species. Surface reactions: The radicals interact with surface-bound precursor ligands through Eley–Rideal or Langmuir–Hinshelwood mechanisms. For example, hydrogen radicals can reduce metal precursors like $Cu(acac)_2$ to deposit metallic copper films at temperatures as low as 140 °C. Self-limiting behavior: REALD retains ALD's cyclic nature. Each cycle includes precursor adsorption, purge, radical exposure, and a second purge. The radicals selectively remove ligands and regenerate reactive surface sites, maintaining atomic-scale control. Material scope: REALD has been successfully applied to deposit metals (e.g., Cu, Ag) and oxides (e.g., Al_2O_3, TiO_2, and Ta_2O_5). The films exhibit low resistivity, high conformality, and reduced impurity levels [55].

C.1.A. Neutral radical activation: REALD relies exclusively on chemically reactive neutral radicals, such as atomic hydrogen ($H^•$), oxygen ($O^•$), and nitrogen ($N^•$), to drive surface reactions. These radicals are typically generated via remote plasma sources or thermal dissociation, enabling low-temperature processing without the risk of ion-induced damage. The radical activation in REALD involves surface reactions driven by non-ionized reactive species via remote plasma or thermal cracking units. These species are highly chemically reactive but lack sufficient kinetic energy, minimizing substrate damage during deposition. The radical flux (τ_{rad}) governs reaction rates and must be optimized to ensure complete ligand abstraction. To restate, these surface reactions typically follow either the Eley–Rideal or the Langmuir–Hinshelwood mechanism, depending on the dynamics of radical–ligand interactions. For example, oxygen radicals react with TMA to form aluminum oxide [56] via eq. (5.12):

$$Al(CH_3)_3{}^* + 3O^• \rightarrow Al_2O_3 + 3CH_3{}^• \rightarrow CO/CO_2 + H_2O \quad (5.12)$$

Similarly, hydrogen radicals reduce $Cu(acac)_2$ to metallic copper via eq. (5.13):

$$Cu(acac)_2{}^* + 2H^• \rightarrow Cu + 2H(acac) \quad (5.13)$$

These reactions regenerate surface functional groups, maintaining self-limiting behavior and enabling atomic-scale control over film growth. Neutral radical activation in REALD uses chemically reactive species such as $H^•$, $O^•$, and $N^•$ to drive surface reactions without ion bombardment. These radicals, generated via remote plasma or thermal dissociation, selectively remove ligands and facilitate bond formation at low temperatures. This mechanism preserves substrate integrity while enabling self-limiting, conformal film growth, making REALD ideal for sensitive materials and high-aspect-ratio structures [57].

C.1.B. Self-limiting surface chemistry: Self-limiting surface chemistry is a defining feature of REALD, enabling precise, layer-by-layer control over film growth. Each deposition cycle initiates with precursor adsorption, during which molecules chemisorb onto reactive surface sites until saturation is reached. This adsorption behavior adheres to Langmuir kinetics, as described by eq. (5.4), ensuring uniform monolayer formation. Following a purge step, chemically reactive neutral radicals facilitate ligand abstraction through either Eley-Rideal mechanisms (eq. 5.5), depending on the interaction between gas-phase radicals and surface-bound species. These reactions complete the formation of the desired chemical bonds while maintaining the self-limiting nature of the process. Subsequent regeneration of surface functional groups, such as hydroxyl (–OH) or amine (–NH_x) terminations, restores the substrate's reactivity, preparing it for the next cycle. This cyclic regeneration preserves atomic-scale reproducibility and conformality, even on substrates with complex geometries or high aspect ratios, making REALD a robust and scalable technique for high-performance thin-film applications [58].

C.1.C. Radical-driven reaction pathways: In REALD, surface reactions are governed by radical-driven mechanisms that facilitate ligand abstraction and bond formation without ion bombardment. These reactions proceed via either Langmuir–Hinshelwood (eq. (5.4)) or Eley–Rideal (eq. (5.5)) pathways, depending on the interaction dynamics between the reactive species and surface-bound precursors. The Langmuir–Hinshelwood mechanism involves both adsorbed radicals and precursor molecules, requiring lateral diffusion and spatial proximity for reaction to occur. Conversely, the Eley-Rideal mechanism involves gas-phase radicals directly colliding with adsorbed ligands, thereby initiating rapid and selective chemical transformations. These pathways enable efficient film growth at reduced temperatures, preserving substrate integrity and supporting atomic-scale precision. For example, hydrogen radicals ($H^•$) can reduce copper(II) acetylacetonate ($Cu(acac)_2$), yielding metallic copper films under mild thermal conditions (eq. (5.13)). The use of neutral radicals ensures self-limiting behavior and high conformality, making REALD particularly suitable for depositing high-purity films on complex or temperature-sensitive substrates. This mechanistic versatility underscores REALD's value in advanced materials synthesis and low-temperature device fabrication [59].

C.1.D. Damage-free deposition: Damage-free deposition in REALD is achieved by eliminating ion bombardment during surface activation. Unlike PEALD, which introduces energetic ions, REALD relies solely on neutral radicals such as $H^•$, $O^•$, and $N^•$ for chemical reactivity [60]. These radicals are generated remotely, minimizing kinetic energy transfer to the substrate and preventing lattice disruption or defect formation. The absence of ion flux (γ_{ion}) ensures that the energy-dependent etch yield Y (E_{ion}) remains effectively zero (eq. 5.14). This condition preserves surface morphology and interface integrity, especially for temperature-sensitive materials like polymers and organic semiconductors. The reaction (as shown in eq. (5.13) above) proceeds without substrate damage, enabling conformal, high-purity film growth. REALD's damage-free nature supports reproducible deposition across delicate architectures, making it ideal for flexible electronics and bio-integrated devices [61]:

$$\gamma_{ion} \propto \tau_{ion} \times YE_{ion} \approx 0 \tag{5.14}$$

C.1.E. Versatile material compatibility: Versatile material compatibility in REALD enables the deposition of diverse functional thin films. REALD has been successfully applied to metals such as copper (Cu) and silver (Ag), using precursors like $Cu(acac)_2$ and $Ag(fod)_2$. During copper deposition, hydrogen radicals ($H^{\bullet}$) reduce $Cu(acac)_2$, according to eq. (5.12). This reaction yields conductive, low-resistivity films with minimal contamination and excellent adhesion to various substrates. Oxide films such as aluminum oxide (Al_2O_3) and titanium dioxide (TiO_2) are formed using oxygen radicals ($O^{\bullet}$) and metal-organic precursors. For Al_2O_3, the reaction proceeds as eq. (4.11). REALD also supports nitride deposition, such as titanium nitride (TiN), using $TiCl_4$ and sequential $H^{\bullet}$ and $N^{\bullet}$ exposure. These reactions maintain conformality across high-aspect-ratio structures, with impurity levels minimized by radical-driven ligand abstraction and purge efficiency [62].

C.2. Synthesis variables: REALD relies on a tightly controlled set of synthesis variables to achieve precise, conformal film growth. These variables govern precursor adsorption, radical reactivity, and surface regeneration, all of which are essential for maintaining self-limiting deposition cycles. Optimizing parameters such as radical flux, substrate temperature, and purge duration ensures reproducibility, minimizes defects, and enables compatibility with sensitive or high-aspect-ratio substrates. Eight factors will be discussed in the following: ***C.2.A. Radical flux***, ***C.2.B. precursor exposure time***, ***C.2.C. substrate temperature***, ***C.2.D. purge time***, ***C.2.E. surface coverage***, and ***C.2.F. radical source configuration***.

C.2.A. Radical flux: Radical flux (τ_{rad}) in REALD defines the concentration of reactive neutral species reaching the substrate surface. These radicals, such as atomic hydrogen ($H^{\bullet}$), oxygen ($O^{\bullet}$), and nitrogen ($N^{\bullet}$), are generated via remote plasma or thermal dissociation sources. The plasma power, precursor gas flow rate, and radical-generation efficiency govern the magnitude of τ_{rad}. A higher τ_{rad} enhances ligand abstraction kinetics, improving surface reactivity and accelerating film GPC [63].

C.2.B. Precursor exposure time: Precursor exposure time (t_p) in REALD defines the duration during which precursor molecules interact with the substrate surface. This variable directly influences surface coverage (θ), which must reach saturation to ensure uniform monolayer formation. Adsorption kinetics follow Langmuir behavior, modeled by eq. (5.4), as described above. If the t_p is too short, incomplete coverage leads to nonuniform growth and reduced film quality. Excessive exposure may lead to physisorption or precursor decomposition, thereby compromising self-limiting behavior and introducing impurities. For example, Al_2O_3 deposition was shown using TMA (eq. 5.15). Optimal *t*precursor ensures complete surface reaction while preserving atomic-scale control, reproducibility, and compatibility with high-aspect-ratio structures [64]:

$$Al(CH_3)_3 + OH^* \rightarrow Al(CH_3)_2^* + CH_4 \tag{5.15}$$

C.2.C. Substrate temperature: Substrate temperature (T_{sub}) in REALD plays a critical role in controlling adsorption kinetics, radical reactivity, and film properties. Typically ranging from 25 to 150 °C, T_{sub} must be carefully tuned to balance precursor chemisorption and surface reaction efficiency. Adsorption behavior follows Arrhenius-type kinetics, where the rate constant *k*ads is expressed as eq. (5.16):

$$k_{ads} = A \times e^{-E_a/RT_{sub}} \tag{5.16}$$

Here, A is the pre-exponential factor, E_a is the activation energy, and R is the ideal gas constant (8.314 J/ J/mol•K).

Lower T_{sub} values are preferred for temperature-sensitive substrates like polymers or organic semiconductors. However, insufficient thermal energy may hinder ligand desorption or reduce the rates of radical–surface interactions. For example, in TiO_2 deposition using titanium tetraisopropoxide (TTIP) and oxygen radicals, as shown in eq. (5.17):

$$Ti(OCH(CH_3)_2)_4 + O^* \rightarrow TiO_2 + \text{volatile organics} \tag{5.17}$$

Substrate temperature (T_{sub}) critically influences the kinetics of precursor adsorption, radical reactivity, and surface regeneration in REALD. Elevated T_{sub} enhances reaction rates and ligand desorption, promoting dense, crystalline films but may limit compatibility with thermally sensitive substrates. Lower T_{sub} preserves fragile interfaces while relying on radical-driven chemistry to maintain deposition efficiency and self-limiting behavior. Careful optimization of T_{sub} enables reproducible, conformal growth of high-purity films across metals, oxides, and nitrides on complex device architectures. Optimizing T_{sub} ensures complete surface reactions, minimizes defect formation, and supports conformal, high-purity film growth across diverse material platforms [65].

C.2.D. Purge time: Purge time (t_p) in REALD is essential for removing unreacted precursor molecules and volatile byproducts between deposition steps. This interval prevents undesired gas-phase reactions that could lead to CVD-like growth and compromise film uniformity. Efficient purging maintains the self-limiting nature of REALD by isolating each half-reaction and preserving surface saturation dynamics. The purge process is governed by chamber flow dynamics and precursor volatility, modeled by first-order decay (eq. 5.18):

$$C_{(t)} = C_0 e^{-k_p \times t_p} \tag{5.18}$$

where $C_{(t)}$ is the residual concentration, C_0 is the initial concentration, and k_p is the purge rate constant.

In Al_2O_3 deposition using TMA and oxygen radicals ($O^•$), incomplete purging can leave residual precursor in the chamber. When these unreacted TMA molecules encounter incoming $O^•$ radicals, uncontrolled combustion reactions may occur, producing excess CO_x and H_2O, as discussed in eq. (5.19). This not only disrupts the self-limiting growth mechanism but also introduces carbon contamination and compromises film uniformity and interface integrity (eq. 5.19) [66]:

$$Al(CH_3)_3 + O^\bullet \rightarrow Al_2O_3 + CO_x + H_2O \tag{5.19}$$

Purge time (t_p) directly influences the chemical isolation between precursor and radical exposure steps in REALD cycles. Insufficient purging can lead to unintended gas-phase reactions, compromising self-limiting behavior and introducing defects or non-uniform growth. By optimizing t_p, residual precursor and byproducts are effectively removed, preserving surface saturation and reaction specificity. This control is essential for achieving reproducible, high-purity films with excellent conformality across high-aspect-ratio and temperature-sensitive substrates [67].

C.2.E. Surface coverage: Surface coverage (θ_{sat}) in REALD quantifies the fraction of reactive sites occupied by chemisorbed precursor molecules during adsorption. Achieving optimal θ is essential for maintaining self-limiting growth and ensuring uniform monolayer formation across the substrate. The adsorption process follows Langmuir kinetics, described by eq. (5.4). As θ approaches saturation (θ_{sat}), further adsorption is inhibited, preserving atomic-scale control. As shown in the previous example, Al_2O_3 deposition was carried out using TMA, as in eq. (5.15). Optimizing θ_{sat} ensures that each reactive site is selectively occupied without oversaturation. Controlled precursor dosing and exposure time allow chemisorption to reach saturation while avoiding physisorption or multilayer formation. Accurate modeling of θ_{sat} using Langmuir kinetics supports reproducible GPC and atomic-scale precision. This optimization is critical for achieving conformal, defect-free films across high-aspect-ratio structures and temperature-sensitive substrates [68].

C.2.F. Radical source configuration: Radical source configuration in REALD determines the generation, transport, and delivery of neutral reactive species to the substrate. These radicals, such as $H^•$, $O^•$, and $N^•$, are typically produced via remote plasma or thermal dissociation systems to avoid ion bombardment. The radical flux (τ_{rad}) depends on source geometry, excitation energy, and precursor gas flow rate. Remote plasma sources, like surface-wave microwave systems, dissociate molecules ($H_2 \rightarrow 2\,H^•$) without accelerating ions toward the substrate. Radical generation efficiency (η_{rad}) is modeled as follows (eq. 5.20):

$$\tau_{\mathrm{rad}} = \eta_{\mathrm{rad}} \times \frac{P_{\mathrm{plasma}}}{E_{\mathrm{diss}}} \tag{5.20}$$

where P_{plasma} is plasma power and E_{diss} is dissociation energy. Reactor designs often include spatial separation between precursor and radical zones to prevent premature reactions [69].

An optimized radical source configuration in REALD ensures that reactive species are delivered with high purity and directional control. By spatially separating radical generation from the substrate, ion bombardment is eliminated, preserving surface integrity and minimizing defect formation. Controlled radical flux enhances ligand abstraction efficiency, enabling precise chemical transformations at lower substrate temperatures. This configuration supports reproducible, conformal film growth with atomic-scale fidelity across diverse materials and complex device architectures [70].

C.3. Applications and diagnosis: REALD has emerged as a highly versatile thin-film synthesis technique, offering low-temperature, damage-free, and conformal deposition across a broad spectrum of materials and substrates. By utilizing neutral radicals such as atomic hydrogen ($H^•$), oxygen ($O^•$), and nitrogen ($N^•$), REALD enables precise surface reactions without ion bombardment, making it particularly well-suited for temperature-sensitive and soft-matter platforms. REALD has demonstrated broad applicability in advanced device fabrication, spanning multiple domains. In flexible electronics, it enables deposition on polymer substrates without thermal degradation, supporting barrier layers, dielectrics, and conductive films. For semiconductor interconnects, low-temperature deposition of copper and titanium nitride, with high conformality, enhances nanoscale reliability and performance. In organic electronics, REALD preserves molecular integrity, facilitating encapsulation and interface engineering for OLEDs and organic photovoltaics. Energy devices benefit from REALD-deposited oxides such as Al_2O_3 and TiO_2, which serve as passivation and transport layers in batteries and solar cells. Additionally, REALD's gentle chemistry supports biocompatible coatings on sensors, biomaterials, and implantable electronics without inducing cytotoxicity or delamination [71].

To ensure reproducibility and atomic-scale control, REALD employs both in situ and ex situ diagnostic techniques. QCM monitors mass gain per cycle, enabling real-time analysis of growth rates and saturation behavior. In situ Fourier transform infrared (FTIR) spectroscopy tracks the evolution of surface functional groups, confirming ligand abstraction and radical-driven reaction pathways. Spectroscopic ellipsometry provides precise measurements of film thickness and optical constants, validating conformality across complex geometries. X-ray photoelectron spectroscopy (XPS) assesses chemical composition and impurity levels, particularly carbon and chlorine residues from incomplete purging. Transmission electron microscopy (TEM) reveals the sharpness and crystallinity of interfaces, which are critical for evaluating multilayer film quality. Residual gas analysis (RGA) detects volatile byproducts during purge steps, diagnosing incomplete reactions or precursor decomposition [72]. Together, these diagnostic tools enable closed-loop process optimization, reinforcing REALD's core advantages of self-limiting growth, damage-free deposition, and compatibility with emerging device architectures. As material and integration demands continue to evolve, REALD remains a scalable and tunable platform for next-generation thin-film technologies [73].

D. Spatial ALD (SALD): SALD achieves film growth by separating precursor exposure zones across space rather than alternating them over time. This mechanism allows substrates to move through distinct chemical regions, often via rotation or translation, within a modular slit-based reactor. The continuous nature of SALD enables high-throughput processing, making it ideal for large-area applications and roll-to-roll manufacturing. Industrial display technologies and photovoltaics benefit from SALD's scalability, conformality, and compatibility with temperature-sensitive materials [74].

D.1. Working principle and reaction mechanism of SALD: ***SALD*** enables thin-film growth by delivering precursor and co-reactant gases in physically separated zones across a reactor. Unlike conventional ALD, which relies on time-based pulsing and purging, ***SALD*** maintains continuous reactant flow. At the same time, the substrate moves via rotation, translation, or roll-to-roll transport through distinct chemical regions. Gas curtains or inert barriers between zones prevent intermixing, ensuring surface reactions remain isolated and self-limiting. This spatial separation allows for atmospheric pressure operation, significantly reducing cycle times and enabling high-throughput, scalable processing. ***SALD*** is particularly well-suited for large-area and temperature-sensitive substrates, making it ideal for industrial applications such as display technologies, photovoltaics, and barrier coatings. Its self-limiting surface chemistry ensures precise control over film thickness, composition, and conformality, even on complex geometries. For example, Al_2O_3 deposition via ***SALD*** can reach growth rates of up to 60 nm per minute, far surpassing the throughput of traditional ALD. Furthermore, ***SALD*** supports multi-component material synthesis through co-dosing or super-cycling strategies, offering tunable functionality and seamless integration into production lines. The core synthesis parameters in ***SALD*** serve as a foundational framework for process optimization. Parameters such as precursor flow rate, substrate velocity, deposition gap, and zone geometry collectively influence spatial separation, surface saturation, and overall throughput. Their interdependence determines whether the system preserves self-limiting ALD behavior or transitions toward chemical vapor deposition growth regimes [75].

Table 5.3 provides a structured reference for these variables and their influence on the products via ***SALD***.

D.1.1 *Roll-to-roll ALD (R2R-ALD)*: ***R2R-ALD*** is a scalable thin-film fabrication method designed for continuous processing of flexible substrates in web format. Unlike batch ALD systems, ***R2R-ALD*** enables uninterrupted deposition by transporting the substrate through reactive zones using rollers, belts, or tension-controlled spools. This configuration is particularly well-suited for polymeric films, textiles, and other deformable materials that require low-temperature, conformal coatings. Mechanistically, ***R2R-ALD*** often integrates principles of SALD, where precursor and co-reactant gases are delivered in physically separated zones along the web path. Gas curtains or inert barriers prevent intermixing, allowing surface reactions to proceed self-limitingly as the substrate moves continuously through each zone. This spatial separation eliminates the need for time-

Tab. 5.3: Summary of ***SALD's*** variables and their influence on the products.

Parameter	Description
Precursor flow rate	Controls reactant delivery; must be balanced to avoid intermixing or depletion
Carrier gas flow rate	Maintains gas curtain integrity and prevents cross-contamination between zones
Deposition gap	Distance between injector head and substrate; critical for spatial separation
Substrate velocity	Determines residence time in each zone; affects surface saturation and growth rate
Zone geometry	Slit width, spacing, and head design influence gas confinement and uniformity
Temperature	Affects surface reaction kinetics and precursor volatility; typically 80–250 °C
Ambient pressure	Operates at atmospheric pressure, simplifying integration and scalability
Precursor type	Must exhibit self-limiting surface chemistry; common examples: TMA, DEZ, TTIP
Gas curtain composition	Typically inert (N_2 or Ar); prevents mixing and maintains spatial isolation

based pulsing and purging, significantly increasing throughput while preserving atomic-scale control. Applications of ***R2R-ALD*** span flexible electronics, barrier coatings for packaging, and wearable sensors, where uniformity, low defect density, and substrate compatibility are critical. Materials such as Al_2O_3, ZnO, and TiO_2 are commonly deposited to enhance moisture resistance, dielectric performance, or optical transparency. As demand for lightweight, conformable devices grows, ***R2R-ALD*** offers a platform for industrial-scale production of functional films with precision and repeatability [76].

E.1. Working principle and reaction mechanism of R2R-ALD: ***R2R-ALD*** is a continuous thin-film fabrication method designed for flexible substrates transported in web format. The process integrates spatial ALD principles, where precursor and co-reactant gases are delivered in physically separated zones along the substrate path. As the substrate moves through these zones via rollers or tension-controlled spools, surface reactions occur sequentially in a self-limiting manner, ensuring monolayer precision without gas-phase intermixing. Inert gas curtains or vacuum barriers maintain spatial isolation, eliminating the need for time-based pulsing and purging. The reaction mechanism relies on chemisorption of the precursor onto reactive surface sites, followed by ligand abstraction through exposure to the co-reactant. For example, in Al_2O_3 deposition using TMA and H_2O, methyl ligands are removed via hydrolysis, forming a stable oxide layer [77]. Each reaction saturates the surface, preventing uncontrolled growth and ensuring conformality. This mechanism supports low-temperature operation, making ***R2R-ALD*** ideal for polymeric substrates. The continuous nature of the process enables high-throughput, scalable production of functional coatings for packaging, electronics, and energy devices [78].

E.2. Kinetic modeling and growth rate equations: Kinetic modeling in ***R2R-ALD*** establishes a predictive framework for understanding and controlling thin-film growth under continuous substrate motion. Unlike conventional temporal ALD, which relies on sequential pulsing and purging, ***R2R-ALD*** utilizes spatially separated precursor and co-reactant zones. This configuration demands dynamic modeling approaches that incorporate substrate velocity, zone geometry, precursor flux, and residence time. The GPC is fundamentally governed by surface saturation kinetics, typically described by Langmuir-type adsorption models, as shown in eq. (4.4). In R2R systems, the adequate exposure time, determined by substrate speed and reactive zone length, directly influences saturation efficiency and film uniformity. Optimizing the growth rate requires precise control of precursor delivery, gas-curtain stability, and web tension to preserve self-limiting behavior. Robust kinetic modeling enables real-time process tuning and scalable deposition of oxides, nitrides, and hybrid materials on flexible substrates with high throughput and reproducibility [79].

E.3. Synthesis variables: The key synthesis variables in ***R2R-ALD*** and their functional impact on process performance are summarized in Tab. 5.4. Variables such as substrate velocity, zone length, and precursor flow rate directly influence residence time, surface saturation, and growth uniformity. Carrier gas flow and deposition gap govern spatial separation and gas confinement, which are critical for maintaining self-limiting reaction behavior. Together, these parameters define the operational window for scalable, conformal film deposition on flexible substrates with high throughput and reproducibility [80].

Tab. 5.4: Summary of the key synthesis variables in roll-to-roll atomic layer deposition (***R2R-ALD***) and their functional impact on process performance.

Variables	Description	Influence on process
Substrate velocity	The speed at which the web moves through reactive zones	Controls residence time; affects surface saturation, growth per cycle (GPC), and uniformity
Zone length	Physical length of precursor and co-reactant exposure region	Determines exposure duration; longer zones improve saturation but may reduce throughput
Precursor flow Rate	Volume of precursor delivered per unit time	Affects surface coverage; excessive flow risks cross-contamination, while low flow limits GPC
Carrier gas flow rate	Inert gas is used to maintain gas curtain separation	Ensures spatial isolation; improper flow leads to reactant mixing and non-uniform growth
Deposition gap	Distance between gas injector head and substrate surface	Influences gas confinement and lateral diffusion; smaller gaps improve spatial resolution
Substrate temperature	Temperature of the moving web during deposition	Affects reaction kinetics and film crystallinity; must be optimized for polymer compatibility

Tab. 5.4 (continued)

Variables	Description	Influence on process
Web tension	Mechanical tension is applied to the substrate during transport	Ensures stable motion and flatness; critical for uniform coating and zone alignment
Ambient pressure	Operating pressure within the reactor	Typically atmospheric; affects gas dynamics and precursor delivery efficiency
Gas curtain composition	Type of inert gas used to separate reactive zones	Impacts the stability and reactivity; N_2 or Ar is commonly used to prevent intermixing
Reactor geometry	Design of slit heads, exhaust ports, and zone layout	Determines spatial separation efficiency and scalability for large-area substrates

E.4. Applications and diagnosis: ***R2R-ALD*** is a high-throughput technique designed for conformal coating of flexible substrates, making it well-suited for industrial-scale manufacturing. Its capability to deposit ultra-thin, pinhole-free films at low temperatures supports a broad range of emerging applications. In flexible electronics, R2R-ALD is employed to fabricate dielectric layers, encapsulation coatings, and transparent conductors on polymer substrates used in displays, sensors, and wearable devices. In photovoltaics, barrier layers such as Al_2O_3 enhance moisture resistance and extend the operational lifetime of flexible solar cells. For packaging applications, inorganic coatings deposited via ***R2R-ALD*** provide effective gas and moisture barriers, replacing complex multilayer laminates in food, pharmaceutical, and electronic packaging. In energy storage systems, conformal coatings on battery separators and electrodes improve thermal stability and ionic conductivity. Additionally, functional coatings applied to textiles and membranes impart hydrophobicity, conductivity, or antimicrobial properties [81].

To ensure reproducibility and film quality during continuous motion, ***R2R-ALD*** relies on a suite of diagnostic tools tailored to dynamic processing conditions. QCM is used in stationary trials to calibrate GPC and validate saturation kinetics before web-scale implementation. In situ spectroscopic ellipsometry enables real-time monitoring of film thickness and optical properties, supporting feedback control during deposition. RGA (Residual Gas Analyzer) detects volatile byproducts and confirms spatial separation of precursor zones, helping prevent gas-phase reactions. XPS evaluates chemical composition and impurity levels, particularly carbon and chlorine residues from incomplete responses. Optical profilometry and atomic force microscopy (AFM) assess surface roughness and uniformity across the web, which are critical for barrier and optical applications. Moisture vapor transmission rate (MVTR) testing quantifies barrier performance, especially in packaging and OLED encapsulation films [82]. Together, these diagnostic strategies enable closed-loop process optimization, ensuring that ***R2R-ALD*** maintains its core advantages of atomic-scale precision, scalability, and compatibility with flexible substrates. As demand for lightweight, comfortable devices increases, ***R2R-ALD*** continues to provide a robust and versatile platform for next-generation thin-film technologies [83].

F. Molecular layer deposition (MLD): ***MLD*** is a vapor-phase thin-film technique that extends the principles of ALD to enable the growth of organic and organic–inorganic hybrid materials. Like ALD, ***MLD*** relies on sequential, self-limiting surface reactions, but it incorporates bifunctional organic precursors that can form polymeric or hybrid structures. This enables precise control of film thickness, composition, and architecture at the molecular level. ***MLD*** is particularly suited for applications requiring conformal polymer coatings, functionalized surfaces, and hybrid interfaces, such as in MOFs, flexible electronics, battery interfaces, and barrier layers. By alternating between metal-containing precursors and organic linkers, ***MLD*** enables the fabrication of hybrid films with tailored mechanical, optical, and chemical properties [84].

F.1. Working principle and reaction mechanism of MLD: MLD relies on a set of finely tuned process variables that govern the growth, composition, and functionality of organic and hybrid thin films (Tab. 5.5). These variables must be carefully controlled to preserve the self-limiting nature of surface reactions and ensure reproducibility across cycles. ***Precursor selection***: MLD uses bifunctional organic molecules (e.g., ethylene glycol and diaminobenzene), paired with metal precursors (e.g., TMA and $TiCl_4$). The chemical compatibility and volatility of these precursors determine film composition, growth rate, and thermal stability. ***Pulse and purge timing***: Precise timing ensures complete surface reactions and prevents gas-phase interactions. Longer purge times may be required due to larger molecular size and slower desorption kinetics compared to ALD. ***Substrate temperature***: It is typically maintained between 80–200 °C to balance precursor reactivity and thermal decomposition. Temperature affects film density, crosslinking, and organic retention. ***Surface functionalization***: Initial surface groups (-OH, $-NH_2$) influence nucleation and growth uniformity. Pre-treatment may be necessary to ensure consistent film initiation. ***Growth per cycle (GPC)***: MLD films often exhibit lower GPC than ALD due to extended molecular structures. Typical values range from 0.3 to 2.0 Å/cycle, depending on precursor pair and reaction completeness. ***Reactor pressure and flow dynamics***: Low-pressure operation (typically 10^{-3} to 10^{-6} mbar) facilitates clean gas–solid interactions. Flow uniformity is critical for conformal coverage on high-aspect-ratio substrates. These variables collectively define the operational window for MLD and must be optimized for each material system to achieve the desired film properties such as flexibility, barrier performance, or chemical functionality [85].

F.2. Kinetic modeling and growth rate equations: Kinetic modeling in MLD provides a quantitative framework for understanding the growth behavior of organic and hybrid thin films formed via sequential, self-limiting surface reactions. Unlike conventional ALD, MLD uses larger, more flexible bifunctional organic precursors, which introduce additional complexity into reaction kinetics due to steric hindrance, conformational variability, and the potential for double reactions. These factors influ-

Tab. 5.5: Technical description of the key technical variables in molecular layer deposition (MLD).

Parameters	Role in process
Precursor type	The selection of bifunctional organic molecules and metal precursors defines film chemistry.
Pulse/purge timing	Controls surface saturation and prevents gas-phase reactions.
Substrate temperature	Influences reaction kinetics and film stability; typically 80–200 °C.
Growth per cycle	Typically lower than ALD; ranges from 0.3 to 2.0 Å/cycle, depending on precursor pair.
Surface functionalization	Initial surface groups (e.g., –OH and $–NH_2$) determine nucleation and growth uniformity.
Reactor pressure	Usually, low-pressure (10^{-3} to 10^{-6} mbar); ensures clean gas–solid interactions.

ence the density and accessibility of reactive surface sites, especially during the initial nucleation phase. The GPC in MLD is typically modeled using modified Langmuir adsorption kinetics, as shown in eq. (5.4). In MLD, the evolution of GPC over successive cycles often exhibits a transition from sub-monolayer growth to steady-state behavior, driven by changes in surface site density and precursor accessibility. Steric effects can block active sites, while double reactions, where a bifunctional precursor reacts with two adjacent sites, can reduce the number of available reactive groups for subsequent cycles. Additionally, precursor physisorption and subsurface diffusion may create new reactive sites, further complicating the growth kinetics. These phenomena necessitate dynamic modeling approaches that account for molecular structure, surface chemistry, and deposition conditions. Accurate kinetic models enable predictive control over film thickness, composition, and uniformity, supporting the rational design of MLD processes for applications in flexible electronics, energy storage, and functional coatings. In situ diagnostics such as spectroscopic ellipsometry and QCM are often employed to validate model predictions and refine process parameters in real time [86].

F.3. Synthesis variables: MLD relies on sequential, self-limiting reactions between vapor-phase precursors to build organic or hybrid thin films with molecular precision. Synthesis variables, such as precursor chemistry, pulse/purge timing, substrate temperature, and surface functionalization, critically influence film growth rate, uniformity, and chemical composition. Due to the larger size and flexibility of organic precursors, steric hindrance and double reactions must be considered when optimizing GPC. Careful control of these parameters enables reproducible deposition of polymers, hybrid materials, and functional coatings for electronics, energy, and barrier applications [87]. Key synthesis variables in MLD and their influence on film growth are summarized in Tab. 5.6.

Tab. 5.6: Summary of key synthesis variables in MLD and their influence on film growth.

Parameter	Description	Influence on film growth
Precursor chemistry	Bifunctional organic molecules, paired with metal or inorganic precursors	Defines film structure, reactivity, and thermal stability; impacts GPC and hybrid content
Pulse/purge timing	Duration of precursor exposure and inert gas removal	Controls surface saturation and prevents gas-phase reactions; longer purge needed for large molecules
Substrate temperature	80–200 °C, depending on precursor volatility and film stability	Affects reaction kinetics, organic retention, and film crystallinity
Surface functionalization	Initial surface groups (–OH, $-NH_2$, and –COOH)	Influences nucleation density, film uniformity, and adhesion
Growth per cycle (GPC)	Thickness added per cycle, typically 0.3–2.0 Å	Indicates reaction completeness; lower GPC may result from steric hindrance or double reactions
Reactor pressure	Low-pressure (10^{-3} to 10^{-6} mbar)	Ensures clean gas–solid interactions and minimizes physisorption artifacts
Steric effects	Spatial constraints due to large organic molecules	Can reduce reactive site availability and limit GPC
Double reactions	Occurs when bifunctional precursors react with multiple surface sites	May reduce reactive group density for subsequent cycles; affects film linearity

F.4. Applications and diagnosis: ***MLD*** has emerged as a versatile technique for fabricating organic and hybrid thin films with molecular-level precision. Its ability to form conformal coatings on complex geometries and low-temperature substrates makes it highly attractive for a range of advanced applications. In flexible electronics, ***MLD*** enables the deposition of dielectric and encapsulation layers on polymeric substrates, supporting bendable displays and wearable sensors. In energy storage, hybrid films such as titanicone and zincone improve interfacial stability and ion transport in lithium-ion batteries and supercapacitors. Barrier coatings formed via ***MLD***, such as alucone, offer excellent moisture and oxygen resistance, critical for OLEDs and packaging. Additionally, ***MLD*** is used for surface functionalization of MOFs, enhancing selectivity and chemical stability for catalytic and gas separation applications. Emerging applications include low-k dielectrics, photoresists, and biointerfaces, where tailored organic functionality is essential [88].

To ensure reproducibility and performance, ***MLD*** processes rely on robust diagnostic strategies. QCM is used to quantify GPC and validate self-limiting behavior. In situ spectroscopic ellipsometry monitors film thickness and optical constants in real time, enabling dynamic process control. XPS and time-of-flight secondary ion mass spectrometry provide chemical composition and depth profiling, which are essential for verifying the integrity of hybrid films. FTIR identifies functional groups and reaction completeness, especially in polymeric systems. AFM and scanning electron microscopy assess surface morphology and uniformity, while electrical and barrier testing (e.g., MVTR, dielectric breakdown) confirm application-specific performance. These diagnostics collectively support closed-loop optimization, ensuring that MLD delivers reproducible, high-quality films tailored to next-generation device architectures [89].

A. Thermal ALD: Thermal ALD is the foundational form of atomic layer deposition, relying solely on heat-driven surface reactions between vapor-phase precursors and reactive surface sites. Each cycle consists of alternating pulses of precursor and co-reactant, separated by inert gas purges to prevent gas-phase reactions. The process is inherently self-limiting, producing highly conformal, pinhole-free films with atomic-scale thickness control. Thermal ALD is widely used for depositing oxides (Al_2O_3, HfO_2), nitrides, and sulfides on substrates with complex geometries. Its strengths include excellent uniformity, scalability, and compatibility with high-aspect-ratio structures. However, it requires relatively high temperatures (typically 150–300 °C), which may limit its use with temperature-sensitive materials such as polymers or organics. Reaction kinetics are governed by precursor volatility, surface reactivity, and saturation behavior. Thermal ALD is extensively applied in microelectronics, catalysis, optics, and protective coatings. It serves as the benchmark for evaluating other ALD variants and remains the most mature and widely adopted technique in both research and industrial settings [90].

B. PEALD: PEALD introduces reactive plasma species such as ions, radicals, and excited molecules into the co-reactant step to enhance surface reactivity and reduce deposition temperature. This method retains the self-limiting nature of ALD while expanding the range of materials and substrates that can be processed. PEALD enables the deposition of high-quality films at lower temperatures (often below 100 °C), making it suitable for flexible electronics, polymers, and temperature-sensitive devices. The plasma step improves film density, reduces impurities, and enhances nucleation on inert surfaces. Common materials deposited via PEALD include SiO_2, TiO_2, AlN, and metal oxides, which exhibit improved electrical and mechanical properties. However, plasma exposure can introduce damage or nonuniformity, especially on delicate substrates. Reactor design must ensure uniform plasma distribution and minimize ion bombardment. PEALD is widely used in semiconductor fabrication, barrier coatings, and optical layers. Its ability to tailor film properties through plasma chemistry makes it a powerful extension of thermal ALD, especially for low-temperature and high-performance applications [91].

C. REALD: REALD utilizes neutral radicals such as atomic oxygen or hydrogen generated from remote plasma or thermal dissociation to drive surface reactions. Unlike PEALD, REALD avoids charged species, reducing substrate damage while maintaining enhanced reactivity. This method is particularly effective for low-temperature deposition of oxides, nitrides, and carbon-based films on sensitive substrates. The radicals react with adsorbed precursor molecules to complete the ALD cycle, enabling conformal growth with improved film purity and reduced carbon contamination. REALD is advantageous for organic electronics, flexible displays, and bio-interfaces where plasma-induced damage must be minimized. Key process variables include radical flux, residence time, and precursor saturation. Materials such as Al_2O_3, ZnO, and TiO_2 have been successfully deposited using REALD with excellent uniformity and low defect density. Reactor design must ensure efficient radical delivery and spatial separation from the substrate. REALD offers a balance between the conformality of thermal ALD and the reactivity of PEALD, making it a promising technique for next-generation thin-film applications requiring gentle yet effective processing [92].

D. SALD: SALD decouples the temporal sequence of precursor exposure by physically separating reactant zones in space rather than time. Substrates move through alternating regions of precursor and co-reactant gases, often with inert gas curtains to prevent intermixing. This configuration enables continuous, high-throughput deposition without pulsing or purging, dramatically increasing process speed while preserving self-limiting surface reactions. SALD is ideal for large-area coatings, roll-to-roll systems, and atmospheric-pressure operation. It supports low-temperature deposition and is compatible with flexible substrates. Key synthesis variables include substrate velocity, zone geometry, gas flow rates, and deposition gap. Materials of metal oxides, such as Al_2O_3, ZnO, and TiO_2, are commonly deposited with excellent uniformity and scalability. SALD's throughput advantage makes it attractive for industrial applications in photovoltaics, packaging, and flexible electronics. However, maintaining spatial separation and curtain integrity requires precise reactor design and process control. SALD bridges the gap between batch ALD and continuous manufacturing, offering a scalable platform for conformal coatings in emerging technologies [93].

E. R2R-ALD: R2R-ALD adapts spatial ALD principles to a continuous web format, enabling high-throughput deposition on flexible substrates transported via rollers or spools. This method is designed for industrial-scale manufacturing of thin films on polymeric, textile, or metallic substrates. R2R-ALD maintains self-limiting reaction behavior by passing the substrate through spatially separated precursor zones, often at atmospheric pressure. Key process variables include substrate velocity, zone length, gas curtain stability, and web tension. The technique supports low-temperature deposition, making it ideal for packaging, wearable electronics, and flexible photovoltaics. Materials such as Al_2O_3, ZnO, and hybrid oxides are commonly deposited with excellent conformality and barrier performance. Diagnostic tools like QCM, ellipsometry, and MVTR testing ensure film quality and reproducibility. R2R-ALD offers scalability, cost efficiency, and compatibility with roll-based manufacturing infrastructure. Its integration into production lines enables real-time coating of large-area substrates, positioning it as a key enabler for next-generation flexible and lightweight devices across multiple industries [94].

F. MLD: MLD extends ALD chemistry to organic and hybrid materials by using bifunctional organic precursors in sequential, self-limiting reactions. MLD enables the growth of polymeric and organic–inorganic hybrid films with molecular-level control over thickness and composition. Each cycle involves alternating exposure to metal or inorganic precursors and organic linkers, forming extended molecular chains or hybrid networks. Key synthesis variables include precursor volatility, pulse/purge timing, substrate temperature, and surface functionalization. MLD typically exhibits lower GPC (0.3–2.0 Å) due to steric hindrance and double reactions. Materials such as alucone, titanicone, polyimides, and zincone are deposited for applications in flexible electronics, energy storage, barrier coatings, and MOF functionalization. Diagnostic tools like QCM, FTIR, and XPS validate reaction completeness and film integrity. MLD offers unique advantages in tailoring mechanical, optical, and chemical properties, especially for biointerfaces and hybrid devices. Its compatibility with low-temperature substrates and ability to engineer functional organic layers make it a powerful tool for advanced material design [95].

G. Nanomaterials of oxides: ALD is a precision thin-film technique that enables the synthesis of nanomaterials with atomic-scale control over thickness, composition, and conformality. Its self-limiting surface reactions allow for uniform coatings on complex 3D structures, high-aspect-ratio features, and porous substrates, making it indispensable for nanotechnology. Among all allotropes of carbon, nanomembranes are lightweight films with a thickness of up to 100 nm, exhibiting large surface area, high flexibility, and appealing quantum confinement effects [96]. Therefore, it has been applied in catalysis, sensing, biomedicine, and energy harvesting and conversion [97]. In the lift-off ALD synthesis of nanomembranes, the inorganic film is deposited onto a sacrificial polymer substrate, which can later be removed by dissolving or chemical etching. The key to etching is the high selectivity between the sacrificial layer and the nanomembranes [98]. Although the limited selection of material sets makes the release of free-standing nanomembranes difficult, coupling lift-off ALD with rolling-up techniques offers opportunities to transform 2D nanomembranes into other 3D architectures [99].

In 2008, Mei and coworkers developed a versatile approach to the engineering of tubular structures by precisely releasing and simultaneously rolling nanomembranes onto polymers [100]. The trick works with the strain gradient caused by the thermal expansion under different temperatures. During the deposition stage, the top layer of the substrate was kept at 80 °C, while the top layer was kept at room temperature. Then, the researchers used acetone to dissolve the polymer, releasing the film. After rolling up those films, TiO_2, ZnO, and Al_2O_3 microtubes were successfully obtained (Fig. 5.3). Lee and coworkers chose polyvinyl alcohol as the sacrificial substrate to prepare free-floating nanosheets of TiO_2, ZnO, and Al_2O_3 [101]. In another study, polyvinyl alcohol and polyacrylic acid were used to produce a TiO_2 nanomembrane [102]. Those sacrificial layers were removed by dissolving the hot water, leaving the free-standing nanomembranes [103]. The latter study controlled film thickness by increasing the number of ALD cycles. These two studies reveal that the sacrificial layers influence the size, thickness, and shape of nanomembranes. Considering that the available surface area of planar polymers is limited, only a small amount of nanomembrane can be generated. Then, increasing the surface area of the polymer substrate leads to a high yield of nanomembranes – porous sponge [104], porous carbon [105], and other templates [106]. It has been reported in the ALD synthesis of nanosheets, nanomembranes, nanopillars, and nanotubes.

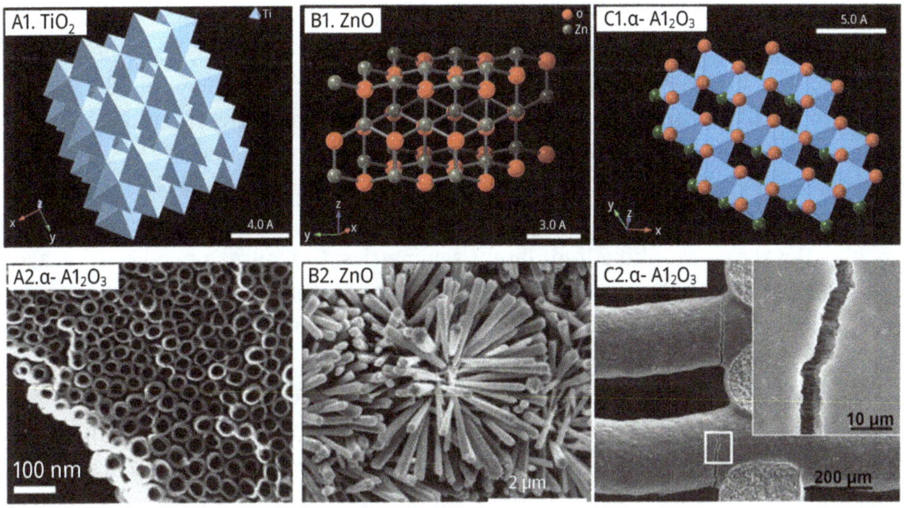

Fig. 5.3: Demonstration of crystallographic structures of metal oxide and their morphology, **A1.** TiO_2, **B1.** ZnO, and **C1.** α-Al_2O_3 microtubes and their morphology, **A2.** TiO_2, **B2.** ZnO and **C2.** α-Al_2O_3. [107].

H. TMDC nanomaterials: TMDCs are 2D nanosheets, developed for their unique photothermal conversion in the semiconductor. The transition metal could be Mo, W, *etc.*, while the chalcogen atom is among S, Se, or Te [108]. TMDCs demonstrate high biocompatibility and biosafety, as well as high photothermal conversion efficiency, making this family a focus of extensive study in energy conversion, drug delivery, and photothermal cancer therapy [109]. Benefiting from the self-limiting feature of ALD, researchers could fabricate nanosheets with controllable thickness, control the number of TMDC layers, manipulate the growth region, and manage the size of the TMDCs to tailor the desired properties [110]. In preparing TMDCs, one of the basic requirements for biomedical uses is the reliable synthesis of ultrathin materials. Here, ALD is superior to conventional CVD due to its precise atomic-level control and its lower defect density. Using a transition metal oxide as the predeposited thin film, researchers observed that the thickness of the TMDC film increased with the number of ALD cycles on a wafer scale [111]. In addition, the successful fabrication of MoS_2, MoO_3, WO_3, Al_2O_3, TiO_2, and ZnO nanomaterials and their applications have been studied [112]. In addition to ongoing efforts to optimize ALD processes and expand the application landscape of TMDC thin films, a rapidly emerging area of research focuses on post-synthetic engineering strategies. These approaches aim to modify, transform, or functionalize pre-deposited TMDC layers through techniques such as thermal annealing, plasma treatment, chemical conversion, and defect modulation. Post-synthetic engineering enables precise control over crystallinity, phase, doping profiles, and interfacial properties, enhancing the performance of TMDCs in electronics, catalysis, and sensing. This paradigm shift enables researchers to decouple deposition constraints from the final material's functionality, offering a versatile route to tailor TMDC behavior beyond the limitations of direct synthesis (Fig. 5.4A). Tungstenite is a rare tungsten sulfide mineral, which is isostructural with Molybdenite. The crystal structure consists of sheets of tungsten atoms, sandwiched between an upper and a lower layer of sulfur atoms (Fig. 5.4B). The resulting "sheet sandwich" is strong, but adjacent sheet sandwiches are only weakly bonded, which gives rise to perfect cleavage parallel to the sheet, along the (001) plane. Note that there are different ways of stacking the sheets, resulting in different symmetries. In nature, a two-layer hexagonal "2 H" polytype is the most common variant, but a three-layer, trigonal polytype also occurs. The high-resolution TEM image (Fig. 5.4C) indicates that the WS_2 nanotubes exhibit a multi-walled cylindrical morphology, with concentric layers forming a hollow core. Each wall consists of stacked WS_2 sheets, likely aligned along the {002} planes, indicating good crystallinity and structural integrity. The uniform diameter and smooth inner cavity suggest controlled synthesis, possibly via sulfurization of tungsten oxide nanowires or template-assisted ALD. This morphology is ideal for applications in nanoelectronics, electrocatalysis, and drug delivery, where high surface area and layered conductivity are advantageous [113].

After preparing the MoO_3 film *via* ALD, Dai and coworkers induced the growth of $MoSe_2$ by selenization in a furnace [115]. This ALD-induced growth of TMDC requires preparing one TMDC as a guide, followed by a modified synthesis using liquid or vapor post-treatment to produce another TMDC. In another study, the ALD-deposited WO_3 film reacted with the gas generated by the sublimation of sulfur powder in an H2S- Ar mixture [116]. The resulting WS_2 nanotubes exhibited a visually uniform appearance. After the successful demonstration of ALD-guided growth of TMDC, other nanomaterials, including MoS_2, Bi_2S_3, and $MoSe_2$, were obtained using a similar strategy [117]. Process control in this ALD-guided growth reveals some significant findings in the growth mechanism. First, the replacement of oxides with sulfides starts around 500–700 °C, then nucleation of metal sulfides grows and completes around 900–1,000 °C. The sulfur atom initiates the conversion, as evident by the decrease of ionic bonds between oxides and metals with increasing temperature when the metal–sulfur bonds start to form [118]. During the growth stage, water is critical to enhancing the transportation of metal oxides. Based on the advances in nanomaterials we surveyed above, ALD shows great promise for the post-processing of TMDC. The resultant materials demonstrate their potential in energy storage, conservation, sensing, and catalysis [119].

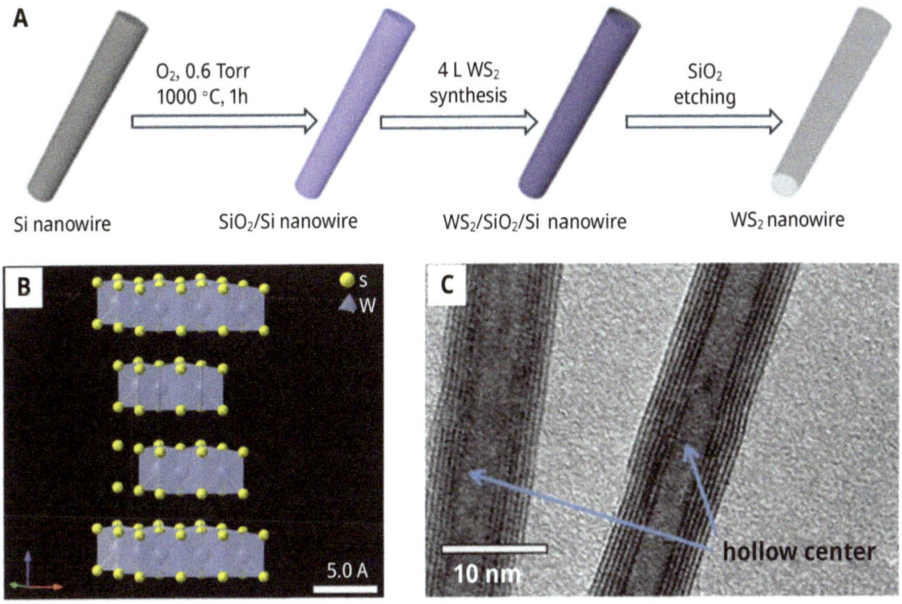

Fig. 5.4: ALD-guided synthesis yields crystalline WS_2 nanotubes with well-defined layered morphology. **A.** Synthetic scheme for the WS_2 nanotubes through ALD-guided growth. Reproduced from Ref. [114] Copyright @ American Chemistry Society 2013, **B.** Crystallographic structure of WS_2 (crystal structure data from: Schutte *et al.* (1987) Crystal structures of tungsten disulfide and diselenide. Journal of Solid State Chemistry 70:207–209) and **C.** Morphology of WS_2 nanotubes.

Chalcogenide-based semiconductors synthesized through ALD exhibit excellent functional properties and strong performance across energy storage systems, energy conversion technologies, and microelectronic devices. Among them, molybdenum disulfide (MoS_2, indirect band gap: ~1.2 eV, direct band gap: ~1.8 eV), bismuth sulfide (Bi_2S_3, direct band gap: ~1.3 eV), and molybdenum diselenide ($MoSe_2$, band gap: 1.1–1.5 eV) stand out as versatile layered materials, whose tunable band structures, anisotropic charge transport behavior, and morphology-dependent characteristics, enabling broad applicability in optoelectronic, catalytic, and electrochemical platforms. Their associated crystal structure parameters are summarized in Tab. 5.7. Their high surface activity, intrinsic structural stability, and compatibility with nanoscale engineering further support advanced applications, ranging from ultrathin transistors and photodetectors to hydrogen-evolution catalysts, supercapacitors, and emerging photovoltaic and sensing technologies.

MoS_2 is a prototypical transition-metal dichalcogenide crystal (Fig. 5.5A) with a layered structure that consists of molybdenum atoms sandwiched between an upper and lower layer of sulfur atoms, forming a robust S-Mo-S trilayer. While each trilayer is strongly bonded internally, adjacent layers interact only through weak van der Waals forces, producing the characteristic perfect cleavage parallel to the basal plane,

Tab. 5.7: Crystal structure parameters for molybdenum disulfide (MoS_2), bismuth sulfide (Bi_2S_3), and molybdenum diselenide ($MoSe_2$), summarizing their lattice constants, symmetry, polytypes, and key structural features.

Crystallographic data	2H-MoS_2	3R-MoS_2	Bi_2S_3	$MoSe_2$
Structure type	Crystal	Crystal	Crystal	Crystal
Number of formula units (Z)	2	3	4	2
Space group	P63/mmc	R3m	Pnma	P63/mmc
Crystal system	Hexagonal	Trigonal	Orthorhombic	Hexagonal
a/*b* (Å)	3.1604	3.1630	11.3160/3.9709	3.2880
c (Å)	12.2950	18.3700	11.1780	12.9000
Cell volume ($Å^3$)	106.351	159.162	502.280	120.777
Asymmetric unit	2 sites	3 sites	5 sites	2 sites
Unit cell	6 sites/cell	9 sites/cell	20 sites/cell	6 sites/cell
Density (g/cm^3)	4.9983	5.0098	6.7996	6.9806
Visible sites	219	156	240	144

commonly indexed as the (001) plane. Variations in how these trilayers stack give rise to distinct polytypes with different symmetries. In naturally occurring MoS_2, the 2H polytype [120], a two-layer, hexagonal stacking sequence, is the most prevalent, although the 3R polytype with a three-layer, trigonal stacking arrangement is also observed (Fig. 5.5B). Molybdenite in the 3R polytype consists of sheets where each molybdenum layer is sandwiched between upper and lower sulfur layers, forming a rigid S-Mo-S trilayer. These trilayers are strongly bonded internally but interact only through weak van der Waals forces, producing perfect cleavage along the basal (001) plane. This layered architecture defines the material's mechanical softness, structural anisotropy, and ease of exfoliation into thin crystalline sheets [120]. The MoS_2 serves as a prototypical transition-metal dichalcogenide, exemplifying the layered S-Mo-S architecture that underpins its mechanical flexibility, electronic anisotropy, and broad technological relevance. This structural anisotropy enables mechanical exfoliation down to monolayer thickness, where MoS_2 undergoes a well-known transition from an indirect band gap in the bulk to a direct band gap of about 1.8 eV. The density of states (DOS) of MoS_2 reveals distinct vibrational modes, with prominent peaks near 9, 17, and 19 THz, indicating vigorous phonon activity at these frequencies (Fig. 5.5C). [121] These features reflect the layered crystal structure and influence MoS_2's electronic and thermal properties, thereby affecting carrier scattering, excitonic transitions, and optoelectronic performance in monolayer and bulk configurations. The spectral distribution underscores its anisotropic vibrational dynamics. This thickness-dependent electronic structure underpins its exceptional optical absorption, excitonic behavior, and suitability for ultrathin optoelectronic devices. The absence of dangling bonds on basal planes minimizes surface scattering, supporting high carrier mobility and low contact resistance attributes that have enabled field-effect transistors with channel lengths approaching the sub-nanometer regime. Catalytically, MoS_2 exhibits highly active edge sites that facilitate the hydrogen (H_2) evolution reaction (Fig. 5.5D), making it a promising non-precious metal electrocatalyst. Its large surface area and tunable adsorption characteristics also support sensitive chemical and biological sensing. Beyond electronics and catalysis, MoS_2 has been incorporated into batteries, solar cells, and spintronic architectures, where its stability, mechanical flexibility, and tunable electronic properties continue to drive broad technological relevance.

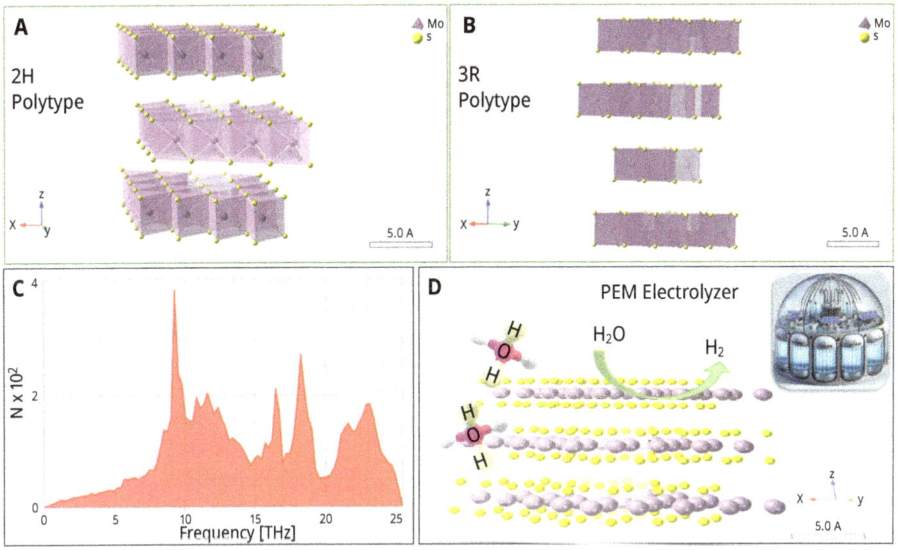

Fig. 5.5: Structural and functional features of molybdenum disulfide (MoS_2), a prototypical transition-metal dichalcogenide. **A.** The 2H polytype exhibits a two-layer, hexagonal stacking sequence; the 3R polytype adopts a three-layer, trigonal arrangement. **B.** Both consist of robust S-Mo-S trilayers held together by weak van der Waals forces, enabling perfect cleavage along the (001) plane. **C.** Density of states (DOS) spectrum reveals distinct vibrational modes, with peaks near 9, 17, and 19 THz, reflecting vigorous phonon activity. **D.** Catalytically active edge sites facilitate hydrogen evolution, highlighting MoS_2's relevance in energy conversion, sensing, and electronic platforms.

Bismuth sulfide (Bi_2S_3) is a direct band gap semiconductor with an optical gap near 1.3 eV, placing it within an ideal range for visible light harvesting and photodetection. Its orthorhombic crystal structure (Fig. 5.6A) is isostructural with stibnite (Sb_2S_3), and the two minerals form a continuous solid-solution series, reflecting their closely related crystal chemistries [121]. The structure of Bi_2S_3 is defined by the linkage of Bi_4S_6 ribbons, which extend parallel to the *y*-axis and form an anisotropic layered framework. Within this architecture, Bi1 atoms (magenta) occupy the apex of trigonal pyramids, coordinated by sulfur, while Bi2 atoms (purple) exhibit five-fold coordination near the basal plane of distorted square pyramids. These polyhedral arrangements produce strongly bonded chains that govern directional charge transport and mechanical anisotropy, while weaker inter-ribbon interactions facilitate cleavage and deformation along preferred crystallographic planes. This anisotropic bonding environment enables the formation of diverse one-dimensional and hierarchical nanostructures, including nanorods, nanowires, nanotubes, and nanoflowers. These morphologies offer tunable surface area, defect density, and carrier pathways, allowing Bi_2S_3 to be engineered for specific optoelectronic and catalytic functions. Compared with cadmium- or lead-based chalcogenides, Bi_2S_3 provides the advantages of lower toxicity and greater elemental abundance, making it attractive for sustainable photovoltaic technologies. Its strong absorption coefficient in the visible and near-infrared regions supports efficient carrier photogeneration. At the same time, its intrinsic n-type conductivity can be modulated toward p-type behavior through doping or defect engineering. These characteristics have enabled Bi_2S_3 to be deployed in thin-film and nanowire solar cells, photodetectors, photocatalytic systems, and electrochemical devices such as supercapacitors and sensors. The combination of structural versatility, favorable band alignment, and environmental compatibility continues to position Bi_2S_3 as a promising material for next-generation optoelectronic and energy-conversion technologies.

Molybdenum diselenide ($MoSe_2$) is a layered transition-metal dichalcogenide that adopts the same structural motif as molybdenite, with molybdenum atoms sandwiched between upper and lower selenium layers to form a robust Se-Mo-Se trilayer (Fig. 5.6B) [122]. In this structure, Mo atoms (blue spheres) occupy trigonal prismatic coordination sites between two Se layers (orange spheres), generating a repeating trilayer arrangement. Strong intralayer bonding stabilizes each Se-Mo-Se unit, while adjacent layers interact only through weak van der Waals forces, enabling mechanical exfoliation and producing perfect cleavage along the (001) basal plane. The 3D axis and 5.0 Å scale bar provide spatial context for interatomic distances and crystallographic orientation. $MoSe_2$ exhibits a thickness-dependent electronic structure, with a band gap tunable from about 1.1 to 1.5 eV, making it suitable for optoelectronic and electrochemical applications. Relative to its sulfur analogue, MoS_2, $MoSe_2$ offers higher intrinsic electrical conductivity and a greater density of catalytically active edge sites, thereby enhancing its performance in hydrogen-evolution and related electrocatalytic reactions. Its stable van der Waals framework supports mechanical flexibility and chemical robustness, facilitating integration into flexible electronics and composite architectures. Multiple accessible oxidation states further enable the formation of hybrid materials with metals, carbon frameworks, and conductive polymers, improving charge storage and transport. These attributes have led to its deployment in supercapacitors, electrochemical sensors, and energy-storage systems, where $MoSe_2$-based electrodes exhibit favorable ion-intercalation behavior and structural stability. Collectively, its tunable band structure, catalytic activity, and compatibility with diverse composite strategies make $MoSe_2$ a versatile material for advanced energy and sensing technologies [123].

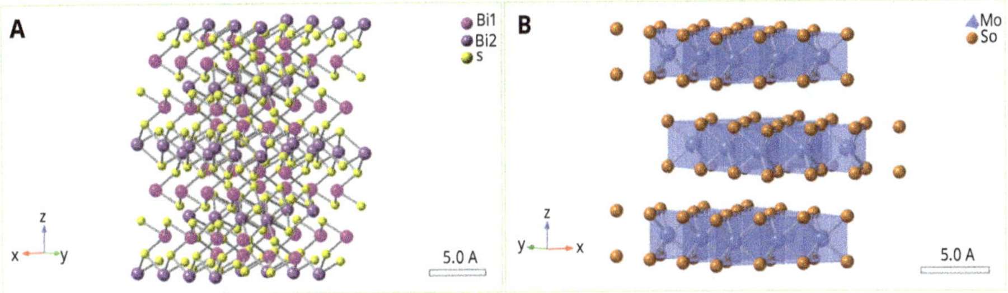

Fig. 5.6: Crystalline structure data for two-layered semiconductors. **A.** Bismuth sulfide (Bi_2S_3) forms Bi_4S_6 ribbons with Bi1 trigonal pyramids and Bi2 five-fold sites, creating an anisotropic framework that supports directional transport and diverse one-dimensional nanostructures. **B.** Molybdenum diselenide ($MoSe_2$) features trigonal-prismatic Mo coordination within Se–Mo–Se trilayers, stabilized by strong intralayer bonding and van der Waals gaps that enable perfect (001) cleavage and mechanical exfoliation.

I. MOFs: Like what we discussed in the case of TMDCs, the flexible integration of layered materials through ALD has also been observed in the 3D architecture of MOF films. Conventional synthesis of MOFs includes the solvothermal and hydrothermal reactions, which were discussed in different chapters. Those synthetic methods produce 3D crystals, so we need alternative ways to make MOF membranes. The Karppinen group developed an ALD synthesis of MOF-2 thin film on various surfaces [124]. This study introduced metal and organic precursors in the gas phase. The thickness of MOF films also depends on the number of ALD cycles. The powder XRD patterns revealed a match between the MOF films and the pristine MOF crystals. Most MOFs take hours to days to form and require higher temperatures, hindering widespread industrial applications. Zhao and his colleagues demonstrated a facile, ultrafast synthesis of MOFs at room temperature via ALD [125]. The ALD-deposited ZnO membrane transformed into a (Zn, Cu) hydroxy double salt intermediate, which is also a layered sheet connected by inorganic and organic anions [126]. The cationic nature enables rapid anion exchange in the linker solution, allowing the growth of MOF HKUST-1 (Hong Kong University of Science and Technology) in the liquid within a minute [127]. They obtained MOFs of Cu-BDC (benzene-1,4-dicarboxylate), ZIF-8 (zeolitic imidazolate frameworks), and IRMOF-3 (isoreticular, IR) by following a similar design. The downside of solvent-based synthesis of MOFs is corrosion and contamination as well as other surface-tension-related problems. [128] Therefore, solvent-free methods serve as alternatives for manufacturing and processing MOFs. Many recent reports discovered ways-dry synthesis or fluid-assisted synthesis to prepare some prototypical MOFs in the presence of trace amounts of solvents and additives, or even without them [129]. Those discoveries support the vapor-phase fabrication of MOFs. The first vapor-phase deposition was to synthesize ZIFs with the ALD-deposited ZnO film as the inducing layer [130]. The vaporized 2-methyl imidazole ligand reacted with the ZnO film in the closed reactor. Although the material was partially converted, the MOF membranes exhibited selectivity for propylene over propane [131]. Those pioneering works open new avenues for research into the fabrication of MOF-based nanomaterials. Other recent advances in MOFs include UiO-66 (Universitetet i Oslo), UiO-66-NH_2, ZIF-6, ZIF-7, Al-PMOF (porphyrin, P), Cu-TCPP (meso-tetra(4-carboxyphenyl)porphyrin), MAF-6 (metal-azolate framework), Fe-BDC, Fe-BTC, Al-BTC, and Al-BDC [132]. Those CVD- and ALD-deposited MOF nanomaterials offer numerous opportunities to tune the nanostructure, composition, and associated optical and electrical properties. However, ALD synthesis methods still suffer from a low effective deposition rate, inevitable impurities in the films, and limited precursor selection. Continuous efforts are needed to optimize film quality, understand the growth mechanism, advance ALD technology, and develop complementary applications [133].

ZIFs are a subclass of MOFs, constructed from tetrahedrally coordinated transition-metal ions bridged by imidazolate (Im) linkers. Their characteristic M-Im-M bond angles closely resemble Si-O-Si angles in aluminosilicate zeolites, enabling the formation of diverse zeolite-type topologies with tunable pore sizes and geometries. ZIFs combine exceptional chemical and thermal stability with the synthetic versatility of organic-inorganic hybrid materials, allowing systematic modification of functionality, aperture dimensions, and framework flexibility. Their porous architectures support high surface areas, selective adsorption, and molecular sieving, while the modular chemistry enables fine control over catalytic, sorption, and transport properties. These attributes have positioned ZIFs as essential materials for gas separation, catalysis, sensing, energy storage, and other technologically relevant applications. ZIF-8 is a ZIF in which imidazolate linkers bridge ZnN4 tetrahedra to generate a robust, 3D network (Fig. 5.7A). The framework adopts the sodalite topology, with an interconnected Zn-Im lattice forming large, symmetric cages that define its characteristic porosity. In structural representations, ZnN_4 tetrahedra appear as grey polyhedra and imidazolate linkers as green pentagons, emphasizing the tetrahedral connectivity that underpins the sodalite-type architecture. The yellow translucent sphere highlights the maximum accessible pore volume, illustrating the spacious cavities and uniform apertures that support selective adsorption and molecular transport. This combination of geometric regularity, high porosity, and exceptional chemical stability has made ZIF-8 a widely studied material for gas separation, catalysis, and membrane-based applications. ZIF-6 is a gismondine-type ZIF constructed from ZnN_4 tetrahedra linked by imidazolate groups to form an open, zeolite-like architecture (Fig. 5.7B). The framework exhibits a characteristic arrangement of cages and channels defined by tetrahedral Zn-Im connectivity, resulting in a porous network with well-defined diffusion pathways. In polyhedral depictions, ZnN_4 units are shown as grey tetrahedra and imidazolate linkers as green pentagons, highlighting the structural motifs that give rise to the gismondine topology. This representation emphasizes the orientation and connectivity of the linkers, which govern pore geometry and framework rigidity. Its resulting structure combines moderate porosity with notable chemical stability, making ZIF-6 relevant for studies of structure–property relationships, sorption behavior, and zeolite-analogue design within the broader ZIF family.

ZIF-zni is a dense zeolitic imidazolate framework composed of ZnN_4 tetrahedra, interconnected by imidazolate linkers and benzene-based ligands, forming a highly ordered 3D network (Fig. 5.7C). Unlike porous ZIFs, the zni topology yields a compact, nonporous structure in which tetrahedral units are tightly packed, thereby conferring exceptional mechanical and thermal stability. In this polyhedral representation, the ZnN_4 tetrahedra appear as grey units, imidazolate linkers as green pentagons, and benzene groups as black hexagons, illustrating the mixed-linker connectivity that defines the zni framework. This structural arrangement provides insight into topological transformations observed in other ZIFs, particularly the conversion of open frameworks into dense polymorphs. ZIF-zni serves as an essential reference for understanding rigidity, framework collapse, and the structural limits of imidazolate-based materials. Another MOF, the zeolite-rho-type ZIF (ZIF-11), is composed of ZnN_4 tetrahedra, linked by imidazolate and benzene-based linkers, to form a spacious, 3D network (Fig. 5.7D). The rho topology generates large cages and interconnected channels, producing a highly porous architecture with tunable adsorption properties. In polyhedral views, ZnN_4 tetrahedra are shown as grey units, imidazolate linkers as green pentagons, and benzene linkers as black hexagons, emphasizing the mixed-linker connectivity that shapes the pore system. This arrangement yields high chemical stability and well-defined pore environments suitable for selective molecular sieving. ZIF-11 demonstrates how imidazolate-bridged metal centers can replicate zeolite-like geometries while offering the synthetic versatility, functional tunability, and structural diversity characteristic of MOFs. In summary, these ZIFs illustrate the remarkable structural diversity achievable through tetrahedral metal-imidazolate connectivity. Despite sharing a common ZnN_4 coordination motif, their distinct topologies, ranging from open sodalite and gismondine architectures to dense zni and spacious rho networks, demonstrate how linker geometry, mixed-linker strategies, and framework packing govern porosity, rigidity, and functional behavior. The polyhedral representations highlight the interplay between inorganic nodes and organic linkers that define pore environments and adsorption pathways. The materials exemplify the versatility, stability, and tunability that make ZIFs an essential platform for studying structure-property relationships and advancing applications in separation, catalysis, and energy-related technologies.

5.5 Summary and outlook

This chapter has presented a comprehensive overview of ALD as a versatile and precise technique for nanomaterial fabrication, emphasizing its unique capability to achieve atomic-scale control through sequential, self-limiting surface reactions. The fundamental principles governing ALD were examined in detail, including precursor chemistries, surface reaction pathways, and the cyclic nature of the deposition process that enables de-

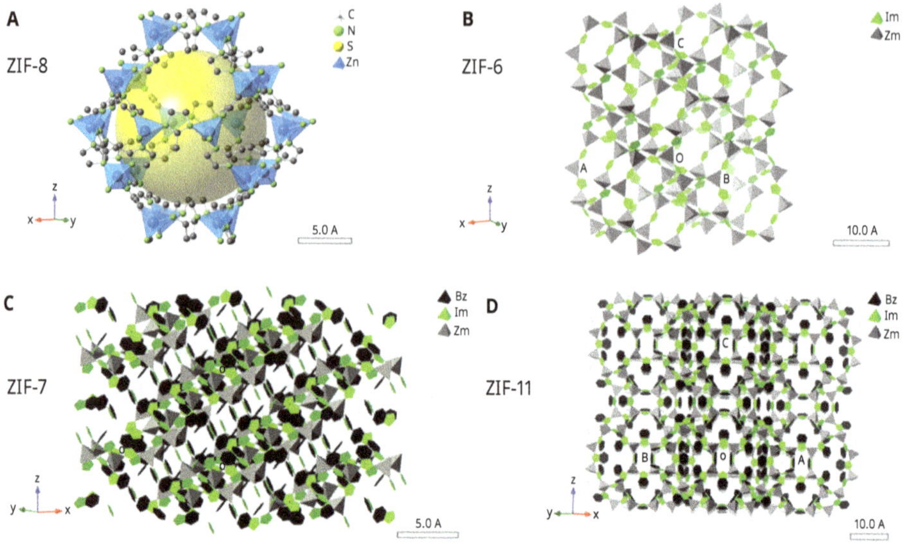

Fig. 5.7: Zeolitic imidazolate frameworks (ZIFs) formed by tetrahedral metal-imidazolate linkages, mimicking zeolite bond angles and enabling diverse, tunable porous architectures. **A.** ZIF-8 adopts a sodalite topology composed of ZnN_4 tetrahedra and imidazolate linkers, forming large cages and uniform pores that enable high porosity and selective molecular transport. **B.** ZIF-6 exhibits a gismondine-type topology with ZnN_4 tetrahedra linked by imidazolate groups, generating an open framework with interconnected channels and moderate, well-defined porosity. **C.** ZIF–zni displays a dense zni topology, composed of tightly packed ZnN_4 tetrahedra and mixed linkers, yielding a nonporous framework with exceptional mechanical and thermal stability. **D.** ZIF-11 adopts a zeolite–rho topology, featuring ZnN_4 tetrahedra, linked by imidazolate and benzene ligands, producing large cages, interconnected channels, and highly tunable adsorption environments.

terministic control over film growth. Particular attention was given to the four-step ALD cycle, precursor exposure, purge, co-reactant exposure, and final purge, which collectively ensure chemical isolation between half-reactions and reproducible layer-by-layer growth. Key synthesis variables were systematically discussed to illustrate how process parameters govern film thickness, uniformity, and material quality. Precursor pulse time and purge duration were shown to directly influence surface saturation, impurity removal, and GPC. At the same time, deposition temperature was highlighted as a critical factor controlling reaction kinetics, precursor stability, and film crystallinity. The roles of precursor reactivity and surface activation were emphasized in determining nucleation behaviors, conformality on high-aspect-ratio structures, and compatibility with chemically inert or temperature-sensitive substrates. Cycle count was identified as the primary variable, enabling precise thickness control, allowing ALD to produce ultrathin films and multilayer heterostructures with sub-nanometer accuracy. The chapter further explored a range of ALD variants, including thermal, plasma-enhanced, radical-enhanced, spatial, roll-to-roll, and molecular layer deposition, demonstrating how modifications to energy input, reactor design, and precursor delivery expand ALD's applicability. These approaches enable lower-temperature processing, higher throughput, and the deposition of hybrid organic–inorganic materials, thereby extending ALD beyond traditional inorganic thin films. The discussion of ALD-derived materials highlighted strong correlations between deposition conditions, structural characteristics such as density and interface quality, and functional performance in electronic, energy storage, catalytic, and biomedical applications.

This chapter also demonstrates the structural characteristics and functional properties of diverse nanomaterials, enabled by ALD, highlighting ALD's versatility across inorganic, chalcogenide, and hybrid material systems. Metal oxide nanomaterials such as TiO_2, ZnO, and Al_2O_3 demonstrate how ALD enables precise thickness control, conformal coating, and crystallographic tuning, supporting applications in catalysis, sensing, energy harvesting, and protective coatings. The lift-off ALD strategy further enables free-standing nanomembranes and tubular architectures, where sacrificial substrates and cycle count govern membrane thickness, morphology, and mechanical flexibility. Transition-metal dichalcogenides synthesized or guided by ALD exhibit thickness-dependent electronic structures, layered crystallinity, and high surface activity. Materials such as MoS_2, Bi_2S_3, and $MoSe_2$ combine tunable band gaps, anisotropic charge transport, and morphology-dependent properties, enabling applications in optoelectronics, electrocatalysis, energy storage, and sensing. ALD-guided post-synthetic conversion of metal oxides into sulfides or selenides provides precise control over the number of layers, crystallinity, and defect density, yielding nanotubes, nanosheets, and hierarchical structures with enhanced catalytic and electronic performance. The chapter also highlights ALD-enabled fabrication of MOFs, such as ZIFs. These materials demonstrate how vapor-phase or ALD-assisted synthesis enables thin-film MOF membranes with controlled thickness, topology, and pore architecture. These ZIFs illustrate the structural diversity achievable through tetrahedral metal–imidazolate connectivity, spanning porous sodalite and gismondine networks to dense ZnI and rho topologies. Their high surface areas, chemical stability, and tunable pore environments support applications in gas separation, catalysis, molecular sieving, and energy-related technologies. Collectively, the materials discussed exemplify how ALD enables deterministic control over composition, structure, and interfaces, linking nanoscale design to functional performance across electronic, catalytic, biomedical, and energy applications.

Looking ahead, continued advances in precursor design, reactor architecture, and in situ diagnostics are expected to enhance further the efficiency, selectivity, and scalability of ALD. Emerging directions include low-temperature and selective-area deposition, integration with two-dimensional materials, porous frameworks, and hybrid organic–inorganic systems, as well as the fabrication of multifunctional heterostructures with precisely engineered interfaces. These developments are enabling the rational design of materials for sustainable energy technologies, including high-performance electrodes for batteries and supercapacitors, catalytic layers for hydrogen evolution and CO_2 conversion, and conformal coatings for photovoltaic and photoelectrochemical devices.

In electronic and optoelectronic applications, ALD continues to play a critical role in producing ultrathin dielectrics, gate oxides, diffusion barriers, and channel materials for advanced transistors, sensors, and flexible electronics, where atomic-scale thickness control and defect minimization are essential. In biomedical and medical research, ALD-derived coatings and nanostructures are increasingly explored for biocompatible implants, drug-delivery platforms, antimicrobial surfaces, and diagnostic devices, owing to their chemical stability, tunable surface chemistry, and conformal coverage of complex geometries. As device architectures become increasingly complex and dimensional tolerances more stringent, ALD is poised to remain a foundational platform for next-generation nanomanufacturing, enabling reliable integration of advanced materials across energy, electronics, and healthcare technologies.

References

[1] George, S. M. (2010). Atomic layer deposition: An overview. *Chemical Reviews*, 110(1), 111–131.
[2] Leskelä, M., & Ritala, M. (2002). Atomic layer deposition (ALD): From precursors to thin film structures. *Thin Solid Films*, 409(1), 138–146.
[3] Puurunen, R. L. (2005). Surface chemistry of atomic layer deposition: A case study for the trimethylaluminum/water process. *Journal of Applied Physics*, 97(12).
[4] Biercuk, M. J., Monsma, D. J., Marcus, C. M., Becker, J. S., & Gordon, R. G. (2003). Low-temperature atomic-layer-deposition lift-off method for microelectronic and nanoelectronic applications. *Applied Physics Letters*, 83(12), 2405–2407.
[5] Poodt, P., Cameron, D. C., Dickey, E., George, S. M., Kuznetsov, V., Parsons, G. N., & Vermeer, A. (2012). Spatial atomic layer deposition: A route towards further industrialization of atomic layer deposition. *Journal of Vacuum Science & Technology A*, 30(1).
[6] Johnson, R. W., Hultqvist, A., & Bent, S. F. (2014). A brief review of atomic layer deposition: From fundamentals to applications. *Materials Today*, 17(5), 236–246.
[7] Knoops, H. C., Potts, S. E., Bol, A. A., & Kessels, W. M. M. (2015). Atomic layer deposition. In *Handbook of Crystal Growth* (pp. 1101–1134). North-Holland.
[8] Leskelä, M., & Ritala, M. (2003). Atomic layer deposition chemistry: Recent developments and future challenges. *Angewandte Chemie International Edition*, 42(45), 5548–5554.
[9] Richey, N. E., De Paula, C., & Bent, S. F. (2020). Understanding chemical and physical mechanisms in atomic layer deposition. *The Journal of Chemical Physics*, 152(4).
[10] Yanguas-Gil, A., & Elam, J. W. (2012). Self-limited reaction-diffusion in nanostructured substrates: Surface coverage dynamics and analytic approximations to ALD saturation times. *Chemical Vapor Deposition*, 18(1-3), 46–52.
[11] Ott, A. W., Klaus, J. W., Johnson, J. M., & George, S. M. (1997). Al_3O_3 thin film growth on Si (100) using binary reaction sequence chemistry. *Thin Solid Films*, 292(1–2), 135–144.
[12] Waldman, R. Z., Mandia, D. J., Yanguas-Gil, A., Martinson, A. B., Elam, J. W., & Darling, S. B. (2019). The chemical physics of sequential infiltration synthesis: A thermodynamic and kinetic perspective. *The Journal of Chemical Physics*, 151(19).
[13] Weisbord, I., Shomrat, N., Azoulay, R., Kaushansky, A., & Segal-Peretz, T. (2020). Understanding and controlling polymer, organometallic precursor interactions in sequential infiltration synthesis. *Chemistry of Materials*, 32(11), 4499–4508.

[14] Park, H. S., Min, J. S., Lim, J. W., & Kang, S. W. (2000). Theoretical evaluation of film growth rate during atomic layer epitaxy. *Applied Surface Science*, 158(1–2), 81–91.

[15] Park, H. K., Yang, B. S., Park, S., Kim, M. S., Shin, J. C., & Heo, J. (2014). Purge-time-dependent growth of ZnO thin films by atomic layer deposition. *Journal of Alloys and Compounds*, 605, 124–130.

[16] Elam, J. W., Groner, M. D., & George, S. M. (2002). Viscous flow reactor with quartz crystal microbalance for thin film growth by atomic layer deposition. *Review of Scientific Instruments*, 73(8), 2981–2987.

[17] Crowell, J. E. (2003). Chemical methods of thin film deposition: Chemical vapor deposition, atomic layer deposition, and related technologies. *Journal of Vacuum Science & Technology A: Vacuum, Surfaces, and Films*, 21(5), S88–S95.

[18] Miikkulainen, V., Leskelä, M., Ritala, M., & Puurunen, R. L. (2013). Crystallinity of inorganic films grown by atomic layer deposition: Overview and general trends. *Journal of Applied Physics*, 113(2).

[19] Van Bui, H., Grillo, F., & Van Ommen, J. R. (2017). Atomic and molecular layer deposition: Off the beaten track. *Chemical Communications*, 53(1), 45–71.

[20] Kim, H. (2003). Atomic layer deposition of metal and nitride thin films: Current research efforts and applications for semiconductor device processing. *Journal of Vacuum Science & Technology B: Microelectronics and Nanometer Structures Processing, Measurement, and Phenomena*, 21(6), 2231–2261.

[21] Oviroh, P. O., Akbarzadeh, R., Pan, D., Coetzee, R. A. M., & Jen, T. C. (2019). New development of atomic layer deposition: Processes, methods, and applications. *Science and Technology of Advanced Materials*, 20(1), 465–496.

[22] Muneshwar, T., Miao, M., Borujeny, E. R., & Cadien, K. (2018). Atomic layer deposition: Fundamentals, practice, and challenges. In *Handbook of thin film deposition* (pp. 359–377). William Andrew Publishing.

[23] Deyu, G. K., Wenskat, M., Díaz-Palacio, I. G., Blick, R. H., Zierold, R., & Hillert, W. (2025). *Recent advances in atomic layer deposition of superconducting thin films: A review. Materials Horizons.*

[24] Oviroh, P. O., Oyinbo, S. T., Karimzadeh, S., Imoisili, P. E., & Jen, T. C. (2024). Introduction to ALD Technologies. In *Emerging Atomic Layer Deposition for Hydrogen Energy* (pp. 1–25). Cham: Springer Nature Switzerland.

[25] Gordon, R. G. (2013). ALD precursors and reaction mechanisms. In *Atomic Layer Deposition for Semiconductors* (pp. 15–46).

[26] Puurunen, R. L. (2003). Growth per cycle in atomic layer deposition: A theoretical model. *Chemical Vapor Deposition*, 9(5), 249–257.

[27] Lim, J. W., Park, H. S., & Kang, S. W. (2001). Kinetic modeling of film growth rate in atomic layer deposition. *Journal of the Electrochemical Society*, 148(6), C403.

[28] Leskelä, M., & Ritala, M. (1999). ALD precursor chemistry: Evolution and future challenges. *Le Journal de Physique IV*, 9(PR8), Pr8–837.

[29] Knez, M., Nielsch, K., & Niinistö, L. (2007). Synthesis and surface engineering of complex nanostructures by atomic layer deposition. *Advanced Materials*, 19(21), 3425–3438.

[30] Hashemi, F. S., Cao, L., Mattelaer, F., Sajavaara, T., Van Ommen, J. R., & Detavernier, C. (2019). Aluminum tri-isopropoxide as an alternative precursor for atomic layer deposition of aluminum oxide thin films. *Journal of Vacuum Science & Technology A*, 37(4), 040901.

[31] Hausmann, D. M., & Gordon, R. G. (2003). Surface morphology and crystallinity control in the atomic layer deposition (ALD) of hafnium and zirconium oxide thin films. *Journal of Crystal Growth*, 249(1–2), 251–261.

[32] Wilk, G. D. (2001). High-k gate dielectrics: Current status and materials properties considerations. *Journal of Applied Physics*, 89(10), 5423–5275.

[33] Matero, R., Rahtu, A., Ritala, M., Leskelä, M., & Sajavaara, T. (2000). Effect of water dose on the atomic layer deposition rate of oxide thin films. *Thin Solid Films*, 368(1), 1–7.

[34] Dingemans, G., & Kessels, W. M. M. (2012). Status and prospects of Al_2O_3-based surface passivation schemes for silicon solar cells. *Journal of Vacuum Science & Technology A*, 30(4).
[35] Ritala, M., Kukli, K., Rahtu, A., Raisanen, P. I., Leskela, M., Sajavaara, T., & Keinonen, J. (2000). Atomic layer deposition of oxide thin films with metal alkoxides as oxygen sources. *Science*, 288(5464), 319–321.
[36] De Los Arcos, T., Awakowicz, P., Benedikt, J., Biskup, B., Böke, M., Boysen, N., & Grundmeier, G. (2024). PECVD and PEALD on polymer substrates (part I): Fundamentals and analysis of plasma activation and thin film growth. *Plasma Processes and Polymers*, 21(2), 2300150.
[37] Groner, M. D., Fabreguette, F. H., Elam, J. W., & George, S. M. (2004). Low-temperature Al2O3 atomic layer deposition. *Chemistry of Materials*, 16(4), 639–645.
[38] Ylilammi, M. (1995). Mass transport in atomic-layer deposition carrier-gas reactors. *Journal of the Electrochemical Society*, 142(7), 2474.
[39] Profijt, H. B., Potts, S. E., Van de Sanden, M. C. M., & Kessels, W. M. M. (2011). Plasma-assisted atomic layer deposition: Basics, opportunities, and challenges. *Journal of Vacuum Science & Technology A*, 29(5).
[40] Zaera, F. (2013). Mechanisms of surface reactions in thin solid film chemical deposition processes. *Coordination Chemistry Reviews*, 257(23–24), 3177–3191.
[41] Tiznado, H., Bouman, M., Kang, B. C., Lee, I., & Zaera, F. (2008). Mechanistic details of atomic layer deposition (ALD) processes for metal nitride film growth. *Journal of Molecular Catalysis A: Chemical*, 281(1–2), 35–43.
[42] De Los Arcos, T., Awakowicz, P., Böke, M., Boysen, N., Brinkmann, R. P., Dahlmann, R., & Grundmeier, G. (2024). PECVD and PEALD on polymer substrates (part II): Understanding and tuning of barrier and membrane properties of thin films. *Plasma Processes and Polymers*, 21(3), 2300186.
[43] Kim, M., Seo, S., Jeon, W. P., Lee, Y., Lee, J., Kim, H., & Bent, S. F. (2025). Anhydrous atomic layer deposition of HfO_2: Mechanistic analysis of the tetrakis (dimethylamido) hafnium (TDMAH)-O_2 Process. *Langmuir*. https://doi.org/10.1021/acs.langmuir.5c05147.
[44] Kukli, K., Kemell, M., Castán, H., Dueñas, S., Stern, R., Heikkilä, M. J., & Leskelä, M. (2020). Magnetic properties and resistive switching in mixture films and nanolaminates consisting of iron and silicon oxides grown by atomic layer deposition. *Journal of Vacuum Science & Technology A*, 38(4).
[45] Fonseca, J., & Lu, J. (2021). Single-atom catalysts designed and prepared by the atomic layer deposition technique. *ACS Catalysis*, 11(12), 7018–7059.
[46] Li, J., Chai, G., & Wang, X. (2023). Atomic layer deposition of thin films: From a chemistry perspective. *International Journal of Extreme Manufacturing*, 5(3), 032003.
[47] Chen, M., Nijboer, M. P., Kovalgin, A. Y., Nijmeijer, A., Roozeboom, F., & Luiten-Olieman, M. W. J. (2023). Atmospheric-pressure atomic layer deposition: Recent applications and new emerging applications in high-porosity/3D materials. *Dalton Transactions*, 52(30), 10254–10277.
[48] Lapteva, M., Beladiya, V., Riese, S., Hanke, P., Otto, F., Fritz, T., & Szeghalmi, A. (2021). Influence of temperature and plasma parameters on the properties of PEALD HfO2. *Optical Materials Express*, 11(7), 1918–1942.
[49] Jeon, J., Park, H., Suh, D. C., & Ko, D. H. (2023). Super-conformal TiN thin film deposition by carrier pulse purge atomic layer deposition system: Chamber design optimization with computational fluid dynamics. *Thin Solid Films*, 771, 139758.
[50] Gangwar, R. K., Pathak, A. K., & Kumar, S. (2023, October). Recent progress in photonic crystal devices and their applications: A review. *Photonics*, 10(11), 1199. MDPI.
[51] Zhang, J., Li, Y., Cao, K., & Chen, R. (2022). Advances in atomic layer deposition. *Nanomanufacturing and Metrology*, 5(3), 191–208.
[52] Von Woedtke, T., Laroussi, M., & Gherardi, M. (2022). Foundations of plasmas for medical applications. *Plasma Sources Science and Technology*, 31(5), 054002.

[53] Chiappim, W., Neto, B. B., Shiotani, M., Karnopp, J., Gonçalves, L., Chaves, J. P., & Pessoa, R. (2022). Plasma-assisted nanofabrication: The potential and challenges in atomic layer deposition and etching. *Nanomaterials*, 12(19), 3497.
[54] Van Ommen, J. R., Goulas, A., & Puurunen, R. L. (2021). Atomic layer deposition. *Kirk-Othmer Encyclopedia of Chemical Technology*, 1–42.
[55] Kaushik, M., Leroy, C., Chen, Z., Gajan, D., Willinger, E., Müller, C. R., & Florian, P. (2021). Atomic-scale structure and its impact on chemical properties of aluminum oxide layers prepared by atomic layer deposition on silica. *Chemistry of Materials*, 33(9), 3335–3348.
[56] Heil, S. B. S., Van Hemmen, J. L., Van De Sanden, M. C. M., & Kessels, W. M. M. (2008). Reaction mechanisms during plasma-assisted atomic layer deposition of metal oxides: A case study for Al_2O_3. *Journal of Applied Physics*, 103(10).
[57] Liu, J., Lu, H., Zhang, D. W., & Nolan, M. (2022). Self-limiting nitrogen/hydrogen plasma radical chemistry in plasma-enhanced atomic layer deposition of cobalt. *Nanoscale*, 14(12), 4712–4725.
[58] Brady-Boyd, A., O'Connor, R., Armini, S., Selvaraju, V., Pasquali, M., Hughes, G., & Bogan, J. (2022). The role of atomic oxygen in the decomposition of self-assembled monolayers during area-selective atomic layer deposition. *Applied Surface Science*, 586, 152679.
[59] Yasmeen, S., Ryu, S. W., Lee, S. H., & Lee, H. B. R. (2023). Atomic layer deposition beyond thin film deposition technology. *Advanced Materials Technologies*, 8(20), 2200876.
[60] Gleason, K. K. (2024). Designing organic and hybrid surfaces and devices with initiated chemical vapor deposition (iCVD). *Advanced Materials*, 36(8), 2306665.
[61] Dufour, T. (2023). From basics to frontiers: A comprehensive review of plasma-modified and plasma-synthesized polymer films. *Polymers*, 15(17), 3607.
[62] Kweon, M., Park, C., Mohapatra, D., Kim, S. B., Bae, J. S., Cheon, T., & Kim, S. H. (2025). Yttrium carbide thin film as an emerging transition metal carbide prepared by plasma-enhanced atomic layer deposition for dual diffusion barrier applications into Cu and Ru metallization. *Applied Surface Science*, 163302.
[63] Han, J. W., Jin, H. S., Kim, Y. J., Heo, J. S., Kim, W. H., Ahn, J. H., & Park, T. J. (2023). Advanced atomic layer deposition (ALD): Controlling the reaction kinetics and nucleation of metal thin films using electric-potential-assisted ALD. *Journal of Materials Chemistry C*, 11(11), 3743–3750.
[64] Nguyen, V. H., Sekkat, A., Jimenez, C., Munoz, D., Bellet, D., & Munoz-Rojas, D. (2021). Impact of precursor exposure on process efficiency and film properties in spatial atomic layer deposition. *Chemical Engineering Journal*, 403, 126234.
[65] Anderson, V. R., Cavanagh, A. S., Abdulagatov, A. I., Gibbs, Z. M., & George, S. M. (2014). Waterless TiO_2 atomic layer deposition using titanium tetrachloride and titanium tetraisopropoxide. *Journal of Vacuum Science & Technology A*, 32(1).
[66] Nwanna, E. C., Coetzee, R. A. M., & Jen, T. C. (2022). A numerical approach on the selection of the purge flow rate in an atomic layer deposition (ALD) process. *Physics of Fluids*, 34(5), 05200.
[67] Jain, H., & Poodt, P. (2021). About the importance of purge time in molecular layer deposition of alucone films. *Dalton Transactions*, 50(17), 5807–5818.
[68] Zaera, F. (2012). The surface chemistry of atomic layer depositions of solid thin films. *The Journal of Physical Chemistry Letters*, 3(10), 1301–1309.
[69] Hori, M. (2022). Radical-controlled plasma processes. *Reviews of Modern Plasma Physics*, 6(1), 36.
[70] Park, S., Park, T., Choi, Y., Jung, C., Kim, B., & Jeon, H. (2022). Radical-induced effect on PEALD SiO_2 films by applying positive DC Bias. *ECS Journal of Solid State Science and Technology*, 11(2), 023007.
[71] Steinfeld, J. I. (1989). Reactions of photogenerated free radicals at surfaces of electronic materials. *Chemical Reviews*, 89(6), 1291–1301.
[72] Knapas, K., & Ritala, M. (2013). In situ studies on reaction mechanisms in atomic layer deposition. *Critical Reviews in Solid State and Materials Sciences*, 38(3), 167–202.

[73] Ramachandran, R. K., Dendooven, J., Poelman, H., & Detavernier, C. (2015). Low temperature atomic layer deposition of crystalline In_2O_3 films. *The Journal of Physical Chemistry C*, 119(21), 11786–11791.

[74] Musselman, K. P., Uzoma, C. F., & Miller, M. S. (2016). Nanomanufacturing: High-throughput, cost-effective deposition of atomic scale thin films via atmospheric pressure spatial atomic layer deposition. *Chemistry of Materials*, 28(23), 8443–8452.

[75] Muñoz-Rojas, D., Nguyen, V. H., de la Huerta, C. M., Jiménez, C., & Bellet, D. (2019). Spatial atomic layer deposition. In *Chemical Vapor Deposition for Nanotechnology* (pp. 1).

[76] Levy, D. H., Nelson, S. F., & Freeman, D. (2009). Oxide electronics by spatial atomic layer deposition. *Journal of Display Technology*, 5(12), 484–494.

[77] Maydannik, P. S., Kääriäinen, T. O., Lahtinen, K., Cameron, D. C., Söderlund, M., Soininen, P., & Zeng, X. (2014). Roll-to-roll atomic layer deposition process for flexible electronics encapsulation applications. *Journal of Vacuum Science & Technology A*, 32(5).

[78] Biyikli, N., & Haider, A. (2017). Atomic layer deposition: An enabling technology for the growth of functional nanoscale semiconductors. *Semiconductor Science and Technology*, 32(9), 093002.

[79] Öğütman, K., Iqbal, N., Gregory, G., Li, M., Haslinger, M., Cornagliotti, E., & Davis, K. O. (2020). Spatial atomic layer deposition of aluminum oxide as a passivating hole contact for silicon solar cells. *Physica Status Solidi (A)*, 217(18), 2000348.

[80] Masse de la Huerta, C., Nguyen, V. H., Dedulle, J. M., Bellet, D., Jiménez, C., & Muñoz-Rojas, D. (2018). Influence of the geometric parameters on the deposition mode in spatial atomic layer deposition: A novel approach to area-selective deposition. *Coatings*, 9(1), 5.

[81] Lahtinen, K., Maydannik, P., Johansson, P., Kääriäinen, T., Cameron, D. C., & Kuusipalo, J. (2011). Utilisation of continuous atomic layer deposition process for barrier enhancement of extrusion-coated paper. *Surface and Coatings Technology*, 205(15), 3916–3922.

[82] Fahlteich, J., Steiner, C., Schiller, N., Miesbauer, O., Noller, K., Deichmann, K. J., & Amberg-Schwab, S. (2017). Roll-to-roll thin-film coating on fluoropolymer webs: Status, challenges, and applications. *Surface and Coatings Technology*, 314, 160–168.

[83] Schumacher, M., Baumann, P. K., & Seidel, T. (2006). AVD and ALD as two complementary technology solutions for next generation dielectric and conductive thin-film processing. *Chemical Vapor Deposition*, 12(2-3), 99–108.

[84] Meng, X. (2017). An overview of molecular layer deposition for organic and organic-inorganic hybrid materials: Mechanisms, growth characteristics, and promising applications. *Journal of Materials Chemistry A*, 5(35), 18326–18378.

[85] Sundberg, P., & Karppinen, M. (2014). Organic and inorganic-organic thin film structures by molecular layer deposition: A review. *Beilstein Journal of Nanotechnology*, 5(1), 1104–1136.

[86] Mitzi, D. B. (2001). Thin-film deposition of organic− inorganic hybrid materials. *Chemistry of Materials*, 13(10), 3283–3298.

[87] George, S. M., Yoon, B., & Dameron, A. A. (2009). Surface chemistry for molecular layer deposition of organic and hybrid organic-inorganic polymers. *Accounts of Chemical Research*, 42(4), 498–508.

[88] Liu, C., Zhu, L., Weng, J. K., Li, Y. C., Chen, Z., Lei, J., & Li, A. D. (2023). Ultraflexible and energy-efficient artificial synapses based on molecular/atomic layer deposited organic-inorganic hybrid thin films. *Advanced Electronic Materials*, 9(2), 2200821.

[89] Plikusiene, I., Maciulis, V., Ramanavicius, A., & Ramanaviciene, A. (2022). Spectroscopic ellipsometry and quartz crystal microbalance with dissipation for the assessment of polymer layers and for the application in biosensing. *Polymers*, 14(5), 1056.

[90] Pakkala, A., & Putkonen, M. (2010). Atomic layer deposition. In *Handbook of deposition technologies for films and coatings* (pp. 364–391). William Andrew Publishing.

[91] Langereis, E., Bouman, M., Keijmel, J., Heil, S., Van de Sanden, M. C., & Kessels, W. M. (2008). Plasma-assisted ALD of Al_2O_3 at low temperatures: Reaction mechanisms and material properties. *ECS Transactions*, 16(4), 247.

[92] Kim, H. (2011). Characteristics and applications of plasma-enhanced atomic layer deposition. *Thin Solid Films*, 519(20), 6639–6644.

[93] Ponraj, J. S., Attolini, G., & Bosi, M. (2013). Review on atomic layer deposition and applications of oxide thin films. *Critical Reviews In Solid State And Materials Sciences*, 38(3), 203–233.

[94] Kääriäinen, T. O., Maydannik, P., Cameron, D. C., Lahtinen, K., Johansson, P., & Kuusipalo, J. (2011). Atomic layer deposition on polymer based flexible packaging materials: Growth characteristics and diffusion barrier properties. *Thin Solid Films*, 519(10), 3146–3154.

[95] Lee, B. H., Yoon, B., Abdulagatov, A. I., Hall, R. A., & George, S. M. (2013). Growth and properties of hybrid organic-inorganic metalcone films using molecular layer deposition techniques. *Advanced Functional Materials*, 23(5), 532–546.

[96] Zhang, Z., Zhao, Y., Zhao, Z., Huang, G., & Mei, Y. (2020). Atomic layer deposition-derived nanomaterials: Oxides, transition metal dichalcogenides, and metal-organic frameworks. *Chemistry of Materials*, 32(21), 9056–9077.

[97] Zhao, Z., Kong, Y., Zhang, Z., Huang, G., & Mei, Y. (2020). Atomic layer-deposited nanostructures and their applications in energy storage and sensing. *Journal of Materials Research*, 35(7), 701–719.

[98] Bouaziz, J., Cancellieri, C., Rheingans, B., Jeurgens, L. P., & La Mattina, F. (2023). Advanced epitaxial lift-off and transfer procedure for the fabrication of high-quality functional oxide membranes. *Advanced Materials Interfaces*, 10(2), 2201458.

[99] Xu, C., Wu, X., Huang, G., & Mei, Y. (2019). Rolled-up nanotechnology: Materials issue and geometry capability. *Advanced Materials Technologies*, 4(1), 1800486.

[100] Mei, Y., Huang, G., Solovev, A. A., Ureña, E. B., Mönch, I., Ding, F., & Schmidt, O. G. (2008). Versatile approach for integrative and functionalized tubes by strain engineering of nanomembranes on polymers. *Advanced Materials*, 20(21), 4085–4090.

[101] Wang, X., & Liu, J. L. (2022). Bottom-up synthesis of nanomaterials. In *Nanochemistry: From Theory to Application for In-Depth Understanding of Nanomaterials* (pp. 239).

[102] Riyanto, E., Martides, E., Junianto, E., Rajani, A., Prawara, B., Gaoshan, H., & Yongfeng, M. (2021, February). A facile route of two-dimensional metal oxide nanosheets fabrication by atomic layer deposition. *Journal of Physics: Conference Series*. 1825(1), 012042. IOP Publishing.

[103] Yang, H. C., Waldman, R. Z., Chen, Z., & Darling, S. B. (2018). Atomic layer deposition for membrane interface engineering. *Nanoscale*, 10(44), 20505–20513.

[104] Deng, S., Kurttepeli, M., Bals, S., Cott, D. J., & Detavernier, C. (2016). Template-assisted synthesis of porous metal oxide and metal nanostructures by ALD. *ECS Transactions*, 75(6), 47.

[105] Sree, S. P., Dendooven, J., Masschaele, K., Hamed, H. M., Deng, S., Bals, S., & Martens, J. A. (2013). Synthesis of uniformly dispersed anatase nanoparticles inside mesoporous silica thin films via controlled breakup and crystallization of amorphous TiO_2 deposited using atomic layer deposition. *Nanoscale*, 5(11), 5001–5008.

[106] Geng, G., Zhang, Z., Li, C., Pan, R., Li, Y., Yang, H., & Li, J. (2022). Atomic layer assembly based on sacrificial templates for 3D nanofabrication. *Micromachines*, 13(6), 856.

[107] Mallick, D., Ghosh, S., Chen, A. H., Liao, J., Yoo, J., Lu, Q., & Moore, R. G. (2025). Next-generation electronics by co-design with chalcogenide materials. *NPJ Spintronics*, 3(1), 41.

[108] Li, X., Shan, J., Zhang, W., Su, S., Yuwen, L., & Wang, L. (2017). Recent advances in synthesis and biomedical applications of two-dimensional transition metal dichalcogenide nanosheets. *Small*, 13(5), 1602660.

[109] Cho, A. J., Ryu, S. H., Yim, J. G., Baek, I. H., Pyeon, J. J., Won, S. O., & Kim, S. K. (2022). Stepwise growth of crystalline MoS_2 in atomic layer deposition. *Journal of Materials Chemistry C*, 10(18), 7031–7038.

[110] Kim, D. H., Park, J. C., Park, J., Cho, D. Y., Kim, W. H., Shong, B., & Park, T. J. (2021). Wafer-scale growth of a MoS_2 monolayer via one cycle of atomic layer deposition: An adsorbate control method. *Chemistry Of Materials*, 33(11), 4099–4105.

[111] Jurca, T., Moody, M. J., Henning, A., Emery, J. D., Wang, B., Tan, J. M., & Marks, T. J. (2017). Low-temperature atomic layer deposition of MoS_2 films. *Angewandte Chemie International Edition*, 56(18), 4991–4995.

[112] Yadgarov, L., & Tenne, R. (2025). Nanotubes from transition metal dichalcogenides: Recent progress in the synthesis, characterization, and electrooptical properties. *Small*, 21(28), 2400503.

[113] Schutte, W. J., De Boer, J. L., & Jellinek, F. (1987). Crystal structures of tungsten disulfide and diselenide. *Journal of Solid State Chemistry*, 70(2), 207–209.

[114] Kim, H., Moon, J. H., Nam, S., Kim, S., Zha, J., & Tan, C. (2025). Scalable fabrication of mid-wavelength and long-wavelength infrared photodetectors based on narrow bandgap semiconductors: Challenges and opportunities. *International Journal of Extreme Manufacturing*, 8, 012012. doi: 10.1088/2631-7990/ae0100.

[115] Romanov, R. I., Kozodaev, M. G., Chernikova, A. G., Zabrosaev, I. V., Chouprik, A. A., Zarubin, S. S., & Markeev, A. M. (2021). Thickness-dependent structural and electrical properties of WS2 nanosheets obtained via the ALD-grown WO_3 sulfurization technique as a channel material for field-effect transistors. *ACS Omega*, 6(50), 34429–34437.

[116] Li, D., Li, Z., Chen, P., Pi, L., Zuo, N., Peng, Q., & Zhai, T. (2021). Two-dimensional metal chalcogenide heterostructures: designed growth and emerging novel applications. *Advanced Materials Interfaces*, 8(15), 2100515.

[117] Kwon, S. G., & Hyeon, T. (2008). Colloidal chemical synthesis and formation kinetics of uniformly sized nanocrystals of metals, oxides, and chalcogenides. *Accounts of Chemical Research*, 41(12), 1696–1709.

[118] Cao, X., Tan, C., Zhang, X., Zhao, W., & Zhang, H. (2016). Solution-processed two-dimensional metal dichalcogenide-based nanomaterials for energy storage and conversion. *Advanced Materials*, 28(29), 6167–6196.

[119] Schonfeld, B., J, H. J., & C, M. S. (1983). Anisotropic mean-square displacements (MSD) in single crystals of 2H- and 3R-MoS2. *Acta Crystallographica*, B39, 404–407.

[120] Rappé, A. K., Casewit, C. J., Colwell, K. S., Goddard, W. A. III, & Skiff, W. M. (1992). UFF, a full periodic table force field for molecular mechanics and molecular dynamics simulations. *Journal of the American Chemical Society*, 114, 10024–10039.

[121] James, P. B., & Lavik, M. T. (1963). The crystal structure of $MoSe^2$. *Acta Crystallographica*, 16, 1183–1183.

[122] Kyono, & Kimata. (2004). Structural variations induced by difference of the inert pair effect in the stibnite-bismuthinite solid solution series $(Sb, Bi)^2S^3$. *American Mineralogist*, 89, 932–940.

[123] Jussila, T., Philip, A., Rubio-Giménez, V., Eklund, K., Vasala, S., Glatzel, P., & Karppinen, M. (2024). Chemical bonding and crystal structure schemes in atomic/molecular layer deposited Fe-terephthalate thin films. *Chemistry of Materials*, 36(13), 6489–6503.

[124] Zhao, J., Gong, B., Nunn, W. T., Lemaire, P. C., Stevens, E. C., Sidi, F. I., & Parsons, G. N. (2015). Conformal and highly adsorptive metal-organic framework thin films via layer-by-layer growth on ALD-coated fiber mats. *Journal of Materials Chemistry A*, 3(4), 1458–1464.

[125] Zhao, J., Nunn, W. T., Lemaire, P. C., Lin, Y., Dickey, M. D., Oldham, C. J., & Parsons, G. N. (2015). Facile conversion of hydroxy double salts to metal-organic frameworks using metal oxide particles and atomic layer deposition thin-film templates. *Journal of the American Chemical Society*, 137(43), 13756–13759.

[126] Han, S., & Mullins, C. B. (2020). Current progress and future directions in gas-phase metal-organic framework thin-film growth. *ChemSusChem*, 13(20), 5433–5442.

[127] Bian, Y., Xiong, N., & Zhu, G. (2018). Technology for the remediation of water pollution: A review on the fabrication of metal organic frameworks. *Processes*, 6(8), 122.

[128] Sun, Q., Duan, Z., Deng, H., Li, R., & Li, J. R. (2025). Green and rapid synthesis of a two-dimensional Zn-MOF with Lewis acid-base sites via a solvent-free strategy. *Journal of Materials Chemistry A*, 13(18), 12911–12916.

[129] Su, P., Tu, M., Ameloot, R., & Li, W. (2021). Vapor-phase processing of metal-organic frameworks. *Accounts of Chemical Research*, 55(2), 186–196.

[130] Ronte, A., Wagle, P., Mahmodi, G., Chevula, M., Dangwal, S., Saeb, M. R., & Kim, S. J. (2023). High-flux ZIF-8 membranes on ZnO-coated supports for propane/propylene separation. *Energy & Fuels*, 37(12), 8456–8464.

[131] Zhong, L., Qian, J., Wang, N., Komarneni, S., & Hu, W. (2023). Metal-organic frameworks on versatile substrates. *Journal of Materials Chemistry A*, 11(38), 20423–20458.

[132] Crivello, C., Sevim, S., Graniel, O., Franco, C., Pané, S., Puigmartí-Luis, J., & Muñoz-Rojas, D. (2021). Advanced technologies for the fabrication of MOF thin films. *Materials Horizons*, 8(1), 168–178.

Telli Alia, Darem Sabrine, Sushesh Srivatsa Palakurthi, Sajid Bashir, Jingbo Louise Liu*

Chapter 6
Sustainable and green routes for nanomaterial synthesis: methods, properties, and applications

6.1 Introduction

The escalating global demand for energy, coupled with the depletion of fossil fuel reserves and the intensifying impacts of environmental pollution, has catalyzed an urgent search for sustainable, carbon-neutral energy solutions. Addressing this challenge requires developing advanced technologies capable of clean energy generation,

Acknowledgments: The research efforts led by Dr Bashir and Dr Liu were supported by the American Chemical Society Petroleum Research Fund (Grant 53,827-UR10) federal agencies, including the Office of Naval Research, the Air Force Office of Scientific Research, the Department of Energy, the National Science Foundation, the Department of Education, the Department of Homeland Security (ORISE), and the National Institutes of Health. The Robert A. Welch Foundation (Departmental Grant AC-0,006) is gratefully acknowledged for its support of student training. The authors also gratefully acknowledge the helpful discussions with colleagues, comments, and reviewers' suggestions, which have improved the presentation.

Authors' contributions: Telli Alia and Darem Sabrine jointly prepared the initial draft of the book chapter. Dr J. Liu substantially revised and rewrote the chapter, incorporating additional figures and data derived from the research and literature contributions of S. Palakurthi, a former graduate student. The final version was edited by Dr Sajid Bashir in his role as co-editor, and he also performed the associated data analyses. All authors reviewed and approved the final manuscript for submission.

***Corresponding author: Jingbo Louise Liu**, Department of Chemistry, Texas A&M University-Kingsville, MSC161, 700 University Blvd., Kingsville, TX 78363, USA; Texas A&M Energy Institute, 1617 Research Parkway, Suite 308, College Station, TX 77843-3372 USA; Louise Nanoinnovation North American, 7534 Beau Terre, Corpus Christi, TX 78414, USA; e-mails: jingbo.liu@tamuk.edu, jingbo.liu@tamu.edu
Telli Alia, Laboratoire de protection des écosystèmes en zone aride et semi-aride, Université de KASDI Merbah, BP 511 la route de Ghardaïa, Ouargla 30000, Algérie, e-mail: telli.alia@univ-ouargla.dz
Darem Sabrine, Laboratoire de sol et développement durable, Université Badji Mokhtar, PB 12, Annaba 23000, Algérie
Sushesh Srivatsa Palakurthi, Department of Pharmaceutical Sciences, Irma Lerma Rangel College of Pharmacy, 1,010 W. Avenue B., Kingsville, TX 78363, USA
Sajid Bashir, Department of Chemistry, Texas A&M University-Kingsville, MSC161, 700 University Blvd., Kingsville, TX 78363, USA; Texas A&M Energy Institute, 1617 Research Parkway, Suite 308, College Station, TX 77843-3372, USA; Louise Nanoinnovation North American, 7534 Beau Terre, Corpus Christi, TX 78414, USA

 | https://doi.org/10.1515/9783111621111-007

efficient harvesting, conversion, and long-term storage. Among these, nanotechnology has emerged as a transformative frontier offering unprecedented control over material properties at the atomic and molecular scale. Within this domain, green nanotechnology represents a convergence of nanoscience with the principles of green chemistry and green engineering. It emphasizes the design and synthesis of nanomaterials through environmentally benign, low-cost, and sustainable methods. These include biogenic synthesis routes that utilize living organisms such as plants, algae, bacteria, yeast, and fungi, as well as naturally derived biomolecules such as phenolic compounds, polysaccharides, and proteins as reducing and stabilizing agents. Such approaches not only minimize the use of toxic reagents and energy-intensive conditions but also enable scalable production of functional nanostructures with tailored properties.

Beyond synthesis, green nanotechnology is increasingly being applied to renewable energy systems, including photocatalytic hydrogen production, solar cells, supercapacitors, and bio-batteries. Nanomaterials fabricated through green routes often exhibit enhanced surface area, tunable bandgaps, and improved charge transport, making them ideal candidates for energy conversion and storage devices. Moreover, their biocompatibility and degradability open new avenues in environmental remediation, smart agriculture, and sustainable biomedical engineering. This chapter explores the diverse methodologies employed in green synthesis of nanomaterials, with a focus on their structural evolution, reaction mechanisms, and integration into renewable energy platforms, as outlined in Fig. 6.1. It also highlights emerging trends such as hybrid bio-inorganic nanostructures, waste-to-nanomaterial conversion, and circular economy frameworks that position green nanotechnology as a cornerstone of future sustainable innovation.

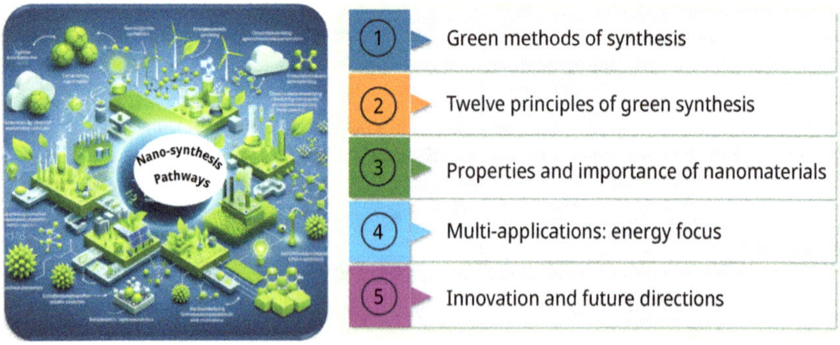

Fig. 6.1: Overview of green nanomaterial synthesis pathways, highlighting green methods of synthesis, properties, and importance of nanomaterials, and their integration into renewable energy applications, innovative trends, and future directions.

The intensifying global energy crisis and environmental degradation resulting from continued reliance on nonrenewable fossil fuels have driven the scientific community to pursue clean, sustainable energy alternatives by adopting advanced green technologies. Among these, nanotechnology has emerged as a pivotal enabler of next-generation energy systems, offering scalable solutions for energy harvesting, conversion, and storage [1]. In response to the ecological and temporal urgency of this challenge, green chemistry has been established as a framework for minimizing the use of hazardous substances and promoting environmentally benign processes [2]. Green nanotechnology, a subset of green technology, integrates the principles of green chemistry and green engineering to reduce energy consumption, minimize waste, and utilize renewable feedstocks throughout the nanomaterial life cycle [3]. It emphasizes low-temperature synthesis (<150 °C), solvent-free or aqueous-phase reactions, and the use of biogenic reducing agents to fabricate functional nanostructures with minimal environmental impact.

Nanomaterial synthesis typically follows two fundamental strategies: top-down and bottom-up. The bottom-up approach is particularly relevant for green nanotechnology and encompasses both physical and chemical techniques (discussed in **Chapters 1–4**). Synthesis methods include physical vapor deposition (PVD), chemical vapor deposition (CVD), molecular beam epitaxy, pulsed laser deposition (PLD), atomic layer deposition (ALD), ion implantation, spray pyrolysis, sputter deposition, and electric arc deposition each offering control over film thickness, crystallinity, and surface morphology [4]. For instance, ALD enables angstrom-level thickness control and conformal coating on high-aspect-ratio substrates. At the same time, PLD can achieve deposition rates of 0.1–10 nm/min, depending on laser fluence and target composition. Chemical bottom-up methods include coprecipitation, sol–gel, chemical reduction, sonochemical and photochemical synthesis, electrochemical deposition, microemulsion, solvothermal, and surface-mediated techniques [5]. These methods allow fine-tuning of nanoparticle (NP) size (typically 5–100 nm), shape (spherical, rod-like, and platelets), and surface functionality. Biogenic synthesis using plants, bacteria, fungi, yeast, algae, and agricultural waste offers a sustainable route to NP fabrication, leveraging natural compounds such as phenolics, flavonoids, and proteins as reducing and stabilizing agents. The selection of an appropriate synthesis method is pivotal in determining the physicochemical characteristics of the resulting NPs, including morphology, crystallinity, particle size distribution, surface charge, and catalytic performance. These attributes are not only critical to the functional behavior of nanomaterials in energy, biomedical, and environmental applications, but also influence their stability, reactivity, and compatibility with target systems. For instance, sol–gel and hydrothermal methods often yield highly crystalline oxide NPs with controlled porosity. At the same time, plasma-assisted techniques can enhance surface activation and defect engineering, thereby improving catalytic performance.

Equally important are the environmental and operational parameters associated with each synthesis route. Reaction temperature, solvent selection, precursor toxicity, energy consumption, and waste generation must be carefully assessed to ensure compliance with the principles of green chemistry and sustainable engineering. Low-temperature aqueous-phase synthesis, for example, reduces energy input and avoids volatile organic solvents, while biogenic methods leverage natural reducing agents and biological templates to minimize hazardous byproducts. These considerations are essential for advancing nanomaterial production toward eco-friendly, scalable, and economically viable platforms. To facilitate a systematic understanding of these approaches, a structured comparative table has been provided that summarizes key nanomaterial synthesis techniques, their underlying mechanisms, and representative materials produced. This framework enables researchers and practitioners to evaluate trade-offs between precision, scalability, environmental impact, and application-specific performance. By aligning synthesis strategies with both technical requirements and sustainability goals, green nanotechnology can serve as a transformative tool in the development of next-generation materials for renewable energy, environmental remediation, and biocompatible systems. A structured comparative table (Tab. 6.1) summarizes key nanomaterial synthesis methods, their mechanisms, typical materials produced, and key parameters.

Tab. 6.1: Summary of key nanomaterial synthesis methods, their mechanisms, and typical materials produced.

Synthesis method	Mechanism	Materials produced	Key parameters
Physical vapor deposition	Evaporation or sputtering of solid targets in a vacuum; the vapor condenses on the substrate	Metal films (Al, Ti, and Cu), nitrides (TiN and CrN), and oxides (Al_2O_3)	Vacuum level (10^{-3}–10^{-6} torr) and substrate temperature (100–500 °C)
Chemical vapor deposition (CVD)	Gas-phase precursors react on a heated substrate to form a solid film	Si, Si_3N_4, graphene, CNTs, MoS_2, WS_2, TiC, and SiC	Temperature (400–1,000 °C), precursor flow rate, and pressure (LPCVD/APCVD)
Atomic layer deposition (ALD)	Sequential, self-limiting surface reactions for layer-by-layer growth	Al_2O_3, ZnO, HfO_2, WS_2, MoS_2, and conformal coatings on 3D structures	Growth per cycle (0.1–1 nm), temperature (150–300 °C), and purge time
Sol–gel method	Hydrolysis and condensation of metal alkoxides in solution	Metal oxides (TiO_2, ZnO, and Fe_2O_3), silica, and ceramics	pH, aging time, solvent type, and calcination temp (300–600 °C)
Coprecipitation	Simultaneous precipitation of multiple ions from solution	Ferrites, mixed oxides, and doped nanoparticles	Precursor ratio, stirring rate, temperature (25–80 °C), and pH control
Hydrothermal/solvothermal	Crystallization under high pressure and temperature in sealed autoclaves	Nanorods, nanowires (ZnO and TiO_2), quantum dots, and metal sulfides	Temp (100–250 °C), pressure (1–10 MPa), and reaction time (6–48 h)

Tab. 6.1 (continued)

Synthesis method	Mechanism	Materials produced	Key parameters
Green/biogenic synthesis	Reduction and stabilization using plant extracts, microbes, or biomolecules	Ag, Au, ZnO, CuO, and Fe_3O_4 nanoparticles with biocompatible surfaces	Extract concentration, reaction temperature (25–80 °C), pH, and time
Pulsed laser deposition (PLD)	Laser ablation of target material; plume condenses on substrate	Oxides (YBCO and ZnO), carbides, nitrides, and thin films with high purity	Laser fluence (1–5 J/cm^2), pulse rate, and substrate temperature (200–700 °C)
Electrochemical deposition	Ion reduction at the electrode surface under applied voltage	Metal films (Cu, Ni, and Ag), conductive polymers, and nanostructured coatings	Voltage (0.1–5 V), electrolyte concentration, and deposition time

Conventional NP synthesis methods often rely on high-temperature reactions, toxic solvents, and hazardous chemical precursors, resulting in the generation of environmentally persistent byproducts and reduced biocompatibility. These practices pose significant risks to both human health and ecological systems, contributing to bioaccumulation, cytotoxicity, and long-term environmental degradation [6–8]. The adverse impacts of traditional synthesis routes, such as elevated energy consumption (>300 °C), the use of volatile organic compounds (VOCs), and the use of nonbiodegradable surfactants, underscore the urgent need for alternative approaches that align with the Sustainable Development Goals. In response, green synthesis has emerged as a reliable, low-cost, and environmentally benign strategy for fabricating a wide range of nanomaterials (NMs), including metal and metal oxide NPs, hybrid nanostructures, and bioinspired materials [9, 142]. This approach leverages naturally occurring reducing and stabilizing agents derived from biological sources such as plant extracts, microbial cultures, and agricultural waste, thereby eliminating the need for toxic reagents and minimizing secondary pollution. Recent studies have demonstrated that green-synthesized NPs exhibit reduced cytotoxicity, enhanced biocompatibility, and functional performance, comparable to that of NPs produced via conventional methods [10]. Key variables influencing green synthesis include the concentration and composition of bioactive compounds (e.g., flavonoids, terpenoids, alkaloids), reaction temperature (typically 25–80 °C), pH (ranging from 4 to 9), and reaction time (from minutes to several hours). The choice of capping agents, reducing agents, and solvent systems plays a critical role in determining NP size (often 10–100 nm), shape (spherical, rod-like, or anisotropic), and surface charge (zeta potential), which in turn affect stability and application-specific functionality [7]. By utilizing renewable inputs and minimizing hazardous outputs, green synthesis directly supports the twelve principles of green chemistry, which emphasize waste prevention, atom economy, safer solvents, and energy efficiency. Its widespread applicability, ease of implementation, and compatibility with diverse biological systems have positioned green nanotechnology as a cornerstone of sustainable innovation. In this context, the present chapter provides a comprehensive overview of green synthesis methodologies for nanomaterials and their integration into renewable energy technologies.

The 12 principles of green chemistry, as articulated by Kharissova et al. [11], provide a foundational framework for designing chemical processes that are inherently safer, more efficient, and environmentally responsible. Among these principles, the concept of a circular economy is emphasized, one in which hazardous byproducts are either avoided entirely or repurposed within the system to minimize waste and resource depletion. This approach (Fig. 6.2) promotes atom economy, real-time pollution prevention, and the use of renewable feedstocks, ensuring that every input and output contributes to a closed-loop sustainable cycle. For example, reactions are encouraged to proceed under ambient temperature and pressure conditions to reduce energy consumption. At the same time, catalytic systems are favored over stoichiometric reagents to enhance selectivity and minimize excess reactants. Additionally, solvent systems are selected based on low toxicity, biodegradability, and recyclability. By integrating these variables, reaction enthalpy, solvent polarity index, E-factor (environmental impact factor), and process yield, green chemistry enables the development of scalable, low-impact technologies that align with both ecological and industrial goals.

Green chemistry, as defined by the U.S. Environmental Protection Agency (EPA), is the design of chemical products and processes that reduce or eliminate the use and generation of hazardous substances. The Department of Energy (DOE) further emphasizes its role in advancing sustainable technologies, particularly in energy conversion, storage, and nanomaterial innovation [12]. When applied to nanotechnology, these principles guide the synthesis of NPs) and nanostructured materials with minimal environmental impact, enhanced safety, and improved life cycle performance. Green nanotechnology integrates renewable feedstocks, low-energy processes, and biocompatible reagents to produce functional materials for catalysis, energy devices, and biomedical applications. Technical parameters such as atom economy, reaction enthalpy, solvent toxicity index, and E-factor are used to quantify sustainability. By aligning nanomaterial synthesis with the 12 principles of green chemistry, researchers can optimize reaction pathways, reduce waste, and design safer, degradable products. This chapter explores each principle in depth, highlighting its relevance to nanomaterial design, process engineering, and environmental stewardship in the context of EPA and DOE frameworks. The key points of the 12 principles of green chemistry were summarized in Tab. 6.2, and discussed in detail as follows.

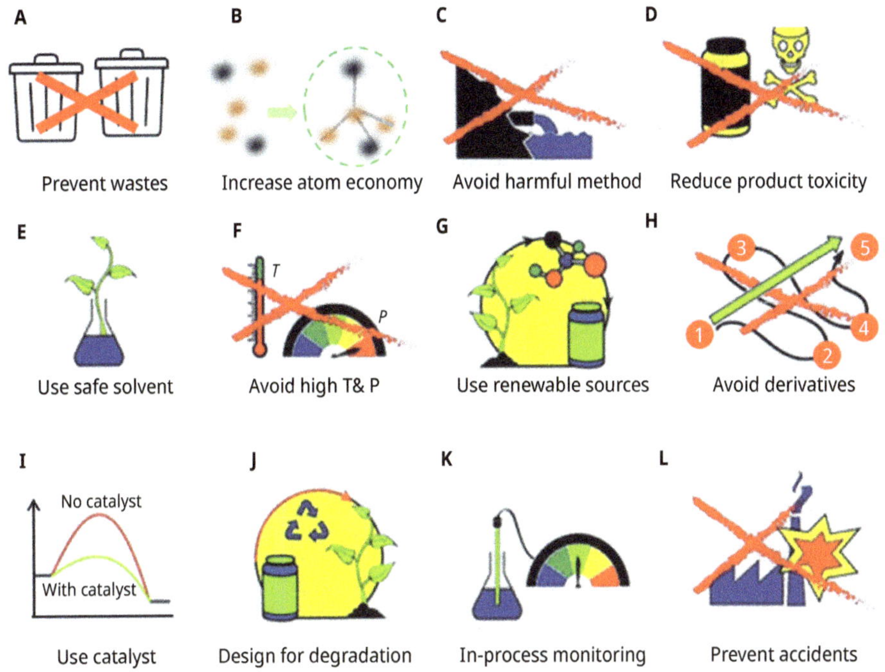

Fig. 6.2: Schematic representation of a green synthesis framework, illustrating how chemical design and process optimization align with the twelve principles of green chemistry. Guided by the U.S. Environmental Protection Agency's definition emphasizing the reduction or elimination of hazardous substances and reinforced by the Department of Energy's focus on sustainable technologies in energy conversion, storage, and nanomaterials, this figure highlights procedural and production-level strategies that advance environmentally responsible chemical manufacturing.

6.1.1 Prevent waste (leave no waste to clean up)

The first principle of green chemistry emphasizes waste prevention at the source rather than post-process remediation. In nanomaterial synthesis, this translates to designing reactions that minimize byproduct formation and maximize conversion efficiency. For example, using stoichiometric ratios of precursors and optimizing reaction kinetics can reduce unreacted residues and the number of purification steps. The EPA advocates for process intensification strategies that reduce solvent volumes and eliminate unnecessary intermediates. In NP fabrication, adopting solvent-free methods, such as mechanochemical synthesis or microwave-assisted reactions, can significantly reduce waste. Quantitative metrics such as the E-factor (mass of waste per mass of product) and process mass intensity are used to assess waste generation. Green nanotechnology also encourages the reuse of reaction media and the valorization of waste streams, for instance, repurposing agricultural waste as a biogenic precursor for

Tab. 6.2: Summary of the key points of the 12 principles of green chemistry.

Principle	Nanotech application	Key technical variables
A. Prevent waste	Design reactions with minimal byproducts	E-factor and process mass intensity (PMI)
B. Maximize atom economy	Use all atoms in the final product	Atom economy % and stoichiometry balance
C. Less hazardous syntheses	Avoid toxic precursors and intermediates	LD_{50}, bioaccumulation index, and reaction enthalpy
D. Safer chemicals and products	Design biocompatible, nontoxic nanomaterials	Zeta potential, ROS generation, and biodegradability
E. Safer solvents and conditions	Use water, ethanol, and ionic liquids	Solvent toxicity index, flash point, and vapor pressure
F. Increase energy efficiency	Low-temperature synthesis and microwave-assisted reactions	Reaction temperature, energy per gram, and carbon footprint
G. Use renewable feedstocks	Plant extracts and microbial metabolites	Biomass yield, carbon content, and extract composition
H. Avoid derivatives	Direct synthesis without protecting groups	Step count, cumulative yield, and atom economy per step
I. Use catalysts	Enzymes, metal NPs, and photocatalysts	TOF, activation energy, and selectivity index
J. Design for degradation	Biodegradable polymers and surface-triggered breakdown	Half-life, hydrolysis rate, and environmental fate modeling
K. Real-time analysis	In situ monitoring of NP growth and surface chemistry	UV–Vis, DLS, FTIR, Raman, and impurity profile
L. Minimize accident potential	Mild conditions, benign reagents, and fail-safe reactor design	Flash point, autoignition temperature, and thermal decomposition threshold

metal oxide NPs. By integrating waste prevention into the design phase, researchers can reduce environmental burden, lower operational costs, and enhance the scalability of sustainable nanomaterial production.

6.1.2 Maximize atom economy (waste few or no atoms)

Atom economy measures the efficiency of a chemical reaction in incorporating all reactant atoms into the final product. In nanotechnology, maximizing atom economy ensures that precursor materials are fully utilized in NP formation, minimizing waste and improving yield. For example, in the sol–gel synthesis of metal oxides, selecting alkoxide precursors that undergo complete hydrolysis and condensation enhances the utilization of atoms. The EPA promotes atom economy as a key metric for sustainable process design, encouraging reactions that avoid the use of sacrificial reagents or excess reactants. In green nanomaterial synthesis, biogenic methods using plant extracts or microbial metabolites often exhibit high atom economy due to the multifunctional nature of bioactive compounds, which act as both reducing and stabilizing agents. Computational modeling and reaction pathway analysis can further optimize atom economy by identifying side reactions and improving selectivity. By prioritizing atom-efficient routes, researchers can reduce material costs, minimize environmental impact, and align nanomaterial production with DOE goals for resource conservation and clean energy innovation.

6.1.3 Design less hazardous chemical syntheses

This principle advocates for the intentional design of chemical reactions that minimize toxicity to human health and the environment. In nanotechnology, this involves selecting precursors and reagents that pose lower risks during synthesis and disposal. For example, replacing metal salts with plant-derived reducing agents or using aqueous solvents instead of organic ones can significantly reduce exposure to carcinogens or mutagens. The EPA encourages the use of predictive toxicology and computational modeling to assess the hazard profiles of synthetic pathways before implementation. DOE-funded research also supports the development of benign-by-design nanomaterials for energy applications, such as nontoxic quantum dots and lead-free perovskites. Reaction conditions such as temperature, pressure, and pH should be optimized to avoid the formation of unstable intermediates or reactive byproducts. Metrics such as LD_{50} (lethal dose), bioaccumulation potential, and environmental persistence are used to assess hazard levels. By designing inherently safer syntheses, researchers can reduce regulatory burdens, improve worker safety, and accelerate the commercialization of green nanotechnologies.

6.1.4 Design safer chemicals and products

Green chemistry emphasizes developing chemical products that are not only effective but also inherently safer for humans and ecosystems. In nanotechnology, this principle guides the design of NPs with reduced cytotoxicity, minimal environmental persistence, and controlled degradation profiles. For instance, surface functionalization with biocompatible ligands, such as polyethylene glycol (PEG), amino acids, or polysaccharides, can reduce immune responses and improve cellular uptake in biomedical applications. The EPA supports the use of structure–activity relationship (SAR) models and high-throughput screening to predict toxicity based on molecular features. DOE initiatives also prioritize developing safe nanomaterials for energy storage and catalysis, such as cobalt-free battery electrodes and nontoxic photocatalysts. Key technical parameters include zeta potential (for colloidal stability), hydrodynamic diameter (for biodistribution), and reactive oxygen species (ROS) generation (for oxidative stress). By designing safer nanomaterials from the outset, researchers can reduce downstream remediation costs, enhance public trust, and align with regulatory frameworks for sustainable innovation.

6.1.5 Use safer solvents and reaction conditions

This principle promotes the selection of solvents and reaction environments that minimize toxicity, flammability, and environmental persistence. In nanomaterial synthesis, replacing volatile organic solvents (e.g., toluene and chloroform) with water, ethanol, or ionic liquids can significantly reduce health risks and improve biodegradability. The EPA encourages solvent selection based on the solvent selection guide, which ranks solvents by safety, environmental impact, and performance. DOE-supported research also explores supercritical CO_2 and deep eutectic solvents for NP dispersion and surface modification. Reaction conditions, such as temperature, pressure, and pH, should be optimized to avoid hazardous extremes and favor ambient or near-ambient conditions, whenever possible. For example, green synthesis of silver NPs using plant extracts can be carried out at 25–80 °C in aqueous media, eliminating the need for high-temperature calcination or toxic surfactants. Technical metrics include flash point, vapor pressure, and biodegradability index. By adopting safer solvents and mild reaction conditions, researchers can reduce operational hazards, improve scalability, and align nanomaterial production with sustainable manufacturing standards.

6.1.6 Increase energy efficiency

Energy efficiency is a cornerstone of sustainable chemical design, aiming to reduce the energy input required for synthesis, purification, and processing. In nanotechnology, this principle encourages the use of low-temperature, low-pressure, and rapid synthesis techniques that conserve energy without compromising material quality. The EPA promotes process intensification strategies such as microwave-assisted synthesis, which can reduce reaction times from hours to minutes while maintaining NP uniformity. DOE initiatives support energy-efficient fabrication of nanostructures for energy storage and conversion, including low-temperature deposition of battery electrodes and solar absorber layers. For example, hydrothermal synthesis of ZnO nanorods at 120 °C for 6 h consumes significantly less energy than conventional sintering at 600 °C. Metrics such as reaction enthalpy, energy per gram of product, and carbon footprint are used to evaluate efficiency. By integrating renewable energy sources (e.g., solar-powered reactors) and minimizing thermal budgets, green nanotechnology can reduce greenhouse gas emissions, lower production costs, and contribute to broader clean energy innovation goals.

6.1.7 Use renewable feedstocks

This principle promotes the use of raw materials derived from renewable sources such as biomass, agricultural waste, and microbial cultures rather than finite petrochemical-based inputs. In nanotechnology, renewable feedstocks are increasingly used to synthesize metal and metal oxide NPs via biogenic routes. For example, plant extracts rich in polyphenols, flavonoids, and terpenoids serve as natural reducing and stabilizing agents in the green synthesis of Ag, ZnO, and CuO NPs. The EPA encourages the valorization of waste biomass and industrial byproducts to reduce reliance on virgin materials. DOE-supported initiatives explore lignocellulosic feedstocks and algae-derived compounds to enable scalable production of nanomaterials for energy systems. Key technical variables include carbon content, functional group availability, and extract yield per gram of biomass. By integrating renewable feedstocks into NP synthesis, researchers can reduce carbon intensity, enhance biodegradability, and support circular economy models. This approach not only aligns with sustainability goals but also opens new pathways for decentralized, low-cost nanomanufacturing in resource-constrained environments.

6.1.8 Avoid chemical derivatives

Avoiding unnecessary chemical derivatization reduces the number of reaction steps, minimizes reagent use, and prevents the formation of transient intermediates that may be hazardous or difficult to isolate. In nanotechnology, this principle discourages the use of protecting groups, activation agents, or post-synthesis modifications that add complexity and waste. For instance, direct synthesis of functionalized NPs using plant-based extracts eliminates the need for surface modification via silanization or thiolation. The EPA promotes reaction simplicity and direct transformation pathways to reduce solvent use and energy input. DOE research supports single-step synthesis of nanostructured catalysts and electrode materials to improve process efficiency and scalability. Technical metrics include step count, atom economy per step, and cumulative yield. Avoiding derivatization also enhances reproducibility and reduces purification burdens, both of which are critical for scaling nanomaterial production. By designing synthesis routes that achieve desired functionality without intermediate transformations, researchers can streamline workflows, reduce environmental impact, and accelerate the deployment of green nanotechnologies.

6.1.9 Use catalysts, not stoichiometric reagents

This principle emphasizes the use of catalytic systems to drive chemical reactions efficiently, selectively, and with minimal waste. In nanotechnology, catalysts are central to both synthesis and application. For example, noble metal NPs (e.g., Pt, Pd, and Au) serve as heterogeneous catalysts in fuel cells and for the hydrogen evolution reaction. During synthesis, enzymes, microbial metabolites, or plant-derived compounds can act as biocatalysts, enabling NP formation under mild conditions. The EPA promotes catalysis as a strategy to reduce excess reagent use and improve atom economy. DOE-funded research focuses on designing nanostructured catalysts with high surface area, tunable active sites, and low activation energy. Technical parameters include turnover frequency (TOF), activation energy (E_a), and selectivity index. Catalysts also enable recyclability and process intensification, reducing energy input and improving scalability. In green nanotechnology, catalytic synthesis routes such as photocatalytic or electrocatalytic NP formation offer cleaner alternatives to conventional redox reactions. By prioritizing catalytic efficiency, researchers can enhance reaction performance while minimizing environmental and economic costs.

6.1.10 Design chemicals and products to degrade after use

This principle advocates for the intentional design of chemical products that break down into nontoxic, environmentally benign components after their functional life. In nanotechnology, this translates to developing biodegradable NPs and coatings that do not persist in ecosystems or accumulate in biological tissues. For instance, polymeric NPs made from polylactic acid (PLA), chitosan, or starch derivatives degrade into natural metabolites under physiological conditions. The EPA supports life cycle analysis and end-of-life planning for chemical products, while DOE initiatives explore degradable materials for energy storage, packaging, and biomedical applications. Key technical variables include degradation rate (e.g., half-life in soil or water), hydrolysis kinetics, and environmental fate modeling. Surface modifications can also be engineered to trigger degradation in response to specific stimuli such as pH, light, or enzymatic activity. By designing nanomaterials with controlled disassembly and minimal ecological footprint, researchers can reduce long-term environmental risks, comply with regulatory standards, and contribute to circular material economies.

6.1.11 Analyze in real time to prevent pollution

This principle emphasizes the importance of real-time monitoring and control of chemical processes to prevent the formation of hazardous substances. In nanotechnology, real-time analysis enables precise control over NP nucleation, growth, and surface modification, reducing batch-to-batch variability and minimizing unintended byproducts. Techniques such as UV–vis spectroscopy, dynamic light scattering (DLS), and in situ Fourier-transform infrared (FTIR) or Raman spectroscopy allow researchers to track reaction progress, particle size evolution, and surface chemistry in real time. The EPA encourages the integration of process analytical technology (PAT) to detect deviations early and adjust parameters dynamically. DOE-supported initiatives leverage machine learning and sensor networks to optimize synthesis conditions for energy materials, such as battery electrodes and fuel cell catalysts. Key metrics include reaction yield, impurity profile, and environmental release potential. By embedding real-time analytics into green nanomaterial synthesis, researchers can reduce waste, improve reproducibility, and ensure compliance with environmental regulations, transforming pollution prevention from a reactive measure into a proactive design strategy.

6.1.12 Minimize the potential for accidents

The final principle of green chemistry focuses on designing chemical processes and materials that inherently reduce the risk of accidents, including explosions, fires, and toxic releases. In nanotechnology, this involves selecting nonvolatile, nonflammable solvents (e.g., water and ethanol), avoiding high-pressure or high-temperature conditions, and using benign precursors. For example, synthesizing NPs at ambient temperature using plant extracts eliminates the need for pressurized reactors or pyrophoric reagents. The EPA promotes inherently safer design principles, which prioritize hazard elimination over mitigation. DOE-funded research supports the development of self-limiting reactions and fail-safe reactor designs for scalable nanomanufacturing. Technical parameters include flash point, autoignition temperature, vapor pressure, and thermal decomposition threshold. Risk assessments and safety audits should be integrated into process development to identify potential failure modes. By minimizing accident potential through thoughtful design and material selection, green nanotechnology enhances laboratory safety, protects workers and communities, and ensures that innovation proceeds responsibly and sustainably.

6.2 Properties and importance of nanomaterials

Nanotechnology enables the precise measurement, visualization, manipulation, and fabrication of materials at the nanoscale, typically ranging from 1 to 100 nm in at least one dimension. At this scale, materials exhibit quantum confinement effects, altered surface energetics, and size-dependent reactivity, all of which are fundamentally distinct from those of their bulk counterparts. One of the most defining attributes of nanomaterials (NMs) is their exceptionally high surface area-to-volume ratio, often exceeding 100 m^2/g, which enhances catalytic activity, adsorption capacity, and interfacial interactions [13]. This characteristic underpins their widespread application across diverse domains, including mechanics, optics, electronics, biotechnology, microbiology, environmental remediation, medicine, and advanced materials engineering. At the nanoscale, materials can demonstrate tunable bandgaps, enhanced photoluminescence (PL), superparamagnetism, and increased mechanical resilience. These emergent properties have driven the development of sophisticated engineering and characterization protocols to resolve atomic-scale features and optimize performance. Techniques such as transmission electron microscopy (TEM), DLS, X-ray diffraction (XRD), and zeta potential analysis are routinely employed to quantify particle size, crystallinity, surface charge, and dispersion stability. A schematic overview of synthesis routes (e.g., sol–gel, hydrothermal, and green synthesis), NP types (e.g., metal, oxide, and hybrid), separation methods (e.g., centrifugation and membrane filtration), and application pathways is provided in Fig. 6.3 to illustrate the integrated life cycle of nanomaterial innovation.

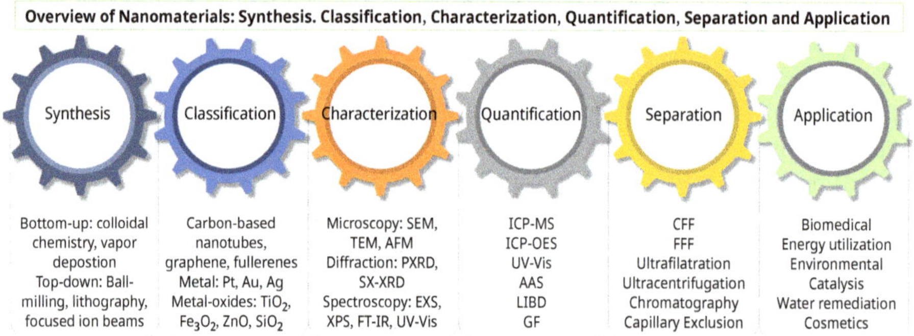

Fig. 6.3: Overview of nanoparticles, synthesis, classification, characterization, quantification, separation, and application.

6.2.1 Properties of nanomaterials

Nanomaterials possess unique physicochemical properties that differ significantly from bulk materials due to their nanoscale dimensions. These properties arise from quantum confinement, enhanced surface energetics, and size-dependent reactivity, which influence their electronic, optical, and catalytic behavior. Additional characteristics such as high surface area, tunable bandgaps, efficient charge transport, biocompatibility, and environmental stability further expand their functional versatility. Together, these traits enable nanomaterials to perform with precision across applications in energy, medicine, and advanced manufacturing. The key properties of nanomaterials are summarized in

Tab. 6.3, and the details are discussed as follows. These properties emerge when particle dimensions approach the 1–100 nm range, where classical mechanics gives way to quantum phenomena. Quantum confinement modifies the electronic band structure, leading to discrete energy levels and tunable optical properties. Altered surface energetics arise from the high fraction of surface atoms, which increases surface energy and modifies thermodynamic stability. Size-dependent reactivity reflects the enhanced chemical activity of NPs due to increased surface area and reduced coordination number of surface atoms. Together, these effects enable nanomaterials to exhibit superior catalytic activity, photonic behavior, and mechanical strength. Applications span from quantum dots in optoelectronics to metal oxide catalysts in environmental remediation. Understanding and controlling these properties is essential for tailoring nanomaterials for specific functions in energy, medicine, and advanced manufacturing, aligning with green chemistry principles and sustainable design frameworks.

Tab. 6.3: Comparison of the properties of nanomaterials and their key perspectives.

Property	Mechanism	Technical variables	Application relevance
A. Quantum confinement	Electron motion is restricted to <10–25 nm, leading to discrete energy levels	Bandgap (E_g), exciton Bohr radius, and quantum yield	Optoelectronics, quantum dots, and solar cells
B. Altered surface energetics	High fraction of surface atoms, surface energy, and reactivity	Surface energy (γ), coordination number, and XPS spectra	Catalysis, sintering control, and adsorption processes
C. Size-dependent reactivity	Increased surface atom exposure enhances chemical activity	TOF, activation energy, and surface coverage	Photocatalysis, fuel cells, and pollutant degradation
D. Surface area	High surface-to-volume ratio boosts interfacial interactions	BET surface area, pore size, and particle diameter	Drug delivery, sensors, and energy storage
E. Bandgap tuning	Particle size and doping alter the electronic band structure	E_g (eV), UV–vis spectra, PL emission wavelength	Photovoltaics, photocatalysis, and LEDs
F. Charge transport	Carrier mobility is affected by defects, grain boundaries, and morphology	Mobility ($cm^2/V \cdot s$), conductivity, and impedance	Transistors, batteries, and biosensors
G. Biocompatibility	Surface chemistry influences cellular uptake and immune response	Zeta potential, cytotoxicity (MTT), and biodistribution	Drug delivery, imaging, and tissue engineering
H. Environmental stability	Resistance to degradation under light, moisture, and temperature	TGA profile, XRD shifts, UV–vis degradation kinetics	Coatings, sensors, and long-term energy devices

6.2.1.1 Quantum confinement effects

Quantum confinement occurs when the dimensions of a material become comparable to the de Broglie wavelength of electrons, typically in the order of 10–25 nm in semiconductors. At this scale, the motion of charge carriers (electrons and holes) is restricted in one or more dimensions, leading to discrete energy levels rather than continuous bands. This confinement alters the bandgap energy; smaller particles exhibit larger bandgaps, resulting in size-dependent optical properties such as blue-shifting of absorption and emission spectra. Quantum dots, for example, demonstrate tunable fluorescence based on particle diameter, making them ideal for bioimaging and photovoltaics. The effect is most pronounced in materials with intrinsic bandgaps, such as CdSe, ZnO, and TiO_2. Technical parameters include exciton Bohr radius, bandgap energy (E_g), and quantum yield. Quantum confinement also influences carrier mobility, dielectric constant, and charge recombination rates. By precisely controlling particle size and shape, researchers can engineer nanomaterials with tailored electronic and photonic properties, enabling breakthroughs in quantum computing, solar energy conversion, and nanoscale sensors.

6.2.1.2 Altered surface energetics

At the nanoscale, a significant fraction of atoms reside on the surface rather than in the bulk, dramatically altering the material's surface energy and thermodynamic behavior. This phenomenon alters surface energetics, resulting in increased surface free energy, reduced coordination number, and enhanced chemical potential of surface atoms. These changes influence melting point depression, phase stability, and adsorption characteristics. For instance, gold NPs exhibit melting points hundreds of degrees lower than those of bulk gold due to surface-atom mobility. Surface energy (γ) becomes a dominant variable, affecting NP aggregation, sintering, and interfacial interactions. Additionally, surface atoms often possess unsaturated bonds, making them more reactive and catalytically active. This is exploited in applications such as heterogeneous catalysis, where high-surface-energy sites accelerate reaction kinetics. Modified surface energetics also impact wettability, zeta potential, and colloidal stability. Techniques like X-ray photoelectron spectroscopy (XPS) and atomic force microscopy (AFM) are used to characterize surface states. By tuning surface chemistry, nanomaterials can be optimized for biomedical interfaces, energy storage, and environmental remediation [14].

6.2.1.3 Size-dependent reactivity

Size-dependent reactivity refers to the enhanced chemical activity of nanomaterials as particle size decreases, primarily due to increased surface area and reduced atomic coordination. As NPs shrink below 100 nm, the proportion of surface atoms rises, exposing more reactive sites and increasing the likelihood of surface-mediated reactions. This leads to higher catalytic efficiency, faster electron transfer, and improved adsorption kinetics. For example, platinum NPs used in fuel cells exhibit superior oxygen reduction activity compared to bulk platinum, owing to their high surface-to-volume ratio and active edge sites. Particle shape, crystallographic facets, and defect density also influence reactivity. Smaller particles often display lower activation energies and altered reaction pathways. Metrics such as TOF, reaction rate constants, and surface coverage are used to quantify reactivity. Size-dependent reactivity is critical in applications like photocatalysis, drug delivery, and pollutant degradation. By engineering particle dimensions and surface morphology, researchers can harness this property to design nanomaterials with precise functionality and minimal environmental impact.

6.2.1.4 Surface area

Nanomaterials possess an exceptionally high surface area-to-volume ratio, often exceeding 100 m^2/g, depending on particle morphology and synthesis method. This property arises from the dominance of surface atoms in nanoscale structures, which dramatically enhances interfacial interactions, adsorption capacity, and catalytic activity. For instance, mesoporous silica NPs with pore diameters of 2–50 nm can achieve surface areas above 800 m^2/g, making them ideal for drug delivery and pollutant capture. Surface area influences reaction kinetics, diffusion rates, and thermal conductivity, and is quantified using Brunauer–Emmett–Teller (BET) analysis. In catalysis, a higher surface area correlates with increased TOF because it provides more active sites. In energy applications, such as supercapacitors and batteries, large surface areas enable rapid ion exchange and charge storage. Surface area also affects colloidal stability, aggregation behavior, and functionalization potential. By tailoring particle size, porosity, and surface roughness, researchers can optimize nanomaterials for specific applications in medicine, environmental remediation, and energy systems, aligning with green chemistry principles of efficiency and minimal waste.

6.2.1.5 Bandgap tuning

Bandgap tuning is a critical property of nanomaterials that enables control over their electronic and optical behavior. At the nanoscale, quantum confinement effects cause the bandgap energy (E_g) to vary inversely with particle size, allowing materials to absorb and emit light at different wavelengths. For example, CdSe quantum dots exhibit size-dependent fluorescence, ranging from red to blue, as their diameter decreases from ~ 6 nm to ~ 2 nm. Bandgap tuning is essential in photovoltaics, photocatalysis, and optoelectronics, where precise control over light absorption and charge separation is required. Techniques such as doping, alloying, and surface modification can further adjust E_g. In ZnO NPs, doping with transition metals (e.g., Mn^{2+} and Co^{2+}) can reduce the bandgap from ~ 3.3 eV to ~ 2.8 eV, enhancing visible-light activity. UV–vis spectroscopy and PL analysis are used to characterize bandgap shifts. By engineering nanomaterials with tailored bandgaps, researchers can optimize energy conversion efficiency, sensor sensitivity, and environmental responsiveness in sustainable technologies.

6.2.1.6 Charge transport

Charge transport in nanomaterials is governed by carrier mobility, conductivity, and interfacial resistance, all of which are influenced by particle size, crystallinity, and surface states. At the nanoscale, reduced dimensionality and high surface-to-volume ratios can enhance or hinder charge mobility, depending on defect density and grain boundaries. For instance, graphene exhibits ultrahigh electron mobility (>200,000 $cm^2/V \cdot s$) due to its delocalized π-electron system, making it ideal for high-speed electronics and transparent conductors. In semiconducting nanowires (e.g., Si and ZnO), charge transport is affected by surface traps and quantum confinement, which can be mitigated through passivation or core–shell architectures. Techniques such as Hall effect measurements, impedance spectroscopy, and field-effect transistor characterization are used to quantify transport properties. In energy devices, efficient charge transport is essential for minimizing recombination losses and maximizing power output. By tuning morphology, doping levels, and surface chemistry, nanomaterials can be engineered to optimize charge transport for applications ranging from solar cells to biosensors and electrochemical capacitors.

6.2.1.7 Biocompatibility

The ability of nanomaterials to interact with biological systems without eliciting adverse effects is a critical property for biomedical applications such as drug delivery, imaging, and tissue engineering. At the nanoscale, surface charge, hydrophobicity, and functional groups dictate cellular uptake, immune response, and biodistribution. For example, NPs with near-neutral zeta potential (~ ± 10 mV) and hydrophilic coatings (e.g., PEG) exhibit reduced protein adsorption and prolonged circulation times. Biocompatibility is assessed through cytotoxicity assays (e.g., MTT and LDH), hemocompatibility tests, and in vivo biodistribution studies. Materials like silica, gold, and iron oxide are generally considered biocompatible when properly surface-modified. However, factors such as particle aggregation, oxidative stress, and unintended bioaccumulation can compromise safety. Surface engineering using biomolecules (e.g., peptides and polysaccharides) enhances targeting and reduces immunogenicity. Regulatory frameworks from the FDA and EMA emphasize rigorous evaluation of nanomaterial biocompatibility. By designing NPs with controlled size, surface chemistry, and degradation profiles, researchers can ensure safe and effective integration into biological environments.

6.2.1.8 Environmental stability

Environmental stability refers to the ability of nanomaterials to maintain their structural, chemical, and functional integrity under ambient or operational conditions. This property is essential for applications in catalysis, sensors, and energy devices, where exposure to light, moisture, oxygen, and temperature fluctuations can degrade performance. For instance, black phosphorus nanosheets degrade rapidly in air due to photooxidation, necessitating surface passivation or encapsulation strategies. Stability is influenced by factors such as crystallinity, surface energy, and protective coatings. Techniques like thermogravimetric analysis, XRD, and UV–Vis spectroscopy are used to monitor degradation kinetics. In photocatalysis, materials like TiO_2 and ZnO exhibit high environmental stability due to their robust oxide frameworks and resistance to photochemical corrosion [15]. Surface modification with polymers, silanes, or inorganic shells can enhance stability by reducing reactivity and shielding against environmental stressors. By engineering nanomaterials with tailored surface chemistry and encapsulation, researchers can extend functional lifetimes, reduce environmental release risks, and support sustainable deployment in real-world conditions [16].

6.2.2 Classification of nanomaterials

Nanomaterials are broadly categorized into organic, inorganic, ceramic, and carbon-based NPs, each exhibiting distinct crystallographic structures and physicochemical behaviors, as summarized in Tab. 6.4. Inorganic NPs are further classified into metal-based (e.g., gold, Au) and metal-oxide-based (e.g., zirconium dioxide, ZrO_2) systems, which are widely utilized in catalysis, biosensing, and photothermal therapy due to their plasmonic resonance, redox activity, and high thermal stability [16]. Carbon-based NPs encompass fullerenes, carbon nanotubes (CNTs), graphene, carbon nanofibers, and carbon black, each offering unique electrical conductivity, mechanical strength, and surface functionalization potential for applications in energy storage, drug delivery, and composite reinforcement [17]. Khan *et al.* (2019) emphasized that NP systems can be classified based on morphology, size, and chemical composition into categories such as carbon-based, metal, ceramic, semiconductor, polymeric, and lipid-based NPs, each tailored for specific biomedical, electronic, or environmental functions [18]. Pandey and Prajapati [19] further refined this classification by chemical composition, identifying four major classes: carbon-based (e.g., CNTs and nanofibers), metal and metal-oxide-based (e.g., Ag and Cu), bio-organic (e.g., liposomes and micelles), and composite-based NPs, which integrate multiple material types for multifunctional performance across industrial and clinical domains.

Tab. 6.4: Summary of representative nanostructured materials and their properties.

Carbon-based nanomaterials	
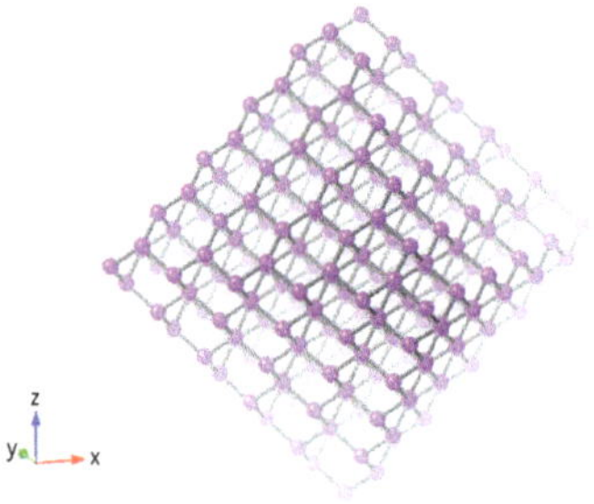	Diamond: the high-pressure form of carbon with an octahedral crystal shape [20] Space group: F*d3m* {origin 1} Origin shift: ¼, ¼, ¼ Crystal system: Cubic *a*: 3.5597 Å Cell volume: 45.107 $Å^3$ Asymmetric unit: 1 site Unit cell: 8 sites/unit cell Density: 3.5371 g/cm^3
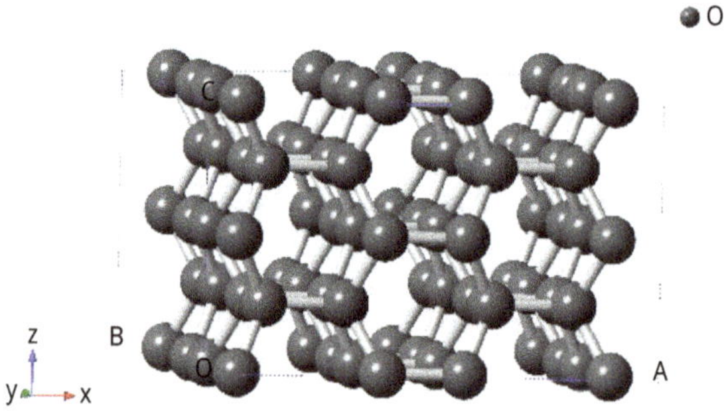	Carbon BCT: intermediate phase between graphite and diamond [21] Space group: *I4/mmm* Crystal system: Tetragonal *a*: 4.3220 Å *c*: 2.4780 Å Cell volume: 46.288 $Å^3$ Asymmetric unit: 1 site Unit cell: 8 sites/unit cell Density: 3.4468 g/cm^3
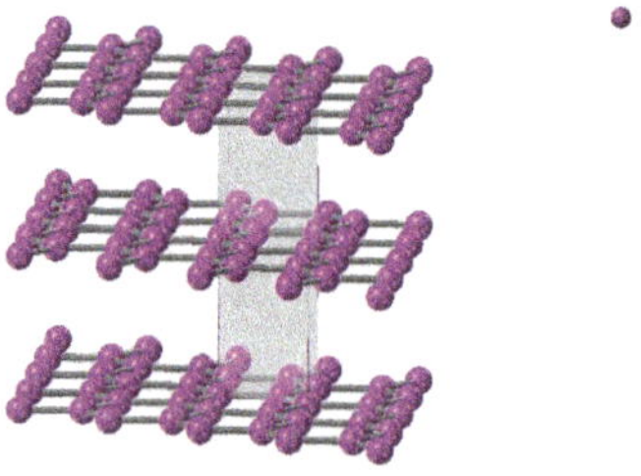	Graphite (the low-pressure/high-temperature polymorph). Three sp^2 bonds form strong networks in the basal (001) plane [22] Space group: *P*3 (chirality; polar) Crystal system: Trigonal *a*: 2.4560 Å *c*: 6.6960 Å Cell volume: 34.979 $Å^3$ Asymmetric unit: 4 sites Unit cell: 4 sites/unit cell Density: 2.2806 g/cm^3

Tab. 6.4 (continued)

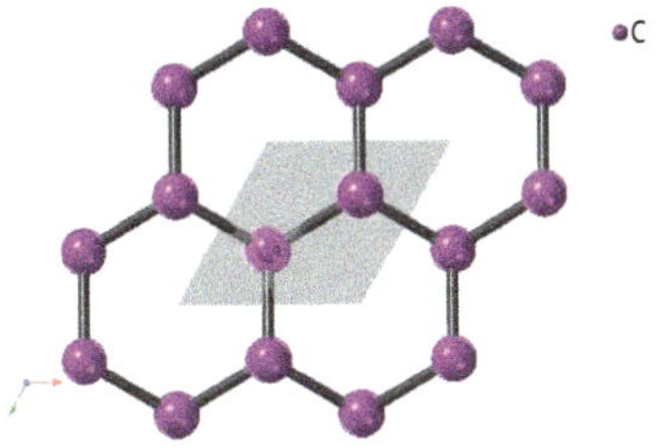	Lonsdaleite – hexagonal carbon Found in microscopic crystals [23] Space group: *P* 63/*mmc* Crystal system: Hexagonal *a*: 2.5200 Å *c*: 4.1200 Å Cell volume: 22.658 $Å^3$ Asymmetric unit: 1 site Unit cell: 4 sites/unit cell Density: 3.5207 g/cm^3
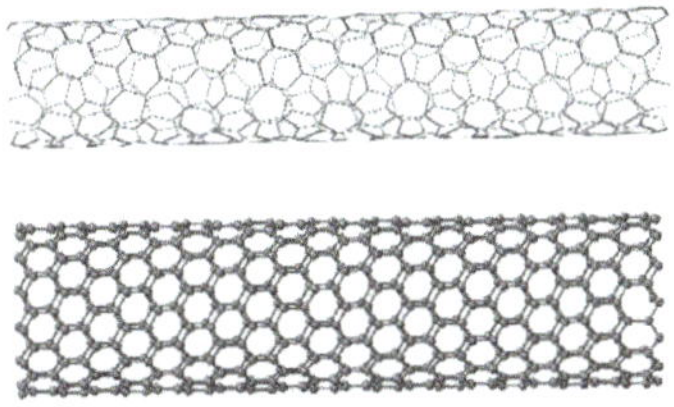	Carbon nanotube, an idealized structure For buckytube [24]
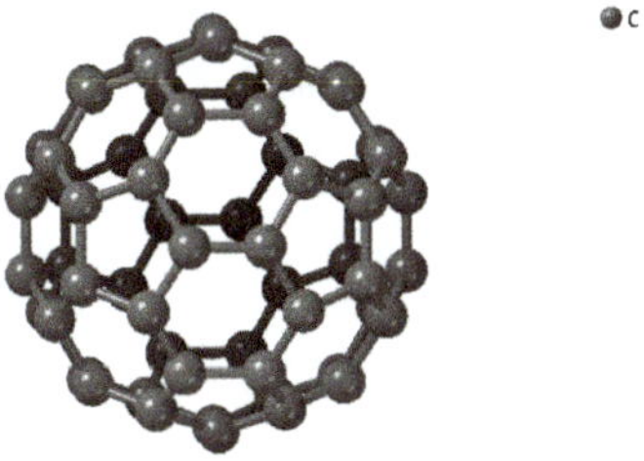	A buckminsterfullerene (C_{60}) molecule [25]
Metal	
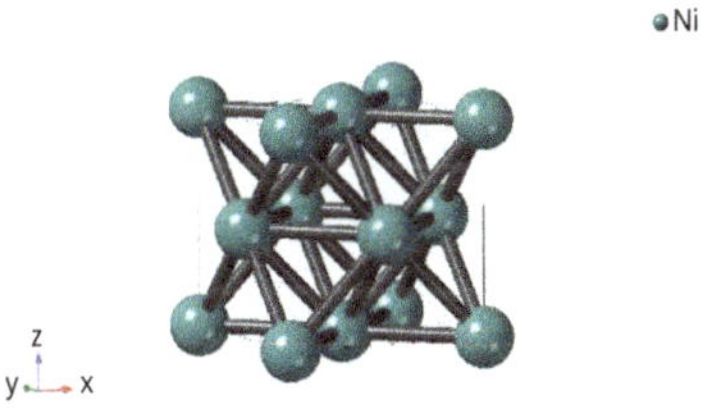	Nickel [26] Space group: *Fm* $\bar{3}$*m* Crystal system: Cubic *a*: 3.5168 Å Cell volume: 43.495 $Å^3$ Asymmetric unit: 1 site Unit cell: 4 sites/unit cell Density: 8.9657 g/cm^3
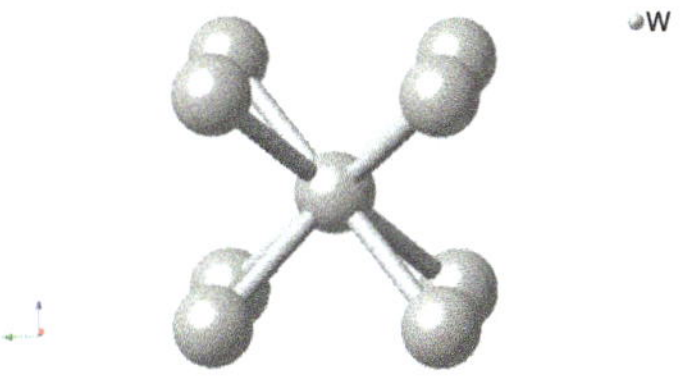	Tungsten (W) [27] The cubic form of tungsten has a bond-centered cubic structure Space group: *I* 1 (chirality; polar) Crystal system: Triclinic $a = b = c$: 3.1647 Å $\alpha = \beta = \gamma$: 90.000 °Cell volume: 31.695 $Å^3$ Asymmetric unit: 1 site Unit cell: 2 sites/unit cell Density: 19.2696 g/cm^3

Tab. 6.4 (continued)

Metal oxides

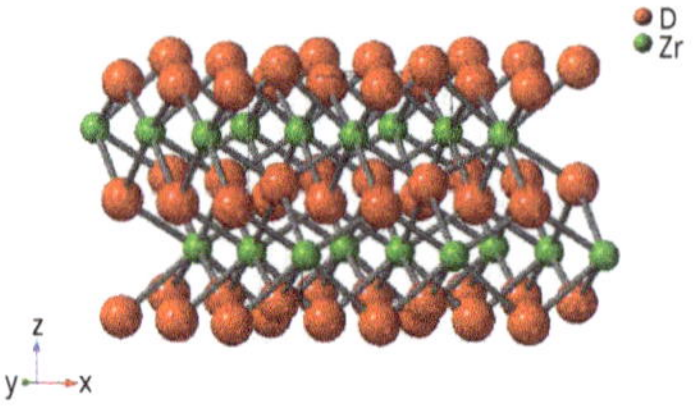

Tetragonal zirconia (ZrO_2) [28]
Space group: *P 4₂/nmc* [origin 2]
Crystal system: Tetragonal
a: 3.6008 Å
c: 5.1793 Å
Cell volume: 67.154 Å^3
Asymmetric unit: 2 sites
Unit cell: 6 sites/unit cell
Density: 6.0939 g/cm^3

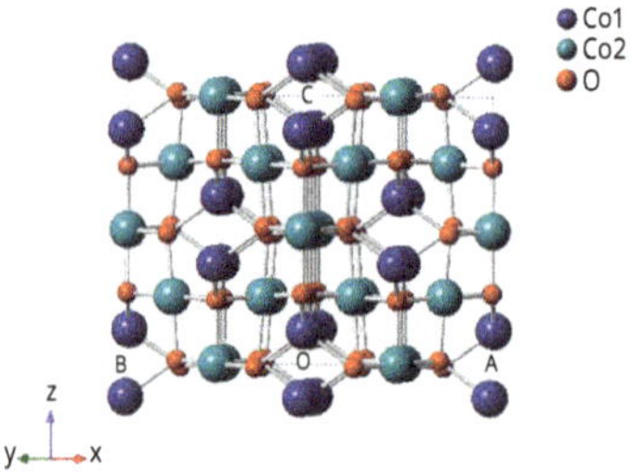

Cobalt oxide (Co_3O_4)
A standard spinel structure at room temperature (measured at 301 K) [29]
Space group: *Fd* $\bar{3}$*m* [origin 2]
Crystal system: Cubic
a: 8.0821 Å
Cell volume: 527.926 Å^3
Asymmetric unit: 3 sites
Unit cell: 56 sites/unit cell
Density: 6.0591 g/cm^3

Metal boride

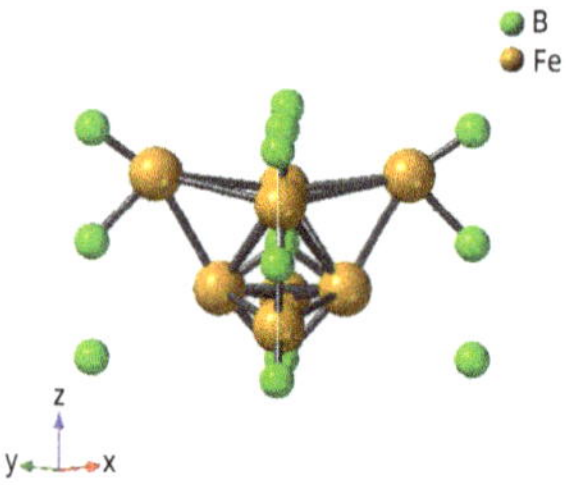

Iron boride (Fe_2B) [30]
Space group: *I*$\bar{4}$*2 m*
Crystal system: Tetragonal
a: 5.0780(50) Å
c: 4.2230(50) Å
Cell volume: 108.895 Å^3
Asymmetric unit: 3 sites
Unit cell: 12 sites/unit cell
Density: 7.4727 g/cm^3

Metal nitride

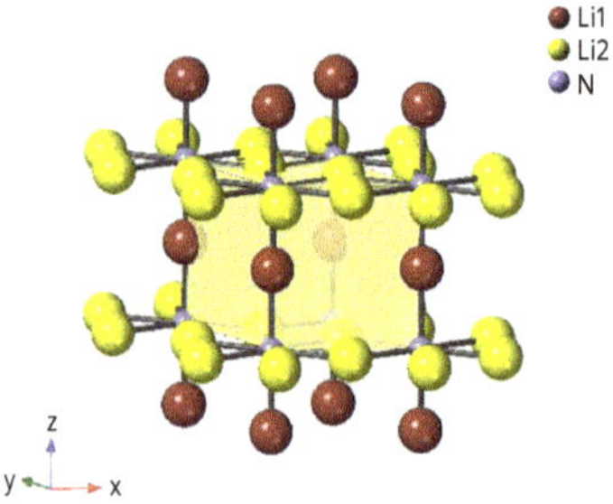

Lithium nitride (Li_3N) [31]
N ions are {8} coordinated by Li, which occupies two sites in this structure
Space group: *P6/mmm*
Crystal system: Hexagonal
a: 3.6520(30) Å
c: 3.8660(20) Å
Cell volume: 44.653 Å^3
Asymmetric unit: 3 sites
Unit cell: 4 sites/unit cell
Density: 1.2954 g/cm^3

Tab. 6.4 (continued)

Silicate	
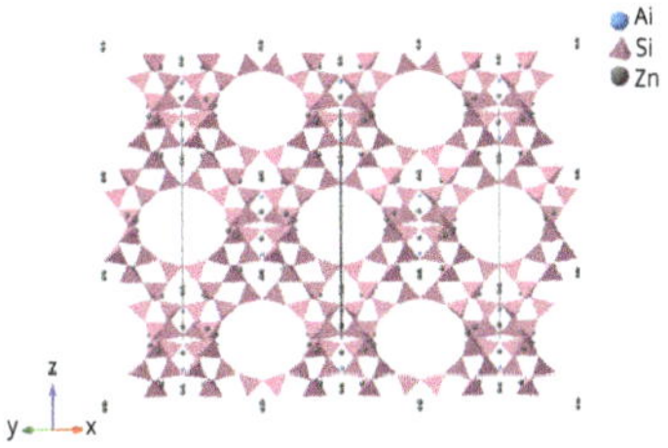 	Zeolite X ($Zn_{54.976}Al_{95.898}Si_{104.0}O_{416.06}$) Based on the faujasite structure comprising a framework of corner-sharing SiO_4 tetrahedra, forming large, 12-fold rings along the {110} direction [32] Space group: $Fd\bar{3}m$ {origin 2} Crystal system: Cubic *a*: 24.7180 Å Cell volume: 15,102.192 Å^3 Asymmetric unit: 10 sites Unit cell: 872 sites/unit cell 0.0444 atoms/Å^3 Density: 1.7329 g/cm^3
Metal hydroxide	
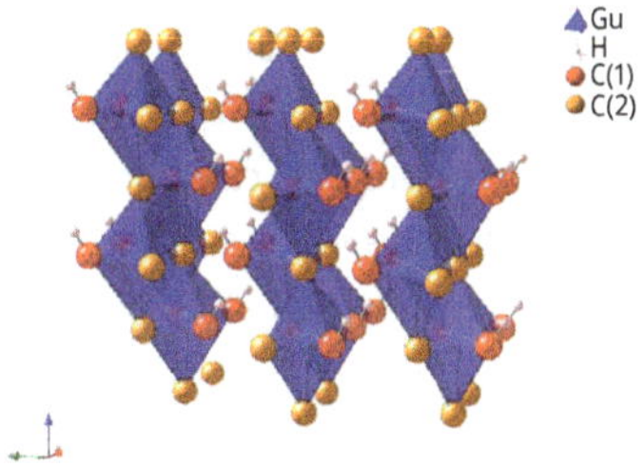 	Copper hydroxide – $Cu(OH)_2$ Copper atoms are octahedrally coordinated by O and O–H groups. The octahedra share edges arranged in slabs parallel to the crystallographic *z*-axis [33] Space group: $Cmc2_1$ (polar) Crystal system: Orthorhombic *a*: 2.9470 Å *b*: 10.5930 Å *c*: 5.2560 Å Cell volume: 164.080 Å^3 Asymmetric unit: 4 sites Unit cell: 16 sites/unit cell Density: 3.9088 g/cm^3

Mr. Palakurthi conducted a comprehensive investigation into the biosynthesis of NPs using a wide range of biological systems, including plants, bacteria, fungi, yeast, algae, and viruses. Among these, plant-mediated synthesis has emerged as a particularly promising route owing to its eco-friendliness, scalability, and inherent biocompatibility. For example, *Pelargonium graveolens* (geranium leaf) extract was employed in the extracellular synthesis of silver NPs (AgNPs), where terpenoids acted as reducing agents to convert Ag^+ ions into metallic silver (Ag^0), as confirmed by FTIR [34].

These AgNPs exhibited high crystallinity, narrow size distribution (10–50 nm), and surface plasmon resonance peaks near 420 nm. Similarly, *Cymbopogon flexuosus* (lemongrass) extract enabled single-step synthesis of gold NPs (AuNPs) at ambient temperature, reducing $AuCl_4$– ions into spherical particles with strong optical absorbance around 530 nm. Beyond plant systems, biological synthesis routes can be broadly categorized into microbial (bacteria, fungi, and yeast), algal, and viral platforms, each offering distinct biochemical mechanisms and material specificity. Bacterial species such as *Bacillus subtilis* and *Pseudomonas aeruginosa* utilize enzymatic pathways and membrane-bound reductases to produce AgNPs and ZnO NPs either intracellularly or extracellularly. Fungi and yeast contribute high biomass yields and secrete extracellular enzymes, enabling controlled NP morphology and scalable production. Algal systems leverage photosynthetic metabolites and metal-binding proteins to synthesize oxide NPs under mild conditions. Viruses, acting as biological templates, facilitate capsid-directed nucleation, thereby enabling precise NP geometry, which is particularly valuable in nanobiotechnology and targeted delivery applications. Biological synthesis of NPs utilizes plants, bacteria, fungi, yeast, algae, and viruses to reduce metal ions through natural metabolites such as terpenoids, enzymes, and proteins. These systems offer distinct synthesis environments, intracellular or extracellular, and produce stable NPs, such as AgNPs, AuNPs, and ZnO NPs, with tunable size and morphology [34]. Each route differs in reaction time, particle size range, and scalability, supporting applications in medicine, energy, and environmental remediation (Khan *et al.*,) [18]. Figure 6.4 was shown for a schematic overview.

6.2.3 Importance of nanomaterials

Nanomaterials are vital to modern science and engineering because properties emerging at 1–100 nm, high surface area-to-volume ratios, tunable band structures, and quantum effects, unlock performance unattainable in bulk materials. In catalysis, precise control of NP composition, morphology, and metal-support interactions boosts activity, selectivity, and durability, aligning with sustainability goals across chemical manufacturing. In electronics, low-dimensional materials and printable nano-inks enable miniaturized, flexible, and 3D architectures with enhanced electrical, optical, and thermal performance. Nanoscale architectures in energy storage reduce the distances traveled by ions and electrons, increase access to redox sites, and enable advanced manufacturing of high-energy, high-power devices. In biomedicine, nanoengineered carriers, sensors, and scaffolds enhance targeting precision, imaging contrast, and regenerative outcomes while demanding rigorous evaluation of biocompatibility and toxicity. These advances are documented in recent reviews and editorials that highlight the system-level impact of nanomaterials, the need for standardization, and translational pathways from synthesis to device integration. Together, nanoscale design principles and metrology are reshaping performance ceilings across energy,

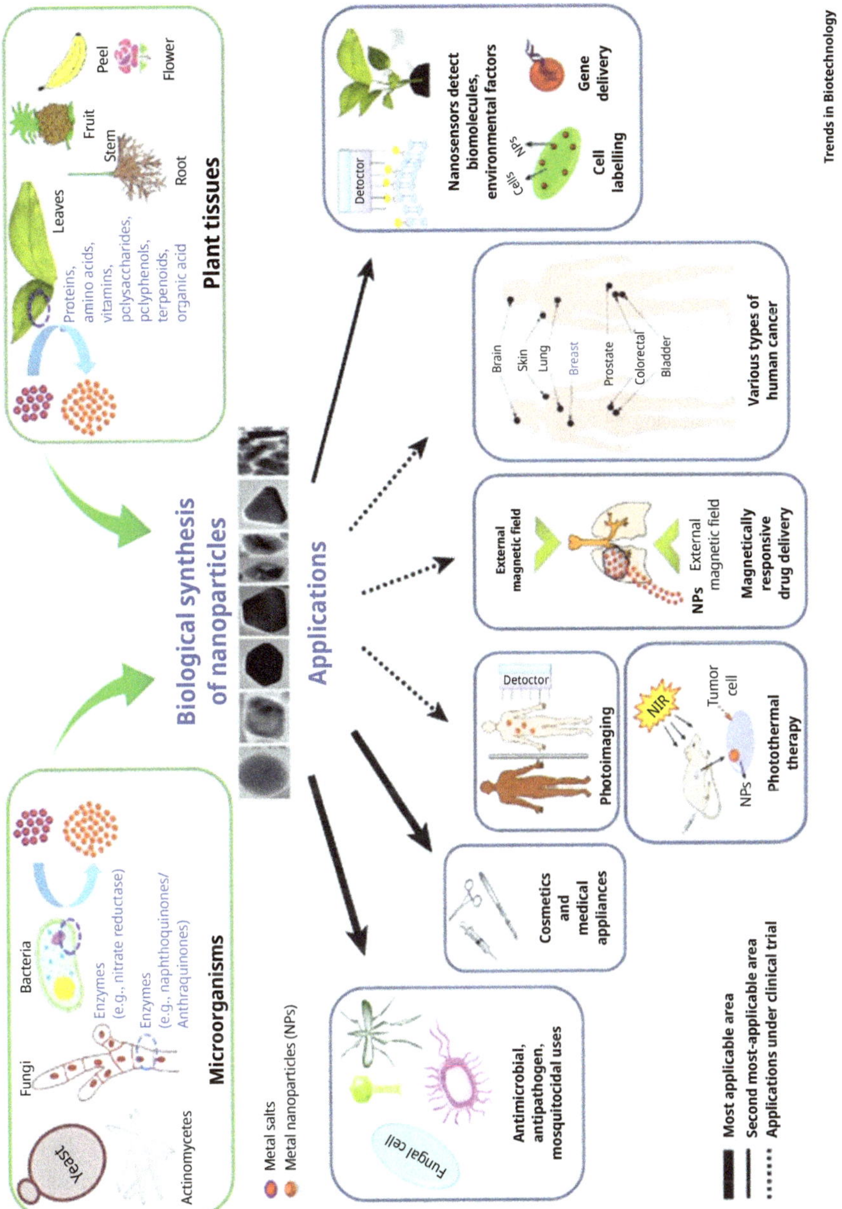

Fig. 6.4: Schematic overview of biological synthesis routes for nanoparticles using plants, bacteria, fungi, yeast, algae, and viruses. The diagram illustrates key reducing agents (e.g., terpenoids, enzymes, and proteins), synthesis environments (intracellular vs. extracellular), and resulting nanoparticle types (e.g., AgNPs, AuNPs, and ZnO NPs). Arrows indicate the flow of ion reduction, stabilization, and particle formation. Comparative metrics such as reaction time, particle size range, and scalability potential are annotated for each biological system. Adapted from [35], and Khan *et al.* [18].

health, climate, and information technologies, accelerating the transition from laboratory phenomena to scalable, application-ready platforms. Table 6.5 summarizes the importance of nanomaterials across four key fields: catalysis, electronics, energy storage, and biomedicine, with emphasis on technical functions, performance enhancements, and translational relevance [35–38].

Tab. 6.5: Comparative importance of nanomaterials across key fields, catalysis, electronics, energy storage, and biomedicine, with emphasis on technical functions, performance enhancements, and translational relevance.

Field	Function of nanomaterials	Performance enhancements	Translational impact
A. Catalysis	Active sites tuning via size, shape, and composition; metal–support interaction optimization	Increased turnover frequency, enhanced selectivity, and lower activation energy	Sustainable chemical manufacturing, CO_2 conversion, hydrogen evolution, and pollutant degradation
B. Electronics	Low-dimensional semiconductors, printable nano-inks, and conductive nanomaterials	High carrier mobility (e.g., graphene: 200k $cm^2/V \cdot s$), flexible 3D architectures, and conductivity	Wearable devices, flexible displays, structural electronics, and MEMS
C. Energy storage	Nanoscale electrode design, ion/electron transport optimization, and redox site accessibility	High specific capacity (>1,000 mAh/g), fast charge/discharge, and enhanced cycle life	Lithium-ion, sodium-ion, and solid-state batteries; supercapacitors; and grid-scale storage
D. Biomedicine	Targeted drug delivery, multimodal imaging agents, and regenerative scaffolds	Controlled release, high imaging contrast, and stimulated tissue regeneration	Cancer theranostics, wound healing, tissue engineering, and personalized medicine

6.2.3.1 Catalysis

Catalysis underpins an estimated 85–90% of manufacturing steps and contributes over 35% of global GDP, making nanocatalysis central to sustainable industry. Nanoscale control of particle size, shape, and electronic structure enables targeted metal–metal and metal–support interactions that tune work function, oxidation state, and adsorption energetics, thereby improving activity, selectivity, and stability. Strategies include facet engineering, defect/vacancy creation, and strong metal–support interactions (e.g., oxygen-vacancy-rich ceria supports that resist sintering and even enable on-stream redispersion of Rh), mitigating particle migration and Ostwald ripening. Advanced operando spectroscopy and microscopy elucidate reaction pathways, while data-driven design accelerates discovery for CO_2 conversion, emissions control, water treatment, and (photo/electro)chemical fuel production. Recent reviews chart progress from historical size effects on Au to high-entropy alloy nanocatalysts, active for hydrogen evolution, underscoring how morphology and compositional complexity couple to kinetic metrics such as TOF and apparent activation energy. With green chemistry pressures rising, nanocatalysis offers precision levers for lifetime, efficiency, and circularity across chemical value chains, as outlined in Fig. 6.5 [39–41]. Catalysis plays a central role in over 85% of industrial chemical processes. It contributes to more than 35% of global GDP, particularly in sectors such as energy, chemicals, and food production. Nanocatalysis enhances catalytic performance by enabling precise control over particle size, morphology, and electronic structure, which directly influences TOF, activation energy, and selectivity[1]. Recent studies have demonstrated that atomically precise metal clusters and single-atom catalysts can achieve near-complete atomic utilization, with high-entropy alloys offering tunable active sites for reactions such as hydrogen evolution (e.g., NiCoP at 238.2 mmol/g h) [42] and CO_2 reduction using heterostructured Ni_2P/CdS systems [43]. Integration of machine learning with operando spectroscopy has accelerated rational catalyst design, enabling predictive modeling of reaction pathways and dynamic surface restructuring[1]. Additive manufacturing and 3D printing now support spatially resolved catalyst deposition, facilitating multifunctional architectures for electronics and energy devices [44]. Nanomaterials such as MXenes, CNTs, and metal oxides provide high surface areas (>200 m^2/g) and conductivity (>10^4 S/m), supporting electrocatalysis and photocatalysis under ambient conditions. As highlighted in *Science*, dynamic restructuring under reaction conditions reveals transient active sites and adaptive behavior, essential for next-generation catalytic systems [38].

6.2.3.2 Electronics

Nanomaterials are revolutionizing electronic device design by enabling high-performance, miniaturized, and multifunctional systems through their integration into 2D semiconductors, conductive carbons, metal oxides, and printable nano-inks. These materials support the fabrication of flexible, conformal, and three-dimensional architectures, where additive manufacturing and printed electronics converge to produce interconnects, antennas, sensors, and energy modules with minimal waste and enhanced functional density, as indicated in Fig. 6.6 [44]. Nanomaterials are reshaping the landscape of electronic device engineering by enabling ultra-compact, high-performance, and multifunctional systems. Through integration of 2D semiconductors, conductive carbons, metal oxides, and printable nano-inks, these materials support the fabrication of flexible, conformal, and three-dimensional architectures. Additive manufacturing and printed electronics converge to produce interconnects, antennas, sensors, and energy modules with minimal material waste and enhanced functional density. This convergence allows for scalable production of wearable devi-

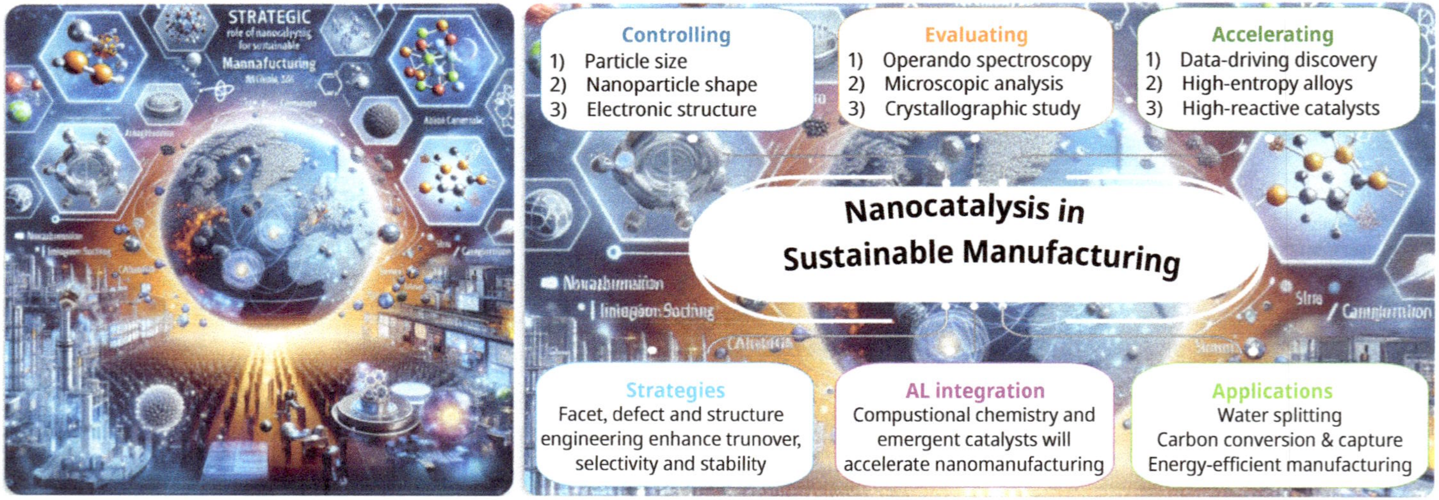

Fig. 6.5: Schematic overview of the strategic role of nanocatalysis in sustainable manufacturing, from atomic-level control to global impact, highlighting six fields: controlling, evaluating, accelerating, strategies, AI integration, and applications.

ces, smart surfaces, and embedded systems that meet the demands of next-generation electronics. The diagram illustrates how each nanomaterial class contributes to specific device components, highlighting their synergistic role in advancing miniaturization, adaptability, and energy efficiency.

Graphene, with its ultrahigh carrier mobility (up to 200,000 $cm^2/V \cdot s$), and CNTs, known for their tensile strength and percolation behavior, are widely used in transistors, transparent conductors, and wearable electronics due to their electrical, mechanical, and thermal advantages. Metal oxides such as ZnO and TiO_2 offer tunable bandgaps (3.3 eV and 3.0 eV, respectively) and high dielectric constants, making them suitable for optoelectronic applications, including displays, photovoltaics, and capacitive sensors. Advances in ALD, inkjet printing, and aerosol jet printing have enabled nanoscale patterning on both planar and curved substrates, facilitating the development of 3D structural electronics, MEMS, and integrated photonic circuits [44]. These techniques allow precise control over film thickness, composition, and interface quality, which are critical for device reliability and performance. Moreover, the emergence of hybrid and multimaterial printing strategies has opened pathways to 4D electronics, in which time-dependent functionality is embedded into spatially resolved nanostructures [44]. Despite these technological breakthroughs, challenges remain in scaling production, ensuring reproducibility, and minimizing environmental impact [45-47]. Variability in NP synthesis, ink formulation, and substrate compatibility can affect device yield and long-term stability. As highlighted in recent literature, the push toward greener synthesis routes, such as aqueous-phase processing, low-temperature sintering, and biodegradable substrates, is essential for sustainable electronics manufacturing [48]. Standardized metrology frameworks are also being developed to assess electrical, mechanical, and thermal properties across nanomaterial interfaces, enabling consistent device qualification and regulatory compliance. Together, these innovations position nanomaterials as foundational elements in next-generation electronics, bridging nanoscale functionality with macroscale integration for applications in healthcare, energy, communication, and environmental monitoring.

6.2.3.3 Energy storage

In batteries and supercapacitors, nanomaterials shorten ion/electron pathways, increase active surface, and tailor interfacial chemistry, enabling simultaneous gains in energy and power density. Reviews document libraries of nanoscale building blocks from 0D to 3D, assembled into electrodes that exploit intercalation, pseudocapacitance, and surface adsorption (details will be discussed in Section 6.3). MXenes deliver conductivities exceeding those of conventional carbons, supporting current-collector-free architectures; carbon–silicon and carbon–sulfur hybrids buffer volume changes

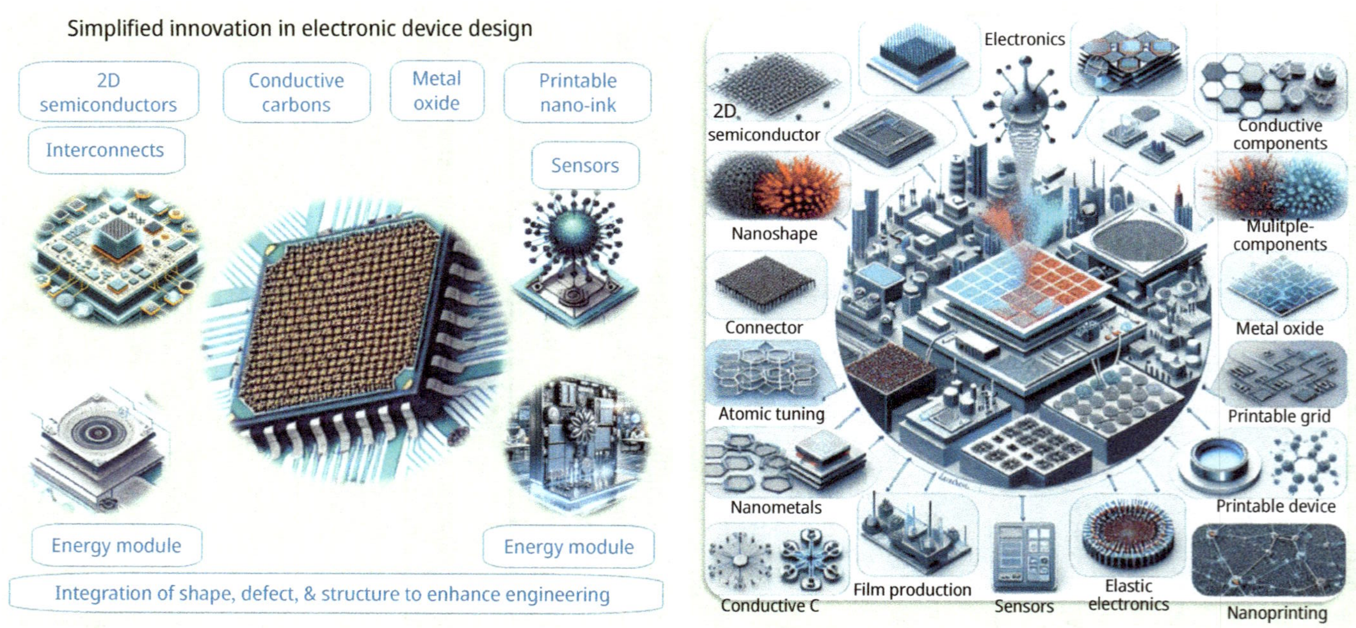

Fig. 6.6: Nanomaterials drive innovation in electronic device design, from 2D semiconductors to printed sensors and energy modules.

and enhance kinetics. Graphene-aerogel–nanowire hybrids increase electrode–electrolyte interfacial area, elevating specific capacity and charge–discharge rate, while carbon nanomaterials and transition-metal oxides/sulfides underpin high-performance supercapacitors. Manufacturing advances, printing, spray coating, roll-to-roll, and self-assembly enable flexible, structural, and wearable devices, bridging laboratory materials to practical formats. Persistent issues include first-cycle irreversibility, parasitic reactions from high surface area, and agglomeration; thus, smart nanoarchitectures and protective coatings are emphasized. Comprehensive surveys anticipate solid-state platforms and AI-guided optimization, positioning nanoscale electrodes as key to grid integration of renewables and fast-charging mobility solutions.

Despite these technological breakthroughs, challenges remain in scaling production, ensuring reproducibility, and minimizing environmental impact. Variability in NP synthesis, ink formulation, and substrate compatibility can affect device yield and long-term stability. As highlighted in recent literature, the push toward greener synthesis routes, such as aqueous-phase processing, low-temperature sintering, and biodegradable substrates, is essential for sustainable electronics manufacturing [49]. Standardized metrology frameworks are also being developed to assess electrical, mechanical, and thermal properties across nanomaterial interfaces, enabling consistent device qualification and regulatory compliance. Together, these innovations position nanomaterials as foundational elements in next-generation electronics, bridging nanoscale functionality with macroscale integration for applications in healthcare, energy, communication, and environmental monitoring. Nanomaterials are reshaping energy storage technologies by enhancing charge transport, surface reactivity, and structural stability at the electrode–electrolyte interface. According to Farooq *et al.* (2024), carbon-based nanomaterials such as graphene, CNTs, and mesoporous carbon exhibit high surface areas (>1,000 m^2/g) and excellent electrical conductivity, making them ideal for supercapacitor electrodes and lithium-ion battery (LiB) anodes [50]. Mohammed et al. [51] highlight that nanostructured materials improve lithium-ion diffusion and reduce interfacial resistance, enabling faster charging and higher capacity retention in LiBs, sodium–sulfur batteries, and redox flow systems. For example, graphene–nanowire hybrids have demonstrated enhanced specific capacities (>1,000 mAh/g) and robust cycling performance due to synergistic effects between high conductivity and increased active surface area. As reported in *Science*, dynamic restructuring of catalytic nanomaterials under operating conditions reveals transient active sites that enhance reaction kinetics and energy-conversion efficiency [46]. These findings underscore the importance of nanoscale design in achieving high-performance, scalable, and sustainable energy storage systems. Emerging trends include solid-state configurations and AI-guided optimization, positioning nanomaterials at the forefront of next-generation power technologies.

6.2.3.4 Biomedicine

Nanomaterials are redefining the biomedical landscape by enabling four interdependent capabilities: precision-targeted delivery, multimodal imaging, integrated theranostics, and regenerative scaffolding [52]. Their nanoscale dimensions (1–100 nm), high surface-area-to-volume ratios, and tunable surface chemistries facilitate selective interaction with biological systems at the molecular level, minimizing off-target effects while enhancing therapeutic efficacy. In wound healing applications, smart nanomaterials such as silver, gold, and zinc oxide NPs exhibit antimicrobial, anti-inflammatory, and pro-angiogenic properties. These particles respond to physiological stimuli, including pH fluctuations, temperature changes, and ROS, thereby triggering controlled drug release and accelerating tissue regeneration. Notably, silver colloid-based dressings have demonstrated up to 85.7% healing efficacy in diabetic foot ulcers, significantly outperforming conventional treatments, which report approximately 41.7% effectiveness. This chapter explores four foundational domains of biomedical nanomaterial application (Fig. 6.7): ***D.1 targeted delivery***: site-specific transport of therapeutic agents to diseased tissues, reducing systemic exposure and improving pharmacokinetics; ***D.2 multimodal imaging***: enhanced diagnostic resolution through integration of optical, magnetic, and nuclear imaging modalities; ***D.3 theranostics***: unified platforms that combine diagnostic and therapeutic functions, enabling real-time monitoring and adaptive treatment strategies; and ***D.4 regenerative scaffolds***: extracellular matrix (ECM)-mimetic structures that support cellular adhesion, proliferation, and tissue repair. Together, these advances position nanomaterials as central to the evolution of personalized and regenerative medicine, offering modular, responsive, and clinically translatable solutions across acute and chronic conditions.

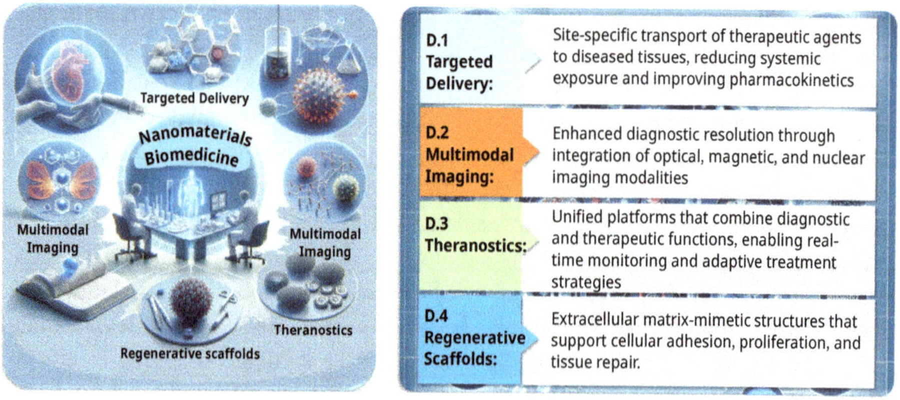

Fig. 6.7: A visual map of how nanomaterials transform biomedicine across targeted delivery, imaging, theranostics, and regenerative scaffolds, all grounded in their nanoscale precision and surface chemistry.

D.1. Targeted delivery: Nanomaterials enable precise drug delivery by leveraging their tunable size (10–200 nm), surface charge, and ligand functionalization to accumulate at disease sites while minimizing systemic toxicity selectively. Engineered NPs, such as liposomes, dendrimers, and polymeric micelles, can be modified with targeting moieties, such as antibodies, aptamers, or peptides, to recognize specific cellular receptors. This approach enhances pharmacokinetics and bioavailability, especially in cancer and cardiovascular therapies. For instance, silver and gold NPs functionalized with folate or transferrin have shown enhanced uptake in tumor cells via receptor-mediated endocytosis. Recent advances in innovative nanomaterials have introduced stimuli-responsive systems that release therapeutic agents in response to environmental cues such as pH, temperature, light, or enzymatic activity. These dynamic carriers, often built from polymers such as poly(*N*-isopropylacrylamide) (PNIPAAm) or PEGylated dendrimers, enable on-demand drug release, thereby reducing dosing frequency and improving patient adherence. Additionally, shape-memory and self-healing nanomaterials offer structural adaptability and resilience in complex biological environments, while bioresponsive platforms can detect and respond to disease-specific biomarkers. Together, these innovations are driving the development of multifunctional, personalized drug delivery systems that integrate sensing, targeting, and controlled release for optimized therapeutic outcomes [52].

Stimuli-responsive systems, activated by pH, temperature, enzymatic activity, or redox gradients, enable precise spatiotemporal control over drug release at pathological sites, minimizing systemic toxicity and enhancing therapeutic efficacy. These platforms are particularly effective at targeting microenvironments characterized by inflammation, hypoxia, or acidosis, which are common in tumors, vascular lesions, and infected tissues. Synthetic biology-inspired frameworks such as *SynBioNanoDesign* integrate programmable genetic circuits into nanocarriers, allowing dynamic sensing and logic-based actuation of therapeutic payloads. These systems can respond to intracellular cues or extracellular biomarkers, triggering drug release, gene silencing, or immunomodulation in real time. Intravascular targeting strategies, such as ligand-functionalized NPs and biomimetic carriers, have demonstrated reduced vascular inflammation, enhanced endothelial repair, and improved delivery of anti-thrombotic or anti-atherosclerotic agents. Extravascular systems, delivered via oral, inhalation, or subcutaneous routes, exploit enhanced permeability and retention (EPR) effects or receptor-mediated transcytosis to reach deep tissue compartments. These approaches support localized treatment of tumors, autoimmune disorders, and neurological diseases. Collectively, these innovations position nanomaterials as foundational to next-generation drug delivery systems, offering modularity, adaptability, and precision across diverse therapeutic landscapes [53].

D.2. Multimodal imaging: Nanomaterials enhance multimodal imaging by serving as multifunctional contrast agents across MRI, CT, PET, SPECT, and optical imaging platforms. Their high surface area and tunable surface chemistry enable conjugation with isotopes, fluorophores, and magnetic domains, allowing simultaneous anatomical and functional visualization. Gold NPs, quantum dots, and iron oxide nanocrystals have demonstrated improved resolution and sensitivity in cancer diagnostics, neurological mapping, and cardiovascular monitoring. For instance, quantum dots radiolabeled with ^{64}Cu and functionalized with RGD peptides have shown high tumor uptake and dual PET/NIRF imaging capability in preclinical models. Radiolabeled NPs offer distinct advantages over conventional tracers, including enhanced signal amplification, prolonged circulation time, and tumor-specific accumulation via the EPR effect. Chelator-free radiolabeling strategies preserve NP integrity and pharmacokinetics, enabling stable conjugation of radionuclides like ^{89}Zr, ^{68}Ga, and $^{99}{}_{m}$Tc for PET/SPECT imaging. In clinical studies, ^{89}Zr-labeled HDL NPs demonstrated safe administration and selective tumor accumulation in patients with esophageal cancer. These multifunctional nanoplatforms also support theranostic applications, integrating imaging and therapy for real-time monitoring and personalized treatment strategies [54].

Advances in NP design, particularly core–shell architectures and hybrid composites, have significantly enhanced the functionality of nanomaterials in biomedical imaging and therapy. These structures enable signal amplification by concentrating imaging agents within the core while functionalizing the shell for targeted delivery, biocompatibility, and controlled release. Hybrid composites, which integrate organic and inorganic components, allow simultaneous loading of diagnostic tracers and therapeutic molecules, facilitating co-localization at disease sites. Such multifunctional platforms support image-guided surgery, early disease detection, and longitudinal monitoring of treatment response with high spatial and molecular resolution. Recent studies have demonstrated that gold–silica core–shell NPs, conjugated with fluorescent dyes and chemotherapeutics, enable dual-mode imaging (e.g., NIRF and CT) and photothermal therapy, achieving tumor ablation with minimal off-target effects. Similarly, iron oxide-based composites functionalized with PEG and targeting ligands have shown enhanced MRI contrast and efficient clearance via renal or hepatic pathways, addressing concerns around long-term toxicity and biodistribution. These nanomaterials exhibit favorable pharmacokinetics, low immunogenicity, and scalable synthesis, positioning them as unified platforms for precision diagnostics and adaptive theranostics in both preclinical and clinical applications [55].

D.3. Theranostics: Theranostics merges diagnostics and therapy into a single nanomaterial-based platform, enabling real-time monitoring and targeted treatment. NPs, such as mesoporous silica, gold, and carbon-based systems, are engineered to carry both imaging agents and therapeutic payloads, allowing simultaneous disease detection and intervention [56]. These "smart" nanomaterials respond to external stimuli, such as light, pH, magnetic fields, or enzymatic activity, to release therapeutic agents selectively at the disease site, thereby minimizing systemic toxicity and enhancing treatment efficacy. Among them, gold nanoshells have gained prominence due to their tunable optical properties and biocompatibility. When functionalized with targeting antibodies and loaded with chemotherapeutics, these nanoshells can accumulate in tumor tissues via the EPR effect. Upon near-infrared (NIR) irradiation, they generate localized heat through surface plasmon resonance, triggering drug release and inducing photothermal ablation of cancer cells. As demonstrated in the referenced study, antibody-conjugated gold nanoshells achieved significant tumor suppression in vivo, with controlled release kinetics and minimal damage to surrounding healthy tissues. The system exhibited high stability, efficient cellular uptake, and precise thermal responsiveness, confirming its potential for integrated cancer theranostics. Such platforms exemplify the convergence of nanotechnology and personalized medicine, where stimuli-responsive behavior, molecular targeting, and multifunctional payloads enable adaptive, site-specific therapies with real-time monitoring capabilities [57]. Metal-based nanomedicines also support multimodal imaging and therapy, including photodynamic, sonodynamic, and immunotherapeutic strategies, by leveraging their unique optical, magnetic, and catalytic properties to enhance both diagnostic precision and therapeutic efficacy. These nanomaterials, ranging from metal oxides and alloys to metal–organic frameworks (MOFs), can be engineered to carry photosensitizers or sonosensitizers, enabling site-specific activation under light or ultrasound exposure. For example, gold and iron oxide NPs exhibit strong NIR absorption and localized surface plasmon resonance, facilitating photothermal and photodynamic cancer ablation. Additionally, metal-based systems such as MnO_2 and TiO_2 exhibit Fenton-like catalytic activity, generating ROS that mediate oxidative stress-induced tumor cell death [58]. Surface modifications improve biocompatibility, circulation time, and tumor penetration, while AI-guided design is emerging to optimize payload ratios and release kinetics. Theranostic nanomaterials are particularly impactful in oncology, where they enable personalized treatment and reduce the risk of recurrence. Despite translational challenges, including toxicity and manufacturing scalability, theranostics represents a paradigm shift toward integrated, adaptive, and patient-specific care.

D.4. Regenerative scaffolds: Nanomaterials are revolutionizing regenerative medicine by serving as bioactive scaffolds that mimic the ECM, promoting cell adhesion, proliferation, and differentiation. Materials such as graphene oxide (GO), CNTs, nanocellulose, and magnetic NPs (MNPs) are increasingly incorporated into 3D scaffolds to regenerate bone, cartilage, nerve, and skin tissues. These nanomaterials offer tunable mechanical strength, electrical conductivity, and surface chemistry, enabling scaffolds to match the physicochemical properties of native tissues. For instance, GO-enhanced hydrogels have demonstrated improved osteoblast adhesion and bone matrix formation, while CNT-reinforced composites support neural stem cell differentiation and electrical stimulation for nerve repair. Nanocellulose scaffolds, with their high porosity and biocompatibility, facilitate vascularization and collagen deposition in skin regeneration. MNPs embedded in scaffolds can be externally actuated to guide stem cell alignment or stimulate mechanotransduction pathways. Advanced fabrication techniques such as electrospinning, 3D bioprinting, and self-assembly enable precise control over scaffold architecture, porosity, and the distribution of bioactive molecules. In vivo studies confirm accelerated healing, reduced inflammation, and functional tissue integration, positioning nanomaterial-based scaffolds as a cornerstone of next-generation regenerative therapies [59].

These scaffolds exhibit tunable mechanical strength, porosity, and surface chemistry, enabling controlled release of growth factors and signaling molecules essential for tissue regeneration. Nanocellulose-based scaffolds, for example, provide a highly porous and biocompatible matrix that enhances osteoblast adhesion, proliferation, and bone matrix deposition. Their hydrophilic nature and nanoscale fiber architecture mimic the ECM, facilitating nutrient diffusion and cellular integration. When reinforced with MNPs, these scaffolds gain additional functionality, responding to external electromagnetic fields to stimulate stem cell differentiation and modulate cellular behavior. As reviewed in recent studies, MNPs such as superparamagnetic iron oxide NPs can activate mechanotransduction pathways and osteogenic signaling cascades, including Wnt/β-catenin, BMP/Smad, and PI3K/Akt, thereby promoting mesenchymal stem cell differentiation into osteoblasts. Magnetic stimulation also enhances scaffold–cell interactions, increases alkaline phosphatase activity, and upregulates osteogenic markers like Runx2 and collagen type I. These dynamic magnetic nanocomposite scaffolds not only improve mechanical integrity and cell viability but also enable spatial control over tissue formation, making them powerful platforms for bone repair, spinal fusion, and orthopedic implant integration [60].

Electrospinning, 3D bioprinting, and self-assembly techniques allow precise fabrication of hierarchical structures that replicate native tissue architecture, offering unprecedented control over scaffold geometry, porosity, and biofunctionality. Electrospinning produces nanofiber matrices with high surface-area-to-volume ratios that mimic the fibrous ECM and promote cell adhesion, proliferation, and differentiation. Multiaxial electrospinning enables the creation of porous, mechanically resilient scaffolds with tunable fiber alignment and bioactive surface modifications. 3D bioprinting, particularly extrusion, inkjet, and light-curing methods, allows spatially resolved deposition of bioinks containing cells, growth factors, and nanomaterials. These printed scaffolds can be customized to match patient-specific anatomical features and mechanical requirements. Emerging 4D bioprinting integrates stimuli-responsive materials that adapt over time, simulating dynamic tissue environments. Self-assembly techniques complement these approaches by enabling bottom-up construction of nanostructures through molecular interactions, forming organized layers and compartments that support tissue regeneration. When combined, these fabrication strategies facilitate the integration of osteoinductive agents, vascular networks, and stem cell niches, accelerating bone healing and functional recovery. Their synergy is critical for advancing regenerative medicine toward clinically viable, nanomaterial-enhanced tissue scaffolds [61]. Integration with stem cells and gene delivery systems further enhances regenerative outcomes. In vivo studies show accelerated healing and reduced inflammation, with promising results in orthopedic, dental, and neural applications. Challenges include biodegradation control, immune response modulation, and regulatory approval, but nanomaterial-based scaffolds are poised to redefine tissue engineering with precision, adaptability, and clinical relevance [62–63].

6.3 Renewable energy and nanotechnology

Renewable energy relies on sources continually replenished by nature, such as solar radiation, wind, hydrological cycles, geothermal heat, and biomass [64]. These fuels are converted into electricity, heat, chemicals, or mechanical power through technologies that have evolved significantly since early deployments [65]. In recent decades, nanomaterials (NMs) have emerged as transformative agents across energy platforms, offering enhanced conversion efficiency, reduced cost, and improved environmental compatibility. Nanomaterials possess high surface-area-to-volume ratios, tunable bandgaps, and catalytic activity, enabling precise control over charge transport, reaction kinetics, and interfacial dynamics. Their integration into energy systems from photovoltaics to fuel cells has led to measurable gains in performance, scalability, and sustainability. The following sections detail four key fields where nanotechnology is driving innovation. These fields include ***solar energy***, ***wind energy***, ***sustainable biofuels***, and ***hydrogen fuel cells*** [66], as shown in Fig. 6.8.

Nanomaterials improve solar energy conversion by enhancing spectral absorption, charge transport, and device flexibility. Quantum dots and perovskites enable tunable bandgaps and high carrier mobility, while TiO_2 and plasmonic structures boost dye-sensitized cell performance. Published results show 10–15% efficiency gains and $9\times$ absorption enhancement, positioning nanotechnology as central to next-generation solar platforms [67]. Nanomaterials enhance wind energy systems via nanogenerators and blade composites. ZnO-based devices convert mechanical energy efficiently, while carbon nanocomposites improve blade strength and thermal resilience. Nanolubricants reduce wear and friction, boosting turbine longevity. These innovations support decentralized energy production and smart grid integration, with published results confirming high durability and mechanical performance. Nanomaterials optimize biofuel production by enhancing catalytic activity and microbial conversion. Fe_3O_4 and CNTs reduce activation energy and boost biodiesel yield, while nanoscale biochar improves biogas generation. Published results show up to a 40% increase in methane yield and a 35% gain in biodiesel efficiency, confirming nanotechnology's role in scalable, clean bioenergy systems [68]. Nanomaterials enhance fuel cell performance by improving catalyst stability and membrane conductivity. Gallium-doped PtNi catalysts retain 96% ORR activity over 12,000 cycles, while NP-modified membranes reduce fuel crossover and boost proton transport. These advances support long-life, high-efficiency fuel cells for transport and stationary power applications [69]. Nanostructured adsorbents, such as activated carbon, MOFs, and metal oxides, enable CO_2 capture via physisorption and chemisorption. Moisture-swing DAC systems using Al_2O_3 and nanographite show fast kinetics and high capacity (up to 3.2 mmol/g CO_2). Pore size tuning and surface functionalization improve selectivity and regeneration. Electrochemical conversion using graphene-based membranes achieves > 85% Faradaic efficiency for CO_2-to-CO transformation [70]. NPs (e.g., SiO_2, Al_2O_3, and ZrO_2) reduce interfacial tension (IFT by 30–50%, alter wettability, and improve oil recovery. Optimal concentrations (0.1–0.5 wt%) enhance polymer/surfactant synergy. Field trials in Colombian reservoirs using Ni–Pd-doped Al_2O_3 nanofluids doubled oil output (from 15,502 to 30,682 bbl), improved API gravity (11.6° to 13.4°), and cut steam costs by 48%. Nanomaterials improve oil recovery by modifying IFT, wettability, and fluid dynamics. SiO_2 and Al_2O_3 NPs enhance chemical EOR, while field trials show a 98% increase in production and a 48% reduction in cost. These results validate nanotechnology's role in extending reservoir life and optimizing hydrocarbon extraction [71].

6.3.1 Solar energy

Solar energy remains one of the most promising renewable energy sources due to its abundance, scalability, and environmental compatibility. Conversion of sunlight into electricity is primarily achieved through two mechanisms: photovoltaics (PV) and concentrated solar power (CSP). In PV systems, solar cells, semiconductor devices, convert solar radiation directly into electrical energy via the photovoltaic effect. However, conventional solar cells face two significant limitations: relatively low efficiency (~10%) and high manufacturing costs [71]. Moreover, the levelized cost of electricity for solar technologies remains economically uncompetitive with fossil fuel-based systems in the absence of subsidies. To address these challenges, researchers continue to identify and develop more efficient and cost-effective technologies for solar energy harvesting and storage. Nanotechnology offers a transformative pathway to produce, store, and convert solar energy through clean, scalable, and high-performance approaches [72]. The integration of nanomaterials into solar cells has catalyzed the emergence of next-generation devices with enhanced functionality.

Among these, perovskite solar cells (PSCs) have garnered significant attention due to their tunable bandgap, high carrier mobility, and favorable film formation kinetics. Two-dimensional (2D) layered materials have been widely applied in PSCs to improve charge transport, interface stability, and overall device performance [73]. The combination of low-cost precursors and solution-based fabrication techniques position PSCs as strong contenders for future solar markets. Since their inception in 2009 with a power conversion efficiency (PCE) of 3.8%, PSCs have achieved laboratory-scale efficiencies of up to 23.3% [74] (Han *et al.*, 2018), rivaling commercial silicon, CIGS, and CdTe thin-film technologies. Nanomaterials used in solar applications exhibit desirable properties, such as high catalytic activity, aqueous stability, ease of synthesis, and material economy. Nonetheless, challenges remain in photocatalytic and photoelectrochemical systems, including long-term stability and scalability. Moore and Wei [75] highlighted the rapid evolution of metal halide-based PSCs, particularly those employing methylammonium lead halides ($CH_3NH_3PbX_3$) as leading candidates due to their record-breaking efficiencies and low production costs. In parallel, nanofluids engineered by suspending NPs in base fluids, have demonstrated enhanced thermal con-

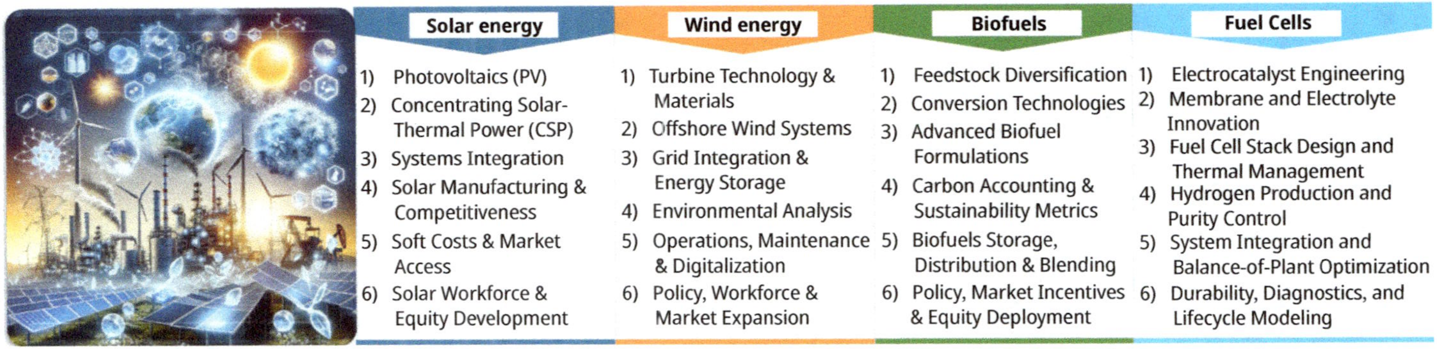

Fig. 6.8: Summary of the diverse applications of nanomaterials across renewable and transitional energy systems, emphasizing their role in sustainable energy conversion, storage, and environmental remediation.

ductivity and solar absorption. This improvement is attributed to the increased surface-to-volume ratio of NPs, which facilitates more efficient heat transfer and energy capture [76]. Recent reviews underscore the role of nanotechnology in boosting solar panel efficiency by up to 15% through innovations in quantum dots, nanowires, and perovskite structures [77]. Additionally, nanomaterials have been shown to enhance absorption of solar radiation by up to a factor of 9, resulting in a 10% improvement in solar collector efficiency compared to conventional designs. These findings affirm the potential of nanotechnology to overcome current limitations and accelerate the global transition to sustainable solar energy systems.

The U.S. DOE's Solar Energy Technologies Office (SETO) continues to play a pivotal role in advancing next-generation photovoltaic technologies, with a strategic focus on hybrid organic–inorganic PSCs. SETO's research and development portfolio targets four interrelated challenges: efficiency, operational stability, scalable manufacturing, and bankability, each of which must be addressed to transition perovskite photovoltaics from laboratory prototypes to commercially viable energy solutions. Recent SETO-supported projects have emphasized the need for robust materials architecture, long-term device durability, and standardized characterization protocols. These efforts are foundational for industry adaptation and certification. In particular, methylammonium lead halides ($CH_3NH_3PbX_3$, MALHs) have been extensively studied due to their high power conversion efficiencies (PCEs), often exceeding 19% in single-junction configurations (Fig. 6.9A). However, their susceptibility to degradation under moisture, heat, and UV exposure has prompted a shift toward more stable formulations, such as formamidinium-based and mixed-cation perovskites. SETO's technical roadmap includes developing defect-tolerant absorber layers, interface engineering to suppress ion migration, and encapsulation strategies to extend device lifetimes. For example, the Photovoltaic Accelerator for Commercializing Technologies (PACT), funded by SETO, has demonstrated minimodules (~25 cm^2) with aperture-area PCEs of 15–18% that retained over 80% of initial performance after five months of outdoor exposure. These results mark a significant improvement over early perovskite devices, which degraded within hours, and align with SETO's long-term goal of achieving an operational lifetime of 20–30 years for grid-scale deployment.

In parallel, SETO's Small Innovative Projects in Solar (SIPS) program has funded exploratory research into rollable PSCs, ITO-free transparent electrodes, and advanced metrology for defect quantification under environmental stressors. These initiatives aim to reduce manufacturing costs, improve mechanical flexibility, and accelerate reliability testing, key factors for commercial bankability and integration into diverse energy markets. In short, SETO's multipronged approach combines materials innovation, device engineering, and technoeconomic validation to reduce the risk of perovskite technologies. The office's support for high-throughput processing, tandem cell architectures, and predictive modeling underscores its commitment to achieving cost-effective, durable, and scalable solar energy solutions. With continued investment and cross-sector collaboration, hybrid PSCs are poised to become a cornerstone of the U.S. clean energy transition. This family of halides is a solid compound with a perovskite structure, with a space group: *P*1 (allows chirality; polar), crystal system: triclinic; *a*: 17.6784 Å, *b*: 8.8392 Å, *c*: 12.6948 Å, $\alpha = \beta = \gamma = 90.000°$, cell volume: 1,983.726 $Å^3$. A simplified schematic of a tandem PSC solar cell is shown in Fig. 6.9B. Fig. 6.9C shows performance evaluation of "a V-shaped perovskite/silicon tandem device based on a bifacial heterojunction silicon cell." The plots in (a) and (b) are the "*J–V* curves of the front side and rear side of the HIT SSC under illumination with different light intensities," respectively. Fig. 6.9C showed the "wavelength-dependent external quantum efficiencies and integrated current densities of the HIT SSC." Fig. 6.9C measured the "VOC vs. Light intensity curves of the HIT SSC. As of January 26, 2022, the efficiency records for perovskite PV cells are the current records: "26.7% for single-junction perovskite devices" and "29.8% for tandem perovskite-silicon devices, compared to other PV technologies, as shown in Fig. 6.9D. [78].

6.3.2 Wind energy

Wind energy arises from the movement of air masses driven by atmospheric pressure gradients, with velocities proportional to the difference between high- and low-pressure zones. Diurnal temperature variations, particularly the cooling of air masses over oceans, seas, and lakes relative to adjacent land, create predictable wind patterns that can be harnessed for power generation. Wind energy has matured into a reliable and cost-competitive technology, with global deployment accelerating in both onshore and offshore sectors. Alternative energy sources such as nuclear have been installed since 2005, contributing to the diversification of the energy mix [79-81]. Electricity generation in wind turbines is achieved through the aerodynamic force exerted on rotor blades, which drives the rotation of a shaft connected to a generator. The resulting electrical energy is then distributed to end-users (Fig. 6.10A). The turbine system comprises four primary components: the generator, tower, gearbox, and blades. Among these, the blade remains the most expensive and structurally demanding component due to its exposure to dynamic loads and environmental stressors. Recent advances in nanotechnology have enabled the development of high-performance blade materials with superior strength-to-mass ratios, improved fatigue resistance, and enhanced thermal stability. Muzammil *et al.* (2019) demonstrated that integrating nanocomposite materials into blade design, specifically CNT-reinforced polymers, can increase power ecoefficiency by 2.8 × when configured as a cross-axis wind turbine with a 45° deflector, compared to conventional vertical-axis designs [82]. Their study also reported a 70% increase in rotor rotational speed, indicating significant gains in converting kinetic energy to electrical energy. To further enhance energy harvesting under variable wind conditions, Li et al. [83] proposed a triboelectric–electromagnetic hybrid strategy. This approach combines the high sensitivity of triboelectric nanogen-

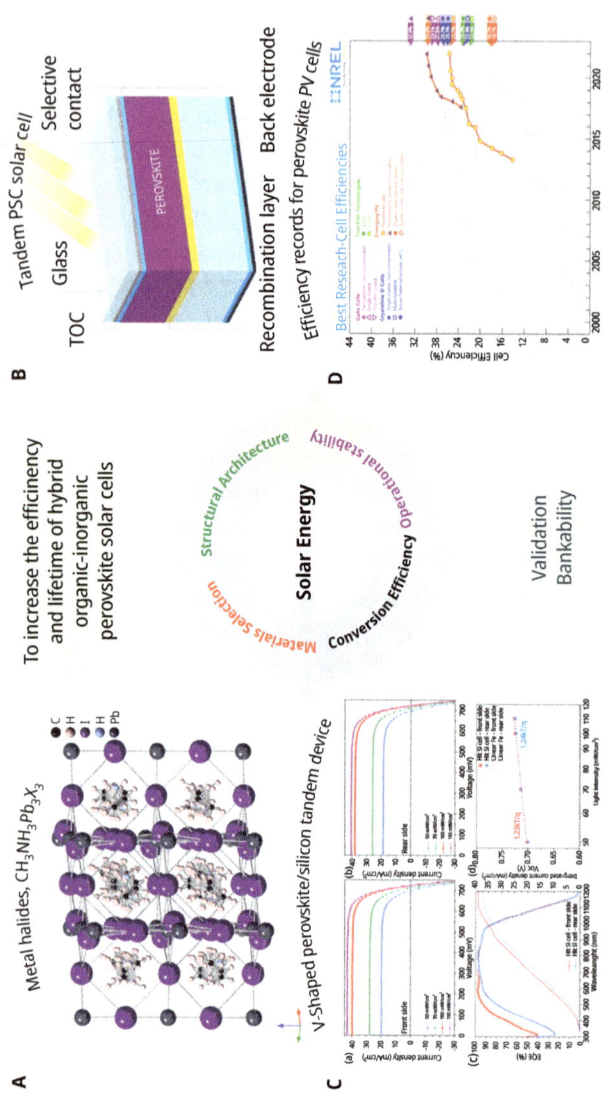

Fig. 6.9: The research and development suggested the efforts to address the challenges in stability, efficiency, manufacturing, validation, and bankability. Plots from C are adapted with permission from [79], and D from NREL.

erators (TENGs, Fig. 6.10B) with the robustness of electromagnetic generators (EMGs), allowing adaptive response to fluctuating wind speeds. The cooperative mechanism enables dynamic modulation of output voltage and current, improving energy capture efficiency under low-speed and turbulent wind regimes (Fig. 6.10C, D).

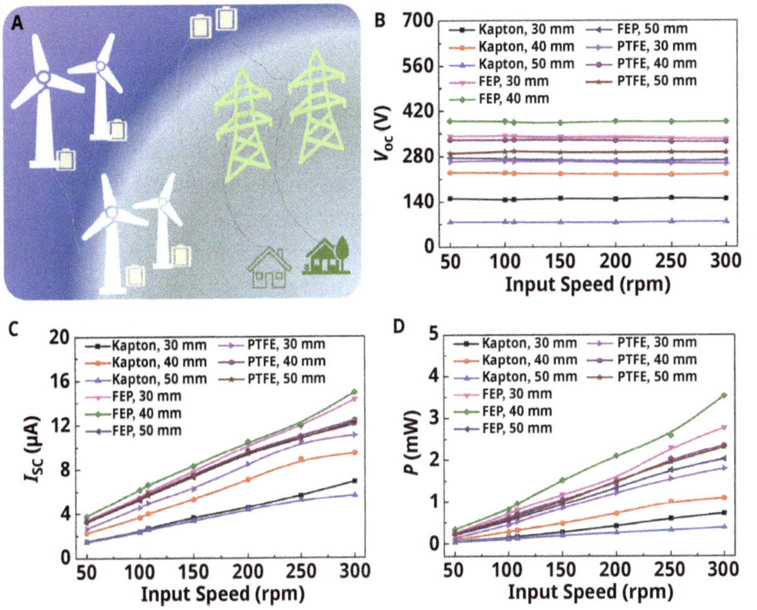

Fig. 6.10: The nanotechnology used to enhance wind energy harvesting and output ability of the triboelectric nanogenerator (TENG) module with different materials and different lengths of dielectric layers (40 mm): **A.** electricity generation by a wind turbine and the distribution of electricity; **B.** the open-circuit voltage of TENG; **C.** the short-circuit current; and **D.** the average power of output. Images of **B–D** were adapted with permission from [83] .

In 2024–2025, the U.S. Department of Energy's Wind Energy Technologies Office (WETO) launched a series of strategic funding initiatives to accelerate innovation in turbine design, materials science, and grid integration. These efforts are part of a broader push under the Energy Efficiency and Renewable Energy (EERE) portfolio, which received $3.460 billion in FY2025 appropriations to support decarbonization, electrification, and grid modernization. A key WETO initiative allocated $6.25 million to aerodynamic research on large-scale turbines, with a focus on blade optimization, wake-interaction modeling, and computational fluid dynamics (CFD) simulations. These studies aim to reduce wake losses by up to 20%, improve tip-speed ratios (λ) beyond 7.5, and extend operational lifetimes under turbulent wind conditions. Blade design improvements are supported by high-fidelity modeling of Reynolds numbers ($Re > 10^6$) and lift-to-drag ratios ($L/D > 100$), enabling more efficient energy capture across variable wind regimes [84]. In parallel, WETO invested $20 million in wind turbine recycling technologies to address end-of-life challenges and promote circularity. Research teams are developing thermoset-free blade composites and recyclable epoxy matrices reinforced with glass fiber, aiming for > 90% material recovery and $a > 50\%$ reduction in landfill waste. These efforts align with DOE's sustainability goals and support the development of a domestic recycling supply chain for wind infrastructure. Nanotechnology plays a critical role in enhancing turbine reliability and performance. Nanolubricants formulated with MoS_2, graphene, and hexagonal boron nitride (h-BN) exhibit ultra-low friction coefficients < 0.05), high load-bearing capacity ($P > 1$ GPa), and resistance to tribo-corrosion. These properties extend gearbox lifespans by 30–40%, reduce maintenance intervals, and improve energy efficiency [85]. Additionally, nanocoatings applied to blade surfaces offer advanced anti-icing and hydrophobic functionalities, reducing ice adhesion forces by > 80% and enabling continuous operation in subzero environments. WETO's support also extends to distributed wind systems, offshore wind R&D, and grid integration technologies, including real-time forecasting, power electronics, and hybrid energy storage. These innovations position nanotechnology as a cornerstone of next-generation wind energy systems, enabling higher efficiency, durability, and adaptability across diverse operating environments. A structured overview of strategic research, development, and deployment milestones supported by the U.S. DOE's Wind Energy Technologies Office, highlighting technical targets, funding priorities, and commercialization pathways across turbine design, materials innovation, grid integration, and offshore systems, as shown in Tab. 6.6.

Tab. 6.6: Multiyear program plan for wind energy technologies office: R&D and deployment roadmap (2024–2030).

Phase	Focus area	Technical targets	Deployment milestones
2024–2025	Materials and aerodynamics	Blade optimization ($L/D > 100$), wake loss reduction (≥20%), and CFD-based design validation	Launch of WETO aerodynamic R&D program, and field testing of recyclable blade prototypes
2025–2026	Reliability and sustainability	Nanolubricant deployment ($\mu < 0.05$), blade de-icing nanocoatings, and recycling recovery rate ≥ 90%	Pilot-scale recycling facilities and integration of nanomaterials in commercial turbines
2026–2028	Grid integration and forecasting	Real-time wind prediction (RMSE < 10%), hybrid storage coupling, and advanced inverter deployment	Regional grid modernization pilots and distributed wind hybrid systems
2028–2030	Offshore and high-capacity systems	Floating platform stability, 20 + MW turbine deployment, blade length > 120 m	Commercial offshore wind farms and DOE certification of next-gen platforms

6.3.3 Sustainable biofuels

Sustainable biofuels are low-carbon energy carriers produced from organic biomass such as agricultural residues, waste oils, algae, and forestry byproducts. They offer scalable alternatives to fossil fuels across transportation, industrial, and power sectors. The U.S. DOE, through its Bioenergy Technologies Office (BETO), leads national efforts to develop and commercialize advanced conversion pathways, including hydroprocessed esters and fatty acids (HEFA), alcohol-to-jet (AtJ), and Fischer–Tropsch synthesis, tailored for aviation, marine, and heavy-duty applications (IEA Bioenergy, 2024). BETO's portfolio supports integrated biorefineries, feedstock logistics, and catalytic upgrading technologies that transform biomass into fuels, chemicals, and electricity. These initiatives align with DOE's broader decarbonization strategy, emphasizing reductions in life cycle emissions, rural economic development, and energy resilience. In 2024, bioenergy contributed approximately 8.6% of the total U.S. energy supply, accounting for nearly half of all renewable energy generation. By advancing sustainable biofuel systems, BETO enables the transition to a circular bioeconomy, one that leverages domestic resources, reduces greenhouse gas emissions, and supports long-term energy security. Classification in sustainable bioenergy will be discussed as follows: ***Solid bioenergy***: Solid bioenergy uses feedstocks such as wood pellets and agricultural residues, converted through combustion, gasification, or pyrolysis to produce heat and power for buildings and industry. Technological improvements, such as enhanced pelletization and coal co-firing, have boosted energy density and cut CO_2 emissions by up to 35%, with global supply projected to reach 73 EJ by 2025, according to the International Energy Agency [86]. ***Liquid biofuels***: Liquid biofuels include biodiesel, bioethanol, and Sustainable Aviation Fuel (SAF), which are produced through advanced conversion pathways such as HEFA, AtJ, and Fischer–Tropsch synthesis for use in the transportation sector. By mid-2024, U.S. SAF output reached 52 million gallons, with DOE's SAF Grand Challenge aiming for 3 billion gallons annually by 2030, as AtJ fuels show over 70% life cycle GHG reductions compared to fossil jet fuels. ***Gaseous bioenergy***: Gaseous bioenergy, including biogas, biomethane, and syngas, is produced from organic feedstocks such as manure and food waste via anaerobic digestion and gasification, and is used for electricity generation, heating, transportation, and grid injection. By 2025, global biomethane output reached 10 EJ, with DOE-supported digesters boosting methane yields by 40% and upgraded systems achieving over 95% purity for pipeline distribution. ***Advanced biofuels***: Advanced biofuels are produced from unconventional feedstocks such as algae and lignocellulosic biomass using enzymatic, catalytic, and hybrid conversion methods to produce drop-in fuels compatible with existing infrastructure. Recent DOE-supported biorefineries have achieved over 50% carbon conversion efficiency. In comparison, algal pilot plants report lipid yields above 30% and net energy ratios exceeding 1.5, supporting IEA projections that advanced biofuels will comprise 75% of liquid bioenergy by 2050. In summary, four major classifications of sustainable bioenergy are tabulated in Tab. 6.7, highlighting that solid bioenergy is dominant in stationary heat and power, with pelletization and co-firing improving efficiency and emissions. Liquid biofuels, especially SAF, are central to transport decarbonization, with DOE's SAF Grand Challenge driving scale-up. Gaseous bioenergy offers flexible grid and fuel solutions, with upgraded biogas systems meeting pipeline standards. Advanced biofuels promise long-term scalability and compatibility with existing infrastructure, especially in the aviation and marine sectors.

Tab. 6.7: Comparative summary of the four major classifications of sustainable bioenergy, highlighting their feedstocks, conversion technologies, applications, and key technical parameters based on recent data (2020–2025).

Bioenergy type	Feedstocks	Conversion technologies	Primary applications	Technical parameters and results
Solid bioenergy	Wood pellets, agricultural residues, forestry byproducts, and energy crops	Combustion, gasification, and pyrolysis	Heat and power for buildings, industry, and district heating	Pellet energy density ↑ 15–20%, co-firing CO_2 emissions ↓ up to 35%, global supply: 73 EJ (IEA, 2025)
Liquid biofuels	Waste oils, corn stover, sugarcane, and algae	HEFA, alcohol-to-jet (AtJ), and Fischer–Tropsch synthesis	Road transport, aviation (SAF), and marine	SAF US output: 52 M gal (2024), DOE target: 3B gal/year by 2030, GHG reduction: >70% (AtJ fuels)
Gaseous bioenergy	Manure, food waste, wastewater sludge, and crop residues	Anaerobic digestion and thermal gasification	Electricity, heating, vehicle fuel, and grid injection	Biomethane purity:>95%, global output: 10 EJ (2025), methane yield ↑ 40% (DOE digesters)
Advanced biofuels	Algae, lignocellulosic biomass, and wet organic waste	Enzymatic hydrolysis, catalytic upgrading, and hybrid platforms	Drop-in fuels for existing infrastructure	Carbon conversion efficiency: >50%, algal lipid yield: >30% dry weight, net energy ratio: >1.5

6.3.3.1 Solid bioenergy

Solid bioenergy refers to the thermal conversion of biomass feedstocks, including wood pellets, agricultural residues, forestry byproducts, and dedicated energy crops, into usable heat and power. These feedstocks are processed through combustion, gasification, or pyrolysis to serve residential, industrial, and district heating systems. Recent innovations in pelletization, such as torrefaction and steam-explosion pretreatment, have increased energy density by 15–20%, reduced moisture content to below 10%, and improved grindability, thereby enhancing fuel logistics and combustion efficiency [87]. Co-firing solid bioenergy with coal in retrofitted power plants has demonstrated significant environmental benefits, with life cycle CO_2 emissions reduced by up to 35%, depending on the blend ratio and feedstock type. Studies published in confirm that optimized co-firing configurations can maintain boiler efficiency while reducing particulate and sulfur oxide emissions [88-90]. Additionally, advanced gasification systems operating at temperatures above 850 °C have achieved cold gas efficiencies exceeding 75%, enabling syngas production for combined heat and power (CHP) systems. According to the International Energy Agency [87], global solid bioenergy supply is projected to reach 73 exajoules (EJ) by 2025, with modern bioenergy systems increasingly replacing traditional biomass combustion in low-efficiency stoves and open fires. The World Bioenergy Association's 2023 report further highlights that solid biomass accounts for over 85% of global renewable heat production, with Asia contributing 40% of the global supply and Europe leading in wood pellet production [91]. Emerging research also explores the integration of solid bioenergy into circular economy frameworks, including ash valorization for soil amendment and carbon-negative pathways using biochar. These developments position solid bioenergy as a scalable, regionally adaptable solution for decarbonizing thermal energy systems while supporting rural economies and forest management [92].

6.3.3.2 Liquid biofuels

Liquid biofuels represent a critical class of renewable energy carriers designed to replace petroleum-based fuels across transportation sectors. This category includes biodiesel, bioethanol, and Sustainable Aviation Fuel (SAF), each produced through distinct conversion pathways such as HEFA, AtJ, and Fischer–Tropsch synthesis. These technologies enable the transformation of diverse feedstocks, including waste oils, lignocellulosic biomass, and microalgae, into drop-in fuels compatible with existing infrastructure, offering substantial reductions in life cycle greenhouse gas (GHG) emissions. SAF has emerged as a strategic priority in aviation decarbonization. By mid-2024, domestic SAF production in the United States reached approximately 52 million gallons, driven by the U.S. DOE's SAF Grand Challenge [93], which aims to scale annual output to 3 billion gallons by 2030. AtJ fuels have demonstrated life cycle GHG emissions reductions of more than 70% compared to conventional jet fuels, as validated by techno-economic assessments published in *Renewable Energy* [94] and *Nature* [95]. HEFA-derived SAF, produced from waste lipids, has achieved blend ratios of up to 50% in commercial aviation trials, with ASTM D7566 certification enabling seamless integration into existing fuel systems. Second-generation bioethanol, synthesized from lignocellulosic biomass, has shown conversion yields exceeding 300 liters per dry metric ton through consolidated bioprocessing and microbial strain engineering, as reported in *Biotechnology for Biofuels* [96]. Biodiesel, produced via transesterification of waste cooking oil and microalgae, has achieved cetane numbers above 55, enhancing combustion efficiency and cold-flow performance in marine and heavy-duty engines. These advancements reflect a broader shift toward feedstock diversification and process intensification in liquid biofuel production [97].

According to the International Energy Agency [98], global liquid biofuel output is projected to surpass 130 billion liters by 2030 under the Net Zero Emissions scenario, with advanced biofuels, including SAF and renewable diesel, expected to comprise over 60% of total production. Fig. 6.11 illustrates the roadmap of development and deployment of SAF, a low-carbon alternative to conventional jet fuel, produced from renewable feedstocks such as waste oils, agricultural residues, and biomass. It is compatible with existing aircraft and fueling infrastructure, enabling immediate emissions reductions without requiring fleet modifications. SAF can reduce life cycle greenhouse gas emissions by over 70%, with U.S. production reaching 52 million gallons by mid-2024 and targeted to scale to 3 billion gallons annually, by 2030, under the DOE's SAF Grand Challenge. Ongoing research and deployment efforts focus on improving conversion efficiency, expanding feedstock availability, and achieving cost parity with fossil-based aviation fuels. The DOE's Bioenergy Technologies Office (BETO) continues to support this trajectory through investments in integrated biorefineries, feedstock logistics optimization, and carbon capture-enabled pathways that enhance fuel sustainability and market readiness. Collectively, these developments position liquid biofuels as a cornerstone of low-carbon transportation strategies, particularly in sectors where electrification remains technically or economically constrained. Their scalability, infrastructure compatibility, and emissions performance make them indispensable to national and global efforts to achieve climate targets and energy resilience.

6.3.3.3 Gaseous bioenergy

Gaseous bioenergy encompasses biogas, biomethane, and syngas, renewable energy carriers derived from organic feedstocks such as livestock manure, food waste, wastewater sludge, and crop residues. These feedstocks are converted through anaerobic digestion and thermal gasification, producing gases that serve diverse applications, including electricity generation, residential and industrial heating, vehicle fuel, and direct injection into natural gas grids [99]. Anaerobic digestion remains the dominant pathway for biogas production, with co-digestion strategies that combine multiple feedstocks yielding up to 40% higher methane output, as demonstrated in DOE-funded pilot systems across the U.S. Midwest. Microbial enhancement techniques, including tailored inoculum blends and real-time pH control, have further improved process stability and volumetric productivity. Upgraded biogas systems now routinely achieve methane purities exceeding 95%, meeting pipeline injection standards and enabling biomethane to function as a drop-in substitute for fossil natural gas. Thermal gasification of dry biomass and organic waste at temperatures above 850 °C has enabled syngas production with cold-gas efficiencies exceeding 75%, supporting com-

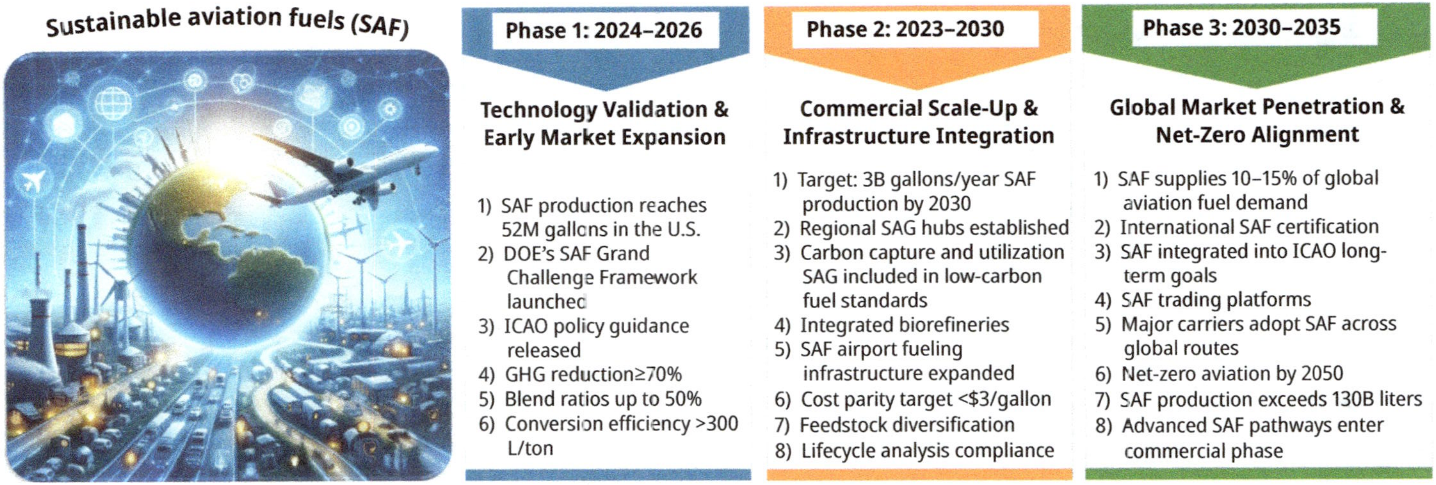

Fig. 6.11: Visual timeline of the Sustainable Aviation Fuel (SAF) Roadmap (2024–2035), illustrating key phases in technology validation, commercial scale-up, and global deployment aligned with decarbonization targets.

bined heat and power (CHP) configurations and synthetic fuel synthesis. According to the International Energy Agency's Outlook for Biogas and Biomethane (2025), global biomethane production reached 10 exajoules (EJ), with sustainable production potential estimated at 1,000 billion cubic meters equivalent (bcme), nearly a quarter of global natural gas demand [99]. Geospatial analyses published by the IEA reveal that 80% of the sustainable biogas potential lies in emerging economies, particularly India, China, and Brazil, where feedstock density and proximity to infrastructure favor decentralized deployment. In advanced economies, biomethane production costs range from USD 10 to USD 30 per gigajoule, with the lowest costs observed in Asia due to high availability of organic waste and favorable climate conditions. These findings underscore the strategic role of gaseous bioenergy in circular economy frameworks, energy security, and climate mitigation. Its ability to valorize organic waste while displacing fossil fuels positions it as a scalable, adaptable solution for low-carbon energy transitions.

6.3.3.4 Advanced biofuels

Advanced biofuels represent a transformative class of renewable energy carriers, synthesized from nontraditional feedstocks such as microalgae, lignocellulosic biomass, and wet organic waste. Unlike conventional biofuels derived from food crops, these fuels offer a sustainable and scalable alternative that aligns with global decarbonization goals and avoids competition with agricultural resources. Engineered as drop-in replacements, advanced biofuels are compatible with existing fuel infrastructure, enabling seamless integration into transportation, industrial, and energy systems.

Feedstock innovation and conversion technologies: The diversity of feedstocks used in advanced biofuel production reflects a strategic shift toward resource efficiency and waste valorization. Microalgae, for instance, offer high lipid content and rapid growth rates, while lignocellulosic biomass comprising agricultural residues, forestry byproducts, and energy crops provides abundant carbon-rich material for biochemical conversion. Wet organic waste streams, including food waste and wastewater sludge, further expand the feedstock base and support circular economy models. Conversion technologies have evolved to accommodate this complexity. Enzymatic hydrolysis enables the breakdown of cellulose and hemicellulose into fermentable sugars, while catalytic upgrading transforms intermediate bio-oils into hydrocarbon fuels. Hybrid biochemical-thermochemical platforms integrate biological and thermal processes to maximize carbon recovery and energy yield. Recent studies from the European Commission's Joint Research Centre (JRC) report carbon conversion efficiencies exceeding 52% in lignin-rich biomass using advanced enzymatic protocols, with integrated biorefineries demonstrating scalable throughput and reduced process energy intensity.

Algal biofuels and energy return metrics: Algal biofuels have gained prominence for their high productivity and low land requirements. Pilot-scale systems have achieved lipid yields exceeding 30% of dry biomass, with net energy ratios exceeding 1.5, indicating a favorable energy return on investment. These metrics are particularly promising in coastal regions and wastewater-integrated operations, where nutrient recycling and co-location strategies enhance system efficiency. Research published in *Algal Research* [100] and *BioEnergy Research* [101] underscores the potential of genetically optimized strains and photobioreactor designs to improve yield and scalability further.

Market dynamics and strategic outlook: The global advanced biofuels market is undergoing rapid expansion. According to the Advanced Biofuels Global Strategic Industry Report (2025), the sector was valued at $136.9 billion in 2024 and is projected to reach $965.1 billion by 2030, reflecting a compound annual growth rate (CAGR) of 38.5% [102]. This growth is driven by rising demand in aviation, maritime shipping, and long-haul transport sectors, where electrification remains technically constrained. Regulatory incentives, carbon pricing mechanisms, and international sustainability standards have further accelerated investment and deployment. The International Energy Agency [103] projects that advanced biofuels will comprise 75% of global liquid bioenergy by 2050, underscoring their central role in achieving net-zero emissions targets. National strategies, such as the U.S. DOE's Bioenergy Technologies Office [104] roadmap and the European Union's RED III directive [105], continue to support feedstock diversification, infrastructure integration, and life cycle emissions accounting.

Emerging technologies and sustainability pathways: Beyond current commercial platforms, a suite of emerging technologies is reshaping the future of advanced biofuels. Consolidated bioprocessing (CBP) integrates enzyme production, hydrolysis, and fermentation into a single step, reducing capital and operational costs. Electrochemical upgrading offers precise control over molecular transformations, enabling the synthesis of designer fuels with tailored combustion profiles. Carbon-negative pathways, such as biochar coproduction and CO_2 capture from fermentation, enhance the climate benefits of biofuel systems and support negative emissions strategies. These innovations position advanced biofuels as a cornerstone of next-generation energy systems. Their ability to deliver high-performance, infrastructure-compatible fuels with minimal environmental footprint makes them indispensable to global efforts in climate mitigation, energy resilience, and sustainable development.

6.3.4 Hydrogen fuel cells

Hydrogen fuel cells are electrochemical devices that convert hydrogen and oxygen into electricity, producing only water and heat as byproducts. Unlike combustion engines, fuel cells operate quietly and efficiently, with energy conversion rates reaching up to 60% for electricity alone and even higher when integrated with cogeneration systems. In polymer electrolyte membrane (PEM) fuel cells, the most widely used type for transportation, hydrogen is separated into protons and electrons at the anode; the electrons travel through an external circuit to generate power, while protons pass through the membrane to recombine with oxygen at the cathode [106]. Hydrogen fuel cells are used in a range of applications, from vehicles and portable electronics to backup power and distributed energy systems. Their efficiency far exceeds that of gasoline engines, which typically convert less than 20% of fuel energy into motion. As hydrogen can be produced from renewable sources such as solar and biomass, fuel cells offer a clean, scalable solution for decarbonizing energy systems across sectors, with growing relevance in climate policy and energy transition strategies [107].

Hydrogen (H_2) is an ideal alternative energy source from an ecological and environmental perspective. Hydrogen is one of the most abundant elements in the universe (also see ***hydrogen production***) (Zhang *et al.*, 2022), and it is stored in water (H_2O), hydrocarbons (e.g., methane, CH_4), and other organic matter [108]. Since the 1990s, the United States has begun promoting hydrogen fuel cell vehicles (FCVs) through legislation [108-109]. In 2014, Toyota showcased the mid-size FCV, the Toyota Mirai, in Los Angeles, which helped increase global FCV annual sales [110]. However, the hydrogen refueling station and its availability (Fig. 6.12A) hindered the utilization of HFCVs and their further commercialization.

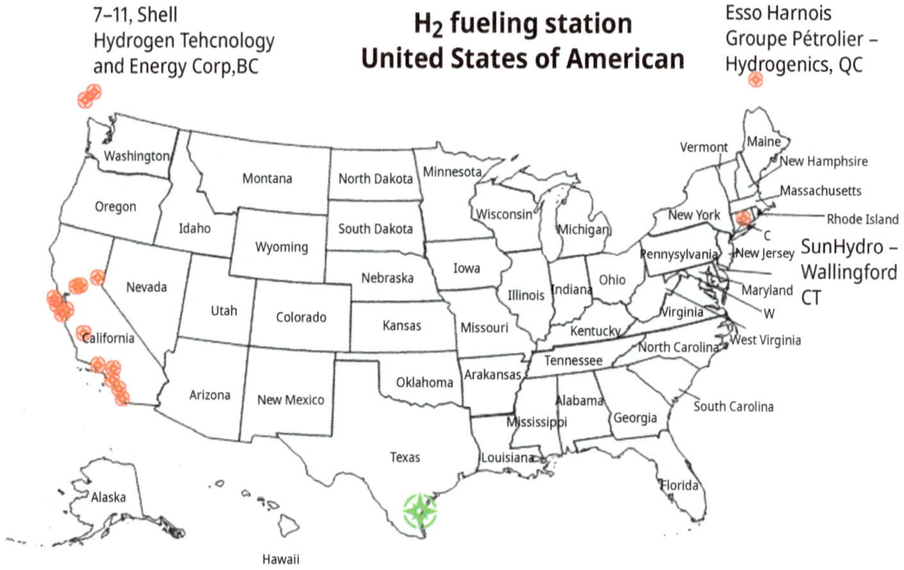

Fig. 6.12: A. The hydrogen fueling stations in the United States and Canada [106]. B. Five separate datasets to guide the priority of hydrogen facilities [108] (Zhang *et al.*, 2022).

By May 2020, some "operating hydrogenation fueling stations" had been established in several provinces of China, where hydrogen availability is expected to increase shortly [111]. Hydrogen technology has been developed in other countries, including England, France, and Canada, to achieve carbon neutrality. The failure of nuclear energy management is a driving force behind Japan's efforts to advance clean energy sources, with a focus on hydrogen technologies. The Japanese government issued the "Basic Hydrogen Energy Strategy" in December 2017 to promote the application of hydrogen energy [112]. This strategy includes ambitious goals to further spread about "800,000 FCVs by 2030, 160 hydrogen refueling stations by 2020, and 320 by 2025." Currently, Japan has become the first country to publish a "policy of subsidy for hydrogen vehicles" (Kongarchapatara *et al.*, 2020) [113]. Chen et al. applied multiple datasets for "data fusion and computation" to achieve two goals, "computation and analysis on the hotspot of hydrogen refueling station" and "location of hydrogen depot," as shown in Fig. 6.12A.

Five distinct datasets (Fig. 6.12B) were integrated to guide and evaluate hydrogen infrastructure development and deployment (Zhang *et al.*, 2020) [114]: (1) government policy data, including zoning regulations, energy incentives, and transportation mandates; (2) large-scale GPS trajectory datasets capturing real-time mobility patterns and vehicle flow; (3) geospatial land use data identifying industrial, residential, and renewable resource zones; (4) detailed road network data for routing, accessibility, and refueling logistics; and (5) spatial distribution of points of interest (POIs), including commercial, transit, and energy demand centers. This multisource framework aligns with the U.S. National Hydrogen Strategy and Roadmap, which emphasizes data-driven planning for hydrogen production, distribution, and end-use across sectors [114].

Government policy dataset: This dataset includes regulatory frameworks, zoning ordinances, transportation policies, and urban development plans sourced from municipal and national agencies. It provides temporal and spatial constraints that influence mobility behavior and infrastructure deployment. Policy layers were encoded using geospatial tags and temporal validity ranges to support simulation modeling. Integration with other datasets enabled scenario analysis under varying regulatory conditions, including congestion pricing and land-use reclassification.

Massive GPS trajectory: Dataset is collected from anonymized mobile devices and vehicle tracking systems, comprising billions of timestamped location points across urban and peri-urban regions. Each trajectory includes latitude, longitude, speed, and heading, enabling fine-grained analysis of travel patterns, modal shifts, and congestion dynamics. Data preprocessing involved map-matching algorithms and noise filtering to ensure spatial accuracy. Temporal resolution ranged from 1 to 5 s, supporting dynamic modeling of route choice and traffic flow.

Geographical land use data: This dataset categorizes spatial regions into residential, commercial, industrial, recreational, and mixed-use zones using satellite imagery and cadastral records. Land-use classifications were standardized using the CORINE and NLCD schemes, with a spatial resolution of 30 m. The data support analysis of origin–destination relationships, trip generation rates, and accessibility metrics. Temporal overlays allowed tracking of land-use transitions and urban sprawl over 10 years.

Geographical road network: Data derived from OpenStreetMap and municipal GIS layers, this dataset includes road geometry, hierarchy (e.g., arterial, collector, and local), speed limits, and intersection topology. Each road segment was encoded with connectivity attributes and traffic control metadata, enabling network-based routing and flow simulations. The dataset supports graph-based algorithms for shortest path, centrality, and bottleneck identification. Updates were synchronized quarterly to reflect construction, closures, and infrastructure upgrades.

POI: This dataset maps the spatial distribution of commercial, cultural, educational, and healthcare facilities using crowd-sourced and institutional data. Each POI entry includes geolocation, category, operating hours, and service capacity. POIs were clustered using DBSCAN and K-means algorithms to identify activity centers and mobility attractors. The dataset informs trip-purpose classification, pedestrian-flow modeling, and accessibility scoring across demographic groups.

Hydrogen is a high-energy-density fuel with a higher heating value of 141.8 MJ/kg at 298 K, nearly three times that of gasoline, making it highly attractive for energy-intensive applications [115-116]. Its combustion produces only water vapor, resulting in zero direct CO_2 emissions, positioning hydrogen as a cornerstone of clean energy transitions. Global hydrogen production exceeds 120 million tons annually, with approximately two-thirds classified as pure hydrogen. However, 95% of current production is derived from fossil sources, primarily steam methane reforming (SMR) and coal gasification, while only 5% originates from electrolysis, often linked to chlorine manufacturing [117]. Hydrogen can be produced from both fossil-based and renewable feedstocks, including natural gas, biomass, solar, and wind-powered electrolysis [118]. Emerging technologies such as photoelectrochemical water splitting and photocatalytic hydrogen generation utilize semiconductor nanomaterials to convert solar energy directly into hydrogen, achieving solar-to-hydrogen efficiencies of 5–10% in lab-scale systems. For storage, solid-state hydrogen carriers, including metal hydrides and porous nanomaterials, offer volumetric densities exceeding 50 kg H_2/m^3, with enhanced safety and reversibility. In end-use systems, proton exchange membrane fuel cells (PEMFCs) fabricated with nanomaterial-based catalysts (e.g., Pt alloys and graphene supports) have demonstrated power densities above 1 W/cm^2 and operational lifetimes exceeding 5,000 h [119]. These advances support the deployment of hydrogen across transportation, stationary power, and portable electronics, in alignment with national strategies for decarbonization and energy resilience.

In summary, energy, food, and water systems are deeply interdependent, forming a dynamic and reciprocal framework known as the energy–food–water (EWF) nexus. This nexus offers a systems-level perspective on resource management, where interventions in one domain, such as energy infrastructure, can significantly influence water availability and food production. Nanomaterials (NMs) have emerged as critical enablers in enhancing the performance, efficiency, and resilience of the EWF nexus, particularly in arid and resource-constrained regions [120]. In solar energy development, NMs, such as perovskite quantum dots and plasmonic NPs, have improved light harvesting and thermal regulation, enabling dual-use systems, such as agrovoltaics, that optimize land for both energy and crop yields (water-energy-food.org, 2025). In wind energy, nanostructured coatings and ZnO-based nanogenerators have enhanced blade durability and micro-scale energy harvesting, thereby supporting decentralized water pumping and irrigation systems [121]. Sustainable biofuels benefit from NM-enabled enzymatic catalysis and biomass conversion platforms, which reduce water intensity and improve fuel yields from agricultural residues [122]. In hydrogen fuel cell development, NM-engineered catalysts and membranes have increased power density and reduced water consumption in electrolysis and fuel cell operation, aligning with low-water-intensity energy goals.

6.4 Green methods of synthesis

NP synthesis is broadly categorized into two foundational approaches: the bottom-up approach, which relies on atomic or molecular self-assembly through physical processes such as vapor deposition, sol–gel techniques, and laser ablation; and the top-down approach, which involves the mechanical or chemical breakdown of bulk materials into nanoscale structures, often using milling, lithography, or etching methods [123]. These approaches differ not only in mechanism but also in scalability, surface control, and energy intensity. Capping agents play a pivotal role in NP fabrication, acting as stabilizers and functional modifiers. By selectively binding to NP surfaces, these agents regulate morphology, particle size distribution, and surface charge, thereby preventing aggregation and enhancing colloidal stability [124]. Common capping agents include thiols, amines, and polymers, which can be tailored to impart biocompatibility, catalytic activity, or optical properties, depending on the application. Nanomaterial (NM) synthesis is further classified into traditional and green methods. Traditional methods often involve toxic solvents, high-temperature reactions, and hazardous reagents, posing risks to both the environment and human health. These techniques may yield high-purity NMs but generate significant waste and require extensive post-processing. In contrast, green synthesis leverages biological systems, such as plant extracts, microbial cultures, and enzymes, to produce NMs under ambient conditions, minimizing energy input and eliminating the need for external stabilizers or harsh chemicals [125]. Recent studies have reinforced the advantages of green synthesis. For instance, Anbalagan et al. [126] demonstrated the use of *Azadirachta indica* leaf extract to produce silver NPs with antimicrobial properties, achieving high yield and stability without synthetic additives. A 2025 review by Kirubakaran et al. [127] emphasized the biomedical and environmental benefits of green-synthesized NPs, highlighting their enhanced biocompatibility and reduced toxicity profiles. Moreover, Baig et al. [128] reported scalable protocols for green synthesis using algae-derived polysaccharides, achieving consistent particle size and functionalization for drug delivery applications. These advancements underscore the transformative potential of green nanotechnology in achieving sustainable, application-specific NP production.

To enhance functional properties and achieve controlled morphology, sol–gel-based green synthesis has emerged as a versatile and environmentally benign method. This approach enables low-temperature processing, uniform particle formation, and minimal use of hazardous reagents. When coupled with physical activation techniques such as microwave irradiation, ultrasound-assisted dispersion, and pulsed laser ablation, sol–gel synthesis can produce NPs with tailored crystallinity, porosity, and surface area [130–131]. For example, Nguyen *et al.* (129) demonstrated that ultrasound-assisted sol–gel synthesis of ZnO NPs yielded hexagonal wurtzite structures with particle sizes ranging from 22 to 30 nm and enhanced photocatalytic activity under UV light [132]. Similarly, González-Lavín et al. [133] reported that microwave-assisted sol–gel synthesis of iron-based aerogels achieved homogeneous heating and improved mechanical integrity, with vessel geometry playing a critical role in reaction uniformity and scale-up feasibility. Comprehensive characterization of synthesized NPs is essential for validating their structural, optical, and surface properties. Techniques commonly employed include DLS for hydrodynamic size and zeta potential; scanning electron microscopy (SEM) and TEM for morphology and lattice resolution; energy-dispersive X-ray spectroscopy (EDS) for elemental composition; and UV-Vis spectroscopy for optical bandgap estimation. Crystallographic analysis is performed using XRD, while FTIR identifies functional groups and bonding interactions. Advanced surface and electronic structure assessments utilize XPS, AFM, high-angle annular dark field (HAADF) imaging, and atomic absorption spectroscopy (AAS) [17]. These tools collectively enable precise correlation between synthesis parameters and NP performance in catalytic, biomedical, and energy applications.

6.4.1 Physical and chemical methods

Recent advancements in NP synthesis have significantly shifted toward safer, more sustainable, and economically viable methodologies. Traditional physical and chemical approaches, such as CVD, sol–gel processing, and thermal decomposition, often require high temperatures, toxic reagents, and extended reaction times, leading to hazardous byproducts and increased energy consumption. These limitations have prompted a transition toward green synthesis frameworks that integrate green chemistry principles, aiming to minimize environmental impact while maintaining control over NP morphology, crystallinity, and functionality. In physical methods, innovations such as microwave-assisted synthesis, ultrasound irradiation, and pulsed laser ablation in liquids (PLAL) have demonstrated rapid reaction kinetics, reduced solvent usage, and enhanced product uniformity. For example, microwave-assisted techniques enable volumetric heating and dielectric coupling, producing metal oxide NPs with particle sizes below 50 nm in less than 10 min, while PLAL offers solvent-free fabrication of high-purity colloidal dispersions without surfactants or stabilizers.

Chemical green synthesis has increasingly adopted biogenic routes utilizing plant extracts, microbial cultures, and biopolymers as reducing and capping agents. These methods operate under ambient conditions and eliminate the need for hazardous chemicals, yielding biocompatible NPs suitable for biomedical and environmental applications. As highlighted in recent reviews [134], biosynthesized NPs exhibit enhanced surface functionality and reduced toxicity, with promising results in drug delivery, antimicrobial coatings, and catalytic systems. Despite these advances, challenges remain in scaling up green synthesis protocols, ensuring reproducibility, and optimizing reaction mechanisms. Ongoing research emphasizes the integration of renewable feedstocks, real-time monitoring, and life cycle assessment to support the development of a nanocircular economy where NP production, application, and disposal are harmonized with ecological and economic goals. As the field matures, green synthesis is poised to become a cornerstone of sustainable nanotechnology, bridging innovation with environmental stewardship.

Among these, microwave-assisted synthesis (Section 6.4.1.1) has emerged as a leading technique due to its ability to deliver rapid, volumetric heating through dielectric interaction, enabling uniform nucleation and accelerated reaction kinetics. This method reduces energy consumption and minimizes thermal gradients, which are common in conventional convective heating systems. Studies have demonstrated that microwave-assisted synthesis can produce metal oxide NPs with controlled crystallinity and particle sizes ranging from 20 to 80 nm in under 10 min, with reduced agglomeration and improved surface functionality [135]. In parallel, PLAL (Section 6.4.1.2) has gained traction as a solvent-free, contamination-resistant method for NP fabrication. PLAL operates under ambient conditions and does not require chemical precursors or stabilizers, making it ideal for producing high-purity colloidal dispersions. The ultra-short laser pulses (femtosecond to nanosecond range) induce localized plasma formation and rapid material ejection from the target surface, followed by quenching in the surrounding liquid. This process enables the synthesis of spherical and low-dimensional nanostructures with tunable size distributions (typically 10–100 nm), depending on pulse energy, duration, and liquid composition [136-137]. Recent work has shown that PLAL can produce TiO_2-, Au-, and graphene-based NPs with distinct optical and catalytic properties, suitable for biomedical and energy applications.

6.4.1.1 Microwave-enhanced synthesis

Microwaves are electromagnetic waves composed of pure radiative energy, typically propagating at the speed of light in vacuum. However, in condensed media, propagation speed decreases due to refraction, governed by the material's refractive index and dielectric properties. This interaction enables localized heating and selective activation of reactants, which is particularly advantageous for NP formation. Microwave synthesis parameters, especially power intensity, irradiation time, and reaction vessel geometry, critically influence NP morphology, crystallinity, and yield. Studies have shown that adjusting microwave power from 300 to 1,200 W and exposure times from 30 s to 10 min can produce NPs with controlled size distributions ranging from 10 to 100 nm, depending on precursor concentration and solvent polarity. Additionally, microwaves are frequently coupled with hydrothermal synthesis apparatus, forming microwave-assisted solvothermal systems that operate under elevated pressure and temperature, enabling the synthesis of complex nanostructures with high phase purity. The most commonly used microwave frequencies for NP synthesis range from 2.45 to 45 GHz, where dielectric loss, penetration depth, and temperature-dependent permittivity play key roles in energy absorption and heat generation [138]. For example, Schmidt et al. [139-140] demonstrated that microwave-assisted hydrothermal synthesis of TiO_2 NPs yielded anatase-phase crystals with enhanced photocatalytic activity and a narrow particle size distribution of 18–25 nm, outperforming conventional hydrothermal methods. Similarly, González-Lavín et al. [133] reported that microwave-assisted synthesis of iron-based aerogels yielded uniform porosity and mechanical stability, with surface areas exceeding 600 m^2/g, making them suitable for energy storage and catalysis. Dr Liu and her team successfully synthesized MOF motifs using microwave-assisted solvothermal techniques, achieving high crystallinity, uniform pore distribution, and distinctive structural features (Fig. 6.13A–C). Their approach demonstrated reaction times reduced by over 80% compared to conventional solvothermal methods, with BET surface areas exceeding 1200 m^2/g and tunable pore sizes optimized for gas adsorption and separation applications. These results underscore the efficacy of microwave-assisted synthesis in producing advanced nanomaterials with tailored properties for energy, environmental, and biomedical applications [141-142].

6.4.1.2 Pulsed laser ablation in liquids

The PLAL method, a clean, versatile, and solvent-mediated technique, allows the fabrication of NPs without byproducts by synthesizing NPs without the use of chemical precursors or surfactants, thereby eliminating secondary byproducts. In ***PLAL***, ultrashort laser pulses, ranging from femtoseconds to nanoseconds, irradiate a solid target submerged in a liquid medium, generating localized high-temperature plasma and inducing rapid material ejection (Fig. 6.14A). This process yields high-purity colloidal NPs with tunable size and surface properties, suitable for post-synthetic functionalization [143, Shabalina *et al.*, 2021). Dell'Aglio and De Giacomo (2020) demonstrated the synthesis of silver NPs (AgNPs) using a nanosecond pulsed laser system (Quanta system PILS-GIANT, 6 ns pulse duration) within a stainless steel chamber. The system op-

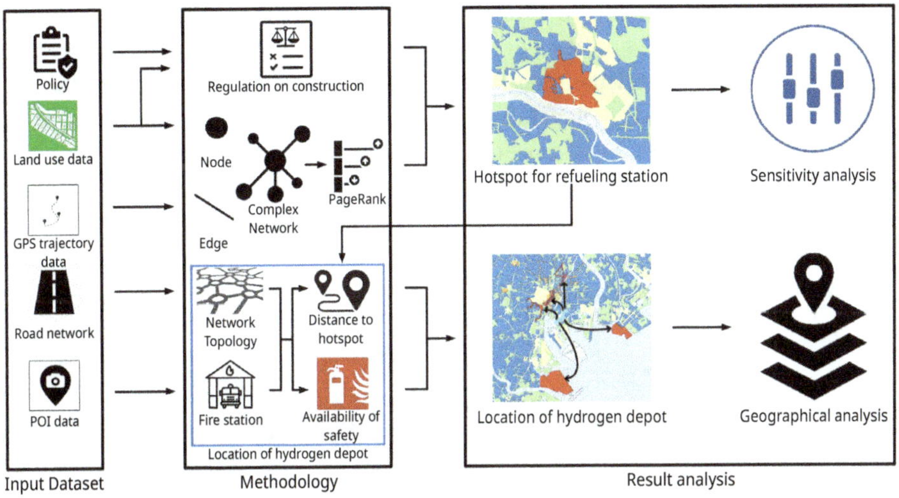

Fig. 6.13: The microwave-assisted solvothermal chemistry to prepare Co-MOFs used as disinfectants: **A.** tetragonal space group *P42₁m* with dimetric μ_2-H_2O-centered basic carboxylate cluster of $[Co_2(H_2O)(O_2CR)_4(H_2O)_4]$; **B.** TEM image of Co-MOF motifs; and **C.** zoomed-in TEM image of Co-MOF motifs [108].

erated at the second harmonic wavelength (532 nm), with a repetition rate adjustable between 1 and 10 Hz and pulse energies ranging from 50 to 217 mJ, depending on the experimental configuration (Fig. 6.14B). The target was immersed in 400 mL of deionized water and mounted on a rotating holder to ensure uniform ablation. A 4.0 cm focal lens was positioned within a spacer tube opposite the target to concentrate the beam. Under these conditions, AgNPs with diameters between 15 and 40 nm were obtained, exhibiting narrow size distributions and high colloidal stability (Fig. 6.14C). Spectroscopic analysis confirmed the absence of residual contaminants, and TEM imaging revealed spherical morphology with crystalline lattice fringes [144].

Recent studies have expanded ***PLAL***'s capabilities to include low-dimensional nanomaterials such as quantum dots, nanorods, and nanosheets. For example, Ye and Musselman [138] synthesized GO quantum dots via femtosecond ***PLAL*** in ethanol, achieving particle sizes below 10 nm with strong PL and oxygen-rich surface functionalities suitable for bioimaging and catalysis. Naderi-Samani and Razavi [137] reported that varying the liquid medium (e.g., acetone, SDS, or polymer solutions) significantly influenced NP morphology and dispersibility, with dielectric constant and viscosity acting as key modulators of plasma–liquid interactions. In parallel, ultrasound-assisted NP synthesis has gained traction due to its ability to operate under mild conditions while achieving high yields and uniform particle dispersion. The core physical mechanisms, acoustic cavitation and nebulization, involve the formation, growth, and implosive collapse of microbubbles, generating localized hotspots (>5,000 K) and pressures (>1,000 atm) that facilitate rapid nucleation and fragmentation [130, 145]. These microdroplets act as transient nanoreactors, enabling the synthesis of non-aggregated

NPs with controlled morphology. Manjamadha and Muthukumar [146] demonstrated that ultrasound-assisted synthesis of ZnO NPs reduced reaction time by over 60% compared to conventional methods, yielding particles with average diameters of 20–30 nm and enhanced photocatalytic activity. Together, ***PLAL*** and ultrasound-assisted synthesis represent two powerful, green nanofabrication strategies that offer precise control over NP characteristics while minimizing environmental impact. Their scalability, reproducibility, and compatibility with diverse target materials and solvents make them ideal for applications in catalysis, biomedicine, energy storage, and environmental remediation.

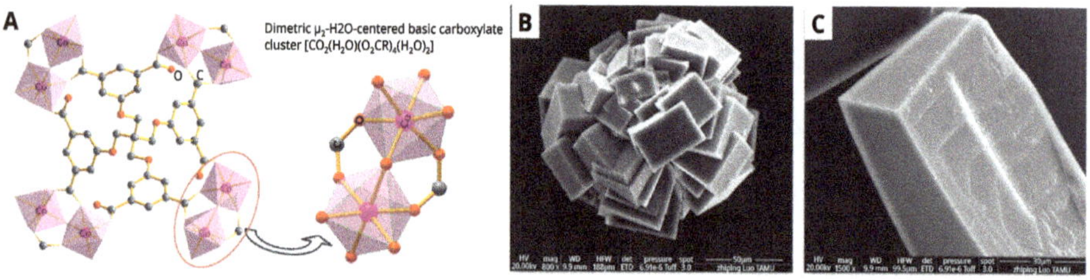

Fig. 6.14: A. Pulsed laser ablation in liquids (PLAL) was used to produce nanostructured silver metal. Experimental setup. **B.** Temporal resolved shadowgraph images of the laser-induced bubble on Ag target immersed in water, obtained at different laser energies, 217, 156, 108, and 60 mJ, respectively. **C.** Surface plasmon resonance spectra of AgNPs colloidal solutions produced with different laser energies. With permission from [141].

6.4.2 Biological methods

Biological entities, including plants, bacteria, fungi, and enzymes, are increasingly recognized for their ability to synthesize NPs through both extracellular and intracellular pathways. These living systems leverage intrinsic biochemical mechanisms to remodel inorganic metal ions into nanoscale structures, opening new frontiers in bio-inspired nanotechnology and molecular-level catalysis [141]. The biosynthesis process typically follows a bottom-up approach, involving either reduction or oxidation reactions mediated by biomolecules such as proteins, polysaccharides, and secondary metabolites [147]. This "green method" of NP production is inherently safe, nontoxic, and environmentally benign, avoiding the use of hazardous solvents, high temperatures, or external stabilizers. It utilizes natural sources such as microbial cultures, plant extracts, and purified enzymes as both reducing and capping agents, enabling the synthesis of stable, functionalized NPs with controlled morphology and surface chemistry [148]. For example, biosynthesized AgNPs using *Azadirachta indica* leaf extract have demonstrated particle sizes ranging from 10 to 40 nm, with potent antimicrobial activity and minimal cytotoxicity.

6.4.2.1 Using plants

Plants, in particular, serve as efficient and low-maintenance biofactories for NP synthesis. Their phytochemical diversity enables selective reduction of metal ions and surface passivation. At the same time, their natural detoxification pathways enable the accumulation and transformation of heavy metals, even at trace concentrations that are toxic to most organisms [148]. This property has been exploited in environmental remediation, where plant-mediated NPs facilitate the removal of lead, cadmium, and arsenic from contaminated water and soil. Numerous plant species have been documented for their NP synthesis capabilities, including *Aloe vera*, *Moringa oleifera*, *Camellia sinensis*, and *Ocimum sanctum*. These systems offer rapid, eco-friendly, and nonpathogenic protocols that often require only a single step, typically mixing a plant extract with a metal salt solution under ambient conditions [149]. The resulting NPs exhibit high stability and functional versatility, suitable for applications in catalysis, drug delivery, biosensing, and agriculture. The bioactive compounds responsible for NP synthesis include flavonoids, alkaloids, phenols, saponins, carbohydrates, proteins, quinine, glycosides, tannins, and steroids, which act synergistically to reduce metal ions and stabilize the resulting nanostructures [150, 152-153]. Recent studies have shown that adjusting parameters such as pH (optimal range: 6.5–8.0), temperature (25–60 °C), and extract concentration can fine-tune NP size, shape, and dispersity. For instance, a 2025 study published in *BioNanoScience* demonstrated that *Moringa oleifera*-mediated synthesis of ZnO NPs yielded hexagonal particles with an average diameter of 28 nm, exhibiting enhanced photocatalytic degradation of methylene blue under UV irradiation [150]. As the field advances, biological synthesis methods are being integrated with real-time monitoring, high-throughput screening, and machine learning models to optimize yield, reproducibility, and scalability. These innovations position green biosynthesis as a cornerstone of sustainable nanotechnology, bridging ecological responsibility with industrial relevance. The majority of active chemical constituents of the above natural products are tabulated in Tab. 6.8. In addition, the reaction rate is much faster than that of the microorganism-based methods. The functional groups in these compounds are actively involved in bonding or interacting with the reactants of the targeted products. They also act as surfactants, creating spatial hindrances that prevent particle growth into large clusters. More importantly, the richness of plant extracts by molecules is known for their antioxidant power, which participates in the process of reduction and stabilization of NPs during biosynthesis due to the presence of some functional groups (like –O–H, –C–O–C, –C=O, –C=C) [151]. Different parts of plants are employed in the bioformation of NPs (leaf, stem, bark, flower, root, seed, and fruit) owing to the exceptional metabolites they produce. The leaves are the most widely used part, as they are the site of photosynthesis and are rich in various metabolites [154]. The biochemical and molecular mechanisms of NP biosynthesis remain poorly characterized and should be investigated to optimize further the process [155].

Tab. 6.8: A summary of plant water extracts used to synthesize nanomaterials.

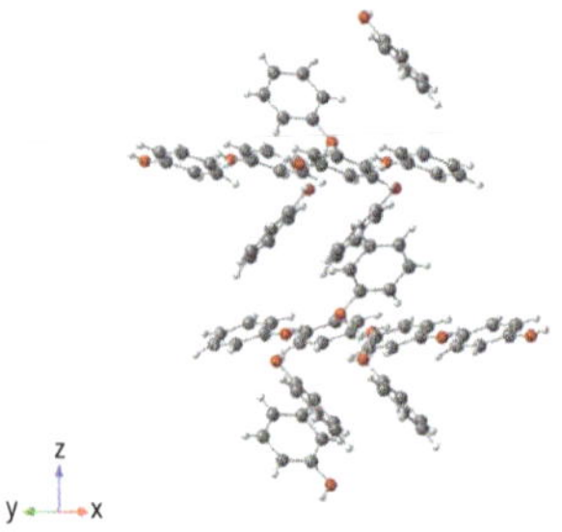	Phenol, C_6H_5OH Space group: $P112_1$ (chirality; polar) Crystal system: Monoclinic [156] a: 6.0500 Å b: 8.9250 Å c: 14.5940 Å β: 90.000 °Cell volume: 788.021 $Å^3$ Asymmetric unit: 39 sites Unit cell: 78 sites/unit cell Density: 1.1898 g/cm^3
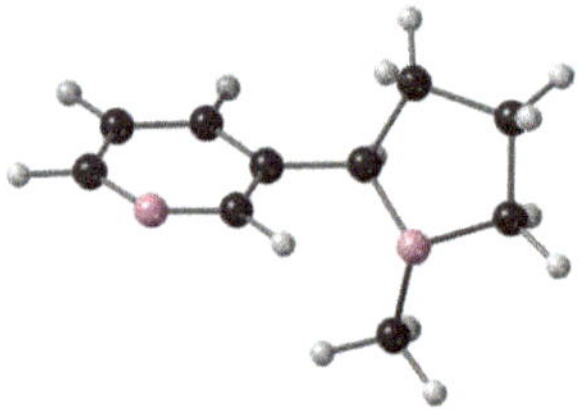	Nicotine (alkaloids) [157] A naturally produced alkaloid in the nightshade family of plants; constituting approximately 0.6–3.0% of the dry weight of tobacco; Highly addictive unless used in slow-release forms; using it as a tool for quitting smoking has a good safety history.
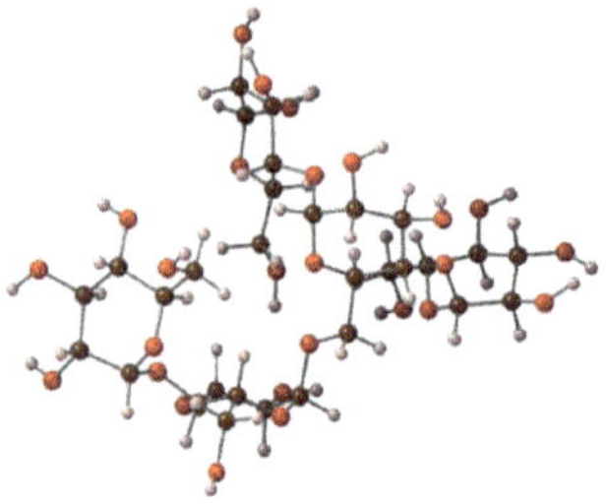	Glycogen (a polysaccharide carbohydrate) The multibranched polysaccharide of glucose, one of two forms of energy reserves, is stored in the body, particularly in the muscles and liver. Stored in skeletal muscle, serves as a form of energy storage for the muscle itself, and is the analog of starch, a glucose polymer [158].
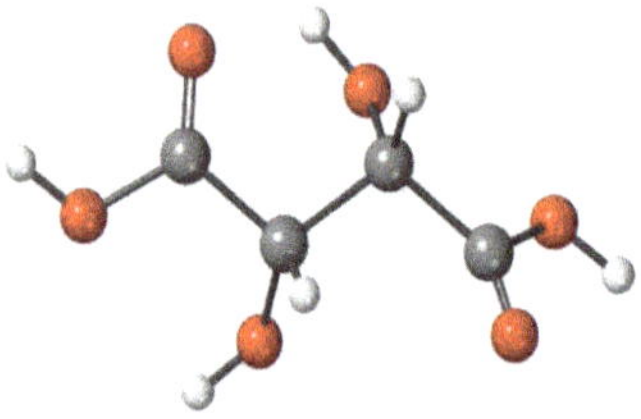	Tartaric acid (2,3-dihydroxy-butanedioic acid, $C_4H_6O_6$) is a chiral molecule with two forms: L and D, of which the D form is the naturally occurring form and is illustrated here [159].

Tab. 6.8 (continued)

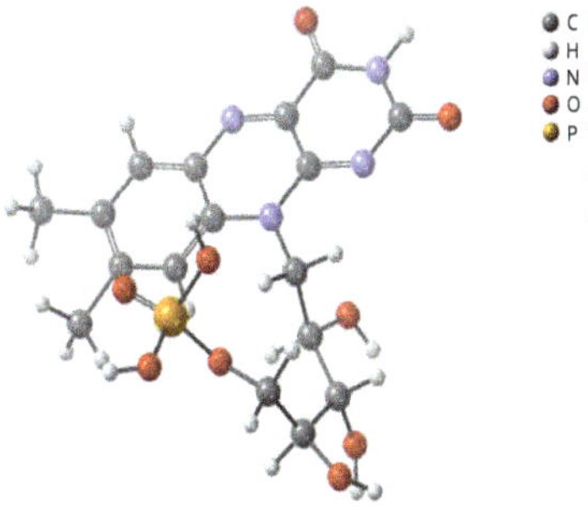	Flavin mononucleotide (E101a, $C_{17}H_{21}N_4O_9P$) is a flavin mononucleotide that is an orange-red food color additive. It is found in many jams, milk, and sugar products. Other food products designed for small children and babies may also contain flavin mononucleotide [160].
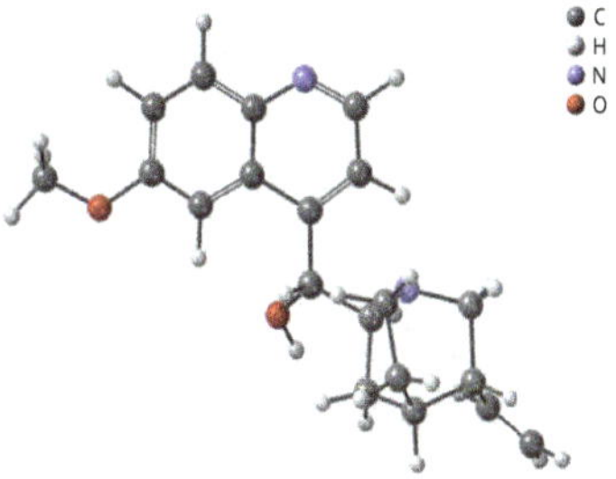	Quinine ($C_{20}H_{24}N_2O_2$) is a medication used to treat malaria. It is also used in tonic water to give its bitter taste [161].
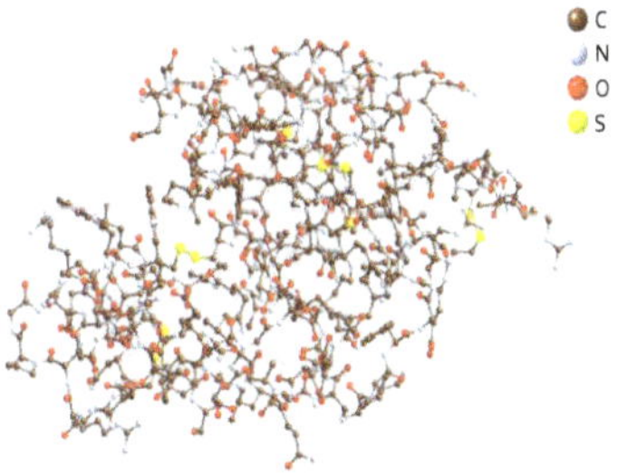	Lysozyme ($C_{613}\ N_{19}3O_{277}S_{10}$) is found in egg-white lysozyme and tears and acts as an antimicrobial agent [162].
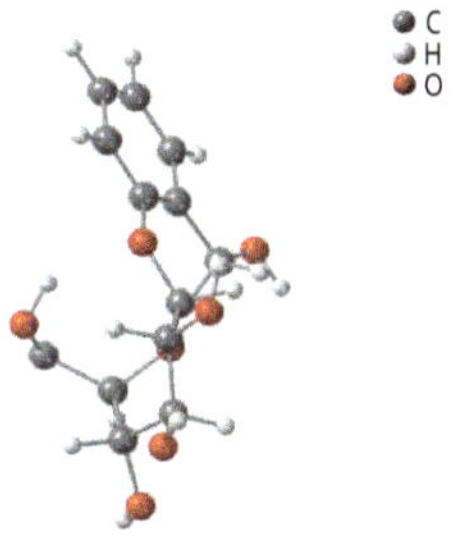	Salicin ($C_{13}H_{16}O_7$) is produced in willow bark and acts as an anti-inflammatory. It is the chemical from which aspirin was first developed [163].

Tab. 6.8 (continued)

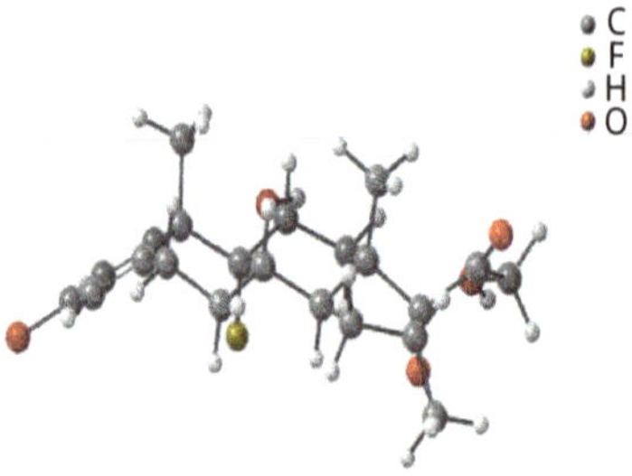	Dexamethasone ($C_{22}H_{29}FO_5$) is a corticosteroid medication used to treat many conditions, including rheumatic problems, skin diseases, severe allergies, asthma, chronic obstructive lung disease, croup, and brain swelling [164].

The plant drying method and extraction method significantly influence the antioxidant content and, consequently, the rate of NPs synthesis [165]. Indeed, Farooqui et al. [50] found that the method of plant drying (fresh, sun-dried, and hot air oven-dried leaf) has a significant impact on the size of Ag NPs produced from the leaf extract of *Clerodendrum inerme*. These results showed that the methods used for drying plants are promising for developing Ag NPs of different sizes tailored for specific applications. The NPs of the smallest size were obtained with the extract of fresh leaf, while the biggest one was observed in the extract of the sun-dried leaf. However, the processing conditions need to be optimized to obtain spherical, monodisperse NPs. Freeze-drying is a drying method used to preserve bioactive molecules in plants. Freeze-drying (lyophilization) is a drying process in which the solvent (usually water) or the suspension medium is crystallized at a low temperature and, after that, sublimated from the solid state directly into the vapor phase [166]. Freeze-drying plant parts is the most effective method for preserving plant's active ingredients [167]. The primary extraction methods for the synthesis of AuNPs and AgNPs are (a) solvent-based extraction, (b) microwave-assisted extraction, and (c) maceration extraction. The most commonly used solvents for extraction are water, methanol, and ethanol. Water is preferred as a green solvent for extracting biomolecules of plant species [2].

Various parameters can affect the characteristic features, production rate, and quantity of NPs synthesized from plant extracts, including extract concentration, metal salt concentration, pH, temperature, time, and extraction solvent. These parameters need to be optimized to improve the NPs biosynthesis and make it competitive with traditional synthesis approaches [169]. The concentration of the plant extract, pH, and reaction temperature affect the morphology, size, and shape of the obtained NPs [170]. Bezares et al. [171] showed that a higher extract concentration implies a greater amount of reducing agents and stabilizing molecules in the reaction mixture, thereby leading to a higher reaction rate and smaller particle size. The study by Kredy [172] on the biosynthesis of silver NPs using *Lawsonia inermis* leaves extract showed an increase in the rate of silver NPs formation with increasing temperature.

Monodisperse silver NPs were obtained at pH 9. [173] also demonstrated the synthesis of Ag NPs from silver nitrate solution and seed extracts of different vegetable seeds of the Brassicaceae family. The obtained results indicated that the optimal conditions of Ag NPs synthesis were a concentration of 20 mL, a temperature of 80 °C, and a pH of 8.5.

Many active ingredients in plant extracts may help reduce and stabilize NPs. Primary metabolites (proteins and polysaccharides) and secondary metabolites in the extract (terpenoids, flavonoids, phenols, alkaloids, steroids, saponins, tannins) operate as reducing agents [174–177]. The essential oils can be utilized as reducing and capping agents, however; the study carried out by Dzimitrowicz et al. [178] on the green synthesis of AuNPs by using aqueous extracts and essential oils of *Eucalyptus globulus* and *Rosmarinus officinalis* exhibited that the aqueous extract of these two species led to obtaining small NPs, in comparison with those synthesized by essential oils, possibly because of the higher amounts of possible reducing and stabilizing agents in the crude extracts compared with the corresponding essential oils [177]. The structure of salicin-coated cubic Cu metal was generated in CrystalMaker as a guide for preparing zero-valence NPs (Fig. 6.15). The simulated structure of salicin-coated cubic copper (Cu^0) NPs represents a biofunctionalized metallic system designed to stabilize zero-valence copper at the nanoscale. Salicin, a naturally occurring glucoside with phenolic and hydroxyl groups, serves as both a reducing and capping agent, facilitating the formation and stabilization of Cu^0 NPs through specific intermolecular interactions. In the initial phase of NP formation, salicin molecules orient themselves around the surface of the cubic Cu^0 crystal lattice (Fig. 6.15A). The oxygen atoms in salicin, particularly those from hydroxyl and ether groups, exhibit partial negative charges due to their electronegativity. These oxygen atoms interact with the partially positive surface of zero-valence copper atoms via dipole-ion interactions. This electrostatic attraction helps anchor salicin to the Cu surface, forming a stabilizing shell that prevents oxidation and aggregation. Density functional theory (DFT) simulations suggest that the Cu–O interaction energy ranges from −0.45 to −0.65 eV, depending on the coordination site and salicin conformation. Intermolecular forces between salicin and two Cu crystal units were illustrated in Fig. 6.15B. Beyond single-point interactions, salicin molecules can bridge adjacent Cu^0 crystal units through multisite coordination. The flexible molecular backbone of salicin allows simultaneous binding of its oxygen-containing functional groups to two separate Cu surfaces. This results in interparticle cross-linking, which contributes to colloidal stability and spatial organization of the NPs. A combination of hydrogen bonding, van der Waals forces, and chelation-like coordination primarily governs these interactions, forming a quasi-network that maintains dispersion in aqueous or polar media. Molecular dynamics simulations indicate that salicin-mediated bridging can reduce NP aggregation by up to 70%, with interparticle distances maintained in the 2.5–4.0 nm range. This structural model supports the use of salicin as a green stabilizer for Cu^0 NPs, offering both chemical protection and spatial control through well-defined intermolecular forces.

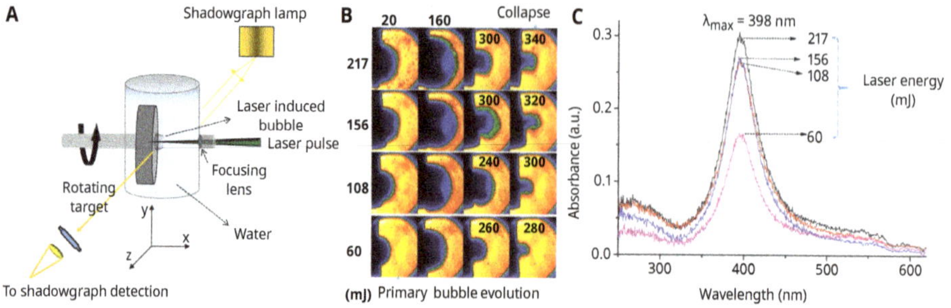

Fig. 6.15: The simulated structure of salicin-coated cubic Cu metal to prepare zero valence nanoparticles: **A**. The anticipated dipole ion intermolecular forces between Cu and O from salicin, and **B**. the intermolecular forces between salicin and two Cu crystal units.

Several works reported the utilization of phyto-assisted synthesis of NPs for manufacturing solar cells. Sutradhar and Saha [179] showed that the fabricated ZnO NPs using green tea leaf extract were further used to prepare ZnO/natural graphite (NG) composite material. The phenolic compound was used as a capping agent to form ZnO compounds, preventing crystal growth (Fig. 6.16). The current–voltage (I–V) characteristics of a ZnO/NG nanocomposite thin film were investigated. The ZnO/NG nanocomposite was measured for short-circuit photocurrent, open-circuit photovoltage, fill factor, and solar cell efficiency. Interestingly, the cell exhibited a high PCE of 3.54% and good stability. *Aloe vera* aqueous extract is employed for the preparation of TiO_2 NPs. The prepared TiO_2 NPs are used to fabricate solar cells on ITO glass substrates, along with TiO_2, ruthenium, graphite, and potassium iodide [180]. Au-doped TiO_2 NPs were successfully synthesized using *Terminalia arjuna* bark extract. The energy conversion efficiency was more important for Au-doped TiO_2 NPs than for pure and Pt-doped TiO_2 NPs. The synthesized NPs were investigated for use in dye-sensitized solar cells. Au noble metal presents a TiO_2 matrix and an improved open-circuit voltage (V_{oc}) of dye-sensitized solar cells [180]. Many NPs are synthesized using green methods that utilize plants. Some papers are available on the applications of water splitting using these green NPs [181]. The leaf extract of *Ocimum sanctum* was used as a reducing agent for the synthesis of platinum NPs from an aqueous chloroplatinic acid ($H_2PtCl_6 \cdot 6H_2O$). The reduced platinum exhibited hydrogen evolution potential and catalytic activity similar to those of pure platinum, as measured by linear scan voltammetry [182]. Silver-deposited titanate nanotube array composite (Ag/TNA-c) was successfully produced using tea leaves and ground coffee as reducing agents. The main reducing agents were catechin and chlorogenic acid. The fabricated Ag/TNA-c showed the best photocatalytic (PC) performance in photocurrent response, electrochemical impedance spectrum, ibuprofen degradation, and hydrogen generation [183].

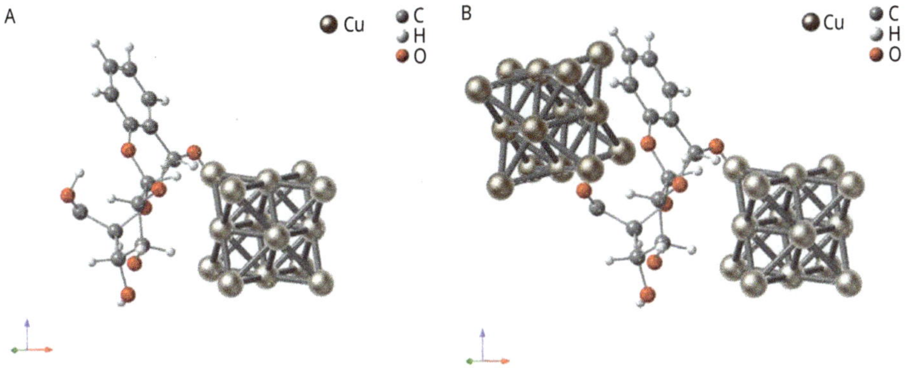

Fig. 6.16: The simulated structure of zinc oxide (ZnO) nanoparticles coated by tea leaf extract: **A.** the anticipated dipole ion intermolecular forces between Zn from ZnO and O from tea leaf extract (phenol as an example), and **B.** intermolecular forces between phenolic compounds with ZnO crystal units inhibit the crystal growth into large clusters.

6.4.2.2 Using microorganisms

Over the past decade, the biosynthesis of NPs using microbial agents, including bacteria, actinomycetes, fungi, yeast, viruses, and marine algae, has gained significant momentum in the field of green nanotechnology due to their intrinsic capabilities for biotransformation, bioaccumulation, and metal ion detoxification [184]. These microorganisms act as natural nanofactories, leveraging enzymatic and metabolic pathways to reduce metal ions into stable nanostructures under ambient conditions. Their ability to secrete extracellular enzymes such as nitrate reductase, hydrogenase, and sulfite reductase is pivotal in facilitating redox reactions that drive NP formation [185]. Recent studies have demonstrated that microbial synthesis can yield metal NPs (such as Ag, Au, ZnO, CuO, and Fe_3O_4) with particle sizes ranging from 10 to 80 nm, depending on the strain, precursor concentration, and culture conditions. For example, *Bacillus subtilis* has been shown to produce spherical AgNPs with an average diameter of 22 nm, exhibiting strong antimicrobial activity and surface plasmon resonance at 420 nm. Similarly, *Aspergillus niger* has been used to synthesize ZnO NPs with hexagonal wurtzite structures and photocatalytic degradation efficiencies exceeding 85% for methylene blue under UV light. Despite these advances, microbial synthesis remains largely confined to terrestrial-origin microorganisms, with limited exploration of extremophiles and marine species. The current scope is predominantly focused on elemental metals, select metal sulfides (e.g., CdS and PbS), and low-complexity oxides, due to constraints in metabolic compatibility and ion transport mechanisms [186].

However, recent work by Campaña et al. [187] has expanded the microbial toolkit to include bimetallic NPs (Au–Ag and Cu–Zn alloys) synthesized via co-reduction pathways in engineered bacterial strains. These hybrid nanostructures exhibit enhanced catalytic and optical properties, opening new avenues for biosensing, energy conversion, and environmental remediation. Moreover, marine algae and cyanobacteria have shown promise in producing NPs with unique morphologies, such as nanoplates, nanowires, and nanobelts, through intracellular compartmentalization and exopolysaccharide-mediated stabilization. Genetic engineering approaches are now being applied to enhance microbial NP synthesis by upregulating metal-binding proteins and optimizing redox enzyme expression, thereby improving yield, monodispersity, and functionalization potential. As microbial nanotechnology continues to evolve, integrating omics-based strain selection, bioreactor optimization and real-time monitoring will be essential to scale production and expand the compositional diversity of biosynthesized nanomaterials. For NP formation, different biological agents react differently with different metal solutions. Numerous microbes produce inorganic materials, either extracellularly or intracellularly, and the mechanism differs from one organism to another, either intra or extracellularly [188]. In intracellular synthesis, metal ions are transported into the microbial cell, and NP forms enzymes inside the cell. In extracellular biosynthesis, it involves trapping metal ions on the surface of microbial cells and reducing them into NPs in the presence of various enzymes (e.g., nitrate reductase) available at the surface [189]. Extracellular synthesis of NPs has an advantage over intracellular synthesis, in that it is less time-consuming, as it does not require any downstream process to collect NPs from the organisms [190]. The precise mechanism for NPs' synthesis using biological agents has not yet been elucidated. Different biological agents react differently with metal ions, leading to the formation of NPs [191]. Biological synthesis using microbes offers an advantage over plants since microbes are easily reproduced. Nonetheless, there are many drawbacks to isolating and screening potential microbes [192]. The microbial synthesis of NPs is characterized by the high cost of microorganism isolation [193]. In addition, the main problem is the cost-effectiveness of the fabrication procedures, as they are time-consuming and use chemicals in the growth medium [194]. In this section, the microbial platforms will be discussed, offering distinct advantages for green nanomaterial synthesis. Their selection depends on the desired NP type, functional properties, and application domain, as outlined in Tab. 6.9.

Bacteria: Bacteria are readily available in the environment, can be cultivated quickly, and adapt to diverse conditions, making them excellent candidates for NP synthesis [195]. Some bacterial species employ specific defense mechanisms to quell stress from heavy metal ions (e.g., *Pseudomonas stutzeri* and *Pseudomonas aeruginosa*). These mechanisms of resistance include efflux pumps, metal efflux systems, inactivation and complexation of metals, impermeability to metals and the lack of specific metal transport systems, alteration of solubility and toxicity by changes in the redox state of the metal ions, extracellular precipitation of metals, and volatilization of toxic metals by enzymatic reactions [196, Beveridge *et al.*, 1997). Numerous bacterial species are applied in the biosynthesis of NPs, such as *Acinetobacter* sp., *Aeromo-*

Tab. 6.9: *A* comparative summary of fungi, yeasts, and algae in the context of green chemistry synthesis of nanomaterials, integrating recent findings from 2020 to 2025.

Microbe	Synthesis mode	Key nanomaterials	Advantages	Applications
Bacteria	Extra-/ intracellular	Ag, Au, Pd, ZnO, and TiO_2	Rapid growth, metal tolerance, and enzyme diversity	Bioremediation, fuel cells, and antimicrobial agents
Fungi	Extra-/ intracellular	Ag, Au, ZnO, and Fe_3O_4	High enzyme activity and scalable biomass	Antimicrobials, catalysis, and soil remediation
Yeasts	Intracellular	Se, CdS, and Au	Genetic tractability and fermentation control	Biosensing, drug delivery, and bioelectronics
Algae	Extracellular	Au, Ag, Cu, and Fe_3O_4	Fast growth and photosynthetic biomass	Energy systems, water purification, and biofuels

nas sp., *Bacillus* species, *Clostridium thermoaceticum*, *Corynebacterium* sp., *Desulfovibrio* species, the species of Enterobacteriaceae family (*E. coli* and *Klebsiella pneumoniae*), *Geobacillus* sp., *Lactobacillus* sp., *Pseudomonas* sp., *Rhodobacter sphaeroides*, and *Rhodopseudomonas capsulata* [197–200]. Microorganisms' biomineralization of metals into NPs depends on environmental parameters such as pH, pO_2, pCO_2, redox potential, and temperature. Optimizing these physiological parameters requires synthesizing NPs with accurate size, morphology, and chemical compositions [191, 201]. Magnetotactic bacteria, known to synthesize magnetic nanocrystals with uniform shapes and sizes under physiological conditions, serve as an inspiration and a source of biological macromolecules used for the biomimetic synthesis of a variety of magnetic materials [202].

Fungi: Fungi are eukaryotic organisms widely distributed across terrestrial and aquatic ecosystems, typically functioning as decomposers within ecological food webs. Although the estimated global fungal diversity exceeds 1.5 million species, only about 70,000 have been formally described and characterized [185, 202]. Their unique cellular architecture and metabolic versatility position fungi as highly effective biofactories for NP synthesis. Notably, fungi secrete a broad spectrum of extracellular enzymes, including reductases, oxidases, and hydrolases, that facilitate large-scale biotransformation of metal ions into stable nanostructures. This enzymatic abundance supports scalable NP production under ambient conditions, with minimal energy input and reduced environmental burden. Fungal-mediated green synthesis of metallic NPs offers several operational advantages: economic viability, ease of biomass cultivation and handling, and compatibility with aqueous systems [203]. According to [204], the fungus *Verticillium* sp. was able to trap, bioreduce, and cap gold ions, forming monodisperse AuNPs with diameters of 8–25 nm. These particles exhibited surface plasmon resonance peaks near 530 nm and showed excellent stability due to protein-mediated capping. The fungal cell wall, rich in chitin, glucans, and glycoproteins, provides high-affinity binding sites for metal ions, thereby enhancing metal uptake and tolerance. This property enables fungi to produce higher NP yields compared to bacterial systems, especially under elevated metal concentrations [185]. Several fungal species have been extensively studied for their ability to biosynthesize NPs. *Fusarium oxysporum, Fusarium solani, Pleurotus sajor-caju, Fusarium semitectum, Alternaria alternata, Fusarium acuminatum, Penicillium* spp., and *Aspergillus* spp. have been employed in the synthesis of silver, gold, zinc oxide, and iron oxide NPs [205]. For instance, *F. oxysporum* has been shown to produce AgNPs with average sizes of 28–35 nm, exhibiting strong antimicrobial activity against *Staphylococcus aureus* (*S. aureus*) and *Escherichia coli* (*E. coli*), with minimum inhibitory concentrations (MICs), as low as 6.5 µg/mL [206]. Recent proteomic studies have revealed that fungal secretomes contain multifunctional proteins, such as FAD-oxidoreductases and glyceraldehyde reductases, which simultaneously mediate metal ion reduction and NP stabilization [206]. Moreover, marine-derived fungi such as *Fusarium equiseti* have demonstrated enhanced biosynthetic performance under saline conditions, producing spherical AgNPs with an average diameter of 50 nm and potent cytotoxicity against MCF-7 breast cancer cells (IC_{50} = 24.38 µg/mL; [207]. These findings underscore the potential of fungal systems not only for eco-friendly NP synthesis but also for biomedical and environmental applications. As research progresses, fungal nanotechnology is expected to expand toward multimetallic NP production, hybrid nanostructures, and genetically engineered strains for precision biosynthesis [208].

Yeasts: Yeasts are eukaryotic, unicellular microorganisms belonging to the kingdom Fungi, widely recognized for their metabolic versatility and industrial relevance. Several strains, particularly those within the family Saccharomycetaceae, play a pivotal role in the food and beverage industries due to their robust sugar fermentation capabilities. To date, approximately 1,500 yeast species have been formally described, with *Saccharomyces cerevisiae* serving as a model organism for both metabolic engineering and biotechnological applications [209]. Yeast cultivation is highly controllable under laboratory conditions, requiring only simple carbon and nitrogen sources, which facilitates rapid biomass accumulation and reproducible experimental outcomes. This ease of handling, combined with their facultative anaerobic metabolism and high tolerance to metal ions, makes yeasts ideal candidates for the biosynthesis of metal NPs. Their ability to secrete biologically active compounds, including reductases, amino acids, organic acids, and vitamins, enables both intracellular and extracellular NP formation through enzymatic reduction and stabilization mechanisms. Recent studies have demonstrated that yeast-mediated synthesis can yield monodisperse NPs with particle sizes ranging from 10 to 80 nm, depending on species, precursor concentration, and reaction conditions. For example, *S. cerevisiae* has been shown to produce AgNPs with average diameters of 25–35 nm, exhibiting vigorous antimicrobial activity and surface plasmon resonance peaks near 420 nm. Similarly, *Yarrowia lipolytica* has been employed to synthesize zinc oxide (ZnO) and selenium NPs with enhanced antioxidant and photocatalytic properties. A diverse array of yeast species has been utilized in NP biosynthesis, including *Y. lipolytica, S. cerevisiae, Trichophyton* spp., *Microsporum canis, Kluyveromyces marxianus, Candida albicans, Candida glabrata, Schizosaccharomyces pombe, Rhodotorula mucilaginosa, Chlorococcum humicola, Aphanothece* sp., *Sargassum muticum*, and *Caulerpa racemosa* [210, 212]. These species have demonstrated the ability to synthesize a wide range of metallic and metal oxide NPs, including Au, Ag, CuO, ZnO, and TiO_2, under mild conditions, often without the need for external stabilizers or toxic reagents. Notably, *Schizosaccharomyces pombe* has been reported to produce palladium and platinum NPs, with catalytic activity suitable for hydrogenation reactions. At the same time, *Candida glabrata* has shown promise in synthesizing selenium NPs with anticancer potential. The biosynthesis process can be modulated by adjusting parameters such as pH (optimal range: 6.0–7.5), temperature (25–37 °C), and incubation time (24–72 h), allowing for fine-tuning of NP size, shape, and surface charge. As research advances, yeast-based nanofabrication is being integrated with omics technologies and bioreactor optimization to enhance yield, reproducibility, and scalability. These developments position yeast-mediated NP synthesis as a cornerstone of green nanotechnology, with applications spanning medicine, agriculture, environmental remediation, and energy systems.

Algae: Algae represent a phylogenetically diverse group of photoautotrophic, eukaryotic organisms that thrive in aquatic environments and exist in both unicellular and multicellular forms. They are broadly classified into three major groups based on their pigment composition and photosynthetic machinery: brown algae (Phaeophyceae), red algae (Rhodophyceae), and green algae (Chlorophyceae). These classifications reflect differences in chlorophyll content, accessory pigments (e.g., fucoxanthin and phycobilins), and metabolic pathways, which also influence their biochemical profiles and suitability for NP biosynthesis. Beyond their well-established roles in commercial sectors, including food, feed, cosmetics, pharmaceuticals, and biofertilizers, algae have emerged as potent biological platforms for green nanomaterial fabrication. Their cellular machinery and metabolic versatility enable the biosynthesis of metal and metal oxide NPs through environmentally benign processes. This is primarily facilitated by secondary metabolites such as polyphenols, flavonoids, terpenoids, alkaloids, and sulfated polysaccharides, which act as reducing, capping, and stabilizing agents during NP formation [213]. Recent studies have demonstrated that algal extracts can produce NPs with controlled size, shape, and surface charge under ambient conditions. For example, *Galaxaura rugosa*, a red alga, was used to synthesize Ag–Fe bimetallic NPs with an average diameter of 32.6 nm, as confirmed by XRD and TEM. These particles exhibited spherical morphology, crystalline structure, and a surface plasmon resonance band at 327 nm, indicating successful alloy formation and optical activity suitable for biomedical and catalytic applications [214]. Similarly, brown algae such as *Fucus vesiculosus* and *Sargassum wightii* have been employed to biosynthesize AuNPs and AgNPs with high colloidal stability and antimicrobial efficacy. Green algae such as *Chlorella vulgaris* and *Spyrogira insignis* have shown promise in producing ZnO and TiO_2 NPs with photocatalytic properties, useful for water purification and environmental remediation [215]. The biosynthesis process can occur intracellularly, where metal ions are absorbed and reduced within algal cells, or extracellularly, where metabolites in the culture medium mediate NP formation. Parameters such as pH (optimal range: 6.5–8.0), temperature (25–40 °C), and extract concentration significantly influence NP morphology and yield. Zeta potential measurements often range from − 15 to − 30 mV, indicating good colloidal stability in the absence of synthetic surfactants. As research advances, algae-mediated NP synthesis is being integrated into phyco-nanotechnology. This subfield combines algal biotechnology with nanoscience to develop sustainable solutions for energy, medicine, and environmental engineering. The scalability, low toxicity, and metabolic richness of algae make them ideal candidates for producing next-generation green nanomaterials.

6.4.2.3 Using biomolecules

A wide array of biomolecules has been investigated for their roles as reducing, capping, and stabilizing agents in the green synthesis of nanomaterials (NMs), particularly metallic and metal oxide NPs. These biomolecules, such as polysaccharides, peptides, proteins, amino acids, and phytochemicals, offer multifunctional capabilities that enable controlled nucleation, growth, and surface passivation of NPs under mild, eco-friendly conditions. Their molecular structures contain functional groups (e.g., hydroxyl, carboxyl, amine, and thiol) that facilitate electron transfer and metal ion chelation, thereby driving reduction and stabilization reactions without the need for toxic reagents or high-energy inputs. Polysaccharides like chitosan, alginate, and starch have demonstrated excellent reducing and stabilizing properties. For instance, chitosan-mediated synthesis of silver NPs yielded spherical particles with diameters of 15–30 nm and a zeta potential of − 28 mV, indicating strong colloidal stability and antimicrobial efficacy against *E. coli* and *S. aureus*. Proteins such as bovine serum albumin (BSA) and casein have been used to fabricate gold and selenium NPs with enhanced biocompatibility and controlled release profiles for drug delivery applications. Amino acids such as cysteine and tyrosine have shown selective reduction capabilities, producing monodisperse NPs with tailored surface chemistry suitable for biosensing and catalytic applications. Phytochemicals, including flavonoids, tannins, alkaloids, and terpenoids extracted from plant sources, have also been widely employed in NP synthesis. A 2024 study demonstrated that *Azadirachta indica* leaf extract, rich in quercetin and catechins, facilitated the synthesis of ZnO NPs with average sizes of 22 nm and photocatalytic degradation efficiencies exceeding 85% for methylene blue under UV light [216]. The fabrication of biologically derived NPs typically follows either a two-step emulsification-based method or a one-step direct synthesis route. The two-step method involves emulsification, followed by solvent evaporation or phase separation, often used for polymeric NPs and drug-loaded nanocarriers. In contrast, the one-step method, comprising nanoprecipitation, desolvation, gelation, and drying, is favored for metallic and oxide NPs due to its simplicity and scalability [217]. For example, nanoprecipitation using curcumin and gelatin yielded iron oxide NPs with particle sizes of 18–25 nm and superparamagnetic behavior, suitable for MRI contrast enhancement.

Polysaccharides: Polysaccharides are natural biopolymers increasingly recognized as versatile and sustainable hosts for the synthesis of MNPs, owing to their exceptional biocompatibility, biodegradability, and functional diversity [218]. Structurally, polysaccharides are composed of repeating monosaccharide units linked via glycosidic bonds, forming linear or branched macromolecules with tunable physicochemical properties. Common examples include chitosan, cellulose, starch, hyaluronic acid, and dextran, each offering unique molecular architectures and reactive sites suitable for NP fabrication [219, 222-228]. The presence of hydroxyl groups, hemiacetal reducing ends, carboxyl, and amine functionalities allows polysaccharides to act as both reducing and stabilizing agents during NP synthesis. These groups facilitate electron transfer to metal ions (e.g., Ag^{+}, Au^{3+}, and Zn^{2+}), initiating nucleation and growth of NPs while simultaneously capping their surfaces to prevent aggregation and oxidation [220]. For instance, chitosan has been shown to produce AgNPs with particle sizes of 20–50 nm and zeta potentials of –25 to –35 mV, indicating strong colloidal stability and antimicrobial efficacy against gram-positive and gram-negative bacteria [221]. Dr Liu and her team employed starch, a polysaccharide composed primarily of amylose and amylopectin, to coat ceria-doped zinc oxide (Ce: ZnO) nanoparticles for dental applications. The synthesis was conducted via colloidal chemistry under controlled pH and temperature conditions, yielding uniformly distributed nanoparticles with average diameters of 18–25 nm. XRD and transmission electron microscopy (TEM) confirmed the coexistence of hexagonal wurtzite ZnO and triclinic fluorite CeO_2 phases. The inherent mismatch in lattice parameters between ZnO ($a = 3.25$ Å, $c = 5.21$ Å) and CeO_2 ($a = 5.41$ Å) was mitigated by the starch matrix, which acted as a molecular scaffold to "glue" the two crystalline domains together, enhancing interfacial compatibility and structural integrity (Fig. 6.17) [229]. The resulting Ce: ZnO@starch nanocomposite exhibited high whitening efficacy, attributed to its strong UV absorption ($\lambda_{max} \approx 370$ nm), generation of ROS, and enamel-safe surface interactions. Spectrophotometric analysis showed an improvement in ΔE^* of 4.8 units after 10 min of application, surpassing commercial peroxide-based formulations. Additionally, FTIR revealed characteristic C–O and O–H stretching vibrations from starch, confirming its role in surface functionalization and phase stabilization. Recent reviews have emphasized the potential of polysaccharide-based nanomaterials in biomedical, environmental, and energy applications, citing their low toxicity, renewable origin, and tunable surface chemistry [218]. As synthesis techniques evolve, such as ionic gelation, nanoprecipitation, and emulsion cross-linking, polysaccharides continue to serve as foundational materials for next-generation nanocomposites with enhanced performance and sustainability.

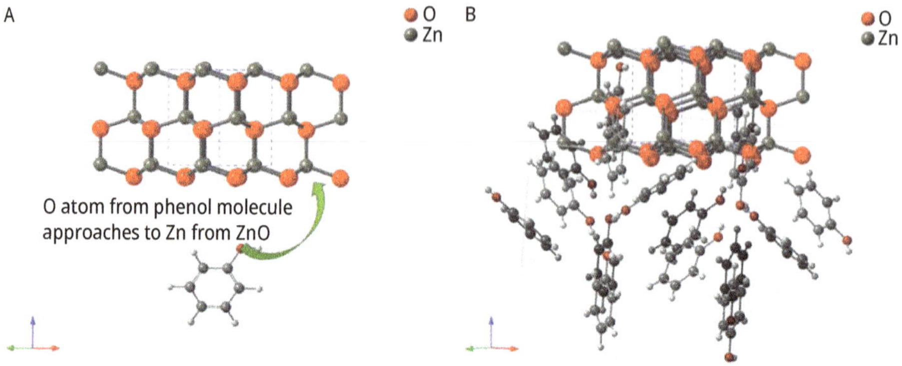

Fig. 6.17: The simulated structure of ceria-doped zinc oxide (ZnO) nanoparticles coated by starch: **A.** the anticipated dipole ion intermolecular forces between Zn from ZnO, Ce from CeO_2, and O from starch and **B.** intermolecular forces between phenolic compounds with ZnO crystal units inhibit the crystal growth into large clusters.

Protein: Proteins are among the earliest and most versatile naturally occurring materials employed in nanoparticle (NP) fabrication, owing to their well-defined primary structures, diverse functional groups, and intrinsic biological compatibility [217, 219, 223]. Their amino acid sequences provide reactive sites, such as carboxyl, amine, thiol, and hydroxyl groups, which enable metal ion chelation, reduction, and surface stabilization. These functionalities make proteins ideal candidates for green synthesis routes, particularly in biomedical, pharmaceutical, and environmental applications. Biological protein-based nanoparticles derived from silk fibroin, keratin, collagen, elastin, corn zein, and soy protein have demonstrated significant advantages, including biodegradability, bioavailability, mechanical tunability, and cost-effectiveness [223]. For example, silk fibroin nanoparticles synthesized via ionic gelation have exhibited particle sizes of 80–150 nm, with encapsulation efficiencies exceeding 85% for hydrophobic drugs such as curcumin. Keratin-based NPs processed with ionic liquids have exhibited enhanced solubility and mechanical strength, with tensile moduli up to 120 MPa, making them suitable for wound-healing and tissue-scaffolding applications [224]. Collagen nanoparticles fabricated via desolvation methods have demonstrated sustained drug-release profiles of over 72 h and high cell viability (> 90%) in fibroblast cultures. Zein and soy protein nanoparticles, often produced via nanoprecipitation, have been used to encapsulate polyphenols and essential oils, yielding spherical particles with diameters between 100 and 250 nm and retaining antioxidant activity above 70% after 30 days of storage [223]. In parallel, phytochemicals, including phenolic compounds, terpenoids, essential oils, alkaloids, and saponins, are increasingly utilized in NP synthesis due to their strong reducing capabilities and bioactive potential [225, 230-232]. These compounds not only facilitate the reduction of metal ions (e.g., Ag^+, Au^{3+}, and Zn^{2+}) but also impart functional properties such as antimicrobial, anti-inflammatory, and anticancer activity to the resulting nanoparticles. For instance, phenolic-rich extracts from *Camellia sinensis* have been used to synthesize gold nanoparticles with average sizes of 20–30 nm and surface plasmon resonance peaks at 530 nm, showing selective cytotoxicity against MCF-7 breast cancer cells ($IC_{50} \approx 18$ μg/mL). Recent studies have also explored hybrid systems, combining proteins and phytochemicals, to enhance NP stability and functionality. A 2024 report demonstrated that soy protein–polyphenol conjugates produced silver nanoparticles with improved dispersion stability (zeta potential −32 mV) and synergistic antimicrobial effects against *Pseudomonas aeruginosa* and *Candida albicans*. These findings underscore the potential of protein–phytochemical systems in developing multifunctional nanomaterials for targeted delivery, regenerative medicine, and biosensing.

6.4.2.4 Using wastes

According to the United Nations Environment Programme, approximately 11.2 billion tons of solid waste is generated globally each year, contributing significantly to environmental degradation, ecosystem disruption, and adverse public health outcomes [226]. In response to this escalating challenge, researchers have increasingly turned to agricultural and industrial food wastes as renewable, low-cost, and nontoxic feedstocks for the green synthesis of nanoparticles (NPs). These waste-derived materials not only reduce the environmental burden but also align with circular economy principles by valorizing biomass into functional nanomaterials. Among agricultural residues, peanut skin has emerged as a promising candidate due to its high availability, low cost, and rich phytochemical content, including polyphenols, flavonoids, and tannins, which serve as natural reducing and stabilizing agents. Liu *et al.* (2020) proposed a mechanistic pathway for NP synthesis that involves the formation of intermediate metal-organic complexes, followed by electron-transfer reactions and the adsorption of nonreducing macromolecules at the solid–liquid interface. Their study demonstrated that peanut skin extract could reduce silver nitrate to form AgNPs with average diameters of 18–25 nm, as confirmed by UV-vis spectroscopy ($\lambda_{max} \approx 420$ nm) and XRD analysis, which showed a face-centered cubic crystalline structure [227]. Similarly, Rajput et al. [228] utilized used tea leaves, an abundant post-consumer waste, to synthesize silver nanoparticles. The biosynthesized AgNPs exhibited particle sizes ranging from 10 to 40 nm, with high crystallinity and surface plasmon resonance peaks at 430 nm, indicating successful reduction and stabilization. TEM imaging revealed spherical morphology, and FTIR spectra confirmed the presence of phenolic and carboxyl groups responsible for capping. These NPs demonstrated potent antimicrobial activity against *E. coli* and *S. aureus*, with minimum inhibitory concentrations (MICs), below 10 μg/mL. In the realm of industrial waste valorization, Zhou et al. [229] developed an approach for converting coal gangue, a by-product of metallurgical and mining operations, into porous zeolite-based nanomaterials for photocatalytic applications (Fig. 6.18). Their process involves alkali activation, hydrothermal treatment, and template-assisted crystallization to transform aluminosilicate-rich gangue into zeolite frameworks, with surface areas exceeding 600 m^2/g and pore diameters in the mesoporous range (2–50 nm). These materials have exhibited high photocatalytic degradation efficiencies (>90%) for organic dyes such as methylene blue and rhodamine B under simulated solar irradiation, with recyclability over five cycles without significant loss of activity.

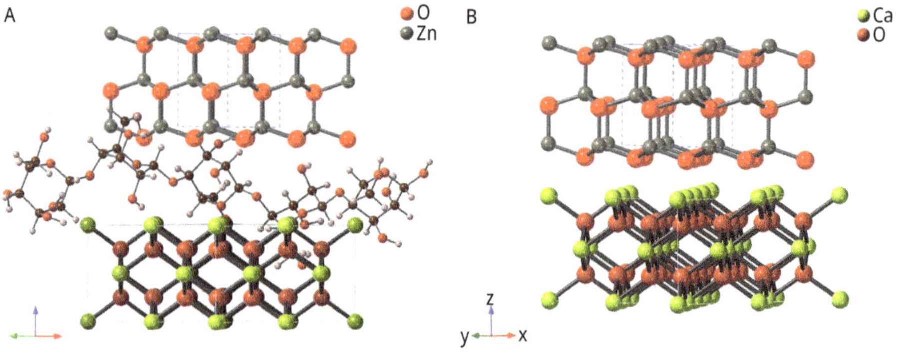

Fig. 6.18: The conversion of waste into value-added products with high porosity, used for water remediation: **A**. scanning electron microscopic images of zeolites, treated by 3 mol/L NaOH solution for different processing time (a) 0.5 h, (b) 1.0 h, and **B**. adsorption curves of Cu^{2+} and Rh-B adsorbed by zeolites treated by 3 mol/L NaOH solution for different processing times: (a) 0.5 h and (b) 1.0 h. Permission from [1, 229].

Recent reviews emphasize that the use of agro-industrial waste in NP synthesis not only reduces production costs but also enhances sustainability by minimizing hazardous reagents and energy consumption. As highlighted in a 2024 *Frontiers in Nanotechnology* review, waste-derived nanomaterials are increasingly being applied in food preservation, water purification, and biomedical coatings [230]. Agricultural residues such as tea leaves, peanut skins, and fruit peels have been successfully converted into metallic and metal oxide nanoparticles using green synthesis methods. These nanoparticles exhibit potent antimicrobial properties, making them ideal for use in active food packaging and edible coatings. For example, silver nanoparticles synthesized from used tea leaves exhibited particle sizes between 10 and 40 nm and inhibited *E. coli* and *S. aureus* at concentrations below 10 µg/mL. In water purification, nanomaterials derived from biochar, eggshells, and coal gangue have shown high adsorption capacities for heavy metals and organic pollutants. Zeolite-based composites synthesized from coal gangue achieved surface areas exceeding 600 m^2/g and removed over 90% of methylene blue under solar irradiation. Biomedical coatings using phytochemical-rich nanoparticles have also shown promise in wound healing and antimicrobial barrier formation, offering biocompatibility and reduced cytotoxicity compared to synthetic alternatives [230]. Moreover, a 2024 *Springer* chapter on agricultural waste valorization [231] underscores the potential of biodegradable residues to serve as sustainable precursors for metallic NPs, provided synthesis scalability and toxicity assessments are addressed. The chapter highlights that crop residues such as rice husk, wheat straw, and sugarcane bagasse contain high levels of cellulose, lignin, and silica. These key components facilitate redox reactions and nanoparticle stabilization. For instance, rice husk ash, rich in SiO_2 (up to 97.3%), has been used to produce nanostructured silica with particle sizes below 50 nm and surface areas exceeding 300 m^2/g. Similarly, fruit peels and vegetable waste, rich in phenolic compounds and flavonoids, have demonstrated strong reducing capabilities for the formation of silver and gold nanoparticles. However, the authors emphasize that while lab-scale synthesis is well-established, industrial scalability remains a challenge due to variability in feedstock composition and lack of standardized protocols. Furthermore, toxicity assessments of both environmental and biological are critical to ensure safe deployment of waste-derived nanomaterials in sectors such as medicine, agriculture, and water treatment.

6.5 Advantages and disadvantages of green methods

The green synthesis of nanomaterials (NMs) using plant extracts, microorganisms, biomolecules, and agricultural or industrial wastes has emerged as a promising alternative to conventional chemical and physical methods. These approaches are widely recognized for being cost-effective, environmentally benign, sustainable, and operationally simple. They align with the principles of green chemistry by minimizing hazardous reagents and energy consumption, making them attractive for applications in medicine, agriculture, and environmental remediation [232]. Despite these advantages, several limitations persist. The reproducibility and scalability of plant-mediated NP synthesis remain constrained by variability in phytochemical composition, seasonal availability, and extract concentration. Similarly, microbial synthesis is challenged by slow growth rates, contamination risks, and the need for precise control over culture conditions. Optimization of parameters such as pH, temperature, and metal salt concentration is still required to ensure consistent nanoparticle morphology and yield. The underlying mechanisms of biosynthesis, particularly the roles of specific metabolites and enzymes in reduction and stabilization, remain only partially elucidated. This gap in mechanistic understanding hampers predictive modeling and process control. Moreover, the cultivation and genetic manipulation of eukaryotic microorganisms to enhance NP production remain technically complex and resource-intensive [233].

Recent studies have increasingly highlighted the need to critically assess the toxicity and environmental persistence of green-synthesized nanoparticles (NPs), despite their natural and biodegradable origins. While these NPs are produced using plant extracts, microbial agents, and agricultural waste – often without hazardous reagents – their physicochemical properties can still pose risks to biological systems. Factors such as particle size, surface charge, aggregation behavior, and surface functionalization significantly influence their interaction with cells and ecosystems. For instance, smaller NPs (<20 nm) with high surface reactivity have been shown to penetrate cellular membranes and induce oxidative stress, DNA damage, and inflammatory responses, even when synthesized via green routes. A 2025 review in *BioNanoScience* emphasized the urgent need for comprehensive toxicological profiling, long-term exposure studies, and life cycle assessments to ensure the safe deployment of green nanomaterials in biomedical and environmental applications [234]. Supporting this, a 2025 study in *Sustainable Environment* reported that green-synthesized TiO_2 and ZnO NPs used in water treatment exhibited bioaccumulation potential and disrupted microbial communities in aquatic systems [235]. Additionally, the review called for standardized protocols to evaluate posttreatment risks, including ecosystem disruption, residual toxicity, and regulatory gaps in nanomaterial disposal and reuse. These findings underscore that while green synthesis reduces upstream hazards, downstream impacts must be rigorously evaluated. Future research should integrate real-time monitoring, ecotoxicological assays, and circular economy frameworks to balance innovation with environmental stewardship [236].

Furthermore, a 2025 study in the *International Journal of Research* Publication and Reviews emphasized that the lack of standardized protocols and regulatory frameworks remains a critical barrier to the industrial adoption of green-synthesized nanoparticles [135]. Despite the growing interest in eco-friendly synthesis using plant extracts, microorganisms, and biowaste, most reported methods are confined to laboratory-scale demonstrations with limited reproducibility and scalability. The absence of harmonized guidelines for synthesis parameters (extract concentration, reaction time, pH, and temperature) leads to inconsistent nanoparticle morphology, yield, and functionality across studies. The review also highlighted that regulatory oversight of green nanomaterials is underdeveloped, particularly regarding environmental release, occupational exposure, and long-term toxicity. Without precise classification and safety thresholds, commercial deployment remains restricted. To bridge this gap, future research must prioritize mechanistic elucidation of biosynthetic pathways, process optimization through kinetic modeling and automation, and comprehensive safety validation using in vitro and in vivo assays. Supporting this, a 2025 Springer review on sustainable nanotechnology called for the development of industry-ready protocols, including ISO-compliant characterization standards and life cycle assessments, to facilitate regulatory approval and market integration [237]. These efforts are essential to transition green synthesis from academic innovation to scalable, compliant, and economically viable industrial practice.

6.6 Conclusion

In recent years, nanotechnology has emerged as a transformative platform for addressing global economic and environmental challenges, particularly through the development of sustainable synthesis routes. Among these, the green synthesis of nanoparticles (NPs) using plant extracts, microorganisms, biomolecules, and agricultural waste has gained significant momentum due to its low cost, minimal ecological footprint, and alignment with circular economy principles. Unlike conventional chemical and physical methods, which often rely on hazardous reagents and energy-intensive conditions, green synthesis reduces environmental pollutants, conserves natural resources, and avoids toxic byproducts. Green and bio-synthesized NPs have demonstrated promising applications across diverse sectors, with relevance to renewable energy technologies. Biosynthesized ZnO, TiO_2, and MnO_2 nanoparticles have shown high photocatalytic efficiency in solar-driven water purification and hydrogen evolution reactions. For instance, green-synthesized MnO_2 NPs have been evaluated as electrode materials in lithium-ion batteries and supercapacitors, achieving specific capacities exceeding 1,000 mAh/g and enhanced cycle stability under ambient conditions. Similarly, iron oxide NPs derived from plant extracts have been integrated into redox-active electrodes for sodium-ion batteries, offering improved ion transport and electrochemical reversibility. Nanomaterials synthesized via green routes also play a critical role in energy conversion systems. In solar energy, bio-derived TiO_2 NPs have been used in dye-sensitized solar cells, enhancing light absorption and electron mobility. In wind energy infrastructure, nanocoatings improve blade durability and reduce friction losses. In biofuel production, enzyme-functionalized NPs accelerate biomass breakdown and fermentation efficiency. For hydrogen fuel cells, green-synthesized platinum and palladium NPs serve as electrocatalysts with reduced toxicity and cost.

Despite these advances, several technical challenges remain. Key areas requiring further research include precise control over nanoparticle morphology, size distribution, and surface functionalization, which directly influence performance, stability, and biocompatibility. Moreover, scaling up green synthesis for industrial production is hindered by variability in biological feedstocks, lack of standardized protocols, and limited mechanistic understanding of biosynthetic pathways. Addressing these limitations through kinetic modeling, real-time monitoring, and regulatory harmonization will be essential to fully realize the commercial potential of green nanotechnology. Guided by the U.S. EPA's definition of green chemistry, which emphasizes the reduction or elimination of hazardous substances, and reinforced by the DOE's focus on sustainable technologies in energy conversion, storage, and nanomaterials, green synthesis frameworks are increasingly being aligned with the 12 principles of green chemistry. These principles support the design of safer materials, energy-efficient processes, and waste-minimizing production strategies. At the nanoscale (1–100 nm), materials exhibit quantum confinement effects, altered surface energetics, and size-dependent reactivity, all of which are fundamentally distinct from those of their bulk counterparts. One of the most defining attributes of nanomaterials is their exceptionally high surface area-to-volume ratio, often exceeding 100 m^2/g, which enhances catalytic activity, adsorption capacity, and interfacial interactions [13]. These properties underpin their widespread application in catalysis, electronics, energy storage, and biomedicine. In energy systems, nanomaterials enable tunable bandgaps, enhanced PL, superparamagnetism, and increased mechanical resilience, critical for optimizing solar cells, battery electrodes, and fuel cell membranes. Advanced characterization techniques such as transmission electron microscopy (TEM), DLS, XRD, and zeta potential analysis are routinely employed to quantify particle size, crystallinity, surface charge, and dispersion stability. These tools support the development of synthesis protocols (e.g., sol–gel, hydrothermal, green synthesis), nanoparticle types (e.g., metal, oxide, and hybrid), and separation methods (e.g., centrifugation and membrane filtration) tailored to specific energy applications. Nanomaterials are broadly categorized into organic, inorganic, ceramic, and carbon-based nanoparticles, each exhibiting distinct crystallographic structures and physicochemical behaviors. As green nanotechnology continues to evolve, its integration into solar energy, wind energy, sustainable biofuels, and hydrogen fuel cells will be pivotal for advancing clean energy transitions. Both physical/chemical methods and biological methods must be optimized to ensure reproducibility, safety, and scalability in real-world energy systems.

6.7 Innovative trends and future directions

Green synthesis of nanomaterials is rapidly evolving from proof-of-concept demonstrations using plant extracts and microbial cultures into a disciplined, data-driven manufacturing paradigm. This transition is being shaped by advances in mechanistic understanding, bioengineering, process intensification, and application-specific material design, particularly in the energy sector.

Mechanistic clarity and data-driven optimization: Recent efforts have focused on mapping redox pathways through advanced spectroscopy and omics technologies. These approaches are helping to identify key metabolites, enzymes, and intermediate complexes that govern nanoparticle nucleation and growth, enabling predictive control over size, shape, and phase. Kinetic modeling, coupled with CFD and discrete element modeling (DEM), is informing reactor design with targets such as polydispersity index (PDI) below 0.15 and batch-to-batch variance under 5%. Artificial intelligence is being integrated into experimental workflows, where Bayesian optimization and active learning reduce experimental iterations by up to 70% while maximizing yield and crystallinity under mild conditions. In situ monitoring tools, including UV–vis, ATR-FTIR, Raman spectroscopy, and DLS, are shifting synthesis from empirical endpoints to specification-driven release criteria.

Bioengineering of production hosts and catalysts: Bioengineering is enhancing the productivity and specificity of biological synthesis platforms. CRISPR-enabled microorganisms are being tailored to overexpress reductases, metal-binding peptides, and export systems, resulting in space-time yield improvements of 2–5 ×. Enzyme nanofactories, composed of immobilized oxidoreductases and multienzyme cascades, offer cleaner nanoparticle populations with reduced endotoxin risk and simplified downstream processing. Phytochemical standardization through chemotype selection and controlled cultivation conditions is reducing seasonal variability and improving reproducibility. Hybrid bio-chemo systems are also emerging, in which biogenic reduction is followed by mild posttreatments, such as photothermal aging or low-temperature annealing, to enhance crystallinity and electronic properties without introducing toxic solvents.

Circular feedstocks and low-impact processing: Green synthesis is increasingly linked to circular economic strategies. Agricultural residues, food processing byproducts, and mining wastes are being repurposed as reductants, supports, and precursors, thereby mitigating waste while producing functional nanomaterials. Green solvents such as water, deep eutectic mixtures, and supercritical CO_2 are replacing VOCs, with solvent recovery rates exceeding 90% and environmental factor (E-factor) values below 5. Process intensification through microfluidics and continuous stirred-tank cascades is improving heat and mass transfer, enabling continuous production with narrow residence time distributions and consistent product quality at kilogram-per-day scales. Low-energy synthesis protocols operating below 60 °C and under ambient pressure are becoming standard for positive life cycle assessments.

Materials frontiers tailored to function: Material innovation is expanding the scope of green-synthesized nanomaterials. Biogenic routes to multimetallic and doped nanoparticles, such as Au–Ag, Pd–Cu, and nitrogen-doped carbon-supported catalysts, are unlocking synergistic catalytic sites for energy conversion reactions. Green-synthesized metal oxides (e.g., TiO_2, ZnO, Fe_3O_4, and MnO_2) and emerging two-dimensional materials (e.g., bio-exfoliated MXenes and graphene derivatives) are being tuned for bandgap, defect density, and conductivity. Biocomposites incorporating proteins, polysaccharides, and lignin serve as scaffolds and capping agents, enhancing nanoparticle stability (zeta potential > 25 mV), dispersibility, and biocompatibility. Form-factor engineering is enabling the transition from colloidal suspensions to functional inks, aerogels, and coatings suitable for integration into electrodes, membranes, and catalytic surfaces.

Energy-focused applications: Green nanomaterials are demonstrating strong potential in energy systems. In solar fuels and photocatalysis, defect-engineered TiO_2 and ZnO nanoparticles, combined with co-catalysts such as Ni and Co, are driving hydrogen evolution and CO_2 reduction, with solar-to-hydrogen efficiencies of 1–3% under simulated AM1.5 G conditions. Biogenic MnO_2, Fe_3O_4, and doped carbon materials derived from biomass are being used as battery and supercapacitor electrodes, achieving high rate capabilities (>200 mAh/g at 2C) and long cycle lives (>10,000 cycles with capacitance retention > 200 F/g). Fuel cell electrocatalysts synthesized via green routes, including bio-reduced Pt, Pd, and non-precious metal catalysts such as Fe–N–C, achieve oxygen reduction reaction (ORR) mass activities above 0.44 A/mgPt and voltage losses below 30 mV at 0.8 A/cm^2 after 30,000 cycles. Thermo- and photoelectrocatalytic upcycling strategies leverage waste-derived supports such as zeolites and biochar to host active nanoparticles for pollutant degradation and VOC abatement, thereby coupling environmental remediation with energy recovery. In membrane technologies, antifouling nanocomposite membranes incorporating green-synthesized nanoparticles increase water flux by over 30% and reduce transmembrane pressure, thereby lowering energy consumption in desalination and wastewater treatment.

Safety, standards, and scale-up: Safety and standardization are critical for industrial translation. Nanoparticle design is increasingly guided by toxicological principles, with surface properties engineered to minimize the generation of ROS and bioaccumulation. Benchmark targets include hemocompatibility above 90% and negative Ames mutagenicity test results. Life cycle assessments are being integrated into phase–gate criteria to evaluate gate-to-gate energy use, carbon intensity, and solvent recovery. Technology readiness level advancement is now tied to ISO-compliant characterization and GLP/GMP alignment. Regulatory convergence is being supported through the development of consensus standard operating procedures, materials passports, hazard banding, and release specifications, which reduce approval friction and enable green certification for procurement. Manufacturing readiness is being achieved through modular skids equipped with inline PAT, automated dosing systems, and closed-loop purification via membranes and centrifugation, delivering GMP-compliant production with process capability indices (C_{pk}) above 1.33 for critical attributes.

Roadmap and actionable targets: To accelerate progress, the field must prioritize several key objectives. Mechanistic mapping of dominant reductants and enzymes across top-performing biosystems should be published alongside open-access kinetic models. Reactor-scale demonstrations of continuous production at 1–10 kg/day with PDI below 0.15 and solvent recovery above 95% are essential. Performance benchmarks must include energy-relevant metrics such as ORR activity, hydrogen evolution reaction (HER) overpotential below 100 mV, and battery capacity retention above 80% after 500 cycles. Safety protocols should be standardized through tiered toxicological and ecotoxicological panels, supported by life cycle reporting packages for regulatory and funding submissions. Finally, circularity targets should mandate sourcing at least 50% of precursors and reductants from certified waste streams without compromising material specifications. Together, these trends signal the emergence of a mature discipline, one that is mechanistically informed, digitally optimized, circular by design, and engineered for energy-relevant performance at an industrial scale.

References (order first cited)

[1] Liu, J., Birss, V., & Hill, J. (2010). Electrochemical performance and microstructure characterization of nickel yttrium-stabilized zirconia anode. *AIChE Journal*, 56(6), 1651–1658.

[2] Castillo-Henríquez, L., Alfaro-Aguilar, K., Ugalde-Álvarez, J., Vega-Fernández, L., Montes de Oca-vásquez, G., & Vega-Baudrit, J. R. (2020). Green synthesis of gold and silver nanoparticles from plant extracts and their possible applications as antimicrobial agents in the agricultural area. *Nanomaterials*, 10(9), 1763.

[3] Verma, A., Gautam, S. P., Bansal, K. K., Prabhakar, N., & Rosenholm, J. M. (2019). Green nanotechnology. *Advancement in Phytoformulation Research. Medicines*, 6(1), 39.

[4] Thunugunta, T., Reddy, A. C., & Reddy, D. C. L. (2015). Green synthesis of nanoparticles: Current prospectus. *Nanotechnology Reviews*, 4(4), 303–323.

[5] Patil, N., Bhaskar, R., Vyavhare, V., Dhadge, R., Khaire, V., & Patil, Y. (2021). Overview on methods of synthesis of nanoparticles. *International Journal of Current Pharmaceutical Research*, 13(2), 11–16.

[6] Hossain, A., Abdallah, Y., Ali, M. A., Masum, M. M. I., Li, B., Sun, G., & An, Q. (2019). Lemon-fruit-based green synthesis of zinc oxide nanoparticles and titanium dioxide nanoparticles against soft rot bacterial pathogen Dickeya Dadantii. *Biomolecules*, 9(12), 863.

[7] Cherian, T., Ali, K., Saquib, Q., Faisal, M., Wahab, R., & Musarrat, J. (2020). Cymbopogon citratus functionalized green synthesis of CuO-nanoparticles: Novel prospects as antibacterial and antibiofilm agents. *Biomolecules*, 10(2), 169.

[8] Srihasam, S., Thyagarajan, K., Korivi, M., Lebaka, V. R., & Mallem, S. P. R. (2020). Phytogenic generation of NiO nanoparticles using Stevia leaf extract and evaluation of their in-vitro antioxidant and antimicrobial properties. *Biomolecules*, 10(1), 89.

[9] Singh, J., Dutta, T., Kim, K. H., Rawat, M., Samddar, P., & Kumar, P. (2018). Green's synthesis of metals and their oxide nanoparticles: Applications for environmental remediation. *Journal of Nanobiotechnology*, 16(1), 84.

[10] Ansari, M. A., Murali, M., Prasad, D., Alzohairy, M. A., Almatroudi, A., Alomary, M. N., & Niranjana, S. R. (2020). Cinnamomum verum bark extract-mediated green synthesis of ZnO nanoparticles and their antibacterial potential. *Biomolecules*, 10(2), 336.

[11] Kharissova, O. V., Kharisov, B. I., Oliva González, C. M., Méndez, Y. P., & López, I. (2019). Greener synthesis of chemical compounds and materials. *Royal Society Open Science*, 6(11), 191378.

[12] EPA (2019) https://www.epa.gov/greenchemistry,https://files.nano.gov/wp-content/uploads/2025/06/EPA.png

[13] Saif, S., Tahir, A., & Chen, Y. (2016). Green synthesis of iron nanoparticles and their environmental applications and implications. *Nanomaterials*, 6(11), 209.

[14] Islam, M. M. (2016). Quantum Confinement-Size effects and properties of nanostructures in Nanoscience and Nanotechnology I: *Engineering and Technology*, https://ebooks.inflibnet.ac.in/msp08/chapter/quantum-confinement-size-effects-and-properties-of-nanostructures/

[15] Lee, M., Han, S. I., Kim, C., Velumani, S., Han, A., Kassiba, A. H., & Castaneda, H. (2022). ZrO_2/ZnO/TiO_2 nanocomposite coatings on stainless steel for improved corrosion resistance, biocompatibility, and antimicrobial activity. *ACS Applied Materials & Interfaces*, 14(11), 13801–13811.

[16] Argyriou, D. N., & Howard, C. J. (1995). Re-investigation of Yttria-Tetragonal Zirconia Polycrystal (Y-TZP) by Neutron Powder Diffraction-a Cautionary Tale. *Applied Crystallography*, 28(2), 206–208.

[17] Ijaz, I., Gilani, E., Nazir, A., & Bukhari, A. (2020). Detailed review on chemical, physical, and green synthesis, classification, characterizations, and applications of nanoparticles. *Green Chemistry Letters and Reviews*, 13(3), 223–245.

[18] Ibrahim Khan, K. S., & Khan, I. (2019). Nanoparticles: Properties, applications, and toxicities. *Arabian Journal of Chemistry*, 12(7), 908–931.

[19] Pandey, R. K., & Prajapati, V. K. (2018). Molecular and immunological toxic effects of nanoparticles. *International Journal of Biological Macromolecules*, 107, 1278–1293.

[20] Wyckoff, R. W. G., & Wyckoff, R. W. (1963). *Crystal Structures* (Vol. 1, pp. 239–444). New York: Interscience Publishers.

[21] Zhou, X. F., Qian, G. R., Dong, X., Zhang, L., Tian, Y., & Wang, H. T. (2010). Ab initio study of the formation of transparent carbon under pressure. *Physical Review B*, 82(13), 134126.

[22] Bundy, F. P., & Kasper, J. S. (1967). Hexagonal diamond - a new form of carbon. *The Journal of Chemical Physics*, 46(9), 3437–3446.

[23] Frondel, C., & Marvin, U. B. (1967). Lonsdaleite is a hexagonal polymorph of diamond. *Nature*, 214(5088), 587–589.

[24] Tersoff, J., & Ruoff, R. S. (1994). Structural properties of a carbon nanotube crystal. *Physical Review Letters*, 73(5), 676.

[25] Kroto, H. W., Heath, J. R., O'Brien, S. C., Curl, R. F., & Smalley, R. E. (1985). C60: Buckminsterfullerene. *Nature*, 318(6042), 162–163.

[26] Jette, E. R., & Foote, F. (1935). Precision determination of lattice constants. *The Journal of Chemical Physics*, 3(10), 605–616.

[27] Movchan, B. A., & Demchishin, A. V. (1969). Structure and properties of thick condensates of nickel, titanium, tungsten, aluminum oxides, and zirconium dioxide in vacuum. *Fizika Metallov I Metallovedenie*. 28: 653–660 (Oct 1969).

[28] Wyckoff, R. W. G. (1960). *Crystal Structures: Chapt. 1-4* (Vol. 1). Interscience Publishers.

[29] Liu, X., & Prewitt, C. T. (1990). High-temperature X-ray diffraction study of Co_3O_4: Transition from normal to disordered spinel. *Physics and Chemistry of Minerals*, 17(2), 168–172.

[30] Wever, F., & Müller, A. (1930). Über das Zweistoffsystem Eisen-Bor und über die Struktur des Eisenborides Fe_4B_2. *Zeitschrift Für Anorganische Und Allgemeine Chemie*, 192(1), 317–336.

[31] Iguchi, M., Umezu, M., Kataoka, M., Nakamura, H., & Ishii, M. (2006). Influence of nitride on sinterability of the composite of lithium aluminum silicate and silicon carbide. *Key Engineering Materials*, 317, 177–180.

[32] Bae, D., Zhen, S., & Seff, K. (1999). Structure of dehydrated Zn^{2+}-exchanged zeolite X. Overexchange, framework dealumination and reorganization, stoichiometric retention of monomeric tetrahedral aluminate. *The Journal of Physical Chemistry B*, 103(27), 5631–5636.

[33] Oswald, H. R., Reller, A., Schmalle, H. W., & Dubler, E. (1990). Structure of copper (II) hydroxide, $Cu(OH)_2$. *Crystal Structure Communications*, 46(12), 2279–2284.

[34] Prakash, P., Gnanaprakasam, P., Emmanuel, R., Arokiyaraj, S., & Saravanan, M. (2013). Green synthesis of silver nanoparticles from leaf extract of Mimusops elengi, Linn. for enhanced antibacterial activity against multidrug-resistant clinical isolates. *Colloids And Surfaces B: Biointerfaces*, 108, 255–259.

[35] Lin, Z., Kim, I. Y., & Personick, M. L. (2025). Nanocatalysis-facing a sustainable future. *Nanoscale*, 17(8), 4172–4174. https://pubs.rsc.org/en/content/articlehtml/2025/nr/d4nr90242d

[36] Singh, P., Kim, Y.-J., Zhang, D., & Yang, D.-C. (2016). Biological synthesis of nanoparticles from plants and microorganisms. Trends in Biotechnology, 34(7), 588 –599

[37] Słoma, M. (2023). 3D printed electronics with nanomaterials. *Nanoscale*, 15(12), 5623–5648. https://pubs.rsc.org/en/content/articlehtml/2023/nr/d2nr06771d

[38] Pomerantseva, E., Bonaccorso, F., Feng, X., Cui, Y., & Gogotsi, Y. (2019). Energy storage: The future enabled by nanomaterials. *Science*, 366(6468), eaan8285. https://www.science.org/doi/10.1126/science.aan8285

[39] Yentekakis, I. V., Gournis, D. P., & Karakassides, M. A. (2023). Nanomaterials in catalysis applications. *Catalysts*, 13(3), 627. https://www.mdpi.com/2073-4344/13/3/627

[40] Lakhani, P., Bhanderi, D., & Modi, C. K. (2024). Nanocatalysis: Recent progress, mechanistic insights, and diverse applications. *Journal of Nanoparticle Research*, 26(7), 148. https://link.springer.com/article/10.1007/s11051-024-06053-9

[41] Pan, Y., Guo, Y., Zhao, M., Li, C., & Yang, B. (2021). Multiple "Hot exciton" channel molecular design in organic electroluminescence materials: A theoretical investigation. *Materials Advances*, 2(4), 1351–1357.

[42] Shaheen, S., Ali, S. A., Mir, U. F., Sadiq, I., & Ahmad, T. (2023). Recent advances in transition metal phosphide nanocatalysts for H2 evolution and CO_2 reduction. *Catalysts*, 13(7), 1046.

[43] Kong, Q., Yang, J., Lu, Q., Li, H., Yuan, Z., Liu, H., & Zhang, X. (2025). Fe-doping engineered dual-active-domain CeO_x-CoP/Ni_2P heterostructure with superwettability interfaces for industrial-current-density alkaline hydrogen production. *Chemical Engineering Journal*, 171272.

[44] Jain, N., & Waidi, Y. O. (2025). The Multifaceted Role of 3D Printed Conducting Polymers in Next-Generation Energy Devices: A Critical Perspective. *JACS Au*, 5(2), 411–425.
[45] De Volder, M. F., Tawfick, S. H., Baughman, R. H., & Hart, A. J. (2013). Carbon nanotubes: Present and future commercial applications. *Science*, 339(6119), 535–539.
[46] Xu, L., Papanikolaou, K. G., Lechner, B. A., Je, L., Somorjai, G. A., Salmeron, M., & Mavrikakis, M. (2023). Formation of active sites on transition metals through reaction-driven migration of surface atoms. *Science*, 380(6640), 70–76.
[47] Corti, D., Misasi, J., Mulangu, S., Stanley, D. A., Kanekiyo, M., Wollen, S., & Sullivan, N. J. (2016). Protective monotherapy against lethal Ebola virus infection by a potently neutralizing antibody. *Science*, 351(6279), 1339–1342.
[48] Nigro, R. L., Fiorenza, P., Pécz, B., & Eriksson, J. (2022). Nanotechnology for electronic materials and devices. *Nanomaterials*, 12(19), 3319. https://www.mdpi.com/journal/nanomaterials/special_issues/nano_electronic_device
[49] Salaudeen, H. D., Akinniranye, R. D., Kolawole, M. I., & Yahaya, S. (2024). Nanomaterials in electronics: Advancements and challenges in high-performance devices. *World Journal of Advanced Research and Reviews*, 24(1), 830–845. https://wjarr.com/sites/default/files/WJARR-2024-3116.pdf
[50] Farooqui, M. A., Chauhan, P. S., Krishnamoorthy, P., & Shaik, J. (2010). Extraction of silver nanoparticles from the leaf extracts of Clerodendrum Inerme. *Digest Journal of Nanomaterials and Biostructures*, 5(1), 43–49.
[51] Mohammed, H., Mia, M. F., Wiggins, J., & Desai, S. (2025a). Nanomaterials for energy storage systems-a review. *Molecules*, 30(4), 883. https://www.mdpi.com/1420-3049/30/4/883
[52] Rani, E. E., Sanjana, D. S., Karthikeyan, E., & Nandhini, J. (2025). Smart Nanomaterials: Current State and Future Prospects in *Drug Delivery and Tissue Engineering. Biomedical Materials & Devices*, 1–28. https://link.springer.com/article/10.1007/s44174-025-00350-0
[53] Cheng, X., Xie, Q., & Sun, Y. (2023). Advances in nanomaterial-based targeted drug delivery systems. *Frontiers in Bioengineering and Biotechnology*, 11, 1177151. https://www.frontiersin.org/journals/bioengineering-and-biotechnology/articles/10.3389/fbioe.2023.1177151/full
[54] Najdian, A., Beiki, D., Abbasi, M., Gholamrezanezhad, A., Ahmadzadehfar, H., Amani, A. M., & Assadi, M. (2024). Exploring innovative strides in radiolabeled nanoparticle progress for multimodality cancer imaging and theranostic applications. *Cancer Imaging*, 24(1), 127. https://link.springer.com/article/10.1186/s40644-024-00762-z
[55] Tiwari, H., Singh, S., Kumar, R., Mandal, A., Pathak, A., Verma, N. K., & Gautam, V. (2025). Novel advancements in nanomaterials-based contrast agents across multimodal imaging and theranostic applications. *Nanoscale Advances*, 7(21), 6753–6773. https://pubs.rsc.org/en/content/articlelanding/2025/na/d5na00596e
[56] Priyanka, P., & Hemalatha, S. (2025). Nanomaterials for Bio-Imaging and Cancer Theranostics. *Current Pharmacology Reports*, 11(1), 1–19. https://link.springer.com/article/10.1007/s40495-025-00428-9
[57] Kashyap, B. K., Singh, V. V., Solanki, M. K., Kumar, A., Ruokolainen, J., & Kesari, K. K. (2023). Smart nanomaterials in cancer theranostics: Challenges and opportunities. *ACS Omega*, 8(16), 14290–14320. https://pubs.acs.org/doi/10.1021/acsomega.2c07840
[58] Yu, H. J., Liu, J. H., Liu, W., Niu, R., Zhang, B., Xiong, Y., & Zhang, H. J. (2025). Metal-based nanomedicines for cancer theranostics. *Military Medical Research*, 12(1), 41. https://link.springer.com/article/10.1186/s40779-025-00627-x
[59] Kandhola, G., Park, S., Lim, J. W., Chivers, C., Song, Y. H., Chung, J. H., & Kim, J. W. (2023). Nanomaterial-based scaffolds for tissue engineering applications: A review on graphene, carbon nanotubes, and nanocellulose. *Tissue Engineering and Regenerative Medicine*, 20(3), 411–433. https://link.springer.com/article/10.1007/s13770-023-00530-3

[60] Sadeghzadeh, H., Dianat-Moghadam, H., Del Bakhshayesh, A. R., Mohammadnejad, D., & Mehdipour, A. (2023). A review on the effect of nanocomposite scaffolds reinforced with magnetic nanoparticles in osteogenesis and healing of bone injuries. *Stem Cell Research & Therapy*, 14(1), 194. https://link.springer.com/article/10.1186/s13287-023-03426-0

[61] Emon, N. U., Zhang, L., Osborne, S. D., Lanoue, M. A., Huang, Y., & Tian, Z. R. (2025). Novel nanomaterials for developing bone scaffolds and tissue regeneration. *Nanomaterials*, 15(15), 1198. https://www.mdpi.com/2079-4991/15/15/1198

[62] Hosny, S., Mohamed, L. Z., Ragab, M. S., Alomoush, Q. K., Abdalla, E. M., & Aly, S. A. (2025). Nanomaterials in biomedical applications: Opportunities and challenges-a review. *Chemical Papers*, 1–22. https://link.springer.com/article/10.1007/s11696-025-03937-5

[63] Goshisht, M. K., Goshisht, A., Bajpai, A., & Bajpai, A. (2025). Recent advances in biomedical applications of smart nanomaterials: A comprehensive review. *RSC Pharmaceutics*. https://pubs.rsc.org/en/content/articlehtml/2025/pm/d5pm00137d

[64] Oh, Y. K., Hwang, K. R., Kim, C., Kim, J. R., & Lee, J. S. (2018). Recent developments and key barriers to advanced biofuels: A short review. *Bioresource Technology*, 257, 320–333.

[65] Energy Information Administration (US)Department of Energy (DOE) (Ed.), . (2010). Annual Energy Outlook 2001: With Projections to 2020. Government Printing Office. DOE/EIA-0383(2001) https://www.eia.gov/outlooks/archive/aeo01/

[66] Goyal, A., Meena, P. K., Shelare, S., & Kumar, S. (2025a). Advancing biofuel production through nanomaterials: A review of efficiency improvements, challenges, and future directions. *Nanocomposites*, 11(1), 151–172. https://www.tandfonline.com/doi/pdf/10.1080/20550324.2025.2518755

[67] Goyal, A., Meena, P. K., & Shelare, S. (2025b). Nanotechnology in Biofuel Production: Enhancing Efficiency and Sustainability Through Nanomaterials. *Journal of Cluster Science*, 36(3), 87. https://link.springer.com/article/10.1007/s10876-025-02797-z

[68] Mohammed, Z. S. (2025b). Advancing Bioenergy and Biofuels: Role of Nanotechnology in Sustainable Energy Systems. *Journal of Environmental Nanotechnology*, 14(2), 82–95. https://nanoient.org/journals/index.php/jent/article/view/2061

[69] Farajollahi, A. (2025). Nanotechnology in solar energy: From active systems to Advanced Solar cells. *Frontiers In Energy Research*, 13, 1560718. https://www.frontiersin.org/journals/energy-research/articles/10.3389/fenrg.2025.1560718/full

[70] Munir, S., Khan, U. U., Ullah, A., Hussain, A., Rehman, U., & Jan, M. F. (2025). Nanomaterials in solar cells: A review of current advances. *Global Scientific Journal*, 13(7), 392. https://www.globalscientificjournal.com/researchpaper/Nanomaterials_In_Solar_Cells_A_Review_Of_Current_Advances.pdf

[71] Jadhav, M. V., Todkar, A. S., Gambhire, V. R., & Sawant, S. Y. (2011). Nanotechnology for powerful solar energy. *International Journal of Advanced Biotechnology and Research*, 2(1), 208–212. https://www.researchgate.net/profile/Supriya-More/publication/315892213_Nanotechnology_for_powerful_solar_energy/links/58ec911aaca272bd287607a5/Nanotechnology-for-powerful-solar-energy.pdf

[72] Pandey, G. (2018). Nanotechnology for achieving green economy through sustainable energy. *RASĀYAN Journal of Chemistry*, 11(3), 942–950. https://www.researchgate.net/profile/Garima_Pandey12/publication/326732293_Nanotechnology_for_achieving_green-economy_through_sustainable_energy/links/60f681a40859317dbdf4f29a/Nanotechnology-for-achieving-green-economy-through-sustainable-energy.pdf

[73] Zhou, Q., Duan, J., Duan, Y., & Tang, Q. (2022). Review on engineering two-dimensional nanomaterials for promoting efficiency and stability of perovskite solar cells. *Journal Of Energy Chemistry*, 68, 154–175.

[74] Rong, Y., Hu, Y., Mei, A., Tan, H., Saidaminov, M. I., Seok, S. I., & Han, H. (2018). Challenges for commercializing perovskite solar cells. *Science*, 361(6408), eaat8235.

[75] Abdel-Mottaleb, M. S. A., Byrne, J. A., & Chakarov, D. (2011). Nanotechnology and solar energy. *International Journal of Photoenergy*, 2011(1), 194146.
[76] Moore, K., & Wei, W. (2021). Applications of carbon nanomaterials in perovskite solar cells for solar energy conversion. *Nano Materials Science*, 3(3), 276–290.
[77] Verma S. K. & Tiwari A. K. (2015). Application of nanoparticles in solar collectors: A review. *Materials Today: Proceedings*, 2(4-5), 3638–3647.
[78] Eduarda. (2024, October 10). Nanotechnology: Boosting solar panel efficiency by 15% in 2025 https://ecoenergystream.com/nanotechnology-boosting-solar-panel-efficiency-by-15-in-2025/
[79] U.S. Department of Energy, DOE/EERE Solar Energy Technologies Office. (2022). Fiscal Year 2022 Photovoltaics Research and Development (PVRD) Funding Program. https://www.energy.gov/eere/solar/seto-fiscal-year-2022-photovoltaics-research-and-development-pvrd-funding-program
[80] Zheng, L., & Xuan, Y. (2021). Performance estimation of a V-shaped perovskite/silicon tandem device: A case study based on a bifacial heterojunction silicon cell. *Applied Energy*, 301, 117496.
[81] Acosta-Silva, Y. D. J., Torres-Pacheco, I., Matsumoto, Y., Toledano-Ayala, M., Soto-Zarazúa, G. M., Zelaya-Ángel, O., & Méndez-López, A. (2019). Applications of solar and wind renewable energy in agriculture: A review. *Science Progress*, 102(2), 127–140.
[82] Muzammil, W. K., Rahman, M. M., Fazlizan, A., Ismail, M. A., Phang, H. K., & Elias, M. A. (2018). Nanotechnology in renewable energy: Critical reviews for wind energy. In *Nanotechnology: Applications In Energy, Drug and Food* (pp. 49–71). Cham: Springer International Publishing.
[83] Li, X., Gao, Q., Cao, Y., Yang, Y., Liu, S., Wang, Z. L., & Cheng, T. (2022). Optimization strategy of wind energy harvesting via triboelectric-electromagnetic flexible cooperation. *Applied Energy*, 307, 118311.
[84] U.S. Department of Energy, Wind Energy Technologies Office. (2025). Wind energy funding opportunities. https://www.energy.gov/eere/wind/wind-energy-funding-opportunities
[85] Congressional Research Service (CRO). (2024). Energy Efficiency and Renewable Energy (EERE) appropriations: FY2025 (CRS In Focus No. IF12710). https://www.congress.gov/crs_external_products/IF/PDF/IF12710/IF12710.4.pdf
[86] International Energy Agency (IEA) Bioenergy . (2024). End-of-term report 2020-2025. https://www.ieabioenergy.com/wp-content/uploads/2025/06/IEA-Bioenergy-Annual-Report-2024.pdf
[87] International Energy Agency (IEA). (2023a). Renewables 2023: Analysis and forecast to 2028. https://www.iea.org/reports/renewables-2023
[88] Dadi, M., Siwale, W., Munalula, F., Ravi, S. N., Mundike, J., Ramasamy, S., & Mulenga, C. (2025). A comprehensive review of advances in bioenergy, including emerging trends and future directions. *Discover Energy*, 5(1), 26. https://link.springer.com/article/10.1007/s43937-025-00095-3
[89] Discover Energy. (2025). Discover Energy Australia. https://www.discoverenergy.com.au/ and https://www.discoverenergy.com.au/blog
[90] Bhoi, P. R., & Sarkar, S. (2025). The Co-Firing of Pine Biomass and Waste Coal in 100 and 600 MW Power Plants: A Sustainable Approach to Reduce GHG Emissions. *Sustainability*, 17(10), 4473.
[91] IEA Bioenergy. (2021). IEA Bioenergy annual report 2020. https://www.ieabioenergy.com/wp-content/uploads/2021/04/IEAB-Annual-Report-2020.pdf
[92] Rakos, C., & Prauhart, P. (2023). Pellet cookstoves: An affordable and sustainable modern clean cooking solution. https://www.worldbioenergy.org/uploads/Pellets%20a%20modern%20cooking%20solution.pdf
[93] U.S. Department of Energy (DOE), Bioenergy Technologies Office. (2024, September 10). (2024a). Fact sheet: Sustainable Aviation Fuel Grand Challenge-Tracking metrics and Mid-2024 dashboard. https://www.energy.gov/eere/bioenergy/articles/sustainable-aviation-fuel-grand-challenge-tracking-metrics-and-mid-2024
[94] Kourkoumpas, D. S., Sagani, A., Hull, A., Hull, A., Karellas, S., & Grammelis, P. (2024). Life cycle assessment of an innovative alcohol-to-jet process: The case for retrofitting a bioethanol plant for sustainable aviation fuel production: *Renewable Energy*, 228, 120512.

[95] Chong, C. T., & Ng, J. H. (2023). Limitations to sustainable renewable jet fuels production attributed to cost than energy-water-food resource availability. *Nature Communications*, 14(1), 8156.
[96] Wang, L. R., Zhang, Z. X., Nong, F. T., Li, J., Huang, P. W., Ma, W., & Sun, X. M. (2022). Engineering the xylose metabolism in Schizochytrium sp. to improve the utilization of lignocellulose. *Biotechnology for Biofuels and Bioproducts*, 15(1), 114.
[97] U.S. Department of Energy (DOE), Bioenergy Technologies Office. (2024, September 11). (2024b). Just announced: Federal interagency sustainable aviation fuel progress metrics and $12 million in funding to support the advancement of integrated biorefinery technologies. https://www.energy.gov/eere/bioenergy/articles/just-announced-federal-interagency-sustainable-aviation-fuel-progress
[98] International Energy Agency (IEA). (2025a). Liquid biofuel demand, main and accelerated case, 2024–2030. https://www.iea.org/data-and-statistics/charts/liquid-biofuel-demand-main-and-accelerated-case-2024-to-2030
[99] International Energy Agency (IEA). (2025b). Outlook for biogas and biomethane: Key findings. https://www.iea.org/reports/outlook-for-biogas-and-biomethane/key-findings
[100] Aratboni, H. A., Rafiei, N., Allaf, M. M., Abedini, S., Rasheed, R. N., Seif, A., & Morones-Ramírez, J. R. (2023). Nanotechnology: An outstanding tool for increasing and better exploitation of microalgae valuable compounds. *Algal Research*, 71, 103019.
[101] Htoo, N. Y. M., Pumas, C., Srinuanpan, S., Kee, P. E., Lan, J. C. W., Lim, J. W., & Khoo, K. S. (2024). Advancements in valorization of food waste resources by microalgae for sustainable biorefinery production. *Bioenergy Research*, 17(4), 2062–2081.
[102] Global Industry Analysts, Inc. (2025). Advanced biofuels: Global strategic industry report, 2025-2030. https://www.globenewswire.com/news-release/2025/06/19/3102123/28124/en/Advanced-Biofuels-Global-Strategic-Industry-Report-2025-2030-Tech-Innovations-Boost-Biofuel-Efficiency-Industry-Eyes-Localized-Supply-Chains.html
[103] International Energy Agency (IEA). (2023b). Bioenergy - Analysis. https://www.iea.org/reports/bioenergy-2
[104] U.S. Department of Energy (DOE, Bioenergy Technologies Office. (2023a). Bioenergy Technologies Office (BETO). https://www.energy.gov/eere/bioenergy/bioenergy-technologies-office
[105] European Parliament & Council of the European Union (EU). (2023). Directive (EU) 2023/2413 on the promotion of energy from renewable sources (RED III). https://eur-lex.europa.eu/eli/dir/2023/2413/oj
[106] U.S. Department of Energy (DOE). (2006). Hydrogen fuel cells: Fact sheet. https://www.californiahydrogen.org/wp-content/uploads/files/doe_fuelcell_factsheet.pdf
[107] U.S. Department of Energy (DOE), Hydrogen and Fuel Cell Technologies Office. (2025). Hydrogen and fuel cell technology basics. https://www.energy.gov/eere/fuelcells/hydrogen-and-fuel-cell-technology-basics
[108] Chen, J., Zhang, Q., Xu, N., Li, W., Yao, Y., Li, P., & Zhang, H. (2022). Roadmap to hydrogen society of Tokyo: Locating priority of hydrogen facilities based on multiple big data fusion. *Applied Energy*, 313, 118688.
[109] Cannon, J. S. (1994). Hydrogen vehicle programs in the USA. *International Journal of Hydrogen Energy*, 19(11), 905–909.
[110] Yoshida, T., & Kojima, K. (2015). Toyota MIRAI fuel cell vehicle and progress toward a future hydrogen society. *The Electrochemical Society Interface*, 24(2), 45.
[111] Meng, X., Gu, A., Wu, X., Zhou, L., Zhou, J., Liu, B., & Mao, Z. (2021). Status quo of China hydrogen strategy in the field of transportation and international comparisons. *International Journal of Hydrogen Energy*, 46(57), 28887–28899.
[112] Iida, S., & Sakata, K. (2019). Hydrogen technologies and developments in Japan. *Clean Energy*, 3(2), 105–113.

[113] Thananusak, T., Punnakitikashem, P., Tanthasith, S., & Kongarchapatara, B. (2020). The development of electric vehicle charging stations in Thailand: Policies, players, and key issues (2015-2020). *World Electric Vehicle Journal*, 12(1), 2.
[114] Page, B., Turan, G., Zapantis, A., Burrows, J., Consoli, C., Erikson, J., & Zhang, T. (2020). The global status of CCS 2020: vital to achieve net zero. https://www.h2knowledgecentre.com/content/researchpaper1679
[115] U.S. Department of Energy (DOE), Hydrogen and Fuel Cell Technologies Office. (2023b). U.S. National Hydrogen Strategy and Roadmap. https://www.hydrogen.energy.gov/library/roadmaps-vision/clean-hydrogen-strategy-roadmap
[116] International Renewable Energy Agency (IRENA). (2019). Hydrogen: A renewable energy perspective. https://www.irena.org/-/media/Files/IRENA/Agency/Publication/2019/Sep/IRENA_Hydrogen_2019.pdf
[117] Osman, A. I., Mehta, N., Elgarahy, A. M., Hefny, M., Al-Hinai, A., Al-Muhtaseb, A. A. H., & Rooney, D. W. (2022). Hydrogen production, storage, utilisation and environmental impacts: A review. *Environmental Chemistry Letters*, 20(1), 153–188.
[118] Mao, S. S., Shen, S., & Guo, L. (2012). Nanomaterials for renewable hydrogen production, storage, and utilization. *Progress in Natural Science: Materials International*, 22(6), 522–534.
[119] Okonkwo, E. C., Abdullatif, Y. M., & Tareq, A. A. (2021). A nanomaterial integrated technology approach to enhance the energy-water-food nexus. *Renewable and Sustainable Energy Reviews*, 145, 111118.
[120] Water, Energy & Food Security Nexus Resource Platform. (2025). The water, energy & food security resource platform. https://www.water-energy-food.org/
[121] Adaikalam, K., Vikraman, D., Karuppasamy, K., & Kim, H. S. (2024). Solar hydrogen production and storage in solid form: Prospects for materials and methods. *Nanomaterials*, 14(19), 1560. www.mdpi.com/journal/nanomaterials/special_issues/QU86U453OD
[122] Shah, T. A., Li, Z., Zhang, A., & Lu, D. (2024). Energy, food, and water nexus, green options. In *Technical and Technological Solutions Towards A Sustainable Society and Circular Economy* (pp. 379–394). Cham: Springer Nature Switzerland. https://link.springer.com/chapter/10.1007/978-3-031-56292-1_31
[123] Aakash, L., Vinduja, B. S., Arunkumar, T., Sivakumar, T., & Gajalakshmi, D. (2020). A systematic review on green synthesis of nanoparticles and its medical applications. *Plant Archives*, 20(2):6069–6076. http://www.researchgate.net/publication/343760340
[124] Sharma, D., Kanchi, S., & Bisetty, K. (2019). Biogenic synthesis of nanoparticles: A review. *Arabian Journal of Chemistry*, 12(8), 3576–3600.
[125] Dikshit, P. K., Kumar, J., Das, A. K., Sadhu, S., Sharma, S., Singh, S., & Kim, B. S. (2021). Green synthesis of metallic nanoparticles: *Applications and limitations. Catalysts*, 11(8), 902.
[126] Anbalagan, K., Kaviya, P., Vijayalakshmi, V., Ananthi, K., Sivakumar, C., & Rani, A. (2024). Effect Of Botanicals in Neem (*Azadiracha Indica*) Leaf On Growth of Field Crops. *Authorea Preprints*. https://www.authorea.com/doi/full/10.22541/au.172900090.00882498
[127] Kirubakaran, D., Wahid, J. B. A., Karmegam, N., Jeevika, R., Sellapillai, L., Rajkumar, M., & SenthilKumar, K. J. (2026). A comprehensive review on the green synthesis of nanoparticles: Advancements in biomedical and environmental applications. *Biomedical Materials & Devices*, 4(1), 388–413. https://link.springer.com/article/10.1007/s44174-025-00295-4
[128] Baig, J. A., Akhtar, K., & Talpur, F. N. (2024). Production of alginate oligosaccharides and their plant growth-promoting activity: Effect of ionizing radiation processing. In *Handbook Natural Polymers* (pp. 377–399). Elsevier.
[129] Liem, L. N., The, N. P., & Nguyen, D. (2019). Microwave-assisted green synthesis of silver nanoparticles using Mulberry leaves extract and silver nitrate solution. *Technologies*, 7(1), 7. https://www.mdpi.com/2227-7080/7/1/7

[130] Deshmukh, A. R., Gupta, A., & Kim, B. S. (2019). Ultrasound-assisted green synthesis of silver and iron oxide nanoparticles using fenugreek seed extract and their enhanced antibacterial and antioxidant activities. *Biomed Research International*, 2019(1), 1714358. https://onlinelibrary.wiley.com/doi/full/10.1155/2019/1714358

[131] Priya, A. K., Krishna, R. S., Kumar, Y. G., Rohini, P., & Dillibabu, S. (2021, August). Green synthesis of silver nanoparticles by pulsed laser ablation using citrus limetta juice extract for clad-modified fiber optic gas sensing application. In *Nanoengineering: Fabrication, Properties, Optics, Thin Films, and Devices* XVIII (Vol. 11802, pp. 184-193). https://www.spiedigitallibrary.org/conference-proceedings-of-spie/11802/1180214/Green-synthesis-of-Silver-Nanoparticles-by-Pulsed-Laser-ablation-using/10.1117/12.2597642.short

[132] Nguyena, N., & Nguyen, V. (2023). Ultrasound-assisted sol-gel synthesis, characterization, and photocatalytic application of ZnO nanoparticles. *Digest Journal of Nanomaterials and Biostructures*, 18, 889–897. https://chalcogen.ro/889_NguyenNT.pdf

[133] González-Lavín, J., Arenillas, A., & Rey-Raap, N. (2025). Scalable Microwave-Assisted Synthesis of Iron Oxide Aerogels: Tuning Morphology, Crystallinity, and Electrocatalytic Performance. *Electrochimica Acta*, 147927.

[134] Abady, M. M., Mohammed, D. M., Soliman, T. N., Shalaby, R. A., & Sakr, F. A. (2025). Sustainable synthesis of nanomaterials using different renewable sources. *Bulletin of the National Research Centre*, 49(1), 24. https://link.springer.com/article/10.1186/s42269-025-01316-4

[135] Gavhane, K. D., Kharat, M. G., Gawali, N. B., Anwane, H., & Mhaske, S. (2025). Green synthesis of nanoparticles: A sustainable approach to nanotechnology. *International Journal of Research Publication and Reviews*, 6(5), 18616–18631. https://ijrpr.com/uploads/V6ISSUE5/IJRPR47348.pdf

[136] Abed, L. H. (2025). Preparation of TiO_2 nanoparticles using microwave and techniques of pulsed laser ablation and their properties. *International Journal of Research and Review*, 12(4), 316–324. https://doi.org/10.52403/ijrr.20250439

[137] Naderi-Samani, H., & Razavi, R. S. (2024). Synthesis of nanoparticles using pulsed laser. *Pulsed Laser Processing of Materials* D. Yang and K. Gibson (ed.), IntechOpen. https://www.intechopen.com/chapters/1179096

[138] Ye, F., & Musselman, K. P. (2024). Synthesis of low-dimensional nanomaterials by pulsed laser ablation in liquid. *APL Materials*, 12(5). https://pubs.aip.org/aip/apm/article/12/5/050602/3293177/

[139] Das, M., & Chatterjee, S. (2019). Green synthesis of metal/metal oxide nanoparticles toward biomedical applications: Boon or bane. In *Green Synthesis, Characterization and Applications of Nanoparticles* (pp. 265–301). Elsevier. https://doi.org/10.1016/B978-0-08-102579-6.00011-3 https://pubs.aip.org/aip/apm/article/12/5/050602/3293177

[140] Schmidt, R., Prado-Gonjal, J., & Morán, E. (2022). Microwave-assisted hydrothermal synthesis of nanoparticles. *Arxiv Preprint*:2203.02394. https://arxiv.org/abs/2203.02394

[141] Zhuang, W., Yuan, D., Li, J. R., Luo, Z., Zhou, H. C., Bashir, S., & Liu, J. (2012). Highly potent bactericidal activity of porous metal-organic frameworks. *Advanced Healthcare Materials*, 1(2), 225–238.

[142] Singh, A., Vihinen, J., Frankberg, E., Hyvärinen, L., Honkanen, M., & Levänen, E. (2016). Pulsed laser ablation-induced green synthesis of TiO_2 nanoparticles and application of novel small-angle X-ray scattering technique for nanoparticle size and size distribution analysis. *Nanoscale Research Letters*, 11(1), 447.

[143] Shabalina, A. V., Svetlichnyi, V. A., & Kulinich, S. A. (2022). Green laser ablation-based synthesis of functional nanomaterials for generation, storage, and detection of hydrogen. *Current Opinion in Green and Sustainable Chemistry*, 33, 100566.

[144] Dell'Aglio, M., Alrifai, R., & De Giacomo, A. (2018). Nanoparticle-enhanced laser-induced breakdown spectroscopy (NELIBS), a first review. *Spectrochimica Acta Part B: Atomic Spectroscopy*, 148, 105–112.

[145] Popov, V., Hinkov, I., Diankov, S., Karsheva, M., & Handzhiyski, Y. (2015). Ultrasound-assisted green synthesis of silver nanoparticles and their incorporation in antibacterial cellulose packaging. *Green Processing and Synthesis*, 4(2), 125–131.
[146] Manjamadha, V. P., & Muthukumar, K. (2016). Ultrasound-assisted green synthesis of silver nanoparticles using a weed plant. *Bioprocess And Biosystems Engineering*, 39(3), 401–411.
[147] Ghotekar, S., Pagar, T., Pansambal, S., & Oza, R. (2020). A review on green synthesis of sulfur nanoparticles via plant extract, characterization, and its applications. *Advanced Journal of Chemistry-Section B*, 2(3), 128–143.
[148] Shahid, M., Dumat, C., Khalid, S., Schreck, E., Xiong, T., & Niazi, N. K. (2017). Foliar heavy metal uptake, toxicity, and detoxification in plants: A comparison of foliar and root metal uptake. *Journal of Hazardous Materials*, 325, 36–58.
[149] Ahmed, S., Ahmad, M., Swami, B. L., & Ikram, S. (2016a. A review on plants extract mediated synthesis of silver nanoparticles for antimicrobial applications: A green expertise. *Journal of Advanced Research*, 7(1), 17–28.
[150] Edo, G. I., Mafe, A. N., Ali, A. B., Akpoghelie, P. O., Yousif, E., Isoje, E. F., & Alamiery, A. A. (2025). Green biosynthesis of nanoparticles using plant extracts: mechanisms, advances, challenges, and applications. *Bionanoscience*, 15(2), 267. https://link.springer.com/article/10.1007/s12668-025-01883-w
[151] Mohamad, N. A. N., Arham, N. A., Jai, J., & Hadi, A. (2014). Plant extract as reducing agent in synthesis of metallic nanoparticles: A review. *Advanced Materials Research*, 832, 350–355.
[152] Verma, M. L., Dhanya, B. S., Thakur, M., Jeslin, J., & Jana, A. K. (2021). Plant-derived nanoparticles and their biotechnological applications. In *Comprehensive Analytical Chemistry* (Vol. 94, pp. 331–362). Elsevier.
[153] Santhoshkumar, J., Rajeshkumar, S., & Kumar, S. V. (2017). Phyto-assisted synthesis, characterization, and applications of gold nanoparticles-A review. *Biochemistry And Biophysics Reports*, 11, 46–57.
[154] Jadoun, S., Arif, R., Jangid, N. K., & Meena, R. K. (2021). Green synthesis of nanoparticles using plant extracts: A review. *Environmental Chemistry Letters*, 19(1), 355–374.
[155] Chung, I. M., Park, I., Seung-Hyun, K., Thiruvengadam, M., & Rajakumar, G. (2016). Plant-mediated synthesis of silver nanoparticles: Their characteristic properties and therapeutic applications. *Nanoscale Research Letters*, 11(1), 40.
[156] Allan, D. R., Clark, S. J., Dawson, A., McGregor, P. A., & Parsons, S. (2002). Pressure-induced polymorphism in phenol. *Structural Science*, 58(6), 1018–1024.
[157] Koo, C. H., & Kim, H. S. (1965). The crystal structure of nicotine dihydroiodide. *Journal of the Korean Chemical Society*, 9(3), 134–141.
[158] Buschiazzo, A., Ugalde, J. E., Guerin, M. E., Shepard, W., Ugalde, R. A., & Alzari, P. M. (2004). Crystal structure of glycogen synthase: Homologous enzymes catalyze glycogen synthesis and degradation. *The EMBO Journal*, 23(16), 3196–3205.
[159] Stern, F. T., & Beevers, C. A. (1950). The crystal structure of tartaric acid. *Acta Crystallographica*, 3(5), 341–346.
[160] Park, F., Gajiwala, K., Noland, B., Wu, L., He, D., Molinari, J., & Buchanan, S. (2004). The 1.59 A resolution crystal structure of TM0096, a flavin mononucleotide binding protein from Thermotoga Maritima. *Proteins*, 55(3), 772–774.
[161] Hisaki, I., Hiraishi, E., Sasaki, T., Orita, H., Tsuzuki, S., Tohnai, N., & Miyata, M. (2012). Crystal structure of quinine: The effects of vinyl and methoxy groups on molecular assemblies of cinchona alkaloids cannot be ignored. *Chemistry-An Asian Journal*, 7(11), 2607–2614.
[162] Kundrot, C. E., & Richards, F. M. (1987). Crystal structure of hen egg-white lysozyme at a hydrostatic pressure of 1000 atmospheres. *Journal of Molecular Biology*, 193(1), 157–170.
[163] Ueno, K. (1984). Structure of salicin, $C_{13}H_{18}O_7$. *Crystal Structure Communications*, 40(10), 1726–1728.

[164] Raynor, J. W., Minor, W., & Chruszcz, M. (2007). Dexamethasone at 119 K. *Structure Reports*, 63(6), o2791–o2793.
[165] Mohamad, N. A. N., Arham, N. A., Jai, J., & Hadi, A. (2014). Plant extract as reducing agent in synthesis of metallic nanoparticles: A review. *Advanced Materials Research*, 832, 350–355.
[166] Ciurzyńska, A., & Lenart, A. (2011). Freeze-drying application in food processing and biotechnology: A review. *Polish Journal of Food and Nutrition Sciences*, 61(3), 165–171.
[167] Mohd Zainol, M. K., Abdul-Hamid, A., Abu Bakar, F., & Pak Dek, S. (2009). Effect of different drying methods on the degradation of selected flavonoids in *Centella asiatica. International Food Research Journal*, 16(4), 531–537.
[168] Christopher, J. G., Saswati, B., & Ezilrani, P. (2015). Optimization of parameters for biosynthesis of silver nanoparticles using leaf extract of *Aegle marmelos. Brazilian Archives of Biology and Technology*, 58(5), 702–710.
[169] Cruz, D., Falé, P. L., Mourato, A., Vaz, P. D., Serralheiro, M. L., & Lino, A. R. L. (2010). Preparation and physicochemical characterization of Ag nanoparticles biosynthesized by *Lippia citriodora* (Lemon Verbena). *Colloids And Surfaces B: Biointerfaces*, 81(1), 67–73.
[170] Pourmortazavi, S. M., Taghdiri, M., Makari, V. & Rahimi-Nasrabadi, M. (2015). Procedure optimization for green synthesis of silver nanoparticles by aqueous extract of *Eucalyptus oleosa. Spectrochimica Acta Part A: Molecular and Biomolecular Spectroscopy*, 136, 1249–1254.
[171] Bezares, B., Jaña, Y., Cottet, L., & Castillo, A. (2018). Effect of pH and leaf extract concentration of Pittosporum undulatum on gold nanoparticles synthesis. *Materials Express*, 8(5), 450–456.
[172] Kredy, H. M. (2018). The effect of pH and temperature on the green synthesis and biochemical activities of silver nanoparticles from *Lawsonia inermis* extract. *Journal of Pharmaceutical Sciences and Research*, 10(8), 2022–2026.
[173] Perveen, R., Shujaat, S., Naz, M., Qureshi, M. Z., Nawaz, S., Shahzad, K., & Ikram, M. (2021). Green synthesis of antimicrobial silver nanoparticles with *Brassicaceae* seeds. *Materials Research Express*, 8(5), 055007.
[174] Kuppusamy, P., Yusoff, M. M., Maniam, G. P., & Govindan, N. (2016). Biosynthesis of metallic nanoparticles using plant derivatives and their new avenues in pharmacological applications-An updated report. *Saudi Pharmaceutical Journal*, 24(4), 473–484.
[175] Villaseñor-Basulto, D. L., Pedavoah, M. M., & Bandala, E. R. (2019). Plant materials for the synthesis of nanomaterials: Greener sources. In *Handbook of Ecomaterials* (pp. 105–121). Springer, Cham.
[176] Naikoo, G. A., Mustaqeem, M., Hassan, I. U., Awan, T., Arshad, F., Salim, H., & Qurashi, A. (2021). Bioinspired and green synthesis of nanoparticles from plant extracts with antiviral and antimicrobial properties: A critical review. *Journal of Saudi Chemical Society*, 25(9), 101304.
[177] Qamer, S., Romli, M. H., Che-Hamzah, F., Misni, N., Joseph, N. M., Al-Haj, N. A., & Amin-Nordin, S. (2021). Systematic review on biosynthesis of silver nanoparticles and antibacterial activities: Application and theoretical perspectives. *Molecules*, 26(16), 5057.
[178] Dzimitrowicz, A., Berent, S., Motyka, A., Jamroz, P., Kurcbach, K., Sledz, W., & Pohl, P. (2019). Comparison of the characteristics of gold nanoparticles synthesized using aqueous plant extracts and natural plant essential oils of *Eucalyptus globulus* and Rosmarinus officinalis. *Arabian Journal of Chemistry*, 12(8), 4795–4805.
[179] Sutradhar, P., & Saha, M. (2015). Synthesis of zinc oxide nanoparticles using tea leaf extract and its application for solar cells. *Bulletin of Materials Science*, 38(3), 653–657.
[180] Gopinath, K., Kumaraguru, S., Bhakyaraj, K., Thirumal, S., & Arumugam, A. (2016). Eco-friendly synthesis of TiO_2, Au, and Pt doped TiO2 nanoparticles for dye sensitized solar cell applications and evaluation of toxicity. *Superlattices and Microstructures*, 92, 100–110.
[181] Basheer A.A., & Ali, I. (2019). Water photo-splitting for green hydrogen energy using green nanoparticles. *International Journal of Hydrogen Energy*, 44(23), 11564–11573.

[182] Soundarrajan, C., Sankari, A., Dhandapani, P., Maruthamuthu, S., Ravichandran, S., Sozhan, G., & Palaniswamy, N. (2012). Rapid biological synthesis of platinum nanoparticles using Ocimum sanctum for water electrolysis applications. *Bioprocess and Biosystems Engineering*, 35(5), 827–833.
[183] Peng, Y. P., Liu, C. C., Chen, K. F., Huang, C. P., & Chen, C. H. (2021). Green synthesis of nano-silver-titanium nanotube array (Ag/TNA) composite for concurrent ibuprofen degradation and hydrogen generation. *Chemosphere*, 264, 128407.
[184] Gahlawat, G., & Choudhury, A. R. (2019). A review on the biosynthesis of metal and metal salt nanoparticles by microbes. *RSC Advances*, 9(23), 12944–12967.
[185] Boroumand Moghaddam, A., Namvar, F., Moniri, M., Md. Tahir, P., Azizi, S., & Mohamad, R. (2015). Nanoparticles biosynthesized by fungi and yeast: A review of their preparation, properties, and medical applications. *Molecules*, 20(9), 16540–16565.
[186] Asmathunisha, N., & Kathiresan, K. (2013). A review on biosynthesis of nanoparticles by marine organisms. *Colloids And Surfaces B: Biointerfaces*, 103, 283–287.
[187] Campaña, A. L., Saragliadis, A., Mikheenko, P., & Linke, D. (2023). Insights into the bacterial synthesis of metal nanoparticles. *Frontiers in Nanotechnology*, 5, 1216921.
[188] Saravanan, A., Kumar, P. S., Karishma, S., Vo, D. V. N., Jeevanantham, S., Yaashikaa, P. R., & George, C. S. (2021). A review on biosynthesis of metal nanoparticles and its environmental applications. *Chemosphere*, 264, 128580.
[189] Farooqui, A., Tabassum, H., Ahmad, A., Mabood, A., Ahmad, A., & Ahmad, I. Z. (2016b). Role of nanoparticles in growth and development of plants: A Review. *International Journal of Pharmaceutical and Biological Sciences*, 7(4), 22–37.
[190] Singh, A., Gautam, P. K., Verma, A., Singh, V., Shivapriya, P. M., Shivalkar, S., & Samanta, S. K. (2020). Green synthesis of metallic nanoparticles as effective alternatives to treat antibiotics resistant bacterial infections: A review. *Biotechnology Reports*, 25, e00427.
[191] Hulkoti, N. I., & Taranath, T. C. (2014). Biosynthesis of nanoparticles using microbes-a review. *Colloids And Surfaces B: Biointerfaces*, 121, 474–483.
[192] Mohd Yusof, H., Mohamad, R., Zaidan, U. H., & Abdul Rahman, N. A. (2019). Microbial synthesis of zinc oxide nanoparticles and their potential application as an antimicrobial agent and a feed supplement in animal industry: A review. *Journal of Animal Science and Biotechnology*, 10(1), 57.
[193] Korde, P., Ghotekar, S., Pagar, T., Pansambal, S., Oza, R., & Mane, D. (2020). Plant extract assisted eco-benevolent synthesis of selenium nanoparticles-a review on plant parts involved, characterization and their recent applications. *Journal of Chemical Reviews*, 2(3), 157–168.
[194] Hashmi, M. Z., & Mughal, A. F. (2025). Microbial and chemically induced reductive dechlorination of polychlorinated biphenyls in the environment. *Environmental Science and Pollution Research*, 32(5), 2167–2181.
[195] Rouch, D. A., Lee, B. T., & Morby, A. P. (1995). Understanding cellular responses to toxic agents: A model for mechanism-choice in bacterial metal resistance. *Journal of Industrial Microbiology and Biotechnology*, 14(2), 132–141.
[196] Beveridge, T. J., Hughes, M. N., Lee, H., Leung, K. T., Poole, R. K., Savvaidis, I., & Trevors, J. T. (1996). Metal-microbe interactions: Contemporary approaches. *Advances in Microbial Physiology*, 38, 177–243.
[197] Iravani, S. (2014). Bacteria in nanoparticle synthesis: Current status and future prospects. *International Scholarly Research Notices*, 2014(1), 359316.
[198] Koul, B., Poonia, A. K., Yadav, D., & Jin, J. O. (2021). Microbe-mediated biosynthesis of nanoparticles: Applications and future prospects. *Biomolecules*, 11(6), 886.
[199] Mughal, B., Zaidi, S. Z. J., Zhang, X., & Hassan, S. U. (2021). Biogenic nanoparticles: Synthesis, characterisation and applications. *Applied Sciences*, 11(6), 2598.
[200] Abhilash, Revati, K., & Pandey, B. D. (2011). Microbial synthesis of iron-based nanomaterials-A review. *Bulletin of Materials Science*, 34(2), 191–198.

[201] Bahrulolum, H., Nooraei, S., Javanshir, N., Tarrahimofrad, H., Mirbagheri, V. S., Easton, A. J., & Ahmadian, G. (2021). Green synthesis of metal nanoparticles using microorganisms and their application in the agrifood sector. *Journal of Nanobiotechnology*, 19(1), 86.

[202] Prozorov, T., Bazylinski, D. A., Mallapragada, S. K., & Prozorov, R. (2013). Novel magnetic nanomaterials inspired by magnetotactic bacteria: *Topical Review. Materials Science and Engineering: R: Reports*, 74(5), 133–172.

[203] Jacob, J. M., Lens, P. N., & Balakrishnan, R. M. (2016). Microbial synthesis of chalcogenide semiconductor nanoparticles: A review. *Microbial Biotechnology*, 9(1), 11–21.

[204] Mukherjee, P., Ahmad, A., Mandal, D., Senapati, S., Sainkar, S. R., Khan, M. I., Ramani, R., Parischa, R., Ajayakumar, P. V., Alam, M., Sastry, M., & Kumar, R. (2001). Bioreduction of AuCl4− ions by the fungus Verticillium sp. and surface trapping of the gold nanoparticles formed. Angewandte Chemie International Edition, 40(19), 3585–3588. https://d1wqtxts1xzle7.cloudfront.net/50044279/Bioreduction_of_AuCl4_Ions_by_the_Fungus20161101-25822-veridf-libre.pdf?1478059407=&response-content-disposition=inline;filename=Bioreduction_of_AuCl4_Ions_by_the_Fungus.pdf&Expires=1767924029&Signature=MRXsoaCsblkhUdcw5N1K9YC3XFac~hr6MyLFSVTPw-y0vkatxMhiPDFe-8fG2JYfUl1nVlZ4-1KSWx-iEzvP2RxTxT1nmW1QFwAiKcjR2k5NiuRYJU269~eApe6oG7mnhZ-Yeyof6UawmErd2lh9F1RGJzCMWTXi8GxyXaJrQIctKqZkfd7oljJHZFsZxVgiRXprvdZdGFvULbqRXbLhlfw~uwblmei~u0QOknI9CenY88IkYm4RUhADUEn5KQ-gML4zgWqzcdbKRH8AUlCY66IyTVZ-EocLd1

[205] Saxena, J., Sharma, M. M., Gupta, S., & Singh, A. (2014). Emerging role of fungi in nanoparticle synthesis and their applications. *World Journal of Pharmaceutical Sciences*, 3(9), 1586–1613.

[206] Ali, S. A., Osman, M. E., & Mohamed, E. T. (2025a). Eco-friendly biosynthesis of silver nanoparticles using marine-derived Fusarium exquisite: Optimization, characterization, and evaluation of antimicrobial, antioxidant, and cytotoxic activities. *World Journal of Microbiology and Biotechnology*, 41(5), 1–18. https://link.springer.com/article/10.1007/s11274-025-04368-w

[207] Da Costa, T. S., Delgado, G. G., Braga, C. B., & Tasic, L. (2025). Insights into the fungal secretomes and their roles in the formation and stabilization of the biogenic silver nanoparticles. *RSC Advances*, 15(9), 6938–6951. https://pubs.rsc.org/en/content/articlehtml/2025/ra/d4ra07962k

[208] Sidhu, A. K., Agrawal, S. B., Verma, N., Kaushal, P., & Sharma, M. (2025). Fungal-mediated synthesis of multimetallic nanoparticles: Mechanisms, unique properties, and potential applications. *Frontiers in Nanotechnology*, 7, 1549713. https://www.frontiersin.org/journals/nanotechnology/articles/10.3389/fnano.2025.1549713/full

[209] Skalickova, S., Baron, M., & Sochor, J. (2017). Nanoparticles biosynthesized by yeast: A review of their application. *Kvasny Prumysl*, 63(6), 290–292.

[210] Roychoudhury, A. (2020). Yeast-mediated green synthesis of nanoparticles for biological applications. *Indian Journal of Pharmaceutical and Biological Research*, 8(03), 26–31.

[211] Banerjee, K., & Ravishankar Rai, V. (2018). A review on mycosynthesis, mechanism, and characterization of silver and gold nanoparticles. *BioNanoScience*, 8(1), 17–31. https://link.springer.com/article/10.1007/s12668-017-0437-8

[212] Kapoor, R. T., Salvadori, M. R., Rafatullah, M., Siddiqui, M. R., Khan, M. A., & Alshareef, S. A. (2021). Exploration of microbial factories for synthesis of nanoparticles-a sustainable approach for bioremediation of environmental contaminants. *Frontiers in Microbiology*, 12, 658294.

[213] Chaudhary, R., Nawaz, K., Khan, A. K., Hano, C., Abbasi, B. H., & Anjum, S. (2020). An overview of the algae-mediated biosynthesis of nanoparticles and their biomedical applications. *Biomolecules*, 10(11), 1498.

[214] Ali, E. M., Elsayed, A., & El Shehawy, A. S. (2025b). A novel green biosynthesis approach and structural characterization of Ag-Fe bimetallic nanoparticles using the red alga *Galaxaura rugosa. Scientific Reports*, 15(1), 31129. https://www.nature.com/articles/s41598-025-16009-1.pdf

[215] Mukherjee, A., Sarkar, D., & Sasmal, S. (2021). A review of green synthesis of metal nanoparticles using algae. *Frontiers in Microbiology*, 12, 693899. https://www.frontiersin.org/journals/microbiology/articles/10.3389/fmicb.2021.693899/full

[216] El-Beltagi, H. S., Ragab, M., Osman, A., El-Masry, R. A., Alwutayd, K. M., Althagafi, H., & El-Saber, M. M. (2024). Biosynthesis of zinc oxide nanoparticles via neem extract and their anticancer and antibacterial activities. *PeerJ*, 12, e17588.

[217] Saallah, S., & Lenggoro, I. W. (2018). Nanoparticles carrying biological molecules: Recent advances and applications. *KONA Powder and Particle Journal*, 35, 89–111.

[218] Wang, C., Gao, X., Chen, Z., Chen, Y., & Chen, H. (2017). Preparation, characterization, and application of polysaccharide-based metallic nanoparticles: A review. *Polymers*, 9(12), 689.

[219] Plucinski, A., Lyu, Z., & Schmidt, B. V. (2021). Nanoparticles: From fabrication to applications. *Journal of Materials Chemistry B*, 9(35), 7030–7062.

[220] Park, F., Gajiwala, K., Noland, B., Wu, L., He, D., Molinari, J., & Buchanan, S. (2004). The 1.59 A resolution crystal structure of TM0096, a flavin mononucleotide binding protein from *Thermotoga maritima. Proteins*, 55(3), 772–774.

[221] Sharma, V., Ambekar, A., & Singh, K. (2025). Exploring chitosan nanoparticles: Preparation methods, cross-linking strategies, and potential therapeutic applications. *Polymer Bulletin*, 1–50. https://link.springer.com/article/10.1007/s00289-025-05979-x

[222] Akdasci, E., Duman, H., Eker, F., Bechelany, M., & Karav, S. (2025). Chitosan and its nanoparticles: A multifaceted approach to antibacterial applications. *Nanomaterials*, 15(2), 126. https://www.mdpi.com/2079-4991/15/2/126

[223] DeFrates, K., Markiewicz, T., Gallo, P., Rack, A., Weyhmiller, A., Jarmusik, B., & Hu, X. (2018). Protein polymer-based nanoparticles: Fabrication and medical applications. *International Journal of Molecular Sciences*, 19(6), 1717. https://www.med.upenn.edu/shorterlab/Papers/MemberPapers/ProteinPolymer-BasedNanoparticles-FabricationandMedicalApplications.pdf

[224] Salama, A., & Guarino, V. (2022). Ionic liquids to process silk fibroin and wool keratin for bio-sustainable and biomedical applications. *Journal of Polymers and the Environment*, 30(12), 4961–4977. https://link.springer.com/article/10.1007/s10924-022-02592-1

[225] Afreen, A., Ahmed, R., Mehboob, S., Tariq, M., Alghamdi, H. A., Zahid, A., & Hasan, A. (2020). Phytochemical-assisted biosynthesis of silver nanoparticles from Ajuga bracteosa for biomedical applications. *Materials Research Express*, 7(7), 075404.

[226] UNEP. (2020). Solid waste management. United Nations Environment Programme. https://www.unep.org/events/publication-launch/launch-global-waste-management-outlook-2024

[227] Kenry, Yeo, T., Manghnani, P. N., Middha, E., Pan, Y., Chen, H., & Liu, B. (2020). Mechanistic understanding of the biological responses to polymeric nanoparticles. *ACS Nano*, 14(4), 4509–4522.

[228] Rajput, D., Paul, S., & Gupta, A. (2020). Green synthesis of silver nanoparticles using waste tea Leaves. *Advanced Nano Research*, 3(1), 1–14.

[229] Zhou, J., Zheng, F., Li, H., Wang, J., Bu, N., Hu, P., & Liu, J. L. (2020). Optimization of post-treatment variables to produce hierarchical porous zeolites from coal gangue to enhance adsorption performance. *Chemical Engineering Journal*, 381, 122698.

[230] Flores-Contreras, E. A., González-González, R. B., Pablo Pizaña-Aranda, J. J., Parra-Arroyo, L., Rodríguez-Aguayo, A. A., Iñiguez-Moreno, M., & Melchor-Martínez, E. M. (2024). Agricultural waste as a sustainable source for nanoparticle synthesis and their antimicrobial properties for food preservation. *Frontiers in Nanotechnology*, 6, 1346069. https://www.researchgate.net/publication/378260007_Agricultural_waste_as_a_sustainable_source_for_nanoparticle_synthesis_and_their_antimicrobial_properties_for_food_preservation

[231] Kimta, N., Dhalaria, R., Kuča, K., Cimler, R., Guleria, V., Guleria, S., & Kumar, H. (2024). Production of metallic nanoparticles from agriculture waste and their applications. In *Transforming Agriculture*

Residues for Sustainable Development: From Waste to Wealth (pp. 131–156). Cham: Springer Nature Switzerland. https://link.springer.com/chapter/10.1007/978-3-031-61133-9_6

[232] Bao, Y., He, J., Song, K., Guo, J., Zhou, X., & Liu, S. (2021). Plant-extract-mediated synthesis of metal nanoparticles. *Journal of Chemistry*, 2021(1), 6562687.

[233] Rauwel, P., Küünal, S., Ferdov, S., & Rauwel, E. (2015). A review on the green synthesis of silver nanoparticles and their morphologies studied via TEM. *Advances in Materials Science and Engineering*, 2015(1), 682749.

[234] Sakthivel, A., Chandrasekaran, R., Balasubramaniam, S., Sathyanarayanan, H., Gnanajothi, K., & T. S. (2025). Nanomaterials as potential plant growth modulators: Applications, mechanism of uptake, and toxicity: A comprehensive review. *BioNanoScience*, 15(1), 53.

[235] Kholopo, M., & Rathebe, P. C. (2025). Lifecycle, toxicology, and exposure pathways of functional nanomaterials in water treatment: Implications for environmental and human Health. *Sustainable Environment*, 11(1), 2558248.

[236] Singh, H., Dhanu, A. S., Joshi, A. S., Mijakovic, I., & Singh, P. (2026). Next-generation nanomaterials for environmental remediation: Smart design, hybrid materials and sustainable use. Frontiers in Chemistry, 14, 1772161. https://doi.org/10.3389/fchem.2026.1772161

[237] Sivasubramaniyan, P., Balakrishnan, P., Safa, M., & Saranya, G. (2025). Sustainable and eco-friendly nanomaterials: Innovations for green healthcare solutions. In Smart Nanomaterials in Biomedical Applications (pp. 423–469). Springer. https://doi.org/10.1007/978-981-96-8023-8_16

Saurabh Vyas, Vivek Anand, Roli Mishra*

Chapter 7
Synthesis and stabilization of metallic nanoparticles

Abstract: The development of varied methodologies for the synthesis of metal nanoparticles (MNPs) has gained tremendous boost because of the unique properties of the nanoparticles as compared to their bulk counterparts. The quantization of electronic effects results in special optical and magnetic properties. Moreover, a high surface-to-volume ratio promotes exceptional mechanical, thermal, and catalytic properties. Therefore, the synthesis and stabilization of MNPs are of paramount importance. Herein, we have highlighted the synthesis, stabilization, and applications of transition metallic nanoparticles like Pd, Ag, Au, Pt, Cu, Co, etc., using different methods, which include the sol–gel method, chemical reduction method, green method (natural resources), colloidal injection method, etc. Furthermore, we have elaborately discussed the different types of stabilizers used to stabilize the metallic nanoparticle. Moreover, the role of stabilizers and their effect on the size of nanoparticles have also been discussed.

Keywords: Metallic nanoparticles, stabilization, nanotechnology, transition metals

7.1 Introduction

Any particle with a diameter ranging between 1 and 100 nm in at least one dimension is defined as a nanoparticle [1]. Moreover, these materials can be zero-, one-, two-, or three-dimensional, depending on the shape. The properties of nanomaterials are

Acknowledgments: The authors RM and SV would like to thank IAR for supporting the research work. The author SV would like to recognize the Gujarat government for the Shodh research fellowship for carrying out the research. The author VA acknowledges Chandigarh University for its support.

Author contribution: The manuscript was written through the contributions of all the authors. All authors have approved the final version of the manuscript.

***Corresponding author: Roli Mishra**, Department of Engineering and Physical Sciences, Institute of Advanced Research, Gandhinagar 382426, Gujarat, India, e-mails: rolimishra2@gmail.com, roli.mishra@iar.ac.in
Saurabh Vyas, Department of Engineering and Physical Sciences, Institute of Advanced Research, Gandhinagar 382426, Gujarat, India, e-mail: saurabhvyas.phd2020@iar.ac.in
Vivek Anand, Department of Chemistry, University Institute of Science, Chandigarh University, Gharuan, Mohali 140413, Punjab, India, e-mail: vivekanandac88@gmail.com

 | https://doi.org/10.1515/9783111621111-008

unique in comparison with bulk materials. This is because of the better surface-to-volume ratio present in nanomaterials [2]. Because of their unique physical and chemical properties, the synthesis of metal nanoparticles (MNPs) exhibiting desired size, shape, morphology, surface area, etc., is significantly important. No surprise, nanomaterials have found their application in various fields such as catalysis [3], drug delivery [4], biosensors [5], paints and coatings [6], agriculture [7], wastewater treatment [8], etc.

Silver and gold have grabbed special attention among the noble metals because of their unparalleled applications in catalysis, anticancer, and antimicrobial properties. These metals in the nanoscale are excellent in exhibiting gram-negative and gram-positive bacteria. Nanoparticles can be synthesized by different methods, such as laser ablation [9], electro-explosion [10], and chemical vapor deposition [11], sol–gel synthesis [12]. The nanoparticles formed through chemical methods are often toxic to the environment. Therefore, biological methods involving living organisms such as bacteria, fungi, or plant extracts are applied for environmentally friendly and less toxic ways of obtaining nanoparticles.

Generally, there are two methods for the synthesis of nanoparticles: top-down and bottom-up methods for the synthesis of MNPs [13]. In the top-down approach, bulk materials are broken down into nano-sized materials. On the other hand, the bottom-up approach includes the synthesis of the nanoparticles atom by atom to nano-sized particles with desired properties. The former is easy to implement and hence more frequently applied. The latter one is environmentally benign and less costly. Thus, both techniques have their pros and cons, and are used according to the requirements. MNPs are very reactive due to their high surface-to-volume ratio. To effectively utilize the properties of MNPs, stabilizers are used so that the agglomeration of MNPs can be restricted [14]. Metallic nanoparticles easily lose their size and shape when subjected to thermal and chemical changes. Therefore, stabilization is very important to maintain the shape and size and gain the maximum output from metallic nanoparticles [15]. MNPs are stabilized by different stabilizers, such as steric stabilization, encapsulation of MNPs, and electrostatic stabilization via alloying [16]. The stability of nanoparticles is a challenge owing to their large surface energy. The utilization of stabilizing agents can prevent the coagulation of the particles to form bulk materials. These agents, through steric repulsion and electrostatic interaction, restrict coagulation. Examples include surfactants, polymers, ligands, etc.

The MNPs are characterized by the phenomenon of surface plasmon resonance (SPR), which is the vibration of the conducting electrons on interaction with the electromagnetic radiation. The noble MNPs show the surface plasmon phenomenon due to their closely located conduction and valence bands and free electrons, which can move freely in these bands. Distinguished color and signature UV-visible peaks characterize the SPR. Changes in particle size, shape, and surrounding environment can affect this surface plasmon phenomenon [17].

7.2 Synthesis and stabilization of transition MNPs

Transition MNPs are a vital class of MNPs with excellent properties, which include organic catalysis, electrocatalysis [18], energy storage, biological activity (antimicrobial, antibacterial), and optoelectronics [19]. The bimetallic nanoparticles are obtained when the transition metal is doped with noble MNPs to enhance its properties. Thus, these bimetallic nanoparticles have found applications in various fields. The synthesis of important MNPs like Cu, Ag, Au, Co, Pd, and Pt is discussed below.

7.2.1 Synthesis and stabilization of palladium nanoparticles (PdNPs)

Although an expensive metal, palladium finds significant importance in organic synthesis. PdNPs have excellent catalytic properties; hence, they are used in reactions like hydrogenation of alkynes or alkenes, petroleum cracking, and carbon–carbon cross-coupling reaction (Suzuki coupling and Heck coupling reaction). PdNPs can be prepared from different synthetic methods, such as sonochemical, chemical reduction, and electrochemical [20]. The most commonly used method is the synthetic method, in which the reduction of palladium salt is accomplished using various reducing agents. The reducing agents used are hydrazine, sodium borohydride, ascorbic acid, alcohol, etc. Because of the high surface area of the PdNPs, they are very reactive and hence need stabilizers such as polymers, dendrimers, micelles, citrate salts, and toluene [21]. Some important synthetic methodologies applied to synthesize palladium nanoparticles are discussed below.

Mallick et al. [22] synthesized PdNPs by reflux and γ-radiolysis, while aniline was used as a nanoparticle stabilizer. To synthesize PdNPs, the mixture of monomers, metal salt, and the solvent was bubbled with nitrogen gas to remove oxygen. Then, the mixture was refluxed for a few hours at a fixed temperature. Aniline was the best stabilizer among the other monomers like *N*-ethylaniline, *N*-methylaniline, *o*-anisidine, and *o*-toluidine (Scheme 7.1). PdNPs showed polydispersity when synthesized by the γ-irradiation method [22].

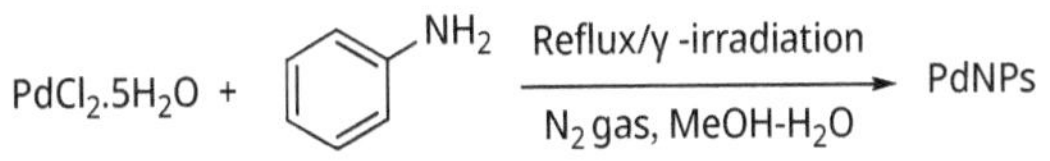

Scheme 7.1: Synthesis of PdNPs using aniline as a stabilizer.

Mallick et al. [23] reported the synthetic route for the metal–polymer uniform composite material, in which MNPs were dispersed in the host polymer. The *o*-methoxyaniline solution was prepared by dissolving it in toluene with stirring. Then, palladium acetate was added drop by drop to obtain the precipitates with a deep yellow color (Scheme 7.2). The PdNPs were stabilized in a polymer matrix under reducing conditions [23].

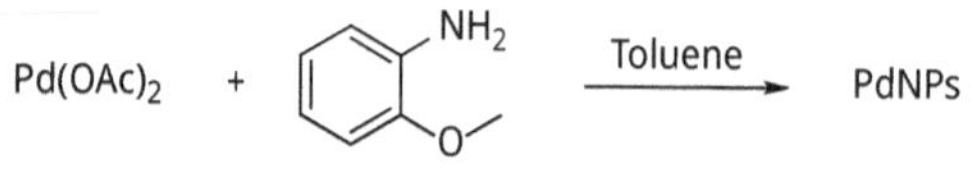

Scheme 7.2: Synthesis of PdNPs using *o*-methoxyaniline as a stabilizer.

Chen et al. [24] reported the synthesis of PdNPs. The H_2PdCl_4 and sodium citrate solutions were mixed, stirred, and then H_2O_2 was added. After the complete dissolution, the gold seed was added. The color of the solution changed from yellow to gray. After the color transition, the solution was centrifuged, and suspensions were washed with Milli-Q water and dried (Scheme 7.3). The size of PdNPs was between 33 and 110 nm [24].

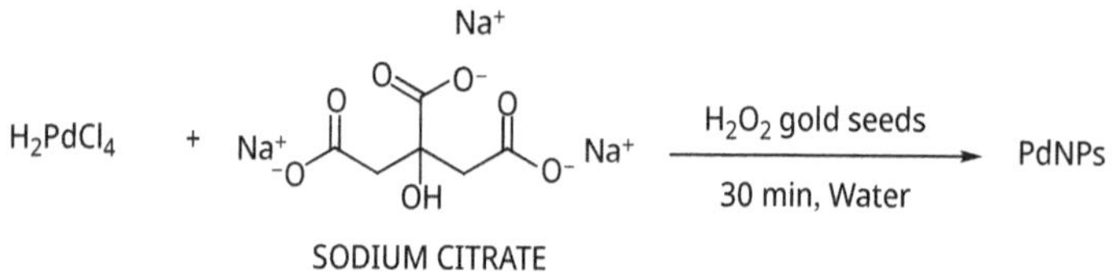

Scheme 7.3: Synthesis of PdNPs using sodium citrate as a stabilizer.

7.2.2 Synthesis and stabilization of gold and silver nanoparticles

Gold nanoparticles (AuNPs) are most commonly obtained by the chemical reduction method. Gold salts are reduced in the presence of a reducing agent, such as reducing gold chloroauric acid in sodium citrate. AuNPs are also predominantly obtained by using thiol ligands. This method (Brust–Schiffrin method) involves two steps. First, the transfer of gold to the organic phase via the tetrabutylammonium bromide salt of gold, then the reduction by the thiol ligand. Finally, the strong reducing agent $NaBH_4$ is added to give AuNPs protection by the thiolate group. During the synthesis of AuNPs, size control, thermal stability, and ease of preparation need to be optimized [25].

One of the most common MNPs synthesized vibrantly is the silver nanoparticles (AgNPs). This is because of many applications of AgNPs in catalysis, antimicrobial, anticancer activity, sensors, etc. One of the popular methods is the reduction of silver salts by citrate anion. Citrate anion plays the dual role of stabilizer and a reducer for Ag^+ to Ag^0. Moreover, the concentration of citrate anion controls the size and shape of the AgNPs. Furthermore, at room temperature, the reduction of Ag^+ to Ag^0 is obtained by gallic acid, which has a reduction potential of 0.5 V. The –OH groups present in the molecular structure of gallic acid are mainly involved in the reduction procedure [26].

Apart from the above-discussed methods, physical and biological methods are used to form AgNPs. For instance, UV irradiation, followed by the addition of PVP (capping agent) and ethylene glycol (reducing agent), leads to the formation of AgNPs. Nowadays, plant extracts are frequently used as reducing agents for synthesis. Moreover, many bacteria and fungi are also used for this purpose. Examples include eukaryotic fungus and prokaryotic bacteria. However, the reaction rates are slower than chemical methods[27].

Moreover, the size, shape, and ease of formation of AgNPs significantly depend on the concentration of the plant extract. Nevertheless, biological methods are economical and eco-friendly. Major synthetic methods used to prepare Ag and Au–NPs are discussed below.

Prabhu Charan et al. [28] highlighted the synthesis of MNPs (MNP = gold and silver nanoparticles) by using polymeric ionic liquid (PIL) as a stabilizer. PIL was dissolved in water, and the aqueous solution of metal salt was added dropwise with continuous stirring. The reaction was kept stirring at room temperature under dark conditions for 12 h. Here, sodium borohydride was used as a reducing agent and added to the reaction mixture with heating at 40 °C for 12 h (Scheme 7.4). The mixture was centrifuged and collected [28].

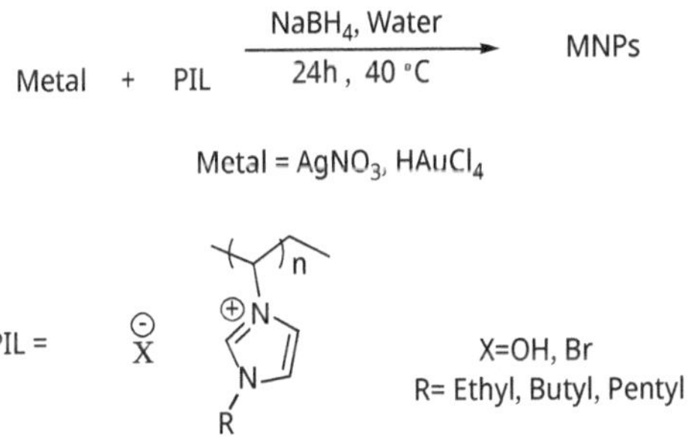

Scheme 7.4: Polyionic liquid-stabilized metal nanoparticles.

Sun et al. [29] proposed the synthesis of AgNPs using tea leaf extract. Solution of tea leaves extract was used as a reducing agent and capping agent. Different amounts of the stock solution of tea leaves extract and silver nitrate solution were stirred (700 rpm) for 120 min. AgNPs were formed and purified by ultrafiltration and washed with Milli-Q water (Scheme 7.5). The maximum AgNP production efficiency was achieved with 5% (v/v) tea extract [29].

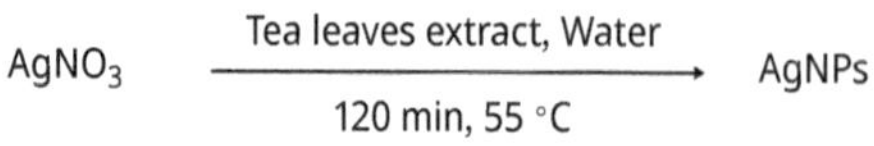

Scheme 7.5: Synthesis of AgNPs using tea leaves extract.

Navaladian et al. [30] prepared the solution of silver oxalate by mixing 50 mL of 0.5 M $AgNO_3$ solution with 30 mL of 0.5 M oxalic acid. Then, a convenient amount of polyvinyl alcohol (PVA) was added to 40 mL of water with stirring. Eventually, 0.05 g of $Ag_2C_2O_4$ was added with stirring, and after 10 min, the mixture was purged with N_2 gas. Then, the mixture was refluxed with a continuous nitrogen gas flow at 100 °C for 3 h in an oil bath till the mixture turned into a yellow-colored colloidal solution. The reaction mixture was then cooled to room temperature under N_2 atmosphere. Finally, the solution was centrifuged to yield Ag nanoparticles (Scheme 7.6). PVA was the capping agent, and it was also involved in the reduction process [30].

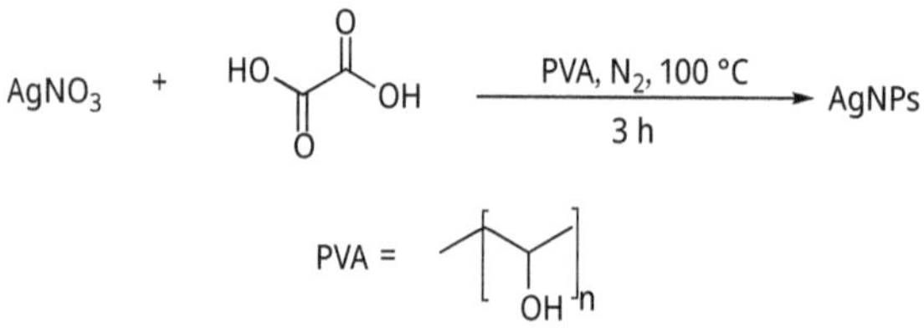

Scheme 7.6: Synthesis of AgNPs using polyvinyl alcohol as a stabilizer.

Wei et al. reported the synthesis of Ag and Au NPs using chitosan as a mediator agent. To synthesize Ag/Au NPs, a metal salt solution was added to the chitosan solution under stirring conditions at room temperature. The reaction mixture was put in the water bath at 45–95 °C for 6–12 h for reduction (Scheme 7.7). The MNPs were characterized by transmission electron microscope (TEM) and Fourier-transform infrared [31].

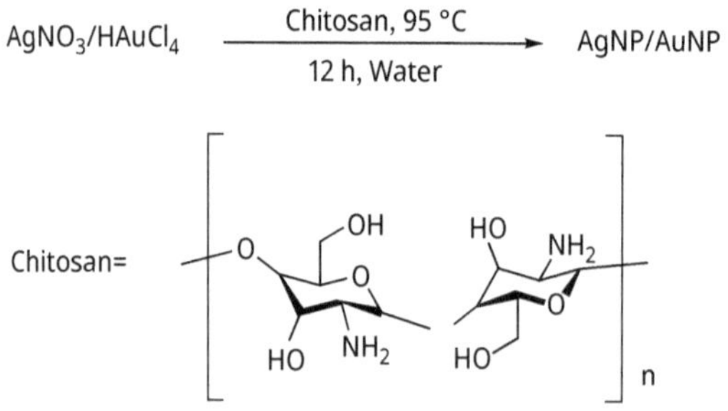

Scheme 7.7: Synthesis of Ag/AuNP using chitosan as a stabilizer.

APP (1-(3-(acetylation)propyl)pyrazin-1-ium) solution was prepared by adding APP into a 10% mixture of methanol and deionized water by Anwar et al. [32]. This solution was added to the aqueous solution of tertachloroauric acid and stirred for 10 min. A freshly prepared $NaBH_4$ solution was added to reduce the reagent (Scheme 7.8). The final reaction mixture was stirred for 2 h at room temperature to reduce the gold salt (if any). The solution was centrifuged to obtain the solid AuNPs, and the suspensions were washed with distilled water and dried [32].

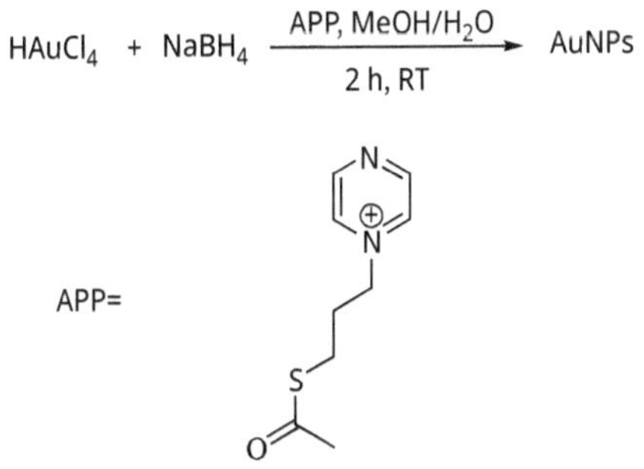

1-(3-(acetylthio)propyl)pyrazin-1-)pyrazin -1-ium

Scheme 7.8: Synthesis and stabilization of AuNPs using APP.

7.2.3 Synthesis and stabilization of copper nanoparticles (CuNPs)

The chemical reduction method is the most popular one for synthesizing copper nanoparticles. Other widely used methods are microwave-assisted green synthesis, sol–gel auto combustion method, mechanochemical synthesis, sonochemical method, and colloidal injection method [33]. CuNPs exhibit unique properties such as very high electrical conductivity, low cost, and low electrochemical migration behavior. To stabilize CuNPs, OLA (oleylamine), TOP (tri-*n*-octyl phosphine), chitosan, and Tween-80 have been frequently used. Functional groups such as amine and alcohol, present in the stabilizers above, are mainly responsible for stabilizing CuNPs. CuNPs show antimicrobial and anti-tumor activity, and it has been reported that amino acid chelated CuNPs give good antimicrobial properties [34]. Moreover, CuNPs are also frequently used for the catalytic reduction of nitroaromatic compounds [35], in bio-sensing [36], and as a disinfectant for wastewater. The different synthetic methodologies adopted for the preparation of CuNPs are discussed below.

The aqueous solution of copper(II) sulfate pentahydrate ($CuSO_4{\cdot}5H_2O$) and orange juice extract were taken in a 1:2 ratio and kept in a microwave system, stirring for 15 min. After 15 min, a dark brown colloidal mixture was obtained (Scheme 7.9), and the resultant copper nanoparticles [37].

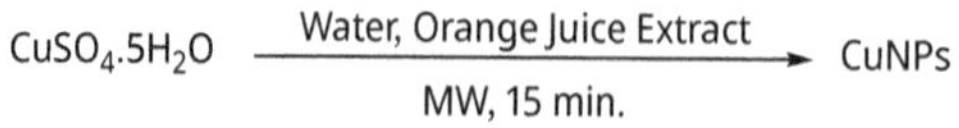

Scheme 7.9: Synthesis of CuNPs using orange juice extract.

The copper(III) formate was dissolved in the mixture of oleylamine and dodecane at 60 °C by [38]. The temperature was increased to 140 °C at a heating rate of 10 °C/min. CuNCs (copper nanocrystals) were extracted from the reaction mixture and purified by repetitive precipitation/resuspension cycles (Scheme 7.10) using toluene as solvent and ethanol or methanol as non-solvents [38].

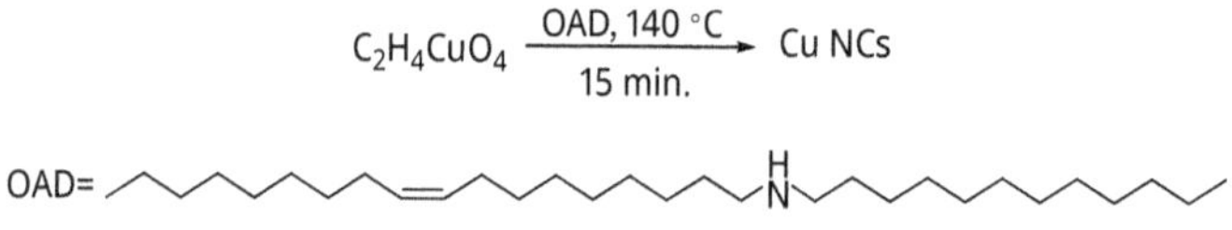

Scheme 7.10: Synthesis of CuNCs using OAD as a stabilizer.

Shu et al. [39] reported the green synthesis of C/Cu–NPs by the sol–gel auto combustion method. Aqueous solutions of $Cu(NO_3)_2{\cdot}3H_2O$ and Tween-80 were mixed, and the pH of the resulting solution was maintained at 7.0 by using NaOH. Then, the g-PGA solution was added to the mixture and heated in a water bath at 60 °C so that the water evaporated and the solution became thick. The viscous product obtained was heated to 150 °C in an oven. After that, it was heated at 400 °C in a muffle furnace for 10 min. Then the reaction mixture was cooled to room temperature, washed with ethanol and distilled water, and finally dried at 100 °C. Thus, the biochar-conjugated copper NPs were obtained (Scheme 7.11). This method uses γ-polyglutamic acid and Tween-80 as carbon sources. Tween-80 is a nonionic hydrophilic surfactant, and γ-PGA was used as a dispersant [39].

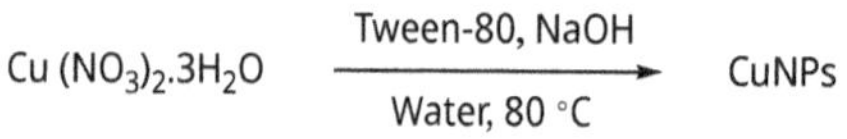

Scheme 7.11: Synthesis of CuNPs using Tween-80 as a stabilizer.

Mbewana-Ntshanka et al. [40] proposed the synthesis of copper oxide, copper sulfide, and copper selenide nanoparticles (Scheme 7.12) by the colloidal hot injection method. Zhou et al. reported the synthesis of CuNPs by using chitosan as a stabilizer and hydrazine as a reducing agent. To produce the CuNPs, chitosan solution was added to $CuCl_2$ solution with continuous stirring while the pH of the solution was maintained basic (9.0) using concentrated ammonia solution. After 10 min of stirring at room temperature, hydrazine hydrate was added to the mixture as a mild reducing agent. The reaction mixture was kept stirring for 5 h. The color change of the solution to reddish-brown confirms the formation of copper nanoparticles (Scheme 7.13). To stabilize the CuNPs, chitosan was used. Chitosan contains NH_2 and OH groups in its structure, which help in stabilizing the nanoparticles by adsorbing on the surface of NPs [41].

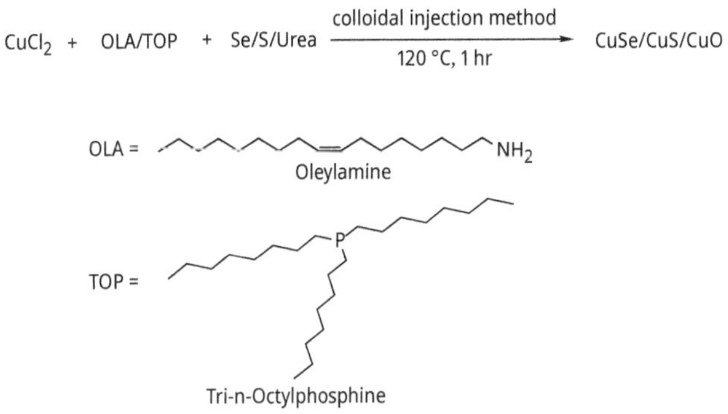

Scheme 7.12: Synthesis of CuONPs using OLA, TOP as a stabilizer.

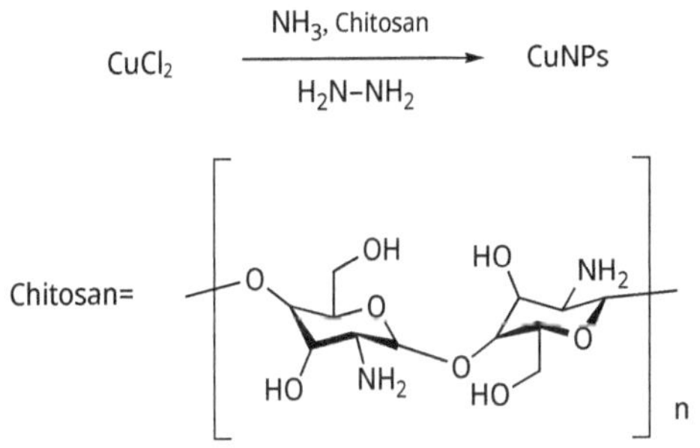

Scheme 7.13: Chitosan-stabilized CuNPs.

7.2.4 Synthesis and stabilization of cobalt nanoparticles (CoNPs)

Cobalt nanoparticles are gaining considerable research interest because of their biomedical and catalytic applications. Other major application includes the asymmetric catalytic hydrogenation of ketones [42]. CoNPs show antimicrobial, anticancer, antifungal, antioxidant, and enzyme inhibition properties in the biomedical field, and they are also used as biomedical sensors [43]. To prepare CoNPs, chemical reduction, thermal decomposition, and precipitation methods are commonly used.

Shao et al. [44] reported the synthesis of CoNPs by using cobalt acetate tetrahydrate through the thermal decomposition method. Trioctylphosphine (TOP), oleylamine, and oleic acid were used as stabilizers to control the size of CoNPs. To synthesize the CoNPs, the cobalt acetate solution was prepared by dissolving it in 2-pyrrolidinone and adding a mixture of oleylamine, oleic acid, and TOP. Then, the mixture was heated to 260 °C for 90 min under an N_2 gas atmosphere to prevent oxidation. The precipitation of CoNPs was achieved using ethanol. After frequent washing with ethanol, CoNPs were dried in a vacuum and characterized by a TEM, X-ray diffractometer (XRD), and a vibrating sample magnetometer (VSM). The synthesized CoNPs were stabilized by the mixture of TOP, oleic acid, and oleylamine (Scheme 7.14). With the addition of the stabilizer, the average particle size was reduced from 200 to 8 nm. The TOP as a stabilizer provides significant steric hindrance owing to the long-chain structure. Hence, the particles were well dispersed with no trace of agglomeration. Oleic acid binds tightly with the metal NP's surface. Hence, the use of the mixture of TOP, OA, and oleylamine enhances the stability of nanoparticles [44].

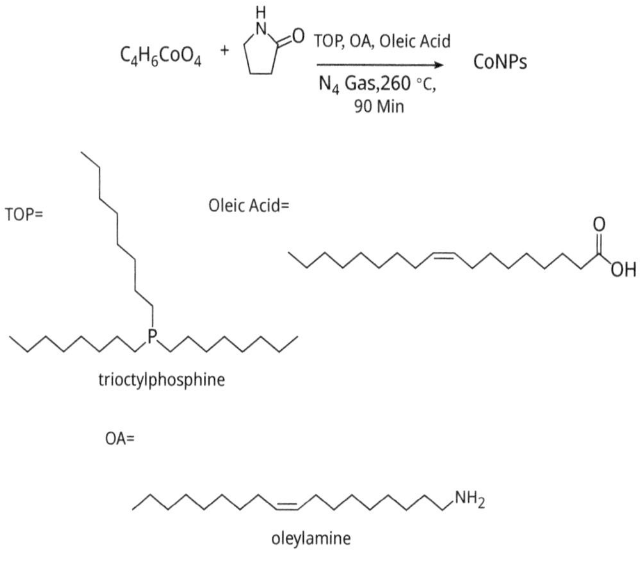

Scheme 7.14: Synthesis of CoNPs using TOP, OA, and oleic acid as a stabilizer.

Santos et al. [45] reported the synthesis of cobalt hydroxide nanoparticles using the precipitation method. For this method, an aqueous solution of $Co(NO_3)_2{\cdot}6H_2O$ was mixed with NaOH solution at room temperature with continuous stirring (2,400 rpm). As a result, black ppt was formed (Scheme 7.15), which was extracted by centrifugation and washed with a sufficient amount of distilled water, and dried at 70 °C [45].

$$Co(NO_3)_2.6H_2O + NaOH \xrightarrow[2400\ rpm]{Centrifuge} (Co(OH)_2)\ NPs$$

Scheme 7.15: Synthesis of CoNPs of oleic acid in the presence of aniline.

In the following method, the KOH solution was added dropwise into the aqueous solution of CoCl2 · $6H_2O$ with continuous stirring. Hydrazine monohydrate and OA (oleic acid) were added with vigorous stirring for 20 min. The mixture was kept in the autoclave. After the complete dissolution of the solid in solution, it was added to a Teflon-lined autoclave and was placed in an electric oven at 160 °C for 1 day (Scheme 7.16). Then, the mixture was cooled at room temperature. To obtain the CoNPs, the solution was centrifuged, and the suspensions were washed with distilled water and dried [46].

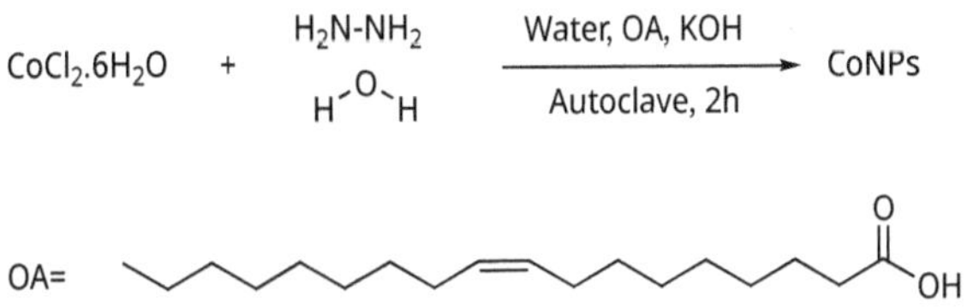

Scheme 7.16: Synthesis of CoNPs using water OA and KOH.

In another study by Ansari et al. [47], the mixture of solutions of sodium citrate dihydrate and aqueous cobalt sulfate was deaerated by bubbling argon gas through the solution. The solution was then heated for around 60–120 min at a fixed temperature (Scheme 7.17). To obtain CoNPs, the solution was centrifuged, and suspensions were washed with distilled water and dried [47].

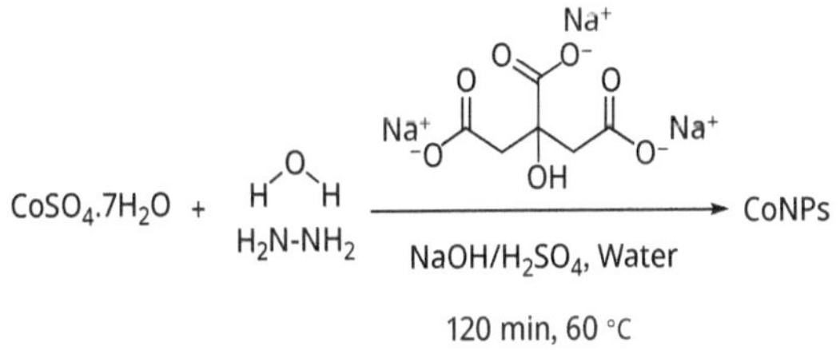

Scheme 7.17: Synthesis of CoNPs using sodium citrate.

Metal salt and glycine (green fuel) were taken in a fixed ratio (1:4.4), and ammonia was used to maintain the basic pH of the solution (Scheme 7.18). The final product was sintered at 600 °C for 5 h [48].

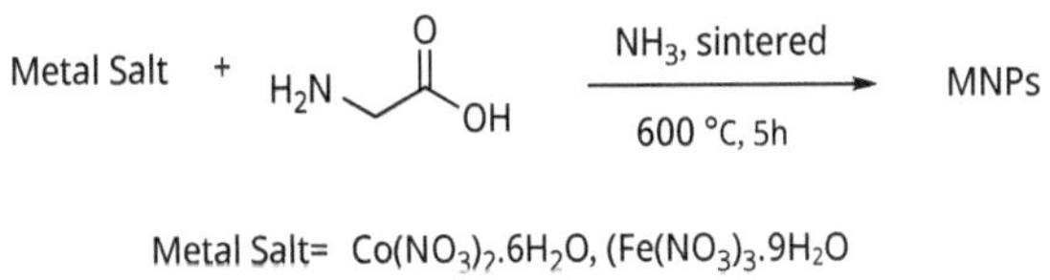

Scheme 7.18: Synthesis of CoNPs, $Fe(NO_3)_3$ NPs.

7.2.5 Synthesis and stabilization of platinum nanoparticles (PtNPs)

In the growing field of nanoscience, nanoparticles of the noble metal platinum have attracted the attention of researchers because of their excellent applications and properties. For instance, PtNPs show excellent catalytic properties for crucial organic transformations, antioxidant capabilities, low toxicity, anticancer, and antimicrobial properties[49]. PtNPs have also found crucial applications in the biomedical field. PtNPs can be synthesized by chemical reduction of Pt salt, green synthesis using natural bioactive compounds from different leaf extracts[50], and the water-in-oil micro-emulsion method. MNPs are very sensitive to reaction conditions. Thus, the reaction time, the concentration of reducing or capping agent, and temperature can affect the size and shape of nanoparticles. To stabilize the PtNPs, different stabilizers such as thiol-containing polymers, natural leaf extracts, and CGA (chlorogenic acid) are frequently used. The various synthetic procedures used for the preparation of PtNPs are discussed below.

Rong Chen et al. [51] reported the synthesis of PtNPs using H_2PtCl_6 as a metal salt precursor and CGA (chlorogenic acid) as a reducing agent and stabilizer (Scheme 7.19). The synthesis of PtNPs was accomplished using a single reaction flask containing the metal salt. A CGA solution was mixed with deionized water at 100 °C, and the reaction continued for 3 h. After that, a solution of H_2PtCl_6 was added to the mixture and incubated at 95 °C in a water bath for 2.5 h. Here, CGA was added as a reducing and capping agent for the preparation of PtNPs. Moreover, for a higher yield, the reaction time was increased [51].

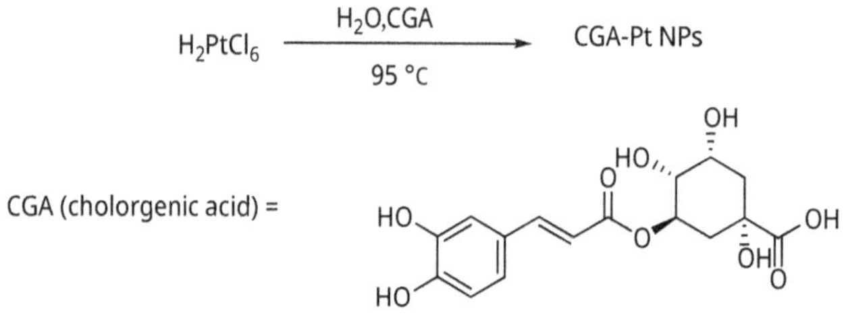

Scheme 7.19: Synthesis of CGA-capped PtNPs.

Roberto A. et al. proposed the synthesis of PtNPs by the water-in-oil micro-emulsion method and investigated the effect of HCl on their surface structure. In this method, the nanoparticles were produced at room temperature, but this method allows only slight modification in the surface structure of the nanoparticles. For preparing PtNPs, H_2PtCl_6 was taken as a metal salt, and sodium borohydride was used to reduce the metal (Scheme 7.20). As a stabilizer, polyethylene glycol dodecyl ether was used. The role of HCl was to control the size and shape of Pt nanoparticles. The optimal percentage was found to be 15–25% of HCl [57].

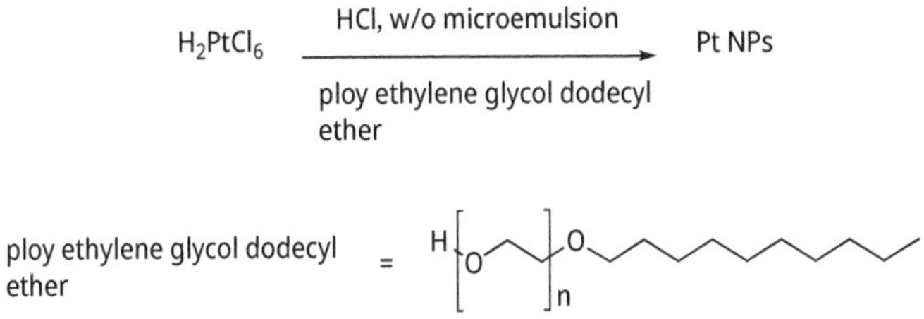

Scheme 7.20: Synthesis of PtNPs using polyethylene glycol dodecyl ether as a stabilizer.

Raj Karthik et al. [52] were the first to report the synthesis of ethylcellulose-supported PtNPs (Scheme 7.21). The role of ethylcellulose was to act as a reducing agent and stabilizer to prevent the aggregation of MNPs. The ion–dipole interaction in cellulose can disperse the MNPs on the entire surface of ethylcellulose. For the synthesis of PtNPs, a solution of H_2PtCl_6 was added to the solution of ethylcellulose in ethanol. The mixture was sonicated for 60 min. Then, the reducing agent ($NaBH_4$) was prepared freshly and added to the mixture till the color of the solution was completely converted to black [52].

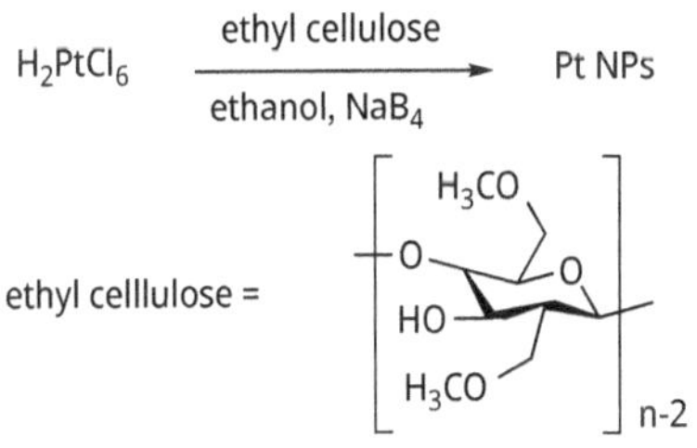

Scheme 7.21: Synthesis of ethylcellulose-supported PtNPs.

Wei Li et al. [27] reported the synthesis of PtNPs and their application in sensing. For the synthesis of PtNPs, dimethylamine borane (DMAB) was used as a reducing agent, and K_2PtCl_4 was used as a metal salt precursor. To stabilize the PtNPs, GSH (glutathione) was used as a stabilizer. Initially, the GSH and PBS buffer (pH 5.0) mixture was incubated with K_2PtCl_4 for 2 h at room temperature. Then, the reducing agent DMAB was freshly prepared and added to the mixture in a ratio (DMAB/ K_2PtCl_4) of 5. The PtNPs were produced after 12 h of reduction at room temperature (Scheme 7.22). GSH-capped PtNPs were utilized to sense heavy metal ions, selective toward Hg^{2+}. The limit of detection (LOD) was found to be 0.25 nm.

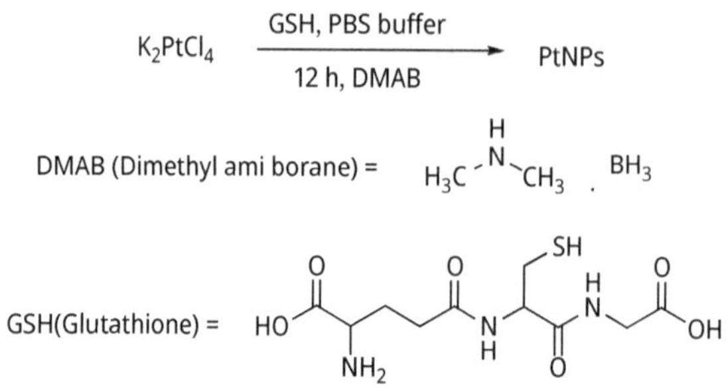

Scheme 7.22: Synthesis of glutathione-stabilized PtNPs.

7.2.6 Synthesis of other important MNPs

To synthesize cerium nanoparticles (CeNPs), cerium(III) nitrate hexahydrate was used as a metal source and added to NaOH. In contrast, choline serinate ionic liquid was used as a stabilizer. The resulting mixture was stirred for 45 min and transferred to a glass tube. Then it was irradiated with an ultrasound bath for 12 h at room temperature (Scheme 7.23). After the centrifugation, the product was separated and washed with ethanol and distilled water, and then kept for drying at 80 °C overnight [53].

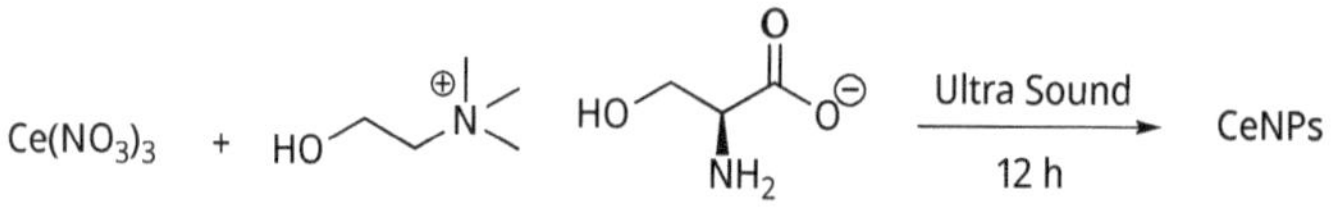

Scheme 7.23: Synthesis of CeNPs using ionic liquid as a stabilizer.

To synthesize nickel nanoparticles (NiNPs), Wu and Chen took $NiCl_2$ as a Ni source and dissolved it in ethylene glycol. To the solution, hydrazine was added, followed by NaOH at 60 °C with stirring (Scheme 7.24). The reaction mixture was kept under the same conditions for 60 min, and finally, NiNPs were formed [54].

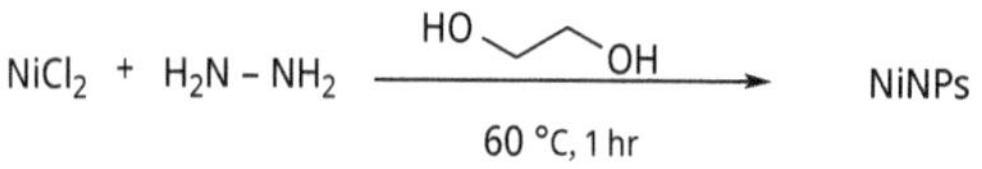

Scheme 7.24: Synthesis of NiNPs in the presence of ethylene glycol.

Hanying et al. reported the synthesis of semiconductor nanoparticles in salt-induced micelles (Scheme 7.25). Polystyrene-block-2-vinyl pyridine (PS-b-P2VP) was dissolved in THF with continuous stirring for about 1 h. Solution of cadmium acetate in the mixture of methanol and THF (tetrahydrofuran) was added to the solution with continuous stirring at room temperature. The blue-colored solution showed the formation of polymer micelles. H_2S gas was bubbled through this solution for 10 min, and yellow cadmium sulfide nanoparticles (CdSNPs) were formed [55].

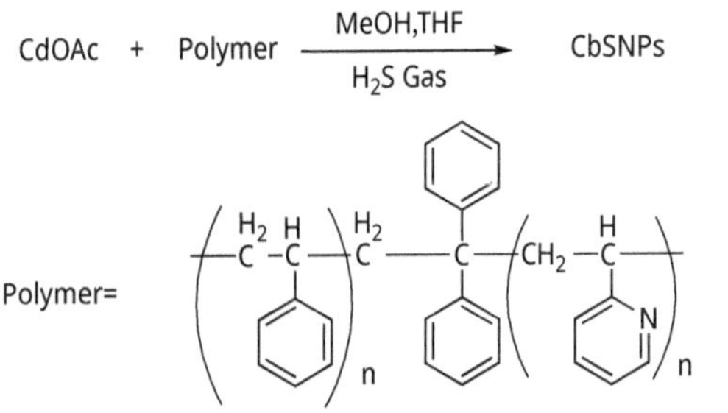

Scheme 7.25: Synthesis and stabilization of CdNPs using PS-b-P2VP.

7.3 Conclusion

MNPs have attracted sustained and widespread interest across the chemical, physical, biological, and engineering research communities due to their remarkable properties and broad technological potential. Noble MNPs, in particular, exhibit unique optical, magnetic, catalytic, and biological behaviors that make them highly suitable for applications ranging from sensing and imaging to catalysis and energy conversion. Among the numerous synthetic strategies available, chemical methods remain the most widely adopted due to their versatility, scalability, and ability to finely tune particle size, morphology, and surface chemistry. Radiolytic and photolytic approaches have also gained prominence, offering advantages such as narrow size distributions, controlled nucleation, and minimal contamination. Techniques such as electron pulse radiolysis and laser-based synthesis further deepen understanding of nanoparticle growth kinetics, enabling researchers to probe transient intermediates and reaction pathways with high temporal resolution.

This chapter provides a detailed overview of chemical synthesis routes for transition metal nanoparticles, focusing on systems based on Pd, Ag, Au, Pt, Cu, and Co. Special emphasis is placed on the critical roles of stabilizers, surfactants, and capping agents, which govern nucleation, prevent aggregation, and impart functional surface characteristics essential for targeted applications. Despite significant progress, challenges remain in translating laboratory-scale methods to industrial production while maintaining uniformity, stability, and performance. Continued innovation in synthetic design, mechanistic understanding, and materials engineering will be essential for expanding the practical utility of MNPs. By compiling diverse methodologies and highlighting key considerations in nanoparticle synthesis, this chapter aims to support researchers in developing more refined, efficient, and application-driven approaches for next-generation nanomaterials.

References

[1] Whitesides, G. M. (2003). The 'right' size in nanobiotechnology. *Nature Biotechnology*, 21, 1161–1165.

[2] Rane, A. V., Kanny, K., Abitha, V. K., & Thomas, S. (2018). Methods for Synthesis of Nanoparticles and Fabrication of Nanocomposites. *Elsevier Ltd*, https://doi.org/10.1016/b978-0-08-101975-7.00005-1.

[3] Taher, A., Nandi, D., Choudhary, M., & Mallick, K. (2015). Suzuki coupling reaction in the presence of polymer-immobilized palladium nanoparticles: A heterogeneous catalytic pathway. *New Journal of Chemistry*, 39, 5589–5596. https://doi.org/10.1039/c5nj00969c.

[4] Kaphle, A., Navya, P. N., Umapathi, A., & Daima, H. K. (2018). Nanomaterials for agriculture, food, and environment: Applications, toxicity, and regulation. *Environmental Chemistry Letters*, 16, 43–58. https://doi.org/10.1007/s10311-017-0662-y.

[5] Li, L., & Yang, Q. (Eds). (2019). Coating materials: Nanotechnology (Part II, pp. 175–233). In *Advanced coating materials*. WileyScrivener. https://onlinelibrary.wiley.com/doi/book/10.1002/9781119407652?msockid=0b479d1851c06266020e8bc150026383.

[6] Masoudi Asil, S., Ahlawat, J., Guillama Barroso, G., & Narayan, M. (2020). Nanomaterial-based drug delivery systems for treating neurodegenerative diseases. *Biomaterials Science*, 8, 4088–4107. https://doi.org/10.1039/d0bm00809e.

[7] Willner, M. R., & Vikesland, P. J. (2018). Nanomaterial-enabled sensors for environmental contaminants Prof Ueli Aebi, Prof Peter Gehr. *Journal of Nanobiotechnology*, 16, 1–16. https://doi.org/10.1186/s12951-018-0419-1.

[8] Thekkudan, V. N., Vaidyanathan, V. K., Ponnusamy, S. K., Charles, C., Sundar, S. L., Vishnu, D. et al. (2017). Review on nano adsorbents: A solution for heavy metal removal from wastewater. *IET Nanobiotechnology*, 11, 213–224. https://doi.org/10.1049/iet-nbt.2015.0114.

[9] Lin, Z., Yue, J., Liang, L., Tang, B., Liu, B., Ren, L. et al. (2020). Rapid synthesis of metallic and alloy micro/nanoparticles by laser ablation towards the water. *Applied Surface Science*, 504, 144461. https://doi.org/10.1016/j.apsusc.2019.144461.

[10] Siwach, O. P., & Sen, P. (2008). Fluorescence properties of Fe nanoparticles prepared by electro-explosion of wires. *Materials Science and Engineering B: Solid-State Materials for Advanced Technology*, 149, 99–104. https://doi.org/10.1016/j.mseb.2007.12.007.

[11] Iravani, S., Korbekandi, H., Mirmohammadi, S. V., & Zolfaghari, B. (2014). Synthesis of silver nanoparticles: Chemical, physical, and biological methods. *Research in Pharmaceutical Sciences*, 9(6), 385–406. *Res Pharm Sci* 2014;9:385–406.

[12] Hasnidawani, J. N., Azlina, H. N., Norita, H., Bonnia, N. N., Ratim, S., & Ali, E. S. (2016). Synthesis of ZnO Nanostructures Using the Sol-Gel Method. *Procedia Chemistry*, 19, 211–216. https://doi.org/10.1016/j.proche.2016.03.095.

[13] Saravanan, A., Kumar, P. S., Karishma, S., Dvn, V., Jeevanantham, S., Yaashikaa, P. R. et al. (2021). A review on the biosynthesis of metal nanoparticles and their environmental applications. *Chemosphere*, 264, 128580. https://doi.org/10.1016/j.chemosphere.2020.128580.

[14] Pachón, L. D., & Rothenberg, G. (2008). Transition-metal nanoparticles: Synthesis, stability and the leaching issue. *Applied Organometallic Chemistry*, 22, 288–299. https://doi.org/10.1002/aoc.1382.

[15] Feng, J., Gao, C., & Yin, Y. (2018). Stabilization of noble metal nanostructures for catalysis and sensing. *Nanoscale*, 10, 20492–20504. https://doi.org/10.1039/c8nr06757k.

[16] Cao, A., Lu, R., & Veser, G. (2010). Stabilizing metal nanoparticles for heterogeneous catalysis. *Physical Chemistry Chemical Physics*, 12, 13499–13510. https://doi.org/10.1039/c0cp00729c.
[17] Baruwati, B., & Varma, R. S. (2009). High-value products from waste: Grape pomace extract- a three-in-one package for synthesizing metal nanoparticles. *ChemSusChem*, 2, 1041–1044. https://doi.org/10.1002/cssc.200900220.
[18] Wang, C., Li, C., Liu, J., & Guo, C. (2021). Engineering transition metal-based nanomaterials for high-performance electrocatalysis. *Materials Reports: Energy*, 1, 100006. https://doi.org/10.1016/j.matre.2021.01.001.
[19] Cid, A., & Simal-Gandara, J. (2020). Synthesis, characterization, and potential applications of transition metal nanoparticles. *Journal of Inorganic and Organometallic Polymers and Materials*, 30, 1011–1032. https://doi.org/10.1007/s10904-019-01331-9.
[20] Yang, X., Li, Q., Wang, H., Huang, J., Lin, L., Wang, W. et al. (2010). Green synthesis of palladium nanoparticles using the broth of Cinnamomum camphora leaf. *Journal of Nanoparticle Research*, 12, 1589–1598. https://doi.org/10.1007/s11051-009-9675-1.
[21] Saldan, I., Semenyuk, Y., Marchuk, I., & Reshetnyak, O. (2015). Chemical synthesis and application of palladium nanoparticles. *Journal of Materials Science*, 50, 2337–2354. https://doi.org/10.1007/s10853-014-8802-2.
[22] Athawale, A. A., Bhagwat, S. V., Katre, P. P., Chandwadkar, A. J., & Karandikar, P. (2003). Aniline as a stabilizer for metal nanoparticles. *Materials Letters*, 57, 3889–3894. https://doi.org/10.1016/S0167-577X(03)00235–0.
[23] Mallick, K., Witcomb, M. J., & Scurrell, M. S. (2006). Formation of palladium nanoparticles in poly (o-methoxyaniline) macromolecule fibers: An *in-situ* chemical synthesis method. *European Physical Journal E*, 19, 149–154. https://doi.org/10.1140/epje/e2006-00027-2.
[24] Chen, H., Wei, G., Ispas, A., Hickey, S. G., & Eychmüller, A. (2010). Synthesis of palladium nanoparticles and their applications for surface-enhanced Raman scattering and electrocatalysis. *Journal of Physical Chemistry C*, 114, 21976–21981. https://doi.org/10.1021/jp106623y.
[25] Venkatesh, N. (2018). Metallic nanoparticle: *A Review. Biomedical Journal of Scientific & Technical Research*, 4, 3765–3775. https://doi.org/10.26717/bjstr.2018.04.0001011.
[26] Ibrahim, K., Saeed, K., & Idrees, K. (2019). Nanoparticles: Properties, applications, and toxicities. *Arabian Journal of Chemistry*, 12, 908–931. https://doi.org/10.1016/j.arabjc.2017.05.011.
[27] Li, X., Xu, H., Chen, Z. S., & Chen, G. (2011). Biosynthesis of nanoparticles by microorganisms and their applications. *Journal of Nanomaterials*, 2011. https://doi.org/10.1155/2011/270974.
[28] Prabhu Charan, K. T., Pothanagandhi, N., Vijayakrishna, K., Sivaramakrishna, A., Mecerreyes, D., & Sreedhar, B. (2014). Poly(ionic liquids) as “smart” stabilizers for the metal nanoparticles. *European Polymer Journal*, 60, 114–122. https://doi.org/10.1016/j.eurpolymj.2014.09.004.
[29] Sun, Q., Cai, X., Li, J., Zheng, M., Chen, Z., & Yu, C. P. (2014). Green synthesis of silver nanoparticles using tea leaf extract and evaluating their stability and antibacterial activity. *Colloids and Surfaces A: Physicochemical and Engineering Aspects*, 444, 226–231. https://doi.org/10.1016/j.colsurfa.2013.12.065.
[30] Navaladian, S., Viswanathan, B., Viswanath, R. P., & Varadarajan, T. K. (2007). Thermal decomposition as a route for silver nanoparticles. *Nanoscale Research Letters*, 2, 44–48. https://doi.org/10.1007/s11671-006-9028-2.
[31] Wei, D., & Qian, W. (2008). Facile synthesis of Ag and Au nanoparticles utilizing chitosan as a mediator agent. *Colloids and Surfaces B: Biointerfaces*, 62(136), 42. https://doi.org/10.1016/j.colsurfb.2007.09.030.
[32] Anwar, A., Minhaz, A., Khan, N. A., Kalantari, K., Afifi, A. B. M., & Shah, M. R. (2018). Synthesis of gold nanoparticles stabilized by a pyrazinium thioacetate ligand: A new colorimetric nanosensor for detecting heavy metal Pd(II). *Sensors and Actuators B: Chemical*, 257, 875–881. https://doi.org/10.1016/j.snb.2017.11.040.

[33] Gawande, M. B., Goswami, A., Asefa, T., Huang, X., Silva, R., Zou, X. et al. (2016). Cu, and Cu-based nanoparticles : synthesis and applications in catalysis 2016. *Chemical Reviews*, 116(6), 3722–3811. https://doi.org/10.1021/acs.chemrev.5b00482.

[34] Dealba-Montero, I., Guajardo-Pacheco, J., Morales-Sánchez, E., Araujo-Martínez, R., Loredo-Becerra, G. M., Martínez-Castañón, G. A. et al. (2017). Antimicrobial properties of copper nanoparticles and amino acid chelated copper nanoparticles produced by using a soya extract. *Bioinorganic Chemistry and Applications*, 2017, 15–17. https://doi.org/10.1155/2017/1064918.

[35] Kaur, R., Giordano, C., Gradzielski, M., & Mehta, S. K. (2014). Synthesis of highly stable, water-dispersible copper nanoparticles as catalysts for nitrobenzene reduction. *Chemistry – An Asian Journal*, 9, 189–198. https://doi.org/10.1002/asia.201300809.

[36] Din, M. I., & Rehan, R. (2017). Synthesis, Characterization, and Applications of Copper Nano-particles. *Analytical Letters*, 50, 50–62. https://doi.org/10.1080/00032719.2016.1172081.

[37] Jahan, I., Erci, F., & Isildak, I. (2021). Facile microwave-mediated green synthesis of non-toxic copper nanoparticles using Citrus sinensis aqueous fruit extract and their antibacterial potentials. *Journal of Drug Delivery Science and Technology*, 61, 102172. https://doi.org/10.1016/j.jddst.2020.102172.

[38] Oliva-Puigdomènech, A., De Roo, J., Van Avermaet, H., De Buysser, K., & Hens, Z. (2020). Scalable Approaches to Copper Nanocrystal Synthesis under Ambient Conditions for Printed Electronics. *ACS Applied Nano Materials*, 3, 3523–3531. https://doi.org/10.1021/acsanm.0c00242.

[39] Shu, X., Feng, J., Liao, J., Zhang, D., Peng, R., Shi, Q. et al. (2020). Amorphous carbon-coated nano-copper particles: Novel synthesis by Sol-Gel and carbothermal reduction method and extensive characterization. *Journal of Alloys and Compounds*, 848, 156556. https://doi.org/10.1016/j.jallcom.2020.156556.

[40] Mbewana-Ntshanka, N. G., Moloto, M. J., & Mubiayi, P. K. (2020). Role of the amine and phosphine groups in oleylamine and trioctylphosphine in the synthesis of copper chalcogenide nanoparticles. *Heliyon*, 6e05130. https://doi.org/10.1016/j.heliyon.2020.e05130.

[41] Hongfeng, Z., El-Kott, A., Ezzat Ahmed, A., & Khames, A. (2021). Synthesis of chitosan-stabilized copper nanoparticles (CS-Cu NPs): Its catalytic activity for C-N and C-O cross-coupling reactions and treatment of bladder cancer. *Arabian Journal of Chemistry*, 14, 103259. https://doi.org/10.1016/j.arabjc.2021.103259.

[42] Michalek, F., Lagunas, A., Jimeno, C., & Pericàs, M. A. (2008). Synthesis of functional cobalt nanoparticles for catalytic applications. Use in asymmetric transfer hydrogenation of ketones. *Journal of Materials Chemistry*, 18, 4692–4697. https://doi.org/10.1039/b808383e.

[43] Iravani, S., & Varma, R. S. (2020). Sustainable synthesis of cobalt and cobalt oxide nanoparticles and their catalytic and biomedical applications. *Green Chemistry*, 22, 2643–2661. https://doi.org/10.1039/d0gc00885k.

[44] Shao, H., Huang, Y., Lee, H. S., Suh, Y. J., & Kim, C. O. (2006). Cobalt nano-particles synthesis from Co(CH3COO)2 by thermal decomposition. *Journal of Magnetism and Magnetic Materials*, 304, 28–30. https://doi.org/10.1016/j.jmmm.2006.02.032.

[45] Santos, G. A., Santos, C. M. B., da Silva, S. W., Urquieta-González, E. A., & Sartoratto, P. P. C. (2012). Sol-gel synthesis of silica-cobalt composites by employing Co_3O_4 colloidal dispersions. *Colloids and Surfaces A: Physicochemical and Engineering Aspects*, 395, 217–224. https://doi.org/10.1016/j.colsurfa.2011.12.033.

[46] Salman, S. A., Usami, T., Kuroda, K., & Okido, M. (2014). Synthesis and characterization of cobalt nanoparticles using hydrazine and citric acid. *Journal of Nanotechnology*, 2014. https://doi.org/10.1155/2014/525193.

[47] Ansari, S. M., Bhor, R. D., Pai, K. R., Sen, D., Mazumder, S., Ghosh, K. et al. (2017). Cobalt nanoparticles for biomedical applications: Facile synthesis, physicochemical characterization, cytotoxicity behavior, and biocompatibility. *Applied Surface Science*, 414, 171–187. https://doi.org/10.1016/j.apsusc.2017.03.002.

[48] Borade, R. M., Kale, S. B., Tekale, S. U., Jadhav, K. M., & Pawar, R. P. (2021). Cobalt ferrite magnetic nanoparticles as highly efficient catalysts for the mechanochemical synthesis of 2-aryl benzimidazoles. *Catalysis Communications*, 159, 106349. https://doi.org/10.1016/j.catcom.2021.106349.

[49] Shiny, P. J., Mukherjee, A., & Chandrasekaran, N. (2014). Haemocompatibility assessment of synthesized platinum nanoparticles and their implications in biology. *Bioprocess and Biosystems Engineering*, 37, 991–997. https://doi.org/10.1007/s00449-013-1069-1.

[50] Zheng, B., Kong, T., Jing, X., Odoom-Wubah, T., Li, X., Sun, D. et al. (2013). Plant-mediated synthesis of platinum nanoparticles and their bioreductive mechanism. *Journal of Colloid and Interface Science*, 396, 138–145. https://doi.org/10.1016/j.jcis.2013.01.021.

[51] Chen, R., Wu, S., & Meng, C. (2021). Size-tunable green synthesis of platinum nanoparticles using chlorogenic acid. *Research on Chemical Intermediates*, 47, 1775–1787. https://doi.org/10.1007/s11164-020-04377-4.

[52] Karthik, R., Karikalan, N., & Chen, S. M. (2017). Rapid synthesis of ethylcellulose-supported platinum nanoparticles for the non-enzymatic determination of H_2O_2. *Carbohydrate Polymers*, 164, 102–108. https://doi.org/10.1016/j.carbpol.2017.01.077.

[53] Inbasekar, C., & Fathima, N. N. (2020). Collagen stabilization using ionic liquid functionalized cerium oxide nanoparticles. *International Journal of Biological Macromolecules*, 147, 24–28. https://doi.org/10.1016/j.ijbiomac.2019.12.271.

[54] Wu, S. H., & Chen, D. H. (2003). Synthesis and characterization of nickel nanoparticles by hydrazine reduction in ethylene glycol. *Ournal of Colloid and Interface Science*, 259, 282–286. https://doi.org/10.1016/S0021-9797(02)00135-2.

[55] Zhao, H., Douglas, E. P., Harrison, B. S., & Schanze, K. S. (2001). Preparation of CdS nanoparticles in salt-induced block copolymer micelles. *Langmuir*, 17, 8428–8433. https://doi.org/10.1021/la011348q.

[56] Li, W., Zhang, H., Zhang, J., & Fu, Y. (2015). Synthesis and sensing application of glutathione-capped platinum nanoparticles. *Analytical Methods*, 7, 4464–4471. https://doi.org/10.1039/c5ay00365b.

[57] Martínez-Rodríguez, R. A., Vidal-Iglesias, F. J., Solla-Gullón, J., Cabrera, C. R., & Feliu, J. M. (2014). Synthesis of Pt nanoparticles in water-in-oil microemulsion: Effect of HCl on their surface structure. Journal of the American Chemical Society, 136(4), 1280–1283. https://doi.org/10.1021/ja411939d.

Sajid Bashir, Jingbo Louise Liu

Postface: Convergent nanochemistry, responsible innovation, and the path forward

Abstract: This postface synthesizes the scientific, technological, and societal themes developed throughout this volume, emphasizing the convergence of nanochemistry, materials engineering, and responsible innovation. Building on the conceptual framework established in the Preface, the chapter integrates advances in the synthesis, characterization, and applications of nanomaterials with emerging considerations of sustainability, ethics, and societal impact. The collective contributions demonstrate how bottom-up and top-down fabrication strategies, atomic-scale control techniques, and data-driven optimization have enabled unprecedented control over material structure and function [1, 2]. At the same time, the deployment of nanotechnologies across energy, environmental remediation, healthcare, and manufacturing underscores the need to embed ethical reasoning, life cycle assessment, and equitable governance into research and development frameworks [8]. This postface highlights cross-cutting scientific insights, identifies shared challenges and opportunities, and outlines future research directions that prioritize safety, scalability, and societal benefit. By framing nanotechnology as both a technical discipline and a sociotechnical enterprise, the chapter articulates a forward-looking vision in which scientific excellence is aligned with environmental stewardship, public trust, and long-term sustainability [6, 8].

Acknowledgments: The research efforts led by Dr Bashir and Dr Liu were supported by the American Chemical Society Petroleum Research Fund (Grant 53827-UR10), federal agencies including the Office of Naval Research, the Air Force Office of Scientific Research, the Department of Energy, the National Science Foundation, the Department of Education, the Department of Homeland Security (ORISE), and the National Institutes of Health. The Robert A. Welch Foundation (Departmental Grant AC-0006) is gratefully acknowledged for its support of student training. Funding agencies supporting the contributions of other authors are acknowledged within their respective chapters to ensure accurate and appropriate recognition of all sources of support. The authors also acknowledge the technical assistance provided by Texas A&M University, Kingsville, and the Texas A&M Energy Institute. We further thank our colleagues and collaborators for their valuable discussions and the reviewers for their constructive comments, which have strengthened the quality of this work.

Authors' contributions: S. Bashir and J. Liu collectively conceived the projects and oversaw their progress. S. Bashir was in charge of the submission processing of all chapters. J. Liu wrote the first draft, which all authors edited. All authors agreed to the publication of this chapter.

 | https://doi.org/10.1515/9783111621111-009

1 Introduction

Nanotechnology has evolved from a conceptual exploration of nanoscale phenomena into a mature scientific and engineering platform that underpins advances across energy, environmental science, medicine, and manufacturing. The defining attributes of nanomaterials arise not merely from dimensional reduction but from the emergence of new physicochemical behaviors governed by size confinement, interfacial dominance, and quantum-mechanical effects; these attributes enable materials to exhibit reactivity, selectivity, and functionality that are inaccessible in their bulk counterparts [3]. The Preface introduced nanochemistry as the intellectual foundation for this transformation, emphasizing the interdependence of synthesis, structure, stability, and function. Subsequent chapters illustrate how this interdependence is realized through diverse fabrication strategies, including sol–gel processing, hydro(solvo)thermal synthesis, microemulsion templating, atomic layer deposition (ALD), and green synthesis methodologies [2, 3]. Collectively, these approaches provide deterministic control over composition, morphology, and surface chemistry, enabling the rational design of nanomaterials tailored for specific applications.

Advances in characterization techniques and computational modeling further reinforce this capability, enabling the elucidation of structure–property relationships with greater precision and predictive power [4, 10]. Beyond technical innovation, the chapters in this volume underscore a critical transition in the field; nanotechnology is no longer confined to laboratory-scale discovery but is increasingly integrated into real-world systems. Applications in sustainable energy, carbon capture and conversion, water remediation, biomedical diagnostics, and advanced manufacturing illustrate both the promise and the responsibility associated with nanoscale engineering. As nanomaterials move toward broader deployment, considerations of environmental persistence, toxicity, life cycle impact, and social equity become inseparable from scientific progress [6, 7].

This postface integrates these scientific and societal dimensions, drawing connections across chapters and identifying unifying themes that define the current state and future trajectory of nanotechnology. By synthesizing technical advances with ethical and sustainability considerations, the chapter articulates a coherent framework for responsible innovation. In doing so, it reinforces the central premise of this volume that the continued success of nanotechnology depends not only on what can be engineered at the nanoscale but also on how such innovations are evaluated, governed, and aligned with societal needs [8].

2 Synthesis of nanomaterial strategies, applications, and challenges

The chapters assembled in this volume collectively illustrate the evolution of nanochemistry from a discipline focused on nanoscale phenomena to an integrated framework for materials design, functional optimization, and application-driven innovation. Despite the diversity of topics, ranging from wet chemical synthesis and ALD to carbon capture, environmental remediation, and societal impact, several unifying scientific themes emerge. These themes reflect a shared emphasis on structure–property relationships, deterministic control of matter at the atomic and molecular scale, and the translation of nanoscale functionality into macroscopic performance [1, 3].

2.1 Synthesis as design

A central convergence across chapters is the recognition that synthesis is not merely a preparatory step but a primary determinant of material behavior. Bottom-up approaches such as sol–gel processing, hydro(solvo)thermal synthesis, microemulsion templating, and green biological routes enable precise control over particle size, morphology, crystallinity, and surface chemistry; these parameters, in turn, govern catalytic activity, adsorption capacity, optical response, and electronic behavior [3, 6]. Top-down strategies, including milling and patterning, complement these methods by enabling scalability and integration with existing manufacturing infrastructures. ALD further bridges these paradigms by offering atomic-scale precision and conformal coating, thereby enabling the functionalization of complex three-dimensional architectures and the engineering of interfaces with sub-nanometer control [2].

2.2 Interfaces, confinement, and active sites

Another unifying theme is the increasing role of interfaces and surfaces as active sites for functionality. Across energy, environmental, and biomedical applications, performance is dictated by interfacial phenomena such as charge transfer, adsorption, and catalytic turnover. Chapters focusing on carbon capture and conversion demonstrate how pore engineering, surface functionalization, and heterojunction formation enhance selectivity and efficiency. Similarly, discussions of noble metal nanomaterials and hybrid composites highlight how electronic coupling between components can be tuned to optimize optical, catalytic, and sensing properties. These insights reinforce the notion that nanoscale interfaces represent both an opportunity and a challenge, requiring careful design to balance activity, stability, and safety [3, 7].

2.3 Characterization and data-driven optimization

The integration of advanced characterization and data-driven methodologies represents a further point of convergence. High-resolution microscopy, spectroscopy, and in situ analytical techniques provide direct insight into growth mechanisms, defect structures, and dynamic behavior under operating conditions [4]. When coupled with computational modeling and machine-learning-assisted optimization, these tools enable predictive design strategies that accelerate discovery while reducing experimental redundancy [10, 11]. Several chapters illustrate how accumulated experimental data can inform algorithmic models, guiding the selection of synthesis parameters and accelerating the development of next-generation nanomaterials.

2.4 Shared challenges

Importantly, the scientific advances described throughout the volume are increasingly contextualized within broader considerations of sustainability, scalability, and societal impact. Green synthesis approaches, life cycle assessment, and safer-by-design principles reflect a growing awareness that material performance must be evaluated alongside environmental and health implications [6]. The convergence of technical rigor with ethical reflection underscores a shift toward responsible innovation, in which scientific progress is aligned with long-term societal benefit [8].

To highlight these cross-cutting themes, Table 1 (in the volume) summarizes principal synthesis strategies, representative material classes, application domains, and shared scientific challenges discussed across the chapters.

3 Translational pathways and device integration

While individual chapters emphasize specific synthesis routes, material classes, or application domains, a unifying insight emerges when these contributions are viewed collectively; the translational success of nanotechnology depends on the seamless integration of synthesis precision, functional validation, and system-level deployment. Across wet chemical methods, vapor-phase deposition, ALD, and hybrid fabrication strategies, the chapters demonstrate that nanoscale control alone is insufficient unless it is coupled with performance metrics relevant to real-world environments [5].

Several contributions illustrate how laboratory-scale advances transition toward applied technologies through iterative feedback between synthesis and evaluation. Nanomaterials designed for catalysis, energy storage, or carbon capture are repeatedly refined through in situ characterization, electrochemical testing, and computational modeling. This iterative loop – spanning synthesis, diagnostics, and performance assessment – represents a translational pathway that bridges fundamental nanochemistry with engineering implementation. Importantly, this pathway is not linear but adaptive, responding to constraints such as scalability, durability, and environmental compatibility [4, 5].

Cross-chapter integration is particularly evident in the convergence of materials platforms. Porous frameworks, metal–oxide heterostructures, and surface-functionalized nanocomposites recur throughout chapters on energy conversion, environmental remediation, and electrochemical systems. These recurring material motifs underscore a shift toward modular nanomaterial design, in which core architectures are tailored through surface chemistry, dopant selection, or hierarchical structuring to meet diverse functional requirements. Such modularity enhances reproducibility and accelerates technology transfer by reducing the need for entirely new material systems for each application [3, 17].

Equally significant is the role of characterization and modeling as translational enablers. Chapters focused on in situ spectroscopy, X-ray absorption techniques, and molecular-scale simulations demonstrate how mechanistic insight informs material optimization under operating conditions. These tools provide the evidentiary foundation required to move beyond proof-of-concept demonstrations toward predictive design and regulatory confidence [4, 10]. Collectively, the volume highlights that translational nanotechnology is not defined by a single breakthrough but by the integration of complementary expertise across synthesis, analysis, and application.

4 Sustainability, ethics, and responsible innovation

As nanotechnology advances from laboratory-scale discovery toward widespread societal deployment, sustainability and ethics emerge as central considerations rather than peripheral constraints. The chapters in this volume collectively demonstrate that nanoscale innovation is inherently a sociotechnical enterprise, in which scientific progress must be evaluated alongside environmental impact, human health, and social equity [6–8]. This convergence reflects a broad shift in materials science toward responsible innovation frameworks that integrate technical excellence with ethical foresight. Such frameworks emphasize transparency, life cycle assessment, and stakeholder engagement to ensure that emerging nanotechnologies deliver long-term societal benefits while minimizing unintended consequences and fostering public trust [8].

A recurring theme across the chapters is the recognition that nanoscale materials, by virtue of their high surface activity and tunable functionality, can deliver transformative benefits while simultaneously introducing new risks. High-aspect-ratio nanostructures, reactive surfaces, and persistent composite materials exemplify this duality. Their enhanced reactivity underpins advances in catalysis, adsorption, and sensing, yet the same properties may contribute to environmental persistence, bioaccumulation, or adverse biological interactions if not carefully managed [7]. Consequently, sustainability in nanotechnology extends beyond resource efficiency to encompass life cycle considerations, including synthesis, deployment, recovery, and end-of-life disposal.

Green synthesis strategies discussed throughout the volume represent a tangible response to these challenges. The use of benign solvents, biological templates, and low-energy processing routes reflects an effort to reduce the environmental footprint of nanomaterial production. Such approaches align with the principles of green chemistry by minimizing the use of hazardous reagents, reducing waste generation, and improving energy efficiency [6]. However, the chapters also highlight that green synthesis alone is insufficient if downstream impacts are not addressed. Materials designed for environmental remediation or energy conversion must themselves be evaluated for long-term stability, recyclability, and potential ecological effects [7].

Ethical considerations further extend to questions of access, equity, and governance. Nanotechnologies deployed in healthcare, energy infrastructure, and environmental management can exacerbate existing social disparities if benefits are unevenly distributed. The societal impact chapter underscores how technological adoption is shaped by public trust, transparency, and inclusive discourse. Historical precedents in industrial chemistry and energy production show that public skepticism often arises when risks are perceived as externalized or inadequately communicated. In this context, ethical governance frameworks serve not only to mitigate harm but also to foster legitimacy and acceptance [8].

Responsible innovation, therefore, requires the integration of ethical reasoning into the earliest stages of research design. Several chapters implicitly adopt this perspective by emphasizing safer-by-design principles, in which material properties are engineered to reduce toxicity, enhance degradability, or limit unintended interactions. Such strategies exemplify a proactive approach to risk management, shifting the focus from post-hoc regulation to anticipatory design. When combined with life cycle assessment and stakeholder engagement, these approaches provide a pathway for aligning nanoscale innovation with societal values.

5 Convergence, automation, and future research directions

The trajectory of nanotechnology outlined in this volume reflects an accelerating convergence of disciplines, methodologies, and technological paradigms. Nanochemistry no longer operates in isolation but increasingly intersects with data science, automation, systems engineering, and sociotechnical governance. This convergence is reshaping how materials are discovered, optimized, and deployed, while simultaneously redefining the role of human expertise in scientific innovation [10, 11].

One of the most significant developments highlighted across chapters is the integration of automation and data-driven methodologies into nanomaterials research. High-throughput synthesis platforms, robotic handling systems, and machine-learning -guided optimization are transforming experimental workflows. These tools enable rapid exploration of multidimensional parameter spaces, allowing the optimization of synthesis conditions, compositional variables, and processing parameters with unprecedented efficiency. When coupled with advanced characterization and real-time feedback, automated platforms reduce experimental redundancy and accelerate the translation of laboratory discoveries into scalable technologies [10, 11].

Artificial intelligence and machine learning further extend this capability by enabling predictive modeling of structure–property relationships. Several chapters demonstrate how accumulated experimental data can inform algorithmic models that guide material design, from catalyst optimization to thin-film deposition. Such approaches shift nanotechnology toward a more deterministic, predictive science, in which hypotheses are refined iteratively through data-driven insights rather than trial-and-error experimentation. However, this shift also raises questions regarding transparency, reproducibility, and the interpretability of algorithmic decision-making, underscoring the need for rigorous validation and open data practices [10].

Convergence is also evident in the integration of nanomaterials into complex systems addressing global challenges. Energy conversion and storage technologies increasingly rely on hybrid architectures that combine nanoscale catalysts, conductive scaffolds, and functional coatings. Environmental remediation strategies leverage multifunctional nanocomposites that can adsorb, catalyze, and regenerate within a single platform. Biomedical applications similarly depend on the convergence of materials science, biology, and medicine to achieve targeted diagnostics and therapeutics. These examples illustrate that future advances will depend not only on individual material innovations but on the orchestration of nanoscale components within larger, system-level designs.

Looking forward, several grand challenges and research priorities emerge from the collective contributions. Scalability remains a persistent barrier, as many synthesis techniques optimized for laboratory conditions must be adapted for industrial throughput without compromising precision or sustainability. Long-term stability and life cycle performance represent another critical frontier, particularly for materials deployed in harsh or dynamic environments. Equally important is the development of standardized testing and characterization protocols that enable meaningful comparison across studies and facilitate regulatory assessment [5].

Human capital and workforce development constitute an additional dimension of future research as automation and artificial intelligence assume a greater role in experimental execution; the human contribution shifts toward conceptual design, ethical oversight, and systems integration. Training the next generation of scientists and engineers, therefore, requires an interdisciplinary education spanning chemistry, materials science, data analytics, and ethics. Such training is essential to ensure that technological convergence enhances, rather than diminishes, human agency and accountability [12].

6 Outlook and concluding remarks

The collective contributions presented in this volume demonstrate that nanotechnology has entered a phase of scientific maturity characterized by convergence, integration, and societal relevance. What began as an exploration of nanoscale phenomena has evolved into a comprehensive framework for engineering matter with atomic and molecular-level precision, enabling transformative advances across energy, environmental remediation, healthcare, and manufacturing. The chapters illustrate that nanochemistry now functions as both a foundational science and an enabling technology, linking synthesis, structure, stability, and function in a unified design paradigm [1, 3].

A central insight emerging from this synthesis is that future progress in nanotechnology will depend on the deliberate alignment of technical innovation with ethical responsibility and sustainability. Advances in synthesis precision, interfacial engineering, and data-driven optimization have expanded the functional landscape of nanomaterials. Yet, these same advances amplify the importance of life cycle assessment, environmental stewardship, and equitable deployment. As nanomaterials transition from laboratory prototypes to components of large-scale systems, their long-term performance, recyclability, and societal impact must be evaluated with the same rigor as their immediate functionality [6, 7].

The convergence of nanochemistry with automation, artificial intelligence, and systems engineering further reshapes the research enterprise. Automated synthesis platforms and predictive modeling tools promise to accelerate discovery and reduce resource consumption, but they also redefine the role of human expertise. In this emerging paradigm, scientists and engineers increasingly serve as architects of conceptual frameworks, ethical stewards, and integrators of complex systems. Preparing the next generation of researchers, therefore, requires interdisciplinary training that encompasses materials science, data analytics, environmental science, and ethics, ensuring that technological capability is matched by societal awareness [12].

Looking forward, the grand challenges facing nanotechnology are not solely technical. Scalability, standardization, and regulatory harmonization remain critical barriers to widespread adoption. Equally important is cultivating public trust through transparency, open communication, and inclusive governance. Historical experience across chemical and energy technologies demonstrates that societal acceptance is earned through demonstrated benefit, responsible risk management, and equitable access. Nanotechnology, by virtue of its invisibility and pervasiveness, demands particular attention to these dimensions [8].

Positioning the second edition within the evolving landscape of nanoscience, this postface closes by reiterating a pragmatic vision: sustained impact will follow from teams that couple ambitious materials design with pragmatic process engineering, rigorous characterization, and transparent governance. When scientific excellence is pursued alongside environmental stewardship and social responsibility, nanotechnology can deliver durable, scalable, and equitable benefits.

References

[1] Anastas, P. T., & Warner, J. C. (2000). *Green chemistry: Theory and practice*. Oxford University Press.

[2] Brinker, C. J., & Scherer, G. W. (1990). *Sol-gel science: The physics and chemistry of sol-gel processing*. Academic Press.

[3] Greeley, J., Jaramillo, T. F., Bonde, J., Chorkendorff, I., & Nørskov, J. K. (2006). Computational high-throughput screening of electrocatalytic materials for hydrogen evolution. *Nature Materials*, 5(11), 909–913. https://doi.org/10.1038/nmat1763.

[4] Jain, A., Ong, S. P., Hautier, G., Chen, W., Richards, W. D., Dacek, S., Cholia, S., Gunter, D., Skinner, D., Ceder, G., & Persson, K. A. (2013). The Materials Project: A materials genome approach to accelerating materials innovation. *APL Materials*, 1(1), 011002. https://doi.org/10.1063/1.4812323.

[5] Puurunen, R. L. (2005). Surface chemistry of atomic layer deposition: A case study for the trimethylaluminum/water process. *Journal of Applied Physics*, 97(12), 121301. https://doi.org/10.1063/1.1940727.

[6] Roduner, E. (2006). Size matters: Why nanomaterials are different. *Chemical Society Reviews*, 35(7), 583–592. https://doi.org/10.1039/b502142k.

[7] Klaine, S. J., Alvarez, P. J. J., Batley, G. E., Fernandes, T. F., Handy, R. D., Lyon, D. Y., Mahendra, S., McLaughlin, M. J., & Lead, J. R. (2008). Nanomaterials in the environment: Behavior, fate, bioavailability, and effects. *Environmental Toxicology and Chemistry*, 27(9), 1825–1851. https://doi.org/10.1897/08-090.1.

[8] Owen, R., Macnaghten, P., & Stilgoe, J. (2012). Responsible research and innovation: From science in society to science for society, with society. *Science and Public Policy*, 39(6), 751–760. https://doi.org/10.1093/scipol/scs093.

[9] Wang, X., Bashir, S., & Liu, J. Eds. (2022). *Nanochemistry: From theory to application for in-depth understanding of nanomaterials* (Vol. 1). Walter de Gruyter GmbH & Co. KG.

[10] Tao, F., & Salmeron, M. (2011). *In situ* studies of chemistry and structure of materials in reactive environments. *Science*, 331(6014), 171–174. https://doi.org/10.1126/science.1195826.

[11] Stankiewicz, A. I., & Moulijn, J. A. (2000). Process intensification: Transforming chemical engineering. *Chemical Engineering Progress*, 96(1), 22–34.

[12] Roco, M. C., Mirkin, C. A., & Hersam, M. C. (2011). Nanotechnology research directions for societal needs in 2020: Summary of international study. *Journal of Nanoparticle Research*, 13(3), 897–919. https://doi.org/10.1007/s11051-011-0223-3.

About the editors

Dr Sajid Bashir is a distinguished chemical scientist, educator, and academic leader with more than two decades of experience in analytical chemistry, interdisciplinary research, and higher-education service. He earned his PhD in analytical chemistry from the University of Warwick, England, in 2001, where his doctoral research focused on advanced mass spectrometry techniques. Following his doctorate, he served as a postgraduate research associate at Cornell University, contributing to research in plant genetics, proteomics, and genomics, further strengthening his interdisciplinary scientific foundation.

He is currently a full professor of chemistry at Texas A&M University-Kingsville (TAMUK) and serves as a faculty fellow with the United States Air Force, reflecting his intense engagement with both academic and federal research communities. His research portfolio spans analytical chemistry, spectroscopy, nanomaterials, sustainable energy, forensic science, and applied materials characterization. Over the course of his career, he has directed and participated in more than 20 externally and internally funded research projects, supported by organizations, including the Welch Foundation, TAMUK, the Texas Workforce Commission, and the U.S. National Institutes of Health.

He has made substantial scholarly contributions, coauthored more than 80 peer-reviewed journal articles and book chapters, and edited and contributed to multiple academic volumes in energy, materials science, and analytical chemistry. His work has influenced both fundamental research and applied technologies, particularly in energy conversion, advanced materials, and chemical detection.

In recognition of his professional standing, he is a Fellow of the Royal Society of Chemistry (FRSC) and holds the designations of Chartered Chemist (CChem) and Chartered Scientist (CSci) from the Science Council. These honors reflect his sustained contributions to chemical science, professional leadership, and international engagement.

As an educator, he has demonstrated a deep commitment to student success and workforce development. During his tenure at TAMUK, he has trained and mentored more than 3,000 undergraduate and graduate students, integrating research-driven instruction and experiential learning into the curriculum. He has also developed online courses and established comprehensive laboratory safety training protocols in collaboration with the university's Risk Management, enhancing both instructional quality and institutional compliance.

He actively engages with the broader community. He currently collaborates with local law enforcement agencies as a scientific consultant, applying analytical and forensic expertise to real-world challenges. Through his combined roles as researcher, educator, and public-sector collaborator, he continues to advance scientific innovation, workforce preparation, and community impact at the regional and national levels.

Dr Jingbo Louise Liu is an internationally recognized materials scientist, educator, and scientific leader whose career spans more than two decades of impactful research, academic leadership, and professional service. She received her PhD in materials science and engineering from the University of Science and Technology Beijing, establishing a strong foundation in materials synthesis, characterization, and structure–property relationships. Her scholarly work has since advanced the frontiers of engineered nanomaterials, energy storage, fuel cells, solid-state electrolytes, and environmental and biological applications of materials science.

She is currently a full professor at TAMUK and a faculty affiliate of the Texas A&M Energy Institute. She has also served as director of the Center for Teaching Effectiveness at TAMUK, where

 | https://doi.org/10.1515/9783111621111-010

she led institution-wide initiatives to enhance instructional quality, faculty development, and evidence-based teaching practices. Her leadership in pedagogy complements her scientific research, positioning her as a rare scholar who bridges cutting-edge materials innovation with transformative education.

Her research contributions are widely recognized for both scientific depth and practical relevance. She is credited with establishing record-high power density in proton exchange membrane fuel cells, significantly advancing fuel-cell performance and durability. She has also pioneered the application of metal–organic frameworks in disinfection science, opening new pathways for environmental remediation, public health protection, and sustainable materials design. Her work integrates nanostructured materials, electrochemistry, and surface science to address pressing challenges in energy sustainability and environmental resilience.

Her scholarly output is extensive and influential. She has authored and coauthored more than 100 peer-reviewed journal articles, books, and book chapters, and has delivered over 150 invited and contributed presentations at national and international conferences. Her publications have garnered more than 21,000 citations, reflecting sustained global impact across materials science, chemistry, and energy research communities. She has also served as editor and guest editor for major scientific publishers, including Elsevier, Springer Nature, and American Chemical Society (ACS) Publications, further shaping the dissemination of knowledge in her field.

She has demonstrated exceptional success in securing research funding and leading programs. She has directed and/or participated in more than 40 funded projects supported by major agencies and organizations, including the National Science Foundation (United States and China), Natural Sciences and Engineering Research Council of Canada (NSERC), ACS Petroleum Research Fund, U.S. Department of Education, and industrial partners. She has served in roles ranging from principal investigator and coprincipal investigator to senior personnel, managing complex, multi-institutional research efforts that integrate fundamental science with applied outcomes.

A defining feature of her career is her deep commitment to education, mentorship, and workforce development. During her decade-plus years of service at TAMUK, she has taught more than 8,700 students and trained approximately 150 undergraduate, graduate, postdoctoral researchers, and visiting scholars in leading-edge research. A significant proportion of her mentees come from underrepresented and minority backgrounds, reflecting her sustained dedication to broadening participation in STEM fields. Her mentorship has produced graduates who have gone on to careers in academia, national laboratories, industry, and public service.

Her leadership and impact have been recognized through numerous prestigious honors and awards. She is an FRSC and holds the professional designations of CSci and CChem. She is also a DEBI Faculty Fellow at the U.S. Air Force Research Laboratory, reflecting her trusted engagement with national defense research programs. Her national leadership in chemistry and materials science has been further recognized through honors from the ACS, including election as an ACS Fellow and selection as an ACS Certified Career Consultant, a role held by only a small fraction of ACS members.

Her contributions to innovation, leadership, and community engagement have earned national recognition from the National Small Business Association, underscoring her impact at the intersection of science, entrepreneurship, and societal benefit. This distinction highlights her ability to translate scientific expertise into broader economic and community value, extending her influence beyond the laboratory and classroom.

She has played a central role in professional service and scientific governance. She currently serves as Division Chair-Elect of the Energy and Fuels Division of the ACS, following prior service as division chair, councilor, and national officer. In these roles, she has guided strategic planning, technical programming, and member engagement for one of ACS's most interdisciplinary divisions. Her leadership during periods of global disruption, including the COVID-19 pandemic, the Science Cafe, and the "ChemFit" Program, was particularly notable for maintaining scientific continuity, supporting members, and delivering public-facing technical outreach, which earned her the IUPAC Women of the Year Award in Chemistry/Chemical Engineering.

In addition to her ACS leadership, she has held appointments and fellowships at major national and international research institutions, including Argonne National Laboratory, Lawrence Berkeley National Laboratory, the University of Tokyo (JSPS Fellow), and the U.S. Air Force Research Laboratory. These experiences reflect her global scientific reach and her ability to collaborate effectively across institutional and national boundaries.

Through her integrated roles as scientist, educator, mentor, and leader, she has made enduring contributions to materials science, sustainable energy, and the chemical sciences community. Her career exemplifies excellence in research, commitment to inclusive education, and visionary leadership that continues to shape the future of energy, materials innovation, and professional service.

About the contributing authors

Ashraf Abedin received his Ph.D. from the Department of Chemical Engineering at Louisiana State University in 2021. His research focuses on the design, synthesis, and characterization of site-specific heterogeneous catalysts to convert fossil fuels into value-added chemicals.

He has published more than 20 peer-reviewed articles, including four book chapters. He has presented more than 25 papers at national and international conferences and received subawards from DOE research grants based on research proposals for clean energy projects.

Dr. **Matthew Alexander**
Associate Professor
Education
Ph.D. in Chemical Engineering, 1990, Purdue University
M.S. in Chemical Engineering, 1986, Georgia Institute of Technology
B.S. in Engineering Science and Chemistry, 1984, Trinity University
Research Interest at Texas A&M University-Kingsville
Renewable and biofuel production processes
Sustainable energy production using chemical/biochemical processes
Remediation of hazardous wastes in soil and groundwater, concentrating on bioremediation processes Industrial wastewater treatment Chemical process simulation/modeling

Dr. **Vivek Anand** has been working as an Assistant Professor of Chemistry in the Department of Chemistry at Chandigarh University since July 2021. He had two years of working experience as an Assistant Professor at the Institute of Advanced Research, Gandhinagar (from August 2019 to July 2021). He received his Ph.D. in Physical and Organic Chemistry from the Department of Chemistry at IIT Madras in July 2018. Have research expertise in the synthesis and characterization of conjugated organic monomers and polymers via various organic synthetic methodologies. Sonogashira coupling, Knoevenagel condensation, and Vilsmeyer-Heck reaction are used to achieve vivid optoelectronic properties, such as aggregation-induced emission, white light generation, and fluorescence sensing. To date, he has published 8 research articles in international journals.

Hassnain Asgar is a Ph.D. candidate in Prof. Gadikota's research group. He obtained his Master's degree in Engineering from Central Michigan University and his undergraduate degree in Metallurgy and Materials Engineering from the University of the Punjab, Lahore. His research in Prof. Gadikota's group focuses on understanding how confinement geometry and chemistry influence the organization of confined fluids for various energy and environmental applications.

 | https://doi.org/10.1515/9783111621111-011

Dr. **Jong-Won Choi** is an Associate Professor in the Department of Civil and Architectural Engineering at Texas A&M University – Kingsville, with extensive experience spanning research, teaching, and academic leadership. His scholarly work focuses on energy and environmental geomechanics by integrating numerical and analytical methods. He has published papers in peer-reviewed journals and contributed to several federal and state grants. His academic service includes serving as a graduate coordinator, developing curricula, mentoring graduate student populations, and advancing collaborative research initiatives.

Sai Raghuveer Chava completed his MS in Chemistry at Texas A&M University-Kingsville and is working in the Biopharmaceutical Industry.

Sabrina Darem received her Master's degree in environmental science from the University of Ghardaia. She is currently a Ph.D. student in the Department of Biology, Laboratory of Soils and Sustainable Development, University of Badji-Mokhtar, Annaba, Algeria. Her research interests include protecting and conserving soil in Algerian Saharan zones using geostatistical modeling, medicinal plants, plant ecophysiology, plant adaptation mechanisms to biotic and abiotic stress, and the interactions among soil, plants, and microorganisms.

Durowoju Emmanuel is a graduate student majoring in Mechanical Engineering at Texas A&M University, where he conducts research under the supervision of Professor Hua Li. His work focuses on improving the performance of solid-state electrolyte lithium batteries through data-driven modeling approaches. He has applied machine-learning algorithms to validate battery performance.

Dr. **Greeshma Gadikota** has the following research interests: developing clean methods for producing clean energy carriers such as hydrogen while simultaneously capturing and converting the carbon dioxide produced into a useful, environmentally harmless solid. One thread of her research uses X-ray scattering, tomography, and spectroscopy to observe chemical reactions as they happen, potentially enabling the development of a method to produce hydrogen without producing greenhouse gases. More broadly, Dr. Gadikota would like to develop a method for observing the reaction environments for any solid-liquid-gas system. Gadikota is also studying the feasibility of pulling acid gases from dilute, heterogeneous gas streams and understanding how reactive fluids and solids behave in extreme subsurface and engineered environments.

Professor **Yousef Haik** served as professor and department chair of the Mechanical and Industrial Engineering department at Texas A & M University-Kingsville. His academic career includes appointments in engineering (Mechanical, Biomedical, and Nanoengineering) and science (Medicinal Chemistry, Physics, and Nanoscience) programs. His academic and administrative career includes department chair, dean, and provost appointments. He is the inventor and co-inventor of several U.S. and international patents. He published and edited nine books and technical proceedings, authored and coauthored over 250 technical papers, and presented many papers at global conferences. He serves as Editor-in-Chief and on the Editorial Board for a dozen International Journals. He serves on the review boards for International Research Foundations. He organized and chaired several international conferences and symposia. He is a member of a dozen professional associations. He is a Fellow of the American Society of Mechanical Engineers and the National Academy of Innovators. His expertise includes nano-energy, nanomaterials, biofuels, diagnostics, recycling, and environmental remediation. He earned his Ph.D. in Mechanical Engineering from Florida State University in 1997, an M.S. from the University of Iowa in 1994, and a B.S. in Mechanical Engineering from Jordan University in 1986.

Dr. Tanveer ul Haq is an Assistant Professor at the University of Sharjah (UoS), a leading research-intensive institution in the United Arab Emirates comprising more than a dozen colleges and over 130 academic programs across science, medicine, engineering, business, and the arts. UoS maintains a strong research ecosystem supported by multiple research institutes, advanced laboratories, core facilities, and interdisciplinary research groups. Dr. ul Haq is actively engaged in green hydrogen production from seawater, a strategic research area aligned with the UAE's national priorities and Vision 2030, and has recently secured approximately USD 1 million in competitive research funding. Trained as a chemist, he earned his Master's degree in 2018 from the Lahore University of Management Sciences (LUMS), where his thesis focused on the development of nanoclusters for electrochemical water splitting. He subsequently completed his Ph.D. at Texas A&M University under the supervision of Prof. Yousef Haik.

Guanjie He is an Associate Professor in Materials Chemistry and Advanced Functional Materials Research Group Lead at the University of Lincoln and an Honorary Lecturer at University College London (UCL). Dr. He received a Ph.D. degree in the Chemistry Department at UCL under his supervision. Dr. He's research focuses on electrochemical energy storage and conversion applications, with a particular emphasis on materials, especially electrode materials and electrocatalysts, in aqueous electrolyte systems.

William Houf was an undergraduate student at Texas A&M University-Kingsville, who was taking a B.S in Chemical Engineering with a minor in Chemistry.

Hsuanyi Huang was a graduate student at Texas A&M University-Kingsville, who graduated with an M.S. in Chemistry.

Zheng Jiang is a Full Professor at Shanghai Advanced Research Institute and Shanghai Institute of Applied Physics and beamline leader of X-ray Absorption Fine Structure (BL14W1). After receiving his Ph.D. degree from the University of Science and Technology of China in 2005, he worked at the Shanghai Synchrotron Radiation Facility (SSRF). In 2007 and 2013, he went to the Australian Synchrotron and BESSY II HZB, respectively, as a senior visiting scholar. His main research interests focus on: 1) the design and construction of X-ray spectroscopy beamlines; 2) developing advanced synchrotron radiation X-ray spectroscopy methods, including high-energy-resolution XES, HERFD-XAFS, RIXS, *operando/in-situ* X-ray spectroscopy, and so on, forming a research system for energy materials from a single atom to a cluster.

Dan Kong is an Associate Professor at the Department of Materials Science and Physics, China University of Mining and Technology, China. She obtained her Ph.D. in Chemical Engineering in 2019 from the University College London, UK, and then served as a postdoctoral researcher at Aalto University, Finland, for one year. She completed her MSc (2014) from Zhejiang University and her BSc (2011) from Wuhan University of Technology, China. Dan's research focuses on semiconductors for solar energy conversion, such as overall water splitting and CO_2 conversion. Dan's interests include efficient microwave-assisted synthetic techniques, photoelectrocatalysis, and mechanism investigation.

John-Ryan C Lawrence is an undergraduate student at Texas A&M University-Kingsville, majoring in Electrical Engineering, minor in Nuclear Engineering and Mathematics, and will return to the master of Science program in Space & Aeronautical Engineering.

Wen Lei received his Ph.D. in Materials Physics and Chemistry from Huazhong University of Science and Technology, China, in 2018. He was a visiting student at the University of Waterloo conducting research in plant genetics. Currently, he is an Associate Professor at Wuhan University of Science and Technology (WUST) and the "Macao Young Scholar" at the University of Macau. His research interests are materials and devices for electrochemical energy storage and conversion. So far, Dr. Lei has published more than 60 research papers, including Energy Environ. Sci., Mater. Today, Adv. Energy Mater., ACS Catal. His papers have been cited more than 2500 times.

Dr. **Hua Li** is a full professor and Industrial Engineering Graduate Coordinator in the Department of Mechanical and Industrial Engineering at Texas A&M University-Kingsville. His research expertise includes renewable energy, simulation, optimization, machine learning, and product design. He has received more than $7M grants as PI and over $10M grants as Co-PI from various federal agencies. He has authored one book chapter and published more than 100 peer-reviewed journal articles and conference papers. He has graduated three doctoral students and over 40 Master's thesis students.

Dr. **Robert Luckett** is an Assistant Professor at Texas A&M University-Kingsville
Education
- Ph. D., Education with a concentration in Higher Education Administration and Leadership, 2017
- Texas A&M University-Kingsville Kingsville
- M.S.W., Social Work with a concentration in Micro Practice, 1999
- University of South Carolina, Columbia
- B.S., Sociology with a minor in Psychology, 1997
- College of Charleston

Areas of interest
- Community college transfer students
- The Human Animal Bond (HAB) and Human Animal Interaction (HAI)
- Changes in state laws that conflict with the NASW Code of Ethics

Bingbao Mei received his Ph.D. from the Shanghai Institute of Applied Physics, Chinese Academy of Sciences, in 2021. He got his B.E. degree from Nanjing University of Aeronautics and Astronautics in 2016. Currently, he is a Research Assistant at the BL14W1 beamline in SSRF. His work focused on operando high-energy-resolution X-ray spectroscopy, including XES, RIXS, and HERFD-XAS, and its application in electrochemistry.

Roli Mishra is working as an Assistant Professor of Chemistry at the Institute of Advanced Research, Gandhinagar (India). She received her Ph.D. (Chemistry) in 2006 from the University of Allahabad (India). She completed postdoctoral studies at the Indian Institute of Science, Bangalore (India), the University of Minnesota, USA, and the Indian Institute of Technology, Delhi (India). She has worked broadly on peptide synthesis, developing a novel protecting group for nucleosides, peptidomimetics, supramolecular chemistry, and thiazolium ionic liquids. Her current research focuses on the design, synthesis, and application of task-specific ionic liquids for organic synthesis, ion recognition, peptide and drug stability, and fluorescence and optoelectronic properties.

Sohaib Mohammed holds an M.S. in chemical engineering from the University of Illinois at Chicago and a B.S. in chemical engineering from Tikrit University in Iraq. Sohaib's research focuses on understanding the properties of confined fluids and phase change, and on colloids' driven and self-assembly at immiscible liquid interfaces. He utilizes computational tools, such as Density Functional Theory and Molecular Dynamics Simulations, and advanced experimental techniques, such as *in-operando* X-ray scattering, to probe the multi-scale morphological, dynamical, and reactivity details of the systems of interest.

Yesirat O Olaseni was a graduate student at Texas A&M University-Kingsville, who graduated with an M.S. in Chemistry.

Sushesh Palakurthi was a graduate student at Texas A&M University-Kingsville, who graduated with an M.S. in Chemistry and now graduated with a PhD in Pharmaceutical Science at Texas A&M University Health Science Center.

Ivan. P. Parkin is the Dean of Mathematical and Physical Sciences (MAPS), Faculty at University College London. He received his Ph.D. from Imperial College London, and his research focused on chemical vapor deposition, Thin films for windows, superhydrophobic surfaces, inorganic materials for electrocatalysis, and zinc-ion batteries. Ivan has published over 600 papers in international journals, and all his work has over 39,000 citations. He has an h-index of 89 (Google Scholar).

Yeshu Tan received his Master's Degree from Soochow University in 2019. He is a Ph.D. student in the Department of Chemistry, University College London. He focused on the synthesis of nanoparticles and their application in electrocatalysis, including the hydrogen evolution reaction and the oxygen evolution reaction.

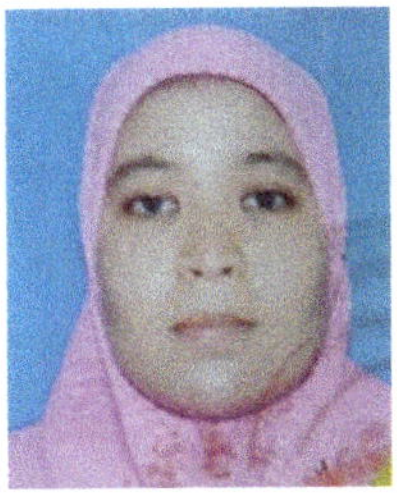

Dr. **Alia Telli** received her graduate degree (2003), Master's degree (2009), and doctorate (2017) in biochemistry from Kasdi Merbah University, Ouargla, Algeria. She is currently working as a lecturer at Kasdi Merbah University. Her research interests include an ethnobotanical survey of medicinal plants, the antioxidant, antidiabetic, and antimicrobial activities of herbal remedies, and the green synthesis of nanoparticles and their applications.

Pete Villarreal was a graduate student at Texas A&M University-Kingsville, who graduated with an M.S. in Chemistry and is now working as an Adjunct Professor of Chemistry, also at Texas A&M University-Kingsville.

Luis F Villanueva was a graduate student at Texas A&M University-Kingsville, who graduated with an M.S. in Chemistry.

Saurabh Vyas did his B.Sc. and M. Sc in Chemistry from the Institute of Advanced Research, Gandhinagar (India). He is currently pursuing his Ph.D. under the supervision of Dr. Roli Mishra at the Institute of Advanced Research, Gandhinagar. His research interests are in organic synthesis, Synthesis of ionic liquids and their based complexes, and applications in organic transformations and optoelectronics.

Dr. **Zhaohui Wang** received his Ph.D. in Electrical and Computer Engineering from the University of Arizona in 2011. He also received a Master of Science in Bioengineering from the University of Toledo and a Master of Science in Mechanical Engineering from the University of Arizona. From 2011, he continued post-doctoral training at the University of Pittsburgh Medical Center. In 2014, he was promoted to Visiting Assistant Professor of Electrical Engineering and Computer Science at Texas A&M University-Kingsville. His current research includes medical imaging, bioinformatics, and nanomaterials. His Research Interests at North Carolina Agricultural and Technical State University Include Data Mining, Embedded System Design, Biomedical Physics, and Nanomaterials.

Haijun Zhang received his Ph.D. in Physical Chemistry in Metallurgy from the University of Science and Technology in Beijing, China, in 1999. After two years as a postdoctoral fellow at the Functional Materials Research Laboratory, he joined the High-Temperature Ceramics Institute at Zhengzhou University in 2001. He was promoted to Associate Professor and then Full Professor there. In 2007, he joined Prof. Toshima's group at Tokyo University of Science, Yamaguchi, as a full-time research scientist. Currently, Prof. Zhang is a Full Professor at The State Key Laboratory of Refractories and Metallurgy, Wuhan University of Science and Technology (WUST, China). His research interests include refractories, high-temperature ceramics, and nanostructured materials. He has published more than 470 papers in refereed journals.

Hao Zhang received his Ph.D. in Nuclear Technology and Application from the Shanghai Institute of Applied Physics, Chinese Academy of Sciences, in 2021. He is currently a Lecturer at the Institute of Functional Nano & Soft Materials (FUNSOM), Soochow University. His research interests include applying Synchrotron-based X-ray spectroscopy in various electrochemical reactions catalyzed by highly dispersed metal sites.

Shaowei Zhang received his Ph.D. in Materials Science and Engineering from Nagoya Institute of Technology, Japan, in 1996. Prof. Zhang is currently a Professor and Royal Society Industry Fellow in the College of Engineering, Mathematics, and Physical Sciences at Exeter, U.K. Before moving to Exeter, he had worked for about 14 years in the Department of Materials Science and Engineering at the University of Sheffield as a Postdoctoral Research Associate, EPSRC Advanced Fellow, Lecturer, and Reader. His main research interests are in the processing, microstructures, and properties of structural and functional materials. Current research topics include low-temperature fabrication of nanomaterials, novel synthesis of 2-D quantum dots, exploration of 2-D nanocatalysts, and preparation of porous ceramics and ceramic matrix composites. Other work includes the development of functional coatings, novel armor materials, and ultrahigh-temperature ceramics.

Index

 | https://doi.org/10.1515/9783111621111-012

www.ingramcontent.com/pod-product-compliance
Lightning Source LLC
LaVergne TN
LVHW060552201026
850618LV00034B/492